BIOCHEMISTRY AND MOLECULAR BIOLOGY IN THE POST GENOMIC ERA

CYCLODEXTRINS

SYNTHESIS, CHEMICAL APPLICATIONS AND ROLE IN DRUG DELIVERY

BIOCHEMISTRY AND MOLECULAR BIOLOGY IN THE POST GENOMIC ERA

Additional books in this series can be found on Nova's website under the Series tab.

Additional e-books in this series can be found on Nova's website under the e-book tab.

BIOCHEMISTRY AND MOLECULAR BIOLOGY IN THE POST GENOMIC ERA

CYCLODEXTRINS

SYNTHESIS, CHEMICAL APPLICATIONS AND ROLE IN DRUG DELIVERY

FRANCIS G. RAMIREZ
EDITOR

nova publishers
New York

NOTICE TO THE READER

The Publisher has taken reasonable care in the preparation of this book, but makes no expressed or implied warranty of any kind and assumes no responsibility for any errors or omissions. No liability is assumed for incidental or consequential damages in connection with or arising out of information contained in this book. The Publisher shall not be liable for any special, consequential, or exemplary damages resulting, in whole or in part, from the readers' use of, or reliance upon, this material. Any parts of this book based on government reports are so indicated and copyright is claimed for those parts to the extent applicable to compilations of such works.

Independent verification should be sought for any data, advice or recommendations contained in this book. In addition, no responsibility is assumed by the publisher for any injury and/or damage to persons or property arising from any methods, products, instructions, ideas or otherwise contained in this publication.

This publication is designed to provide accurate and authoritative information with regard to the subject matter covered herein. It is sold with the clear understanding that the Publisher is not engaged in rendering legal or any other professional services. If legal or any other expert assistance is required, the services of a competent person should be sought. FROM A DECLARATION OF PARTICIPANTS JOINTLY ADOPTED BY A COMMITTEE OF THE AMERICAN BAR ASSOCIATION AND A COMMITTEE OF PUBLISHERS.

Additional color graphics may be available in the e-book version of this book.

Library of Congress Cataloging-in-Publication Data

ISBN: 978-1-63482-788-1
Library of Congress Control Number: 2015939096

Published by Nova Science Publishers, Inc. † New York

CONTENTS

PREFACE

Cyclodextrins (CDs) are capable of interacting with a large variety of guest molecules to form inclusion complexes. The authors of the chapters that form this book discuss the development of ion-selective electrodes based on selected cyclodextrins and their derivatives as selective ionophores for various drugs and their applications in pharmaceutical research; the wide range of applications of CDs in biomedical electrochemistry, focusing both on their behavior in solutions and embedded in thin films formed at the electrode surface; the general features of CDs and the progress done so far in the field of microencapsulation of pesticides in CDs; cyclodextrins as encapsulating agents for bioactive plant molecules in the pharmaceutical field; the effect of cyclodextrins on the photodegradation profile of drugs when they are used in drug formulations; applications of cyclodextrins in stabilizing or solubilizing ingredients of pharmaceutical relevance used in the food industry (commonly called nutraceuticals); the use and importance of cyclodextrins in the pharmaceutical industry; the main aspects of the crystal structures of the dimeric β-CD inclusion complexes; the influence of biologically active inorganic ions on complex formation of cyclodextrins with some aromatic carboxylic acids possessing pharmacological and biological activity; the importance of detailed characterization of the composition of substituted CDs by the example of sulfobutylether cyclodextrins; the several crystal structures of dye inclusion complexes using cyclodextrin, the preparation of DISC (dye inclusion single crystal); the structural details about the dye inclusion complex analyzed by X-ray and the properties of porphyrin DISC.

Chapter 1 – Cyclodextrins (CDs) are capable of interacting with a large variety of guest molecules to form inclusion complexes, for instance with drugs molecules through their lipophilic inner cavities and hydrophilic outer surface. One of the uses of cyclodextrins is their application in potentiometry. Potentiometric studies concerning the use of cyclodextrins as ionophores (in sensor development) are very new and not numerous.

In particular, lipophilic cyclodextrins (with an alkylated hydroxyl group in positions 2, 3, and 6), dihydroxypropyl-β-cyclodextrin, triacetyl-β-cyclodextrin, α-, β-, and γ-cyclodextrin derivatives and heptakis (2,3,6,-tri-O-methyl)-β-cyclodextrin, as well as (2-hydroxypropyl)-γ-cyclodextrin and many other can be incorporated in plasticized PVC membrane graphite electrodes as excellent ionophores.

The derivatives of β-cyclodextrins are used as excipients in oral or external drug formulation. β-CDs can be used to increase the solubility and dissolution rate of poorly water-soluble drugs and can affect drugs stability. β-CDs may be used to develop new dosage forms of drugs in pharmaceuticals or many other medicines. Many pharmaceutical products containing CD-s are already on the market. From a group of non-steroidal anti-inflammatory drugs (NSAIDs), the following are included: indomethacin, piroxicam, meloxicam, nimesulide, tiaprofenic. Therefore, it seemed fully justified to perform investigations into the effects of β-CDs on the performance of ion-selective electrodes which are used for determination of selected drugs.

This review discusses the development of ion-selective electrodes based on selected cyclodextrins and their derivatives as selective ionophores for various drugs and their applications in pharmaceutical research.

Chapter 2 – The supramolecular complex formation between various chemical species and cyclodextrins, significantly influences the physical-chemical features of the guest molecule entrapped inside the cyclodextrins (CD) cavity. The inclusion complexes of many pharmaceutical substances with CDs have been investigated by using various analytical methods, including the electrochemical ones. The decreased peak currents and shifts in the oxidation and reduction potentials are the main changes of the electrochemical behavior of the guest molecules, attributed to the decrease of concentration, smaller diffusion coefficients of the complexes, and to their relatively high stability.

Different techniques, such as: inclusion in carbon paste electrode, entrapment in polymers or in plasticized membranes, immobilization onto screen printed electrodes, nanoparticles, carbon based materials, and nanostructured polymeric films, Langmuir-Blodgett technique, self-assembled monolayers, were used for the development of novel (bio)sensors based on CDs and their derivatives.

The CDs based electrodes represents a promising future for a lot of applications in various fields: agriculture, food, toilet articles, textile manufacturing, environment, pharmaceuticals, and therapy. The CDs specific features recommend them for a wide range of applications in analytical chemistry as catalysts, enantioselectors and molecular receptors for the development of new electrochemical sensors/biosensors.

Chapter 3 – Globally it is estimated that between 7.9 and 15.1 % of agricultural harvests are destroyed by pests, being wheat, rice, maize, potato, soybean and cotton, the most affected crops. With the exponential growth of worldwide population, an optimization in the crops protection against pests is urgently required, and because of that a 15- to 20-fold increasing in the chemical pesticides use has been observed. However, no significant effects on pest control have been observed. Despite of pesticides harmful properties against pests, they can indirectly improve the nutritional value of food. Nevertheless, the increased knowledge on their benefits and risks posed additional safety issues, namely those related to their low water solubility (bioavailability) and easy photodegradation. Several pesticides are easily decomposed by light and in that way a large amount of pesticide is needed to be applied in the crops in order to be effective. Some estimates point to the fact that less than 1 % of applied pesticides effectively reach the target. Furthermore, in some cases the metabolites originated by photodegradation of pesticides display a higher toxicity level. These facts associated with their capacity to act as carcinogenic, mutagenic and teratogenic agents to mankind, have led to a growing concern among scientific community to find a solution aimed to reduce or minimize the mentioned drawbacks and increase pesticide efficiency against pests.

To increase the solubility and photostability as well as to facilitate the handling of pesticides, microencapsulation in nanostructures, namely cyclodextrins (CDs), has been proposed as a putative solution. CDs are cyclic oligosaccharides composed by D-glucose units connected by α-1,4-glycosidic linkages. The CDs structural characteristics, hydrophobic interior cavity and hydrophilic shell, and its ability to alter the physical, chemical, and biological properties of guest molecules has been regarded as a solution to improve pesticide formulations and to attain environmental sustainability.

In this chapter, the authors will give special attention to the main pesticide groups and their inherent health and environmental issues. General features of CDs and the progress done so far in the field of microencapsulation of pesticides in CDs, will also be reported.

Chapter 4 – Since ancient times, plant-derived compounds have been a great source of constituents used in beneficial medical treatments. Plants are a source of a wide range of molecules capable of improve healing: fibre, vitamins, phytosterols, alkaloids, sulphur-containing compounds, carotenoids, organic acids and polyphenolics. Remarkably, many of these constituents possess a scarce bioavailability due to their low water solubility or poor stability.

Cyclodextrins are cyclic oligosaccharides arising from the degradation of starch, which can be a viable option for encapsulation technique to generate aqueous drug solutions without the use of organic co-solvents, surfactants, or lipids, as formulation adjuncts.

The inclusion complexes are reported to have remarkable advantages over conventional formulations of plant actives and extracts, which include enhancement of solubility, protection from toxicity, enhancement of pharmacological activity and *in vivo* stability, sustained delivery, and protection from physical and chemical degradation, consequently increasing bioavailability and bioefficacy of guest molecules.

Therefore, the aim of this chapter is to highlight the use of cyclodextrins as encapsulating agents for bioactive plant molecules in the pharmaceutical field.

Chapter 5 – Most drugs are hydrophobic in nature, and while this is good for their pharmacological activity, it often affects their absorption into the systemic circulation due to poor aqueous solubility. One of the common approaches for the improvement of drug solubility is by complexation with cyclodextrins (CDs). Moreover, it can also make possible the preparation of injectable aqueous formulations of poorly soluble drugs. It has been proven that the presence of CD can significantly increase the aqueous solubility of some drugs, and in addition be a good absorption enhancer. However, cyclodextrins are also known to affect the drug stability, either positively or negatively. The rate of drug photodegradation and distribution of photodegradants may be altered in the presence of CD [1-3]. Therefore, CDs have the potential to control chemical and photochemical reactions. For example, it has been shown that the presence of randomly methylated beta-cyclodextrin (RM-β-CD) slightly decreases the photostability of midazolam. The decreased stability was accompanied by the appearance of a new degradation product, an intermediate, 6-chloro-2-methyl-4-(2-fluorophenyl)-quinazoline and the dimer. In general, photosensitivity reactions have occurred in patients to whom light sensitive drugs have been administered, and have also been associated with the formation of toxic photoproducts. Such drugs decompose to form radical intermediates and highly reactive products, which react with the tissue cells resulting in adverse effects, making the detection and identification of photodegradants important. Thus, there is a need to determine the effect of cyclodextrins on the photodegradation profile of drugs when they are used in drug formulations.

Chapter 6 – The most interesting property of cyclodextrins (CDs) is, undoubtedly, their ability to form inclusion complexes (ICs) through the encapsulation into their central cavity of guest molecules of suitable size, geometry and hydrophobicity. These guest molecules may be apolar, low solubility drugs, volatile compounds warranting stabilization or light-sensitive compounds.

The present chapter deals with applications of cyclodextrins in stabilizing or solubilizing ingredients of pharmaceutical relevance used in the food industry, which, for the combination of these two functions, are commonly called nutraceuticals. This emerging field is gaining relevance and acceptance amidst the consumers, and the use of cyclodextrins as carriers for antioxidant compounds will help meet the growing market demands. Furthermore, the cyclodextrins themselves are also revealing intrinsic nutraceutical properties, as is the case of incorporating α-CD in starchy meals to lower post-prandial glycemia and of the delayed release of glucose by γ-CD. The benefits and challenges of this new field of application for the native cyclodextrins, from the physico-chemical to the formulation standpoints, as well as a thorough revision of the regulatory status, are presented and discussed.

Chapter 7 – Cyclodextrins due to their hydrophobic cavity and hydrophilic surface can form the "host-guest" inclusion complexes with various compounds. The central hydrophobic cavity allows that the cyclodextrins have the role of the "host" molecules. The preparation of inclusion complexes leads to the changes in the physical and chemical properties of "guest" molecules. In this way it is possible to protect the compounds sensitive to the effect of light, heat and oxygen from the air. Also, the aqueous solubility of poorly soluble substances can be increased. Because of these properties and negligible toxicity, the cyclodextrins have found a broad application in the pharmaceutical industry. Today, the drugs based on the cyclodextrins represent about 40% of the total available drugs on the market. Their use in the pharmaceutical formulations is justified, because they improve the aqueous solubility of the hydrophobic drugs without the addition of the organic solvents, surfactants or lipids, increase the dissolution rates, oral bioavailability and stability. The cyclodextrin inclusion complexes reduce the gastrointestinal and ocular irritations, eliminate the unpleasant odors or tastes and prevent the drug-drug or drug-excipients interactions. The controlled release is also one of the important properties of the cyclodextrins which justifies their use in order to improve a drug delivery from a pharmaceutical formulation. Also, the cyclodextrins are used in the preparation of deodorants, lotions, diapers, paper towels, etc. Their role is to form the complexes with smelly substances in order to reduce their volatility.

Chapter 8 – *β*-Cyclodextrins (*β*-CD), consisting of seven *α*-(1→4)-linked D-glucopyranose units, have the tendency to form dimers by linking two molecules through hydrogen bonds relating their O3*n*-H hydroxyl groups. Inside the formed dimeric cavity, a host environment is created where a plethora of guest molecules, disposing an appropriate shape and conformation, can be included forming complexes both in solution and in solid state. The guest molecules are held inside the cavity not by covalent bonds but by weak interactions, like hydrogen bonds, van der Waals interactions, hydrophobic effects, etc. The formed inclusion compounds have wide applications, as the encapsulation of a molecule inside the CD cavity changes beneficially its physicochemical properties enhancing its water solubility and protecting it from hydrolysis, oxidation, photolysis, volatility and chemical reactivity. Moreover, due to the weak non-covalent interactions between the host and the guest molecules, these inclusion complexes are not fixed or permanent but rather in a dynamic equilibrium with their free components in solution, thus offering a slow release of the guests and sustaining their action. The *β*-CD complexes has been investigated by many methods like NMR, FTIR, UV-visible spectroscopy, spectrofluorometric studies, thermal studies, molecular dynamics and finally by X-ray crystallography.

In this chapter, the main aspects of the crystal structures of the dimeric *β*-CD inclusion complexes reported so far are reviewed. During the last four decades several crystals of dimeric *β*-CD complexes have been formed and analyzed by X-ray crystallography, disposing many common features like: the crystallographic space group (the great majority of them being P1, C2, $P2_1$ or $C222_1$), the crystal packing mode (the majority of them being Channel, Screw Channel, Intermediate, or Chess Board), isostructurality, polymorphism etc. The structural studies of these inclusion complexes have also revealed in detail the inclusion geometry, the guest's conformation, the encapsulation of water molecules inside the dimeric interface in the cases of Channel or Screw Channel packing modes usually being hydrogen bonded either with the guest or the host molecules, host-guest CH-π interactions, π-π interactions between the guests disposing phenyl or naphthyl moieties etc.

Chapter 9 – The growing interest in cyclodextrins is determined by their successful use as encapsulating materials and based on the ability of these natural cyclic oligosaccharides to form inclusion complexes. Application of inclusion compounds in medicine, cosmetology and food industry implies that they are affected by physiological medium of the living organisms, content of which is rather complicate. Inorganic salts are the main components of these biological liquids. As it is well known, salts are not always inert in the reaction medium, and equilibria occurring in saline solution are sensitive to the nature and concentration of the electrolytes. Therefore, the aim of this work was to reveal the influence of biologically active inorganic ions on complex formation of cyclodextrins with some aromatic carboxylic acids possessing pharmacological and biological activity.

Complex formation of α- and β-cyclodextrins with benzoic, nicotinic, p-aminobenzoic and m-aminobenzoic acids in water and saline solutions (KCl, KBr, KH_2PO_4 and K_2SO_4) was studied. Salt effects were discussed taking into account the possibility of nonspecific and specific interactions of the inorganic ions with cyclodextrins and acids. Possibility of ternary complex formation CD/acid/ion was also considered. Dependence of the manifestation of salt effects on the nature and properties of inorganic ions, the acid structure and size of cyclodextrin cavity was analyzed.

Chapter 10 – Intermolecular interactions between guest molecules and various sulfobutyl ether cyclodextrins (SBECD) differing in degree of substitution and number of anhydroglucose units were studied. For pharmaceutical purposes a special grade of SBECD has been authorized as excipient (i.e. complying US Pharmacopoeia 37) which is a highly specified product of strictly determined chemical properties. The aim of the study was to reveal how the functional properties would change by alteration from the pharmacopoeial specifications. For the characterization of the composition of this group of substituted cyclodextrins, three independent methods were compared. By NMR limited but undistorted structural information may be obtained, while capillary electrophoresis is able to provide the compositional profile, too. A HPLC method is presented by which information on the compositional profile as well as the quantity of typical impurities may be obtained. Several drug molecules were tested with a sulfobutyl ether cyclodextrin library using capillary electrophoresis. Based on the collected data it has been demonstrated that multiple factors may play significant role in the binding process ensuring some sort of selectivity amongst the members of the cyclodextrin library used. Besides the dimensions of the cavities, the steric effect of the relatively bulky sulfobutyl ether moieties as well as the charge born on the terminal of the substituent chains influence the strength of interaction. These data may serve as examples how the great variety of sulfobutylether cyclodextrins may be exploited in specific fields of application.

Chapter 11 – The dyes which have the large π system easily associate themselves. To inhibit their π–π stacking, host-guest interaction with cyclodextrins are often employed in water medium. Porphyrins and fullerenes are such typical dyes as guest molecules. Still there are many reports for their host-guest interactions, the structures of their inclusion complexes have been discussed only on the basis of the theoretical calculations and the spectral properties in solution, e.g. NMR and absorption spectra.

It is really surprising that, so far, nobody knows the real complex structure determined from single-crystal analysis. In 2011, the authors reported the first successful isolation of a dye inclusion single crystal (DISC) between trimethyl-β-cyclodextrin (TMβCD) and tetrapyridyl porphyrin, and analyzed its X-ray crystal structure. It revealed the real complex structure which has the prospective barrel shape bicapped complex.

Furthermore, the X-ray crystal structure provided several newly information in the molecular conformation. Recently, several crystal structures of dye inclusion complexes using cyclodextrin were reported from other groups. The crystallography of dye inclusion complex should invite the new field of functional materials designed by supramolecular chemistry.

In: Cyclodextrins
Editor: Francis G. Ramirez
ISBN: 978-1-63482-788-1

Chapter 1

CHEMICAL APPLICATION OF CYCLODEXTRINS IN POTENTIOMETRY: REVIEW

Joanna Lenik*
Faculty of Chemistry, Maria Curie-Skłodowska University,
Lublin, Poland

ABSTRACT

Cyclodextrins (CDs) are capable of interacting with a large variety of guest molecules to form inclusion complexes, for instance with drugs molecules through their lipophilic inner cavities and hydrophilic outer surface. One of the uses of cyclodextrins is their application in potentiometry. Potentiometric studies concerning the use of cyclodextrins as ionophores (in sensor development) are very new and not numerous.

In particular, lipophilic cyclodextrins (with an alkylated hydroxyl group in positions 2, 3, and 6), dihydroxypropyl-β-cyclodextrin, triacetyl-β-cyclodextrin, α-, β-, and γ-cyclodextrin derivatives and heptakis (2,3,6,-tri-O-methyl)-β-cyclodextrin, as well as (2-hydroxypropyl)-γ-cyclodextrin and many other can be incorporated in plasticized PVC membrane graphite electrodes as excellent ionophores.

The derivatives of β-cyclodextrins are used as excipients in oral or external drug formulation. β-CDs can be used to increase the solubility and dissolution rate of poorly water-soluble drugs and can affect drugs stability. β-CDs may be used to develop new dosage forms of drugs in pharmaceuticals or many other medicines. Many pharmaceutical products containing CD-s are already on the market. From a group of non-steroidal anti-inflammatory drugs (NSAIDs), the following are included: indomethacin, piroxicam, meloxicam, nimesulide, tiaprofenic. Therefore, it seemed fully justified to perform investigations into the effects of β-CDs on the performance of ion-selective electrodes which are used for determination of selected drugs.

This review discusses the development of ion-selective electrodes based on selected cyclodextrins and their derivatives as selective ionophores for various drugs and their applications in pharmaceutical research.

* E-mail: j.lenik@poczta.umcs.lublin.pl;j.lenik@outlook.com.

Cyclodextrins and Inclusion Complexes

Cyclodextrins belong to the group of cyclical oligosaccharides and they are formed from α-D-glucopyranose units linked together by α-(1,4) acetalbonds. The most common cyclodextrins are α-, β- and γ-cyclodextrins, which consist of 6-, 7-, and 8- glucose units respectively. α-, β- and γ-cyclodextrins occur in the form of white crystal powders of a delicate sweetish taste. They are hygroscopic substances, often soluble in water, and they take the shape of a toroid (conical frustum) in solutions. The wider edge of the toroid is formed by hydroxyls located next to secondary carbon atoms, whereas the narrower one is formed by hydroxyls located next to primary carbon atoms.

They form lipophilic inner cavities and hydrophilic outer surface. The diameter of the toroid's interior in the case of α-, β- and γ-cyclodextrins is 570, 780 and 950 pm respectively [1, 2]. They are capable of interacting with a large variety of guest molecules: organic compounds, ions, and even structures as small as radicals to form inclusion complexes; therefore they are widely used in different branches of industry [3, 4]. The CDs are used in the pharmaceutical, food, clothing, cosmetics, and chemical industries, and they have agricultural applications. In pharmaceutical industry they can be used to increase the solubility and dissolution of poorly water-soluble drugs. In comparison with α- and γ-cyclodextrins, β-cyclodextrin is poorly soluble in water, therefore it has the weakest solubilisation properties among the mentioned cyclodextrins. Many pharmaceutical products containing β-CD are already on the market. From a group of non-steroidal anti-inflammatory drugs (NSAIDs), the following are included: indomethacin, piroxicam, meloxicam, nimesulide, tiaprofenic acid, and diclofenac [5].

The formation of inclusion complexes made of cyclodextrins and other compounds is possible in gas, liquid, and solid states. These reactions are based on numerous weak intermolecular forces such as: hydrogen bonds, Van der Waals forces, dipole-dipole, and other dispersion forces [6]. The whole system aims to achieve the optimum energetic status. The inclusion may also occur as a result of a spatial match between the host and the guest. This is the so-called steric effect which depends on the guest's molecule size as well as on the presence of functional groups. Cyclodextrin hydrophobic gap constitutes the appropriate area to which nonpolar molecules of an applicable size (or only their fragments) can enter the host and form an inclusion complex [7]. The driving force behind the formation of inclusion complexes may also be the displacement of polar molecules - for example, water in hydrophobic gap can be displaced by more hydrophobic molecules of the guest present in the solution, and as a result a more stable state of lower energy is achieved.

The intensity of occurrence of steric effect and the effects related to other intermolecular interactions may depend on particular compounds. Because of the size of lipophilic cavity, β-cyclodextrins prefer bonds with aromatic and heterocyclic groups, α-cyclodextrins – with aromatic compounds, and γ-cyclodextrins – with steroids and macrocycles.

The modification of cyclodextrins, which is based on the alkylation of hydroxyls in position 2,3,6, results in more lipophilic character of cyclodextrins thanks to which they can be used as potential ionophores in membranes of ion-selective electrodes [8] and they can interact with other molecules of compounds, for example drugs. In the case of synthetic CD, intermolecular forces between the host and the guest occur and they include hydrogen bonding, hydrophobic forces, and Van der Waals forces. Natural cyclodextrins are mostly characterised by hydrogen bonding. The size of the guest's molecules and their geometry are the most important factors that have an impact on the intensity of bonding, though these interactions can be strengthened by modification of three groups of hydroxyls in glucose units [9].

In most inclusion complexes, mole ratio of the guest's molecule towards the host's is 1:1 though there are cases in which the ratio is 2:1, 1:2, or 2:2 [3].

The formation of the inclusion complexes by cyclodextrins can alter some physicochemical properties of quests molecules, that is electrochemical properties. They can be used in analytical techniques such as: potentiometry [8], polarography [10], and voltametry [11, 12]. Electroanalytical application of CD-s are based also on chiral molecular recognition and selective preconcentration of analyte on the electrode.

Methods Employed in the Study of Inclusion Complexes

The study of complex compounds with cyclodextrins can employ numerous techniques and methods [13]. The techniques that can be applied in the solid state include: thin layer chromatography (TLC) scanning electron microscopy (SEM), thermo-analytical methods, X-ray diffractiometry, X-ray structure analysis, wettability and dissolution tests, Infra-Red (I R) spectroscopy. A greater number of methods can be applied in the study concerning liquid state. These methods include: polarography, conductivity, polarimetry, solubility studies, nuclear magnetic, resonance (NMR), electron spin resonance (ESR), UV/VIS spectroscopy, fluorescence spectroscopy, circular dichroism (CD) spectroscopy, pH-potentiometric titration, microcalorimetry. Many scientific monographs discuss simultaneous usage of different methods. For example the study described in the article [15] presents the following methods: differential scanning calorimetry, infra-red, nuclear magnetic resonance, phase solubility, powder X-ray diffraction, thermogravimetric analysis, and ultraviolet methods. Other scientific articlesdescribe one or two research methods related to inclusion compounds, among others: X-ray structure analysis [16], nuclear magnetic, resonance (NMR) [17], UV, IR spectroscopy [18] fluorescence spectroscopy [19], chromatography methods [20], differential scanning calorimetry [21], and polarography [22].

One of the methods used in the study of inclusion complexes is potentiometry [23]. In comparison with the methods mentioned above, potentiometry is cheap, fast, and easy to perform. This method can also be used to determine the equilibrium constant of chemical synthesis of drug – cyclodextrin complexes. The chemical synthesis can be formulated as follows:

Drug+ βCD ↔Inclusion Complex

By establishing calibration curves of the electrode sensitive to a drug in the solution of a free drug and in the presence of cyclodextrin in a given concentration, we can observe the changes in the electrode's slope and its linear range. This confirms the formations of an inclusion complex. The equilibrium constant of the inclusion complex formation – K_x– can be determined on the basis of electromotive force (EMF) changes in the cells of reference and indicator electrodes and with the use of Hildebrand equation:

$v^{-1}=(K_s m_1)^{-1} + 1$
v– concentration of inclusion complex over its total concentration
m_1 – concentration of free drug

The chart showing the dependence between v^{-1} and m_1^{-1} refers to the complexation constant. The linear dependence observed in the data proves the stoichiometry of the complex with the ratio 1:1 [23]. ISE sensitive to N-(1,1,2,2-tetrahydroperfluorooctyl)-N,N,N-trimethylammoniumbromide (FTABr) (fluorinated surfactant) was used in another work [24] in order to determine the constant of complex formation with β-cyclodextrins. The electrode was immersed in fluorinated surfactant solution of constant concentration. The concentration of β-cyclodextrin was changing. The decrease of the electromotive force with simultaneous increase of cyclodextrin concentration was caused by the decrease of free ion concentration and formation of the complex. The interpretation of the experimental data used a calculation method that assumes the formation of a complex of 1:1 stoichiometry (β-CDFTABr) between β-CD and surfactant (FTABr). The final form of the equation used for the determination of the complex formation constant has been formulated as follows:

$$\lg (K) = \lg[FTABr]_o - [FTABr] / [\beta\text{-CD}]_o - [FTABr]_o + [FTABr]$$
$$-\lg ([FTABr]) = Y - \lg ([FTABr])$$

The value [FTABr] was determined on the basis of the experimental data of the measurement with the use of the following dependency:

$$[FTABr] = 10^{\ E-\text{const./slope}}$$

Association constant K for β-cyclodextrin/FTABr complex is evaluated to be ~1.26×10^5.

One of the examples of ion-selective electrode (ISE) application to determine complexes stability constants is a selective sensor sensitive to the drug called bupivacaine [25] that has cation characteristics. The electrode has a polymer PVC membrane plasticised with dibutylphthalate that contains bupivacaine phosphotungstate as an active component. The electrode was immersed in solutions of bupivacaine hydrochloride with cyclodextrin and the potential difference was determined on the basis of the concentration of the bupivacaine hydrochloride cation.

The measured potentiometric data processed by both a linear and nonlinear regression corroborated the formation of weak 1:1 bupivacaine cation-cyclodextrin complexes and the corresponding complexation constants K (11) approximately 50-155 M(-1) were evaluated by the non-linear least-squares method.

Among the studied compounds with cyclodextrin features that may potentially form complexes with bupivacaine, the following should be mentioned: alpha-cyclodextrin (alpha-CD), beta-cyclodextrin (beta-CD), as well as their random-substituted derivatives hydroxypropyl-alpha-cyclodextrin (HP-alpha-CD), hydroxypropyl-beta cyclodextrin (HP-beta-CD) and methyl-beta-cyclodextrin (M-beta-CD). The highest K was exhibited by M-beta-CD followed by alpha-CD.

Such studies have their practical applications as they allow to modify the composition of a given drug and to replace bupivacaine hydrochloride with an appropriate cyclodextrin complex.

The formation of inclusion complexes of carbisocaine cation BH+ with α-cyclodextrin (α-CD), β-cyclodextrin (β-CD) and randomly substituted methyl (M-β-CD) and hydroxypropyl (HP-β-CD) has been studied with potentiometric method, with the use of a cation-selective electrode [26]. The potentiometric data corroborated formation of the carbisocaine-cyclodextrin complexes (BH+)CD (1:1) with all the cyclodextrins and the respective complexation constants K11 were estimated using a modified Scatchard method.

Potentiometry - Ion-Selective Electrodes

One of the solutions that allows for effective cost reduction is the use of a well-known potentiometric technique. This technique, which makes use of ion-selective electrodes (ISE), has a competitive advantage over other expensive and complicated methods because it is relatively simple in use, cheap, and at the same time, it enables one to determine the examined compound in a selective, sensitive and precise way. The sensors for determination of selected drugs, developed within the undertaken research project, will help expand the knowledge in the field of chemistry and pharmacy, as far as the possibilities of their use and application are concerned. Classical potentiometry with the use of ion-selective electrodes is an analytical procedure which may be implemented in the future as one of cheap basic techniques in routine quality control of drugs (for example in pharmaceutical laboratories) for quantitative determination of the active medicinal substance.

Potentiometric methods are extensively used in pharmaceutical enantioselective analyses.

In comparison with other methods, potentiometry has great advantages, such as: accuracy, simplicity, non-time consuming nature, and it is easily carried out in the laboratory at a relatively low cost; for example, in order to produce approximately 5-7 electrodes, only 1 mg of ionophore, which is relatively expensive, is necessary.

Other materials, such as casing, electrical conductors, and other membrane components are relatively cheap.

Optimisation of the Membrane Phase

The optimisation of the membrane composition of ion-selective electrodes is a very important and, at the same time, really interesting object of study. The selection of an appropriate ionophore (ion carrier), solvent (plasticizer) and the membrane matrix, undoubtedly influences the analytical parameters of the electrode.

Over the last years, many studies have discussed the use of macrocycles, such as cyclodextrins, as ionophores. Cyclodextrins interaction with particular molecules of drugs is very unique and depends on their spatial structure. As far as these mostly inert compounds are concerned, the stability of the complexes formed in the membrane depends on complexation constants and their stoichiometry.

Stability constants should being enough so as to ensure appropriate stability and lability of complexes. If the values are too high, co-extraction takes place and the electrode loses its selectivity. The reaction rate of complex formation is also important – if the reaction is dominated by diffusion, the electrode will show a longer response time, or even non-theoretical slope of the characteristics [27].

One of the common additions to the membrane is, apart from the inert ionophore, lipophilic salt. The salt provides electrical inertness of the membrane and causes many favourable changes in the membrane parameters.

IONOPHORES WITH CYCLODEXTRINS PROPERTIES

Most drugs have cation properties because of the hydrogenation of the groups (and the formation of hydrochlorides). Drugs from this group are more common than the drugs belonging to anionic group; these are for example barbiturates, and non-steroidal analgesics. That is why the literature on the subject focuses more often on anion cyclodextrins than on cation cyclodextrins used in thin-layer chromatography (TLC), such as capillary electrophoresis. The analysis of pharmaceutical compounds that employs TLC (electrophoresis or chromatography) allows the researchers to predict, select and implement a particular compound, for example cyclodextrin, as an ionophore in the membrane of ion-selective electrodes.

Numerous studies focus on the ion-selective electrodes used for determining chiral compounds [28, 29].

Anionic cyclodextrin derivatives include the following compounds that function as sensitive selectors of cationic drugs:

- Sulfated CDs (S-β-CD), sulfobutylether-β-CD (SBE-β-CD), SEE-β-CD (sulfoetylether-β-CD), SPE-β-CD (sulfopropylether-β-CD),2,3-dimethyl-6-sulfato-β-CD DMS-β-CD),
- carboxylated CDs: carboxymethyl-β-CD (CM-β-CD), carboxyethyl-β-CD (CE-β-CD), DAS-β-CD (2,3-di-O-acetyl- 6-sulfato-β-CD),
- phosphated-CDs (both γ-and β-CD),
- SUCC-β-CD (succinylated-β-CD) TGA-β-CD (6-(2-thioglycolic acid)-6-deoxy-β-CD).

Cationic cyclodextrin derivatives used in selective separation of carboxylic acids or amino acids usually include mono- and poly-substituted amino derivatives of CDs:

AEA-β-CD (6-β-aminoethylamino-6-deoxy-β-CD), DMA-β-CD (6,6-dimethylamino-6-deoxy-β-CD), ED-β-CD (6-ethylenediamino-6-deoxy-β-CD), HA-β-CD (6-hexylamino-6-deoxy-β-CD), HEA-β-CD (6-hydroxyethylamino-6-deoxy-β-CD), Hm-β-CD (6-N-histamino-6-deoxy-β-CD), Im-β-CD ([4-(2-aminoethyl)imidazolyl]-6-deoxy-β-CD), MA-β-CD (6-methylamino-deoxy-β-CD), MEA-β-CD (6-methoxyethylamino-6-deoxy-β-CD), NH2-β-CD (6-amino-6-deoxy-β-CD), QA-β-CD (2-hydroxypropyl-trimethyl ammonium-β-CD).

The most widely used cationic CD is 2-hydroxypropyltrimethylammonio-β-CD, which is the only commercially available cationic(quaternary ammonium substituted) β-CD [28].

Popular cyclodextrins most commonly used in plasticized PVC membranes include:

- native cyclodextrins: α-, β- and γ-cyclodextrins
- randomly substituted cyclodextrins: heptakis (2,6-di-O-methyl)-β-CD, 2,6-didodecyl β-cyclodextrin, 2,6-didodecyl-α-cyclodextrin, 2-hydroxypropyl-β-cyclodextrin, triacetyl-β-cyclodextrin
- persubstitued cyclodextrins: heptakis (2,3,6-tri-O-methyl)-β-CD, heptakis(2,3,6-tri-O-benzoyl)-β-cyclodextrin, 2,3,6-tri-O-ethyl-β-cyclodextrin,

Structural formulas of these compounds are presented in Figure.1

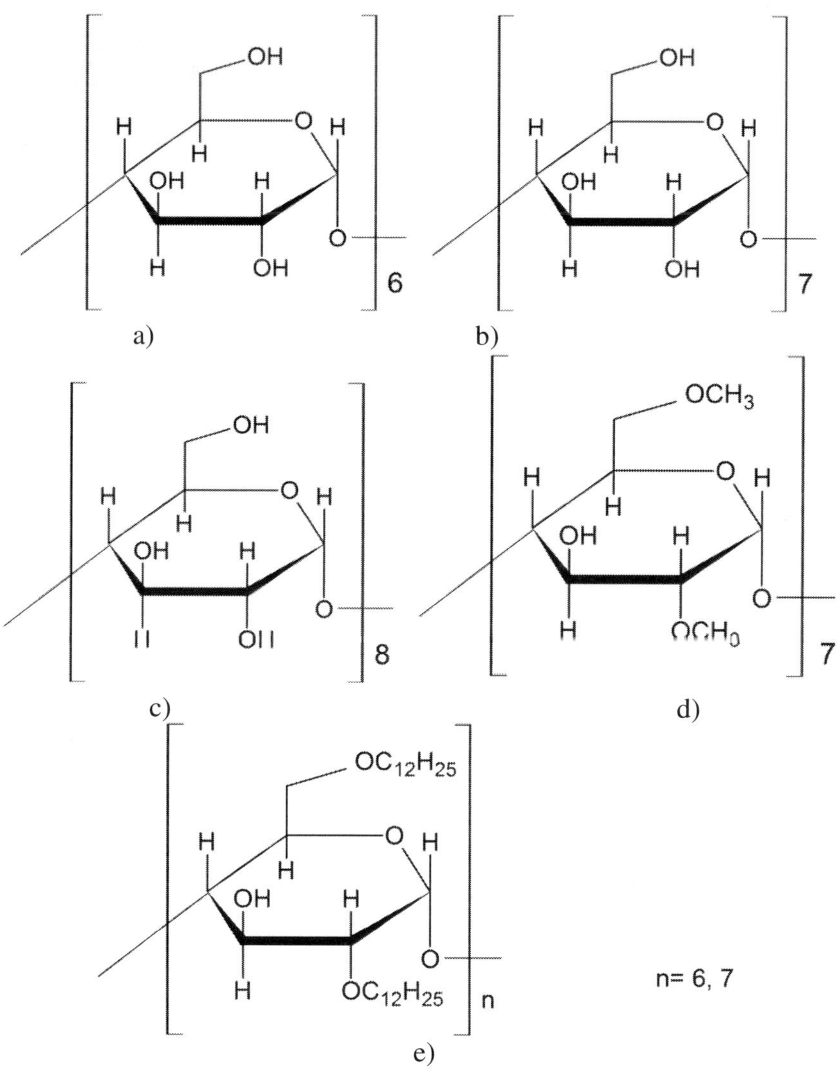

Figure 1. (Continued).

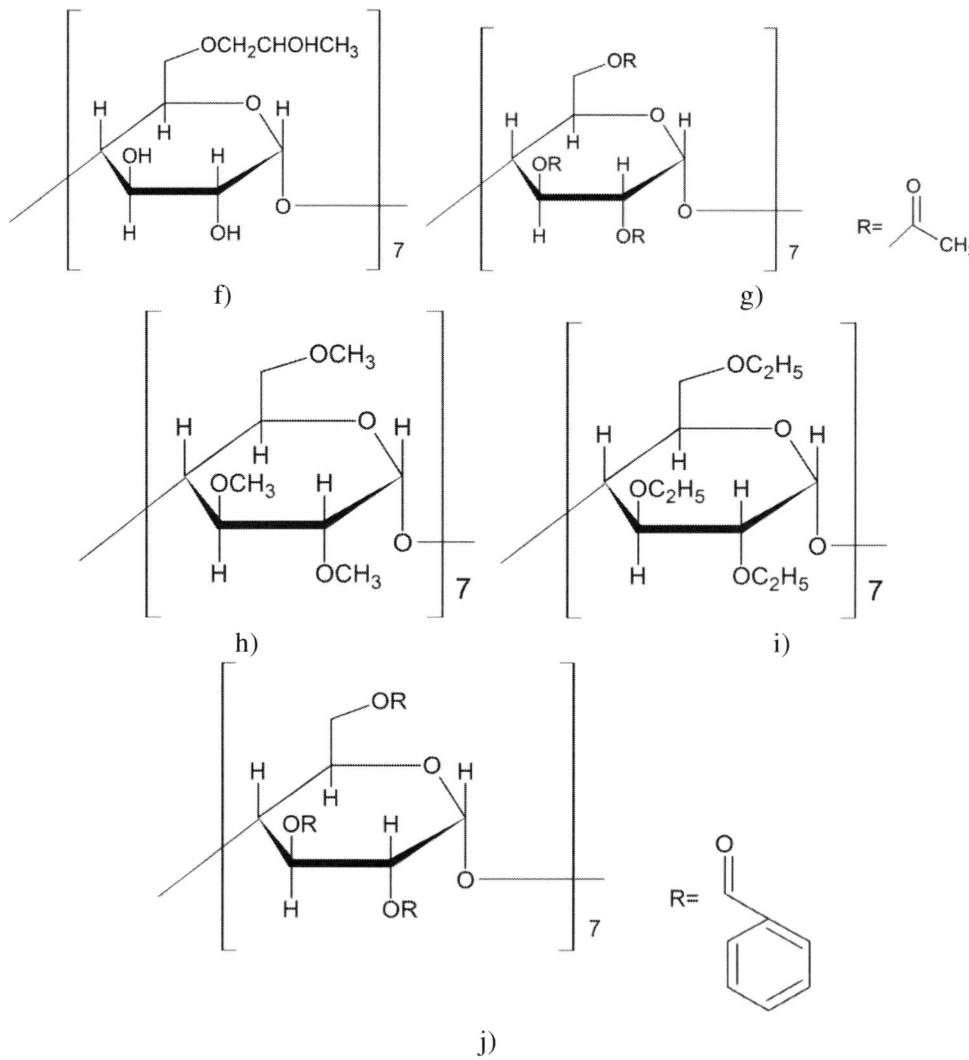

Figure 1.

The literature on the subject offers not many publications that focus on the use of CDs in ion-selective electrodes. The accessible monographs that are worth mentioning include the works by Mohammad Reza Ganjal et al. [30], E. L. Izake [31], and Vinod K. Gupta, et al. [32].

Ion-Selective Electrodes Used in the Analysis of Drug Compounds that Contain Cyclodextrins – Literature on the Subject (1990-2014)

This subchapter presents the review of the literature on ion-selective electrodes sensitive to drug substances which may form inclusion complexes with cyclodextrins. Particular monographs focus on properties of the electrodes that contain cyclodextrins in the membrane (other electrodes with different active compounds in the membrane are not included).

Special attention has been paid to the composition of membranes (since cyclodextrins are found in various membrane composites – PVC, graphite).

The subchapter concludes with information on potential application of such electrodes in different fields.

Electrodes Sensitive to alkyl and Aryl Ammonium Ions of Pharmaceutical Significance Based on α-,β-, γ-cyclodextrins

The first ion-selective electrodes containing cyclodextrins used as chiral selectors were introduced in the 1990s. The research group of D. Parker, R. Kataky and others prepared the electrodes sensitive to ephedrine derivatives, propranolol and amphetamine. Their work described preparation and properties of ephedrine ion-selective electrode (PVC type) based on peroctylated α-cyclodextrin with o-NPOE and BBPA and potassium tetrakis(p-chlorophenyl)borate [33, 34]. Polymer membranes were placed in IS Philips corpuses (561), and hydrochloride ephedrine (Eph·HCl) and NH_4Cl were used as the electrodes' inner solutions of the electrodes.

The electrodes were examined in pure enantiomer solutions: (+), (-) hydrochloride ephedrine, pseudoephedrine, norephedrine, and/or in racemic mixture (±), hydrochloride ephedrine, pseudoephedrine, and norephedrine. The selectivity was studied on the basis of the determined selectivity ratios with the use of mixed solutions method.

The electrode containing o-nitrooctyphenyl ether (o-NPOE) in the membrane behaves in a slightly different way. In a pure dextrorotatory enantiomer solution and with the addition of interfering ion, the electrode exhibits linear response with Nernstian slope of 59.0 mV decade^{-1}, and limit of detection –log c 4.80. The changes of the electrode sensitivity in enantiomer solution (-) depend on the following aspects: concentration, time of measurement, as well as on the presence of added interfering ions. Taking into consideration the fact that o-nitrophenol forms inclusion complexes with a-cyclodextrins in water solutions, the authors concluded that the changes in electrode sensitivity might be attributed to the interaction of NPOE with the molecules of peroctylated CD. Such interactions will not take place in the case of ephedrine hydrochloride enantiomer (+) because of the small differences in enthalpies of both enantiomers formation. In this particular case the aggregation of cyclodextrin and plasticizer seems to be the most probable and it depends on ionic strength and the concentration of the solution.

The study [35] by P. S. Bates et al. discussed the electrodes sensitive to drug substances, such as acetylcholine chloride, dopamine hydrochloride, and a substance of surfactant properties– myristyltrimethylammonium bromide. The authors used functionalized, lipophilic α-, β- and γ-cyclodextrins as membrane active substances. They also studied the electrode sensitivity to simple alkyl- and aryl-ammonium ions.

The obtained membranes containing peroctylated β-cyclodextrin with o-NPOE as plasticiser that have very good sensitivity (60 mV decade^{-1}), a limit of detection, -log a of 5.8, and selectivity for alkylammonium ions, (-logK for Na^+, 3.8 and for K^+, 3.2, and for NH_4^+, 3.5) were treated as optimal membranes.

As far as arylammonium ions are concerned, both α- and β-cyclodextrins dissolved in o-NPOE or BBPA exhibit good sensitivity and minimal interferences towards alkali metal ions.

The electrodes based on peracetylated α- and β-CDs exhibit Nernstian response (60-61 mV decade^{-1}) to catecholamine dopamine ion, and the limit of detection –log c =5.0. The selectivity of the electrodes is very good in comparison with Ca^{2+} and Mg^{2+} ions (-log K=3.3, 3.2). Slightly worse selectivity is exhibited for K^+ ions (-log K =1.5) and Na^+ ions (-log K=1.8). In this case, the type of plasticizer does not affect the parameters of the obtained electrodes because the aryl part is probably included inside the cyclodextrin cavity.

In the above-mentioned work, the authors also compared the so-called 'empty' electrodes which did not contain any cyclodextrin. These electrodes exhibited longer conditioning time to obtain stable potential (approx. 60 hours) in comparison to the electrodes with cyclodextrins that needed only 12 hours to achieve the same result. The electrodes slope was always sub-Nernstian, the limit of detection approx. –log c=2.0. Placing cyclodextrin in the membrane significantly improved their functionality and minimized interferences.

The study discussed also other methods used in complex formation, for instance nuclear magnetic resonance measurement, and electroscopic mass spectrometry.

Applying a similar measuring system, that is the same electrode construction and study methods, the same team developed other selective electrodes sensitive to different compounds using various cyclodextrins in that process, for example, the electrode sensitive to guanidine including 2,6-didodecyl-β-cyclodextrin [36] or 2,6-didodecyl-α- and -β-cyclodextrins, 2,3,6-tri-O-ethyl-β-cyclodextrin [37], creatinine hydrochloride, and acetylcholine chloride including peroctylated and 2,6 didodecyl β-CD [38].

More electrodes sensitive to local anaesthetics (procaine, prilocaine, lidocaine and bupivacaine hydrochlorides) were developed [39]. In the membrane phase, the electrodes contain inert cyclodextrins such as 2,3,6-trioctyl and 2,6-didodecyl-β-cyclodextrins and they exhibit very good analytical parameters, theoretical slope, very low limit of detection -log c=5.0, and advantageous selection ratios for Na^+, Ca^{2+} K^+ ions (–log K > 4.0). Moreover, in the presence of organic interferents, such as serum albumin, α_1 acid glycoprotein, Nernstian slope is retained thanks to a more biocompatible polyurethane membrane (Tecoflex SG80) than in the case of PVC.

The authors of the work mentioned above, developed electrodes sensitive to tricyclic antidepressants imipramine, desipramine and trimipramine as their hydrochloride salts [40] with the use of alkylated α-, β- and γ-cyclodextrins. Nernstian response and excellent potentiometric limits of detection were observed with each of the three examined alkylated cyclodextrins. Interference from 'serum' levels of Na^+, K^+ and Ca^{2+} ions was very low. Interference from organic molecules such as glycine, L-histidine and nicotinamide was also low in most cases.

The Electrodes Sensitive to Organic Amines of Pharmaceutical Significance Based on Derivative of β-cyclodextrins

Ion-selective electrodes with polymer membrane containing active compounds in the form of a long alkyl chain derivative of β-cyclodextrin(6A,6B,6C,6D,6E,6F,6G-hepta-S-dodecyl-6A,6B,6C,6D,6E,6F,6G-heptathio-β-cyclodextrin tetradecaacetate [41] were used to determine organic amines. Several membrane solvents were used for the membrane processing, among which 2-fluoro-2′-nitrodiphenyl ether (FNDPE) was found to be by far the most effective for potentiometric responses by the cyclodextrin host. The changes of potential on the electrodes were determined in protonic solutions of organic amines of pH = 5.0 with different types of nonpolar regions. The magnitude of the guest-induced potential change was as follows: 2-phenylethylamine > 1-adamantanamine >benzylamine>cyclohexanamine> dopamine. The electrodes were also responsive to the positional isomers of di- and trimethoxybenzylamines, whereas the reaction to mono-methoxybenzylamines isomers was slight. The differences in selectivity of polymer electrodes and electrodes with liquid membrane containing no particular sensory element were also observed. When potassium tetrakis(p-chlorophenyl)borate was added to the membrane as anionic carrier, no changes were observed regarding the electrode behaviour. Such electrode behaviour may suggest the ability (geometric match) and accessibility of the complexes forming between the molecules of cyclodextrin and amines, which are stable in the near membrane layer.

Electrodes Sensitive to Heavy Ions Metals Based on Cyclodextrins

The next work [42] discussed the use of electrodes operating in the electrochemical cell. The technique of chemical grafting of cyclodextrin molecules was used in a solid polysiloxane gel matrix. The use of FTIR, ^{13}C and ^{29}Si solid state NMR methods confirmed the reaction between cyclodextrin and polymethylhydrosiloxane (PMHS).

In order to compare the parameters, the electrodes based on PVC were prepared and as a result, the membranes sensitive to heavy ions metals, such as Pb with β-CD (sensitivity 29 mV $decade^{-1}$) and Cd with α-CD (sensitivity30 mV $decade^{-1}$) were obtained. The membrane sensitivity to cadmium was confirmed by the measurement of capacitance.

Lidocaine Modified Ion-Selective Electrodes (ISEs) Based on 2,6-didodecyl-β-cyclodextrin

The authors of the above mentioned work prepared modified ion-selective electrodes for anaesthetic lidocaine hydrochloride (lignocaine), using 2,6-didodecyl-β-cyclodextrin (2,6ddβCD) as an ionophore. They were used in discrete solutions and inflow injection analysis manifold for monitoring the drug in the presence of endogenous cations and proteins.

Two types of membrane matrices were used: poly (vinyl chloride) (PVC) or polyurethane (Tecoflex SG 80). The calibration curves obtained in the presence of bovine serum albumin (BSA), α1 acid glycoprotein (AAG) and human serum (HS) demonstrated that the Tecoflex-based membrane matrices are preferable to PVC as the drift induced by proteins on baseline lines and peak height reproducibility was considerably reduced in the former.

A very high rate of selectivity -log K_{ij}^{pot}) (overall) ≥ 4.0) for ions from 'serum' levels of sodium, potassium and calcium (145 mmol dm^{-3} Na^+, 1.26 mmol dm^{-3} Ca^{2+}, 4.30 mmol dm^{-3} K^+) was reached. The major organic interferents were molecules of similar size and structure, which caused reduced slope and drift in baseline potentials at equimolar concentration levels. A reduction in interferent concentration by a factor of 10 negated these effects.

Ammonium (Onium) Ions, Choline, Acetylcholine, and Acetyl-β-methylcholine Potentiometric Sensor Based on Cyclodextrins

The next study [44] used novel cyclodextrin-based ISEs to determine activity coefficients of some onium chlorides in aqueous solution. Lipophilic cyclodextrins, rendered by alkylation and dissolved in an appropriate polymer membrane were used as ionophores. The electrodes showed sensitivity to ammonium compounds, that is choline, acetylcholine, and acetyl-β-methylcholine. Potentiometric sensors based on cyclodextrins were used to determine activity coefficients of individual solutions choosing one solution as reference and using a theoretical model (for example Pitzer equations).

This study was one of the first that employed potentiometry to measure the activity coefficients, which was important as far as the biological approach to ions is concerned.

Diclofenac Potentiometric Electrode Based on α-cyclodextrin

A new ion-selective electrode with polymer membrane phase and diclofenac function containing α-cyclodextrin as an electrode active substance was prepared [45]. The polymer membrane consisted of: 1.2% w/w of α-cyclodextrin, 0.4% w/w tetradodecylammonium bromide, 65.6% w/w 2-nitrophenyloctyl ether, and 32.8% w/w of PVC. The prepared electrode was used to measure the concentration of diclofenac in a classic potentiometric system and in an automated analytical system, based on the concept of sequential injection analysis, providing real-time assessment of results quality.

The general electrode characteristics presented a wide range of linear response $4x10^{-6}$ – $1x10^{-2}$ M with an anionic slope of 57.5 mV $decade^{-1}$, a practical limit of detection 1 $x10^{-6}$ M, good reproducibility 0.4 mV per day, and rapid response time (<20 s). The pH range of that electrode response is practically constant – between 8.2-12.7. Under flow conditions the characteristics of electrode are similar: linear working ranges from 5×10^{-6} to 1×10^{-2} M, slope of characteristic 56.6 mV decade $^{-1}$, a practical limit of detection $2x10^{-6}$ M, and quadratic correlation coefficient 0.9992. Relative standard errors of 0.93 and 0.53% were obtained after 10 consecutive injections of a 5×10^{-5} M diclofenac standard solution. The sampling rate was about 33 samples h^{-1}.

The selectivity of the prepared electrodes is very good. The selectivity coefficients for ions (-log K) are the following: citrate, phosphate, tartrate ~4.0, borate ~3.5, aminobenzoate~3.0, chloride, acetate~2.5, aminosalicilate, nitrate~2.0, salicilate~1.5. For pharmaceutical excipients such as: lactose, glucose, fructose, mannitol, starch, carboxymethyl cellulose no interferences in quantities up to 100-fold excess were detected. The prepared electrode was used to determine active substance in different diclofenac products and the results obtained were in accordance with the provided reference method for a 95% confidence level.

Propranolol Enantiomers Potentiometric Sensor Based on Lipophilic 2-hydroxy-3-trimethylammoniopropyl-β-cyclodextrin

The authors of the next study [46] focused on the preparation of an electrode sensitive to enantiomers of ropranolol containing lipophilic cyclodextrin 2-hydroxy-3-trimethylammoniopropyl-β-cyclodextrin as chiral selector. They used the classic type of ion-selective electrode with PVC membrane. Different plasticizers, (o-NPOE) and (BBPA), and inner filling solutions were studied. The membrane composition, the measurement conditions for both electrodes named A and B respectively, as well as the parameters and selectivity coefficients are presented in Table 1.

Table 1. Composition of propranolol electrodes and their parameters

Electrode	A	B
Plasticizer	o-NPOE	BBPA
Inner solution	1.0×10^{-3} M R(+)-enantiomer	1.0×10^{-3} M NH_4Cl
Conditioning solution	1.0×10^{-3} M R(+)-enantiomer	1.0×10^{-3} M NH_4Cl
Slope [mV decade^{-1}]	59.1	60.0
Linear range [M]	$2.2\times10^{-5} – 1.0\times10^{-2}$	$1.0\times10^{-5} – 1.0\times10^{-2}$
Limit of detection [-logc]	4.92	5.28
Response time [s]	30	30
Selectivity coefficient log K Na^+, K^+, Ca^{2+}, Zn^{2+}	2.13 2.14 3.62 3.08	2.18 2.38 4.01 3.78
racemic salbutamol sulphate	2.02	1.91
R-sabutamolsulfate	1.73	1.91

The enantioselectivity coefficient ($K\pm^{pot}$), determined in 1.0×10^{-3} M appropriate enantiomer solutions, was 0.20.

S-Perindopril Ion Selective Electrode Based on α-, β- and γ-cyclodextrins

Three S-perindopril selective carbon paste electrodes were investigated with α-, β- and γ-cyclodextrins used as ionophore, and paraffin oil and graphite powder as electrode matrices [47].

Response characteristics of enantioselective, potentiometric membrane electrodes with α-, β- and γ-cyclodextrins were obtained. The slope (mV decade^{-1}) intercept, E^o (mV) linear concentration range (M) detection limit (M) for these electrodes were: 56.64 −106.79 10^{-7} to 10^{-3} 10^{-8} (for α-CD); 50.71 −101.00 10^{-7} to 10^{-3} 1.5×10^{-8} (for β-CD), 63.00 −119.33 10^{-7} to 10^{-3} 10 (for γ-CD). The responses of the electrodes do not depend on the pH ranges 4.30–7.1 (α-CD), 3.78–8.1 (β-CD) and 3.0–5.0 (γ-CD).

Mixed solution method was used for the investigation of selectivity. Enantioselectivity was checked against R-perindopril and d-proline. The enantioselectivity decreases with the increasing cavity size, and the best enantioselectivity was shown by α-CD based electrode. Other conclusions concerning electrode selectivity can be formulated in the following way:

Polyvinylpyrolidone (PVP) did not interfere. L-proline is the only interferent for the three electrodes. Inorganic ions such as Na^+,K^+ and Ca^{2+} did not interfere with the analysis of S-perindopril. The electrodes were used to assay S-perindopril in pills. The remaining electrodes had the average values of recoveries (n = 10) of 100 ± 0.02% (I), 99.70 ± 0.03% (II) and 98.50 ± 0.08% (III).

L-proline ion Selective Electrode Based on α-, β- and γ-cyclodextrins

The authors of the work [48] prepared a similarly constructed enantiometric sensor for assaying l-proline. The research focused on various α-, β- and γ-cyclodextrins. The best parameters were obtained by the electrode with γ-cyclodextrin in the membrane: slope of characteristic −59.50 (mV decade^{-1}), intercept 586.39, (mV) linear concentration range 10^{-8}–10^{-3} (M), and detection limit 10^{-10} (M). The enantioselectivity of the electrodes was investigated over d-proline, using mixed solutions method. The values of pK^{pot} were 2.96, 2.52 and 2.05 for (α-CD), (β-CD) and (γ-CD), respectively. The enantioselectivity decreases with the increasing cavity size, the best enantioselectivity was shown by α-cyclodextrin based electrode. Inorganic ions such as Na^+, K^+ and Ca^{2+} did not interfere with the analysis of l-proline.

The electrodes can be successfully used for the assay of l-proline in the presence of d-proline. The enantioselectivity is good for all the three electrodes proposed, the best response is recorded when α-cyclodextrin is used as a chiral selector.

L- and D-2-Hydroxyglutaric Acid Ion Selective Electrode Based on α-, β- and γ-cyclodextrins

Three carbon paste enantioselective potentiometric membrane electrodes (EPMEs) based on immobilization of β-, γ-cyclodextrin (CD) or 2-hydroxy-3- trimethylammoniopropyl-β-cyclodextrin (as chloride salt) (β-CD-derivative) were prepared [49]. The optimal electrodes were prepared on the basis of β-CD and β-CD-derivative for the determination of L-2-hydroxyglutaric acid (L-2-HGA), in the concentration range of 10^{-8}- 10^{-6} and 10^{-7}-10^{-5} M. The electrode that proved more useful in assaying D-2-hydroxyglutaric acid (D-2-HGA) was the electrode based on γ-CD, whose measurement range was 10^{-6}- 10^{-4} M. The β-CD showed the lowest detection limit (1×10^{-9} M). The enantioselectivity and selectivity of the proposed electrodes for the assay of L-2-HGA and D-2-HGA, respectively, were determined over D-2-HGA/L-2-HGA, creatine, and creatinine. The proposed EPMEs can be applied to the enantioanalysis of 2-hydroxyglutaric acid in urine samples.

Tobramycin Potentiometric Electrode Based on β-cyclodextrin

Another polymeric electrode for determination of the amino glycosidic antibiotic, tobramycin was described [50]. The polymer membrane contained β-cyclodextrin as ionophore carboxylated poly (vinyl chloride) as matrix polymer, tetrakis-fluorophenyl borate as fixed anionic site and NPOE as plasticizer. The electrode shows 10^{-2} to 10^{-5} M tobramycin with cationic slope of mV decade^{-1} and accuracy of concentration range of 99.8 ± 2.14%. The proposed method was successfully applied to the determination of tobramycin uptake, by 100 μl E. coli suspension in the presence of up to 10 μg ml^{-1} of EDTA solution, with average recovery of 99.3 ± 2.77%.

Vesamicol Potentiometric Electrode Based on α-, β-, γ- and 2-hydroxy-3-trimethylammoniopropyl-β-cyclodextrin

The electrodes described in the article [51] were based on cyclodextrins (α-, β-, γ-) and 2-hydroxy-3-trimethylammoniopropyl-β-cyclodextrin (as chlorine salt).

The linear concentration ranges for the proposed EPMEs based on cyclodextrins were from 10^{-9} to 10^{-7}mol L^{-1} for those electrodes based on α-, β- and β-derivative-cyclodextrins with detection limits of 5.7×10^{-10}, 2.42×10^{-10} and 4.77×10^{-10} M, respectively. Relatively less favourable parameters were obtained by γ-cyclodextrin-based electrode which is characterized by linear concentration range 10^{-5} to 10^{-2} M with detection limit of 7.86×10^{-6} M.

The electrodes show near-Nernstian slopes of 52-59 mV decade^{-1} of l-vesamicol in the range of concentration of 4×10^{-6} - 10^{-2} M. The electrodes displayed near-Nernstian slopes of 52-59 mV decade^{-1}of l-vesamicol concentration. The electrodes were used for the direct assay of l-vesamicol in serum samples and proved to be reliable, reproducible and stable. The surface of the electrodes can be simply renewed by polishing on an alumina paper to obtain a fresh surface ready for use in a new test.

Hyoscinebutylbromide Potentiometric Electrode Based on β-cyclodextrin

Another sensor prepared was the potentiometric sensor for determining cationic compound - hyoscinebutylbromide [52]. The polymer membrane consisted of the following compounds in appropriate ratio: 0.04g β-CD, 0.4g NPOE, 0.01 g ammonium reineckate, and 0.18g PVC. The performance characteristic of this sensor was evaluated according to IUPAC recommendations, and revealed a fast, stable and linear response for hyoscine over the concentration range of 10^{-6}-10^{-2}M with slope of 51.51 mV decade^{-1}. Correlation coefficient was R=0.9992. The electrode is characterized by a very low limit of quantitation 4.9 10^{-7} M, the response time of 20-30 s, and pH range of 7-9. The duration of constant performance of the electrode is 5-6 weeks.

The direct potentiometric determination of hyoscinebutylbromide using the proposed sensors gave average recoveries % of 99.87 ± 1.39. The sensor is used for determination of hyoscinebutylbromide in laboratory prepared mixtures, pharmaceutical formulations in combination with ketoprofen and in plasma. A very thorough validation of the method was conducted and it consisted in the measurements of the measurement range. Lower limit of detection, accuracy, (recovery), trueness, precision, repeatability (inter- and intra-day variability), linearity and sensitivity, show suitability of the proposed sensors for the use in the quality control assessment of hyoscinebutylbromide. The developed method was found to be simple, accurate and precise when compared with a reported HPLC method.

Midazolam and Diazepam Potentiometric Electrode Based on α- β-, γ-cyclodextrin

A solid-contact ion selective electrode based on α-cyclodextrin, β-cyclodextrin and (2-hydroxypropyl)-γ-cyclodextrin as ionophores, 2-fluorophenyl 2-nitrophenyl ether as plasticizer, PVC-COOH and potassium tetrakis (p-chlorophenyl) borate as ionic additive was investigated for its response to midazolam and diazepam [53].

Also implemented were electrodes with membranes without cyclodextrins and with an additive without cyclodextrins. Solid-contact electrodes were prepared in the process of dispensing the membrane cocktail on the conductive element (epoxy or powdered graphite) and then allowing them to dry at room temperature for two days.

The near Nernstian slope of characteristic of 61.9 ± 1.3 mV decade^{-1}, a lower limit of linear response LLLR of 5.7 ± 2.7 × 10-4 g L^{-1} and pH range of 2.6-5.4 were obtained for midazolam electrodes. These parameters were achieved for electrodes with optimal ionophore, that is β-CD with the lipophilic additive (1:4 ratio). Electrodes with α-CD, obtained higher limit of detection.

Moreover, the electrodes with the same exchanger showed irreproducible potential. The addition of inert cyclodextrin improved also the shape of the curve – from the S-shaped to a more linear – as the flow of ions to the centre of the membrane was limited. The lifetime of the proposed midazolam electrode was longer than a year without any particular conditioning procedure.

Diazepam electrodes containing α-CD and β-CD respectively, obtained reproducible super-Nernstian slope of characteristic. High lipophilicity of diazepam allowed for the observation that the size of the inclusive cavity and the 2-hydroxypropyl)-γ-cyclodextrin have influence on the process, as it improves the characteristics of the electrode and the inclination of 67.6 ± 3.0 mV decade^{-1}, LLLR 4.9 ± 1.5 × 10-2 g L^{-1} and 1.9-2.7 pH units, respectively.

Miniature electrodes were used in sequential-injection lab-on-valve (LOV) flow system. Their lifespan was shortened and amounted to 15 days for the electrodes working continuously 8 hours a day and the electrodes working non stop.

The optimized flow conditions were achieved for sample injection volumes of 20 μL propelled towards the detection cell at the flow rate of 16 μL s^{-1} during 80 s.

The selectivity coefficients for optimal electrodes towards inorganic species in pharmaceutical formulations were established with the use of MPM method and the values were less than 0.3. The drugs in pharmaceutical samples were determined with the use of injection analysis and the results were satisfactory: recovery 97-104% for diazepam, and 100.7% for midazolam.

Vitamin B1 and Vitamin B6 Potentiometric Sensor Based on β-cyclodextrins

This study focuses on the construction, and analytical parameters of potentiometric sensors sensitive to vitamin B1 and vitamin B6 [54]. The three solid contact electrodes were produced using β-cyclodextrins as ionophores in a carboxylated poly(vinyl chloride) support matrix. The first sensor for vitamin B1 contained 0.04g of the ionophore (2-hydroxypropyl-β-cyclodextrin) 0.4g of mediator solvent (FNDPE) and 0.18g PVC-COOH. The sensors for vitamin B6, contained in the membrane either β-cyclodextrin (sensor 2) or more lipophilic 2-

hydroxypropyl –β-cyclodextrin (sensor 3) dissolved in components of the same quality and quantity as the components of membrane 1.They also contained addition of lipophilic salt – potassium tetrakis(p-chlorophenyl) borate. The most important conclusions concerning the characteristics of the study include:

1. Sensor 1 parameters are least favourable, the characteristic slope- 51 mV decade^{-1}, the narrowest linear range 1.0×10^{-4} to 1.0×10^{-1} M;
2. Electrodes sensitive to vitamin B6 do not show any considerable differences as far as the parameters are concerned. Near Nernstian slopes (mV decade^{-1}) of 60.6 ± 0.6 and 61.1 ± 1.4, within the intervals (M) of 5.8×10^{-5} to 1.0×10^{-1} and 4.3×10^{-5} to 1.0×10^{-1} were obtained, for thiamine and pyridoxine I and II prepared membranes, respectively;
3. A pH operational range of 6.5-8.5 for thiamine and 2-4.5 for pyridoxine electrodes was found;
4. Selectivity coefficients toward a large number of inorganic and organic cations usually present in multivitamin formulations vitamin B5, inositol, vitamin C, nicotinamide, riboflavin, cyanocobalamin and biotin revealed good performance;
5. Analysis of vitamins B1 and B6 in complex multivitamin drugs was achieved with recoveries within the intervals of 95.1-99.6% for thiamine and 95.1-102% for pyridoxine.
6. The use of cyclodextrins as ionophores for construction of the proposed electrodes proved the assessment of good quality ISEs when compared with classical ion exchange materials especially concerning slope, reproducibility and selectivity. This ionophore allowed obtaining electrodes for vitamin B1 with slopes corresponding to monoprotonated or diprotonated species, rendering their use in samples with different pH.

1,4-nitrophenylethyl Amine, Ephedrine Potentiometric Electrodes Based on permethoxylated-β-CD or peroctyl-α-cyclodextrins

The study [55] focused on the influence of different parameters of ion-selective electrodes on enantioselectivity of selected compounds. 1,4-nitrophenylethyl amine was selected as the model chiral compound. The research analysed permethoxylated β-cyclodextrin as a chiral selector. The parameters were defined in a classic ion-selective electrode that contained 32.8% weight of PVC, 65.6% of a relevant plasticizer (NPOE or DOS), 1.2 cyclodextrin) and 4.0% of lipophilic tetrakis(4-chlorophenyl)borate. After primary conditioning in an appropriate enantiomer solution, the electrodes were calibrated and the selectivity coefficients established with the use of solutions separate for the second enantiomer and other free ions of chloride, nitrate and nitrite salts. The analysis led to the following findings: the obtained electrodes showed Nernst slope of the calibration curve; NPOE and DOS plasticized electrodes obtained relatively high enantioselectivity; log $K_{R,S}$=2.11 and 1.27, K for univalent ions were of value 10^{-5} and for divalent ions – of 10^{-2}. The next stage of the research focused on the obtaining the enantioselectivity for ephedrine with the use of carboxymethyl-β-cyclodextrin or peroctyl-α-cyclodextrins. As far as the first cyclodextrin is concerned, the Nernstian response was obtained – the electrode stability in time; the parameters for the second ionophore were less positive. Selectivity factors concerning enantiometers were similar for both types of electrodes, with the value of log $K_{R,S}$= -0.31, -0.39.

Thiamine and Pyridoxine Potentiometric Sensor Based on β-cyclodextrin

The authors of the next study [56] presented the research concerning a PVC tubular ion-selective electrode without internal reference to accomplish the simultaneous determination of thiamine and pyridoxine in multivitamin products. Composition of the membrane phase was: 6.35% (w/w) of β-cyclodextrin as ionophore, 1.59% (w/w) of potassium tetrakis (p-chlorophenyl) borate as lipophilic additive, and 63.5% (w/w) of 2-fluorophenyl-2-nitrophenyl ether as mediator solvent.

The electrode responds selectively to thiamine but presents high initial diffusion potential contribution from pyridoxine. This fact was exploited under flow conditions for the quantitative assessment of both species. A multitask flow system was optimized to dilute appropriately real samples and prepare in-line the set of calibrating solutions.

The potentiometric signal was modelled by partial least squares regression, enabling the prediction of the content of each vitamin in the range of 1×10^{-3} to 1×10^{-4} M. Absolute errors of prediction inferior to 0.06×10^{-4} M and 0.16×10^{-4}mol L^{-1} for pyridoxine and thiamine respectively were obtained when compared with the chromatographic method.

Pyridostigmine Bromide (PB) Selective Electrodes Based on 2-hydroxy propyl β-cyclodextrin

In order to determine pyridostigmine bromide (PB), polymeric selective electrodes were investigated with 2-hydroxypropyl β-cyclodextrin as an ionophore [57].

The matrix was made from PVC-COOH plasticized with 2-nitrophenyl octyl ether. In order to demonstrate ionophore properties of cyclodextrin the study employed the analysis of the sensor in the membrane that did not contain inert cyclodextrin but only PVC-COOH, and replaced ionophore due to the reaction between dissociated COOH groups and cations of the drug extracted from the solution.

The sensor showed linear responses of PB within a concentration range of 10^{-3}-10^{-2} with slopes of 51.9 ± 0.31 mV decade^{-1}. The utility of 2-hydroxypropyl β-cyclodextrin as an ionophore has a significant influence on increasing both membrane sensitivity and selectivity of sensor 2 in comparison with sensor 1. Linear responses 10^{-5}-10^{-2} M, with slopes of 56.7 ± 0.40 mV/decade over pH range of 5-10 were obtained.

The proposed sensor 2 was applied for determination of PB in tablets with average recoveries of 100.22 ± 0.62, and in plasma with average recoveries of 99.14 ± 1.19. The methods were also used to determine the intact drug in the presence of its degradate, and thus they could be used as stability-indicating methods.

Cetrizine Potentiometric Electrode Based on β-cyclodextrin

Ion-selective electrodes with cationic functions sensitive to cetrizine were developed with the use of two types of electrodes: classic with the PVC membrane with inner solution (CE) and with carbon paste electrode (CPE) [58].

Table 2. Membrane composition, analytical parameters and potentiometric selectivity coefficient for cetrizine sensors

Electrode	CE	CPE
Plasticizer [%wt]	65.3 o-NPOE	
Ionophore	2.0 β-CD	2.0β-CD
Lipophilic salt	0.1 KTFPB potassium tetrakis[3,5-bis(trifluoro-methyl) phenyl]borate	0.1 KTFPB
Polymer/ other material	33 PVC	Carbon powder, paraffin oil
Inner solution	Cetirizine + KCl 1.0×10^{-2} M	-
Conditioning solution	1.0×10^{-2} M cetirizine	1.0×10^{-2} M cetirizine
Slope [mV decade^{-1}]	60.2	57.4
Linear range [M]	1×10^{-1}-5×10^{-6}	1×10^{-1}-7×10^{-6}
Lower Limit of Detection [M]	$5x10^{-6}$	$7x10^{-6}$
Response time [s]	10±2	30±2
Recovery time for 10^{-3} M	25±2	10±2
Life span (weeks)	6	6
pH range	2-4	2-4
Selectivity coefficients for selected ions [K]		
Na^+, K^+,	$2.2x10^{-6}$ $1.3x10^{-6}$	$2.3x10^{-6}$ $1.4x10^{-6}$
NH_4^+,Mg^{2+}	$8.2x10^{-5}$ $2.7x10^{-6}$	$7.2x10^{-6}$ $3.1x10^{-6}$
Glucose, Fructose,	$7.1x10^{-3}$ $3.8x10^{-4}$	$7.4x10^{-3}$ $4.2x10^{-4}$
Glycine, Cisteine	$5.4x10^{-4}$ $4.7x10^{-3}$	$3.4x10^{-5}$ $7.8x10^{-3}$
5-nitrobarbiturate	$7.5x10^{-5}$	$6.4x10^{-3}$
Catecholate	$7.8x10^{-4}$	$3.4x10^{-3}$
p-Nitroaniline	$1.1x10^{-5}$	$3.1x10^{-3}$

The membrane composition and characteristics of electrodes are presented in Table 2.

The prepared electrodes were used to assay cetrizine in pharmaceuticals and in spiked human urine. The results show recoveries in pure samples and pharmaceuticals 99.5±0.4 and 99.8±0.3% for conventional electrode and carbon paste electrode. The determination of cetrizine in urine samples was conducted with the use of standard addition method; the recoveries obtained were 100.9, 99.0% for conventional electrode and carbon paste electrode respectively. The mean standard deviation was SD ± 0.4.

Tetracycline Membranes Ion- Selective Electrodes Based on β- cyclodextrin

Tetracyclines form inclusive complexes with β-cyclodextrin through placing the widest part of β- cyclodextrin against aromatic part of tetracycline. The analysis of such complexes allowed the authors to conduct further research using cyclodextrin as potential ionophore of ion-selective electrodes [59].

The ion-selective membranes of electrodes were created in order to determine the tetracycline type of antibiotics such as: chlortetracycline, oxytetracycline, and doxycycline. The research employed solid-contact electrodes based on conductive graphite/epoxy support filling the distal interior tip of a long wide Perspex tube. An additional layer was coated in the process of electrochemical coating with the layer of conductive polymer, in order to ensure a better conductivity between the solid contact and polymeric layer – polypyrrole containing potassium heksacyanoferrate (II)/(III). Depending on the membrane composition, different compound compositions have been analysed and they are presented in the Table 3. below.

Table 3. Composition of tetracycline membrane electrode

Polymeric membrane		
DBP + βCD		FNDP + βCD
PVC	PVC-COOH	PVC

Electrodes with the best responses were based on plastic membranes containing 31% (w/w) PVC-COOH, 68% (w/w) of dibutylphthalate as plasticizer and 1% (w/w) of β-cyclodextrin as ionophore. The electrodes were responsive for over 6 months to tetracycline, oxytetracycline, doxycycline and chlortetracycline in glycine buffer solution (pH= 2), in the dynamic range 2×10^{-5}-10^{-2} M with a constant slope of about 55 mV $decade^{-1}$.

The procedure – a sequential-injection analysis system – was used for determination of oxytetracycline in pharmaceutical formulation and in wastewater analysis. The sample throughput of 51 samples h^{-1} was enabled by the system with a total volume of waste of about 6 mL per determination as well as results that favourably agree with those provided by chromatographic analysis. The relative errors were inferior to 4.5%.

Acetylcholines Potentiometric Sensor Based on heptakis (2, 3, 6-tri-o-methyl)-β-CD

Acetylcholines belong to the group of drugs of cationic characteristic and their derivatives include: butylcholine (BCh), acetylmethylcholine (AmCh) and acetylthiocholine (AtCh). The authors of the study [60] prepared electrodes sensitive to these chemical compounds. The preliminary research attempted to analyse different α-, β-, and γ-cyclodextrins as ionophores sensitive to acetylcholine derivatives. It was concluded that only β-cyclodextrins and their derivatives demonstrate the abilities to form inclusive compounds, while the remaining cyclodextrins do not have compatible cavities to form such complexes. Electrodes incorporated with heptakis (2, 3, 6-tri-o-methyl)-β-CD as sensing ionophore showed stable Nernst response (3 weeks).

Further research employed two types of electrodes: classic with inner r-rem and PVC membrane, and carbon paste electrode (CPE). The research focused on the influence of various components of membrane cocktails on the sensitivity and stability of the electrodes. The analysed possible combinations of membrane components and short descriptions of their application in electrodes are presented below:

Type of electrode: PVC, CPE, better parameters were shown by PVC electrodes,

Types of β-cyclodextrins derivatives: β-CD, heptakis (2,6-di-o-methyl)-β-CD, heptakis 2, 3, 6-tri-o-methyl)-β-CD; the best electrodes were based on heptakis 2, 3, 6-tri-o-methyl)-β-CD,

Types of plasticizers: NPOE, DBP, DOP, DOS, FPNPE (2-fluorophenyl, 2-nitrophenyloctyl ether), the highest sensitivity was shown by the electrodes with the most polar plasticizers, that is NPOE – 56.3, and FPNPE- 57,4 whose aromatic parts may strongly affect the solubility of cyclodextrin molecule. Optimal plasticizing properties were established for NPOE, in which hydrogen or Lewis acid and base interactions with cyclodextrin molecule probably occurred (this has improved solvation, and finally, the solubility of the ionophore).

Types of lipophilic salts: sodium tetraphenyl borate, sodium tetrakis (4-fluorophenyl) borate, potassium tetrakis (4-chlorophenyl) borate, phosphotungstic acid, phosphomolybdic acid sodium dodecyl sulphate, reineckate ammonium salt. Tetraphenylborates demonstrate the highest lipophilic compounds among the analysed salts and they have best applications as lipophilic additives in sensitive ion-selective membranes.

The selected optimal electrodes sensitive to acetylcholine worked satisfactorily in the concentration range from 10-6 to 10-2 mol L-1 with detection limit reaching 8.3×10^{-7} M and fast response time of 2s. The selectivity coefficients for acetylcholine derivatives towards various ions K^+, Li^+, NH_4^+, (-log K ~3.8) and molecules: choline citrate, starch glycine (-log K ~2,5), caffeine (-log K ~2.5) maltose (-log K ~2.4) sucrose (-log K ~2.0) were determined.

The sensors were successfully applied in the potentiometric determination of acetylcholines and its derivatives under flow injection analysis (FIA) and potentiometric titration conditions.

Butyrylcholinesterase (BuChE) Potentiometric Electrode Based on (2,3,6-tri-o-methyl)-β- cyclodextrin

The electrodes sensitive to butyrylcholinesterase were prepared with the use of disposable screen-printed butyrylcholine (BuCh) potentiometric sensor, based on other β-cyclodextrins: native β-CD, heptakis (2,6-di-o-methyl)- β-CD, heptakis (2,3,6-tri-o-methyl)-β-cyclodextrin and 2-hydroxypropyl-β-CD as ionophores [61]. Other components of ion-selective membrane were analysed to achieve the best electroanalytical characteristics. The analysis employed different plasticizers, including: NPOE, 2-fluorophenyl-2-nitrophenylether (FPNPE), dibutylphthalate (DBP), dioctylphthalate (DOP), dioctylsebacate (DOS) and tricresylphosphate (TCP) and different lipophilic additives, that is: sodium tetraphenylborate (NaTPB), sodium tetrakis (4-fluorophenyl) borate (NaTFPB), potassium tetrakis (4-

chlorophenyl) borate (KTCPB), phosphotungstic acid (PTA), phosphomolybdic acid (PMA), sodium dodecylsulfate (SDS) and reineckate ammonium salt (RAS).

Screen printed electrodes show many advantages: their mass production is inexpensive, they are reproducible, universal, used in pharmaceutical and biomedical analyses, as well as in the protection of the environment. Among the analysed electrodes, the optimum membrane was found in heptakis (2,3,6-tri-o-methyl)-β-cyclodextrin, KTCPB as an anion additive, high polar aromatic plasticizers; o-NPOE and FPNPE. Regarding these plasticizers, the electrodes show the highest sensitivity, 61.3±0.9 and 58.7±1.4 mV decade^{-1} respectively. The employment of some less polar plasticizers causes the decrease of the electrodes sensitivity, which may be explained by a lower degree of ionophore solvation. The lack of ionophore, on the other hand, decreases the electrodes sensitivity to about 17.5±1.0 mV decade^{-1}. The addition of this compound to the membrane causes a stable slope of the electrode S=59.6±1.4mV decade^{-1}. The electrode is characterized by the linear range of 10^{-6} to 10^{-2} M with detection limit of 8×10^{-7} M, fast response time (1.6 s) and adequate shelf-life (6 months).

The developed cholinesterase sensors with low selectivity for choline chloride were used to test the enzymatic reaction. They have been successfully applied for real-time intoxication monitoring through assaying cholinesterases (ChEs) activity in human serum. Determination of organophosphate pesticide was conducted by measuring their inhibition of BuChE with successful assaying of malathion in insecticide samples (Malathion Coupra, Prioderm lotion) with high accuracy and precision.

Sulfadiazine Potentiometric Electrode Based on α- β-or γ-cyclodextrin

The electrodes for the selective screening of sulfadiazine (SDZ) in aquaculture waters were reported [62]. The sensors used an epoxy-graphite matrix as a solid contact. The membranes were made from PVC doped with tetraphenylporphyrin-manganese(III) chloride as charged ionophore, α-cyclodextrin, β-cyclodextrin, or γ-cyclodextrin as neutral ionophores that were dispersed in NPOE as plasticizer.

In the case of charged ionophore, the electrode showed a near-Nernstian response with slopes of 52 mV decade^{-1} and detection limits of 3.91×10^{-5} M. Much better parameters were shown by the electrode after the tetraoctyl ammonium bromide was added to the membrane, slope of characteristic of 59.7-62.0 mV decade^{-1} and wider linear ranges.

The applicability of different cyclodextrins containing 6,7 and 8 α-(1,4 glycosyl) units was analysed.

The electrodes demonstrated different calibration curves depending on the quantitative weight ratio of the components and the liophilic additive.

In general, better parameters were demonstrated by the electrodes with positively charged additives, when compared with electrodes without any additives. The electrode with β-cyclodextrin showed theoretical slope of the calibration curve, regardless of the amount of the lipophilic additive. As far as the remaining sensors are concerned, the increase in the amount of the liophilic additive, generally increases the sensitivity of the sensors, broadens the working range, and lowers LLLR and LOD.

As far as the influence of the two analysed cationic lipophilic additives – tetraoctylammonium bromide and dimethyldioctylammonium bromide is concerned, the former significantly influences the increase of the sensors' slope in sensors containing cyclodextrin with a bigger cavity, and the latter influences the electrodes slope of electrodes containing cyclodextrins with a smaller cavity.

The electrodes can work independently, regardless of the pH, in the sulfadizine solution in the range 2-7. All of them have a very short response time – less than 15 seconds. The proper time of electrode functioning, stored in 10^{-3} M with reproducible slope of the curve in the range ±3% amounts to 2 months. For sulfadiazine electrodes, the determined selectivity coefficients can be arranged according to the following scheme (near Hofmeister behavior): nitrate>nitrite> chloride> fluoride~ hydrogen hydrogen carbonate> carbonate.

The sensors were successfully applied to the fast screening of SDZ in real water samples from aquaculture fish farms. The method offered the advantages of simplicity, accuracy, precision (recoveries ranged 93.6 – 102.9%) and automation feasibility. The sensing membrane may contribute to the development of small devices allowing in locus measurements of sulfadiazine or parent-drugs.

Dextromethorphan Potentiometric Electrode Based on β-cyclodextrin Derivatives

Simple rapid and accurate potentiometric methods for the determination of dextromethorphan (DXM) in pharmaceutical preparation were described [63]. The three types of electrodes successively named silver coated wire electrode (CWE), coated graphite electrode (CGE) and the conventional PVC membrane electrodes (CE) were described. The quantitative composition of the analysed membranes was as follows:

- ionophores β-CDs derivatives including; heptakis (2,6-di-o-methyl)-β-CD heptakis (2,3,6-tri-o-methyl)-β-CD, 2-hydroxypropyl-β-CD and native β-CD,
- lipophilic additive sodium tetraphenylborate or sodium tetrakis (4-fluorophenyl) borate or potassium tetrakis (4-chlorophenyl) borate
- plasticizers o-nitrophenyloctylether or dibutylphthalate or dioctylphthalate dioctylsebacate and tricresylphosphate (TCP, Fluka)
- matrix polyvinyl chloride (PVC).

The preliminary research concerning the selection of the best group of cyclodextrins allowed for drawing a conclusion that the most compatible group of these macrocyclic particles for determining dextromethorphan are β-cyclodextrins. It is assumed that they form inclusive compounds with the analysed drug in the 1:1 ratio. Lipophilic cavity incorporates methoxy groups and the part of the adjoining aromatic ring of dextromethorphan.

The electrodes based on β-cyclodextrins ionophores show a considerable selectivity towards DXM with Nernstian slopes depending on the type of the electrode and the type of β-cyclodextrins.

The criteria for the selection of the best ionophore included the increase of the slope of the calibration curve as well as the potential reproducibility.

Generally, the potentiometric response of the fabricated sensors prepared using heptakis (2,3,6-tri-o-methyl)-β-CD was the best among other tested ionophores as the slope were

62.7 ± 4.0, 59.0 ± 2.8 and 62.0 ± 5.2 mV decade^{-1} for CWE, CGE and CEs, over the period of two weeks respectively. From the lipophilic additives and plasticizers, the optimal one proved to be sodium tetrakis (4-fluorophenyl) borate (NaTFPB) and o-nitrophenyloctylether (o-NPOE). The analytical parameters of the optimum CWE electrode are presented in Table 4.

The developed electrodes have been successfully applied for the potentiometric determination of DXM in pharmaceutical formulation under batch and flow injection analysis (FIA) conditions. FIA allows the analysis of 90 samples h $^{-1}$, the average peak heights were 187 ± 2.0 mV

The potentiometric titration was conducted with the use of titrant tetraphenylborate sodium. Potential jump between 274 and 380 mV was noted in the ideal conditions (when the two curves are symmetrical), determining 0.8 mg of dextromethorphan. The proposed potentiometric methods offer the advantage of simplicity, accuracy and automation feasibility.

Table 4. Analytical parameters of heptakis (2,3,6-tri-o-methyl)-β-CD based dextrometorphan sensor (Type CWE)

Concentration range (mol L^{-1})	10^{-7}–10^{-2}
Slope (mV decade^{-1})	59.0 ± 0.2
R	0.9997
LOD (M)	7.0×10^{-8}
Response time (s)	4
Lifetime (Day)	30
Working pH range	4–7

Metronidazole and Spiramycin Potentiometric Electrodes Based on β-cyclodextrin

Two microsized graphite-design sensors containing β-cyclodextrin (β-CD), polyvinyl chloride carboxylated (PVC-COOH) and dioctyl phthalate have been investigated for their response to metronidazole and spiramycin [64].

The microsized graphite rods were coated with thin films of PVC-COOH-β-CD-dioctylphthalate (DOP), and used as potentiometric sensor. These sensors were soaked in 10^{-3}M of metronidazole solution for 12 hours and in 10^{-2} M spiramycin solution for 24 hoursrespectively and they remained in the solutions between the measurements.

The study determined analytical parameters of the electrodes made in accordance with IUPAC recommendations [65]. The findings are presented in Table 5.

Moreover, the potentiometric coefficients of selectivity, value of K~10^{-5}, were defined.

The developed electrodes have been successively applied for the determination of the two drugs in their pharmaceutical formulations Spirazole® tablets. Validation of the method according to the quality assurance standards shows suitability of the proposed electrodes for the use in the quality control assessment of these drugs. The recovery percentages for the determination of metronidazole and spiramycin by the two proposed selective electrodes are 99.86 ± 0.249% and 99.69 ± 0.856% respectively.

Table 5. Analytical parameters of metronidazole (MZ), and spiramycin (SP) coated graphite electrodes

Parameters	MZ electrode	SP electrode
Slope mV decade^{-1}	58	58.3
Intercept (mV)	105.5	112.7
R	0.9999	0.9996
LOD (M)	6.8x10^{-7}	5.9x10^{-6}
Linear range (M)	1x10^{-5} - 1x10^{-3}	1x10^{-5}-1x10^{-2}
Response time (s)	45	30
Working pH range	5.5-7.5	5.5-7.5
Stability (weeks)	5	6

Cyclobenzaprine Hydrochloride Potentiometric Electrode Based on 2-hydroxy Propyl β-cyclodextrin

The authors of the study [66] prepared ion-selective graphite electrode sensitive to cyclobenzaprine hydrochloride containing 2-hydroxypropyl β-cyclodextrin immersed in PVC - dibutylsebacate - ammonium reineckate membrane. This electroactive substance has a significant influence on the increase of the membrane sensitivity and selectivity of sensor. The developed electrode can work properly in the concentration range of 1x10^{-4} – 1x10^{-2} M with slopes of 55.5 mV decade^{-1}, over the pH range 2-4. The proposed method displayed useful analytical characteristics for determination of CZ in its pure powder form with average recoveries 99.61 ± 0.34% and in plasma with good recoveries. The sensor was also used to determine the intact drug in the presence of its degradate and, thus, could be used as stability-indicating method. Results obtained with the electrode revealed its performance characteristic, which was evaluated according to International Union of Pure and Applied Chemistry recommendations.

Pramoxine HCl Potentiometric Electrode Based on 2-hydroxypropyl β-cyclodextrin

The study [67] discusses the use of cyclic compounds, that is cyclodextrins and calixarene, as active compounds of polymer membranes. These cyclic compounds demonstrate better sensitivity and selectiveness for pramoxine HCl in comparison with ordinary cation exchanger – sodium-tetraphenylborate. The electrodes were constructed with the use of classic body of electrode with inner solution of KCl and drug. Polymer membranes based on PVC were made through the evaporation of the THF solvent. NPOE was used as plasticizer. The parameters of electrodes with 2-hydroxy propyl β-cyclodextrin ionophore and the determination results in pharmaceuticals and plasma are presented in Table 5.

Better electrodes parameters and determination results in organic samples were achieved for sensors with 4-sulfocalix-6-arene and 4-sulfocalix-8-arene in the membrane. The authors' explanation is a type of host-guest interaction. A positively charged hydrogenated pramoxine particle prefers different donor sites (OH-groups and sulphonic acid) of 2-hydroxy propyl-β-cyclodextrin and calixarene structures). The presence of OH groups, derivatives of cyclodextrin, does not suffice to form stable complexes, which manifests itself in the slope of the electrode by only 54.3 mV decade^{-1}. In the case of calixarene, the complexes are more stable due to the electrostatic interactions with anionic sulfonic groups, and the sensitivity of the electrodes increases to 56-59 mV decade^{-1}

The proposed sensors displayed useful analytical characteristics for the determination of PAM in bulk powder, pharmaceutical formulation, and in biological fluids. Validation of the method showed the suitability of the proposed electrodes for the use in the quality control assessment of the drug (Table 6).

Table 6. Electrochemical response and determination results of pramoxine using ion-selective electrode

Parameters	Values	Statistical analysis(n=5)	Values in pure sample	Values in pharmaceutical formulation	Values in human plasma
Slope mV decade^{-1}	54.3 ± 0.8,	Average accuracy (%)	99.6	98.6(10^{-3}) 100.1 (10^{-4})	99.9 (10^{-3}) 100.2 (10^{-4}) 101.2 (10^{-5})
Intercept (mV)	211	Standard deviation	0.5	-	-
Correlation coefficient (r)	1.0	RSD (%)	0.5	0.5 (10^{-3}) 0.4 (10^{-4})	0.4 (10^{-3}) 0.2 (10^{-4}) 0.7 (10^{-5})
LOD (M)	3.6×10^{-5}	Repeatability (%)	0.5		
Concentration Range (M)	1.0×10^{-5} to 1.0×10^{-2}	Reproducibility (%)	0.8		
Response time (s)	45	Variance	0.3	-	
Working pH range	3-6	F test	1.3		
Stability (days)	28	Student's test	1.2		

Sulfamethoxazole Potentiometric Electrode Based on 2-hydroxy propyl β-cyclodextrin

Sulfamethoxazole (SMX) is one of the antibiotics employed in aquaculture for prophylactic and therapeutic reasons. Environmental and food spread may be prevented by controlling its levels in several stages of fish farming. The present work [68] proposes for this purpose new SMX selective electrodes for the potentiometric determination of this sulphonamide in water. The selective membranes were made of polyvinyl chloride (PVC) with tetraphenylporphyrin manganese (III) chloride or cyclodextrin-based acting as ionophores. 2-nitrophenyl octyl ether was employed as plasticizer and tetraoctylammonium, dimethyldioctadecylammonium bromide or potassium tetrakis (4-chlorophenyl) borate was used as anionic or cationic additive.

The best analytical performance was reported for ISEs of tetraphenylporphyrin manganese (III) chloride with 50% mol of potassium tetrakis (4-chlorophenyl) borate in comparison with ionophore. Nernstian behaviour was observed from 4.0×10^{-5} to 1.0×10^{-2} M (10.0 to 2500 μg/mL), and the limit of detection was 1.2×10^{-5} M (3.0 μg/mL). In general, the electrodes displayed steady potentials in the pH range of 6 to 9. Emf equilibrium was reached in under 15 s in all concentration levels.

The electrodes revealed good discriminating ability in environmental samples. The analytical application to contaminated waters showed recoveries from 96 to 106%.

Differential Dynamic Potentiometry (DDP) Based on β-CDion Selective Electrode for Different Drugs

This paper [69] discusses the differential dynamic potentiometry (DDP) for determination of different drugs such as procaine, lidocaine, quinidine, quinine, procainamide, papaverine, tetracaine, bupivacaine, clomipramine and chlorpromazine. This novel potentiometric technique consists in recording the dynamic potential difference between two ion-selective electrodes (ISEs). This was done by measuring the DDP response between a membrane electrode containing PVC, plasticizer, ionic additive and β-CD, and another with a blank membrane without β-CD, when both electrodes were simultaneously exposed to different drugs. The influence of various plasticisers on the response of the electrodes was analysed, and as far as the reproducibility of the DDP responses is concerned, the membranes plasticized with TCP tricresyl phosphate were selected to be the best out of the eight analysed membrane compositions.

If the dynamic response of the electrodes for different drugs is to be compared, it is necessary to achieve a reproducible initial stage of the corresponding membrane systems. Interestingly, when this was done and the electrodes were calibrated in serial calibration mode, the DDP responses obtained for all the drugs showed unusual shapes characterized by a potential inversion. Moreover, a different shape for different assayed drugs was obtained. These signals may be treated as drug fingerprints that may be useful for drug identification purposes. Also, when discrete calibration mode was followed, the non-monotonic DDP responses obtained for each concentration were used for quantitative analysis.

Piroxicam Potentiometric Sensor Based on β-cyclodextrin

The novelty in the work [70] was the employment of the new potentiometric electrode, not yet discussed in scientific literature. This carbon paste electrode contains functionalized multi wallet carbon nanotubes in the membrane cocktail. The modification included oxidation of the carbon nanotubes in the acidic environment (groups –COOH were incorporated).Then, the tubes have been triturated with alcohol and β-cyclodextrin. The occurrence of the reaction between nanotubes and cyclodextrin was proven by the presence of the IR spectra with characteristic bands of 3021, 2778, 2442 cm^{-1}.

The optimisation and the selection of the best piroxicam electrode consisted in the comparison of analytical parameters of the electrodes of three different matrix compositions and achieving relevant, most important conclusions and findings:

1. Carbon paste electrodes (CPE) containing different cyclodextrins; from the α, β, γ – CDs, β-cyclodextrins proved to be optimal, and among different derivatives of β-CDs, (heptakis, methyl, hydroxypropyl), β-cyclodextrin proved optimal properties as

ionophore. The best plasticising properties of graphite powder were shown by 2 - fluorophenyl ether (f-NPE), hyamine was used as the ion additive to the inert cyclodextrin. The electrode with such composition showed the near Nernstian sensitivity -61.7 mV decade^{-1} over the linear range of $1x10^{-5}$ - $1x10^{-2}$ M and lower limit of detection $1.0x10^{-5}$ M.

2. Multi-wallet carbon nanotubes carbon paste electrodes (MWCNTP-CPE), electrodes containing multi-wallet nanotubes in the graphite matrix, enlarged the surface of the electrode, and their aim was to capture ions on the paste surface. Electrodes of this type improved the parameters and they achieved broader linear range $1x10^{-6}$ - $1x10^{-2}$ M and lower limit of detection $1.0x10^{-6}$ M, and also lower slope of the characteristic –51.1 mV decade^{-1}
3. Functionalized multi-walled carbon nanotubes (FMWCNTs/β-CD-CPE). This type of electrodes contained FMWCNTs/β-CD composite, that replaced free cyclodextrin placed on the graphite paste, and hyamine (Hy) and 2-fluorophenyl 2-nitrophenyl ether (f-NPE) as electrode plasticizer. Such electrodes have shown optimal parameters achieving the concentration range from 10^{-6} to 10^{-2} M with Nernstain compliance (58.7±0.9 mV decade^{-1}) with fast response time of about 2 s and exhibit adequate operational lifetime (16 weeks).

The developed electrodes have been applied for the potentiometric determination of PXM in pharmaceutical formulation under batch and flow injection analysis (FIA). FIA allows for the analysis of 120 samples h^{-1} with the advantage of simplicity, accuracy and automation feasibility.

Amilsupride Potentiometric Electrode Based on 2-hydroxypropyl β-cyclodextrin (β-CD)

This study [71] examined the preparation of amilsupride membrane sensors based on other ionophores: ion exchangers type and β-cyclodextrin characteristic, selectivity and response time. Several plasticized poly(vinyl chloride) membranes compositions were examined and the best electrode based on 2-hydroxypropyl β-cyclodextrin (β-CD) with membrane plasticized with dibutylsebacate (DBS) was selected.

The electrode shows the highest sensitivity near Nernstian 31.1 mV decade^{-1} over the linear range of $1x10^{-6}$ - $1x10^{-2}$ M. The present electrode exhibits clear discrimination of amilsupride ions from several inorganic, organic, some common drug excipients and amilsupride degradation products. This electrode has a response time of 10-20 s and can be used in the pH range 4.0-7.0 in the period of 4-6 months.

The sensors were used for determination of amisulpride, in pure form, in presence of its degradation products, in tablets and in plasma gave average recoveries 99.94±0.87 for sensor with β-cyclodextrin. Validation of the method shows suitability of the proposed sensors for use in the quality control assessment of amisulpride and for routine analysis as stability indicating method.

Metformin Potentiometric Electrode Based on β-cyclodextrin (β-CD)

In this work [72] the miniaturized coated wire electrodes (CWEs) and coated graphite (CGE) electrodes for metformin were prepared. Various commercially available cyclodextrins heptakis (2,6-di-o-methyl)-β-CD heptakis (2,3,6-tri-o-methyl)-β-CD, 2-hydroxypropyl-β-CD and native β-CD, as active substances of the membrane phase (instead of commonly applied ion-pair complexes) were used. The influence of ionic sites and polarity of plasticizers on the calibration curves parameters and the electrodes selectivity was examined. Basic parameters of the optimum electrode were assigned. The authors comprehensively examined also inclusive complexes of the available cyclodextrins with metformin. The structures were confirmed by the infrared spectroscopy and gravimetric methods – DSC. It was also established that the degree of inclusion (complexation constant) with the examined drug increased according to the following sequence: heptakis (2,3,6-tri-o-methyl)-β-CD >heptakis (2,6-di-o-methyl)-β-CD> 2-hydroxypropyl-β-CD and native β-CD. Methylated cyclodextrins became more lipophilic, and hence provided better penetration of the lipophilic particle of the drug inside the inclusion cavity in comparison with natural hydroxy cyclodextrin (inert, with no substituents). Moreover, the presence of hydroxypropyl substituents in 2-hydroxypropyl-β-CD, hindered the formation of the complex with the guest's molecule due to the steric hindrance effect. Therefore, CWEs modified with heptakis(2,3,6-tri-o-methyl)-β-CD, sodium tetrakis(4-fluorophenyl)borate (NaTFPB) as lipophilic site and 2-fluorophenyl 2-nitrophenyl ether (f-NPE), work satisfactorily.

For the best electrode the following basic parameters were investigated: measurement range 10^{-6} to 10^{-1} M, slope of characteristic 55.7 ± 0.4 mVdecade^{-1} activity, limit of detection 8×10^{-7} M, 6-month lifetime, dependence of the electrode potential on pH (4.0-7.0), reproducibility of potential (±04mV per day) and selectivity coefficients in relation to some organic and inorganic anions.

The electrode was applied for determination of metformin under batch and flow injection analysis (FIA). FIA allowed for the analysis of 90 samples per hour offering the advantages of simplicity, accuracy and automation feasibility. The dissolution profile for metformin pharmaceutical samples (Cidophage®) was monitored with the employment of the proposed electrode.

Midodrine Potentiometric Electrode Based on 2-hydroxypropyl-β-CD

Midodrine PVC membrane sensors based on different plasticizers DOP and NPOE and active substance of phosphotungestate (PT) as cation exchanger, and β-cyclodextrin (β-CD)-based technique with PT as a fixed anionic site were prepared [73]. The electrode's basic parameters, such as the slope of characteristics, selectivity response time lifetime, and the influence of pH on the electrode's potential were established. The electrode with DOP and β-cyclodextrin response to midodrine hydrochloride has the near Nernstian sensitivity (56±1.0 mV decade^{-1}) over the linear range of $5x10^{-5}$ - $1x10^{-1}$ M. The present electrodes show clear discrimination of midodrine cation, with respect to urea starch, sucrose, lactose, talc, NaCl, KCl, $CaCl_2$, glycine, and deglymidodrine. The electrode has a response time of 15 s and can be used in the pH range of 5.0 - 7.0. The notable properties and attractive qualities of the indomethacin sensor are low cost and comfortable application. Sensor can be used for two weeks without any measurable change in sensitivity. The suggested electrodes succeeded to determine intact MD in the presence of up to 10% of its degradation product and in pharmaceutical products without prior extraction.

Ibuprofen Potentiometric Electrodes Based on Other Cyclodextrins

The aim of the research was to create an electrode with a polymeric membrane phase with ibuprofen function on the basis of PVC as well as to establish its analytical parameters [8]. The optimisation of the electrode consisted in the selection of appropriate cyclodextrin from the commonly available α-, β-cyclodextrin, FNDP 2-fluoro-2nitrodiphenyl ether or o-NPOE plasticizer and lipophilic additive (ammonium bromide). The preliminary analysis investigated α-CD, β-CD and hydroxypropyl-β-cyclodextrin with the membrane optimisation. The last cyclodextrin is less soluble in polar solvent – FNDP, and the fourth examined membrane was the membrane prepared in the following way: first the components were dissolved in ethanol, and next - in tetrahydrofuran. From the four electrodes, the optimal characteristics were observed for the first and fourth, with the slope about 50-55 mV decade^{-1}. Further research on optimisation with the inclusion of other components led to preparation of the best electrode sensitive to ibuprofen that contained 1.2 wt.% α-cyclodextrin, 65.6 wt.% o-nitrophenyloctylether plasticizer, 42% mol (relative to the molar concentration of the ionophore) of tetradodecylammonium bromide and 32.8 wt.% PVC. This electrode response to ibuprofen has the Nernstian sensitivity -59.0 over the linear range of $5x10^{-5}$ - $1x10^{-2}$ M, 3.87×10^{-6} to 10^{-2} M, for samples with pH adjusted to 9. The response time was fast <15 s and reproducible (±0.8 mV) with constant performance over more than 6 months. The practical limit of detection was of $(3.34 \pm 0.03) \times 10^{-6}$ M ibuprofen.

The present electrode show clear discrimination of ibuprofen ions from several inorganic and organic excipients – acetate, aminosalicylate, benzoate, borate, Br-, carboxyibuprofen, Cl-, creatinine, citrate, F^-, glycerol, methyparaben, NO_3^-, NO_2^-, phosphate, polysorbate, salicylate, sorbitol, SO_4^{2-}, tartrate, urea, and uric acid. The response time was fast <15 s and reproducible (±0.8 mV) with constant performance over more than 6 months.

The analytical usefulness of electrode was examined by determining ibuprofen in pharmaceuticals and several river water samples with the use of the Gran plot method. The recovery from four replicate measurements for the determination of ibuprofen in water solutions was determined to be 95.±5 – 102±5%. For pharmaceutical formulations variation coefficients less than 1% were obtained.

Ephedrine Potentiometric Electrode Based on Triacetyl-β-cyclodextrin [(triacetyl-β-CD)]

This paper describes the results of research leading to creating an electrode with a polymeric membrane phase on the basis of PVC carboxylated poly(vinyl chloride) [(PVC-COOH)] of ephedrine function [74]. The triacetyl-β-cyclodextrin [(triacetyl-β-CD)] as a neutral ionophore dissolved in a plasticizer (DOS, DOP or 2-nitrophenyl-phenylether (2-NPPE) is the active substance of this electrode. Potassium tetraphenylborate was used as the ion exchanger.

The advantages of electrodes with membrane phase of the following composition: 32.7 % wt. PVC, 1.0 % wt. of triacetyl-β-CD, 65.8 % wt. DOS, and 0.5% TPB^- are near Nernstian slope of characteristic 57.0±0.3 in the range $3.0x10^{-5}$ - $8.0x10^{-3}$ M and low detection limits of $5.7x10^{-6}$ M. The working pH range is 3-8 and the response time is 10-20 s. The analytical usefulness of the created electrodes was tested by determining ephedrine in pharmaceuticals and biological fluids with the use of the direct potentiometry method. They were also used as tubular detectors in a double channel flow injection system. The intrinsic characteristics of the detectors in a low dispersion manifold under hydrodynamic mode of operation were determined and compared with data obtained under batch mode of operation. The developed sensor for ephedrine is characterized by long life span – 5 weeks, good selectivity for EPD+ over a wide variety of other organic compounds, long-term stability, high reproducibility acceptable accuracy (99.5) and precision (CV=0.4%).

Moxifloxacin Ion-Selective Electrode Based on 2-hydroxypropyl β-cyclodextrin

Another classic electrode with inner solution and polymer membrane to identify moxifloxacin was prepared [75]. In order to make the membrane, 1,400 mg NPOE was mixed with 50 mg TpClPB and 190 mg PVC. The mixture was dissolved in THF and then 50.0 mg of 2-hydroxypropyl β-cyclodextrin was added. The electrode was filled with an internal solution of 10^{-2} M moxifloxacin and 10^{-2} M KCl. Ag/AgCl wire was used as an internal reference electrode. The results reveal that, among others, 2-hydroxypropyl β-cyclodextrin provides high stability to the complexes formed with cationic drug present in the solution; thus, the membrane selectivity and sensitivity are substantially enhanced.

The performance of characteristic of obtained electrode is presented in Table7.

Table 7. Analytical parameters of moxifloxacin electrode

Electrode	Cyclodextrin based sensor for moxifloxacin
Slope [mV $decade^{-1}$]	51.50
Intercept [mV]	315.35
Limit of detection	6.8×10^{-6}
Linear range [M]	10^{-5} to 10^{-2}
Response time [s]	15
Working pH range	2–5
Stability (days)	45
Correlation coefficient	0.9999
Average recovery (%) ± S.D. (n=5)	99.97±0.42

The selectivity coefficients of the cyclodextrin based sensor indicated excellent selectivity for moxifloxacin. Satisfactory values were achieved in the case of the following ions: degradate, Na^+ K^+ NH_4^+ Ca^{2+} Mg^{2+} Ca^{2+} Fe^{3+} glycine, glucose, lactose, starch, urea, K~1-5 $x10^{-3}$.

The proposed sensor was applied for the analysis of moxifloxacin in pharmaceutical formulation. The results prove the applicability of the three sensors for the determination in bulk powder, pharmaceutical formulation, and biological fluids (plasma) and in the presence of its alkaline degradation product, and thus could be used for stability-indicating studies.

Rivastigmine Ion Selective Electrode Based on 2-hydroxypropyl Beta Cyclodextrin

The scientific novelty of study [9] was the incorporation of calixarene and 2-hydroxyl propyl beta cyclodextrin ionophores in the PVC membrane sensitive to rivastigmine and the achievement of very good analytical parameters for ion-selective electrodes. Moreover, the kinetic analysis of the rivastigmine degradation was conducted with the use of the prepared electrode. Polymer membranes based on o-nitrophenyloctyl ether, cyclodextrin (and/or PVC)were prepared in the process of evaporation of the solvent (THF). Then, the membranes were placed at the end of a glass tube filled with electrolyte solution with 10^{-2} M of RIV solution. The created electrodes are characterized by a higher sensitivity (54,6-56,8 mV $decade^{-1}$) in comparison with the electrodes with typical ion exchangers, low detection limit 10^{-7} - 10^{-8} M, and long stability – around 50 days. The electrode containing calixareneis characterized by considerably higher selectivity than the electrode with cyclodextrin, which may be caused by higher interaction preference for rivastigmine cations with highly polarcalixarene sulfonic groups.

Both sensors were used to determine rivastigmine in pharmaceutical preparations and biological fluids: plasma and celebrospinal fluid. This compound was also tagged in the presence of the alkaline degradation products (>80%). The achieved determination results were very good, for instance for the sensor with cyclodextrin (recovery % 99,4-100.73).The created electrodes may be successfully used to determine rivastigmine in the quality control processes in various laboratories.

Diclofenac Potentiometric Electrode Based on Heptakis(2,3,6-tri-O-benzoyl)-β-cyclodextrin

This paper [76] reports the preparation of diclofenac-selective membrane electrodes with solid contact, incorporating β-cyclodextrins: (2-hydroxypropyl) – β-cyclodextrin, heptakis (2,3,6-tri-O-methyl) - β-cyclodextrin, and heptakis (2,3,6-tri-O-benzoyl) - β-cyclodextrin. Several plasticized poly (vinyl chloride) membranes of different compositions were tested with the best electrode being the one incorporating heptakis(2,3,6-tri-O-benzoyl)-β-cyclodextrin with the membrane plasticized with 2-nitrophenyloctyl ether.

The membrane phase of the electrode consists of two layers: the inner layer containing plasticized PVC in which the Ag/AgCl electrode is placed. The potential of the internal reference Ag/AgCl electrode is stable and determined by Cl^- ions originating from PVC degradation, as well as from dissolution dissociation of AgCl in TBP.

The outer layer consist of: 32,8% wt. PVC, 65,6% wt. of mixed plasticizers (NPOE+TBP), 1,2% of appropriate cyclodextrin, and 0,4% of tetraoctylammonium chloride. The best electrode function for the determination of diclofenac was found for electrodes 2, 5,

6, and 7 with HSMβCD+DIBP,HSMβCD+o-NPOE, HSBβCD+o-NPOE, HPβCD+o-NPOE respectively.

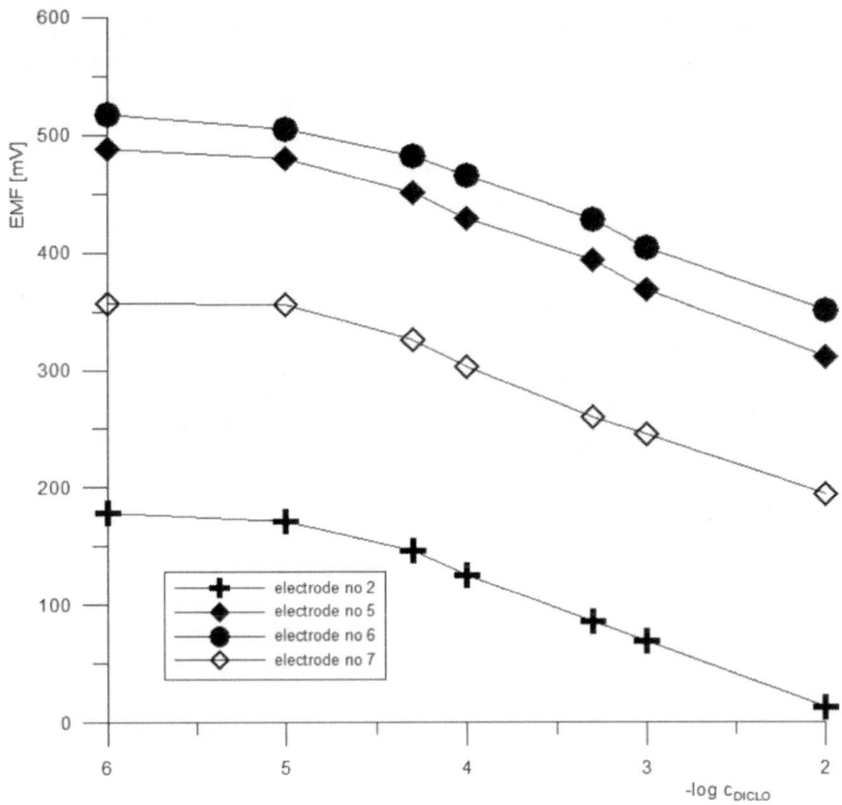

Figure 2.

The overall good characteristics (Figure 2) including a wide linear range (5.0×10^{-5}–1×10^{-2} M), a low limit of detection (1.0×10^{-5}–5.0×10^{-5} M), and a reproducible Nernstian slope were obtained. The proposed electrodes can easily discriminate diclofenac ions from several inorganic and organic interferents and some common drug excipients.

For electrode 6, different lipophilic patterns were obtained: $I^- > C_6H_4(OH)COO^- > H_2PO_4^- > NO_3^- > Br^- > CH_3COO^- > Cl^-$

In particular, the incorporation of heptakis(2,3,6-tri-O-benzoyl)-β-cyclodextrin resulted in excellent electrodes (no. 2 and 6) with a short response time and long lifetime (8–10 months).

The electrical properties of the membrane electrode were studied by impedance spectroscopy. For the studied electrode, bulk resistance was 945 kΩ and geometric capacitance was 39.3 pF.

The notable advantages of the diclofenac-selective electrode include its high sensitivity, selectivity, cost-effectiveness, and comfortable application in drug and urine analysis. Recovery from 6 replicate diclofenac measurements was found to be 94.0–100.5% (RSD 2.2–3.4%) for clear solutions, 96.1–98.8% (RSD 2.5–4.5%) for pharmaceuticals, and 99.0–99.7% (RSD 2.7–3.1%) for urine solutions.

Trigonelline Potentiometric Electrode Based on 2-hydroxy Propyl β-cyclodextrin

The study concerns the preparation of the electrode sensitive to trigonelline, a bioactive marker for determination of the quality of crude drugs and the formulations [77]. The authors employed a classic electrode with polymer membrane construction. The particular elements of the membrane – 0.19 g of PVC and 0.4 g of NPOE were mixed with 2-hydroxypropyl β-cyclodextrin; then the mixture was dissolved in THF. A master membrane of 0.1 mm thickness was obtained. The electrodes were preconditioned by immersing in 10^{-2} M trigonelline hydrochloride solution for 24 hours and stored in the same solution between the tests.

The proposed sensors showed Nernstian slopes of 59 mV decade^{-1} in the concentration over pH range 4-9, the concentration range covered 10^{-3} - 10^{-5} M. Very short response time 15 s, and 5-week stability for electrode were obtained. The response of the electrodes was monitored as a function of temperature and the calibration graphs obtained at different temperatures in the range 25 – 45°C at 5°C interval were parallel.

The slope, response time and limit of detection did not significantly vary with variation of temperature up to 35° C, indicating reasonable thermal stability. The proposed sensors offer the advantages of fast response and moderate stability time. The selectivity coefficients of the developed sensors indicated excellent selectivity for trigonelline towards: Isoleucine, Leucine, Lysine, Vitamin B2, Vitamin B3, K^+, Na^+, Zn^{+2}, Mg^{+2}, (K = 1- $6x10^{-3}$).

The proposed electrode was successfully applied for direct determination of trigonelline in Trigonellafoenum-graecumseeds extract without prior purification or pretreatment steps, and in plasma. Medium yield and medium standard deviation in the plasma samples were ± SD 100.06 ± 0.55 99.82 ± 1.10 (n=3)

Dopamine Potentiometric Electrode Based on β-cyclodextrin

The study [78] presents a very interesting, different construction of the solid-contact potentiometric sensor with the gold surface to determine dopamine.

The sensors were fabricated electrodes utilizing glass substrates coated with gold then followed by the growth of ZNRs (ZnO nanorods). ZNRs were grown on a gold-coated glass substrate by a low temperature aqueous chemical growth (ACG) method. The polar nature of dopamine leads to a change of surface charges on the ZNR surface via metal ligand bond formation which results in a measurable electrical signal. Polymeric membranes consisted of PVC (0.18 g), β-CD as ionophore (0.04 g), potassium tetrakis(4-chlorophenyl) borate as ionic additive (0.01 g), and 2-fluoro-2-nitrodiphenyl ether (0.4 g). The membrane was immobilized on the ZNR surface. The fabricated electrodes were characterized by X-ray diffraction (XRD) and scanning electron microscopy (SEM) techniques. The grown ZNRs were well aligned and exhibited good crystal quality. The electrode prepared to measure dopamine can be used after 72 hours from its preparation, as the whole structure should dry at room temperature.

The electrode shows sensitivity of 49 mV/decade, in the concentration range $1 \times 10^{-6} \times 10^{-1}$ M in 10^{-2} M acetic acid/sodium acetate buffer solution at pH 5.45. The electrode is characterized by short response time of approx. 6 s, lifetime – 8 weeks, and excellent repeatability.

Selectivity of the proposed sensor was described in the presence of common interferents such as ascorbic acid, uric acid, urea and glucose. It was discovered that the sensor may be usedto determine dopamine in plasma samples in the presence of the mentioned interferents of low concentration. However, if the interferences are of high concentration (over 1×10^{-2} M), the measurement may be distorted.

Cooper Potentiometric Electrode Based on β-cyclodextrin

Novel chemically modified carbon paste ion-selective electrodes (CMCPEs) for copper (II) ions determination based on β-cyclodextrin (β-CD) and β-cyclodextrin/1,4-bis(6-bromohexyloxy)benzene (β-CD/BHOB) as new ionophores were prepared and studied.

Preparation of the electrodes required the use of graphite paste consisting of pure graphite powder and (β-CD or β-CD/BHOB) ionophore and plasticizer (0.2 mL of DOP, TCP, DBP, DOS or o-NPOE). The modified paste is filled in electrode body and kept in distilled water for 24 hours before use.

A fresh surface was obtained by gently pushing the stainless-steel screw forward and polishing the new carbon-paste surface with filter paper to obtain a shiny new surface.

The optimisation of the electrodes was conducted (the influence of different quantities of cyclodextrins, the influence of the type of plasticizer) with the use of titration method and led to the achievement of the biggest jump. The best parameters were ascribed to the electrodes III and VI that contained: 10 mg β-CD, and 12.5 mg β-CD/BHOB plasticized with polar. A potential jump amounts to 125 and 214 mV for electrode (III) and electrode (VI), respectively. These sensors have wide linear dynamic range from 3.1×10^{-7} -1×10^{-2} and 1×10^{-7}-1×10^{-2} M with a Nernstian slope of 28.90 ± 0.92 and 29.94 ± 0.12 mV decade^{-1} and low detection limit of 3.1×10^{-7} and 1×10^{-7} M for electrode (III) and electrode (VI), respectively. The response time is up to 5 s. The electrode can be used in a pH range of 6-9. The -log K selectivity coefficients are: 2-2.9 for acetates, 1.8-2.2 for chlorates, 0.7-1.0 for nitrates and 0.6-1.3 for acetosalicylates. The electrodes can be used for about 3 to 4 months without any considerable divergence in potential for electrode (III) and electrode (VI), respectively. The frequently used Cu(II) solution of analytical and technical grade as well as Cu(II) ion in different water samples has been successfully titrated with RSD % 0.5-1.6 and recovery % was 97.2 – 100.8 for electrode III and RSD % 0.3-1.1, and 98.4 – 101.0 for electrode VI.

The Influence of Cyclodextrins on Potentiometric Measurements

The ability to form inclusive complexes by drugs with cyclodextrin may be estimated by direct potentiometric measurement that examines the influence of a given cyclodextrin of different concentration on the characteristics of the electrode sensitive to a given drug. The research was included in the study [80] that discussed preparation of ion-selective electrodes sensitive to propranolol and the analysis of the influence of different compounds, that is surfactants and cyclodextrin, on the electrode's parameters.

The effect of 1 × 10^{-6} to 1 × 10^{-2} M of β-cyclodextrin on the calibration curve of propranolol polyvinyl chloride selective electrode based on tetraphenylborate (PRO-TPB) ion-exchanger as an electroactive material was investigated. It was observed that the electrode potential of the calibration curves decreased gradually departing from Nernestian value 59 mV as β-cyclodextrin concentration increased. When the concentration of cyclodextrin reaches the 10^{-2}M, the change in the characteristic's slope of the electrode is highly visible. It may be caused by the formation of the complex. The hydrophobic cavity of cyclodextrin will eagerly incorporate lipophilic particles of propranolol; therefore the decreasing concentration of the drug in the solution will result in lower sensitivity of the electrode, and the potential decrease in the case of cationic drug.

The following article discusses the research of preparation of an indomethacin electrode and the changes of analytical parameters in the presence of 2-hydroxypropyl-β-cyclodextrin [81]. The research has been inspired by numerous scientific studies on the application of β-CD in pharmaceuticals of indomethacin and many other medicines that may be found in developing new dosage forms of drugs.

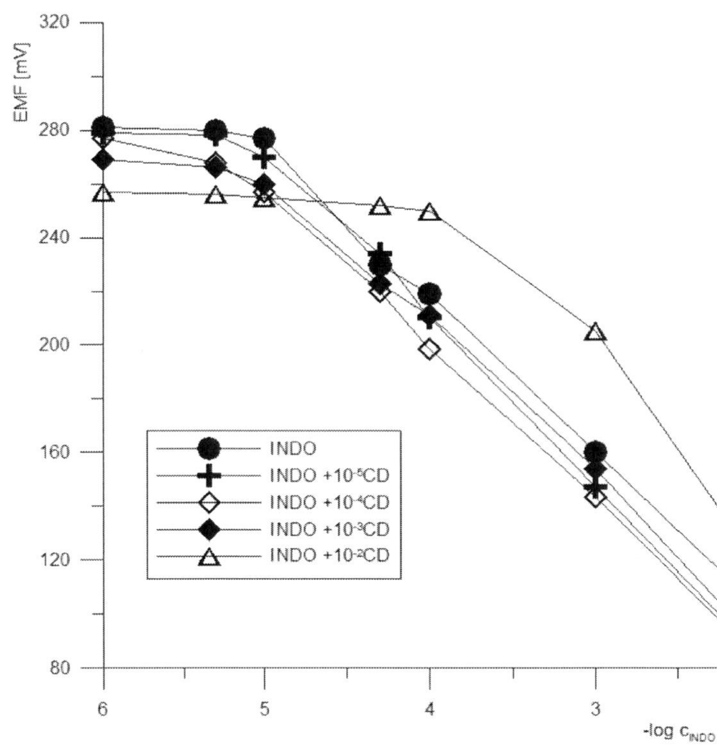

Figure 3.

Approaching cyclodextrins as potential interferents in drug quality control processes, for instance potentiometry, it seemed fully justified to perform investigations into the effect of HPβCD on the course of ion-selective calibration of indomethacin electrode. In the course of the research the calibration curves presented in Figure 3 were obtained for DBP plasticized electrode containing tetraoctylammonium 1-(p-chlorobenzoyl)5-methoxy-2-methyl-3-indolylacetate as ionophore.

The characteristic slope of electrode 4 remains constant to the value -63.1 mV $decade^{-1}$(in $1.0x10^{-5}$ M CD), -59.7mV $decade^{-1}$(in $1.0x10^{-4}$, $1.0x10^{-3}$ M CD). The concentration of HP β CD higher than $1.0x10^{-2}$ M causes the contraction of the linearity range (to $1.0x10^{-3}$ – $1.0x10^{-2}$ M), increase of detection limit ($3.16x10^{-4}$ M), and increase of potential. This effect might be attributed to the formation of inclusion complexes with the indomethacin in solution with HP β CD. This in turn may result in an increase in the electrode potential.

To conclude, the constructed electrode can be used even in the presence of β-cyclodextrin at a concentration below 1×10^{-2} M.

CONCLUSION

In recent years, growing interests in cyclodextrins have been observed in the field of chemistry, pharmaceutics, and numerous branches of industry. Cyclodextrins are becoming more and more popular and widely used in different analytical techniques, especially in potentiometry. Cyclodextrins' popularity is related to the possibility to create new inclusive complexes with a broad range of compounds as far as the construction and properties are concerned. The unique physicochemical properties of cyclodextrins and the modification possibilities of various hydroxyl groups allow for the synthesis of new selective receptors based on hydrophobic interactions, Van der Waals interactions, and hydrogen bonds. Thanks to that, the included molecule changes its properties, such as solubility, reactivity, and it affects spectrum.

Potentiometry uses cyclodextrins mainly as selective ionophores of electrodes sensitive to various organic compounds, especially drug substances. Other electrodes can also be used in the studies of complex forming reactions, establishing solidity constants of a drug – cyclodextrin complexes. The review of the literature on the subject and the progress in the preparation of ion-selective electrodes over the years, prove that the research constitutes an interesting and precious input into the development of chemistry and will be a growing trend in the years to come.

REFERENCES

[1] Janicki S., Fiebig A.: Farmacja stosowana, Wydawnictwo Lekarskie PZWL, 1998.

[2] http://www.naukowy.pl/encyklopedia/Cyklodekstryny

[3] Dodziuk H. Cyclodextrins and their complexes, Wiley-VCh Verlag GmbhH & Co. KGaA, Weinheim, 2006.

[4] Śliwa W. Cyclodextrins, Wydawnictwo Częstochowa, 2001.

[5] Brewster M. E., Loftsson T., Cyclodextrins as pharmaceutical solubilizers, Advanced Drug Delivery Reviews, 59 (2007) 645–666.

[6] Frömming K. H., Szejtli J.: *Cyclodextrins in pharmacy.* Dordrecht: Kluwer Academic Publishers, 1994.

[7] Loftsson T, Brewster M. Pharmaceutical applications of cyclodextrins. 1. Drug solubilization and stabilization, *J. Pharm. Sci.* (1996) 85, 1017-1025.

[8] Sousa, T.F.A., Amorim, C.G. , Montenegro, M.C.B.S.M., Araújo, A.N. Cyclodextrin based potentiometric sensor for determination of ibuprofen in pharmaceuticals and waters , Sensors and Actuators, B: *Chemical* (2013) 176, 660-666.

[9] El-Sayed, M.A. Advantages of the incorporation of 2-hydroxyl propyl beta cyclodextrin and calixarene as ionophores in potentiometric ion-selective electrodes for rivastigmine with a kinetic study of its alkaline degradation, Sensors and Actuators, B: *Chemical* (2014) 190, 101-110.

[10] Caballero J., Zamora C., Aguayo D., Yañez C., González-Nilo F.D., Study of the interaction between progesterone and β-cyclodextrin by electrochemical techniques and steered molecular dynamics, *Journal of Physical Chemistry* B (2008) 112, 10194-10201.

[11] Chen, M., Wei, X., Qian, H., Diao, G. Fabrication of GNPs/CDSH-Fc/nafion modified electrode for the detection of dopamine in the presence of ascorbic acid, *Materials Science and Engineering* C (2011) 31, 1271-1277.

[12] Izaoumen N., Cubillana-Aguilera L.M., Naranjo-Rodríguez I., de Cisneros J.L.H.-H, Bouchta D., Temsamani K.R., Palacios-Santander J.M., β-Sonogel-Carbon electrodes: A new alternative for the electrochemical determination of catecholamines, *Talanta* (2009) 78, 370-376.

[13] Ramnik S. Nitin B., Jyotsana M.,. Hiremath S. N., Characterization of Cyclodextrin Inclusion Complexes – A Review *Journal of Pharmaceutical Science and Technolog*y (2010) 2(3), 171-183.

[14] Mura, P. Analytical techniques for characterization of cyclodextrin complexes in aqueous solution: A review *Journal of Pharmaceutical and Biomedical Analysis* (2014) 101, 238-250.

[15] Jambhekar, S., Casella, R. Maher, T. The physicochemical characteristics and bioavailability of indomethacin from β-cyclodextrin, hydroxyethyl-β-cyclodextrin, and hydroxypropyl-β-cyclodextrin complexes, *International Journal of Pharmaceutics* (2004) 270 (1-2) 149-166.

[16] Brown, G.R., Caira, M.R., Nassimbeni, L.R., Van Oudtshoorn, B. Inclusion of ibuprofen by Heptakis(2,3,6-tri-O-methyl)-β-cyclodextrin: An X-ray diffraction and thermal analysis study, *Journal of Inclusion Phenomena and Molecular Recognition in Chem*istry, (1996) 26(4) 281-294.

[17] Tárkányi, G., Németh, K., Mizsei, R., Todoubleacuteke, O., Visy, J., Simonyi, M., Jicsinszky, L., Szemán, J., Szente, L., Structure and stability of warfarin-sodium inclusion complexes formed with permethylated monoamino-β-cyclodextrin *Journal of Pharmaceutical and Biomedical Analysis* (2013) 72, 292-298.

[18] Jambhekar, S., Casella, R., Maher, T. The physicochemical characteristics and bioavailability of indomethacin from β-cyclodextrin, hydroxyethyl-β-cyclodextrin, and hydroxypropyl-β-cyclodextrin complexes *International Journal of Pharmaceutics* (2004) 270 (1-2), 149-166.

[19] Poór, M., Kunsági-Máté, S., Szente, L., Matisz, G., Secenji, G., Czibulya, Z., Koszegi, T. Interaction of ochratoxin A with quaternary ammonium beta-cyclodextrin *Food Chemistry* (2015) 172(1) , 143-149.

[20] Sbârcea, L., Udrescu, L., Drăgan, L., Trandafirescu, C., Szabadai, Z., Bojiţă, M Thin-layer chromatography analysis for cyclodextrins inclusion complexes of fosinopril and zofenopril, *Farmacia* (2010) 58 (4), 478-484.

[21] Grandelli, H.E., Stickle, B., Whittington, A., Kiran, E. Inclusion complex formation of β-cyclodextrin and Naproxen: A study on exothermic complex formation by differential scanning calorimetry *Journal of Inclusion Phenomena and Macrocyclic Chemistry* (2013) 77 (1-4), 269-277.

[22] Jones, S.P., Parr, G.D. Polarographic detection of beta-cyclodextrin inclusion complexes *International Journal of Pharmaceutics*, (1986) 33 (1-3), 105-114.

[23] Soleymanpour, A., Ghasemian, M. Chemically modified carbon paste sensor for the potentiometric determination of carvedilol in pharmaceutical and biological media Measurement: *Journal of the International Measurement Confederation Volume* (2015) 59, 14-20.

[24] Patil, S.R., Turmine, M., Peyre, V., Durand, G., Pucci, B. Study of β-cyclodextrin/fluorinated trimethyl ammonium bromide surfactant inclusion complex by fluorinated surfactant ion selective electrode Talanta (2007) 74(1), 72-77.

[25] Kopecký F, Vojteková M, Kaclík P, Demko M, Bieliková Z. Bupivacaine hydrochloride complexation with some alpha- and beta-cyclodextrins studied by potentiometry with membrane electrodes. *J Pharm. Pharmacol.* (2004) 56, 581-7.

[26] Kopecký, F., Vojteková, M., Vrana, M., Čížová, K. Potentiometric study of carbisocaine micellization and inclusion complexation with α-cyclodextrin, β-cyclodextrin, methyl-β-cyclodextrin, and (hydroxypropyl)-β-cyclodextrin, *Collection of Czechoslovak Chemical Communications* (2002) 67(2), 245-256.

[27] Gląb S., Hulanicki A. Elektrody potencjometryczne jonoselektywne i uczulane. Nałęcz M. Biocybernetyka i inżynieria biomedyczna vol. 2, *Biopomiary, Exit, Warszawa* (2001) 513-543.

[28] T. de Boer, R.A. de Zeeuw, G.J. de Jong and K. Ensing. The Use of Charged Cyclodextrins in Capillary Electrophoresis for Chiral Separations in Pharmaceutical Analysis, Chapter 7 was published as a review in: *Electrophoresis*, 21 (2000) 3220-3239.

[29] De Boer, T., De Zeeuw, R.A., De Jong, G.J., Ensing, K. Recent innovations in the use of charged cyclodextrins in capillary electrophoresis for chiral separations in pharmaceutical analysis *Electrophoresis* (2000) 21(15), 3220-3239.

[30] Ganjali M. R. , Norouzi P., Rezapour M. , Faridbod F., Pourjavid M. R. , Supramolecular *Based Membrane Sensors, Sensors,* (2006) 6, 1018-1086.

[31] Izake E. L. Chiral discrimination and enantioselective analysis of drugs: An overview (Review) *Journal of Pharmaceutical Sciences* (2007) 96(7) 1659-1676.

[32] Recent Advances on Potentiometric Membrane Sensors for Pharmaceutical Analysis Combinatorial Chemistry & High Throughput Screening, Volume 14, Number 4, May 2011, pp. 284-302.

[33] [Kataky, R., Bates, P.S., Parker, D Functionalized α-cyclodextrins as potentiometric chiral sensors The Analyst.1992, 117 (8), 1313-1317

[34] Bates, P.S., Kataky, R., Parker, D A chiral sensor based on a peroctylated α-cyclodextrin, *Journal of the Chemical Society, Chemical Communications*, 1992, 2, 1313-131

[35] Bates, P.S., Kataky, R., Parker, D. Functionalized cyclodextrins as potentiometric sensors for onium *ions The Analyst* 119, (2) 1994, 181-186.

[36] Kataky, R.a, Kelly, P.M.a, Parker, D.a, Patti, A.F. Selective binding and sensing of guanidinium ions by lipophilic cyclodextrins *Journal of the Chemical Society*, 12, 1994, 2381-2382.

[37] Kelly, P.M., Kataky, R., Parker, D., Patti, A.F. Selective sensing of guanidinium and tetraalkylammonium ions using lipophilic cyclodextrins *Journal of the Chemical Society*, 11, (1995) 1955-1963.

[38] Kataky, R., Parker, D., Kelly, P.M. Potentiometric, enantioselective sensors for alkyl and aryl ammonium ions of pharmaceutical significance, based on lipophilic cyclodextrins *Scandinavian Journal of Clinical and Laboratory Investigation,* 55,(5) 1995, Pages 409-419.

[39] Kataky, R., Palmer, S. Local Anesthetics Measured by Lipophilic β-Cyclodextrin-Based Ion-Selective Electrodes, *Electroanalysis,* 8(6), (1996) 585-590.

[40] Kataky, R., Palmer, S., Parker, D., Spurling, D. Alkylated Cyclodextrin-Based Potentiometric and Amperometric Electrodes Applied to the Measurement of Tricyclic Antidepressants *Electroanalysis* 9(16) (1997)1267-1272.

[41] Odashima, K., Hashimoto, H., Umezawa, Y Potentiometric discrimination of organic amines by liquid membrane electrodes based on a long alkyl chain derivative of β-cyclodextrin , *Mikrochimica Acta,* 113 (3-6) (1994) 223-238.

[42] Ben Ali, M., Kalfat, R., Sfihi, H., Ben Ouada, H. , Chovelon, J.M., Jaffrezic-Renault, N. Cyclodextrin-polymethylhydrosiloxane gel as sensitive membrane for heavy ion sensors, *Materials Science and Engineering* C, 6(1) (1998) 53-58.

[43] Kataky, R., Toth, K., Palmer, S., Feher, Z. A study of the effect of proteins and endogenous cations on a lipophilic β-cyclodextrin-based potentiometric lidocaine sensor using discrete solution and flow-injection analysis *Talanta* 50(5) (1999) 939-946.

[44] Barriada, J.L., Covington, A.K., Kataky, R. New cyclodextrin-based ion-selective electrodes for determining activity coefficients of biologically important onium ions *Journal of Solution Chemistry*, 28 (5) (1999) 555-565.

[45] Pimenta, A.M., Araújo, A.N., Montenegro, M.C.B.S.M. Simultaneous potentiometric and fluorimetric determination of diclofenac in a sequential injection analysis system *Analytica Chimica Acta* 470(2) (2002) 185-194.

[46] Cheng, J.S., Xian, X.S., Aboul-Enein, H.YPotentiometric, enantioselective electrode responsive to propranolol enantiomers based on a lipophilic β-cyclodextrin . *Analytical Letters* 371(11) (2004) 2259-2266.

[47] Ozoemena, K.I., Stefan, R.-I. , Staden, J.F.V., Aboul-Enein, H.Y. Enantioanalysis of S-perindopril using different cyclodextrin-based potentiometric sensors , Sensors and Actuators, B: *Chemical,* 105(2) (2005) 425-429.

[48] Ozoemena, K.I., Stefan, R.-I. Enantioselective potentiometric membrane electrodes based on α-, β- And γ-cyclodextrins as chiral selectors for the assay of l-proline *Talanta* 66(2) (2005) 501-504.

[49] Staden, R.-I.S.-V., Nejem, R.M. Enantioselective, potentiometric membrane electrodes based on cyclodextrins for the assay of Analytical Letters 38 (12) (2005) 1847-1855.

[50] El-Kosasy, A.M. Potentiometric assessment of Gram-negative bacterial permeabilization of tobramycin *Journal of Pharmaceutical and Biomedical Analysis* 42 (3) (2006) 389-394.

[51] Stefan-van Staden, R.-I., Nejem, R.M. Cyclodextrins-based enantioselective, potentiometric membrane electrodes for l-vesamicol assay in serum samples Sensors and Actuators, B: *Chemical* 117(1) (2006) 123-127.

[52] El-Saharty, Y.S., Metwaly, F.H., Refaat, M., El-Khateeb, S.Z. Development of membrane electrodes for the selective determination of hyoscine butylbromide *Talanta* 72 (2) (2007) 675-681.

[53] Amorim, C.G., Araújo, A.N., Montenegro, M.C.B.S.M. , Silva, V.L. Cyclodextrin-based potentiometric sensors for midazolam and diazepam *Journal of Pharmaceutical and Biomedical Analysis* 48(4) (2008) 1064-1069.

[54] Pires, A.R., Araújo, A.N., Montenegro, M.C.B.S.M. , Chocholous, P., Solich, P.

[55] New ionophores for vitamin B1 and vitamin B6 potentiometric sensors for multivitaminic control *Journal of Pharmaceutical and Biomedical Analysis,* 46, (4) (2008) 683-691.

[56] Kaniewska, M., Sikora, T., Kataky, R., Trojanowicz, M. Enantioselectivity of potentiometric sensors with application of different mechanisms of chiral discrimination. *Journal of biochemical and biophysical methods* 70 (6) (2008) 1261-1267.

[57] Pires, A.R., Araújo, A.N., Lopes, J.A., Montenegro, M.C.B.S.M. Simultaneous potentiometric determination of thiamine and pyridoxine in multivitamins using a single cyclodextrin-based thiamine-selective electrode Analytical Letters 42 (13) (2009) 1923-1939.

[58] El-Kosasy, A.M., Salem, M.Y., El-Bardicy, M.G., El-Rahman, M.K.A. Membrane electrodes for the determination of pyridostigmine bromide *Journal of AOAC International* 92 (6) (2009) 1631-1638.

[59] Rizk, N.M.H., Abbas, S.S., El-Sayed, F.A., Abo-Bakr, A. Novel ionophore for the potentiometric determination of cetirizine hydrochloride in pharmaceutical formulations and human urine *International Journal of Electrochemical Science* 4 (3) (2009) 396-406.

[60] Cunha, C.O., Silva, R.C.R., Amorim, C.G., Júnior, S.A., Araújo, A.N., Montenegro, M.C.B.S.M., Silva, V.L. Tetracycline Potentiometric Sensor Based on Cyclodextrin for Pharmaceuticals and Waste Water Analysis Electroanalysis 22 (24) (2010) 2967-2972.

[61] Khaled, E., Hassan, H.N.A., Mohamed, G.G., Ragab, F.A., Seleim, A.E.A. β-Cyclodextrin-based potentiometric sensors for flow-injection determination of acetylcholines *International Journal of Electrochemical Science* 5 (4) (2010) 448-458.

[62] Khaled, E., Hassan, H.N.A., Mohamed, G.G., Ragab, F.A., Seleim, A.E.A. Disposable potentiometric sensors for monitoring cholinesterase activity *Talanta* 83 (2) (2010) 357-363.

[63] Almeida, S.A.A., Heitor, A.M., Montenegro, M.C.B.S.M., Sales, M.G.F. Sulfadiazine-selective determination in aquaculture environment: Selective potentiometric transduction by neutral or charged ionophores *Talanta* 85 (3) (2011) 1508-1516.

[64] Khaled, E., Kamel, M.S., Hassan, H.N.A., Aboul-Enein, H.Y. Cyclodextrin-based dextromethorphan potentiometric sensors *Journal of Electroanalytical Chemistry* 661(1) (2011) 239-244.

[65] Khattab, F.I., Ramadan, N.K., Hegazy, M.A., Ghoniem, N.S. Microsized graphite sensors for potentiometric determination of metronidazole and spiramycin *Portugaliae Electrochimica Acta* 29 (2) (2011) 79-90.

[66] Y. Umezawa, P. Bühlmann, K. Umezawa, K. Tohda, S. Amemiya, "Potentiometric Selectivity Coefficients of Ion-Selective Electrodes Part I. Inorganic Cations (Technical Report)," *Pure Appl. Chem.* 72 (2000) 1851-2082.

[67] Ramadan, N.K., Zaazaa, H.E., Merey, H.A. Microsized graphite sensors for potentiometric determination of cyclobenzaprine hydrochloride in pure powder, tablets, and plasma *Journal of AOAC International*, 94 (6) (2011) 1807-1814.

[68] Ramadan, N.K., Merey, H.A. Ion-selective electrodes for the potentiometric determination of pramoxine HCl using different ionophores *Acta Chimica Slovenica* 59(4) (2012) 870-878.

[69] Almeida, S.A.A., Heitor, A.M., Sá, L.C., Barbosa, J., Da Conceição, M., Montenegro, B.S.M., Sales, M.G.F. Solid contact PVC membrane electrodes based on neutral or charged carriers for the selective reading of anionic sulfamethoxazole and their application to the analysis of aquaculture water, *International Journal of Environmental Analytical Chemistry*, 92(4) (2012) 479-495.

[70] Cuartero, M., Ortuño, J.A., García, M.S., Martínez-Ortiz, F. Differential dynamic potentiometry with ion selective electrodes: A tool for drug fingerprinting *Electrochimica Acta* 69, 1 (2012) 152-159.

[71] Khaled, E., Kamel, M.S., Hassan, H.N.A., Haroun, A.A., Youssef, A.M., Aboul-Enein, H.Y. Novel multi walled carbon nanotubes/β-cyclodextrin based carbon paste electrode for flow injection potentiometric determination of piroxicam *Talanta* 97 (15) (2012) 96-102.

[72] AbdElHamid, G., ElGhobashy, M.R., ElZanfaly, E.S., Amer, S.M., ElKhateeb, S.Z. Ion selective membrane electrodes for stability-indicating determination of amisulpride Analytical and Bioanalytical *Electrochemistry* 4 (6) (2012) 579-592.

[73] Khaled, E. , Kamel, M.S., Hassan, H.N., Abd El-Alim, S.H., Aboul-Enein, H.Y. Miniaturized ionophore-based potentiometric sensors for the flow-injection determination of metformin in pharmaceutical formulations and biological fluids Analyst 137 (23) (2012) 5680-5687]

[74] Elzanfaly, E.S., Zaazaa, H.E., Merey, H.A. Ion selective phosphotungestate and β-cyclodextrin based membrane electrodes for stability-indicating determination of midodrine hydrochloride Acta *Chimica Slovenica* 60, (2) (2013) 256-262.

[75] Hassan, S.S.M., Kamel, A.H. , El-Naby, H.A. New potentiometric sensors based on selective recognition sites for determination of ephedrine in some pharmaceuticals and biological fluids , *Talanta* 103 (15) (2013) 330-336.

[76] Elghobashy, M.R., Rezk, M.R. Comparative study of different ionophores in ion selective electrodes for stability indicating determination of moxifloxacin *Analytical and Bioanalytical Electrochemistry* 6 (4) (2014) 461-474.

[77] Lenik, J. A new potentiometric electrode incorporating functionalized β-cyclodextrins for diclofenac determination *Materials Science and Engineering C* 45 (2014) 109-116.

[78] Abdel Ghany, M.F., Hussein, L.A., Yamani, H.Z. Comparative study of calix-6-arene, 2-hydroxy propyl beta cyclodextrin and 18-crown-6 as ionophores in potentiometric ion- selective electrodes for determination of trigonelline in Trigonella foenum-graecum seeds extract and plasma *Journal of Applied Pharmaceutical Science* 4 (7) (2014) 32-37.

[79] Elhag, S., Ibupoto, Z.H., Nur, O., Willander, M. Incorporating β-Cyclodextrin with ZnO nanorods: A potentiometric strategy for selectivity and detection of dopamine, Sensors (Switzerland) 14 (1) (2014) 1654-1664.

[80] Ali, T.A., Eldidamony, A.M., Mohamed, G.G., Elatfy, D.M. Construction of chemically modified electrode for the selective determination of copper(II) ions in polluted water samples based on new β-cyclodextrine and 1,4-bis(6-bromohexyloxy)benzene ionophores *International Journal of Electrochemical Science* 9 (5) (2014) 2420-2434.

[81] El-Sherbiny, D., Elewady, G.Y. , Shalabi, K., El-Askalany, A. Effect of different organized media on ion-selective electrode (ISE) determination of propranolol in pharmaceutical dosage forms *Journal of Food and Drug Analysis,* 17 (2) (2009) 85-92.

[82] Lenik, J., Wardak, C. Studies on influence of (2-Hydroxypropyl)-β-cyclodextrin on properties of a new indomethacin electrode, *IEEE Sensors Journal* 13(12) (2013) 4638-4647.

In: Cyclodextrins
Editor: Francis G. Ramirez

ISBN: 978-1-63482-788-1

Chapter 2

APPLICATIONS OF CYCLODEXTRINS IN BIOMEDICAL ELECTROCHEMISTRY

Luminiţa Fritea, Mihaela Tertiş and Robert Săndulescu
Analytical Chemistry Department,"Iuliu Haţieganu"
University of Medicine and Pharmacy,
Cluj-Napoca, Romania

ABSTRACT

The supramolecular complex formation between various chemical species and cyclodextrins, significantly influences the physical-chemical features of the guest molecule entrapped inside the cyclodextrins (CD) cavity. The inclusion complexes of many pharmaceutical substances with CDs have been investigated by using various analytical methods, including the electrochemical ones. The decreased peak currents and shifts in the oxidation and reduction potentials are the main changes of the electrochemical behavior of the guest molecules, attributed to the decrease of concentration, smaller diffusion coefficients of the complexes, and to their relatively high stability.

Different techniques, such as: inclusion in carbon paste electrode, entrapment in polymers or in plasticized membranes, immobilization onto screen printed electrodes, nanoparticles, carbon based materials, and nanostructured polymeric films, Langmuir-Blodgett technique, self-assembled monolayers, were used for the development of novel (bio)sensors based on CDs and their derivatives.

The CDs based electrodes represents a promising future for a lot of applications in various fields: agriculture, food, toilet articles, textile manufacturing, environment, pharmaceuticals, and therapy. The CDs specific features recommend them for a wide range of applications in analytical chemistry as catalysts, enantioselectors and molecular receptors for the development of new electrochemical sensors/biosensors.

INTRODUCTION - SUPRAMOLECULAR COMPLEXATION VIA "HOST-GUEST" INTERACTION

Cyclodextrins (CDs) are non-reducing cyclic oligosaccharides consisting of six (α-cyclodextrin), seven (β-cyclodextrin), eight (γ-cyclodextrin) or more D(+)-glucopyranose units (chair conformation) linked by α-bonds (1,4-bonds), with relatively hydrophobic cavity compared to water and with a hydrophilic outside described as a 'micro heterogeneous environment' [1-8]. The molecular lipophilicity models (MLM) of the cyclodextrin surface, achieved by molecular modeling with MOLCAD software, allowed the view of the surface regions with hydrophilic and hydrophobic behavior, respectively. To be easier to examine, they were represented with different colors: hydrophilic regions, with shades of blue, and the hydrophobic ones from yellow to brown, depending on the strength of hydrophobicity (Figure 1) [9].

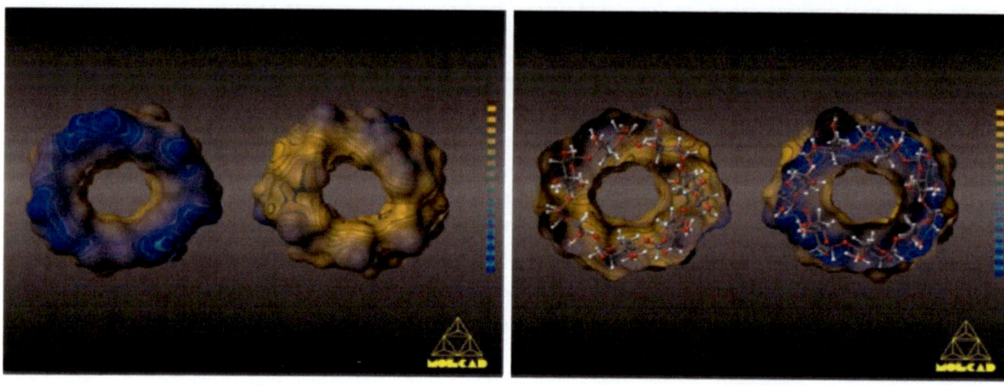

A B

Figure 1. MLM of the contact surfaces of (A) α-cyclodextrin (upside view with 2-OH/3-OH groups (left) and downside view with 6-CH_2OH (right) of the toroidal structure, models were rotated 180° around vertical axis); (B) β-cyclodextrin (half of the molecule has been removed to allow viewing of the interior of the structural model).

The most notable characteristic of CDs is the inclusion complexes formation with various organic and inorganic molecules (solid, liquid and gaseous compounds) in aqueous, nonaqueous and mixed media. The complex formation consist in a dimensional fit between host cavity and guest molecule when the high enthalpy water molecules from the host cavity are displaced by the more hydrophobic guest molecules leading to a nonpolar-nonpolar association and a decrease of cyclodextrin ring strain with a more stable lower energy state.

The main driving forces for the inclusion complexes include Van der Waals interactions, hydrophobic interactions between the guest hydrophobic moiety and the CD cavity, hydrogen bonding between the polar groups of the guest molecules and the CDs hydroxyl groups and electrostatic interactions for ionic guests.

The inclusion complexation leads to a significant change of the physico-chemical properties of the guest (solubility, volatility, taste, stability, reactivity, electrochemical and spectral properties) entrapped inside the CD cavity. The main analytical techniques which are used for the characterization of inclusion complexes include: spectroscopic techniques, electroanalytical techniques, separation techniques, polarimetry, isothermal titration calorimetry, scanning electron microscopy (SEM), X-ray crystallography [10, 11].

Their unique behavior is responsible for the widespread applications of CDs in agriculture, food, toilet articles, textile industry, environment monitoring, pharmaceutical and medical areas [12]. CDs are widely used in analytical chemistry as catalysts, enantioselectors and molecular receptors. They were employed for the determination of certain analytes and many results concerning electrochemical sensing devices were published [1-3].

2. Cyclodextrins Applications in Electroanalysis

Cyclodextrins (in their native form or chemically modified) are the subject of extended research in electrochemistry and this is mainly due to their properties, such as: variable size and hydrophobic cavities. The research interest is focused both on their behavior in solutions and embedded in thin films formed at the electrode surface.

2.1. Electrochemical Behavior of Cyclodextrins in Solution

A great variety of compounds were used as guest molecules within CD cavity and among them there are also many substances presenting pharmacological effects. The most important such complexes were obtained in the case of: sodium barbitone, phenylhydrazine hydrochloride, 1,4-benzoquinone, 9,10-anthraquinone, phenothiazine, indomethacin, nalidixic acid, irisquinone, catecholamines (adrenaline, dopamine), hexamethylenetetramine, lumazine, rutin, *β*-lapachone, steroids (estrone, estradiol), methyl violet and ethyl violet, gliclazide, Pb, chloramine T, chlorpromazine, vitamin K_3 and ascorbic acid [2].

Among the various analytical methods employed for the study of interactions between the guest molecules and the host CD we address only to electrochemical ones, such as: potentiometry, cyclic voltammetry (CV), differential pulse stripping voltammetry (DPSV), differential pulse voltammetry (DPV) by using different electrodes. The formation of supramolecular complex between the drug molecule and CD significantly modifies the electrochemical behavior of the drug [2].

In the case of CV, the main effect consists in the decrease of the peak currents and shifts in the oxidation and reduction potentials due to the relatively high stability and the smaller diffusion coefficients of the complexes compared with the guest alone [2]. It was also observed that in aqueous solutions CDs inhibited and even greatly suppressed the electrochemical oxidation of the guest molecules [13, 14] and this behavior can be used for the protection of the easily oxidative molecules.

The experimental results suggest that both the dimension and polarity of the guest and host molecules influence the inclusion complex stability. It was also observed that the stability of the indomethacin inclusion complexes presented a higher affinity for *β*-CD because the guest-host interaction was the most exergonic [15]. The same observation was noticed for nalidixic acid revealing 2 recognition elements in the binding process: hydrophobic interaction between 2 methylpyridine sides and the *β*-CD cavity, and a hydrogen bond between the carboxyl group which remained outside the cavity and the secondary hydroxyl group of β-CD [16]. Lumazine had a greater affinity for α-CD than for β-CD [17].

The influence of β-CD on the electrochemical behavior of ascorbic acid (AA), uric acid (UA), caffeine, caffeine citrate, theophylline, aminophylline and acetaminophen (AC) was studied by using carbon paste electrodes. It can be noticed that the potential is anodicaly shifted and the current intensity is decreasing after the addition of β-CD in the solution. The decrease in current intensity was attributed to the decrease in concentration of free analyte molecules in the solution due to the entrapment in the β-CD host cavity that hinder the oxidation for the encapsulated molecules. The positive shift of the E_p indicates that the oxidation center of the analyte is included in the β-CD cavity, leading to a more difficult oxidation. The formation of the inclusion complexes was also investigated by FTIR, which revealed some important modifications of the complexes spectra, consisting in shift, attenuation and disappearance of some vibration bands in comparison with the analyte and β-CD spectra [18].

2.2. Cyclodextrin Based (Bio)Sensors in Pharmaceutical and Biomedical Analysis

The CDs applications in electrochemistry are based on the inclusion complexes formation, molecular recognition and selective preconcentration of the analyte at the electrode surface leading to the development of highly selective sensors and biosensors. Many techniques can be used for designing the molecular architecture of the electrode/solution interface which supplies various and attractive methods for the preparation of CD based electrochemical sensors.

The most significant modification methods that involve CD are: *Langmuir-Blodgett and Self-Assembly method*; the *incorporation in plasticized membranes*; the *deposition of a film at the surface of solid electrodes*; the *inclusion in carbon paste*; the use *of CD for the decoration of metal particles*; and *CD embedded in different polymer films*.

2.2.1. Langmuir-Blodgett Technique and Self-Assembly Method

The Langmuir-Blodgett monolayers are obtained in the presence of the amphiphilic CD molecules that promote the monolayers transfer from the air-water interface onto solid substrates [2]. Highly organized monolayers (self-assembled monolayers, SAMs) are formed onto metal electrodes (especially gold) by CDs modified with organo-sulfur compounds (thiols, disulfides, silanes). These substances present chemisorption capacity and the number of the thiol groups and the spacer length, between the thiol group and the CD cavity, highly influence the density and permeability of CD films. Thus, the film obtained with monothiolated CDs is less permeable than multithiolated films, the packing density of the CD cavities increases with spacer length [19]. The potential-controlled adsorption of β-CD on Au(III) surfaces lead to the formation of "nanotube" structures [2].

The preparation of CD monolayers is used to obtain modified electrodes responsive towards different guest molecules such as: ferrocene, ibuprofen, methylene blue, dopamine, menadione, etc. [20]. The inclusion into cyclodextrin cavities is monitored in order to evidence the mediating role of the immobilized electroactive molecules towards the electroinactive species from the solution. The guest molecule included in the cavity of the CD attached to the electrode surface may act as a mediator, providing electrical contact and the electrons transfer between the electrode and the solution species. This function of the CD complex is especially important in case of electro and bioelectrocatalysis and was observed for methylene blue and dopamine [21]. The CD monolayer separates mechanically the biomolecules (e.g. redox enzymes) from the electrode surface, thus preventing their denaturation or conformational changes. CD attached to the electrode surface does not cover the electrode uniformly and bare regions of surface are still exposed to the solution. On the other hand, multi anchoring guest is not a dynamic structure and the anchoring does not reorganize the formed monolayer. Therefore the concept of a sealing component responsible for covering the surface between the CD molecules was introduced. The CD cavities have to be earlier blocked by a guest so that the sealing molecule does not enter the cavities during the self-assembly process [22].

These types of CDs modified electrodes have many applications in pharmaceutical and medical analysis. Many sensors based on the formation of CD-SAMs were thus developed. A highly sensitive, reproducible and stable sensor based onSAM of thiolated *β*-CD on a gold electrode was elaborated for the specific determination of ultratrace lead concentrations in blood with a limit of detection (LOD) of 7.1 nM [23].

A novel and easy to regenerate sensor platform based on supramolecular interactions has been developed, based on the self-assembly of bifunctionalized polymers bearing adamantane and DNA onto CD surfaces. The sensor presented a LOD of 0.08 nM and high selectivity, clearly differentiating between complementary and non-complementary DNA sequences in real samples [24]. Another approach based on the patterning of gold electrodes with a mixed SAM of perthiolated *β*-CD and 1-octanethiol was reported for the sensitive electrochemical determination of transglutaminase activity. The *β*-CD functionalized with primary amino groups act as an amino donor substrate for transglutaminase and provides molecular nanopores for the enzyme-controlled diffusion of $Fe(CN)_6^{3-/4-}$ to the electrode surface [25].

The construction of nanoporous assemblies based on single SAM of mono-(6-deoxy-6-mercapto)-*β*-cyclodextrin (*β*-CDSH) and two-component monolayers of *β*-CDSH and 11-mercaptoundecanoic acid was also described and was applied for the investigation of the tricyclic antidepressants drugs, chlorprothixene and imipramine [26]. A chemically modified electrode based on a thiolate *β*-CD derivative in the form of SAM was elaborated and was used for the evaluation of trimethyilcetylammonium ion and ursodeoxycholic acid. This sensor can work as channel sensor being used for the detection of non electroactive analytes forming a strong inclusion complex with *β*-CD in the presence of a suitable marker able to permeate its cavity [27].

A molecular imprinting method involving a three-step sequential self-assembly procedure was applied to prepare gold electrodes responsive towards ibuprofen, based on the competition of ferrocene and ibuprofen for the cavities of CDmolecules arranged in monolayer [28].

β-CD coated Au electrodes were also used for a reversible immobilization of adamantane modified phenylalanine dehydrogenase by a supramolecular association consisting in the formation of inclusion complexes between CD and adamantane. The biosensor showed a linear amperometric response up to 3 mM *L*-phenylalanine with a LOD of 15 μM [29, 30]. Another two sensors based on Au electrode modified with a *β*-CD/thioctic acid mixed monolayer and a self-assembled *D,L*-cysteine grafted *β*-CD sulfonic acid monolayer respectively, were developed for selective dopamine detection in the presence of AA as a common interfering compound [31, 32]. Dopamine was included in the cavities of *α*-CD molecules, anchored to the gold electrode surface in a SAM that acts as a mediator for the dopamine molecules diffusion from the bulk solution to the electrode [33]. The inclusion complexes between cortisol and cortisone and a SAM of *β*-CD have been evaluated by using both, permeable and impermeable, electroactive species for the *β*-CD cavity [34].

2.2.2. Cyclodextrins in Plasticized Membranes

The lipophilic CDs can be incorporated in plasticized, polyvinyl chloride (PVC) membranes and used as ionophores in the fabrication of potentiometric ion-selective electrodes for the detection of a wide range of pharmaceutical and biomedical analytes (ephedrinium ions, tetraalkylammonium ions, dopaminium ion, local anesthetics, tricyclic antidepressants as imipramine, desipramine, trimipramin, propranolol, ephedrine, amphetamine, propanolol, guanidinium ions as metformin, phenformin, guanidine, creatinine, choline, acethylcholine, lidocaine); and also for the enantioselective determination of ephedrinium ions and (+) propanolol. Voltametric and amperometric sensors for dopamine and tricyclic antidepressant drugs were elaborated by depositing alkylated CDs embedded in plasticized polymer matrix at the surface of screen-printed electrodes. The highly selective detection of acetylcholine was performed at subpicomolar levels in the presence of AA, dopamine and atropine as interfering species by using a screen-printed electrode modified with layers of ferrocene-based redox mediator, enzymes (horseradish peroxidase, choline oxidase and acetylcholine esterase) and alkylated-*β*-CD, embedded in a plasticized polyurethane membrane [35].

Two polymeric membranes were fabricated by incorporating *β*-CD and (2-hydroxipropyl)-*γ*-CD as ionophores, 2-fluorophenyl 2-nitrophenyl ether as plasticizer and potassium tetrakis (*p*-chlorophenyl) borate as ionic additive and were used for the development of ion-selective electrodes for diazepam and midazolam analysis in pharmaceuticals [36].

Three novel potentiometric sensors for dextromethorphan: silver coated wire electrode, coated graphite electrode and the conventional PVC membrane electrode, were developed based on CD. The presence of CDs ionophores determines the considerable increase in selectivity towards dextromethorphan and the Nernstian slopes are related to the type of the electrode and the nature of the used ionophore. The best analytical performances were obtained on silver coated wire electrode incorporated with heptakis (2,3,6-tri-O-methyl)-*β*-CD and sodium tetrakis(4-fluorophenyl) borate with *o*-nitrophenyloctylether as membrane plasticizer (Nernstian slope of 59.0±0.22 mV decade^{-1} and a LOD of 0.07 μM) [37].

Good analytical performances for ephedrine (LOD of 5.7 μM; fast response: 10 s; wide pH range; better selectivity than previously reported ephedrinesensors) were obtained by using potentiometric ephedrine-selective membrane sensors using triacetyl-*β*-CD as neutral ionophore, dioctylsebacate as plasticizer, Na tetraphenyl borate as anionic excluder and PVC as polymeric matrix [38].

Another potentiometric sensor was developed for ibuprofen monitoring in pharmaceuticals and water based on *α*-CD as ionophore, *o*-nitrophenyloctylether as plasticizer, tetradodecylammonium bromide and PVC. The sensor presented a constant sensitivity of -59.0 mV decade^{-1} in the concentration range of $3.87 \cdot 10^{-6}$to 10^{-2}M at pH 9 and a LOD of 3.34±0.03 μM for ibuprofen [39].The preparation of diclofenac-selective membrane electrodes incorporating 2-hydroxypropyl)-β-CD, heptakis(2,3,6-tri-*o*-methyl)-β-CD and heptakis(2,3,6-tri-*o*-benzoyl)-β-CD as ionophores was also reported. After optimization, heptakis(2,3,6-tri-*o*-benzoyl)-β-CD based membrane electrode provided the best analytical results (near-Nernstian response slope of -60.0 mV decade^{-1} and a LOD of 14μM). The proposed electrode can easily discriminate diclofenac ions in the presence of several inorganic and organic interferents and some common drug excipients [40].

A high sensitive dopamine potentiometric sensor was also elaborated using ultrathin cobalt oxide (Co_3O_4) nanowires grown on Au coated glass substrates, PVC as plasticized polymer and β-CD as ionophore [41].

2.2.3. Cyclodextrins Modified Screen-Printed Electrodes

Screen printed electrodes (SPEs) are miniaturized devices, based on different layers of inks printed on an inert plastic or ceramic substrate suitable for the development of disposable sensors. The design and operation of analytical determinations by using SPEs is easier and in accordance with the requirements for a decentralized assay [42]. Recently, the SPEs have demonstrated a great interest especially in the development of rapid single use analytical analysis and biosensor's fabrication. Moreover, regarding their simplicity, they are simply mass-produced at low costs, having at the same time superior characteristics in comparison with the commonly used electrode materials.

Carbon based SPEs were modified with thin films of the β-CD condensation polymers or carboxymethylated β-CD for the determination of tricyclic antidepressive drugs from serum samples, by using a simple two-step procedure, consisting in preconcentration, followed by DPV quantification. The accumulation propertiesof the carboxymethylated β-CD film appeared to be favorable as compared to those of the β-CD film [43].

An α-CD modified graphite based SPE was obtained by the integration of CD directly inthe graphite ink used for the printing of the working electrode. This electrode was applied for the simple and sensitive determination of phenolic compounds in aqueous solution and a LOD of 30 nM was obtained by using SWV [44].

Dopamine selective quantification was also achieved in the presence of interfering agents such as AA and UA, using a SPE modified by the electropolymerization of β-CD with glucose oxidase (GOx) and multi-walled carbon nanotubes (MWCNT). This biosensor displayed a LOD of 0.48 μA and a sensitivity of 0.0302 μA μM^{-1} for dopamine [45].

2.2.4. Cyclodextrins Modified Carbon Paste Electrodes

Carbon paste electrodes (CPEs) were intensively used for a wide range of electrochemical applications as a simple, fast and low cost manner to regenerate the electrode surface. Carbon paste is obtained by simply mixing graphite microparticles with paraffin, silicon or other and more important, itcan be easily modified with different substances (including cyclodextrins) that can be displayed on the electrode surface, embedded in the carbon paste or in polymer films.

A selective and sensitive hydroxypropyl-β-CD modified CPE was elaborated for the quantification of (+) catechin in some commercial drinks (tea, cocoa, and coffee) and urine samples and good analytical parameters were obtained both for anodic and cathodic voltammetry [46, 47].

Nitrendipine determination was achieved with a CPE modified with β-CD by using DPV with a significant increase in the reduction current peak, probably due to the formation of an inclusion complex with β-CD [48]. A β-CD modified CPE presenting significantly increased sensitivity and selectivity for donepezil was applied with a LOD of 1.2·nM [49]. α- and β-CD electroactive films were electrogenerated on CPEs in $HClO_4$and the obtained sensors were applied for the determination of Pb(II) ions in solution [50].

β-CD-ferrocene inclusion complex was obtained at the surface of a CPE and used for AA electrocatalytic oxidation. A LOD of 0.1 μM was obtained for AA by using this modified CPE [51]. The electrochemical determination ofthe tricyclic antidepressants (imipramine, trimipramine and thioridazine) [52] and dopamine [53] in pharmaceuticals was investigated on β-CD modified CPE.

The inclusion complexes formation between the quinone groups of the drugs: sparfloxacin, ofloxacin, norfloxacin, gatifloxacin and lomefloxacin and β-CD at the surface of a CD modified CPE were described. These complexes wereresponsible for the increase in the reduction peak currents and the modified electrode exhibited good sensitivity and stability, which allowed its application to the determination in drugs and spiked human urine [54, 55]. Another type of inclusion complexes were formed between the corticosteroids ketone functional group and β-CD modified CPE, presenting increased sensitivity and accuracyforprednisolone, dexamethasone and hydrocortisone determination [56].

A CPE modified with a β-CD-azomethine-H inclusion complex, where azomethine-H plays the role of molecular recognition agent, was developed for the potentiometric determination of H_3BO_3 in aqueous solution. Glutaraldehide and Nafion were used to avoid the loss of the recognition agent [57]. A β-CD modified CPE was elaborated and exhibited an increased sensitivity and selectivity for anagrelide in drugs, spiked serum sample and urine [58].

Some fluoroquinolones (ciprofloxacin, ofloxacin, norfloxacin and gatifloxacin) were analysed by using an electrochemical sensor based on the polymerization of β-CD and *l*-arginine (*l*-arg) at CPE surface. Electrochemical impedance spectroscopy (EIS) indicated that the β-CD/*l*-arg copolymer efficiently decreased the charge transfer resistance value of the electrode and improved the electron transfer kinetic. This sensor was successfully applied to detect the above mentioned fluoroquinolonesin pharmaceuticals and human serum with good reproducibility, long-term stability and fast current response [59].

The tenoxicam oxidation was studied at a CPE modified with a β-CD polymer. The process that occurs is controlled by the adsorption of the inclusion complex formed with the β-CD molecules from the surface of the polymer [60]. Electroanalytic parameters for the determination of nifedipine using β-CD and CNTs as modifiers in a CPE were established. The sensor exhibit high sensitivity, low detection limit and excellent reproducibility and was applied for thenifedipine detection and quantification in pharmaceutical formulations, urine and human serum [61].

A CPE modified with an electrochemical generated conducting polymer based on β-CD, drastically influenced the electrochemical response of dopamine, increasing the peak current density and diminishing the peak potential as a result of the supramolecular interactions between dopamine and the poly-β-CD. At the bare CPE, the dopamine redox process was diffusion-controlled, while at the CPE/poly-β-CD it changed to an adsorption-controlled one. This sensor was tested for the dopamine detection in real samples with excellent recovery and low interference in the presence of AA [62].

Two different strategies using CPEs were reported by Fritea *et al* [18]. The first one consisted in the β-CD powder embedded in carbon paste, while the second one involved the formation of a homogenuous paste with a 10^{-2} M β-CD aqueous solution. The two types of β-CD modified CPEs were employed for AA and UA determinations. The peak intensity was significantly increased (three times) in the case of AA electrochemical oxidation on both β-CD modified CPEs, while the potentials were strongly shifted to more negative values (with 0.134 V and 0.091 V respectively) indicating the formation of inclusion complexes between the β-CD and the guest molecules. This complexation leads to a preconcentration of the analyte at the electrode surface, suggesting that higher complexation efficiency is obtained with the paste prepared with β-CD in solution due to the higher homogeneity of the carbon paste, than the one prepared with β-CD added as powder. In the case of UA, the peak intensity is higher only at the β-CD/CPE obtained with the addition of β-CD solution and the potential shift is much smaller than in the case of AA determination [18].

2.2.5. Nanoparticles Decorated with Cyclodextrins

The recent developments of gold nanoparticles (AuNPs) based biosensors are accomplished due to the unique properties of this material: immobilization of biomolecules with the complete retention of their biological activity and the obtaining of efficient conducting interfaces with electrocatalytic ability. These properties generate an intensive used tool for the electrochemical biosensors construction [63, 64]. Other type of metal nanoparticles, such as Co, Fe, Zn, Ti, Cu, Ag with excellent electric conductivity were also functionalized with CDs resulting in new composite materials with multiple applications in the field of biosensors, especially for biomedical applications, due to their biocompatibility and to the special inclusion ability of the CD molecules.

β-CD molecules can be used as coatings for metallic or composite nanoparticles improving their mechanical and thermal stability, their optical and magnetic properties. The presence of CD at the surface of some nanocomposite materials enhances the electrochemical response and analytical performances. Therefore, AuNPs coated with β-CD were used for the co-immobilization of superoxide dismutase and catalase, via supramolecular association, increasing the pH range and thermal stability of catalase, while superoxide dismutase was 90-fold more resistant to inactivation by H_2O_2 [65].

An electrochemical sensor based on β-CD-cobalt ferrite nanocomposite showed an enhanced electrochemical response and excellent analytical performances for catechol detection, due to the high catechol-loading capacity at the electrode surface, mediated by CD, and to the great electric conductivity of cobalt ferrite nanoparticles [66].

Magnetic particle of Fe_3O_4 encapsulated within a shell of SiO_2 and functionalized with folate ligand, fluorescence tag and CD was able to perform a quadruple of activities: magnetic manipulation, bioimaging, cancer cell-targeting and drug storing-and-releasing [67]. Monochlorotriazinyl-β-CD was grafted onto the surface of ZnO nanoparticles [68], whileFe_3O_4 nanoparticles were modified with mono-6-deoxy-6-(*p*-tolylsulfonyl)-CD. The new material exhibited excellent molecular recognition ability and high electrochemical response due to the combination of individual advantages of magnetic nanoparticles and CDs and were employed for detecting UA [69].

Highly fluorescent, water-soluble CdTe quantum dots with unique optical properties and excellent molecular recognition ability were obtained by using β-CD as surface-coating agent

[70] and were used as fluorescent nanosensor for acetylsalicylic acid determination in the presence of its metabolites and AA [71].

In the case of the TiO_2 nanoparticles, the dehydration of CD molecules from the coating layer in solution provides the force that drives the whole self-assembly process [72]. A surrounding layer of β-CD at the Ag nanoparticles surface shows a promising bactericidal activity against the microorganism *Escherichia coli* [73].

AuNPs embedded in amine functionalized silicate and β-CD composite film were prepared using a single step method applied for the GCE modification. β-CD played a major role both in the synthesis of smaller size NPs and electrocatalytic reduction and sensing of nitroaromatics [74].

The use of β-CD-AuNPs as an intermediate layer for the specific anchoring of adamantane tagged enzyme has led to efficient and sensitive amperometric biosensors for glucose. The β-CD-modified AuNPs were anchored onto the poly(adamantane-pyrrole) functionalized SWCNTs coatings and the 3D composite configuration offered a high-specific surface with excellent accessibility for the enzymes immobilization via affinity interactions and represents an attractive way to develop multilayered enzyme assemblies [75].

The main steps involved in nanoparticle based sensor elaboration, beginning with adamantane immobilization onto CNT's surface, followed by the adamantane/CNT deposition at the electrode and the final decoration with AuNPs (modified with CD molecules), are illustrated in Figure 2. The last step is based on the host-guest specific interaction between adamantane and the CD cavity. Another manner may be the previous deposition of CNTs at the electrode surface, followed by their functionalization with adamantane and AuNP/β-CD immobilization.

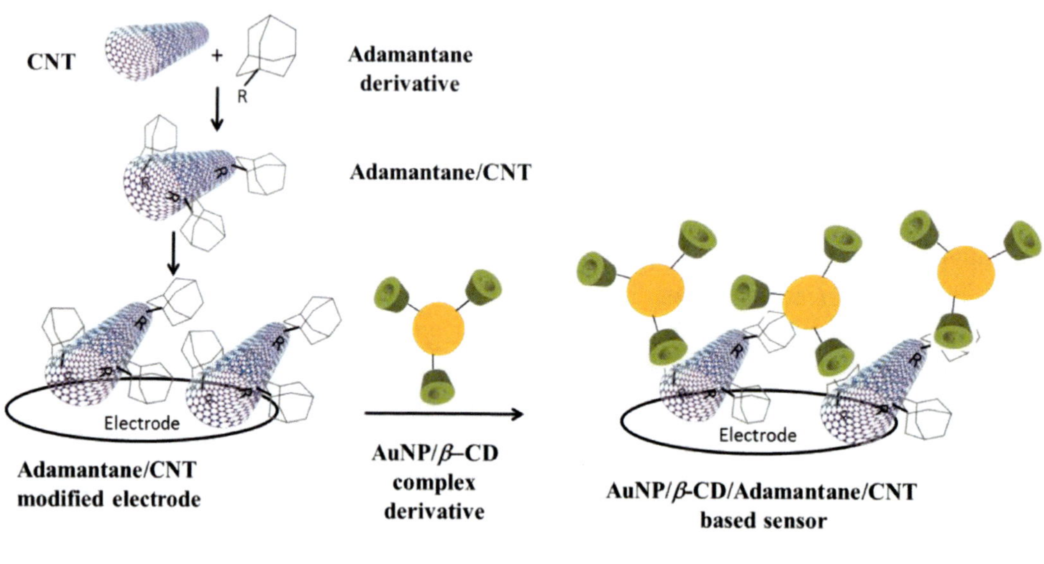

Figure 2. Adamantane derivative immobilization on CNTs and further decoration with β-CD modified AuNPs.

AuNPs with adjustable sizes (10-50 nm) were synthesized in alkaline aqueous solution by reducing hydrochloroauric acid with β-CD and were used as effective catalyst to activate the reduction of 4-nitrophenol [76]. The synthesis of nanometer-sized organic hollow spheres was achieved by the oxidation of AuNPs protected by thiolated *β*-CD molecules. These water-soluble polyCD nanocapsules held together by S-S bonds are very promising materials with multiple possible applications in many domains, such as drug delivery, extraction, as nanoreactors etc. [77].

An amperometric glucose sensor, based on the inclusion complex between mono-6-thio-*β*-CD/ferrocene capped onto AuNPs and GOx, was elaborated and its excellent sensitivity was attributed to the composite film that provided a convenient electron tunneling between the protein and the transducer.

The biosensor showed good stability and analytical performances for glucose detection: fast response time of 5 s; LOD of 15 μM; a high sensitivity of 18.2mA M^{-1} cm^{-2}; a linear range of 0.08-11.5 mM and lack of interferences [78].

GCE was used as support material for the deposition of graphene decorated with AuNPs and CDs and the obtained sensor presented good stability, sensitivity and selectivity for the simultaneous determination of DA and UA in the presence of AA [79].

A superparamagnetic nanomaterial with molecular receptor properties prepared by attaching CD moieties at the surface of silica core-shell Fe_3O_4-NPs was employed as support for the supramolecular immobilization of two different adamantane-modified enzymes: tyrosinase and xanthine oxidase, through host-guest interactions. These biosensors showed excellent electroanalytical behaviors for cathecol and xanthine: sensitivities of 12 mA/M and 130 mA/M, and LODs of 0.022 μM and 2.0 μM, respectively [80].

β-CD molecules were used both for the functionalization of Cu@Ag(Cu@Ag-CD) core-shell nanoparticles as labels and for the functionalization of graphene nanosheet as sensor platform for the development of a nonenzymatic immunoassay applied for the ultrasensitive detection of carcinoembryonic antigen (CEA) (wide linear range: 0.0001-20 ng/mL and LOD of 20 fg/mL) showing good sensitivity, reproducibility, stability and potentially application for CEA in clinical diagnosis [81].

Gold nanorods and AuNPs functionalization with *β*-CD was employed for the elaboration of a sandwich-type immunosensor for *β*-fetoprotein detection. The *in situ* gold superstructures assembly showed excellent performance for the signal amplification and provides a LOD of 0.032 pg mL^{-1} [82].

2.2.6. Cyclodextrins in Polymeric Films

2.2.6.1. Conducting Polymers

CDs can form polymers by their own electropolimerization or polycondensation with dialdehyde or can be attached to other polymers: poly(*N*-acetylaniline), polypyrrole, poly(*N*-methylpyrrole), polyaniline, poly-3-methylthiophene; polyethylenimine, chitosan, Nafion (in which CDs are whether covalent bonded to the monomer, or only entrapped in the polymer matrix). The inclusion of *β*-CD in a poly(*N*-acetylaniline) (PAA) film electrogenerated onto GCE allows the sensitive indirect detection of cinchonine, with 1,4-hydroquinone as redox probe [83].

Pyrrole was electropolymerized at a Pt electrode surface in the presence of *β*-CD allowing the simple and sensitive determination of rifampicin in pharmaceuticals and biological samples with no interferences from other tuberculostatic drugs [84]. The selective and quantitative electrochemical determination of some polyhydroxyphenyls and neurotransmitters, derived from pyrogallol and catechol, with LODs of 0.4 and 1.8 μM for catechol and pyrogallol, respectively, was achieved with apolypyrrole-β-CD film, prepared by electropolymerization [85]. The selective electrocatalytic activity toward the electrochemical oxidation of dopamine and norepinephrine was also investigated with a GCE modified with an electropolymerized film of pyrrole and β-CD (1:1). For this sensor, the selectivity was achieved even at a concentration 20 times higher than the above mentioned neurotransmitters [86]. This composite polymeric film electrogenerated on a gold substrate provided a very good selectivity for some neurotransmitters electrochemical detection towards common interferences such as AA [87], as well as the detection of the muscle relaxant, rocuronium bromide with a LOD of 1 ppm [88].

Carboxymethylated β-CD polymer film was used for the fabrication of a highly selective and sensitive electrochemical sensor for the determination of dopamine in the presence of AA. The analytical performance of the sensor was based on both electrostatic and inclusion interaction between the film and the target analyte [89]. A multi-walled carbon nanotube (MWCNT) with incorporated β-CD and a polyaniline (PANI) film-modified GCE has been successfully developed for determination of dopamine in the presence of AA. The PANI film showed no interference and good stability and the MWCNTs functionalized with carboxyl groups assured the sensitive detection of dopamine, due to the preconcentrationobtained after β-CD and dopamine, supramolecular complex formation [90].

Conducting polypyrrole (PPy) and poly(pyrrole-2,6-dimethyl-β-CD) films were prepared onto a gold electrode, both in aqueous and nonaqueous electrolyte solutions, an increased oxidation potential being observed for theCD modified pyrrole monomer. The same results were obtained for poly(*N*-methylpyrrole) and poly(*N*-methylpyrrole-cyclodextrin) films [91, 92].

A biocompatible hybrid film of β-CD and ionic liquids (ILs), 1-butyl-3-methylimidazolium tetrafluoroborate was used for the development of an electrochemical biosensor by immobilizing the horseradish peroxidase (HRP). β-CD provided a biocompatible microenvironment for HRP and ILs accelerated the electron transfer between the enzyme and the transducer. The biosensor exhibited low operating potential, fast response, high sensitivity, good selectivity and submicromolar LOD [93].

The γ-CD stabilized the poly-3-methylthiophene film electropolymerized onto gold electrode, the resulting sensor exhibited high sensitivity and selectivity toward electrochemical qualitative and quantitative detection of dopamine and l-dopa and chlorpromazine with LODs of 0.2, 1, and 0.1 μM, respectively [94].Simultaneous quantification of serotonin and dopamine using a β-CD/poly(*N*-acetylaniline)/CNT composite modified CPE was achieved in plasma samples with acceptable recoveries [95].

Insoluble β-CD polymer was synthesized by using epichlorohydrin as crosslinking agent under basic conditions. The polymer presented excellent thermal stability, insolubility and catalytic activity for the oxidation of benzyl alcohol to benzaldehyde [96].

Polyethyleneimine (PEI) was used for the β-CD entrapment at GCE surface and the obtained sensors were used for AA and UA voltametric determination. After the optimization of experimental parameters for the PEI and β-CD concentration, the AA current intensity was significantly increased for β-CD-PEI/GCE compared to bare GCE. This behavior was attributed to the formation of inclusion complexes between AA and β-CD, complexation which seems to be more effective for AA than for UA. The negative shift of peak potentials can be explained by the mediator effect of CD, whose cavity acts like a "tunnel", which directs the electrons to the electrode surface for the redox process. The increase of the current intensity is due to the preconcentration of the analyte at the electrode surface and to the complexation with CD, process which retains more molecules at the electrode surface [97]. Microscopic images, RAMAN spectroscopy and electrochemical impedance spectroscopy successfully confirmed the modification of the electrode surface with β-CD-PEI film. In order to overcome the interference between ascorbic and uric acid in electrochemical analysis and to separate their electrochemical signals, GCEs modified with PEI and β-CD films were used. The electrochemical behavior of different concentrations of AA and UA was studied with β-CD+PEI/GCE, two well separated oxidation peaks being obtained (Figure 3 A).

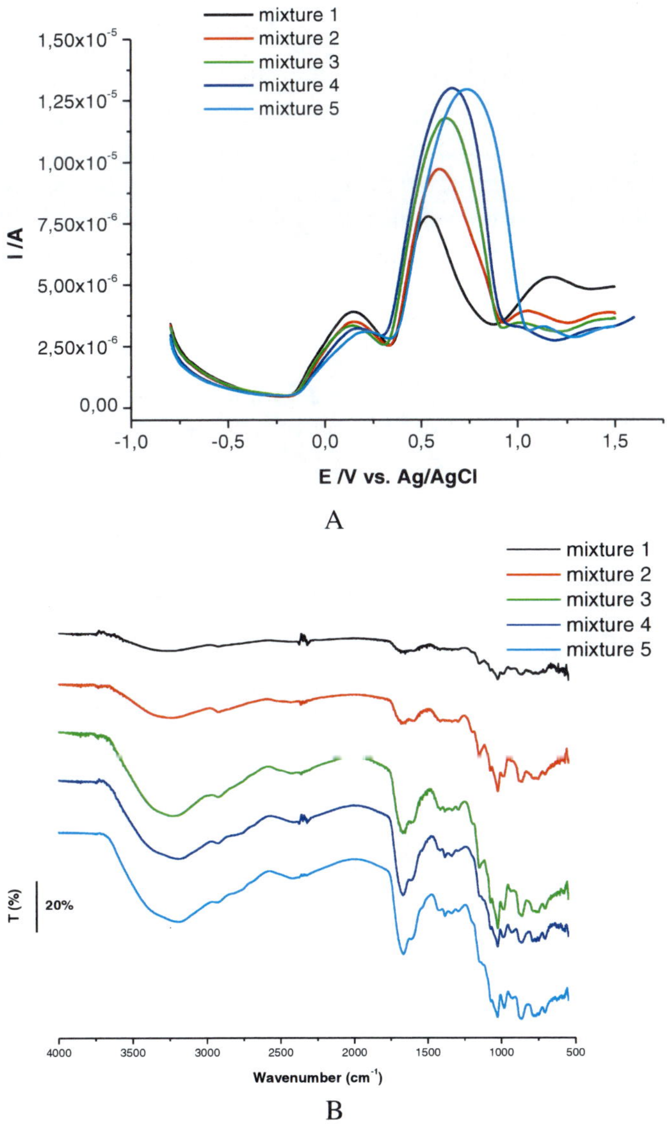

Figure 3. (A) SWVs of different concentrations of UA and 10^{-3}M AA at 1% β-CD-PEI (5mg/ml)/GCE: (mixture 1) 10^{-3}M UA, (mixture 2) $2 \cdot 10^{-3}$M UA, (mixture 3) $4 \cdot 10^{-3}$M UA, (mixture 4) $6 \cdot 10^{-3}$M UA, (mixture 5) $8 \cdot 10^{-3}$M UA; (B) FTIR spectra of: (mixture 1) β-CD:AA:UA, 1:1:1, (mixture 2) β-CD:AA:UA, 1:1:2, (mixture 3) β-CD:AA:UA, 1:1:4, (mixture 4) β-CD:AA:UA, 1:1:6, (mixture 5) β-CD:AA:UA, 1:1:8 [97].

The FTIR spectra of these mixtures revealed that the increasing concentration of AA determined the growth of the band from 1590 cm^{-1}. A possible explanation could be the fact that AA removes UA from the inclusion complex, being preferentially allowed to enter in the cavity of the CD. The bands of AA molecules inside the cavity are screened, meanwhile the UA molecules which were left outside present bands which become stronger with the AA increasing concentration (Figure 3 B) [97].

Confocal microscopy determination were performed for both electrodes modified with thin films, PEI/GCE and β-CD-PEI/GCE (Figure 3), showing that the β-CD structures were uniformly and homogenously distributed in the PEI film covering the entire electrode surface (Figure 4 B) [97].

The optimized β-CD-PEI based sensors were applied for the assay of AA in two commercial drug formulations, with recoveries between 97.29% and 101.02% and a LOD of 0.22 μM (S/N=3) over the concentration range of 9.99-215.26 μM. The detection limit value was comparable with the lowest LOD reported in literature, being 50 times lower than other methods employing electrodes modified with CDs [97]. The sensor was also applied for ascorbic acid monitoring in urine samples, during a *7 days* treatment. The simultaneous detection of AA and UA was achieved due to the obvious anodic peak separation of the two substances (ΔE=0.22V), with excellent recovery of AA in spiked urine samples [98].

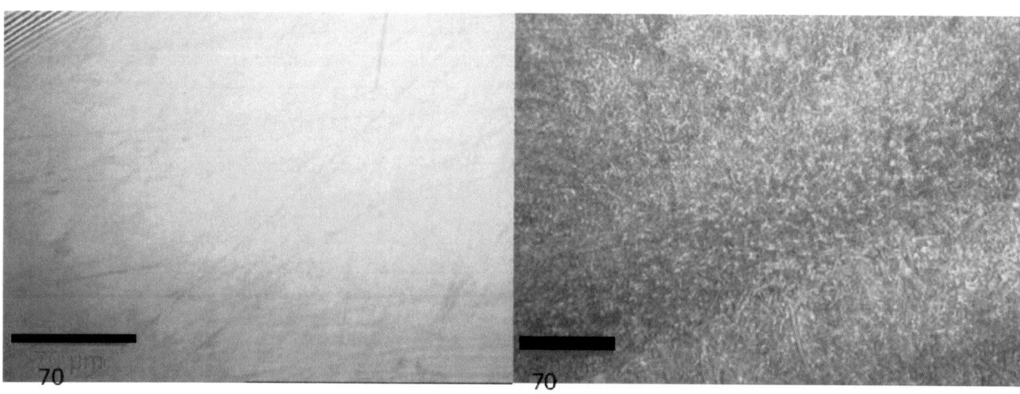

Figure 4. Microscopic images for: (A) PEI/GCE and (B) β-CD-PEI/GCE [97].

2.2.6. Carbon Nanomaterials Modified with Cyclodextrins

2.2.6.1. Carbon Nanotubes with Cyclodextrins

Multiwalled carbon nanotubes (MWCNTs) were sonicated together with β-CD forming β-CD-MWCNTs nanocomposite, which was assembled onto GCE using layer-by-layer (LBL) method based on electrostatic interaction of positively charged biopolymer chitosan and negatively charged MWCNTs. The assembled MWCNTs with β-CD onto GCE exhibited lower capacitive background current and higher electrocatalytic activity towards dopamine than the assembled MWCNTs without β-CD, results which were ascribed to the relatively compact three-dimensional structure of the MWCNTs multilayer films with β-CD and to the excellent molecule recognition function of β-CD [99].

GC, Au and ITO electrodes were covered by a conductive composite film containing functionalized MWCNTs and AuNPs decorated with hydroxypropyl-β-CD (HPβCD) as catalysts. The presence of AuNPs produced an increase in the sensitivity of the modified electrodes and the presence of MWCNTs and HPβCD in the composite film enhanced the active surface coverage concentration of Au by 397.0 %. These modified electrodes permitted individual and simultaneous electrochemical determination of tyrosine, guanine, adenine and thymine [100]. Other electrodes with marked electrocatalytic activity based on β-CD decorated CNTs were proposed for the simultaneous or individual determination of guanine and adenine with LODs of 0.1 μg mL^{-1} and 0.2 μg ml^{-1}, respectively [101]. ITO electrodes coated with β-CD modified MWCNTs showed enhanced and selective electrocatalytic effect towards sodium cholate reduction, while electrochemical response of deoxycholate was not observed [102].

The simultaneous electrochemical oxidation of norepinephrine and AA was studied with β-CD incorporated CNT modified electrode and was successfully achieved due to a separation peak potential of about 264 mV [103]. An α-CD incorporated carbon nanotube-coated electrode with strong catalytic effects for the simultaneous determination of dopamine and epinephrine with a peak potential separation of 390 mV was also reported [104]. Another MWCNTs/β-cyclodextrin based sensor, with increased analytical performances compared to the bare electrode, was elaborated for rutin determination [105]. The selectivity to quercetin and rutin was obtained on succinyl-β-CD-modified MWCNT coated GCE and the most probable complexation occurred between succinyl-β-CD and quercetin [106].

MWCNTs were ultrasound-suspended in β-CD solution and used to modify the GCE elaborating a dopamine (DA) electrochemical sensor. β-CD as molecular receptor and MWCNT as enhancer of electron transfer improved the analytical performances and the catalytic oxidation of DA [107]. Aligned CNTs were functionalized with β-CD diazonium salt through diazotization reaction and the resulting sensor was used to detect DNA hybridization in homogeneous solution based on host-guest recognition. The DNA probe had both ends modified: one with 4-((4-(dimethylamino)phenyl)azo) benzoic acid) as guest molecule for the modified electrode capture, and the other end was labeled with a CdS nanoparticle as an electrochemical tag to indicate the occurrence of DNA hybridization. This electrochemical biosensor has high sensitivity and specificity with a LOD of $5.0 \cdot 10^{-13}$ M for the determination of sequence-specific DNA [108].

Carbon nanomaterials (including fullerenes, SWCNTs, MWCNTs and graphene sheets) were functionalized with β-cyclodextrin polymer (CDP) obtaining CDP-carbon nanomaterials whit high solubility and stability in water. The integrity of carbon nanomaterials was significantly preserved by this noncovalent modification. Furthermore *p*-aminothiophenol used as a guest was trapped in the CDP-modified carbon nanomaterials, after that these complexes interacted with Au or Pt NPs by the formation of thio-Au or thio-Pt bond allowing the highly efficient nanoparticle assembly onto the surface of MWCNTs. Pt-decorated CDP-carbon nanomaterials can be applied in methanol fuel cells. These carbon nanomaterial-based hybrid materials presented not only the good electrical and large surface area properties of carbon nanomaterials, but also displayed high supramolecular recognition and enriching properties of CDs [109].

AuNPs-poly(luminol) (Plu-AuNPs) hybrid film and MWCNTs decorated with β-CD were explored to modify GCE for simultaneous determination of dopamine and UA exhibiting good selectivity and sensitivity with well-separated peaks. AuNPs were dispersed into the porous polymer matrix of poly(luminol) generating additional electrocatalytic sites; *β*-CD was used as a dispersing reagent for CNTs. The DPV response of the modified electrode was about 8-fold greater compared to that without CD and MWCNT. The sensor was tested on real samples after the efficient elimination of the AA and UA interferences [110].

MWCNTs were suspended in a mixed aqueous solution of *β*-CD and *β*-CD prepolymer (pre-CDP) obtaining an insoluble conducting composite film of polycyclodextrin (CDP) and CNTs. Glucose was the target analyte, thus GOx was immobilized on the film and was used for biosensing testing purpose [111].

2.2.6.2. Graphene with Cyclodextrins

Hydroxypropyl-*β*-CD-reduced graphene oxide hybrid nanosheets (HP-*β*-CD-RGO) were synthesized by microwave irradiation. Nafion and in situ deposited bismuth film were used as the sensing material for the simultaneous determination of Pb^{2+} and Cd^{2+} with high sensitivity, good stability and reproducibility. It successfully combines the host-guest recognition and enrichment properties of HP-*β*-CD, the large 2-D electrical conductivity and large surface area properties of RGO, the ion-exchange capacity of Nafion and the ability of bismuth film to form easily alloys with heavy metals in a synergistic effect which improves the analytical performances of the sensor [112].

Hydroxypropyl-*β*-CD was covalently grafted onto the surface of graphene oxides via ester bonds by nucleophilic addition. The electrochemical interactions between hemoglobin and tetraphenyl-porphyrin were investigated by CV with a LOD of 5 nM for hemoglobin, indicating high biorecognition ability [113]. Another *β*-CD@CRG nanocomposite prepared via a simple sonication-induced assembly, in the presence of Nafion, was used for the development of a sensitive electrochemical sensor for rutin detection. Its excellent analytical performance was attributed to the high rutin loading capacity at the electrode surface and the outstanding electric conductivity of graphene [114].

The same simple wet-chemical strategy was used for preparing cyclodextrin-graphene (CD-Gr) nanocomposite which was employed for the fabrication of enzyme electrodes via host-guest supramolecular interactions between adamantane-modified horseradish peroxidase (HRP-ADA) and CD-Gr nanocomposites. The biosensor presents good reproducibility, high sensitivity for H_2O_2 and a LOD of 0.1 mM, revealing that such an electrochemical platform not only preserved the native structure of the immobilized enzyme, but also exhibited good electron transfer properties for HRP [115].

The modification of the graphene sheets with *β*-CD, HRP and adamantane is presented in Figure 5. The graphene oxide is reduced (electrochemically or with a reducing agent) and functionalized with *β*-CD, followed by the HRP/adamantane immobilization at the reduced graphene oxide surface, via the host-guest interaction between *β*-CD and adamantane.

The electrochemically reversible behavior of DA was observed with the modified *β*-CD/graphene sheets compared to bare GCE, revealing a faster electron transfer process. A LOD of 5.0 nM was obtained for DA in the presence of large excess of ascorbic acid. The sensor also showed a good ability to suppress the background current [116]. A simple method for water-phase synthesis of *β*-CD modified RGO reduced by *l*-AA at room temperature was reported and the *β*-CD-RGO was used as enhanced material for sensitivity determination of diethylstilbestrol.

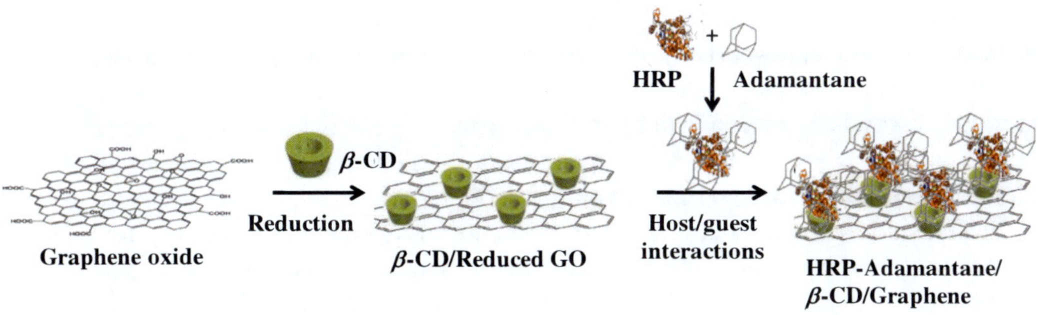

Figure 5. Fabrication of the HRP-Adamantane/β-CD/Graphene complex.

The developed electrochemical sensor exhibited high selectivity, good stability and reproducibility, performances due to the excellent electronic properties of RGO sheets and also to the high supramolecular recognition capability of β-CD [117].

A highly sensitive electrochemical sensor for quercetin was fabricated by using combined potentiostatic and voltammetric techniques for the electrodeposited graphene/polymerized β-CD (E-Gr/P-βCD) nanocomposite film preparation. A LOD of 1 nM was obtained due to the synergistic effect of E-Gr and P-β-CD [118]. The electrochemical behaviors of 2-chlorophenol (2-CP) and 3-chlorophenol (3-CP) were investigated with β-CD functionalized graphene modified CPE. This sensor may provide a promising new application to detect other isomers [119]. Monodisperse AuNPs were anchored onto carboxylic graphene nanosheets (CGS). A dual β-CD functionalized Au@CGS nanohybrid was then fabricated: the NH_2-β-CD was covalently bonded to Au@CGS by combining the amine group of NH_2-β-CD and the carboxyl group of CGS, with the aid of an EDC/NHS coupling agent and then the NH_2-β-CD was strongly attached to the surface of the Au@CGS. The performances of this nanohybrid for the simultaneous determination of two phenol compounds, hydroquinone and *p*-nitrophenol, were tested by DPV. The proposed electrochemical sensor showed a low detection limit and high selectivity caused by the large surface area, the excellent conductivity, and the high host-guest molecular recognition capability of the β-CD-Au@CGS nanohybrids [120].

A β-CD functionalized graphene nanosheets (GN-CD) assembly was prepared by *in situ* thermal reduction of graphene oxide with hydrazine hydrate in the presence of β-CD, then *cytochrome c* (Cyt c) was intercalated to the GN-CD assembly, to form a layered self-assembled structure, through electrostatic interaction. This assembly combined the specific properties of the modifiers, such as: high conductivity of graphene nanosheets, the selectively binding properties and electronegativity of CDs and the electropositivity of Cyt c. The sensor displayed improved electron transfer rate and high supramolecular recognition capability towards six probe amino acids, among which glycine was chosen as a representative analyte [121]. An enhanced electrochemical sensing platform based on the integrating properties of graphene and 2,6-dimethyl-β-CD (DM-β-CD) was elaborated for the determination of isoquercitrin and baicalin in real serum samples. The obtained LODs of 4 nM and 10 nM are related with the characteristic properties of graphene (large surface area and high conductivity) and of DM-β-CD (high supramolecular recognition and enrichment capability), that improved the nanocomposite electrochemical features [122].

A non-enzymatic electrochemical sensor for micromolar detection of cholesterol was obtained using graphene-β-CD hybrid system as the sensing matrix and methylene blue as redox probe. The graphene oxide was treated with β-CD in presence of ammonia and NaOH, the complex was then reduced with hydrazine, forming Gr-β-CD [123].

2.2.6.3. Fullerenes with Cyclodextrins

The chemical and physical features of fullerenes make them very attractive for biological applications. Their well-known lack of solubility in polar solvents can be solved by chemical or supramolecular approaches. The last technique, supramolecular complexation, employed the fullerene entrapment in γ-CD cavity in a molar ratio of 2:1 (CD:fullerene) to form water-soluble fullerene derivatives. The CD ability to mask the fullerene sphere is very important in enhancing solubility in polar solvents and in decreasing aggregation [124, 125].

C70 solubility was improved by complexation with δ-CD using the ball-milling method based on a solid-solid mechano-chemical reaction; the stoichiometric ratio of the complex was 1:2/C70:δ-CD) [126].

PEG/CD-modified dendrimers have been synthetized and applied for the solubilization of C60 fullerene, obtaining an aqueous solution of fullerene with a concentration of 2.8 μM. It has to be mentioned that the solubility of C60 in pure water was reported as $2 \cdot 10^{-24}$ M [127]. Stable solutions of fullerene C60 with human serum albumin (C60/HSA) were prepared via the formation of C60/HP-β-CD nanoparticles, in particular by transferring C60 molecules from HP-β-CD to HSA molecules. The resulting C60/HSA presented an excellent stability and antioxidant activity. It presented also substantial phototoxicity, propertie wich enables HSA to be used as a carrier of C60 in drug delivery systems for photodynamic therapy [128].

A synthesized supramolecular inclusion complex of C60 and β-CD (1:2) interacted with bovine serum albumin (BSA) giving a $(\beta\text{-CD})_2$/C60-BSA water-soluble complex which was characterized by spectroscopic and electrochemical methods in aqueous solutions [129]. Triazole-methoxypyridyl γ-CD was used as a successful hydrosolubilizing reagent for C60 resulting in hydrosoluble inclusion complex of C60 with a concentration of more than 70 mM which is approximately 90 times greater than that with non-substituted γ-CD [130].C60 was immobilized through host-guest interaction into the SAMs of bis-pyridine-substituted γ-CD preserving its electronic properties and stabilising the fullerene against photopolymerisation and aggregation [131]. The isolation of the different regioisomers of fullerene derivatives by using host-guest complexes has been reported. Bis-*N*-methyl-fulleropyrrolidines were separated by complexation with γ-CD by controlling the relative amounts of γ-CD and dimethyl sulfoxide (DMSO). Using small amount of γ-CD, the trans-1 and trans-2·γ-CD complexes were separated from the trans-3·γ-CD complex. The explanation for the observed regioselective separation was a competition between the relative stabilities and solubilities of the complexes in water and water-DMSO mixture [132]. It was proved that host-guest complexes of C60 and its conjugates maintained the electrochemical activity of the fullerene moiety. The determination of redox potentials allowed drawing conclusions on the formation of complexes and determining the binding enhancement due to the redox change on the C60 core [133].

CD-fullerene conjugates were also synthesized by covalent functionalization. In particular, C60 with a notable electron acceptor capability, it was conjugated with azido-CDs via 1,3-dipolar cycloaddition in chlorobenzene, resulting the corresponding azafulleroid with 40-49 % yields. These CD-C60 conjugates are soluble in all the common organic solvents and in THF/H_2O mixtures with up to 90% of water. The redox properties were determined in an *o*-dichlorobenzene/acetonitrile mixture (4:1, v:v) and remained unaltered after ferrocene addition, as internal reference. The four reduction processes corresponding to the four quasi-reversible reductions of the fullerene sphere were displayed in the azafulleroid voltammograms [134, 135]. A bridged bis(β-CD)-fullerene conjugate was also synthesized and characterized by UV-Vis, FTIR, NMR, fluorescence and circular dichroism spectroscopy, techniques which showed that the bis(β-CD)-fullerene conjugate displayed an intramolecular capsule type conformation in aqueous solution. Beyond its satisfactory water solubility at physiological temperature and DNA-cleaving ability under visible light irradiation, a specific advantage of the capsule type CD-fullerene conjugate is that the CD cavity can selectively form complexes with certain fragments of the biological substrates, enabling the site specific interaction between the fullerene and target substrate [136].

Another method for synthesizing a water soluble CD-fullerene conjugate was based on the Diels-Alder reaction between anthryl-CD and fullerene in DMF/toluene solution. The CD-C60 conjugate exhibited water solubility up to 2.5 g L^{-1} at 25°C and remained stable for several weeks, when stored at -10°C [137]. A water-soluble complex of C60 fullerene and β-CD was found to act as the mediator for the electrocatalytic conversion of dinitrogen to ammonia, the rate determining step being the reduction of the fullerene-CD complex [138]. *p*-nitrophenol detection was performed by using a GCE modified with CD-C60 (synthesized by 1,3 cycloaddition), a LOD of 1.2 nM being calculated. The high electrocatalytic performance of the sensor was due to synergetic properties of C60 (microelectrodes) and β-CD (acceptor-host system) [139]. A brief review of the main electroanalytical performances of the CDs based or modified (bio)sensors developed and reported in scientific literature, in the last decade, is summarized in Table 1.

CDs were intensively employed for the (bio)sensors elaboration by using various techniques, such as: the incorporation in membranes and in polymer films, the inclusion in carbon paste, the deposition at the electrode surface or at different micro or nano materials or particles. The obtained (bio)sensors were successfully applied for the determination of a wide variety of target molecules, such as: heavy metals, glucose, phenols, steroids, antibiotics, DNA, neurotransmitters, anti-inflamatory drugs, etc. Among the electrochemical methods applied for the testing and assessment of the electroanalytical performances of the newly developed devices, amperometry, voltammetry, potentiometry and impedance spectroscopy, can be mentioned. The use of CDs and carbon based nanomaterials, especially CNTs or graphene, in (bio)sensors configuration, resulted in a remarkable improvement of the analytical parameters, as it can be noticed in Table 1.

3. Cyclodextrins As Biomimetic Receptors

The bioreceptor, consisting in the biological recognition system is the most important part of a biosensor. Many bioelements, such as: enzymes, DNA, RNA, antibodies, affibodies, aptamers, microRNA, microorganisms, and living cells, were used in biosensor construction, based on different types of interactions between the bioreceptor and the analyte.

Table 1. Electroanalytical performances of CDs based (bio)sensors

Electrode configuration	Analyte	Detection method	LOD(μM)/ Sensitivity	Linear range (μM)	Reference
Electrodes with CD obtained via Langmuir-Blodgett technique and Self-Assembly Method					
(MEA-β-CD)SAMs/Au[1]	Pb(II)	DPSV	0.0071	0.017-0.93	[23]
(β-CDPSH/ ADA-CMC-DNA)SAMs/Au[2]	HLA-DQA1*0201	Amperometry	0.08 nM	0-0.002	[24]
(β-CD/1-octanethiol)SAMs/Au[3]	Transglutaminase	CV	1.42 nA μL U^{-1} s^{-1}	1.9–37 mU/mL	[25]
Electrodes with CD included in plasticized membranes					
β-CD and (2-hydroxiproyl)-γ-CD-FNFE-PTCPH[4]	Midazolam	Potentiometry	0.0012 g/L	0.0023-0.44 g/L	[36]
	Diazepam		0.068 g/L	0.12-1.7 g/L	
β-CD-NaTFPB-*o*-NPOE[5]	Dextromethorphan	Potentiometry	0.07	0.1-10000	[37]
I: triacetyl-β-CD-PVC[6] II: triacetyl-β-CD-PVC-COOH[6]	Ephedrine	Potentiometry	5.70 6.20	30-8000	[38]
α-CD-o-NPOE-TDDAB/PVC[7]	Ibuprofen	Potentiometry	3.34	38.7-1000	[39]
heptakis(2,3,6-tri-*O*-benzoyl)-β-CD-PVC[8]	Diclofenac ions	Potentiometry	14	50-100	[40]
β-CD-PVC-KT- (4-CPB)/Co_3O_4 nanowires-Au/Glass[9]	Dopamine	Potentiometry	-	0.001-10000	[41]
Screen-printed electrodes modified with CD					
CBM-β-CD/SPE[10]	Azepine Phenothiazine	DPV	0.04 0.01	0.196-2.75	[43]
α-CD/SPE[11]	Phenols	SWV	0.03	0.03-50	[44]
		DPV	0.50	0.5-25	[44]
GOx-β-CD-MWCNT/SPE[12]	Dopamine	Amperometry	0.48	10-50	[45]
Carbon paste electrodes modified with CD					
HP-β-CD/CPE[13]	(+) Catechin	Anodic voltammetry Chatodic voltammetry	0.12 0.30 ng/ml	7.20-4.20 g/mL	[46]

Electrode configuration	Analyte	Detection method	LOD(μM)/ Sensitivity	Linear range (μM)	Reference
β-CD/CPE[14]	(+) Catechin	SWV cathodic SWV anodic	1.35 mg /mL	>70.0 mg/mL >45.0 mg/mL	[47]
β-CD/CPE[14]	Nitrendipine	DPV	0.172	1-10	[48]
β-CD/CPE[14]	Donepezil	DPV	0.0012	0.0032-0.042	[49]
β-CD-Fc/CPE[15]	Ascorbic acid	Amperometry	0.0001	0.5-1000	[51]
β-CD/CPE[15]	Sparfloxacin	DPV	0.04	0.043-60	[54]
β-CD/CPE[15]	Ofloxacin Norfloxacin Gatifloxacin Lomefloxacin	DPV	0.024	0.032-20	[55]
β-CD/CPE[15]	Prednisolone, Dexamethasone Hydrocortisone	DPV	0.48 0.36 0.37	0.56-20 0.41-20 0.42-25	[56]
β-CD/CPE[15]	Anagrelide	DPV	0.0022	0.001-1.6	[58]
P-β-CD-*L*-arg/CPE[16]	Ciprofloxacin	DPV	0.01	0.05-100	[59]
P-β-CD/CPE[17]	Tenoxicam	LSV	-	7-56	[60]
β-CD/MWCNT-PE[18]	Nifedipine	DPSV	0.0148	0.0477-20	[61]
P-β-CD/CPE[17]	Dopamine	DPV	4.1	10-130	[62]
Electrodes with nanoparticles decorated with CD					
β-CD-CoFeNC[19]	Catechol	DPV	0.12	1-200	[66]
β-CD-11-ECTUDA-MSACdTe/QD[20]	Ascorbic acid	Fluorescence	0.0085	0-1mg/l	[71]
β-CD-Si sol-gel/AuNp[21]	Nitrobenzene	SWV	-	10-80	[72]
M-6-T-β-CD-Fc/AuNP/GOx[22]	Glucose	DPV	15	80-1150	[78]
AuNPs–β-CD–Gra/GCE[23]	*L*-ascorbic acid Dopamine Uric acid	SWV	10 0.15 0.21	30-2000 0.5-150 0.5-60	[79]
M-6F-β-CD-3-APTES/Fe_3O_4NP-ADA-(Ty and XO)/CPE[24]	Cathecol Xanthine	Amperometry	0.022 2	0.1-12 5-120	[80]

Table 1. (Continued)

Electrode configuration	Analyte	Detection method	LOD(μM)/ Sensitivity	Linear range (μM)	Reference
CEA- Cu@Ag-*β*-CD-CD-GN[25]	Carcino-embryonic antigen	Amperometry	20 fg/mL	0.0001-20 ng/mL	[81]
CD in polymeric films					
β-CD-poly(*N*-acetylaniline)/GCE[26]	Cinchonine	CV	2	4-80	[83]
Ppy-*β*-CD/Pt[27]	Rifampicin	Amperometry	0.032 pg mL^{-1}	0.5 pg/mL-0.5 ng/mL	[84]
Ppy-*β*-CD/GCE[28]	Catechol pyrogallol	CV	0.4 1.8	0.15-8 1-10	[85]
Ppy-*β*-CD/GCE[28]	Dopamine Norepinephrine	CV	6 0.8	2-10 0.4-2	[86]
Ppy-*β*-CD/Au[29]	Dopamine *L*-dopa,	CV	0.6 60	2-20	[87]
Ppy-*β*-CD/Au[29]	Rocuronium bromide	CV	1	1-10000	[88]
β-CD-MWCNT-PANI/GCE[30]	Dopamine	CV	0.012	0.1-1000	[90]
HRP-*β*-CD-ILs/GCE[31]	H_2O_2	LSV	2.65	4-84	[93]
P-3-MT-*γ*-CD/Au[32]	Dopamine L-dopa Chlorpromazine	SWV	0.2 1 0.1	0.1-50 2-50	[94]
β-CD-poly(*N*-acetylaniline)-CNT/CPE[33]	Serotonin (5-HT) Dopamine	CV	<1	4-200	[95]
β-CD-PEI/GCE[34]	Ascorbic acid	SWV	0.22		[97]
Carbon nanomaterials based sensors with CD					
β-CD-MWCNT-chitosan/GCE[35]	Dopamine	DPV	0.15	0.25-20	[99]
fMWCNT-AuNP-HP-*β*-CD/GC, Au and ITO[36]	Tyrosine Guanine Adenine Thymine	DPV	660 90 7780 3500	300-1800 3600-13200 600-3000 1200-7200	[100]
β-CD-CNTs/GCE[37]	Guanine Adenine	DPV	0.1 μg mL^{-1} 0.2 μg ml^{-1}	0.2-20	[101]
β-CD-CNTs/GCE[37]	Norepinephrine	CV	0.5	1-300	[103]

Electrode configuration	Analyte	Detection method	LOD(μM)/ Sensitivity	Linear range (μM)	Reference
α- CD-CNT/GCE[38]	Dopamine Epinephrine	CV	1 0.5	2-1000 1-1000	[104]
MWCNT-β-CD/GCE[39]	Rutin	CV	0.2	0.4-1000	[105]
MWCNT-β-CD/GCE[39]	Dopamine	Amperometry	6.7	0-80	[107]
β-CD-ACNTs/GCE[40]	Sequence –specific DNA	DPV	$5.0\cdot10^{-7}$	10^{-4}-1	[108]
Plu-AuNPs/β-CD-MWCNT/GCE[41]	Dopamine	DPV	0.19	1-5.6	[110]
pre-CDP-MWCNT-GOx[42]	Glucose	Amperometry	3.5	0.004-3230 4.26-10000	[111]
HDP- β-CD-RGO-Nafion/Bi/GCE[43]	Pb^{2+} Cd^{2+}	CV	$9.42\ 10^{-5}$- $6.73\ 10^{-5}$	10^{-4}-$9\ 10^{-3}$ $5\ 10^{-4}$-10^{-3}	[112]
HDP-β-CD-GO/GCE[44]	Hb	CV	0.005	-	[113]
β-CD@CRG-Nafion[45]	Rutin	DPV	0.002	0.006-10	[114]
β-CD-Gr-ADA-HRP[46]	H_2O_2	DPV	0.1	0.7-35	[115]
β-CD-Gr/GCE[47]	Dopamine	Amperometry	0.005	0.9-200	[116]
β-CD-CRGO/GCE[48]	Diethylstilbestrol	DPV	0.004	0.01-13	[117]
E-Gr/P-β-CD[49]	Quercetin	DPV	0.001	0.005-20	[118]
β-CD-fGr/CPE[50]	2-chlorophenol 3-chlorophenol	DPV	0.2 0.09	0.5-40 0.4-77	[119]
AuNPs-CGS-NH_2-β-CD[51]	Hydroquinone *p*-nitrophenol	DPV	0.0065 0.0038	0.01-200	[120]
DM-β-CD-fGr/GCE[52]	Isoquercitrin Baicalin	DPV	0.004 0.010	0.01-3 0.04-3	[122]
CRC-β-CD[53]	Cholesterol	DPV	<1	1-100	[123]

[1](MEA-β-CD)SAMs/Au = a self-assembled monolayer of thiolated β-cyclodextrin on a gold electrode;
[2](β-CDPSH/ ADA-CMC-DNA)SAMs/Au = a SAM of β-CDPSH and adamantine carboxylate sodium salt on bare gold electrodes;
[3](β-CD/1-octanethiol)SAMs/Au = a SAM of β-CD/1-octanethiol on the gold disk electrode;
[4]β-CD and (2-hydroxiproyl)-γ-CD-FNFE-PTCPH = polymeric membranes incorporating β-CD and (2-hydroxiproyl)-γ-CD as ionophores, 2-fluorophenyl 2-nitrophenyl ether as plasticizer and potassium tetrakis (**p**-chlorophenyl) borate as ionic additive; (Continued …)

[5]*β*-CD-NaTFPB-**o**-NPOE = 2,3,6-tri-*o*-methyl-*β*-CD (ionophore)+sodium tetrakis (4-fluorophenyl) borate (anionic site)+**o**-nitrophenyloctylether (membrane plasticizer);
[6]triacetyl-*β*-CD-PVC/PVC-COOH = triacetyl-*β*-CD (neutral ionophore) embedded in a plasticized poly vinyl chloride matrix (sensor I) and carboxylated polyvinyl chloride (simultaneous plastic matrix and ion exchanger) (sensor II);
[7]*α*-CD-o-NPOE-TDDAB/PVC = *α*-CD+*o*-nitrophenyloctylether (plasticizer) tetradodecylammonium bromide and PVC/a conductive polymeric resin coated with the membrane (a low impedance solid contact)
[8]heptakis(2,3,6-tri-*O*-benzoyl)-**β**-CD-PVC = heptakis(2,3,6-tri-*O*-benzoyl)-**β**-CD-plasticized poly(vinyl chloride) membrane based on 2-nitrophenyloctyl ether;
[9]*β*-CD-PVC-KT- (4-CPB)/Co_3O_4 nanowires-Au/Glass = Ultra-thin cobalt oxide nanowires grown on gold coated glass substrates entrapped by using a polymeric membrane containing polyvinyl chloride (plasticized), β-CD (ionophore), and potassium tetrakis (4-chlorophenyl) borate (ionic additive);
[10]CBM-*β*-CD/SPE = carboxymethylated *β*-CD modified screen printed electrode;
[11]*α*-CD/SPE = *α*-CD modified SPE;
[12]GOx-*β*-CD-MWCNT/SPE = glucoseoxidase+*β*-CD decorated multiwalled carbon nanotubes modified SPE;
[13]HP-*β*-CD/CPE = hydroxypropyl+*β*-CD modified carbon paste electrode;
[14]*β*-CD/CPE = *β*-CD modified CPE;
[1\5]*β*-CD-Fc/CPE = inclusion complex of ferrocene with *β*-CD modified CPE;
[16]P-*β*-CD-*L*-arg/CPE = poly-*β*-CD and *L*-arginine modified CPE;
[17]P-*β*-CD/CPE = poly-*β*-CD modified CPE;
[18]*β*-CD/MWCNT-PE = *β*-CD modified MWCNT paste;
[19]*β*-CD-CoFeNC = β-CD-cobalt ferrite nanocomposite;
[20]*β*-CD-11-ECTUDA-MSACdTe/QD = *β*-CD modified with 11-[(ethoxycarbonyl)thio]undecanoic acid and mercaptosuccinic acid CdTe quantum dots;
[21]*β*-CD-Si sol-gel/AuNp = *β*-CD-silicate sol–gel composite embedded gold nanoparticles;
[22]M-6-T-*β*-CD-Fc/AuNP/GOx = inclusion complex of mono-6-thio-*β*-CD/ferrocene capped on gold nanoparticles and glucose oxidase;
[23]AuNPs–*β*-CD–Gra/GCE = Graphene decorated with gold nanoparticles modified GCE;
[24]M-6F-*β*-CD-3-APTES/Fe3O4NP-ADA-(Ty and XO)/CPE = Mono-6-formyl- *β*-CD + 3-aminopropyl)triethoxysilane superparamagnetic Fe_3O_4 nanoparticles reduced with $NaBH_3CN$ + adamantane-modified enzymes (tyrosinase and xanthine oxidase) magnetically modify carbon paste electrodes;
[25]CEA- Cu@Ag-*β*-CD-CD-GN = carcino-embryonic antigen using *β*-CD functionalized core–shell nanoparticles as labels and *β*-CD functionalized graphene nanosheet;
[26]*β*-CD-poly(*N*-acetylaniline)/GCE = *β*-CD-poly(*N*-acetylaniline) modified/GCE;
[27]PPy-*β*-CD/Pt = polypyrrole and*β*-CD modified Pt electrode;
[28]PPy-*β*-CD/GCE = polypyrrole and*β*-CD modified GCE;
[29]Ppy-*β*-CD/Au = polypyrrole and*β*-CD modified gold electrode;
[30]*β*-CD-MWCNT-PANI/GCE = *β*-CD and MWCNT in polyaniline film on GCE;

[31]HRP-β-CD-ILs/GCE = 1-butyl-3-methylimidazolium tetrafluoroborate modified β-CD HRP and ionic liquids modified GCE;
[32]P-3-MT- γ-CD/Au = poly(3-methylthiophene)/γ-cyclodextrin film on a gold electrode;
[33]β-CD-poly(*N*-acetylaniline)-CNT/CPE = β-CD+poly(*N*-acetylaniline)+CNT modified CPE;
[34]β-CD-PEI/GCE = β-CD in polyethileneimine film on GCE;
[35]β-CD-MWCNT-chitosan/GCE = β-CD incorporated MWCNT and chitosan modified GCE;
[36]fMWCNT-AuNP-HP-β-CD/GC, Au and ITO = functionalized MWCNTs and AuNPs decorated with hydroxypropyl-β-CD synthesized on glassy carbon, gold and on indium tin oxide electrodes;
[37]β-CD-CNTs/GCE = β-CD incorporated CNTs-modified GCE;
[38]α- CD-CNT/GCE = α- CD-CNT coated GCE;
[39]MWCNT-β-CD/GCE = multi-walled carbon nanotubes and β- CD film on GCE;
[40]β-CD-ACNTs/GCE = aligned carbon nanotubes functionalized with β-CD derivative through diazotization reaction;
[41]Plu-AuNPs/β-CD-MWCNT/GCE = Gold nanoparticles-poly(luminol) hybrid film and multi-walled carbon nanotubes with incorporated β-CD modified GCE;
[42]pre-CDP-MWCNT-GOx = Carbon nanotubes in mixed solution of CD and CD prepolymer chemical modified electrode;
[43]HDP- β-CD-RGO-Nafion/Bi/GCE = Hydroxypropyl-β-CD-reduced graphene oxide hybrid nanosheets combining Nafion and in situ deposited Bi film on GCE;
[44]HDP-β-CD-GO/GCE = Hydroxypropyl-β-CD-graphene oxide modified GCE;
[45]β-CD@CRG-Nafion = β-CD and chemically reduced graphene-Nafion modified electrode;
[46]β-CD-Gr-ADA-HRP = CD-graphene nanocomposite employed for the fabrication of adamantane-modified horseradish peroxidase nanocomposite electrodes;
[47]β-CD-Gr/GCE = β-CD-graphene sheets modified GCE;
[48]β-CD-CRGO/GCE = β-CD modified RGO (chemically reduced by *l*-AA) modified GCE;
[49]E-Gr/P-β-CD = electrogenerated graphene/polymerized β-CD nanocomposite film modified electrode;
[50]β-CD-fGr/CPE = β-CD functionalized graphene modified CPE;
[51]AuNPs-CGS-NH_2-β-CD = Au NPs were anchored onto carboxylic graphene nanosheets functionalized with NH_2-β-CD;
[52]DM-β-CD-fGr/GCE = 2,6-dimethyl-β-CD functionalized graphene nanosheets modified GCE;
[53]CRC-β-CD = Chemicaly reduced Graphene-β-CD hybrid system.

The major disadvantage of all these biomolecules, represented by the poor chemical and physical stability, was the main reason for developing an alternative approach which involves the use of biomimetic receptor systems, capable of binding target molecules with affinities and specificities comparable to those of natural receptors. These chiral macrocyclic receptors are also able of stereoselective molecular recognition. Among the biomimetic receptors the most used are crown ethers, calixarenes, CDs and MIPs, representing a topic of current interest in supramolecular chemistry [140, 141].

CDs with their particular structural properties have been widely used for the inclusion complexes formation with a variety of guest molecules that fitted in their cavity. The immobilization of enzymes through supramolecular complexation between *β*-CD and adamantane or pyrene tagged molecules was reported for the first time by Villalonga *et al.* [142] starting the great evolution of the sensors based on this affinity system [143-145]. It is very well known that adamantane derivates form 1:1 inclusion complexes with *β*-CD with high stability (association equilibrium constant about10^4-10^5 M^{-1}). The association equilibrium constant of $K_{ass}=5.2\cdot10^4$ M^{-1} for 1:1 inclusion complex between *β*-CD and adamantane labelled with a fluorescent probe (Ada-A488) was calculated by fluorescence correlation spectroscopy [154]. The pyrene-*β*-CD complex has an apparent association constant of 2.3 10^4 M^{-1} [143, 146].Biotin is also a suitable guest for CD forming host-guest complexes with an association constant of 300 M^{-1} used in order to immobilize biomolecules. Due to their high stability the host-guest complexes have found several important applications in supramolecular chemistry and in biomedical fields [147-150].

Cosnier et al. synthetized glucose oxidase conjugated with *β*-CD which showed affinity towards adamantane-polypyrrole/pyrene, thus mimicking the biological avidin-biotin interactions. This approach led to the construction of high-performance amperometric biosensors [75, 143, 151].

Recently, considerable attentions have been paid to molecular imprinting technology, which provides a promising alternative to create highly specific recognition sites within a synthetic polymer network via the template polymerization process. Molecular imprinting technology became available as a technique for the creation of such receptor-like binding sites with selectivity comparable to that of affinity biomacromolecules. Molecularly imprinted polymers (MIPs) can be successfully used in applications relying on selective molecular binding imitating the receptor-ligand, antibody-antigen, or enzyme-substrate biorecognition [152], and CDs have been incorporated in the MIPs structure improving their selective recognition ability. Formation of MIPs with CD has a major advantage over other MIP polymers due to the benefits brought by the CDs: binding capacity and binding selectivity. Thus, artificial receptors for versatile nanometer scaled guests can be obtained.

Polymeric receptors for cholesterol and stigmasterol were synthesized by crosslinking *β*-CD and 2,6-di-*o*-methyl-*β*-CD with hexamethylene diisocyanate or toluene 2,4-diisocyanate in dimethyl sulfoxide. These templates promoted the formation of dimers and trimers of *β*-CD [153, 154]. Another MIP was developed for tryptophan selective detection using bonded *β*-CD and acrylamide by developing a method for chiral amino acid separation and purification [155]. A surface plasmon resonance (SPR) sensor for dextromethorphan detection was elaborated based on thin layer imprinted *β*-CD polymer [156], while the analytical determination of *N*-phenyl-1-naphthylamine was successfully achieved on a MIP using *β*-CD as the building units in the development of an optical sensor receptor [157].

A MIP was prepared using allyl-β-CD and methacrylic acid as the binary functional monomers, with good selective adsorption performance for dipenthylphthalate from cow milk [158]. Maleic anhydride modified β-CD and acrylic acid as functional monomers were used for developing a*l*-phenylalanine sensor [159]. Pseudo-polyrotaxanes protein was used together with CD for a MIP elaboration, demonstrating a highly effective recognition capacity relied on the spatial configuration constructed by the used template [160].

A MIP based on β-CD-MWNTs composite material with imprinted sol-gel film on a PANI layer modified carbon electrode was elaborated for the selective determination of *l*-phenylalanine in blood samples. The response mechanism of the MIP sensor was based both on β-CD inclusion complex formation and the molecular recognition capacity of the MIP film for *l*-phenylalanine. The interconnected β-CD-MWNTs composites film and the PANI layer possessed superior conductivity, high stability, excellent electrocatalytic ability and inclusion interaction for the determination of *l*-phenylalanine obtaining a LOD of 1 nM. This MIP based electrochemical sensor presented excellent sensitivity, selectivity, stability, reproducibility and recovery [161].

The molecularly imprinting technique was used for the development of an electrochemical sensor based on gold electrode decorated by β-CD, MWCNT, AuNPs-polyamide amine dendrimer nanocomposites and chitosan derivative being used for selective and convenient determination of chlortetracycline with a LOD of 0.0495 μM [162].

CONCLUSION

The inclusion complexes formed by CDs with a wide variety of chemical species, both in solid state or in solution (aqueous and non-aqueous media) leads to significant changes in the physical-chemical properties of the guest molecules retained inside the relatively hydrophobic CD cavity, through many types of interactions. Therefore, CDs found a lot of applications in various fields, such as: agriculture, food, toilet articles, textile industry, environment, pharmaceuticals, therapy and analytical chemistry as catalysts, enantioselectors and molecular receptors.

The electrochemical investigations of CDs in solution are based on the formation of inclusion complexes, the binding constants being determined by using potentiometric titrations and cyclic voltammetry. The influence of the CD upon the electrochemical behavior of the guest molecule mainly consists in decreased peak currents and shifts in the oxidation and/or reduction potentials, due to the decrease of the free drug concentration, smaller diffusion coefficients of the formed inclusion complexes, and to their relatively high stability.

Natural receptors are able of specific *host-guest* interaction, but their fabrication is very difficult and very expensive. Therefore, besides other affinity based bioelements (DNA, antibodies or aptamers), the development of artificial biomimetic receptors focused the attention of many research groups. A large variety of cyclodextrins and derivatives, calixarenes, various ligands (valinomycine, crown-ethers, etc.), and molecularly imprinted polymers were studied and a great number of related papers were published in the last decade.

Cyclodextrins (natural or chemically modified) with their variable size hydrophobic cavities are the subject of expanded electrochemical research, including both their behavior in homogeneous solutions and in thin films attached to the electrode surfaces. The CDs applications in electrochemistry are based on the inclusion complexes formation, molecular recognition and selective preconcentration of the analyte at the electrode surface leading to the development of highly selective sensors and biosensors.

The most significant modification methods to obtain CD based electrochemical sensors, reported in literature, are the CDs entrapment in polymers or in plasticized membranes, their inclusion in carbon paste, CD immobilization at screen printed electrodes, CDs based self-assembled monolayers, nanoparticles *(gold, silver, cobalt ferrite, magnetic silica, ZnO,* Fe_3O_4*, nanoparticles, CdTe quantum dots)*, carbon based materials (*CNTs, graphenes, fullerenes*), and nanostructured polymeric films *(poly N-acetylaniline, polypyrrole, polyaniline; molecularly imprinted polymers)* functionalized with CD.

The CD modified electrodes with enhanced analytical performances represent an attractive promise for the future development of electrochemical (bio)sensors and remain a very active field of research for a wide range of applications in many areas, including pharmaceutical and biomedical analysis.

References

[1] Martin Del Valle, EM. Cyclodextrins and their uses. *Process Biochemistry,* 2004 39, 1033-1046.

[2] Abd-Elgawad, R; Shimaa, E. Electrochemistry of Cyclodextrin Inclusion Complexes of Pharmaceutical Compounds. *The Open Chemical and Biomedical Methods Journal,* 2010 3, 74-85.

[3] Dodziuk, H. *Cyclodextrins and Their Complexes.* Wiley-VCH Verlag GmbH & Co. KGaA. Weinheim: Helena Dodziuk; 2006.

[4] Szejtli, J; Atwood, JL; Lehn, J-M; et al. *Comprehensive Supramolecular Chemistry. Vol. 3. Cyclodextrins.* 1st Edition, New York: Pergamon; 1996.

[5] Szejtli, J. Past, present and future of cyclodextrin research. *Pure and Applied Chemistry,* 2004 10, 1825-1845.

[6] Szejtli, J. Introduction and General Overview of Cyclodextrin Chemistry. *Chemical Reviews,* 1998 98, 1743-1753.

[7] Connors, KA. The Stability of Cyclodextrin Complexes in Solution. *Chemical Reviews,* 1997 97, 1325-1357.

[8] Brewster, MS; Loftsson, T. Cyclodextrins as pharmaceutical solubilizers. *Advanced Drug Delivery Reviews,* 2007 59, 645-666.

[9] Lichtenthaler, FW; Immel, S. The Lipophilicity Patterns of Cyclodextrins and of Non-glucose Cyclooligosaccharides. *Journal of Inclusion Phenomena and Molecular Recognition in Chemistry,* 1996, 25, 3-16.

[10] Mura, P. Analytical techniques for characterization of cyclodextrin complexes in aqueous solution: A review. *Journal of Pharmaceutical and Biomedical Analysis,* 2014 101, 238–250.

[11] Singh, R; Bharti, N; Madan, J; Hiremath, SN. Characterization of Cyclodextrin Inclusion Complexes - A Review. *Journal of Pharmaceutical Science and Technology,* 2010 2, 171-183.

[12] Singh, M; Sharman, R; Banerjee, UC. Biotechnological applications of cyclodextrins. *Biotechnology Advances,* 2002 20, 341-359.

[13] Gao, Z-N; Li, H-L. Spectroscopic and electrochemical studies of phenylhydrazine-β-cyclodextrin inclusion complex and its adsorption behaviors. *Acta Physyco-Chimica Sinica,* 1999 15, 1005-1010.

[14] Gao, ZN; Wen, XL; Li, HL. Study of the inclusion complexes of catecholamines with β-cyclodextrin by cyclic voltammetry. *Polish Journal of Chemistry,* 2002 76, 1001-1007.

[15] Shehatta, IS; Ibrahim, MS. Binding of anti-inflammatory drug indomethacin with cyclodextrin and DNA: Solubility, spectroscopic, and voltammetric studies. *Canadian Journal of Chemistry,* 2001 79, 1431-1438.

[16] Shehatta, IS; Ibrahim, MS; Sultan, MR. The antimicrobial nalidixic acid as a probe for molecular recognition of α- and β-cyclodextrins. *Canadian Journal of Chemistry,* 2002 80, 1313-1320.

[17] Ibrahim, MS; Shehatta, IS; Al-Nayeli, AA. Voltammetric studies of the interaction of lumazine with cyclodextrins and DNA. *Journal of Pharmaceutical and Biomedical Analysis,* 2002 28, 217-225.

[18] Fritea, L; Tertiş, M; Topală, TL; Săndulescu, R. Electrochemical and spectral study of cyclodextrins interactions with some farmaceutical compounds, *Farmacia,* 2013 6, 1054-1068.

[19] Chen, Y; Zhang, YM; Liu, Y. Multidimensional nanoarchitectures based on cyclodextrins. *Chemical Communications,* 2010 46, 5622-5633.

[20] Chmurski, K; Majewska, UE; Bilewicz, R. Analytical applications of gold electrodes modified with monolayers of thiolated cyclodextrins, *Journal of Inclusion Phenomena Macrocyclic Chemistry,* 2007 57, 385–389.

[21] Chmurski, K; Koralewska, A; Temeriusz, A; Bilewicz, R. Catalytic Au Electrodes Based on SAMs of per(6-deoxy-6- thio-2,3-di-O-methyl)- b-Cyclodextrin. *Electroanalysis,* 2004 16, 1407-1412.

[22] Rojas, MT; Koniger, R; Stoddart, JF; Kaifer, AE. Supported monolayers containing preformed binding sites. Synthesis and interfacial binding properties of a thiolated β-Cyclodextrin derivative. *Journal of the American Chemical Society,* 1995 117, 336-343.

[23] Li, W; Jin, GY; Chen, H; Kong, JL. Highly sensitive and reproducible cyclodextrin-modified gold electrodes for probing trace lead in blood. *Talanta,* 2009 78, 717–722.

[24] Ortiz, M; Torréns, M; Alakulppi, N; Strömbom, L; Fragoso, AK; O'Sullivan, C. Amperometric supramolecular genosensor self-assembled on cyclodextrin-modified surfaces. *Electrochemistry Communications,* 2011 13, 578–581.

[25] Fernández, I; Araque, E; Martínez-Ruiz, P; Di Pierro, P; Villalonga, R; Pingarrón, JM. Gold surface patterned with cyclodextrin-based molecular nanopores for electrochemical assay of transglutaminase activity. *Electrochemistry Communications,* 2014 40, 13–16.

[26] Damos, FS; Luz, RCS; Kubota, LT. Electrochemical properties of self-assembled monolayer based on mono-(6-deoxy-6-mercapto)-β-cyclodextrin toward controlled molecular recognition. *Electrochimica Acta,* 2007 53, 1945–1953.

[27] D'Annibale, A; Regoli, R; Sangiorgio, P; Ferri, T. Preparation and electrochemical characterization of a β-Cyclodextrin-based chemically modified electrode. *Electroanalysis,* 1999 11, 505-510.

[28] Chmurski, K; Temeriusz, A; Bilewicz, R. Measurement of ibuprofen binding to mixed monolayers containing β-Cyclodextrin active sites. *Journal of Inclusion Phenomena Macrocyclic Chemistry,* 2004 49, 187–191.

[29] Villalonga, R; Tachibana, S; Cao, R; Ortiz, PD; Gomez, L; Asano, Y. Supramolecular-mediated immobilisation of L-phenylalanine dehydrogenase on β-cyclodextrin-modified gold nanospheres. *Journal of Experimental Nanoscience,* 2006 1, 249–260.

[30] Villalonga, R; Fujii, A; Shinohara, H; Asano, Y; Cao, R; Tachibana, S; Ortiz, P. Supramolecular-mediated immobilization of L-phenylalanine dehydrogenase on cyclodextrin-coated Au electrodes for biosensor applications. *Biotechnology Letters,* 2007 29, 447-452.

[31] Fragoso, A; Almirall, E; Cao, R; Echegoyen, L; González-Jonte, R. A supramolecular approach to the selective detection of dopamine in the presence of ascorbate. *Chemical Communications,* 2004 10, 2230-2231.

[32] Li, J; Wu, X; Yu, Y; Le, S. Self-assembled sulfonated β-Cyclodextrin layer on gold electrode for the selective electroanalysis of dopamine. *Journal of Solid State Electrochemistry,* 2009 13, 1811-1818.

[33] Majewska, UE; Chmurski, K; Biesiada, K; Olszyna, AR; Bilewicz, R. Dopamine oxidation at per(6-deoxy-6-thio)-α-cyclodextrin monolayer modified gold electrodes. *Electroanalysis,* 2006 18, 1463-1470.

[34] Frasconi, M; Mazzei, F. Electrochemical and surface plasmon resonance characterization of β-cyclodextrin-based self-assembled monolayers and evaluation of their inclusion complexes with glucocorticoids. *Nanotechnology,* 2009 20, 285502, doi: 10.1088/0957-4484/20/28/285502.

[35] Ferancová, A; Labuda, J. Cyclodextrins as electrode modifiers. *Fresenius Journal of Analytical Chemistry,* 2001 370, 1-10.

[36] Amorim, CG; Araujo, AN; Montenegro, MCBSM; Silva, VL. Cyclodextrin-based potentiometric sensors for midazolam and diazepam. *Journal of Pharmaceutical and Biomedical Analysis,* 2008 48, 1064–1069.

[37] Khaled, E; Kamel, MS; Hassan, HNA; Aboul-Enein, HY. Cyclodextrin-based dextromethorphan potentiometric sensors. *Journal of Electroanalytical Chemistry,* 2011 661, 239–244.

[38] Hassan, SSM; Kamel, AH; Abd El-Naby, H. New potentiometric sensors based on selective recognition sites for determination of ephedrine in some pharmaceuticals and biological fluids. *Talanta,* 2013 103, 330–336.

[39] Sousa, TFA; Amorim, CG; Montenegro, MCBSM; Araújo, AN. Cyclodextrin based potentiometric sensor for determination of ibuprofen in pharmaceuticals and waters. *Sensors and Actuators B,* 2013 176, 660-666.

[40] Lenik, J. A new potentiometric electrode incorporating functionalized β-cyclodextrins for diclofenac determination. *Materials Science and Engineering C,* 2014 45, 109–116.

[41] Elhag, S; Ibupoto, ZH; Liu, XJ; Nur, O; Willander, M. Dopamine wide range detection sensor based on modified Co_3O_4 nanowires electrode. *Sensors and Actuators B,* 2014 203, 543–549.

[42] Wang, J. Decentralized electrochemical monitoring of trace metals: From disposable strips to remote electrodes, *Analyst,* 1994 119, 763-766.

[43] Ferancová, A; Korgová, E; Buzinkaiová, T; Kutner, W; Stepanek, I; Labuda, J. Electrochemical sensors using screen-printed carbon electrode assemblies modified

with the β-cyclodextrin or carboxymethylated β-cyclodextrin polymer films for determination of tricyclic antidepressive drugs. *Analytica Chimica Acta,* 2001 447, 47–54.

[44] Kim, H; Chang, SC; Shim, YB. α-Cyclodextrin Modified Screen Printed Graphite Electrodes for Detection of Phenols. *Bulletin of the Korean Chemical Society,* 2002 23, 427-431.

[45] Alarcón-Ángeles, G; Guix, M; Silva, WC; Ramírez-Silvad, MT; Palomar-Pardavé, M; Romero-Romoe, M; Merkoci, A. Enzyme entrapment by β-cyclodextrin electropolymerization onto a carbon nanotubes-modified screen-printed electrode. *Biosensors and Bioelectronics,* 2010 26, 1768–1773.

[46] Abd El-Hady, D. Selective and sensitive hydroxypropyl-β-cyclodextrin based sensor for simple monitoring of (+)-catechin in some commercial drinks and biological fluids. *Analytica Chimica Acta,* 2007 593,178–187.

[47] El-Hady, D; El-Maali, N. Selective square wave voltammetricdetermination of (+)-catechin in commercial tea samples using β-cyclodextrin modified carbon paste electrode. *Microchimica Acta,* 2008 161, 225-231.

[48] Yanez, C ; Nunez-Vergara, LJ ; Squella, JA. Determination of nitrendipine with β-Cyclodextrin modified carbon paste electrode. *Electroanalysis,* 2002 14, 559-562.

[49] Rao, CN; Subbarayudu, K; Rao, CCN; Venkateswarlu, P. Electrochemical reduction behaviour of donepezil at β-Cyclodextrin modified carbon paste electrode. *Portugaliae Electrochimica Acta,* 2010 28, 349-357.

[50] Morales, RG; Ramirez, T; Galicia, SL. Carbon paste electrodes electrochemically modified with cyclodextrins. *Journal of Solid State Electrochemistry,* 2003 7, 355–360.

[51] Guorong, Z; Xiaolei, W; Xingwang, S; Tianling, S. β-cyclodextrin–ferrocene inclusion complex modified carbon paste electrode for amperometric determination of ascorbic acid. *Talanta,* 2000 51, 1019-1025.

[52] Ferancová, A; Korgová, E; Mikó, R; Labuda, J. Determination of tricyclic antidepressants using a carbon paste electrode modified with β-cyclodextrin. *Journal of Electroanalytical Chemistry,* 2000 492, 74-77.

[53] Ramírez-Silva, MT; Corona-Avendaño, S; Alarcón-Ángeles, G; Rojas-Hernández, A; Romero-Romo, M; Palomar-Pardavé, M. Dopamine electrochemical behavior onto an electrode modified with a β-cyclodextrin polymer. *ECS Transactions,* 2009 20, 151-157.

[54] Reddy, TM; Sreedhar, M; Reddy, SJ. Electrochemical determination of sparfloxacin in pharmaceutical formulations and urine samples using a β-cyclodextrin modified carbon paste electrode. *Analytical Letters,* 2003 36, 1365-1379.

[55] Reddy, TM; Balaji, K; Reddy, SJ. Voltammetric behavior of some fluorinated quinolone antibacterial agents and their differential pulse voltammetric determination in drug formulations and urine samples using a β-cyclodextrin-modified carbon paste electrode. *Journal of Analytical Chemistry,* 2007 62, 168-175.

[56] Balaji, K; Reddy, GVR; Reddy, TM; Reddy, SJ. Determination of prednisolone, dexamethasone and hydrocortisone in pharmaceutical formulations and biological fluid samples by voltammetric techniques using β-cyclodextrin modified carbon paste electrode. *African Journal of Pharmacy and Pharmacology,* 2008 2, 157-166.

[57] Ortega, RD; Silva, RMT; Pardavé, PM; Alarcón-Ángeles, G; Hernández, RA; Romero, RM. Development a Boron potentiometric determination methodology using a carbon

paste electrode modified with a β-Cyclodextrine- Azomethine-H inclusion complex. *ECS Transactions,* 2009 20, 13-19.

[58] Narasimha, RC; Madhavi, G; Venkateswarlu, P. Electrochemical behavior of anagrelide and its determination. *Oriental Journal of Chemistry,* 2009 25, 425-428.

[59] Zhang, FF; Gu, SQ; Ding, YP; Zhang, Z; Li, L. A novel sensor based on electropolymerization of β-cyclodextrin and l-arginine on carbon paste electrode for determination of fluoroquinolones. *Analytica Chimica Acta,* 2013 770, 53– 61.

[60] Guzmán-Hernández, DS; Palomar-Pardavé, M; Rojas-Hernández, A; Corona-Avendano, S; Romero-Romo M, Ramírez-Silva, MT. Electrochemical quantification of the thermodynamic equilibrium constant of the tenoxicam- β-cyclodextrin inclusion complex formedon the surface of a poly-β-cyclodextrin-modified carbon paste electrode. *Electrochimica Acta,* 2014 140, 535-540.

[61] Gaichore, RR; Srivastava, AK. Voltammetric determination of nifedipine using a β-cyclodextrin modified multi-walled carbon nanotube paste electrode. *Sensors and Actuators B,* 2013 188, 1328-1337.

[62] Palomar-Pardavé, M; Corona-Avendaño, S; Romero-Romo, M; Alarcón-Angeles, G; Merkoçi, A; Ramírez-Silva, MT. Supramolecular interaction of dopamine with β-cyclodextrin: An experimental and theoretical electrochemical study. *Journal of Electroanalytical Chemistry,* 2014 717-718, 103-109.

[63] Pingarron, JM; Yanez-Sedeno, P; Gonzalez-Cortes, A, Gold nanoparticle-based electrochemical biosensors. *Electrochimica Acta,* 2008 53, 5848-5866.

[64] Guo, SJ; Wang, EK. Synthesis and electrochemical applications of gold nanoparticles. *Analytica Chimica Acta,* 2007 598, 181–192.

[65] Villalonga, R; Cao, R; Fragoso, A; Damia, AE; Ortiz, PD; Caballero, J. Supramolecular assembly of β-cyclodextrin-modified gold nanoparticles and Cu, Zn-superoxide dismutase on catalase. *Journal of Molecular Catalysis B: Enzymatic,* 2005 35, 79–85.

[66] Han, JT; Huang, KJ; Li, J; Liu, YM; Yu, M. β-cyclodextrin-cobalt ferrite nanocomposite as enhanced sensing platform for catechol determination. *Colloids and Surfaces B: Biointerfaces,* 2012 98, 58-62.

[67] Badruddoza, AZM; Rahman, MT; Ghosh, S; Hossain, MZ; Shi, JZ; Hidajat, K; Uddin, MS. β-cyclodextrin conjugated magnetic, fluorescent silica core–shell nanoparticles for biomedical applications. *Carbohydrate Polymers,* 2013 95, 449-457.

[68] Abdolmaleki, A; Mallakpour, S; Borandeh, S. Tailored functionalization of ZnO nanoparticle via reactivecyclodextrin and its bionanocomposite synthesis. *Carbohydrate Polymers,* 2014 103, 32-37.

[69] Zhu, J; He, J; Du, XY; Lu, RH; Huang, LZ; Ge, X. A facile and flexible process of b cyclodextrin grafted on Fe3O4 magnetic nanoparticles and host–guest inclusion studies. *Applied Surface Science,* 2011 257, 9056-9062.

[70] Chen, XF; Zhou, M; Chang, YP; Ren, CL; Chen, HL; Chen, XG. Novel synthesis of β-cyclodextrin functionalized CdTe quantum dots as luminescent probes. *Applied Surface Science,* 2012 263, 491-496.

[71] Algarra, M; Campos, BB; Aguiar, FR; Rodriguez-Borges, JE; Esteves da Silva, JCG. Novel β-cyclodextrin modified CdTe quantum dots as fluorescence nanosensor for acetylsalicylic acid and metabolites. *Materials Science and Engineering: C,* 2012 32, 799-803.

[72] Yoon, S; Nichols, WT. Cyclodextrin directed self-assembly of TiO_2 nanoparticles. *Applied Surface Science,* 2013 285, 517-523.

[73] Andrade, PF; Fonseca de Faria, A; Soares da Silva, D; Bonacin, AJ, do Carmo, GM. Structural and morphological investigations of β-cyclodextrin-coated silver nanoparticles, *Colloids and Surfaces B: Biointerfaces,* 2014 118, 289–297.

[74] Manivannan, S; Ramaraj, R. Synthesis of cyclodextrin-silicate sol–gel composite embedded gold nanoparticles and its electrocatalytic application. *Chemical Engineering Journal,* 2012 210, 195–202.

[75] Holzinger, M; Bouffier, L; Villalonga, R; Cosnier, S. Adamantane/β-cyclodextrin affinity biosensors based on single-walled carbon nanotubes. *Biosensors and Bioelectronics,* 2009 24, 1128–1134.

[76] Huang, T; Meng, F; Qi, LM. Facile synthesis and one-dimensional assembly of cyclodextrin-capped gold nanoparticles and their applications in catalysis and Surface-Enhanced Raman Scattering. *Journal of Physical Chemistry C,* 2009 113, 13636–13642.

[77] Sun, L; Crook, RM; Chechik, V. Preparation of polycyclodextrin hollow spheres by templating gold nanoparticles. *Chemical Communications,* 2001 4, 359–360.

[78] Chen, M; Diao, GW. Electrochemical study of mono-6-thio-β-cyclodextrin/ferrocene capped on goldnanoparticles: Characterization and application to the design of glucoseamperometric biosensor. *Talanta,* 2009 80, 815–820.

[79] Tian, XQ; Cheng, CM; Yuan, HY; Du, JA; Xiao, D; Xie, SP; Choi, MMF. Simultaneous determination of l-ascorbic acid, dopamine and uric acid with gold nanoparticles–β-cyclodextrin–graphene-modified electrode by square wave voltammetry. *Talanta,* 2012 93, 79-85.

[80] Díez, P; Villalonga, R; Villalonga, ML; Pingarrón, JM. Supramolecular immobilization of redox enzymes on cyclodextrin-coated magnetic nanoparticles for biosensing applications. *Journal of Colloid and Interface Science,* 2012 386, 181–188.

[81] Gao, JA; Guo, ZK; Su, FJ; Gao, L; Pang, XH; Cao, W; Du, B; Wei, Q. Ultrasensitive electrochemical immunoassay for CEA through host–guest interaction of β-cyclodextrin functionalized graphene and Cu@Ag core–shell nanoparticles with adamantine-modified antibody. *Biosensors and Bioelectronics,* 2015 63, 465–471.

[82] Lin, DJ; Wu, J; Ju, HX; Yan, F. Signal amplification for electrochemical immunosensing by in situ assembly of host–guest linked gold nanorod superstructure on immunocomplex. *Biosensors and Bioelectronics,* 2013 45, 195–200.

[83] Yu, Y; Xin, Y ; Feng, YH ; Min, LZ ; Li, LY ; Li, SG ; Qin, YR. Electrochemical sensor for cinchonine based on a competitive host–guest complexation. *Analytica Chimica Acta,* 2005 528, 135–142.

[84] Lomillo, MAA; Renedo, OD; Martinez, MJA. Optimization of a cyclodextrin-based sensor for rifampicin monitoring. *Electrochimica Acta,* 2005 50, 1807–1811.

[85] Izaoumen, N; Bouchta, D; Zejli, H; El Kaoutit, M; Stalcup, AM; Temsamani, KR. Electrosynthesis and analytical performances of functionalized poly(pyrrole/β-cyclodextrin) films. *Talanta,* 2006 66, 111–117.

[86] Izaoumen, N ; Bouchta, D ; Zejli, H ; El Kaoutit, M ; Temsamani, KR. The electrochemical behavior of neurotransmitters at a poly (pyrrole-β-cyclodextrin) modified glassy carbon electrode. *Analytical Letters,* 2005 38, 1869-1885.

[87] Bouchta, D; Izaoumen, N; Zejli, H; El Kaoutit, M; Temsamani, KR. Electroanalytical properties of a novel PPY/γ-cyclodextrin coated electrode. *Analytical Letters,* 2005 38, 1019-1036.

[88] Luk, HN; Jheng, RF; Tsai, TC; Wu, RJ. Novel electrochemical impedance detection for rocuronium by using polypyrrole and cyclodextrin as the sensing material. *Sensor Letters,* 2009 7, 1093-1099.

[89] Yang, Y; Lei, CX; Liu, ZM; Liu, YL; Shen, GL; Yu, RQ. Highly selective dopamine determination by using carboxymethylated β-cyclodextrin polymer film modified electrode. *Analytical Letters,* 2004 37, 2267-2282.

[90] Tanji, Y; Wanzhi, W; Jinxiang, Z. Selective detection of dopamine in the presence of ascorbic acid by use of glassy-carbon electrodes modified with both polyaniline film and multi-walled carbon nanotubes with incorporated β-cyclodextrin. *Analytical and Bioanalytical Chemistry,* 2006 386, 2087–2094.

[91] Arjomandi, J; Holze, R. Spectroelectrochemistry of conducting polypyrrole and poly(pyrrole–cyclodextrin) prepared in aqueous and nonaqueous solvents. *Journal of Solid State Electrochemistry,* 2007 11, 1093–1100.

[92] Arjomandi, J; Holze, R. In situ characterization of N-methylpyrrole and(N-methylpyrrole-cyclodextrin) polymers on goldelectrodes in aqueous and nonaqueous solution. *Synthetic Metals,* 2007 157, 1021–1028.

[93] Cao, AP; Ai, HH; Ding, YL; Dai, CY; Fei, JJ. Biocompatible hybrid film of β-cyclodextrin and ionic liquids: A novel platform for electrochemical biosensing. *Sensors and Actuators B,* 2011 155, 632–638.

[94] Bouchta, D; Izaoumen, N; Zejli, H; El Kaoutit, M; Temsamani, KR. A novel electrochemical synthesis of poly-3-methylthiophene-y-cyclodextrin film. Application for the analysis of chlorpromazine and some neurotransmitters. *Biosensors and Bioelectronics,* 2005 20, 2228–2235.

[95] Abbaspour, A; Noori, A. A cyclodextrin host–guest recognition approach to an electrochemical sensor for simultaneous quantification of serotonin and dopamine. *Biosensors and Bioelectronics,* 2011 26, 4674-4680.

[96] Jiang, HG; Yang, ZJ; Zhou, XT; Fang, YX; Ji, HB. Immobilization of β-Cyclodextrin as insoluble β-Cyclodextrin polymer and its catalytic performance. *Chinese Journal of Chemica Engineering,* 2012 20, 784-792.

[97] Fritea, L; Tertiş, M; Cristea, C; Săndulescu, R. New β-Cyclodextrin entrapped in polyethyleneimine film modified electrodes for pharmaceutical compounds determination. *Sensors,* 2013 13, 16312-16329.

[98] Fritea, L; Tertiş, M; Cristea, C; Cosnier, S; Săndulescu, R. Simultaneous determination of ascorbic and uric acids in urine using an innovative electrochemical sensor based on β-cyclodextrin. *Analitycal Letters,* 2015 48, 88-99.

[99] Kong, B; Zeng, J; Luo, G; Luo, S; Wei, W; Li, J. Layer-by-layer assembled carbon nanotube films with molecule recognition function and lower capacitive background current. *Bioelectrochemistry,* 2009 74, 289–294.

[100] Yogeswaran, U; Thiagarajan, S; Chen, SM. Pinecone shape hydroxypropyl-β-cyclodextrin on a film of multi-walled carbon nanotubes coated with gold particles for the simultaneous determination of tyrosine, guanine, adenine and thymine. *Carbon,* 2007 45, 2783–2796.

[101] Wang, Z; Xiao, S; Chen, Y. β-Cyclodextrin incorporated carbon nanotubes-modified electrodes for simultaneous determination of adenine and guanine. *Journal of Electroanalytical Chemistry,* 2006 589, 237-242.

[102] Kang, SZ; Cui, Z; Mu, J. Electrochemical behavior of sodium cholate and deoxycholate on an electrode modified with multi-walled carbon nanotubes (MWNTs) linked up with cyclodextrin. *Diamond & Related Materials,* 2007 16, 12–15.

[103] Wang, G; Liu, X; Yu, B; Luo, G. Electrocatalytic response of norepinephrine at a β-cyclodextrin incorporated carbon nanotube modified electrode. *Journal of Electroanalytical Chemistry,* 2004 567, 227-231.

[104] Wang, GY; Liu, XJ; Luo, GA; Wang, ZH. α-Cyclodextrin incorporated carbon nanotube-coated electrode for the simultaneous determination of dopamine and epinephrine. *Chinese Journal of Chemistry,* 2005 23, 297-302.

[105] He, JL; Yang, Y; Yang, X; Liu, YL; Liu, ZH; Yu, RQ. *β*-Cyclodextrin incorporated carbon nanotube-modified electrode as an electrochemical sensor for rutin. *Sensors and Actuators B: Chemical,* 2006 114, 94-100.

[106] Joon-Hyung, J; Hyunmyung, K; Seunho, J. Electrochemical selectivity enhancement by using monosuccinyl *β*-cyclodextrin as a dopant for multi-wall carbon nanotube-modified glassy carbon electrode in simultaneous determination of quercetin and rutin, *Biotechnology Letters,* 2009 31, 1739–1744.

[107] Alarcon-Angeles, G; Perez-Lopez, B; Palomar-Pardave, M; Ramirez-Silva, MT; Alegret, S; Merkoci, A. Enhanced host–guest electrochemical recognition of dopamine using cyclodextrin in the presence of carbon nanotubes, *Carbon,* 2008 46, 898-906.

[108] Yang, LZ; Xu, Y; Wang, XH; Zhu, J; Zhang, RY; He, PA; Fang, YZ. The application of *β*-cyclodextrin derivative functionalized aligned carbon nanotubes for electrochemically DNA sensing via host–guest recognition. *Analytica Chimica Acta,* 2011 689, 39–46.

[109] Zhang, W; Chen, M; Gong, XD; Diao, GW. Universal water-soluble cyclodextrin polymer–carbon nanomaterials with supramolecular recognition. *Carbon,* 2013 61, 154-163.

[110] Jia, D; Dai, JY; Yuan, HY; Lei, L; Xiao, D. Selective detection of dopamine in the presence of uric acid using a gold nanoparticles-poly(luminol) hybrid film and multi-walled carbon nanotubes with incorporated *β*-cyclodextrin modified glassy carbon electrode, *Talanta,* 2011 85, 2344– 2351.

[111] Yang, HP; Zhu, YF; Chen, DC; Li, CH; Chen, SG; Ge, ZC. Electrochemical biosensing platforms using poly-cyclodextrin and carbon nanotube composite. *Biosensors and Bioelectronics,* 2010 26, 295-298.

[112] Lv, MJ; Wang, XB; Li, J; Yang, XY; Zhang, CA; Yang, J; Hu, H. Cyclodextrin-reduced graphene oxide hybrid nanosheets for thesimultaneous determination of lead(II) and cadmium(II) using square-wave anodic stripping voltammetry. *Electrochimica Acta,* 2013 108, 412-420.

[113] Xu, CH; Wang, XB; Wang, JC; Hu, HT; Wan, L. Synthesis and photoelectrical properties of β-Cyclodextrin functionalized graphene materials with high bio-recognition capability. *Chemical Physics Letters,* 2010 498, 162–167.

[114] Liu, K; Wei, J; Wang, C. Sensitive detection of rutin based on β-cyclodextrin@chemically reduced graphene/Nafion composite film. *Electrochimica Acta,* 2011 56, 5189–5194.

[115] Lu, LM; Qiu, XL; Zhang, XB; Shen, GL; Tan, WH; Yu, RQ. Supramolecular assembly of enzyme on functionalized graphene for electrochemical biosensing. *Biosensors and Bioelectronics,* 2013 45, 102–107.

[116] Tan, L; Zhou, KG; Zhang, YH; Wang, HX; Wang, XD; Guo, YF; Zhang, HL. Nanomolar detection of dopamine in the presence of ascorbic acid at β-cyclodextrin/graphene nanocomposite platform. *Electrochemistry Communications,* 2010 12, 557–560.

[117] Lu, DB; Lin, SX; Wang, LT; Shi, XZ; Wang, CM; Zhang, Y. Synthesis of cyclodextrin-reduced graphene oxide hybrid nanosheets for sensitivity enhanced electrochemical determination of diethylstilbestrol. *Electrochimica Acta,* 2012 85, 131-138.

[118] Zhang, Z; Gu, SQ; Ding, AP; Shen, MJ; Jiang, L. Mild and novel electrochemical preparation of β-cyclodextrin/graphene nanocomposite film for super-sensitive sensing of quercetin. *Biosensors and Bioelectronics,* 2014 57, 239–244.

[119] Wei, MC; Tian, D; Liu, S; Zheng, XL; Duan, S; Zhou, CL. β-Cyclodextrin functionalized graphene material: A novel electrochemical sensor for simultaneous determination of 2-chlorophenol and 3-chlorophenol. *Sensors and Actuators B,* 2014 195, 452–458.

[120] Yang, L; Zhao, H; Li, YC; Li, CP. Electrochemical simultaneous determination of hydroquinone and p-nitrophenol based on host–guest molecular recognition capability of dual β-cyclodextrin functionalized Au@graphene nanohybrids. *Sensors and Actuators B,* 2015 207, 1-8.

[121] Gong, CB; Guo, CC; Jiang, D; Tang, Q; Liu, CH; Ma, XB. Graphene–cyclodextrin–cytochrome c layered assembly with improved electron transfer rate and high supramolecular recognition capability. *Materials Science and Engineering: C,* 2014 39, 281–287.

[122] Liu, ZG; Zhang, A; Guo, YJ; Dong, CA. Electrochemical sensor for ultra sensitive determination of isoquercitrin and baicalin based on DM-β-cyclodextrin functionalized graphene nanosheets. *Biosensors and Bioelectronics,* 2014 58, 242–248.

[123] Agnihotri, N; Chowdhury, AD; De A. Non-enzymatic electrochemical detection of cholesterol using β-cyclodextrin functionalized grapheme. *Biosensors and Bioelectronics,* 2015 63, 212–217.

[124] Bosi, S; Da Ros, T; Spalluto, G; Prato, M. Fullerene derivatives: an attractive tool for biological applications. *European Journal of Medicinal Chemistry,* 2003 38, 913-923.

[125] Kato, S; Aoshima, H; Saitoh, Y; Miwa, N. Highly hydroxylated or γ-cyclodextrin-bicapped water-soluble derivative of fullerene: The antioxidant ability assessed by electron spin resonance method and β-carotene bleaching assay. *Bioorganic & Medicinal Chemistry Letters,* 2009 19, 5293–5296.

[126] Furuishi, T; Fukami, T; Nagase, H; Suzuki, T; Endo, T; Ueda, H; Tomono, K. Improvement of solubility of C70 by complexation with cyclomaltononaose(δ-cyclodextrin). *Pharmazie,* 2008 63, 54-57.

[127] Kojima, C; Toi, Y; Harada, A; Kono, K. Aqueous solubilization of fullerenes using poly(amidoamine) dendrimers bearing cyclodextrin and poly(ethyleneglycol). *Bioconjugate Chemistry,* 2008 19, 2280–2284.

[128] Abdulmalik, A; Hibah, A; Zainy, BM; Makoto, A; Daisuke, I; Masaki, O; Kaneto, U; Fumitoshi, H. Preparation of soluble stable C60/human serum albumin nanoparticles

via cyclodextrin complexation and their reactive oxygen production characteristics. *Life Sciences,* 2013 93, 277–282.

[129] Zhang, MF; Fu, L; Wang, J; Xu, ZQ; Jiang, FL; Liu, Y. Spectroscopic and electrochemical studies on the interaction of an inclusion complex of β-cyclodextrin/fullerene with bovine serum albumin in aqueous solution. *Journal of Photochemistry and Photobiology A: Chemistry,* 2012 228, 28–37.

[130] Nobusawa, K; Payra, D; Naito, M. Pyridyl-cyclodextrin for ultra-hydrosolubilizationof [60]fullerene. *Chemical Communications,* 2014 50, 8339-8342.

[131] McNally, A; Forster, RJ; Keyes, TE. Interfacial supramolecular cyclodextrin-fullerene assemblies: host reorientation and guest stabilization. *Physical Chemistry Chemical Physics,* 2009 11, 848–856.

[132] Ikeda, A; Ishikawa, M; Aono, R; Kikuchi, J; Akiyama, M; Shinoda, W. Regioselective recognition of a [60]Fullerene-bisadduct by cyclodextrin. *The Journal of Organic Chemistry,* 2013 78, 2534–2541.

[133] Pospišil, L; Hromadová, M; Gál, M; Buličková, J; Sokolová, R; Filippone, S; Yang, J; Guan, Z; Rassat, A; Zhang, YM. Redox potentials and binding enhancement of fullerene and fullerene–cyclodextrin systems in water and dimethylsulfoxide. *Carbon,* 2010 48, 153-162.

[134] Giacalone, F; D'Anna, F; Giacalone, R; Gruttadauria, M; Riela, S; Noto, R. Cyclodextrin-[60]fullerene conjugates: synthesis, characterization, and electrochemical behavior. *Tetrahedron Letters,* 2006 47, 8105–8108.

[135] Yuan, DQ; Koga, K; Kourogi, Y; Fujita, K. Synthesis of fullerene-cyclodextrin conjugates. *Tetrahedron Letters,* 2001 42, 6727–6729.

[136] Liu, Y; Liang, P; Chen, Y; Zhao, YL; Ding, F; Yu, A. Spectrophotometric study of fluorescence sensing and selective binding of biochemical substrates by 2,2'-bridged bis(β-cyclodextrin) and its water-soluble fullerene conjugate. *The Journal of Physical Chemistry B,* 2005 109, 23739-23744.

[137] Liu, Y; Zhao, YL; Chen, Y; Liang, P; Li, L. A water-soluble β-cyclodextrin derivative possessing a fullerene tether as an efficient photodriven DNA-cleavage reagent. *Tetrahedron Letters,* 2005 46, 2507–2511.

[138] Pospišil, L; Hromadová, M; Gál, M; Buličková, J; Sokolová, R; Fanelli, N. Electrochemical impedance of nitrogen fixation mediated by fullerene–cyclodextrin complex. *Electrochimica Acta,* 2008 53, 7445–7450.

[139] Rather, JA; Debnath, P; De Wael, K. Fullerene–β–cyclodextrin conjugate based electrochemical sensing device for ultrasensitive detection of *p*–nitrophenol. *Electroanalysis,* 2013 25, 2145–2150.

[140] Dinh, TV; Cullum, B. Biosensors and biochips: advances in biological and medical diagnostics. *Fresenius Journal of Analytical Chemistry,* 2000 366, 540–551.

[141] Shahgaldian, P; Pieles, U. Cyclodextrin derivatives as chiral supramolecular receptors for enantioselective sensing. *Sensors,* 2006 6, 593-615.

[142] Villalonga, R; Camacho, C; Cao, R; Hernandez, J; Matias, UC. Amperometric biosensor for xanthine with supramolecular architecture. *Chemical Communications,* 2007 9, 942–944.

[143] Le Goff, A; Gorgy, K; Holzinger, M; Haddad, R; Zimmerman, M; Cosnier, S. Biofunctionalization of carbon nanotubes via tris(bispyrene-bipyridine)iron(II): A

supramolecular bridge for the π-stacking and pyrene/β-Cyclodextrin host–guest interactions, *Chemistry: A European Journal,* 2011 17, 10216-10221.

[144] Granadero, D; Bordello, J; Perez-Alvite, MJ; Novo, M; Al-Soufi, W. Host-guest complexation studied by fluorescence correlation spectroscopy: adamantane-cyclodextrin inclusion. *International Journal of Molecular Science,* 2010 11, 173–188.

[145] Holzinger, M; Bouffier, L; Villalonga, R; Cosnier, S. Adamantane/beta-cyclodextrin affinity biosensors based on single-walled carbon nanotubes. *Biosensors and Bioelectronics,* 2009 24, 1128–1134.

[146] Shown, I; Ujihara, M; Imae, T. Sensitizing of pyrene fluorescence by β-cyclodextrin-modified TiO_2 nanoparticles, *Journal of Colloid and Interface Science,* 2010 352, 232–237.

[147] Holzinger, M; Le Goff, A; Cosnier, S. Supramolecular immobilization of bio-entities forbioelectrochemical applications, *New Journal of Chemistry,* 2014 38, 5173-5180.

[148] Ogoshi, T; Harada, A. Chemical sensors based on cyclodextrin derivatives, *Sensors,* 2008 8, 4961-4982.

[149] Davis, ME; Brewster, ME. Cyclodextrin-based pharmaceutics: past, present and future, *Nature Reviews, Drug discovery*, 2004 3, 1023-1035.

[150] Ortiz, M; Fragoso, A; O'Sullivan, CK. Detection of antigliadin autoantibodies in celiac Patient samples using a cyclodextrin-based supramolecular biosensor. *Analytical Chemistry,* 2011 83, 2931–2938.

[151] Haddad, R; Holzinger, M; Villalonga, R; Neumann, A; Roots, J; Maaref, A; Cosnier, S. Pyrene-adamantane-β-cyclodextrin: An efficient host–guest system for the biofunctionalization of SWCNT electrodes, *Carbon,* 2011 49, 2571–257.

[152] Sharma, PS; Pietrzyk-Le, A; D'Souza, F; Kutner, W. Electrochemically synthesized polymers in molecular imprinting for chemical sensing. *Analytical and Bioanalytical Chemistry,* 2012 402, 3177-3204.

[153] Asanuma, H; Kakazu, M; Shibata, M; Hishiya, T, Komiyamas, M. Synthesis of molecularly imprinted polymer of β-cyclodextrin for the efficient recognition ofcholesterol. *Supramolecular Science,* 1998 5, 417-421.

[154] Hishiya, T; Asanuma, H; Komiyama, M. Spectroscopic anatomy of molecular-imprinting of cyclodextrin. Evidence for preferential formation of ordered cyclodextrin assemblies. *Journal of the American Chemical Society,* 2002 124, 570-575.

[155] Qin, L; Hea, XW; Li, WY; Zhang, YK. Molecularly imprinted polymer prepared with bonded β-cyclodextrinand acrylamide on functionalized silica gel for selectiverecognition of tryptophan in aqueous media. *Journal of Chromatography A,* 2008 1187, 94–102.

[156] Roche, PJR; Ng, SM; Narayanaswamy, R; Goddard, N; Page, KM. Multiple surface plasmon resonance quantification of dextromethorphan using a molecularly imprinted β-cyclodextrin polymer: A potential probefor drug–drug interactions. *Sensors and Actuators B,* 2009 139, 22–29.

[157] Ng, SM; Narayanaswamy, R. Molecularly imprinted β-cyclodextrin polymer as potential optical receptor for the detection of organic compound. *Sensors and Actuators B,* 2009 139, 156–165.

[158] Kang, YF; Duan, WP; Li, YN; Kang, JX; Xie, J. Molecularly imprinted polymers of allyl-*β*-cyclodextrin and methacrylic acid for the solid-phase extraction of phthalate. *Carbohydrate Polymers,* 2012 88, 459-464.

[159] Liu, XY; Fang, HX; Yu, LP. Molecularly imprinted photonic polymer based on β-cyclodextrin for amino acid sensing. *Talanta,* 2013 116, 283–289.

[160] Guo, MJ; Hu, X; Fan, Z; Liu, J; Wang, XC; Wang, Y; Mi, HF. Imprinted polymers with cyclodextrin pseudo-polyrotaxanes as pseudo-supports for protein recognition. *Talanta,* 2013 105, 409-416.

[161] Hu, YF; Zhang, ZH; Zhang, HB; Luo, LJ; Yao, SZ. Electrochemical determination of l-phenylalanine at polyaniline modified carbon electrode based on *β*-cyclodextrin incorporated carbon nanotube composite material and imprinted sol–gel film. *Talanta,* 2011 84, 305–313.

[162] Lian, WJ; Huang, JD; Yu, JH; Zhang, XM; Lin, Q; He, XR; Xing, XR; Liu, S. A molecularly imprinted sensor based on *β*-cyclodextrin incorporated multiwalled carbon nanotube and gold nanoparticles-polyamide amine dendrimer nanocomposites combining with water-soluble chitosan derivative for the detection of chlortetracycline. *Food Control,* 2012 26, 620-627.

In: Cyclodextrins
Editor: Francis G. Ramirez

ISBN: 978-1-63482-788-1

Chapter 3

CYCLODEXTRINS: A VALUABLE RESOURCE TO ATTAIN ENVIRONMENTAL SUSTAINABILITY

***Carlos Fernandes*[1], *Igor Encarnação*[1], *André Fonseca*[1], *Jorge Garrido*[1,2], *Fernanda Borges*[1] *and E. Manuela Garrido*[1,2]**
[1]CIQUP/Department of Chemistry and Biochemistry, Faculty of Sciences, University of Porto, Porto, Portugal
[2]Department of Chemical Engineering, School of Engineering (ISEP), Polytechnic of Porto, Porto, Portugal

ABSTRACT

Globally it is estimated that between 7.9 and 15.1 % of agricultural harvests are destroyed by pests, being wheat, rice, maize, potato, soybean and cotton, the most affected crops. With the exponential growth of worldwide population, an optimization in the crops protection against pests is urgently required, and because of that a 15- to 20-fold increasing in the chemical pesticides use has been observed. However, no significant effects on pest control have been observed. Despite of pesticides harmful properties against pests, they can indirectly improve the nutritional value of food. Nevertheless, the increased knowledge on their benefits and risks posed additional safety issues, namely those related to their low water solubility (bioavailability) and easy photodegradation. Several pesticides are easily decomposed by light and in that way a large amount of pesticide is needed to be applied in the crops in order to be effective. Some estimates point to the fact that less than 1 % of applied pesticides effectively reach the target. Furthermore, in some cases the metabolites originated by photodegradation of pesticides display a higher toxicity level. These facts associated with their capacity to act as carcinogenic, mutagenic and teratogenic agents to mankind, have led to a growing concern among scientific community to find a solution aimed to reduce or minimize the mentioned drawbacks and increase pesticide efficiency against pests.

To increase the solubility and photostability as well as to facilitate the handling of pesticides, microencapsulation in nanostructures, namely cyclodextrins (CDs), has been proposed as a putative solution. CDs are cyclic oligosaccharides composed by D-glucose units connected by α-1,4-glycosidic linkages. The CDs structural characteristics, hydrophobic interior cavity and hydrophilic shell, and its ability to alter the physical, chemical, and biological properties of guest molecules has been regarded as a solution to improve pesticide formulations and to attain environmental sustainability.

In this chapter, we will give special attention to the main pesticide groups and their inherent health and environmental issues. General features of CDs and the progress done so far in the field of microencapsulation of pesticides in CDs, will also be reported.

1. Introduction

At present it is estimated that around 10 % of the agriculture profits are destroyed by plagues before the harvest, in spite of intensive measures applied to protect the plantations [1]. From the protection measures taken so far, the utilization of natural, including the use of animals, and/or artificial methods, namely the use of pesticides, can be pointed out as the most relevant [2, 3].

A pesticide is any substance or mixture of substances intended for preventing, destroying, repelling, or mitigating any pest. The term pesticide is generally applied to chemical compounds that are biologically active and interfere with the biological processes of live organisms that are classified as plagues, namely insects, fungi, herbs or toxic plants. Pesticides should be effective, selective and safe, and the benefits of the plague control should outweigh economic, environment and health costs.

Several benefits to society have been regularly mentioned when talking about pesticides, mainly at economic level [4]. Actually, the non-control of the plagues could cause a high loss of food supplies leading consequently to a rise of their price [4]. Pesticides can also be used for controlling vegetation growth near electricity posts and freeways thus allowing infrastructure maintenance and driver safety, respectively, which is important in terms of human well-being. Their use can also avoid disease spread by living beings and so can protect endangered species. For the reasons that have been highlighted, among others, the application of pesticides rose 15 to 20 times in the last 40 years. However, it continues to observe a significant increase in crop losses due to plagues [2]. Although pesticides have reduced crop losses, their use have often led to an increase and unnecessary pest outbreaks. Actually, additional crop losses have been observed due to the inadvertent destruction of pests' natural enemies and the emergence of pest resistance.

The recognition that chemical pesticides are harmful to human health and to the environment result in a tight of the regulatory requirements, focused on toxicity or environmental issues, for the pesticide fabrication and use. The studies carry out so far have shown that the paramount degree of human exposure to pesticides occurs during the mixing and loading processes prior to pesticides application. As a result, intensive research has been performed to discover new technologies that can limit the potential harmful effects from these processes. Recent developments in formulation technology, such as the development of water-soluble sachets, tablets, dust-free granules, and microencapsulated active ingredients, are believed to offer a significant reduction in exposure and so a safer handling of chemical pesticides [5].

Taken into account the need to balance the risks and benefits of pesticides, this chapter presents an overview of the currently used pesticides, their environmental and public health risks, as well as the application of cyclodextrins (CDs) as a solution for environmental sustainability. In this context, some case studies on encapsulation of pesticides in CDs will be highlighted.

2. Pesticides: An Overview

Three millennia have passed since Homer point out that sulfur repelled plagues and over a century since the synthesis of 4,6-dinitro-*ortho*-cresol as the first organic insecticide [6]. With the introduction of the insecticide dichlorodiphenyltrichloroethane (DDT), the herbicide 2,4-dichlorophenoxyacetic acid (2,4-D) and the fungicide thiram in the 1930's and 40's aroused great interest in the scientific community in the search for new compounds and formulations, as well as efforts to understand the interactions between pesticides and living beings [7]. With the developed research was intended to find the ideal pesticide concentration necessary to counteract pests and reduce collateral damage to the environment and health.

Pesticides can be classified based upon their chemical structure, the type of plagues with which they interact, and how they interact (Table 1). It should be stressed that sometimes pesticides with the same chemical structure can be effective against different type of plagues [8, 9]. In general, a pesticide must connect or interact with a specific target interacting with biochemical processes (*e.g.* a receptor, a channel, a protein or membrane) that can be lethal to the plague (a kill effect) or can produce an outbreak (an inactivation effect) turning the plague harmless to the crop. To understand how pesticides work it is important to understand the mechanisms of action to prevent development of pesticide resistance, a problem that can be avoided or delayed by rotating plague control chemicals that work through different modes of action. However, this is a difficult task as diverse pesticides can target the same enzyme, receptor, channel, protein or membrane. For instance, insecticides, fungicides and herbicides can share around 4 to 6 targets with which they interact [6].

In spite of the difficulties found in gathering consensus in the selection of a pesticide by its mechanism(s) of action it is consensual in the area the importance of its selectivity. For instance, insecticides must be toxic for harmful insects and relatively safe for those harmless or beneficial to the plants, like the case of bees. Herbicides, should damage herbs but not the crops, and fungicides should be able to prevent the spread of the disease, like grape disease, and at the same time allow the proliferation of the fungi needed to complete the fermentation wine process. So, at the end one can recognize the hard mission of pesticide industry.

In the past three decades the amount needed to obtain significant results against plagues is only 1 to 10 % of the amount needed before that time. These numbers shows a marked improvement over the years, but at the same time emphasize the emerging need to improve the efficacy of pesticides in plague control.

2.1. Environmental Risks

Besides the growing research interest to potentiate the pesticides capability, the toxicity outline ascribed to these compounds is still a worrying problem either for its formulation or application.

Table 1. Examples of organic pesticides, corresponding target plague and probable mechanism of action

Classification (Chemical family)	Chemical structure	Classification (plague)	Mechanism of action example(s)
Acetamide	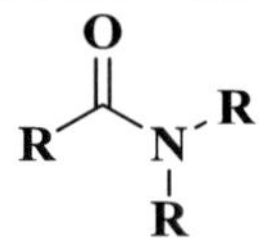	Herbicide	Mitosis and cell division inhibitors
Phenoxyalkanoic acid	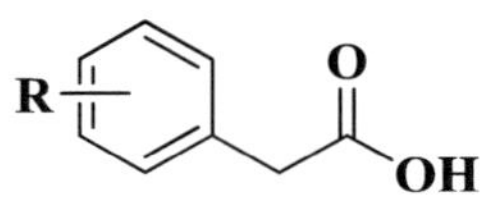	Herbicide	Acetyl CoA carboxylase inhibitors Nucleic acid inhibitors not fully characterized
Carbamate	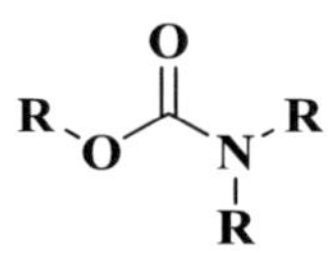	Acaricide Insecticide Fungicide Herbicide	Photosystem II inhibitors Fatty acid and lipid biosynthesis inhibitors Dihydropteroate synthetase inhibitors Inhibition of mitosis and cell division
Aromatic amines (aniline derivatives)	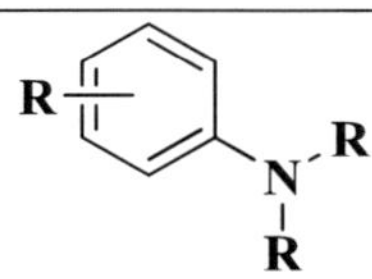	Fungicide Insecticide Herbicide	Mitochondrial respiratory functions
Dicarboximide	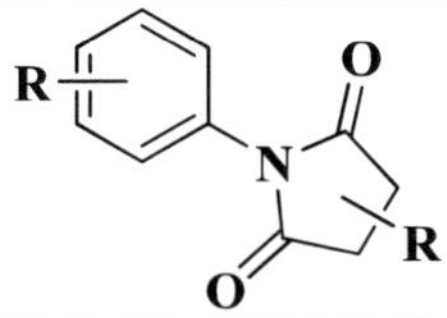	Fungicide	Not fully characterized Affect osmotic regulation
Organochloride	R—Cl	Fungicide Insecticide	Axonic poison
Organophosphorus	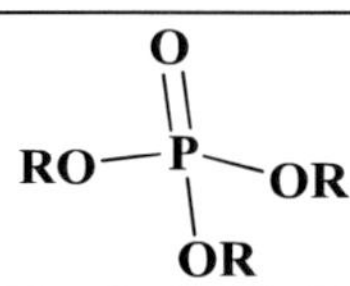	Acaricide Fungicide Rodenticide	Inhibition of acetyl cholinesterase
Pyrethroid	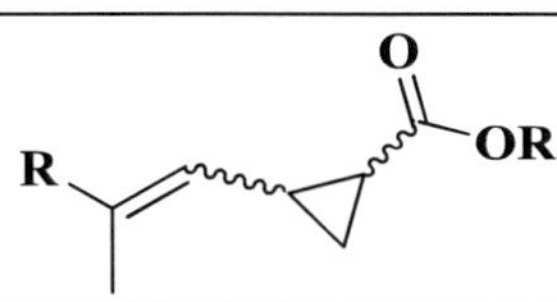	Insecticide	Axonic poison
Urea	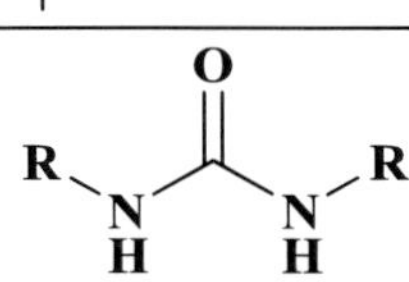	Insecticide	Electron transfer and oxidative mechanisms Inhibition of mitosis and cell division

Ideally, the applied pesticides should only be toxic to the target organisms, should be biodegradable and eco-friendly to some extent [10]. Unfortunately, this is rarely the case as most of the pesticides are non-specific and may kill the organisms that are harmless or useful to the ecosystem. Furthermore, it has been estimated that only about 0.1 % of the pesticides reach the target organisms and, consequently, the remaining bulk contaminates the surrounding environment [11]. The nature of the pesticide and in what kind of medium it is transported in will determine the area it will move and how long it will reside in the environment [12]. Pesticides can be present anywhere in soil, air and water. Actually, transportation of pesticides after they are applied to land involve several simultaneous or successive processes, including emission, wash-off, degradation, sorption/desorption, volatilization, leaching, runoff and plant uptake (Figure 1) [12]. Consequently, bioaccumulation of these compounds can occur in the soil, underground water or in animals and non-target plants [13].

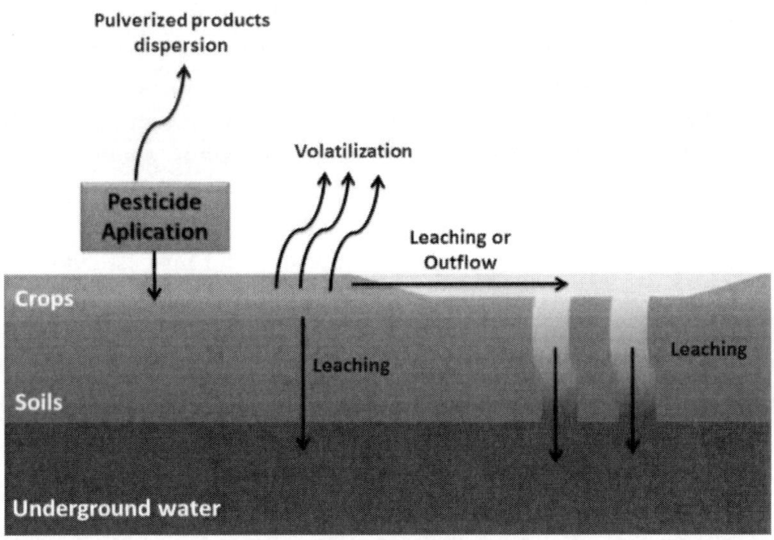

Figure 1. Schematic representation of pesticides fate in the environment.

A major fraction of the pesticides that are used for agriculture and other purposes accumulates in the soil. Several factors such as soil properties and soil micro-flora determine the fate of applied pesticides, owing to which it undergoes a variety of degradation, transport, and adsorption/desorption processes [14].

Pesticides, regardless of the medium that they are applied in, all have the potential to be transported by air. The presence of pesticides in air can be caused by number of factors including spray drift, volatilization from the treated surfaces, and aerial application of pesticides. Extent of drift depends on: droplet size and wind speed. The volatility or semi-volatility nature of the pesticide compounds similarly constitutes an important risk of atmospheric pollution [14].

Pesticide residues in water are a major concern as they pose a serious threat to biological communities including humans. There are different ways by which pesticides can get into water such as accidental spillage, industrial effluent, surface runoff and transport from pesticide treated soils, washing of spray equipments after spray operation, drift into ponds, lakes, streams and river water, aerial spray to control water-inhibiting pests [14, 15].

If pesticides are used in the recommended dosage and applied using modern techniques according the good agricultural practices, the risk of environment contamination is small.

However, the tiniest spill can originate a high concentration in the soil, since some drops of formulated pesticide can include around 1 g of active substance [15]. A spill in about 1 dm^2 of area of soil implies a final pesticide concentration of 1 g dm^{-2}, which is equivalent to 1 ton ha^{-1}. At the present, the normal dosages of pesticides in modern crops are in the range of kg ha^{-1} or g ha^{-1}.

2.2. Public Health Risks

Human poisonings and their related illnesses are clearly the highest price paid for pesticide use. Human exposure to agricultural waste can happen through occupational exposure in the case of farm workers in open fields and greenhouses, workers in pesticide industry and house plague exterminators [16-18]. Obviously, workers that handle with the mixture, transportation and application are usually considered to be a risk group due to a higher exposure potential and consequently have a greater probability of acute poisoning [19]. In addition, the incorrect handling of material and compounds could lead to spills and possible contaminations. Worker exposure increases if basic security measures like using protection equipment, when working with dangerous material, are not implemented.

Another factor that may affect the extent of exposure during pesticide handling is the form of formulation of pesticide products. In solution, pesticides can drip resulting in direct contact to the skin or indirect contact through clothing. On the other hand, pesticides in solid state can generate dust clouds while being loaded into tanks exposing the eyes and respiratory tract of the worker.

In addition, weather conditions at the moment of application can influence human exposure to pesticides, since factors like temperature and air humidity can affect the chemical properties of the product (e.g. volatility) and the human body transpiration rate [20, 21]. The wind velocity has a direct effect on the worker exposure risk, given that it can drag pesticide particles during pulverization.

The pesticide exposure of the general population happens mainly through contaminated food and water ingestion.

A greater exposure can occur if a population is living close to a pesticide production facility or a farm [22]. This exposure is even more alarming because the long term effects are for now unknown and depend on several factors such as different exposition time and levels, the possibility of combining pesticides and meteorological and geographic characteristics of the fields where the pesticides are implemented [23].

Independently from the difficulties in evaluating pesticides utilization risks, the authorization for pesticide commercialization in Europe requires data of potential negative effects of the active substances on human health. These data are generally obtained from different tests focused on metabolism pathways, acute and sub-acute toxicity, chronic and sub-chronic toxicity, genotoxicity, carcinogenicity, teratogenicity and also irritation assays performed using animal models [24].

The United States Environmental Protection Agency (EPA) requires extensive test data from pesticide producers that demonstrate pesticide products can be used without posing harm to human health. EPA uses a four-step process for human health risk assessment: hazard identification, dose-response assessment, exposure assessment and risk characterization [25].

Toxicity tests are also required by EPA for human health risk assessments. The acute toxicity tests are necessary for the calculation of the median lethal dose (LD_{50}), which is the pesticide dose that is required to kill half of the tested animals when entering the body by a particular route [18]. These LD_{50} are used for EPA and World Health Organization (WHO) toxicity classifications of pesticides shown in tables 2 and 3 [18].

As tables 2 and 3 shows, the oral LD_{50} is generally lower than the dermal LD_{50} as pesticides can enter the bloodstream more easily than through the skin [18].

Table 2. Acute toxicity of pesticides according to the EPA classification (adapted from [18])

Class	Warning	Acute toxicity to rat		
		LD_{50} oral (mg/kg)	LD_{50} dermal (mg/kg)	LC_{50} by inhalation (mg/L)
I	Danger	< 50	< 200	< 0.2
II	Warning	50-500	200-2000	0.2-2.0
III	Caution	500-5000	2000-20000	2.0-20
IV	Caution (optional)	> 5000	> 20000	> 20

Table 3. Acute toxicity of pesticides according to WHO classification (adapted from [18])

Class	Classification	LD_{50} for the rat (mg/kg)			
		Oral		Dermal	
		Solids	Liquids	Solids	Liquids
Ia	Extremely hazardous	< 5	< 20	< 10	< 40
Ib	Highly hazardous	5-50	20-200	10-100	40-400
II	Moderately hazardous	50-500	200-2000	100-1000	400-4000
III	Slightly hazardous	> 501	> 2001	> 1001	> 4001
U	Unlike to present acute hazard	> 2000	> 3000	-	-

The LD_{50} values given in the WHO classification are for the active ingredient, whereas these LD_{50} values must be modified to take account of the concentration of the pesticide formulation actually used. This is because the actual toxicity of a commercial pesticide product is significantly affected by the formulation [18].

Pesticides are additionally classified (Table 4) according to the principles of the International Agency for Research on Cancer (IARC) [26]. The classification of a pesticide in each category reflects the strength of the evidence derived from epidemiological studies in humans, from experiments with animals, and from mechanistic and other relevant data [18].

The results on toxicity characterization (based on EPA, WHO and IARC) indicate that from the 276 active substances legally commercialized in Europe, 32 out of 76 fungicides, 25 out of 87 herbicides and 24 out of 66 insecticides are related to at least one health effect (for instance, carcinogenic, endocrine disruptor, acute toxicity, reproductive and development toxicity).

Table 4. IARC classification of pesticides

Group	Warning
1	Carcinogenic to humans
2A	Probably carcinogenic to humans
2B	Possibly carcinogenic to humans
3	Not classifiable as to its carcinogenicity to humans
4	Probably not carcinogenic to humans

3. Cyclodextrins As a Solution for Environmental Sustainability

Reconciling agricultural productivity with other components of sustainability remains one of the greatest challenges for agriculture [27]. A key issue will be to achieve substantial reductions in the level of pesticide use for environmental and health reasons [27]. To achieve more sustainable use of pesticides, in 2009 the European Parliament and the Council of the European Union approved Directive 128 for the sustainable use of pesticides which provides measures to establish and reduce the overall risk that pesticides pose to human health and to the environment during their use [28]. The Directive includes requirements for the training of distributors and users of pesticides, the implementation of the principles for integrated pest management, information and awareness-raising, inspection and monitoring of spraying equipment, the prohibition of aerial spraying, specific measures to protect the aquatic environment and drinking water and the establishment of an EU indicator for plant protection products.

Registration of a pesticide in today's regulatory climate requires environmental friendliness in addition to efficacy against the target pest. Most recent pesticides have rapid environmental degradation rates to meet this former constraint. Efficacy is often correlated with the product of environmental concentration (C) and contact time (T) as an indirect measure of the exposure an organism experiences (C×T). If the pesticide degrades rapidly within the environment, then the time window (contact time) for pest control/efficacy can be small. To compensate this fact normally great amounts of agrochemicals are used which over a long period of time leads to crop damage and environmental contamination [29].

Recently, controlled release technology has emerged as an approach with potential for solving the problems associated with the applications of conventional agrochemicals. Actually, controlled release of agrochemicals (pesticides, herbicides, nutrients) can be used to maintain the local concentration of active ingredients in the soil and to reduce losses due to runoff. Controlled release systems for pesticides involve advanced pesticide delivery technologies, highlighting new means to reduce toxicity, increase efficacy, lessen the environmental impact from pesticides and pesticide applications, reduce potential transportation hazards, as well as facilitating new product development [29, 30].

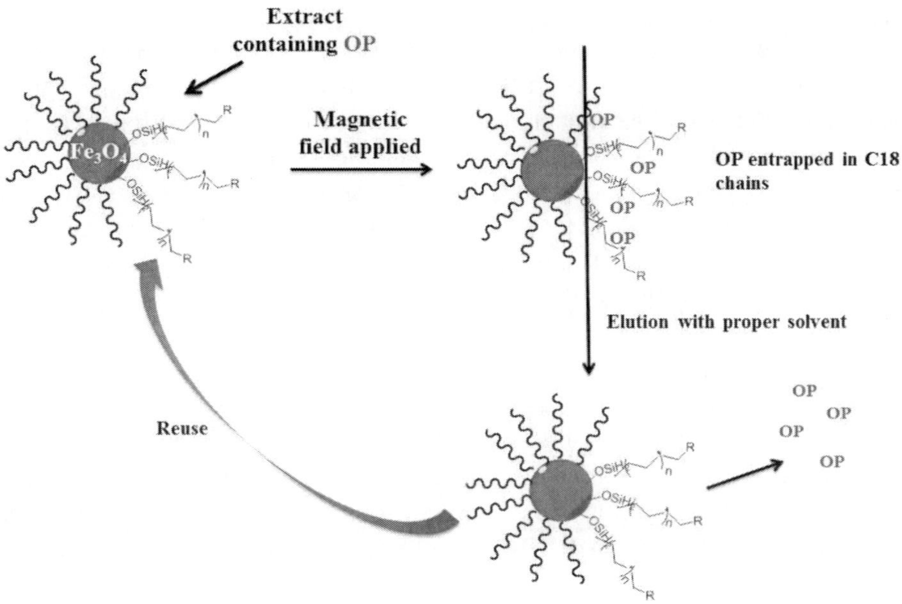

Figure 2. Organophosphorus pesticide removal strategy of contaminated soils using magnetic nanoparticles functionalized with C18 chains.

Rapid advances have been made in the development and use of controlled release systems for agrochemicals [31-33]. As an example, magnetic adsorbents have emerged as a new class of materials with the aim of selectively adsorb to the surface of the desired target compounds. The potential of the magnetic nanoparticles functionalized with C18 compounds for the removal of organophosphorus (OP) pesticides is illustrated in figure 2 [32]. After the application of magnetic field, the pesticides are captured while the remaining sample components are eluted. Removing the magnetic field and after the addition of a suitable eluent pesticides trapped in magnetic nanoparticles are released, getting the nanoparticles ready for a new use.

Cyclodextrins (CDs) and their derivatives have a wide variety of practical applications namely in pharmacy, medicine, foods, cosmetics, toiletries, chemistry, catalysis, biotechnology, textile industry, etc. The vast majority of these chemical and biological applications are based on their ability to form inclusion complexes.

Similarly to CDs applications in other industries, the most common benefits of CDs applications in agriculture include, among others, modification of the solubility of the pesticide, stabilization against the effects of light or biochemical degradation, and a reduction of volatility. Their ability to alter the physical, chemical, and biological properties of guest molecules has been used for the preparation of new formulations of pesticides. In fact, CDs form complexes with a wide variety of agricultural chemicals including herbicides, insecticides, fungicides, repellents, pheromones, and growth regulators.

This section of the chapter will focus the main characteristics of CDs and present some recent studies involving pesticides' encapsulation in CDs as a mean for achieving environmental sustainability.

3.1. Cyclodextrins General Concepts

Cyclodextrins (CDs) are cyclic carbohydrates composed by D-glucose units (α-D-glucopyranose) bonded by glucosidic linkages type α-(1,4) obtained from the enzymatic degradation of one of the most essential polysaccharides, starch. CDs were accidentally discovered in 1891 by the French scientist Villiers, who was able to isolate a crystalline product after bacterial digestion of the starch [34]. Due to the chair conformation of the glucopyranose units, the CDs are shaped like a truncated cone rather than perfect cylinders (Figure 3A). The hydroxyl functions are orientated to the cone exterior with the primary hydroxyl groups of the sugar residues at the narrow edge of the cone and the secondary hydroxyl groups at the wider edge. The central cavity is lined with the skeletal carbons and ethereal oxygens of the glucose residues, which gives it a lipophilic character. CDs cavity plays an important role for the formation of inclusion complexes, since provides a lipophilic microenvironment into which suitably sized drug molecules may enter and be included [34-37]. The main driving forces behind the formation of these so-called host-guest inclusion complexes are hydrophobic and van der Waals interactions, although other factors also play a role, including release of CD ring strain, changes in solvent-surface tensions, and hydrogen bonding with CD's hydroxyl groups [38].

The most common CDs are α-CD, β-CD, and γ-CD which consist of six, seven, and eight glucopyranose units, respectively (Figure 3B). Due to steric factors, CDs having fewer than six glucopyranose units cannot exist [27].

The natural CDs, in particular β-CD, are of limited aqueous solubility meaning that complexes resulting from interaction of lipophiles with these CD can be of limited solubility resulting in precipitation of solid CD complexes from water and other aqueous systems. Substitution of any of the hydrogen bond forming hydroxyl groups results in dramatic improvement in their aqueous solubility. Actually, as the three main native CDs bear 18 (α-CD), 21 (β-CD), or 24 (γ-CD) substitutable hydroxyl groups, the possible number of derivatives is practically unlimited [39].

Water-soluble CD-derivatives of commercial interest include the hydroxypropyl derivatives of β-CD (HP-β-CD) and γ-CD, the randomly methylated β-cyclodextrin (RM-β-CD), and sulfobutylether β-cyclodextrin sodium salt (SBE-β-CD).

As was already mentioned, it is known and widely reported in the literature that CDs and their derivatives have a wide variety of practical applications [40-43].In fact, over the past decade, CDs have aroused considerable interest in many fields (agriculture, nanocomposite technology, chromatography, biotechnology, etc.), because of the low cost of production. Many synthetic pesticides can form inclusion complexes with CDs, often resulting in improvements in their chemical and physical properties including enhanced solubility and bioavailability [44, 45], increased stability of photodegradable and/or unstable pesticides [46, 47], catalytic effects on the degradation of pesticides [48], diminution of unpleasant odor, a reduction of human toxicity [49], improvements in pesticide adsorption, and inhibition of their leaching in soils [50, 51]. In addition, CDs are natural, biodegradable, nontoxic compounds, harmless to microorganisms, and hence, not noxious for the environment [52].

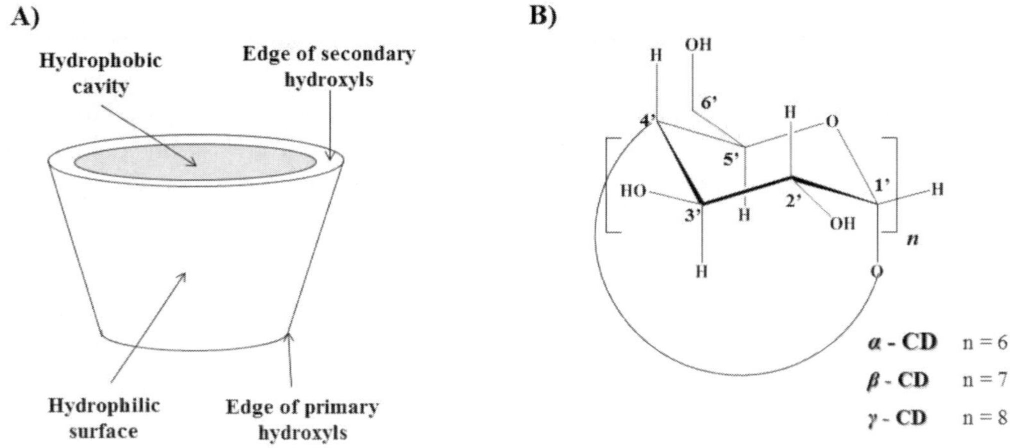

Figure 3. Three-dimensional representation of (A) CDs structure and (B) chemical structure of α-, β-, and γ-CDs.

3.2. Case Studies on Encapsulation of Pesticides in Cyclodextrins

Generally it is considered that an efficient pesticide formulation should not only show an effective biological activity, it must also be user friendly and environmentally friendly. With this goal, different pesticides delivery systems have recently been developed, being microencapsulation perhaps one of the most efficient methods. Actually, encapsulation through CDs inclusion complex formation is one of the most attractive methods, as they are cheap, natural, environmentally safe and water-soluble compounds and enables the formation of a complex in which the physicochemical properties of the guest molecule can change significantly.

In this section, examples of studies performed so far related with the application of different CDs as encapsulation agents in order to achieve environmental sustainability are presented.

3.2.1. Organophosphorus Pesticides

Organophosphorus (OP) pesticides are used extensively to control agricultural, household and structural pests. Their application is still the most effective and accepted means for the protection of plants, and has contributed significantly to enhanced agricultural productivity and crop yields [23]. These pesticides constitute a diverse group of chemical structures exhibiting a wide range of physicochemical properties, with their primary toxicological action arising from inhibition of the enzyme acetylcholinesterase (AChE) [53]. The extensive use of OP pesticides makes them common environmental pollutants in aqueous environments including ground water, surface water, and drinking water. There is now an increasing environmental concern with regard to these compounds because of their potential side effects on the aquatic organisms and humans [54].

Kamiya and co-workers investigated how and to what extent photodegradation rates of the OP pesticides parathion and paraoxon (Figure 4) were influenced by the inclusion effects of α-, β-, γ-CDs and several methylated CDs (hexa-2,6-dimethyl-α-CD and hepta-2,6-

dimethyl-β-CD) [46]. The results found showed that α-CD exerts a slightly promotive effect on the initial decay rates of both parathion and paraoxon. On the contrary, the γ-CD exerts a considerably inhibitive effect on the photodecay rate of both pesticides. The β-CD exerts a markedly inhibitive effect on the parathion photodecay rate, but a markedly promotive effect on the paraoxon photodecay rate. Methylation of the CD hydroxyl groups tends to decrease the CD inclusion effects, whether these are inhibitive or promotive to the photodecay rates of the included pesticides.

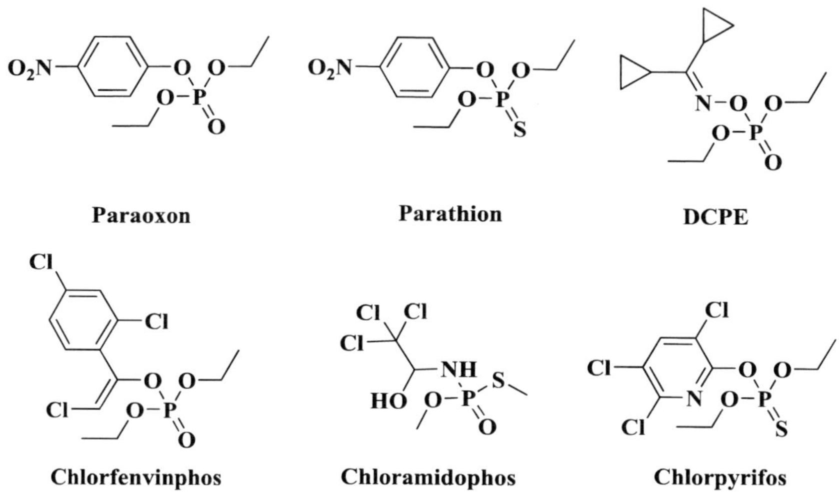

Figure 4. Chemical structures of organophosphorus pesticides.

Basic properties of the CD inclusion effects were explained in terms of the difference in the degree of proximity between the catalytic sites of the CD host cavities and the reaction centers of the included pesticide guests. Actually, the remarkable difference in the inclusion effects of β-CD found for the photolysis rates of parathion and paraoxon was attributed to the fact of thiophosphate group of parathion being considered more hydrophobic in nature or higher in affinity to the hydrophobic CD cavity than the phosphate group of paraoxon [46].

Loukas and colleagues have studied the inclusion complex formation of the insecticide DCPE (Figure 4) and 2-hydroxypropyl-β-CD (HP-β-CD) [49]. Additionally, the authors compared the stability of the free and complexed DCPE with α-CD, β-CD and HP-β-CD. The data showed that for complexes with α-CD and β-CD the degradation rate remained the same. However, for HP-β-CD a decelerating effect on the degradation rate is observed, which shows that this CD has a pronounced stabilizing effect on DCPE. A possible explanation is that HP-β-CD may be able to include more the DCPE molecule or its reactive group within the cavity than can α- and β-CD.

Zhou and his team studied the encapsulation of chloramidophos (CP) (Figure 4) in β-CD in an attempt to gain a more stable but equally efficacious formulation compared with CP alone [55]. The structural identification, thermal stability, and biological assays of the CP/β-CD complex were carried out. The results showed that the degradation rate of CP, in 14-day incubation, was reduced by a factor of 3.6 when it was complexed with β-CD and that both the activity and toxicity of CP were not affected by the encapsulation process. As a result, the encapsulation process greatly improved the thermal stability of the pesticide with no adverse effects on bioefficacy compared to that of CP [55].

Chlorpyrifos is a broadly active nonsystemic insecticide used to control a variety of soil insects and arthropods in a wide range of crops, including citrus fruits, peaches, nectarines, vegetables, cereals, maize, and tobacco [56]. It is easily soluble in organic solvents but shows poor water solubility. Thus, the aqueous solubility of chlorpyrifos (Figure 4) was studied using different types of CDs (*α*-, *β*-, *γ*-, HP-*β*-, and G2-*β*-CDs) [56]. Chlorpyrifos solubility only increased using G2-*β*-CDs, indicating that G2-*β*-CDs trap the insecticide within their hydrophobic cavity. The inclusion of chlorpyrifos in CDs has improved its water solubility, bioavailability, and insecticidal activity and helps prevent overdosing, leading to more cost-effective and more environmentally friendly agricultural practices [56].

Hartnik et al. have developed a new biomimetic extraction method using HP-*β*-CD to estimate uptake of the two insecticides chlorfenvinphos and *α*-cypermethrin (pyrethroid group) in the earthworm Eisenia fetida [57]. The study revealed that HP-*β*-CD is a suitable surrogate for estimating the bioaccessibility of hydrophobic chemicals in soil. Using a 3.5 times higher amount of HP-*β*-CD than soil, effective and reproducible extractions were achieved. At these conditions, inclusion of dissolved chemicals by HP-*β*-CD mimics uptake of a given compound into earthworms and takes into account sorption-related aspects that control biological uptake [7].

A number of OP pesticides are chiral, and their enzyme inhibition activity, toxicity, and biodegradability depend on the absolute configuration of their phosphorus, carbon, or sulfur chiral atoms [58]. In a few cases, such as fenamiphos (Figure 5), one enantiomer is much more toxic than the other, so it would be desirable to use a single enantiomer in the commercial formulation [58].

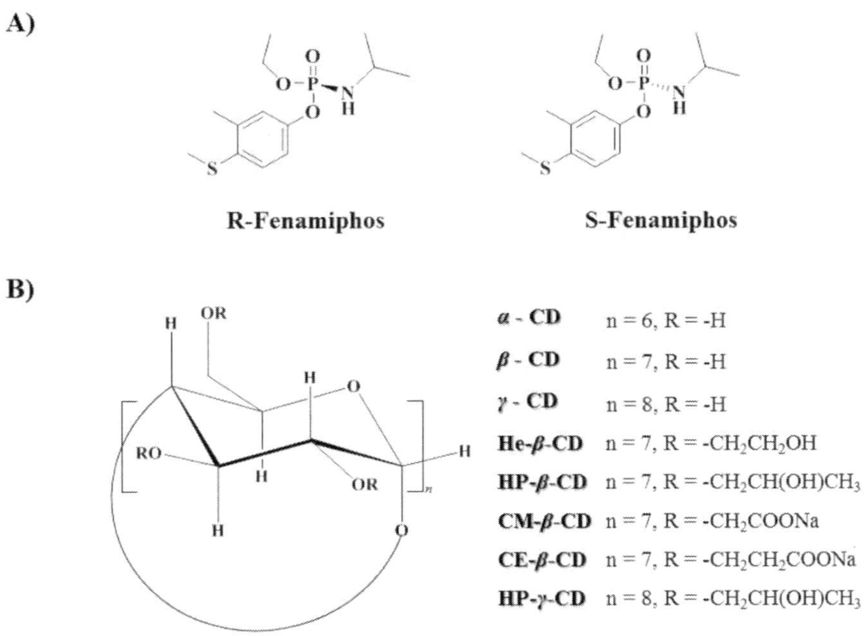

Figure 5. Structures of (A) two enantiomers of Fenamiphos and (B) CDs used as chiral selectors.

Molaabasi et al. used different CDs (α-, β-, RM-β-, hydroxyethyl-β-, HP-β-, HP-γ-, carboxymethyl-β- and carboxyethyl-β-CD) as chiral selectors for the enantiomeric discrimination of fenamiphos (Figure 5) [58]. The CDs were chosen according to the differences in the size of their inner cavities (α-, β-, and γ-CDs) and rim substitution (RM-, HP-, hydroxyethyl-, carboxymethyl-, and carboxyethyl-). The results found showed that β-CD and HP-β-CD were the best chiral solvating agents for the fenamiphos enantiomeric discrimination.

3.2.2. Phenoxyalkanoic Acid Pesticides

Following World War II, chlorinated phenoxyacetic acids, especially 2,4-D, MCPA, and 2,4,5-T (Figure 6), were introduced as selective herbicides against broadleaf weeds [59]. They remain among the most commonly applied weedkillers, although 2,4,5-T international trade was restricted by the Rotterdam Convention. Phenoxyalkanoic acids are considered as hazardous compounds and due to their high water solubility, especially of the dissociated forms, are amenable to transport to the water environment via leaching and runoff processes depending upon the sorption and degradation kinetics [60]. Sorption on soil components depends on a combination of parameters relevant to the chemical composition of minerals and the mechanical and textural characteristics of soils. Phenoxyalkanoic acid herbicides show moderate toxicity, however, some chlorinated metabolites can be toxic to human and aquatic organisms [61-63].

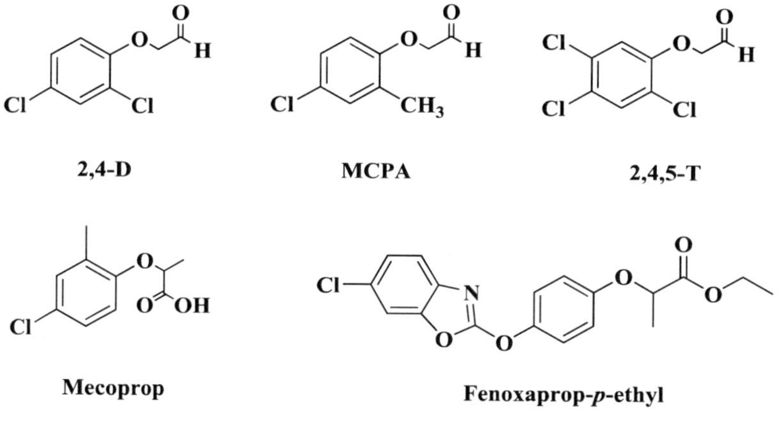

Figure 6. Chemical structures of phenoxyalkanoic pesticides.

MCPA is in the top five of the most used pesticide active ingredients in the home and garden sector in the US and herbicide active ingredients in the EU [64]. The influence of the pattern of substitution of different β-CDs on the stability of MCPA inclusion complexes was evaluated by Garrido and his team in order to assess the possibility of their use in commercial pesticide formulations and for environmental protection [65]. The results showed that the stability constants for HP-β-CD and RM-β-CD systems were two fold greater than that obtained for the native β-CD indicating that these complexes were more stable. The data found suggested that the higher association constant values were related with the presence in CDs of less polar groups, 2-hydroxypropyl or methyl, which seems to stabilize the complex with MCPA.

In another report, Costa and co-workers evaluated the influence of HP-β-CD in the photochemical degradation of the herbicides MCPA and mecoprop (Figure 6) in different aqueous media [66]. The results obtained showed that the half-lives of MCPA and mecoprop inclusion complexes were increased approximately by a factor of three related to the free molecules. Furthermore, it has been shown that the photodegradation of MCPA and mecoprop is influenced by their structural features. Actually, it was found that the existence of a methyl group in the side chain of mecoprop molecular structure has a positive influence on the stabilization of the radical intermediate formed in the first stage of photodegradation [66]. As a result, MCPA and mecoprop form inclusion complexes with higher photostability compared to free herbicides indicating that HP-β-CD may serve as a valuable ingredient in these herbicide formulations.

With the objective of investigate the possibility of improving the aqueous solubility and dissolution properties of 2,4-D via inclusion or complexation with CDs, Ginés and co-workers studied the encapsulation of 2,4-D in different CDs (α-, β- and RM-β-CD) [51, 67, 68]. The aqueous solubility of the pesticide 2,4-D was improved by inclusion complexation into the CDs. Dissolution rate studies of the non-complexed and complexed pesticide were also carried out. Dissolution rate of the complexed 2,4-D was considerably improved as compared with the uncomplexed herbicide. According to authors, these findings would allow a more rational application of 2,4-D in agriculture, and thus diminishing its risks as potential pollutant of ground water and soils [69]. This finding would allow a more rational application of 2,4-D in agriculture, and thus diminishing its risks as potential pollutant of ground water and soils.

Inclusion complex formation of 2,4-D with CDs (β-, HP-β-CD) has been studied as a way of modifying the behaviour of the pesticide in the soil environment [68]. Experiments were performed to study the adsorption-desorption of 2,4-D on soils and the influence of the CDs over these processes. The results found indicated that the complex may have great utility as a rapid way of dissolving the pesticide and that CDs could increase the desorption of 2,4-D previously adsorbed on soils.

The ability of CD to modify the risk-components, such as mobility, water solubility, and biological availability of the contaminants makes possible its use in the environmental risk assessment and risk reduction technologies [70]. Actually, CDs may influence bioavailability of the contaminants during biodegradation and also toxicity of the pollutant on soil microbes and plants. The ability of different native and derivatized CDs (α-, β-, γ-, HP-β-, RM-β-, sulfated-β-, and carboxymethyl-β-CD) to extract 2,4-D and several other commonly used pesticides from contaminated soil was assessed by Flaherty and colleagues [71]. All CDs showed significant binding interactions with each of the pesticides investigated, with the exception of sulfated-β-CD. A general correlation between the pesticide/CD binding constants and the percent extraction enhancements was found. HP-β-CD and RM-β-CD generally displayed the greatest levels of extraction enhancement. The results of this study may, according to authors, be used for future novel methods of contaminated soil remediation, which overcome the disadvantages of organic solvent and surfactant use, and can also be applicable toward the development of pesticide/CD formulations [71].

Fenoxaprop-p-ethyl (FE) (Figure 6), a systemic post-emergence broad spectrum herbicide firstly registered in Germany by Hoechst AG in 1970, was encapsulated in β-CD and two derivatives (RM-β- and HP-β-CD) in an attempt to assess their potential as new formulation additives for more efficient FE delivery and better environmental approaches [72]. The solubility of FE was enhanced in the presence of the CDs due to the formation of inclusion complexes, with RM-β-CD being six times more effective than the other two. Furthermore, the dissolution of the FE/RM-β-CD complex was much faster than other complexes and FE alone. These results indicated that RM-β-CD was a better inclusion complexation agent for FE in terms of both solubility enhancement and dissolution rate. Authors suggested that RM-β-CD may thus be used to improve FE delivery and to mobilize FE in soil for bioremediation. The inclusion inside the β-CD derivatives allowed a 4-fold increase of FE solubility, when compared with the herbicide alone. The results showed that, with regard to solubility and dissolution rate, RM-β-CD is a better complexing agent than HP-β-CD [72].

3.2.3. Urea Pesticides

Substituted ureas are an important group of pesticides that are used as herbicides (phenylureas and sulfonylureas) and insecticides (benzoylureas). The substituted urea herbicides are used for the control of many annual and perennial weeds, for brush control, and for weed control in irrigation and drainage ditches [73]. Likewise the benzoylurea insecticides are widely used on a large number of crops [73]. Several of these compounds are highly persistent in the environment and misuse, carelessness, and leaching can contaminate surface waters used for crop irrigation or for a potable water supply [73]. Yet, toxicity of urea pesticides is not completely understood. Studies showed that wastewater bacteria are able to transform urea derivative compounds in mono or dichloroanilines [74, 75]. This process deserves special attention since chloroanilines were pointed as probable carcinogens by the EPA due to their association with bladder cancer [25].

In order to assess the potential effect of CDs on the transformation of pyrazosulfuron ethyl (PE) and halosulfuron methyl (HM) (Figure 7) in aquatic environments, Zheng and colleagues studied the encapsulation of PE and HM in α- and β-CDs [76]. Results found showed that the presence of α- or β-CD did not promote or decelerate the dissipation of either PE or HM in buffer solutions. Additionally, the data obtained showed that CDs would not react with the sulfonylureas in unbuffered deionized water. According to authors, the hydrolysis of HM and PE was relatively unaffected by the presence of CDs, indicating that natural OH-containing organic compounds occurring in aquatic environments may have little impact on the transformation of these sulfonylurea herbicides [76].

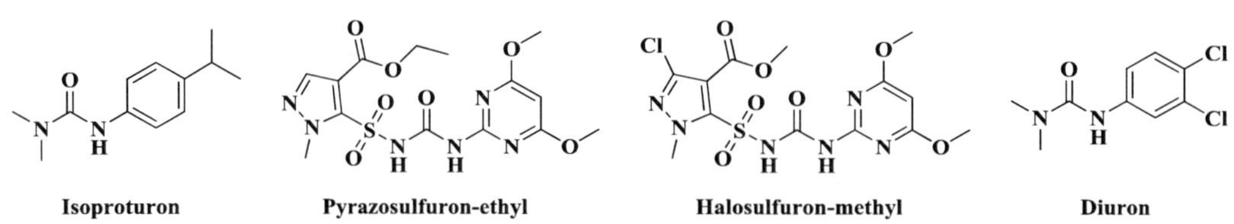

Figure 7. Chemical structures of urea pesticides.

Manolikar and Sawant studied the interaction between herbicide isoproturon (Figure 7) and β-CD as a mean to improve its efficacy [77]. The results found showed that the interaction with the CD increase the solubility and the dissolution rate of isoproturon. Authors considered that these findings would allow the rational use of isoproturon due to the increase in its efficacy [77].

The phenylurea herbicide diuron (Figure 7) is widely used in a broad range of herbicide formulations and, is a biologically active pollutant present in soil, water, and sediments. Diuron has the unfortunate combination of being strongly adsorbed by soil organic matter particles and, hence, slowly degraded in the environment due to its reduced bioavailability. Thus, Villaverde and colleagues performed some studies in order to find procedures that might result in an increase in the bioavailability of diuron in a contaminated soil, through solubility enhancement using biodegradable molecules [78]. A CD-based bioremediation technology involving a bacterial consortium, was evaluated as an effective diuron mineralization system thanks to the increased bioavailability. A double system composed of HP-β-CD, which is capable of forming inclusion complexes in solution, and specific chemical bacterium degraders was used. As a result, an environmentally friendly CD-based bioremediation technology was developed that enabled an almost complete mineralization of diuron in a soil system [78].

3.2.4. Dicarboximide Pesticides

Dicarboximide pesticides are widely used as preventive and persistent fungicides against *Botrytis*, *Cochliobolus*, *Helminthosporium*, *Monilia* and *Sclerotinia spp.* and against several other fungi [79]. These fungicides are applied on field crops, fruits, vegetables, and particularly on vineyards.

As CD molecules are used, in some countries, for retardation of seed germination and therefore can be present in soil together with applied pesticides, Hromadová and colleagues studied the influence of CDs (α-, β-, γ-CD) both in the release and degradation of fungicide vinclozolin (Figure 8) [62]. The data found suggested that CD molecules do not simply act as proton donors in a nonaqueous environment, but also protect parts of the molecule included within the cavity and steer the degradation process of vinclozolin [62].

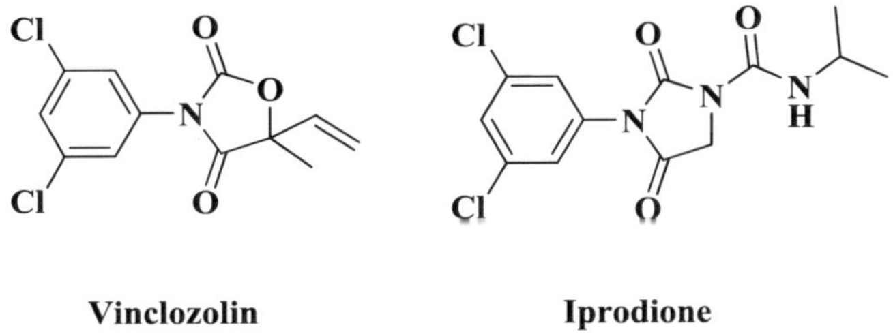

Figure 8. Chemical structures of dicarboximide pesticides.

Fungicide iprodione (Figure 8) has been extensively used to control a variety of crop diseases for decades. However, iprodione has a poor water solubility and transportation property [63]. Therefore, significant increases in resistance against iprodione were observed in a range of important plant pathogens after a long-time field application. Accordingly, Zhu and colleagues investigated the possibility of obtaining inclusion complexes of iprodione with β-CD, as a first step to obtain formulations that supply a more rational use of this fungicide, improving its water solubility, bioavailability, and fungicidal activity [63]. The complexation of iprodione with β-CD significantly improved the bioavailability of iprodione and, therefore, resulted in over two-fold increase of the fungicidal activity. According to authors, the significant increase of the bioactivity of iprodione in the presence of β-CD provides an effective approach for a more rational application of iprodione, diminishing the use of organic solvents and the amount of iprodione and increasing its efficacy [63].

3.2.5. Pesticides of Other Chemical Classes

In this section are included several pesticides with different chemical structures and targets and not belonging to any of the above mentioned class of pesticides.

Pyrimethanil (Figure 9) is an anilino-pyrimidine fungicide particularly active against gray mold (*Botrytis cinerea*) and pear scab (*Venturia spp.*) on fruits, vegetables, and ornamental plants. Although pyrimethanil has no apparent mutagenic, genotoxic, or carcinogenic potential, the concentration of pyrimethanil has been limited in food commodities by the European Commission [80]. Furthermore, this fungicide is degraded by sunlight decreasing drastically its efficiency [81]. As a mean to decrease the fungicide photodegradation, Fernandes and co-workers studied the influence of microencapsulation within HP-β-CD on the environmental photostability of pyrimethanil [82].

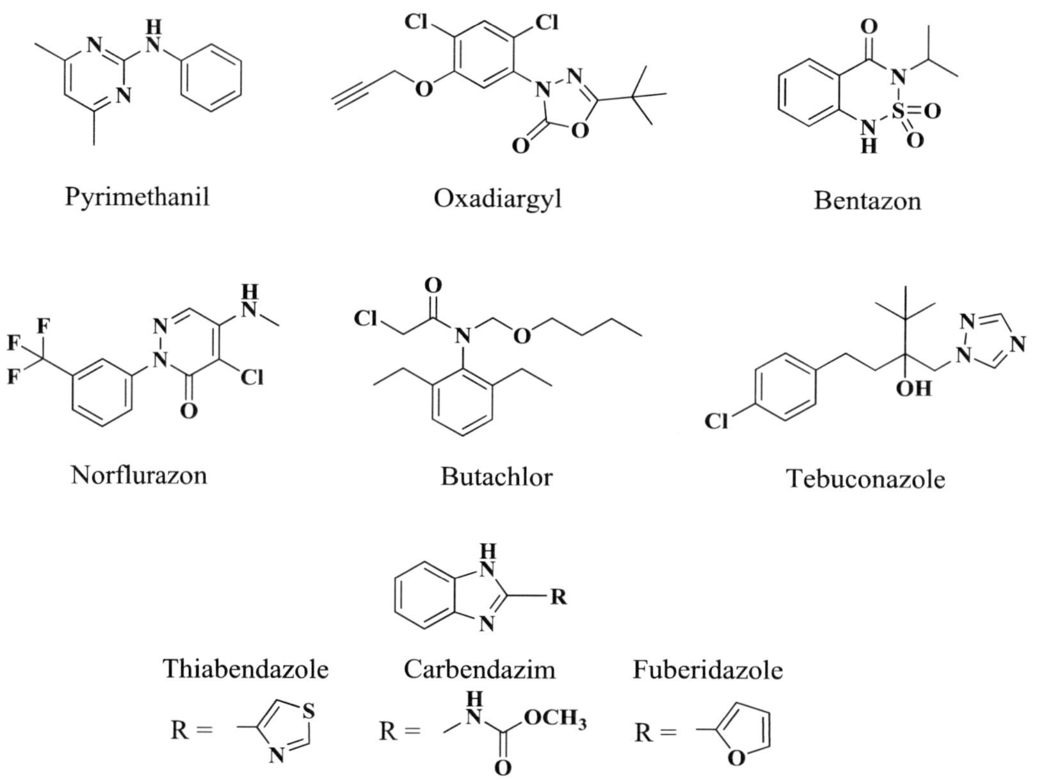

Figure 9. Chemical structures of pesticides.

The results obtained showed that the formation pyrimethanil/HP-β-CD inclusion complex increases significantly the photostability of the fungicide in aqueous solutions, especially in natural water. Actually, the data found suggested that the microencapsulation in HP-β-CD seems to act as a physical barrier, through which the encapsulated pyrimethanil is protected, against the action of light and/or natural photocatalysts present in the water [82].

Oxadiargyl (Figure 9) is an herbicide developed for the control of annual grasses, sedges, and broad-leaved weeds in rice [83]. However, its low aqueous solubility leads to a facile accumulation in water or soil and to a rapid poisoning of environment. In order to surpass the mentioned drawback, Benfeito and colleagues studied the microencapsulation of the oxadiargyl in HP-β-CD [84]. The results obtained showed that HP-β-CD enhance the water solubility of oxadiargyl. According to authors, the results gathered can be regarded as an important step for its removal from industrial effluents and/or to increase the stabilizing action, encapsulation, and adsorption in water treatment plants [82].

Bentazon (Btz, Figure 9) is a selective benzothiadiazole post-emergence herbicide that can provide effective control of broadleaf weeds. The effectiveness of Btz is often improved by adding adjuvants that increase herbicide activity but also its retention and adsorption [85]. Yáñez and co-workers have prepared and characterized some inclusion complexes formed by Btz with native β-CD and two derivatives (HP- and SBE-β-CD) with the aim to contribute to the future applications of CDs to solid samples in agriculture field [85]. The data obtained showed that the formation of inclusion complexes with the three CDs tested improved slightly the solubility of Btz, increased the dissolution rate, decreased the surface tension and greatly increased the photostability of Btz in aqueous solutions. Therefore, it was concluded that the formation of inclusion complexes with CDs may to improve the properties of Btz formulations, increasing the herbicide effectiveness for future applications [85].

Norflurazon (NFL, Figure 9), a fluorinated pyridazinone, is a selective pre-emergent herbicide widely used to control germinating annual grasses and broadleaf weeds in fruits, vegetables, nuts, cotton, peanuts and soybeans [86]. NFL tends to be retained in soils with high organic matter and iron oxide content and therefore can present a high risk of long-term contamination [86]. Thus, Villaverde et al. has investigated the possibility of obtaining inclusion complexes for NFL with α- and γ-CDs in order to increase its aqueous solubility and bioavailability in soils via complexation [86]. Additionally, a study about the effect of different CDs (α-, β-, γ-, RM- and HP-β-CD), as potential photosensitizers, on the photodegradation of the herbicide NFL in aqueous suspensions of different soil components was also accomplished [47, 86]. The results found evidenced a high extraction capacity of α- and γ-CD towards NFL herbicide adsorbed on soils due to the formation of water-soluble inclusion complexes [86]. Moreover, the irradiation of NFL aqueous solutions in the presence of CDs showed that the higher the formation constant of NFL/CD complexes and their solubility, the higher their photocatalytic effects, following the order: RM > HP- > β- > α- > γ-CD. The data gathered pointed out that CDs can be employed as an effective and alternative tool for in situ remediation of pesticide-contaminated waters and soils [47, 86].

Butachlor (BTC, Figure 9) is commonly used as a pre-emergence or early post emergence herbicide in crop fields (wheat, barley, beet, cotton, rape, vegetable and peanut) in Asia, South America and Africa. BTC is highly toxic to some aquatic organisms and influences soil microbial populations and enzyme activities as well as does harm to earthworms [87]. Due to its low water solubility and high hydrophobicity butachlor presents high adsorption in soils with medium-high organic matter. This behavior of butachlor not only may lead to its inactivation, with the consequent result that higher amounts of herbicide have to be applied to obtain the desired control, but also could reduce its bioavailability and thereby increase its environmental persistence [87]. With the aim of investigate the effect of CD complexation on the adsorption of butachlor in relation with its soil mobility and environmental availability, Bian and co-workers prepared and characterized BTC/β-CD inclusion complexes [87]. The results showed that by complexation, not only is the solubility of butachlor greatly improved, but also the butachlor aqueous dissolution rate was remarkably enhanced. Moreover, the complexation significantly reduced the sorption of butachlor by soil. According to authors, the data found indicate that β-CD may be used as an additive to improve the solubility of butachlor, reduce its adsorption on soil, and increase the availability of butachlor for weeds. In addition, β-CD may also be used as a remediation agent to enhance the mobility of butachlor and thus the bioavailability of BTC in contaminated soil [87].

Tebuconazole (TBC, Figure 9) is a triazole fungicide used as a seed dressing and spray. In cereals, the use of seed dressing is a crucial treatment for the control of soilborne diseases caused by several species of filamentous fungi. Balmas et al. tested the suitability of TBC to be included in β-CD and evaluated the fungicidal efficacy of the TBC/β-CD complex, applied as seed dressing, for controlling foot and crown rot of durum wheat [88]. This study confirmed the aptness of the β-CD inclusion complex in agricultural practices and particularly in fungicide seed-coating formulations. In fact, the TBC/β-CD inclusion complex, applied as seed dressing, reduced the disease incidence caused by *Fusarium culmorum* and improved grain yield, showing effects that were generally comparable to those observed upon application of a commercial formulation of TBC. Furthermore, given the very poor water solubility of TBC, complexation in β-CD provided increased aqueous solubility to the fungicide and therefore better bioavailability [88]. Authors suggest that these results prove that by the use of the TBC/β-CD complex it is possible to obtain release of TBC and bioavailabilty of the fungicide without compromising its effectiveness.

Benzimidazole fungicides are used for prevention and treatment of parasitic infections in agriculture and some of them have also found applications as pre- or post-harvest fungicides for control of a wide range of pathogens. However, their low solubility in water is a limiting factor for their fungicidal action. Thus, the potential increase of water solubility of three benzimidazole-type fungicides (thiabendazole, carbendazim, and fuberidazole, Figure 9) by complexation with α- and β-CDs was investigated by Lezcano and colleagues [45]. Phase-solubility diagrams confirmed the formation of inclusion complexes between the fungicides and β-CD and showed significant increases of their solubilities due to complexation [45].

Conclusion

CDs are nowadays used in diverse fields due to their propensity to form inclusion complexes and recognize analytes selectively. Their ability to alter the physical, chemical, and biological properties of guest molecules has been used for the preparation of new formulations of pesticides. The most common benefits of CDs applications in agriculture include, among others, alterations of the solubility of the pesticide, stabilization against the effects of light or biochemical degradation, and a reduction of volatility. Additionally, the molecular inclusion of pesticides offers several advantages over conventional agrochemical formulations, namely the release of a constant level of the active ingredient, providing enhanced efficacy, and reduced mammalian toxicity and phytotoxicity. Promising results and applications have already being described including new formulations of pesticides with enhanced biological activity and reduced or even absent irritating or toxic effects. CDs were also applied to delay germination of seed, in water treatment to increase the stabilising action, encapsulation and adsorption of contaminants and in soil remediation.

Thus, there is a scope for an exciting future in the research and development of CD based technologies as a valuable resource to attain environmental sustainability.

REFERENCES

[1] Oerke, E.-C. *J. Agric. Sci.*, 2006, *144*, 31.
[2] Martin, E. A.; Reineking, B.; Seo, B.; Steffan-Dewenter, I. *Proc. Natl. Acad. Sci. U.S.A.*, 2013, *110*, 5534.
[3] Fantke, P.; Juraske, R. *Environ. Sci. Technol.*, 2013, *47*, 3548.
[4] Cooper, J.; Dobson, H. *Crop Prot.*, 2007, *26*, 1337.
[5] M. Yudelman, A. Ratta, and D. Nygaard, *Pest Management and Food Production: Looking to the Future*; International Food Policy Research Institute, 1998, USA.
[6] Casida, J. E. *Chem. Res. Toxicol.*, 2009, *22*, 609.
[7] Casida, J. E.; Quistad, G. B. *Ann. Rev. Entomol.*, 1998, *43*, 1.
[8] Reedy, R. C.; Scanlon, B. R.; Nicot, J.-P.; Tachovsky, J. A. *Environ. Sci. Technol.*, 2007, *41*, 6914.
[9] Sierra-Alvarez, R.; Yenal, U.; Field, J. A.; Kopplin, M.; Gandolfi, A. J.; Garbarino, J. R. *J. Agric. Food Chem.*, 2006, *54*, 3959.
[10] Rosell, G.; Quero, C.; Coll, J.; Guerrero, A. *J. Pestic. Sci.*, 2008, *33*, 103.
[11] Pimentel, D.; Levitan, L. *BioScience* 1986, *36*, 86.
[12] Gavrilescu, M. *Eng. Life Sci.*, 2005, *5*, 497.
[13] Arias-Estévez, M.; López-Periago, E.; Martínez-Carballo, E.; Simal-Gándara, J.; Mejuto, J.-C.; García-Río, L. *Agr. Ecosyst. Environ.*, 2008, *123*, 247.
[14] Gill, H. K.; Garg, H. *Pesticides: Environmental Impacts and Management Strategies*; Intech, 2014, Croatia.
[15] Castillo, M. d. P.; Torstensson, L.; Stenström, J. *J. Agric. Food Chem.*, 2008, *56*, 6206.
[16] Pimentel, D. *In Integrated Pest Management: Innovation-Development Process*; Eds. Peshin, R., Dhawan, A., 2009, Netherlands.
[17] Atreya, K. *Soc. Sci. Med.*, 2008, *67*, 511.
[18] Damalas, C. A.; Eleftherohorinos, I. G. *Int. J. Environ. Res. Publ. Health.*, 2011, *8*, 1402.
[19] Fenske, R. A.; Day, E. W. *Assessment of Exposure for Pesticide Handlers in Agricultural, Residential and Institutional Environments*; Eds. Franklin, C. A., Worgan, J. P., 2005, UK.
[20] Gil, Y.; Sinfort, C.; Guillaume, S.; Brunet, Y.; Palagos, B. *Biosyst. Eng.*, 2008, *100*, 184.
[21] Jindal, T.; Singh, D. K.; Agarwal, H. C. *J. Environ. Sci. Heal. B*, 2007, *42*, 367.
[22] Davis, J. R.; Brownson, R. C.; Garcia, R. *Arch. Environ. Contam. Toxicol.*, 1992, *22*, 260.
[23] Bolognesi, C. *Mutat. Res-Rev. Mutat.*, 2003, *543*, 251.
[24] Matthews, G. *Pesticides: Health, Safety and the Environment*; John Wiley & Sons, 2008, India.

[25] http://www.epa.gov/pesticides/factsheets/riskassess.htm (accessed on 12 March 2015)
[26] http://monographs.iarc.fr/ENG/Classification/ (accessed on 12 March 2015)
[27] Lechenet, M.; Bretagnolle, V.; Bockstaller, C.; Boissinot, F.; Petit, M.-S.; Petit, S.; Munier-Jolain, N. M. *PLOS ONE*, 2014, *9*, 97922.
[28] Directive 2009/128/EC of the European Parliament and of the Council of 21 October 2009 establishing a framework for Community action to achieve the sustainable use of pesticides. Official Journal of the European Union, L. 309/71, 24 November (2009).
[29] Cryer, S. A. *In Pesticides - Formulations, Effects, Fate*; Stoytcheva, P. M., Ed.; 2011, USA.
[30] Aouada, F. A.; Regina de Moura, M.; Mattoso, L. H. C. *In Pesticides - Formulations, Effects, Fate*; Stoytcheva, P. M., Ed.; 2011; USA.
[31] Pan, J.; Yao, H.; Xu, L.; Ou, H.; Huo, P.; Li, X.; Yan, Y. *J. Phys. Chem. C,* 2011, *115*, 5440.
[32] Liu, B.; Han, M.; Guan, G.; Wang, S.; Liu, R.; Zhang, Z. *J. Phys. Chem. C,* 2011, *115*, 17320.
[33] Das, S. K.; Das, A. R.; Guha, A. K. *Langmuir*, 2009, *25*, 8192.
[34] Loftsson, T.; Brewster, M. E. *J. Pharm. Sci.,* 1996, *85*, 1017.
[35] Shimada, K. A.; Kawano, K. I.; Ishii, J. U.; Nakamura, T. A. *J. Food Sci.,* 1992, *57*, 655.
[36] Saenger, W. *Angew. Chem. Int. Ed.,* 1980, *19*, 344.
[37] Szejtli, J. *Starch – Starke,* 1985, *37*, 382.
[38] Roy, M. N.; Ekka, D.; Saha, S.; Chandra Roy, M. *R. Soc. Chem. Adv.,* 2014, *4*, 42383.
[39] Khan, A. R.; Forgo, P.; Stine, K. J.; D'Souza, V. T. *Chem. Rev*., 1998, *98*, 1977.
[40] Guo, R.; Wilson, L. D. *Curr. Org. Chem*., 2013, *8*, 14.
[41] K. Goyal, A.; S. Johal, E.; Rath, G. *Curr. Nanosci*., 2011, *7*, 640.
[42] Davis, F.; Higson, S. *Macrocycles: Construction, Chemistry and Nanotechnology Applications*; Wiley, 2011, UK.
[43] Chilajwar, S. V.; Pednekar, P. P.; Jadhav, K. R.; Gupta, G. J.; Kadam, V. J. *Expert Opin. Drug Deliv*., 2014, *11*, 111.
[44] Dailey, O. D.; Bland, J. M.; Trask-Morrell, B. J. *J. Agric. Food Chem.,* 1993, *41*, 1767.
[45] Lezcano, M.; Al-Soufi, W.; Novo, M.; Rodríguez-Núñez, E.; Tato, J. V. *J. Agric. Food Chem.,* 2002, *50*, 108.
[46] Kamiya, M.; Nakamura, K. *Environ. Int*., 1995, *21*, 299.
[47] Villaverde, J.; Maqueda, C.; Undabeytia, T.; Morillo, E. *Chemosphere*, 2007, *69*, 575.
[48] Ishiwata, S.; Kamiya, M. *Chemosphere*, 1999, *39*, 1595.
[49] Loukas, Y. L.; Antoniadou-Vyza, E.; Papadaki-Valiraki, A.; Machera, K. G. *J. Agric. Food Chem.,* 1994, *42*, 944.
[50] Morillo, E.; Pérez-Martínez, J. I.; Ginés, J. M. *Chemosphere*, 2001, *44*, 1065.
[51] Pérez-Martínez, J. I.; Ginés, J. M.; Morillo, E.; González-Rodríguez, M. L.; Moyano Méndez, J. R. *Pest Manag. Sci*., 2000, *56*, 425.
[52] Bardi, L.; Mattei, A.; Steffan, S.; Marzona, M. *Enzyme Microb. Technol*., 2000, *27*, 709.
[53] Guo, X.; Jans, U. *Environ. Sci. Technol.,* 2005, *40*, 900.
[54] Aker, W. G.; Hu, X.; Wang, P.; Hwang, H.-M. *Environ. Toxicol*., 2008, *23*, 548.
[55] Zhou, S.; Wang, L.; Zhang, A.; Lin, K.; Liu, W. *J. Agric. Food Chem.,* 2008, *56*, 2708.

[56] Lucas-Abellán, C.; Gabaldón-Hernández, J. A.; Penalva, J.; Fortea, M. I.; Núñez-Delicado, E. *J. Agric. Food Chem.*, 2008, *56*, 8081.
[57] Hartnik, T.; Jensen, J.; Hermens, J. L. M. *Environ. Sci. Technol.*, 2008, *42*, 8419.
[58] Molaabasi, F.; Talebpour, Z. *J. Agric. Food Chem.*, 2011, *59*, 803.
[59] Gintautas, P. A.; Daniel, S. R.; Macalady, D. L. *Environ. Sci. Technol.*, 1992, *26*, 517.
[60] Patsias, J.; Papadakis, E. N.; Papadopoulou-Mourkidou, E. *J. Chromatogr. A*, 2002, *959*, 153.
[61] Cserháti, T.; Forgács, E. *J. Chromatogr. B*, 1998, *717*, 157.
[62] Hromadová, M.; Pospíšil, L.; Fanelli, N.; Giannarelli, S. *Langmuir*, 2005, *21*, 1923.
[63] Zhu, X.-L.; Wang, H.-B.; Chen, Q.; Yang, W.-C.; Yang, G.-F. *J. Agric. Food Chem.*, 2007, *55*, 3535.
[64] Best, G. A.; Ruthven, A. D. *Pesticides: Developments, Impacts and Controls*; Royal Soc. Chem. Ed., 1995, UK.
[65] Garrido, J.; Cagide, F.; Melle-Franco, M.; Borges, F.; Garrido, E. M. *J. Mol. Struct.*, 2014, *1061*, 76.
[66] Costa, C.; Maia, S.; Silva, P.; Garrido, J.; Borges, F.; Garrido, E. M. *Int. J. Photoenergy*, 2013, *2013*, 8.
[67] Pérez-Martínez, J. I.; Arias, M. J.; Ginés, J. M.; Moyano, J. R.; Morillo, E.; Sánchez-Soto, P. J.; Novák, C. *J. Therm. Anal.*, 1998, *51*, 965.
[68] Pérez-Martínez, J. I.; Ginés, J. M.; Morillo, E.; Rodríguez, M. L. G.; Moyano, J. R. *Environ. Technol.*, 2000, *21*, 209.
[69] Arias Blanco, M. J.; Ginés, J. M.; Morillo González, E.; Moyano Méndez, J. R.; Pérez-Martínez, J. I.; Ruiz Conde, A.; Sánchez-Soto, P. J. *Chemosphere*, 1996, *33*, 321.
[70] Gruiz, K.; Molnár, M.; Fenyvesi, E.; Hajdu, C.; Atkári, Á.; Barkács, K. *J. Incl. Phenom. Macrocycl. Chem.*, 2011, *70*, 299.
[71] Flaherty, R. J.; Nshime, B.; DeLaMarre, M.; DeJong, S.; Scott, P.; Lantz, A. W. *Chemosphere*, 2013, *91*, 912.
[72] Zhang, A.; Liu, W.; Wang, L.; Wen, Y. *J. Agric. Food Chem.*, 2005, *53*, 7193.
[73] Dommarco, R.; Santilio, A.; Fornarelli, L.; Rubbiani, M. *J. Chromatogr. A*, 1998, *825*, 200.
[74] Gledhill, W. E. *Water Res.*, 1975, *9*, 649.
[75] Di Corcia, A.; Costantino, A.; Crescenzi, C.; Samperi, R. *J. Chromatogr. A*, 1999, *852*, 465.
[76] Zheng, W.; Yates, S. R.; Papiernik, S. K. *J. Agric. Food Chem.*, 2008, *56*, 7367.
[77] Manolikar, M. K.; Sawant, M. R. *Chemosphere*, 2003, *51*, 811.
[78] Villaverde, J.; Posada-Baquero, R.; Rubio-Bellido, M.; Laiz, L.; Saiz-Jimenez, C.; Sanchez-Trujillo, M. a. A.; Morillo, E. *J. Agric. Food Chem.*, 2012, *60*, 9941.
[79] Worthing, C. R.; Hance, R. J. *The Pesticide manual: a world compendium*; Blackwell Science, 1991, USA.
[80] Organization, W. H. *Pesticide Residues in Food 2007: Toxicological Evaluations*; World Health Organization, 2010.
[81] Kinani, A.; Rifai, A.; Bourcier, S.; Jaber, F.; Bouchonnet, S. *J. Mass Spectrom.*, 2013, *48*, 983.
[82] Fernandes, C.; Encarnação, I.; Gaspar, A.; Garrido, J.; Borges, F.; Garrido, E. M. *Int. J. Photoenergy*, 2014, *2014*, 8.
[83] Gitsopoulos, T. K.; Froud-Williams, R. J. *Weed Res.*, 2004, *44*, 329.

[84] Benfeito, S.; Rodrigues, T.; Garrido, J.; Borges, F.; Garrido, E. M. *Scientific World J.*, 2013, *2013*, 6.

[85] Yáñez, C.; Cañete-Rosales, P.; Castillo, J. P.; Catalán, N.; Undabeytia, T.; Morillo, E. *PLOS ONE*, 2012, *7*, 41072.

[86] Villaverde, J.; Pérez-Martínez, J. I.; Maqueda, C.; Ginés, J. M.; Morillo, E. *Chemosphere*, 2005, *60*, 656.

[87] Bian, H.; Chen, J.; Cai, X.; Liu, P.; Liu, H.; Qiao, X.; Huang, L. *J. Agric. Food Chem.*, 2009, *57*, 7453.

[88] Balmas, V.; Delogu, G.; Sposito, S.; Rau, D.; Migheli, Q. *J. Agric. Food Chem.*, 2006, *54*, 480.

In: Cyclodextrins
Editor: Francis G. Ramirez

ISBN: 978-1-63482-788-1

Chapter 4

CYCLODEXTRINS SUPRAMOLECULAR COMPLEXES WITH BOTANICAL COMPOUNDS

Maria Camilla Bergonzi * ***and Anna Rita Bilia***
Department of Chemistry, Building ex Pharmaceutical Sciences,
University of Florence, Florence, Italy

ABSTRACT

Since ancient times, plant-derived compounds have been a great source of constituents used in beneficial medical treatments. Plants are a source of a wide range of molecules capable of improve healing: fibre, vitamins, phytosterols, alkaloids, sulphur-containing compounds, carotenoids, organic acids and polyphenolics. Remarkably, many of these constituents possess a scarce bioavailability due to their low water solubility or poor stability.

Cyclodextrins are cyclic oligosaccharides arising from the degradation of starch, which can be a viable option for encapsulation technique to generate aqueous drug solutions without the use of organic co-solvents, surfactants, or lipids, as formulation adjuncts.

The inclusion complexes are reported to have remarkable advantages over conventional formulations of plant actives and extracts, which include enhancement of solubility, protection from toxicity, enhancement of pharmacological activity and *in vivo* stability, sustained delivery, and protection from physical and chemical degradation, consequently increasing bioavailability and bioefficacy of guest molecules.

Therefore, the aim of this chapter is to highlight the use of cyclodextrins as encapsulating agents for bioactive plant molecules in the pharmaceutical field.

---

* Corresponding author: Department of Chemistry, Building of Pharmaceutical Sciences, University of Florence, Via Ugo Schiff 6, 50019 Sesto Fiorentino (FI), Florence, Italy; Tel. 0039-055-4573678; email: mc.bergonzi@unifi.it.

1. INTRODUCTION

Cyclodextrins are natural macrocyclic oligosaccharides arising from the degradation of starch, having toroid-shaped structures with a rigid lipophilic cavity and a hydrophilic outer surface insuring good dissolution properties of the complex in an aqueous environment. A true molecular inclusion is achieved when an highly hydrophobic molecule is included in the hydrophobic cavity, constituting a classical suparmolecular complex (Dodziuk, 2006; Duchêne, 1999).

Cyclodextrins are inexpensive, friendly to humans, and also capable of improving the biological, chemical and physical properties of bioactive molecules.

The major advantages of the use of CD-complexation in pharmaceutical applications, foods, cosmetics and toiletries are: enhancement of aqueous drug solubility without the use of organic co-solvents, surfactants, or lipids; as formulation adjuncts protection of the active ingredients against oxidation, light induced reactions, decomposition and thermal decomposition, loss by evaporation and sublimation; elimination or reduction of undesired tastes/odours, to reduce or prevent gastric-intestinal irritation (mainly due to anti-inflammatory drugs) or ocular disturbances, prevent drug-drug or drug-additive interactions, or even to convert oils and liquid drugs into microcrystalline or amorphous powders, to reduce microbiological contamination, fibres or the elimination of other undesired components and hygroscopicity (Rubistein, 1989). Among the other advantages of the use of CDs as carrier, one of the most interesting is their ability to eliminate irritant or toxicological effects of the included molecules, replacing some excipients, such as pH regulators, solubilising agents and organic solvents (Cal, 2008; Marques, 2010; Pinho, 2014).

Moreover, CDs can modify the behaviour of the encapsulated molecule, specifically by modulating the delivery rate (i.e. retarding or prolonging it (Cal, 2008; Aqil, 2013).

There are three main types of natural CDs: (i) α- cyclodextrins, corresponding to 6 glucopyranose units linked by α-(1,4) bonds, (ii) β-cyclodextrins, corresponding to 7 units and (iii) γ-cyclodextrins, corresponding to 8 units. The dimensions of the internal cavity are 0.5-0.8 nm and are crucial for the 'encapsulation' of guest molecules. Only molecules, which fit physically into the cavity can be incorporated (Duchêne, 1999).

In the last years, physicochemical properties and, consequently, the inclusion capacity of the native CDs have been improved by chemical modification of their hydroxyl groups (Matsuda, 1999). Besides natural cyclodextrins, the growing number of semi-synthetic derivatives and co-polymers has been prepared and are already commercially available. Some derivatives are used as structural and chiral selectors, with new properties given by the type and number of substituents. The semi-synthetic derivatives of cyclodextrins can have a better solubility in water, can optimise the dissolution rate, the inclusion capacity and also decrease of the side effects of some molecules.

Based on the studies reported in the literature, especially β-cyclodextrins and its derivatives, in particular the HP-β-CD, are generally selected to improve water solubility and stability of natural compounds, increasing their bioavailability and bioefficacy.

Natural products, either as pure compounds or as standardized extracts, are defined as secondary plant metabolites, usually with very complex and unique carbon skeleton structures and many of them of high interest to the pharmaceutical and chemical industry.

Many of these secondary metabolites have been widely used as "bioactive" constituents by food, cosmetic and pharmaceutical industries for their health benefits and medicinal properties such as anticarcinogenic, antihypertensive, antimutagenic, and antimicrobial activities.

Conversely, in order to have biological activity, molecules need to be able of reaching the action site without losing integrity and to be able to cross the lipophilic membrane. Many plants bioactive compounds possess limited water solubility, poor bioavailability, and can be easily modified by environmental factors such as temperature, pH and light. Therefore, in order to preserve bioactive molecules structural integrity, they need to be protected by a finishing formulation with the capacity to deliver them to the physiological targets without losing any bioactivity (Fang, 2010; Munin, 2011).

Additionally, in the food industry encapsulation is being used for many purposes such as flavours carrier and to impart of protection against evaporation of volatile compounds, reaction, or migration in a food; thus, it may be used for preparation of several fortified foods and functional foods.

These problems can be solved by applying the microencapsulation technology, which provides the required technique for conversion of the liquid extract to an effective functional ingredient. The intake of natural bioactive compound through food is of great interest, but difficulties associated with the susceptibility of those compounds to adverse external effects, or detrimental food processing conditions and their chemical instability have provided many efforts to improve oral bioavailability, among them microencapsulation, which represents a promising concept. The application of microencapsulated bioactive compounds as functional ingredients in various food and beverage applications exhibits a significant potential, since it could enable the enrichment of various food products with natural antioxidants (Diplock et al., 1999).

Various encapsulation systems are used to entrap the desired compounds, among which some novel materials have been used such as cyclodextrins, used for the encapsulation single natural substance or extracts (Belščak-Cvitanovič, 2011).

2. Plant-Derived Bioactive Agents

Botanical bioactive constituents, either as pure compounds or as standardized extracts, provide unlimited opportunities for new drug leads because of the unmatched availability of chemical diversity, according to an increasing demand for chemical diversity in screening programs, seeking therapeutic drugs from natural products.

These constituents are produced within the plants besides the primary biosynthetic and metabolic routes and are regarded as products of biochemical "side tracks" in the plant cells. Several of them are found to hold various types of important functions in the living plants such as protection against microorganisms, insects and herbivores, attraction or signalling.

These compounds are an extremely diverse group of natural products synthesized by plants, fungi, bacteria, algae, and animals. Most of secondary metabolites, such as terpenes, phenolic compounds and alkaloids are classified based on their biosynthetic origin. A simple classification of secondary metabolites includes tree main groups: terpenes (such as plant volatiles, cardiac glycosides, carotenoids and sterols), phenolics (such as phenolic acids, coumarins, lignans, stilbenes, flavonoids, tannins and lignin), nitrogen, and sulphur-containing compounds (such as alkaloids and glucosinolates).

Simplified illustration of different pathways for the production of major groups of plant bioactive compounds are shown in Figure 1 (Azmir, 2013).

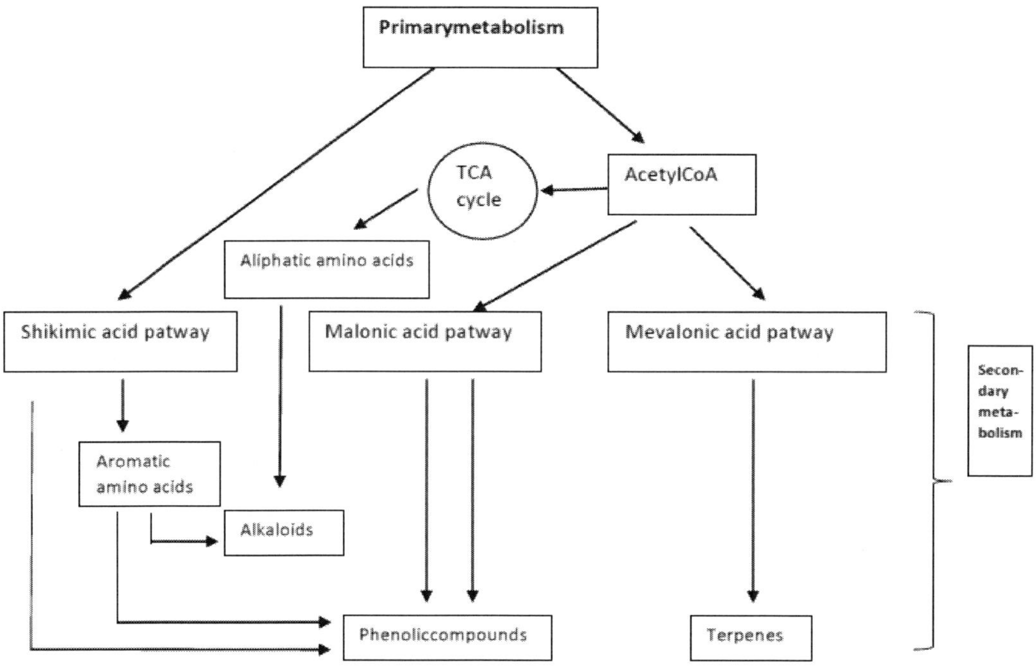

Figure 1. A simplified view of pathways for production of major groups of plant bioactive compounds (adapted from Azmir, 2013).

Terpenoids are the largest and most diverse family of natural products (ca. 29,000 different constituents), ranging in structure from linear to polycyclic molecules and in size from the five-carbon hemiterpenes to natural rubber, comprising thousands of isoprene units. All terpenoids are synthesized through the condensation of isoprene units (C5) and are classified by the number of five-carbon units present in the core structure. The C5 precursor isopentenyl diphosphate (IPP) is generated from mevalonic acid (3,5-dihydroxy-3-methyl valeric acid) pyrophosphate by decarboxylation and dehydration. The C10 compound geranyl pyrophosphate is formed by condensation of isopentnyl pyrophosphate with dimethyl allyl pyrophosphate and it represents the precursor of monoterpenes. Many flavor and aromatic molecules, such as menthol, linalool, geraniol and caryophyllene are formed by monoterpenes (C10), with two isoprene units, and sesquiterpenes (C15), with three isoprene units. Further, C6 units are added by participation of more isopentnyl pyrophosphate units and produce sesquiterpenes. From geranyl and farnesyl pyrophosphates various isoprenoid structures are synthesized, namely diterpenes and triterpenes, steroids, saponines respectively. Many bioactive compounds, such as diterpenes (C20), triterpenes (C30) and tetraterpenes (C40) show very special health and medicinal properties.

The shikmic acid pathway is a key intermediate from carbohydrate for the biosynthesis of C6-C3 units (phenyl propane derivative). Besides serving as precursor for the biosynthesis of amino acids shikmic acid is also an intermediate in production of tannins, flavones, coumarins and vanillin. The shikimic acid pathway converts simple carbohydrate precursors derived from glycolysis and the pentose phosphate pathway to the aromatic amino acids. The shikimic acid pathway is present in plants, fungi, and bacteria but is not found in animals. Animals have no way to synthesize the three aromatic amino acids-phenylalanine, tyrosine, and tryptophan, which are therefore essential nutrients in animal diets.

Phenolic compounds, widely distributed in nature, mainly represent phenylpropane derivatives. Their chemical structures may vary greatly, including simple phenols (C6), such as hydrobenzoic acid derivatives and catechols, as well as long chain polymers with high molecular weight, such as catechol melanins (C6)6, lignins (C6-C3)n and condensed tannins (C6-C3-C6)n. Stilbenes (C6-C2-C6) and flavonoids (C6-C3-C6) are phenolic compounds with intermediate molecular weight that present many pharmacological and biological activities. Flavonoids, including anthocyanins, flavonols (such as quercetin and myricetin), isoflavones (such as daidzein and genistein) and others are formed by multiple biosynthetic branches that originate from chalcone. More than 8,000 phenolics have been actually isolated and more than 2,000 are represented by flavonoids.

Alkaloids are defined as basic compounds synthesized by living organisms containing one or more heterocyclic nitrogen atoms, derived from amino acids (with some exceptions) and pharmacologically active. The class name is directly related to the fact that nearly all alkaloids are basic (alkaline) compounds. Alkaloids constitute a very large group of secondary metabolites, with more than 12,000 substances isolated. A huge variety of structural formulas, coming from different biosynthetic pathways and presenting very diverse pharmacological activities are characteristic of the group. Cyanogenic glycosides (ca. 100) and glucosinolates (ca. 100) are also generated from aminoacids (Dewick, 2009).

Plants bioactive molecules include polyphenolics, alkaloids and polysaccharides, and they all have well-documented pharmacological properties. However, polyphenolics are currently the major group of interest in view of their anti-inflammatory, anti-microbial and antioxidant properties, as well as their availability inthe human diet (Aridogan, 2002; Gou, 2011).

2.2. Flavonoids

Flavonoids are among the biggest group of polyphenols, widely distributed in plant-based foods and mostly of herbal remedies. A plethora of evidence supports the health benefits and value of flavonoids can play in the hysiological function treatment and in the prevention of disease particularly in the prevention of degenerative conditions including cancers, cardiovascular and neurodegenerative diseases. Hence, flavonoids represent the active constituents of many dietary supplements and herbal remedies, as well as there is an increasing interest in this class of polyphenols as functional ingredients of beverages, food grains and dairy products. Conversely, various studies have also shown that flavonoids have some drawbacks after oral administration such as stability, bioavailability and bioefficacy.

Structurally, flavonoids are low-molecular mass phenylchromane derivatives with a basic carbon-based skeleton in C6-C3-C6, a general structural backbone in which the two C6 units (ring A and ring B) are of phenolic nature and according to the variations in the heterocyclic ring C, flavonoids can be divided in different classes (Harborne, 2000). Flavonoids exhibit a wide spectrum of biological activities, either as antioxidants or modulators of cell signaling, and the influence of their metabolism on these properties is key to the evaluation of these potent biomolecules as antioxidants and anti-inflammatory agents, cardioprotectants, and inhibitors of neurodegeneration. Moreover, they have the capacity to inhibit the growth of a wide range of microorganisms and viruses (Ross, 2002; Yao, 2004; Heim, 2002).

Despite the health benefits, the therapeutic outcome is still dependent by the improvement of the pharmacokinetic profile of these compounds after oral administration. Flavonoids have limited water solubility even if they are in the form of glycosides, poor bioavailability, and can be easily modified by environmental factors such as temperature, pH and light. The mechanisms of gastrointestinal absorption of flavonoids are complex, which are poorly absorbed, in their natural form in the intestine. It is believed that flavonoids are extensively degraded by intestinal microorganisms and/or enzymes, producing different metabolites. If absorbed, these metabolites, are subjected to the hepatic enzymatic system and new metabolites can be formed varying in bioactivity (Manach, 2005; Stahl, 2002).

After the hydrolysis of the sugar moieties in the small intestine or due to bacterial activity in the colon, aglycones are generated and further metabolized before reaching the systemic circulation. Briefly, numerous factors could play a role in limiting glucuronidated or sulfated form Manach and Stahl (Manach, 2005; Stahl, 2002). As a consequence, flavonoids result in poor bioavailability, poor permeability, instability and extensive first-pass bioavailability of flavonoids, such as low water solubility, instability to gastric and colonic pH, metabolism by gut microflora, absorption across the intestinal wall, active efflux mechanism and first-pass metabolic effects. Micro and nanocarriers could be useful to enhance biovailability and bioefficacy of flavonoids because they can increase solubilization potential, alter absorption pathways, and prevent the metabolic degradation within the gastrointestinal tract.

In recent years, cyclodextrin complexation has been successfully used to improve solubility, chemical stability, and bioavailability of a number of poorly soluble molecules.

The flavonoids analysed can interact with CDs by (1) directing the B-ring towards the secondary rim of the CD or (2) heading the A-ring towards the secondary rim of the CD (Kim, 2009; Pinho, 2014). The majority of the publications concern the encapsulation of flavonoids with β-CD and its derivatives, in particular heptakis (2,6-di-O-methyl)-β-cyclodextrin (DM-β-CD) and 2-hydroxypropyl-β-cyclodextrin (HP-β-CD).

2.2.1. Flavones

Baicalein (Figure 2) is an active ingredient of the herbal medicine extracted from *Scutellaria baicalensis*. Baicalein has attracted considerable attention in the recent years primarily because of its various interesting activities, such as anti-inflammatory, anti-HIV, anti-tumor, anti-fibrosis, anti-anaphylaxis, protecting neurons, scavenging free radical and anti-oxidation (Gao, 2010; Shen, 2003; Kim, 2007). However, the clinical application of baicalein has been seriously influenced by its low solubility and instability.

Zhou investigated the solubility of baicalein with the addition of modified cyclodextrins using phase solubility method (Zhu, 2013).

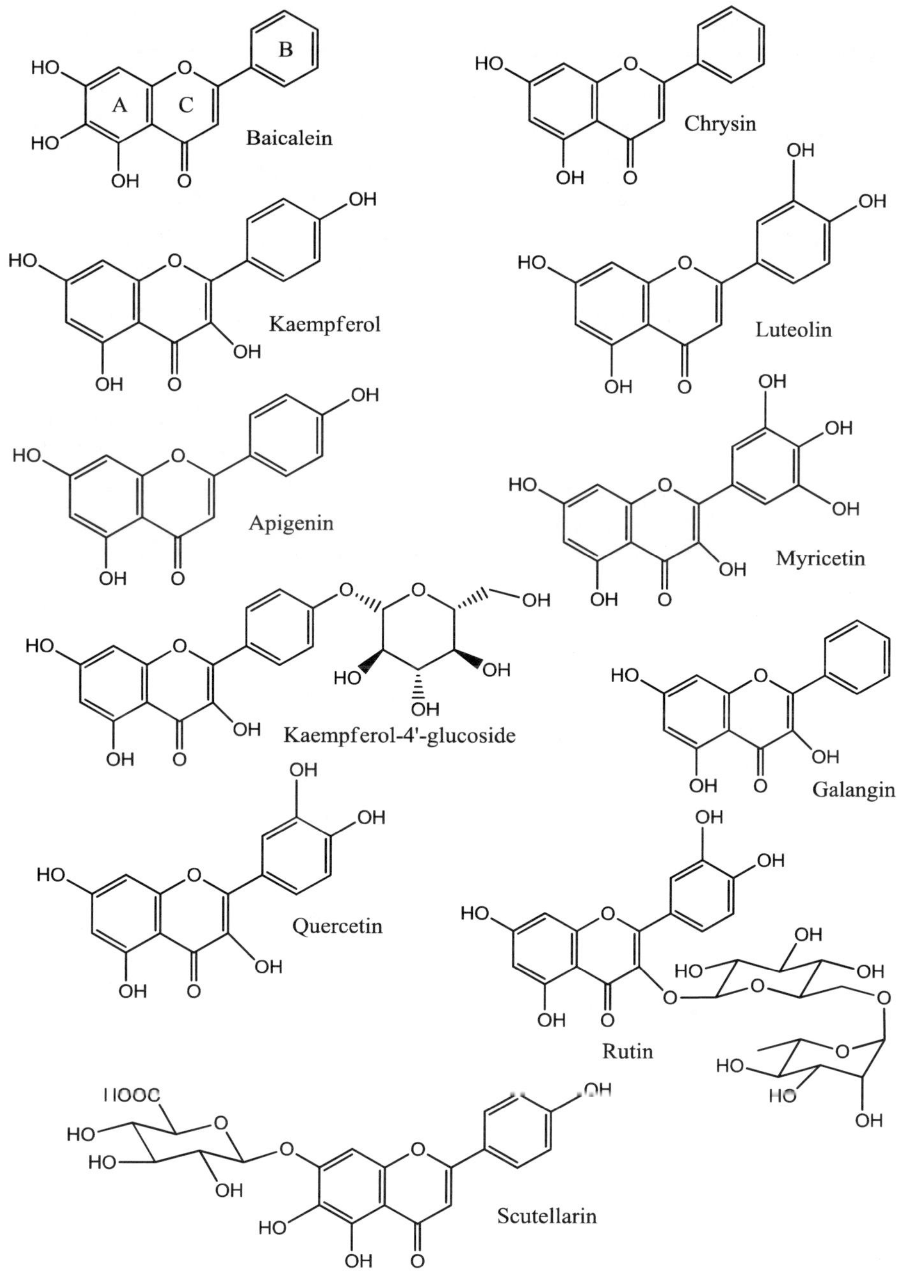

Figure 2. Chemical structure of flavones and flavonols.

The solubility in the presence of natural (α-, β-, and γ-CD) cyclodextrins and its derivatives, namely, hydroxypropyl-β-cyclodextrin (HP-β-CD) and (2,6-di-O-methyl)-β-cyclodextrin (DM-β-CD), was higher than that of free flavone. In particular, the stability constants of ICs (Kc) resulted: 312.63 M^{-1} for α-CD, 2586.66 L mol^{-1} for β-CD, 1016.95 L mol^{-1} for γ-CD, 10771.49 L mol^{-1} for HP-β-CD and 13672.67 L mol^{-1} for DM-β-CD, with a solubilizing efficiency ranging from 6.05 for α-CD and 180.86 for DM-β-CD. Furthermore, the dissolution rate and thermal stability of the inclusion complex were significantly enhanced compared with the pure drug.

Chrysin (Figure 2) is also a flavone and its pharmaceutical applications are related with its antioxidant, anti-inflammatory and anti-hypertension capacity, however, it has low solubility in water (Chakraborty, 2010). Chrysin IC with β-CD was investigated by Chakrabortym (Chakraborty, 2010). The complex had the stoichiometry of 1:1, with a Kc of 1005 M^{-1}. The complexation improved antioxidant activity of chrysin (Chakraborty, 2010). Kim et al., (2008) studied the effect of β-CD, DM-β-CD and HP-β-CD on the aqueous solubility of chrysin. The authors reported the same stoichiometry (1:1) for all the CD used. The order of interaction with three cyclodestrins resulted: HP-β-CD > DM-β-CD > β-CD and the Kc resulted 11858.5, 1062.1, 975.6 M^{-1}, respectively. In both works, authors concluded that the interaction between the chrysin and the CDs occurs by the A-ring of the flavones.

Luteolin (Figure 2) is a yellow coloured flavone found naturally in a variety of plants, especially vegetables such as celery, green pepper and perilla leaves. It has stronger antioxidant activities and lower potential prooxidant activity compared to many common flavonoids, such as quercetin and myricetin, and displays potential health benefits for humans (Seelinger, 2008). Numerous studies over the last decade have shown that luteolin exhibits a number of health-promoting functions including anticancer, anti-inflammatory, anti-allergic and cardioprotective properties (Seelinger, 2008; Ong, 2010; Liu, 2013). However, the poor solubility and stability in water severely restricts its applications in pharmaceutical, nutraceutical and functional food products.

The formation of supramolecular inclusion complexes between luteolin and five cyclodextrins namely β-CD, M-β-CD, HE-β-CD, HP-β-CD and glucosyl-β-cyclodextrin (G1-β-CD) was investigated by Liu (Liu, 2013). Results from phase-solubility studies showed that luteolin formed 1:1 stoichiometric inclusion complexes with these cyclodextrins with the G1-β-CD complex displaying the greatest stability constant. The Kc of the luteolin complexes formed with the five different cyclodextrins followed the rank order of G1-β-CD (Kc, 3355) > HP-β-CD (Kc, 2863) > HE-β-CD (Kc, 2374) > M-β-CD (Kc, 1290) > β-CD (Kc, 545).

Results showed clearly the formation of a supramolecular complex in which the guest molecule, luteolin, was entrapped inside the cavity of the host, G1-β-CD. Furthermore, molecular docking study showed that the complex was formed with the B ring of luteolin inserted into the cavity of G1-β-CD.

The effect of CDs on the aqueous solubility of luteolin was also investigated. Solubility enhancements of flavone obtained with three CDs followed the rank order: HP-β-CD > DM-β-CD > β-CD. The Kc were 51385.9, 4461.2, 2328.3 M^{-1}, respectively. NMR spectroscopic analysis suggested that A-C ring, and part of the B ring of luteolin display favorable interaction with the CDs, which was also confirmed by docking studies based on the molecular simulation (Kim, 2008).

In the same publication, the effect of β-CD, DM-β-CD and HP-β-CD on the aqueous solubility of *apigenin* (Figure 2) was also investigated. Also for this flavone the order of interaction with the three cyclodestrins resulted: HP-β-CD > DM-β-CD > β-CD and the Kc were 14511.5, 1038.6, and 827.6 M^{-1}, respectively.

2.2.2. Flavonols

Kaempferol have also been described as efficient radical scavenger (Mercader-Ros, 2009). However, their high hydrophobicity and sensitivity to external agents such as air, light, and oxidative enzymes constitute a serious problem for their bioavailability, formulation, and manipulation in the elaboration of functional foods. To avoid this limitation, the inclusion of kaempferol in CDs has been suggested as a possible solution. β-CD and its derivatives, such as HP-β-CD, DM-β-CD and maltosyl-β-cyclodextrin (G2-β-CD), were evaluated for the IC formation with kaempferol (Figure 2) (Kim, 2009; Jullian, 2011; Mercader-Ros, 2010 a,b).

Additionally, the temperature conditions the Kc values, but IC formation was improved by lower temperatures (Jullian, 2011). NMR experiments provided data indicating that the B-ring of kaempferol is immersed in the apolar cavity with the A- and C-ring protruding from the wider rim for the three cyclodextrins studied. The IC behave as better antioxidants than kaempferol alone.

The same beneficial effect on the antioxidant activity of kaempferol was determined by Kim (2009) and Mercader-Ros (2010 a,b). In particular Kim analysed also the IC with modified cyclodextrins such as DM-β-CD and HP-β-CD. The solubility of flavonol was increased with all carriers. Stability constant was 12952 and 12456 M^{-1} for DM-β-CD and HP-β-CD, respectively. Flavonol can complex with CDs largely by two binding modes. The first one is that B-ring of flavonol is oriented toward secondary rim of CD. The second one is that A-ring of flavonol is oriented toward secondary rim of CD. Whereas only the first mode was observed in DM-β-CD complexes, both the first and the second mode were observed in HP-β-CD complexes in this study.

The complexation of kaempferol with cyclodextrins (β-, G2-β-, and HP-β-CDs) in basic medium was studied by Mercader-ros and coworkers (Mercader-ros, 2010). The Kcs were calculated by using enzymatic, solubility, and fluorometric methods.

In all cases HP-β-CDs showed the highest Kc value (11161 M^{-1} in the solubility study), indicating that they are more efficient in the formation of inclusion complexes with kaempferol, although depending on the method employed the results are not always superimposable.

Zheng (Zheng, 2014) reported the complexation of *kaempferol-4'-glucoside* (Figure 2) with sulfobutyl ether-β-cyclodextrin (SBE-β-CD), HP-β-CD, and methylated-β-cyclodextrin (M-β-CD) in phosphate buffer solutions of different pH. Experimental results showed that kaempferol-4′-glucoside binds with the three β-Cd in the same 1:1 stoichiometry. The rank order of stability constants is SBE-β-CD > HP-β-CD > M-β-CD at the same pH level and pH 6.0 > pH 7.4 > pH 9.0 for the same cyclodextrin. The possible inclusion mode was that in the cavity of β-CD is included the planar benzopyranic-4-one part of the kaempferol-4′-glucoside.

Kim analysed also the ICs of others two flavonols *myricetin* and *galangin* (Figure 2) with modified cyclodextrins, such as DM-β-CD and HP-β-CD, to improve theis solubility. The solubility of flavonols was increased. Based on the NMR results, it can be inferred that the A-ring of flavonol exhibit significant interaction with the DM-β-CD cavity. In the case of HP-β-CD complex, B-ring as well as A-ring of flavonols shows significant interaction with the CD cavity. DM-β-CD (Kc: 12079 M^{-1}) shows higher stability constant than that of HP-β-CD (Kc: 7674 M^{-1}) in galangin, but on the other hand, HP-β-CD (5494 M^{-1}) shows higher stability constants than that of DM-β-CD (4287 M^{-1}) in myricetin which contain more hydroxyl groups than kaempferol. The overall values of stability constant is largely higher than that of α-, β-, and γ-CD complexes, which has been reported for galangin, kaempferol values below 400 M^{-1} (Bergonzi, 2007). These results are reflecting an enhancement of binding and solubility of flavonols with an increase in substitution and hydrophilicity of the CDs.

In the case of natural CDs (Bergonzi 2007), multidimensional NMR experiments in DMSO and water are consistent with dynamic binding processes, dominated by insertion of the B ring of flavonols into the wider rim of the CD cavity. In addition, some smaller interactions exist with protons of ring A and the protons inside the Cd cavity, probably according to a dynamic binding between flavonol and CD systems.

The effect of complexing myricetin with HP-β-CDs on its antioxidant capacity was studied by means of the oxygen radical absorbance capacity-fluorescein (ORAC-FL) assay (Mercader-ros 2010). This complexation phenomenon increased the antioxidant activity of flavonol, which reached a maximum level when myricetin had been complexed in the hydrophobic cavity of CDs. The antioxidant activity increased because of the flavonol was protected against rapid oxidation by free radicals.

As the most part of flavonoids, *quercetin* (Figure 2) is flavonol with antibacterial, antioxidant and antitumor properties but its use in the pharmaceutical field is limited due to it sparingly solubility in water. β-CD was used to improve solubility and photostability of quercetin by several authors (Calabrò, 2004; Jullian, 2007; Sri, 2007; Bergonzi, 2007).

Bergonzi demonstred an increase of solubility of quercetin in the presence of β-CD with a stability constant of 396 M^{-1}. The host-guest interactions were studied, both in solution and in the solid state.

For all three guests, multidimensional NMR experiments are consistent with dynamic binding processes, dominated by insertion of the B ring into the wider rim of the CD cavity. In addition, some smaller interactions exist with protons of ring A and the protons inside the CD cavity, probably according to a dynamic binding between flavonol and CD systems.

The authors reported a 1:1 stochiometric complex. Sri et al. (Sri, 2007) described instead a 1:2 complex. In fact, as reported in the literature, in the same equilibrium, it is possible to find different interactions between CD and the guest molecule (1:1 or 2:1). Moreover, the Kc values, assessed by these authors, rise with the increase of the temperature, at 30°C (Jullian et al. 2007) the IC had the higher stability constant (602 M^{-1}). In the same work, the IC formation with SBE-β-CD and HP-β-CD was also established. The IC for both β-CD derivatives with quercetin was 1:1 and the solubility of quercetin was enhanced, with Kc of 4032 and 1419 M^{-1}, respectively, without diminishing its antioxidant property. Moreover, the CD derivatives were more efficient on the solubilisation of quercetin than the natural CD and showed better antioxidant activity (Jullian et al., 2007). The solubility of quercetin determined with the three CDs followed the rank order SBE-βCD>HP-βCD>βCD, reflecting and enhancement of binding and solubility with an increasing in substitution and hydrophilicity of the CDs.

The HP-β-CD was also used by Sri et al. (2007) and by Mercader-Ros (2010 a;b) to enhance the solubility and antioxidant properties of quercetin. In both situations, the Kc value found was lower (321 and 900 M^{-1}) than that mentioned by Jullian et al., as well as the temperature used (28 and 25°C) (Mercader-Ros, 2010a,b; Sri, 2007), this means that the temperature is a crucial parameter on quercetin/HP-β-CD IC formation. The antioxidant activity of the quercetin was improved by its encapsulation in β-CD derivatives (HP-β-CD, SBE-β-CD, DM-β-CD and M-β-CD) (Jullian, 2007; Yu, 2007; Carlotti, 2010; Mercader-Ros,

2010a,b). Carlotti (2010) also reduced the photodegradation ratio of the quercetin by its complexation with M-β-CD. The quercetin was in an apolar environment inside the CD cavity and the photolytic reaction was reduced.

Rutin (Figure 2) is a natural flavone derivative found in abundance in buckwheat but also present in many other plants, including asparagus, citrus fruits and berries. Structurally, rutin is a glycoside of the common flavonoid quercetin and the disaccharide rutinose. Therefore, rutin has emerged as a major bioactive ingredient with potential applications in many functional food, nutraceutical and pharmaceutical products. However, a major problem related to this phytochemical is its poor water solubility, which severely restricts its applications. Thus, improving the solubility of rutin would greatly enhance the application potential of this flavonoid.

The encapsulation of rutin with α-CD, β-CD, HP-β-CD and DM-β-CD was described by different authors in order to improve rutin solubility, and consequently pharmacological activity (Shuang, 1997; Haiyun, 2003; Sri, 2007; Yu, 2007).

For the CDs used, each molecule of cyclodextrin was capable of interacting with one molecule of rutin. Regarding HP-β-CD, the IC formed was more stable since the constants described were higher than the one for the rutin/β-CD (260 M^{-1}). However, the DM-β-CD was the CD with higher Kc (3217.62 M^{-1}) (Yu et al., 2007) and the β-CD with lower capacity to complex with rutin probably related to the CDs' cavity size.

The interaction of β-CD, HP-β-CD and γ-CD with the rutin has been investigated by using fluorescence spectroscopy. The stoichiometry of the complexes and their apparent formation constants have been estimated. The results show that the formation constants of rutin with several cyclodextrins exhibit the order HP-β-CD > β-CD > γ-CD with the same stoichiometry of the complexes 1:1. Rutin has no measurable fluoresence enhancement in α-CD system. The complexing ability of HP-β-CD is remarkble stronger than β-CD and γ-CD (Shuang, 2007).

The formation of IC improved rutin solubility and consequently its antioxidant activity and bioavailability (Calabrò, 2005). The enhancement of the oral availability by the capsulation of rutin by HP-β-CD was proved by Miyake and co-workers after administration of the ICs to beagle dogs (Miyake, 2000).

The formation of supramolecular inclusion complexes between rutin and four CDs, namely β-CD, HP-α-CD, HP-β-CD and HP-γ-CD, and the effects of the complexation on the stability and antioxidant activity of rutin were investigated also by Nguyen (Nguyen, 2013). Results from phase-solubility studies showed that rutin formed 1:1 stoichiometric inclusion complexes with HP-α-CD, β-CD, HP-β-CD and HP-γ-CD; the complexes formed with HP-γ-CD and HP-β-CD had the greatest Kc, followed by β-CD and HP-α-CD. The HP groups enhanced the stability of interaction between CD with rutin and the NMR analysis showed that the A ring was inside the cavity of HP-β-CD. Formation of such an inclusion complex conferred moderate degrees of protection to rutin from degradation by heat and UV radiation during storage, and significantly enhanced its antioxidant capacity.

Scutellarin, (Figure 2) a known flavone glycoside, is a primary active ingredient extracted from Chinese herb *Erigerin breviscapus* (Vant.) Hand.-Mazz. Scutellarin have the neuroprotective action and the anticoagulation effect. Moreover, scutellarin can also induce cell death in the human colon cancer cell line and protect against cerebral ischemia-reperfusion injury by many pathways of action (Ge, 2003; Ju, 2006). However, it has been reported that the absolute bioavailability of scutellarin oral preparations was very low due to the poor solubility and hydrophobicity of scutellarin, and the short residence time of scutellarin in the circulation.

A series of cyclodextrin/scutellarin ICs were prepared with α-CD, β-CD and HP-β-CD, and their inclusion complexation behaviors were investigated in order to improve the absolute bioavailability of scutellarin, prolong the duration of the drug in the circulation, effectively maintain the therapeutic drug levels in the blood for a long time, reduce the frequency of injection administration and therefore afford patient compliance (Yang, 2009).

The results showed that scutellarin could be efficiently encapsulated in the CD cavity in aqueous solution to produce complexes that were more soluble than free molecule.

The 1:1 inclusion complexation between host and guest was found. Kcs were calculated to be 210 M^{-1} for α-CD, 634 M^{-1} for β-CD and 1925 M^{-1} for HP-β-CD. From NMR data scutellarin was be included in the β-CD and HP-β-CD cavity with the A-ring and C-ring orienting towards the secondary hydroxyl group. In contrast, α-CD was only weakly associate with scutellarin to form the inclusion complex. The results show that the water solubility of β-CD/scutellarin, HP-β-CD/scutellarin and α-CD/scutellarin complexes comparing with that of scutellarin (ca. 160 μg/mL), is dramatically increased to approximately 9.0, 10.3 and 8.2 mg/mL (calculated as scutellarin residue), respectively. No interactions were observed in the case of γ-CD and heptakis(2,3,6-tri-O-methyl)-β-CD (TM-β-CD).

2.2.3. Flavan-3-ols

The *catechin, epicatechin and epigallocatechin-gallate* (Figure 3) are molecules representative of the flavan-3-ols subgroup. These compounds present antidiabetic and antiobesity properties, besides the antioxidant action (Haidong, Fang, 2011).

Catechin from grape seed has demonstrated high potential to protect against a myriad of diseases such as anti-hyperglycemic effects (Pinent, 2004), anti-inflammatory effects, enhancement of postprandial lipemia, improvement of insulin sensitivity and prevention of hypertriglyceridemia, cardiovascular disease (Al-Awwadi, 2005), prevention of in vitro LDL oxidation, modulation of the expression of antioxidant enzyme systems.

Catechin isolated from grape seed was successfully complexed with β-CD with a 1:1 stoichiometry (Krishnaswamy, 2012). The interactions were evidenced by differential scanning calorimeter, raman laser spectroscopy, scanning electron microscopy and X-ray diffractometer. The results obtained strongly confirmed supramolecular complex formation.

Yan performed a molecular modeling study to elucidate the structural aspects for the complexation of (+)-catechin and (-)-epicatechin (an enantiomer) to β-CD. In the β-CD/catechin inclusion complex, the orientation in which the aromatic A-ring of catechin projects onto the 2-OH/3-OH face of β-CD, and the B-ring projects from the 6-OH face is preferred in the binding energy. In contrast, the inclusion of the B-ring of epicatechin from either the secondary hydroxyl group side or the primary hydroxyl group side gives rise to the two most probable complexes. Epicatechin forms a more stable complex with β-CD than the corresponding catechin (Yan, 2007).

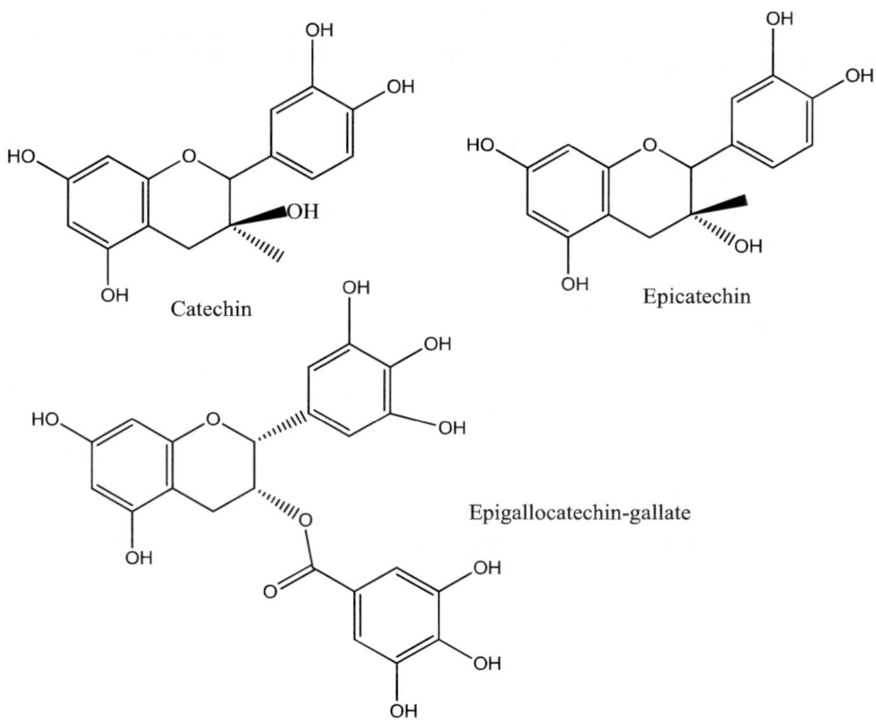

Figure 3. Chemical structure of flavan-3-ols.

Another study (Jullian, 2007) reported the inclusion properties and the factors affecting the complexation selectivity and stabilization of catechin into β-CD, DM-β-CD and HP-β-CD. Analysis of the proton shift change confirms the formation of a 1:1 stoichiometric complex for catechin and the different CDs, in aqueous medium. The formations indicated the following trend upon complex formation: β-CD (21800 M^{-1}) > HP-β-CD (13580 M^{-1}) > DM-β-CD (3500 M^{-1}). The detailed spatial configuration proposed confirmed that when complex is formed, the catechol moiety in the complex is oriented toward the primary rim; however, when CD is derivatized to HP-β-CD and DM-β-CD this ring is oriented toward the secondary rim.

Moreover, the solubility profile of *epigallocatechin* after encapsulation with β-CD, HP-β-CD and DM-β-CD was described by Folch-Cano (Folch-Cano, 2013). He observed that the temperature had a different influence on the Kc, dependent on the CD used. In the case of the native CD, the Kc rise with temperature increases, the opposite was observed for the CD derivatives used in this study. Hence, the DM-β-CD was the most suitable CD for the complexation of epigallocatechin. The authors also described that all the IC had similar geometries, but the difference resides on the exposition degree of the antioxidant rings of EGCG inside the CD cavity (Folch-Cano, 2013).

2.2.4. Dihydroflavonols

Astilbin (Figure 4) is a dihydroflavonol that is widely distributed in many plants, such as *Engelhardia roxburghiana*, *Hypericum perforatum*, *Smilax genera* plants, and grape. It is the dominant compound of *Rhizoma Smilacis Glabrae* and of processed functional food, turtle jelly.

The most interesting pharmacological potential of astilbin is its novel immunosuppressive activity (Xu, 1999). The complexation of astilbin with α-, β-, and γ-cyclodextrin (CD) was studied by Zhang to improve solubility and stability of this compound (Zang, 2013).

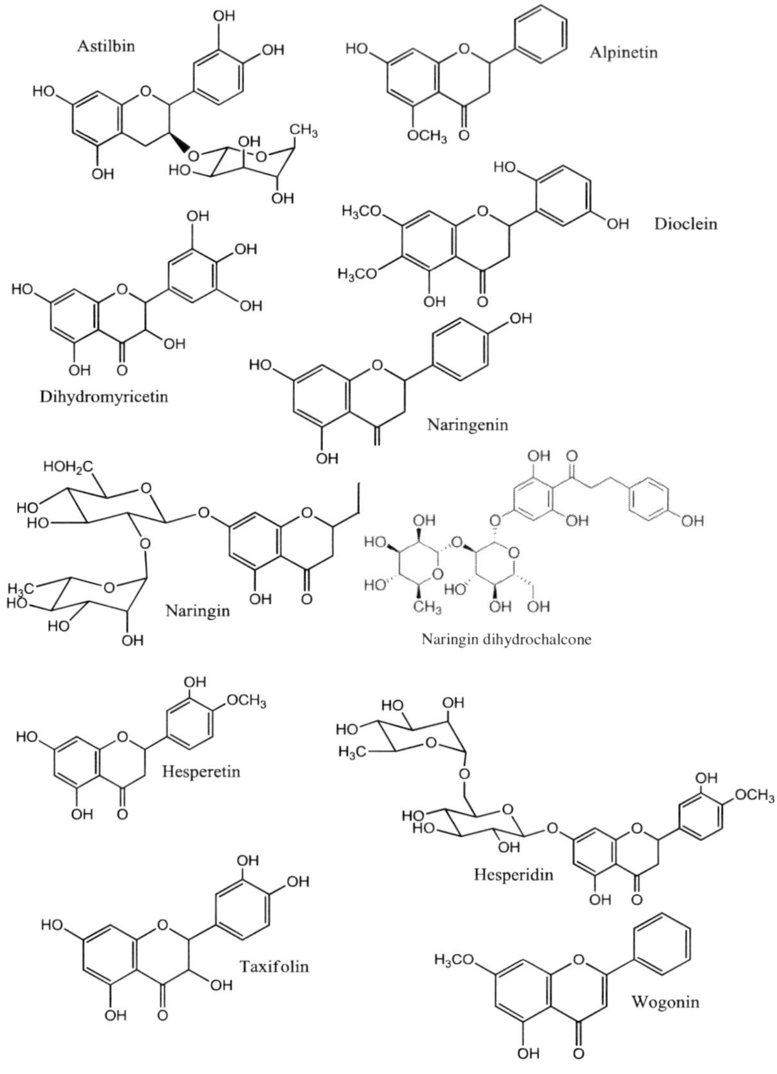

Figure 4. Chemical structure of dihydroflavonols and flavanones.

Temperature studies showed that the Kc value decreased along with the rise of temperature indicating that the affinity between CDs and astilbin decreased. The stoichiometry of astilbin with β- and γ-CD was 1:1. As shown, the Kc values of astilbin with β- and γ-CD were 2305.9 M^{-1} and 995.2 M^{-1} in acetate buffer at pH 3.5, respectively. In alkaline medium isomerization and decomposition of astilbin were found; however, the addition of CDs significantly improved its stability through complexation. The results suggested that the interaction of CDs with ionized astilbin was weaker than that of un-ionized species. Similar results have been observed in many other CD/guest complexation studies. It was also noted that the stability of astilbin inclusion complexes was in the order β-CD > γ-CD > α-CD attributed to the CDs' cavity size.

Liu investigated the inclusion complexation of (2-hydroxypropyl)-cyclodextrins with *dihydromyricetin* (Figure 4) (Liu, 2012). Inclusion complexes were prepared with HP–α–CD, HP-β-CD with different molecular weight and HP-γ-CD. The stability constants of the complexes followed the rank order: HP-β-CD (MW 1540) > HP-β-CD (MW 1460) > HP-β-CD (MW 1380) > β-CD > HP-γ-CD > HP-α-CD. The Kc were highest for the complex formed with the three HP-β-CDs, followed by β-CD and HP-γ-CD, while those formed with HP-α-CD had the lowest stability constants. Kc were 78 M^{-1} for HP-α-CD, 279 M^{-1} for β-CD, 530 M^{-1} for HP-β-CD (MW 1540) and 264 M^{-1} for HP-γ-CD.

2.2.5. Flavanones

The flavonoid *alpinetin* (Figure 4) is widely distributed in seeds of *Alpinia katsumadai Hayata* or *Amomum alnus* and in roots of the genus *Populus*, *Polygonum* or *Scuiellaria*. Several of the plants described as possessing alpinetin are used in traditional and clinical medicine; some are ingested in the human diet and others are being studied for their potential use in drug development. Alpinetin has long been known to show many pharmacological activities including antiemetic, depriving dampness, and expelling cold. It is increasingly being recognized as possessing a broad spectrum of biological activities and important therapeutic applications, such as significant microbial resistance, anti-hemostasi, antitumor activity (Zhang, 2010), anti-hepatotoxic activity, anti-inflammatory and stomachic properties (Yamamoto, 2011). However, the use of alpinetin as herbal medicines is greatly limited by its low water solubility and bioavailability. Therefore, the search for an efficient and nontoxic carrier for alpinetin has become important in order to further its clinical applications.

The inclusion complexation behavior, characterization and binding ability of alpinetin with HP-β-CD has been investigated in both solution and the solid state (Ma, 2012). The results showed that the water solubility and stability of alpinetin were obviously increased in the IC. The stoichiometry of the complex was 1:1 with Kc of 783 M^{-1}. The possible inclusion modes for the alpinetin/HP- β-CD complex was the C-ring of flavanone inserted into the HP-β-CD cavity, towards the secondary rim.

Dioclein (Figure 4) is a flavonoid present in the roots of *Dioclea grandiflora* Mart. Ex Benth. Several beneficial effects of dioclein in the cardiovascular system include vasodilation (Almeida, 2002) and antioxidant and antiarrythmogenic effects (Vianna, 2006). One problem with the use of dioclein as a therapeutic agent, in spite of the wide spectrum of pharmacological properties, is its lack of oral therapeutic efficacy and low water solubility.

To circumvent the problem of solubility and possible intestinal degradation, a 1:1 IC between dioclein and β-CD was prepared and characterized and its hypotensive effect after intraperitoneal or oral administration in mice were measured (Bruno, 2009).

The inclusion of dioclein in β-CD increased the water solubility of 44% compared to free dioclein. The IC produced a higher and long lasting change in systolic blood pressure after intraperitoneal administration compared to free dioclein. The complex improves the hypotensive effect of the flavonoid by increasing its bioavailability and enables dioclein to be effective after oral administration. The mechanism underling the increase in bioavailability is probably a consequence of a protective effect of β-CD against *in vivo* biodegradation by enzymes and possibly increased water solubility. The *in vitro* vasodilator effect of dioclein was unchanged by its inclusion in β-CD howing that the IC does not change the interaction between dioclein and its cellular targets.

Among naturally occurring *Citrus* flavonoids, *hesperetin* and its 7-rhamnoglucoside *hesperidin* (Figure 4) have been reported to exert a wide range of pharmacological effects. Hesperidin possesses antioxidant, blood lipid lowering and anti-carcinogenic activities. In addition, they have been found to improve venous tone, enhance microcirculation, assist healing of venous ulcers and are used for the treatment of chronic venous insufficiency, hemorrhoids and the prevention of post-operative thromboembolism (Tommasini, 2005). Dietary hesperidin is deglycosylated into hesperetin by intestinal bacteria prior to absorption. The very poor solubility of these flavanones in water yet imposes considerable limitations to pharmaceutical use.

Ficarra and Tommasini investigated the effect of the β-CD and HP-β-CD ICs on the solubility of hesperetin and hesperidin (Ficarra, 2002; Tommasini, 2005). In the first work, they were complexed with β-CD and improvement on the solubility and chemical stability was demonstrated (Ficarra, 2002). Tommasini et al. (2005) proved that the ICs formed was 1:1, and the better complexation was observed by HP-β-CD and hesperetin, based on the higher Kc values (21000 M^{-1}) when compared to the ones assessed for hesperidin (Kc 90 M^{-1}). This discrepancy can be justified by the size of the flavanones molecules: the hesperetin is smaller and less polar which make it more appropriate to interact with the hydrophobic cavity of the CD derivative. Nevertheless, the solubility of both flavanones was improved and, consequently, their antioxidant activity.

Liu investigated the inclusion complexation of *hesperetin* with HP-CD (Liu, 2012). Inclusion complexes were formed with HP-α-CD, HP-β-CD with different molecular weight and HP-γ-CD. The stability constants followed the rank order: HP-β-CD (MW 1540) > HP-β-CD (MW 1460) > HP-β-CD (MW 1380) > β-CD > HP-γ-CD > HP-α-CD. The stability constants were highest for the complex formed with the three HP-β-CDs, followed by β-CD and HP-γ-CD, while those formed with HP-α-CD had the lowest stability constants. Ks were 34 M^{-1} for HP-α-CD, 471 M^{-1} for β-CD 1343 M^{-1} for HP-β-CD (MW 1540) and 197 M^{-1} for HP-γ-CD.

The flavonoid *naringenin* (Figure 4), is one of the most abundant flavonoids in grapefruits and citrus fruits. Naringenin is mainly ingested from fruit juices. Naringenin plays a key role as an estrogenic substance in humans and as an endogenous regulator in plants, and is known as a safe natural product with various bioactive effects. It has been reported to have anticancer, antimutagenic, anti-inflammatory, antiatherogenic, antifibrogenic and free radical scavenging properties (Jayachitra, 2012). Naringenin induces cytotoxicity and apoptosis in various cancer cell line and its treatment at a similar dose showed no toxic effect on normal cells (Sabarinathan, 2011).

The formation of IC between naringenin and β-CD and its derivatives (HP-β-CD, DM-β-CD, M-β-CD and TM-β-CD) was also analysed by several authors (Ficarra, 2002; Shulman, 2011; Yang, 2013). For the CDs mentioned above the stoichiometry of the ICs was 1:1, and the HP-β-CD was the CD with higher stability constant and increased of 400 times the solubility of flavonoid (Shulman, 2011). Yang et al. demonstrated that the water solubility and thermal stability of the flavonoid increases when encapsulated by β-CD, DM-β-CD or TM-β-CD. Based on NMR analysis, they assumed that the C-ring of naringenin was interacting with the CDs' cavity (Yang, 2013). The biological effect of the IC naringenin/HP-β-CD was reported. Shulman and co-workers proved that the solubility of the flavonoid was increased 400 times when complexed with the HP-β-CD; moreover the naringenin transport across the model of the gut epithelium (Caco-2 cells) was also enhanced of 11-fold as well as its plasma concentration. Complexation with HP-β-CD increased the oral bioavailability of naringenin of 7.4-fold (Shulman, 2011).

Liu investigated the inclusion complexation of (2-hydroxypropyl)-cyclodextrins with naringenin and its 7-O-glycoside *naringin* (Figure 4) (Liu, 2012). The CDs analyzed were: HP-α-CD, HP-β-CD with different molecular weight, HP-γ-CD. For all two flavanones, the stability constants of their complexes formed with different CDs followed the rank order: HP-β-CD (MW 1540) > HP-β-CD (MW 1460) > HP-β-CD (MW 1380) > β-CD > HP-γ-CD > HP-α-CD. The ability of flavanones to form inclusion complex with (2-hydroxypropyl)-cyclodextrins was determined by both steric effect and hydrophobicity of the flavanone: the B ring of the flavanones was most likely involved in hydrogen bonding with the side groups in the cavity of the CDs. The stability constants were highest for the complexes formed with the three HP-β-CDs, followed by β-CD and HP-γ-CD, while those formed with HP-α-CD had the lowest Kc. Naringenin showed higher Kc respect to naringin. This is most likely attributable to the large molecular volume of naringin due to its two sugar moieties, which makes it difficult to enter the cavity of the CDs.

Naringin dihydrochalcone (Figure 4) is an intense sweetener, which is 500–700 times sweeter than sucrose and it is a derivative of natural flavonoid naringin. It possesses many desirable properties such as clean taste, low calorie, innocuity and safety, which allow it to be used in both food and pharmaceutical products. Apart from being used as a sweetener, its intense sweetness is also useful for masking bitterness in foods and medicines. Therefore, naringin DC has the capacity to function both as a non-calorie sweetener and an antioxidant with potential applications in functional foods and beverages, nutraceuticals and pharmaceuticals. However, the poor solubility and stability of naringin in aqueous systems severely restricts its applications in those products. The feasibility of using hydroxypropyl-β-cyclodextrin (HP-β-CD) to form a supramolecular inclusion complex with naringin dihydrochalcone to improve its solubility was investigated. Results showed clearly the formation of a supramolecular complex inwhich the guest molecule, naringin, was entrapped inside the cavity of the host, HP-β-CD. The close association between naringin dihydrochalcone and HP-β-CD resulted in changes in some of the characteristic spectral, phase transitional and morphological properties of naringin DC. Furthermore, ^{1}H-NMR analyses demonstrated that it was the B ring of naringin DC that was inserted into the HP-β-CD cavity to form the supramolecular complex.

The flavonoid *taxifolin* (Figure 4) is widely distributed in barks of the genus *Pinus* or *Larix* and in seeds of the genus Silybum. Several of the plants described as possessing taxifolin are used in traditional and clinical medicine; some are ingested in the human diet; and others are being studied for their potential use in drug development.

Taxifolin significantly dilates blood vessels, improves microcirculation, increases cerebral blood flow, and inhibits platelet aggregation activity. It has been widely used in the treatment of cerebral infarction and sequela, cerebral thrombus, coronary heart disease and angina pectoris (Shiko, 2009). In recent years, taxifolin has become a common antioxidant additive in the food industry, and it has been described as having antimicrobial activity, anti-inflammatory and analgesic properties, anti-adipogenic capacity, radical scavenger activity, and a protective role in plants against pathogens (Bais, 2003; Delporte, 2005). Taxifolin also affects the gastrointestinal tract by possessing antiulcer, antispasmodic, and antidiarrheal activities (Yang, 2011).

The three native CDs and HP-β-CD were used in the IC with taxifolin (Yang, 2011). The authors proved that the ICs enhanced the taxifolin water solubility and thermal stability, highlighting their use on healthcare products. Moreover, β-CD showed better capacity to interact with this flavanonol (Yang, 2011).The stability constant were 1872 M^{-1} for α-CD, 2145 M^{-1} for β-CD and 2908 M^{-1} for γ-CD. The inclusion stoichiometry of the taxifolin/CD complexes resulted 1:1. The possible inclusion modes of taxifolin was the insertion of the C ring of the flavonol into the cavity of the cyclodextrin, on the side of secondary rim.

Wogonin (Figure 4) is a hydrophobic active flavonoid isolated from the widely used herb in traditional oriental medicine called the Scutellaria baicalensis Georgia. It has exhibited many potential biological activities like antioxidant, antiviral, antithrombotic, antifungal, anti-anxiolytic, anti-inflammatory and recently anti-hepatitis B, neuroprotective and anti-cancer activities mainly by inducing cell differentiation, apoptosis and cell cycle arrest (Ma, 2000; Pal 2003).

However, despite the wide spectrum of pharmacological properties, Wogonin is restricted to clinical administration because of its low water solubility. For this reason, the inclusion complex wogonin-β-CD was investigated (Seridi, 2013). The most stable geometry presents a partial inclusion of Wogonin with the aromatic ring totally embedded in β-cyclodextrin hydrophobic cavity, while the di-hydroxy methoxy phenyl moiety is protruded out of the secondary rim. The IC presented a 1:1 stoichiometry.

2.2.6. Isoflavonoids

Isoflavones and isoflavonoids are diphenolic secondary metabolites in plants and are generated largely in root of these herbs (*Glycine max*, soy bean; *Phaseolus vulgaris*, bean; *Medicago sativa*, alfalfa; *Astragalus*, *Trifolium*, clovers; *Ononis*, restharrows). Since isoflavones have several therapeutically useful effects, e.g. estrogen-like, antioxidant, anti-inflammatory effect, platelet aggregation inhibition, etc., it has been not surprising that have been intensively studied since years (Kerry, 1998; Iovine, 2011). Two most investigated isoflavones are *genistein and daidzein* (Figure 5) both having affinity to estrogen receptors, so may play important role in therapy or prevention of different, estrogen related cancers or postmenopausal symptoms. Genistein and daidzein have rather limited water solubility so they need to be solubilized for nutraceutical and pharmaceutical purposes, to provide sufficient concentration of available monomeric forms. These isoflavones have been found suitable for encapsulation in cyclodextrin cavity (Crupi, 2007; Stancanelli, 2007; Cannavà, 2010).

The molecular interaction of isoflavones and cyclodextrin derivatives has already been studied by UV, circular dichroism and FTIR spectroscopic methods by Crupi et al. (Crupi, 2007) and Stancanelli et al. (Stancanelli, 2007). These studies focused on the molecular inclusion between isoflavones and highly soluble, non-aggregating, chemically modified β-cyclodextrins. Inclusion complex formation was proven by spectroscopic and solubility methods, and the existence of 1:1 molar stoichiometry in water was found. The host-guest interactions have been evidenced by monitoring the changes in the physical mixtures and complexes spectra of some guest molecule's bands (Stancanelli, 2007). The molecular interaction of HP-β-CD with genistein and daidzein was studied in aqueous medium by UV-vis and circular dichroism techniques. The stoichiometric ratios of complexes and stability constants were determined by phase-solubility. The reported Kc values (in order of magnitude 10,000 M^{-1}) (Stancanelli, 2007) indicated a very strong interaction between isoflavones and cyclodextrin derivatives.

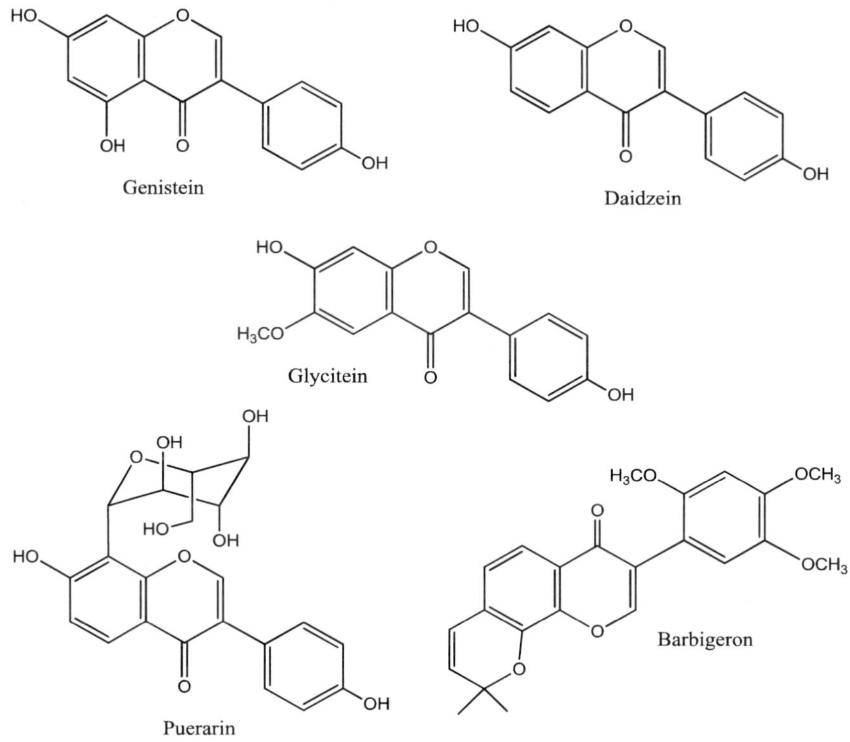

Figure 5. Chemical structure of isoflavanones.

The complexation of genistein with CD can improve its solubility and enhance its biological effects. Crupi and Daruházi (Crupi, 2007; Daruházi, 2013) tested the IC formation between this isoflavonoid and β-CD, γ-CD, HP-β-CD and RM-β-CD. The genistein was capable of interacting 1:1 with the 4 CD, but the CD derivatives induced a higher influence on the solubility of the compounds. The highest value for the stability constant was calculated for the complex with HP-b-CyD (Kc: 60518 M^{-1}), and then, in order, for Me-β-CD *Kc*: 34065 M^{-1}) and β-CD (Kc: 16401 M^{-1}).

The genistein and daizein capacity to cross biological membranes was also improved by the encapsulation (Daruházi, 2013). Determining enhancement in solubility and bioavailability, the author investigated the transport of these inclusion complexes across Caco-2 cell line comparing that of the pure compounds and found significant improving effect of the different CD derivatives on membrane permeation of the two isoflavone aglycons. Transport of genistein was enhanced by γ-CD the best followed by HP-β-CD, RAME-β-CD and β-CD, while in case of daidzein the isoflavone crossed the monolayer in the order of RAME- β-CD, β-CD, HP- β-CD and γ-CD. The microencapsulation of genistein and daidzein unambiguously improved their membrane permeation on Caco-2 monolayer.

Although chemical structures of these isoflavones are so similar, the position of hydroxyl groups has a key role for hydrogen bonds in the CD cavity, so the differences in the order of CDs helping their membrane transport may be caused by structural difference.

Yatsu et al. (2013) assessed the β-CD and HP-β-CD encapsulation capacity of a mixture of daidzein, genistein and *glycitein* (Figure 5) present in an isoflavone enriched fraction. All the isoflavones showed higher affinity to the HP-β-CD and the interaction between the CD and the bioactive molecules occur by introduction of the B-ring into the CD cavity, besides external interactions. In conclusion, the three isoflavones are able to simultaneously complex with HP-β-CD.

The phase-solubility diagrams were studied at 24°C and 37°C. In both temperatures the aqueous solubility of mixture of three isoflavones increased linearly with the concentration of HP-β-CD or β-CD. At 37°C the phase-solubility revealed that, even in wider range of cyclodextrin concentrations, the profile of the phase-solubility diagrams was not altered (AL type). However it resulted in a solubility increase and, on the other hand, a significant reduction of the Kc and complexation efficiency values for both cyclodextrins, especially for genistein. At 24 °C, Kc for HP-β-CD resulted: daidzein 1041 M^{-1}, glycitein 809 M^{-1} and genistein 775 M^{-1}.

Cannavà and Stancanelli evidenced the complexation of genistein with an host nanocarrier as amphiphilic cyclodextrin modified in the upper rim with oligoethylene hydroxyl groups (2-oligo(ethyleneoxide)-6-hexylthio)-β-CD, SC6OH) (Cannavà, 2010; Stancanelli, 2010).

SC6OH is a non-ionic amphiphilic CDs modified with small portions of PEG at the wider rim and hydrophobic chains of intermediate length at the narrower rim. It is highly versatile towards both encapsulation of drug with different polarity and controlled covalent modifications with receptor targeting groups (Quaglia, 2009). The special benefits of using this kind of CD include, among the others, the formation in aqueous medium of stable, potentially low immunogenic, nanoaggregates with an outer hydrophilic surface, that can direct the drug to the site of actions (i.e. solid tumors and cellular membranes).

By monitoring the most significant changes in the shape and position of the absorbance bands of the genistein functional groups, genistein is shown to be entangled in SC6OH nanoaggregates, establishing hydrogen bonding with the hydrophilic PEG chains (Cannavà, 2010). Gen turned out to be complexed starting from 1:1 molar ratio. Genistein turned out to be complexed starting from 1:1 molar ratio. The occurred Gen/SC6OH interaction, which increases isoflavone solubility in water from 3 mM to, at least, 30 mM.

Puerarin (Figure 5) is a natural isoflavone and one of the principal active components found in the traditional Chinese medicinal plant Ge-gen (*Radix puerariae*). Like many other isoflavonoids, puerarin has been found to exhibit strong antioxidant activity with potential beneficial effects to the human body. In recent years, the health effects of puerarin have received a considerable amount of research. Reported health-promoting functions of puerarin include antithrombotic, antihyperglycemic, cholesterol-lowing, anti-inflammatory, anticancer, heptoprotective, and antiosteoporotic effects and the ability to protect pancreatic tissues from oxidative damage and to reduce diabetes-induced damage to retina and improve retinal functions. Therefore, puerarin is emerging as a romising bioactive ingredient with potential applications in pharmaceutical, nutraceutical, and functional food products. However, puerarin is sparingly soluble in water (11 mM at 25°C), which severely restricts its application in medicinal and food formulations. Liu explored the formation of inclusion complex between puerarin and G1-β-CD to improve aqueous solubility (Liu, 2012). Results showed clearly that the process led to the formation of a 1:1 supramolecular complex in which the guest molecule, puerarin, was entrapped inside the cavity of the host, G1-β-CD. The molecule of puerarin enters the cone of G1-β-CD through the wide rim with part of its B ring extruding outside of the cone.

Barbigeron, a naturally occurring isoflavone, was first isolated from the seeds of Tephrosia barbigera in 1979 and could be found in other plants (Figure 5). It has been reported to have diverse pharmaceutical properties, including antioxidant, 15-lipoxygenase inhibitory activity, and anti-plasmodial activity against the malarial parasite *Plasmodium falciparum* (Yenesew, 2003). Barbigerone was found to effectively suppress tumor angiogenesis and tumor growth in vitro and in vivo. However, its aqueous solubility is almost nil, making it difficult for deep research. The complex with HP-β-CD notably increased the solubility: the solubility is only about 0.5 μg/μl in water. In contrast, the water solubility of barbigerone/HP-β-CD inclusion complex was remarkably increased to approximately 6.83 mg/ml. The apparent 1:1 stability constants was alculated to be 20,659 M^{-1}. The phenyl group (D ring) of barbigerone deeply inserted into the cavity of CD. Further more, preliminary in vitro cytotoxicity assay showed that IC still maintain the anticancer activity of barbigerone (Qiu, 2014).

2.2.7. Anthocyanins

Anthocyanins are potentially very interesting as natural food colorants due to their ability to impart vibrant colors. These polyphenolic compounds provide colors ranging from salmon-pink to red and violet, they are water-soluble pigments which simplifies their incorporation in aqueous food systems and do not pose any particular problems in terms of food safety. On the contrary, these compounds are antioxidant and were described to have some putative beneficial health effects (Clifford, 2000). However, the application of anthocyanins as food colorants requires improvement of their chemical stability since they are very unstable and susceptible to be transformed in food matrices during storage. The fact that anthocyanins co-exist in aqueous solution in a complex network of equilibrium species depending on pH is an important factor related to the color displayed by these compounds. The formation of inclusion complexes between cyclodextrins and anthocyanins was found to improve the anthocyanin stability and prevent losses during storage (Chandra, 1993; Mourtzinos, 2008; Tamura, 1998). However, in several studies an anti-copigmentation effect was described, resulting from the better inclusion of the colorless forms into the cyclodextrin cavity (Dangles, 1992; Lewis, 1995; Yamada, 1980).

Yamada (Yamada, 1980) reported a first study on the interation between anthocyanins and CDs. The authors evidenced the IC between α- or β-CD and three anthocyanins: *pelargonidin-3-glucoside* (callistephin), *cyanidin-3-glucoside* (chrysanthemin) and *delphindin-3-(4-(p-coumaroyl)-L-rhamnosyl-(1,6)-glucosido)-5-glucoside* (nasunin) (Figure 6). From absorbance results, nasunin did not interact with cyclodextrins, and other two compounds showed a fading effect more evidenced with β-CD.

Fernandez studied the ability for inclusion complex formation between the different equilibrium forms of *cyanidin-3-O-glucoside* (Figure 6) and β-CD in aqueous solution and the structure of these inclusion complexes (Fernandez, 2014). NMR analysis and molecular dynamics simulation performed in this study have clearly proved the selective inclusion of the anthocyanin hemiketal form into the CD cavity. The A- and C-nucleus of cy-3-glc provide the major contribution to the binding driving force, followed by the B-ring, whilst its glucose residue seems to perturb the cy-3-glc's inclusion in the β-CD and is facing the solvent.

The thermal stability of anthocyanin extract isolated from the dry calyces of *Hibiscus sabdariffa* L. was studied over the temperature range 60-90 °C in aqueous solutions in the presence or absence of β-CD (Mourtzinos, 2008). The extract contains two anthocyanins, *delphinidin-3-sambubioside* and *cyanidin-3-sambubioside*. The presence of β-CD, improved the thermal stability of nutraceutical antioxidants present in *H. sabdariffa* L. extract, both in solution and in solid state. In the presence of β-CD, anthocyanins degraded at a decreased rate. The formation of complexes in solution was confirmed by NMR studies of β-CD solutions in the presence of the extract: the presence of many cross-peaks due to intermolecular interactions of the cavity with the aromatic and double bond hydrogens (ring B and C), whilw ring A is not included. Moreover, differential scanning calorimetry revealed that the inclusion complex of *H. sabdariffa* L. extract with β-CD, in the solid state was more stable against oxidation as compared to the free extract, as the complex remained intact at temperatures 100-250 °C where the free extract was oxidized.

2.3. Polyphenols Non-Flavonoids

Non-flavonoids include all the other polyphenolics from the simplest phenolics and phenolic acids to the phenolic complex tannin. This non-flavonoids group incorporates phenolic acids, quinines, stilbenes, tannins, and coumarins. The non-flavonoids had great importance on the pharmaceutical and cosmetic industry due to their biological properties.

However, their application is limited because these polyphenolics are very susceptible to degradation by environmental factors such as light, temperature and pH. In order to overcome this, some works have been published regarding the inclusion of non-flavonoids into CD.

2.3.1. Phenolic Acids

Although phenolics and phenolic acids are single-substituted phenolic rings, they are capable of numerous biological effects. In fact, some authors have related the number of hydroxyl groups on the aromatic ring with their antimicrobial activity: the higher the number of hydroxyl groups the higher is the level of toxicity exerted to microorganisms. Phenolic acids include derivatives of benzoic acids, i.e. (C_1-C_6) hydroxybenzoic acids, and derivatives of cinnamic acid, i.e. (C_3-C_6)- hydroxycinnamic acids. Gallic, p-hydroxybenzoic, and ellagic acids are representatives of the first group, and frequently occur in the form of glucosides. These compounds are water-soluble and sensitive to temperature, elevated pH values, oxidation and light. Caffeic, ferulic and p-coumaric acids belong to the class of hydroxycinnamic acids, and are rarely found in the free form; indeed, they are usually present in nature as simple esters with hydroxy carboxylic acids or glucoses. The hydroxycinnamic acids are also sensitive to oxidation and high pH values, but they are poorly soluble in water.

Phenolic acids have been described in the literature as molecules with effective antimicrobial activity, particularly as fungicides. Their anti-microbial activity critically depends on their chemical structure, especially on the number and position of the substitution in the benzene ring, and also on the saturated chain length. The microbicidal capacity was found to be increased with augmentation of the alkyl chain length (Cueva, 2010; Sánchez-Maldonado, 2011).

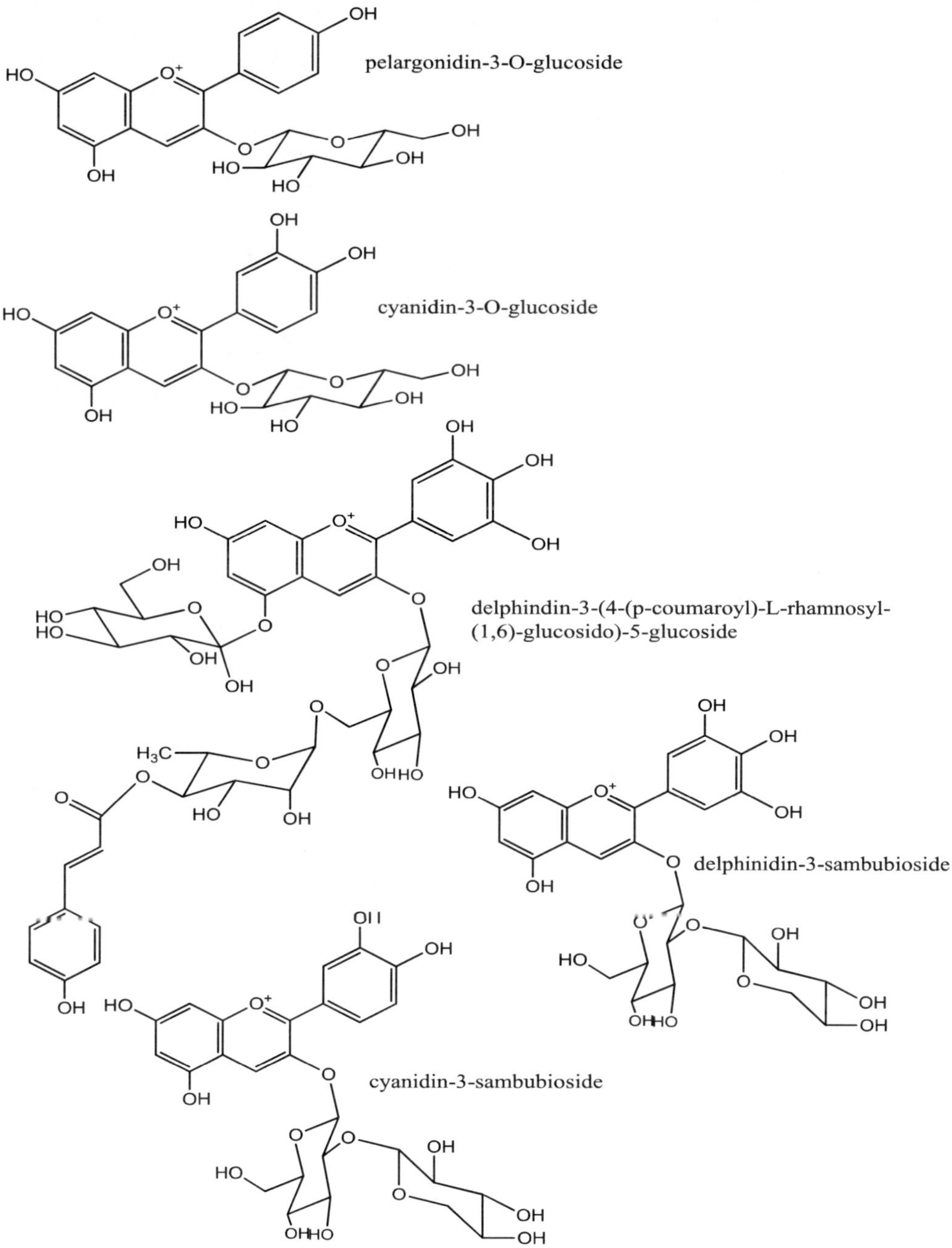

Figure 6. Chemical structure of anthocyanins.

Ferulic acid (Figure 7) is an extremely abundant hydroxycinnamic acid of low toxicity in the plant kingdom, which is synthesised from the shikimate pathway from phenylalanine or L-tyrosine and occurs mainly as a trans isomer. It plays important roles in plant cell walls by cross-linking with polysaccharides, including protection against pathogen invasion, and control of extensibility of cell walls and cell growth. It is one of the effective components of Chinese medicinal herbs such as *Angelica sinensis*, *Cimicifuga heracleifolia* and *Lignstcum chuangxiong*, which has been reported to have physiological functions, including antioxidant, anti-carcinogenic, antimicrobial, anti-thrombosis, anti-inflammatory, and anticancer properties (Ou, 2004). It also protects against coronary disease, lowers cholesterol and increases sperm viability.

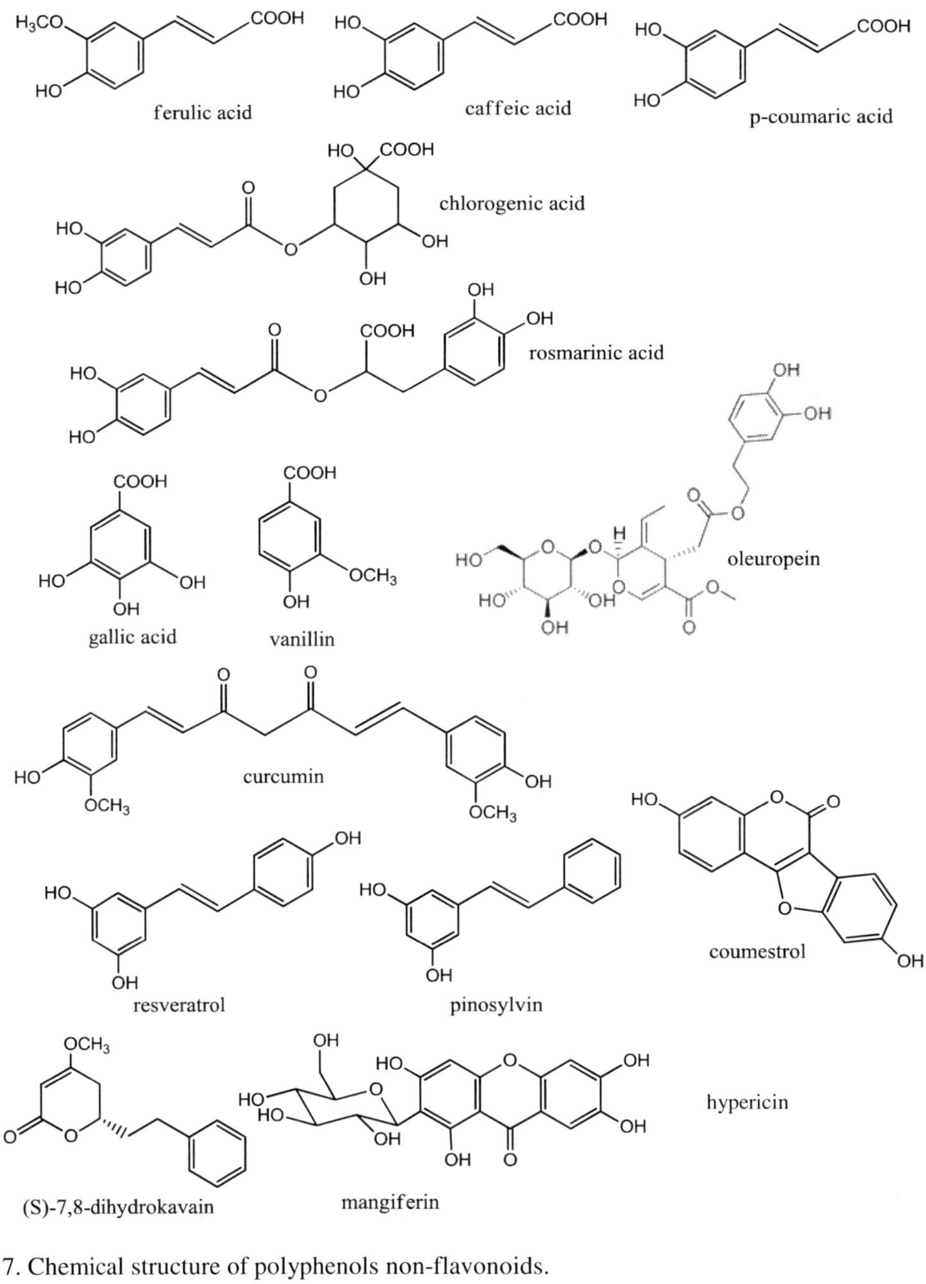

Figure 7. Chemical structure of polyphenols non-flavonoids.

Despite the applicable qualities and biological activities, the therapeutic usefulness of ferulic acid are limited because of its unfavourable physicochemical properties, especially very poor water solubility and low oxidative stability (Piber, 2005).

Anselmi (Anselmi, 2008) investigated the IC formation between ferulic acid and α-CD. Thus, the authors proposed the inclusion of this phenolic acid into the α-CD, with a Kc value of 1162 M^{-1} and equimolecular complexation. Based on their results, the α and β unsaturated part of the ferulic acid and part of its aromatic skeleton were inside the hydrophobic cavity of the CD. The IC increased the ferulic acid resistance to the degradation by UVB and also, decreases its rate release.

The same phenolic acid was used by Casolaro for the complexation with α-CD (Casolaro, 2005). The stoichiometry of the IC was the same and they assumed that the ferulic acid was neutralised inside the CD cavity. The cutaneous permeation and distribution through skin of the IC ferulic acid and α-CD was assessed by Monti et al. (Monti, 2011). They concluded that the IC prevented the formation of the less active cis-isomer of ferulic acid and its degradation by UV light. The ferulic acid/α-CD IC presented lower penetration on the skin which enlarge the skin protection against UV damages, since the ferulic acid remains at the skin surface (Monti, 2011). The encapsulation of ferulic acid with HP-β-CD was also studied (Wang, 2011). Ferulic acid was able to form an inclusion complex with HP-β-CD, and the ferulic acid/HP-β-CD inclusion compounds exhibited a 1:1 stoichiometry. The calculated apparent stability constant was 166.3 M^{-1}, and the water solubility was significantly improved by phase solubility studies, about 15-fold. Moreover, the irradiation-induced decomposition of ferulic acid in aqueous solution was markedly reduced by complexation with HP-β-CD.

Caffeic acid (Figure 7) shows antibacterial and antioxidant activities, but its utilization may be jeopardised by its sensibility to oxidation and lower solubility (Zhang, 2009). Thus, some authors had described its encapsulation with CD to overcome these issues. Górnas, (Górnas, 2009) and Divakar (Divakar, 1997) complexed this phenolic acid with β-CD. In both cases, the experimental results suggest a 1:1 IC with K of 270 and 516 M^{-1}. The hydroxyl groups of the phenolic acid were trapped inside the CD cavity and the carboxyl moiety was projected outwards the CD (Divakar, 1997; Górnas, 2009). Górnas studied the influence of the pH on the IC formation, and concluded that the K decreased with the presence of caffeic acid charged species. The caffeic acid was, also, encapsulated by HP-β-CD to increase solubility (Zhang, 2009). The IC formation was better in acid conditions and the IC ratio was 1:1; also, the caffeic acid solubility increases. The lipophilic aromatic ring and ethylene portion of the caffeic acid was entrapped inside the CD cavity and the polar groups were outside the HP-β-CD cavity (Zhang, 2009).

Chlorogenic acid (Figure 7) is naturally present in many fruits, such as apple and plum. Its highly antioxidant activity has attracted much attention from researchers. Moreover, some in vitro and in vivo pharmacological activities of chlorogenic acid, such as hypoglycaemic, antiviral, hepatoprotective and immunoprotective activities, have also been reported (Perrone, 2008). However, chlorogenic acid is labile when exposing to air.

Cyclodextrin inclusion complex of chlorogenic acid is an interesting attempt to improve the stability under ambient conditions (Zhao, 2010). β-CD was used to form IC with chlorogenic acid. Quinic acid moiety was revealed to be included within the β-CD cavity, while caffeic acid moiety located outside. The storage stability of chlorogenic acid was apparently improved by measurement of absorbance change during 20 weeks. Through determinations of DPPH radical scavenging and antimicrobial activities, no significant difference was observed between chlorogenic acid and its complex.

Rosmarinic acid (Figure 7), a hydroxycinnamic acid with high antioxidant properties and poor solubility, was encapsulated with α-CD, β-CD, HP-β-CD, HE-β-CD and M-β-CD (Celik, 2011) in order to improve both properties. By the observation on the UV–Vis spectrum of the rosmarinc acid and the ICs, the authors assessed the stoichiometry of all ICs (1:1) and the Kc values. They reported that ability forming stables IC was as follows M-β-CD (328 M^{-1}) > HP-β-CD (267 M^{-1}) > HE-β-CD (168 M^{-1}) > β-CD (164 M^{-1}) > α-CD (82 M^{-1}). Additionally, the antioxidant activity of the ICs was higher than the rosmarinic acid alone (Celik, 2011).

Gallic acid (Figure 7) is a hydroxybenzoic acid biosynthesized from precursors of the shikimate pathway by the enzyme dihydroshikimate. It is present in great abundance in the plant kingdom, especially in berries, citrus fruits, cereals, tea, wine and herbs. Gallic acid occurs naturally both in the free form and as an ester or salt. Gallic acid has several biochemical properties, as antioxidant, an antimicrobial agent, and an antihyperglycaemic agent; it also prevents oxidative stress and some kinds of cancer (Ho, 2010). Gallic acid retains these properties even at low amount (between 5 and 20 mg), but it is unstable at extreme temperatures or in the presence of oxygen or light, conditions which are common in food processing and storage. It was encapsulated by lyophilization in β-CD. The encapsulated gallic (encapsulation efficiency 80%) acid showed no loss of antioxidant capacity and different characteristics from the pure gallic acid (da Rosa, 2013).

The formation of ICs of the antioxidants gallic acid, and its hydrophobic derivatives ethyl (EG), propyl (PG) and butyl (BG) gallate with β-CD were reported (Martínez-Alonso, 2015). Results show that addition of increasing amounts of β-CD to solutions of gallic acid and EG do not modify their spectral characteristics, suggesting that the formation of inclusion complexes is negligible or very weak. On the contrary, the spectral properties of propyl and butyl gallates are modified in the presence of cyclodextrin, suggesting the formation of inclusion complexes and their binding constants were determined from the variations of the absorbance at increasing β-CD. The obtained value for BG, Kc: 200 M^{-1}, is higher than that for PG, Kc: 105 M^{-1}, as a consequence of the increase in the alkyl chain of the derivative.

Coffee brew is rich in chlorogenic and cinnamic acids, potent antioxidants. Its interaction with β-CD leads to formation of inclusion complexes, which affect the physicochemical properties of the guest molecule. Górnas investigated spectral changes in coffee-originated phenols, chlorogenic and caffeic acids (caffeic acid, ferulic acid and p-coumaric and o-coumaric acids), due to complexation with β-CD in aqueous solutions and studied thermodynamic properties of the complexes formed. The data indicate the formation of a 1:1 complex between β-CD and chlorogenic or caffeic acids. Inclusion constants at 20°C were 424, 278 and 128 M^{-1} for chlorogenic acid, caffeic acid and coffee-originated phenols, respectively (Górnas, 2009).

Vanillin (Figure 7) is widely used as a flavour and fragrance and it is also used as a food preservative due to its antioxidant properties in food industry (Karathanos, 2007). The source of vanilla is the bean or pod of the tropical Vanilla orchid (principally Vanilla planifolia Andrews, syn. V. fragrans).

However, vanillin has a short shelf-life because of its volatile nature, therefore, the stabilization of vanillin is very important for its prolonged functionality.

Divakar carried out a detailed structural study of the inclusion of vanillin and β-CD. Based on its studies vanillin displays a phenol part within the cavity of CD and the aldehyde projecting outside. The stoichiometry of the complex was found to be 1:1 (Divakar, 1997), as also reported by Karanthanos. Karathanos prepared the complex with β-CD by using freeze-drying method. The complex is protected towards oxidation since it remains intact at temperatures where the free vanillin is oxidizing (Karanthanos, 2007).

A binding constant value of $1.11 \times 10^4\ M^{-1}$ was determined for the 1:1 complex (Divakar, 1990). NMR experiments revealed that vanillin exists with the phenolic end nearer to the narrower end and the aldehyde end to the wider end of the torusshaped β-CD molecule.

Kayaci reports the formation of solid ICs with three types of CDs (α-CD, β-CD, and γ-CD) with the aim to enhance the thermal stability and sustained release of vanillin (Kayaci, 2011). ^1H-NMR study elucidated that the complexation stoichiometry for both vanillin/β-CD IC and vanillin/γ-CD IC was a 1:1 molar ratio, whereas it was 0.625:1 for vanillin/α-CD IC. In the case of vanillin/α-CD IC, the sample was mostly amorphous and some uncomplexed vanillin was present, suggesting that α-CD was not very effective for complexation with vanillin compared to β-CD and γ-CD. The complexation increased the thermal stability of vanillin. The amount of vanillin released with increasing temperature was lowest for vanillin/γ-CD IC and highest for vanillin/α-CD IC, suggesting that the strength of interaction between vanillin and the CD cavity was in the order γ-CD > β-CD > α-CD for solid vanillin/CD ICs.

Inclusion complexes of o-, m- and p-*coumaric acid* (Figure 7) with β-CD were prepared in stoichiometric ratios (1:1) and stability constants and antioxidant activity of the complexes were calculated. The apparent stability constants in aqueous solution of $0.39 \times 10^3\ M^{-1}$, $2.81 \times 10^3\ M^{-1}$ and $49 \times 10^3\ M^{-1}$ for o-, m- and p-CA complexes, respectively, were determined by phase solubility tests (Străzĭsar, 2008).

2.3.2. Other Polyphenols

Curcumin (Figure 7), a hydrophobic yellow-orange polyphenol derived from the rhizome of the herb *Curcuma longa*, is an important natural colourant, used in food and pharmaceutical preparations (Zhang, 2013). In the food industry, curcumin is added as a stabiliser in jellies or is used as a natural colourant, as a substitute for artificial colourants in cheeses, pickles, mustards, cereals, soups, ice creams and yogurts. Curcumin is also a very interesting pharmacological compound due to its innumerous pharmacological applications, including anti-inflammation, anti-human immunodeficiency virus, anti-microbial, anti-oxidant, anti-parasitic, anti-mutagenic and anti-cancer (Mohan, 2012). It is considered safe for human use, even in high doses (Singh, 2010). The applications of curcumin are limited due to its low water solubility and sensitivity to alkaline conditions, thermal treatment, light, metallic ions, enzymes, oxygen and ascorbic acid. Additionally, curcumin is poorly absorbed in the gut, independent of the route of administration, which limits its bioavailability. These factors usually restrict the application of curcumin in the food industry and in pharmaceutical formulations. Thus, an improvement in the stability and solubility of curcumin is necessary.

Curcumin was complexed with β-CD (Mangolim, 2014). The possible complexation had an efficiency of 74% and increased the solubility 31-fold. Curcumin/β-CD complex exhibited a sunlight stability 18% higher than the pure colourant. This material was stable to pH variations and storage at -15 and 4 °C. With an isothermal heating at 100 and 150 °C for 2 h, the material exhibited a colour retention of approximately 99%. The application of curcumin/β-CD complex in vanilla ice creams intensified the colour of the products and produced a great sensorial acceptance.

The encapsulation by the native CD was described by Patro (Patro, 2013). They were able to improve the solubility and oral availability with all the three IC; however the α-CD was the one that showed higher Kc (1124 M^{-1}). López-Tobar et al. (López-Tobar, 2012) also applied to β-CD and γ-CD as curcumin drug carriers. The large cavity of γ-CD was more efficient in the IC formation. Moreover, both CDs were able to form 2:1 IC with this phenolic compound and the molecular interaction proposed was that the aromatic rings and the hydrogen bonds were involved and a change occurs from the curcumin planar ketoenol form to non-planar diketo. The chemical stability and bioavailability may be upgraded by this conformational alteration (López-Tobar, 2012).

The same stoichiometry of the IC β-CD/curcumin was observed by Tang and Rahman, (Tang, 2002; Rahman, 2012). In both situations, the solubility of curcumin was described as well as the same molecular interaction, as referred above. Dandawate (Dandawate, 2012) used a synthetic form of curcumin but accomplished the same stoichiometry and solubility. In this work, the anti-carcinogenic, systemic bioavailability and tissue distribution of the IC β-CD/synthetic curcumin were compared with the synthetic curcumin alone and concluded that they have been improved by the encapsulation (Dandawate, 2012).

CD derivatives were also tested as drug carriers of curcumin, in order to overcome the difficulties of its application as anti-carcinogenic agent. For instance, 2 molecules of HP-β-CD formed stable IC (Kc 5000–62000 M^{-1}) with 1 molecule of curcumin (Tomren , 2007; Ghanghoria, 2012; Mohan, 2012). The transdermal capacity of curcumin was raised by the complexation with HP-β-CD as well as the decrease of skin irritation (Ghanghoria, 2012). Besides, HP-β-CD, HP-α-CD and HP-γ-CD were used with the same goal. Mohan et al. (2012) compared the encapsulation of the three CD derivatives and reported that the encapsulation may occur both in 1:1 and 2:1 stoichiometry and the HP-γ-CD has a better complex-ion capacity. This IC HP-γ-CD/curcumin was capable of reducing cell proliferation and increases the apoptosis of cancer cells by interfering in the protein production (Rocks, 2012).

Tønnesen reported the encapsulation of curcumin with several CD derivatives, namely HP-α-CD, RM-β-CD, HP-β-CD, SBE-β-CD, HTA-β-CD and HP-γ-CD (Tønnesen, 2002). The greater positive change on the stability and solubility of curcumin was achieved by the complexation with the RM-β-CD and HP-γ-CD, probably due to the more hydrophobic environment found inside the cavity of the first CD and large cavity of the second leading to a better accommodation of the bioactive molecule. The increase of the curcumin resistance to hydrolysis under alkaline environments when encapsulated with these CDs, was also proved (Tønnesen, 2002).

Recently, diamide linked γ-cyclodextrin (γ-CD) dimers (Figure 8a and 8b) are used to capture curcumin and suppress its decomposition in in the aqueous environment at pH 7.4 and 37 °C (Harada, 2011). In this study, succinamide and urea linked γ-CD dimers (Figure 8a and 8b) joined through the C6 carbon on each γ-CD are used. The diamide linker is expected to be hydrolyzed enzymatically in the intracellular environment, thereby releasing the encapsulated materials. The γ-CD dimers showed a remarkable ability to suppress the decomposition of curcumin and extend its half-life from less than 30 min to greater than 16 h. The 1:1 association of curcumin with dimers has high stability constants of $8.7x10^6$ M^{-1} and $2.0x10^6$ M^{-1}, for succinimide and urea linker, respectively. The stability was ameliorated and the half-life was increased by at least 180-750-fold.

The linked γ-CD dimers also demonstrated an ability to deliver and release curcumin to membranes and furthermore to the intracellular milieu. In the study, sodium dodecyl sulfate micelles were used as a model membrane system.

The potential efficacy of the same linked γ-CD dimers for intracellular curcumin delivery to cancer cells was evaluated (Harada, 2013). Cellular viability assays demonstrated a dose-dependent antiproliferative effect of curcumin in human prostate cancer cells that was preserved by the succinamide linked γ-CD dimer.

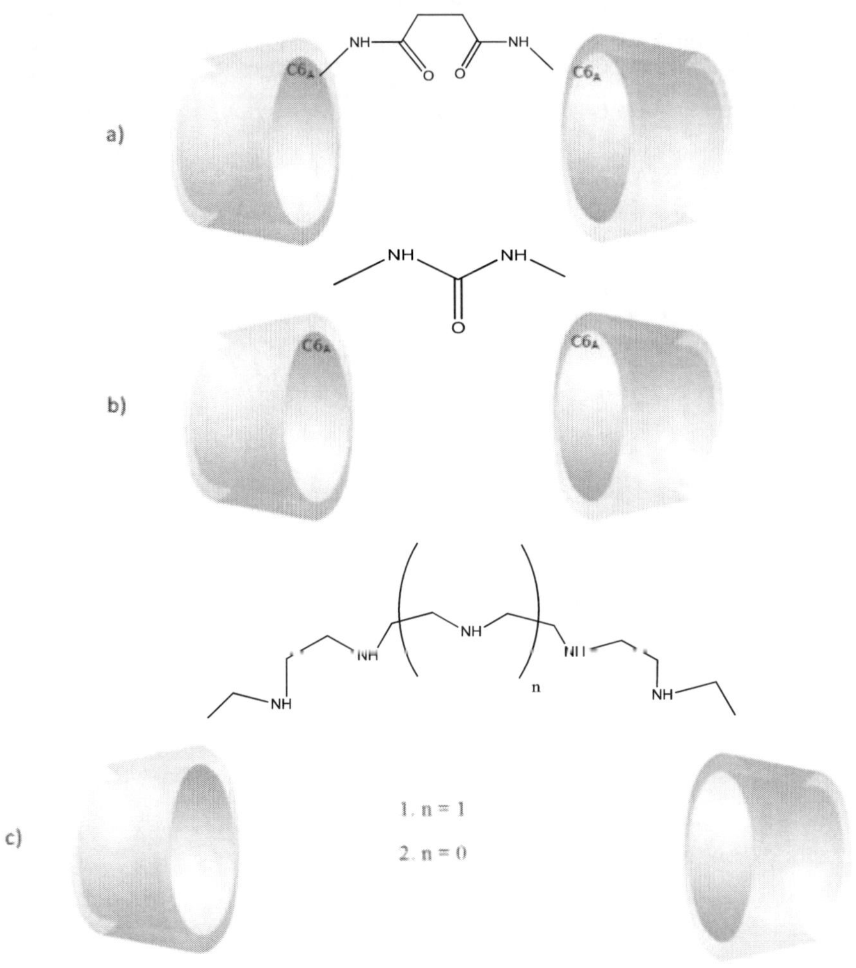

Figure 8. Chemical structures of diamide linked γ-cyclodextrin dimers (a: succinamide linker; b urea linker) and oligo(ethylenediamino) bridged bis(β-cyclodextrin)s 1 and 2(c; adapted from Harada, 2011; Harada, 2013 and Liu, 2004).

In contrast, delivery of curcumin by urea linked γ-CD dimer significantly delayed the antiproliferative effect. These delivery agents offer molecular-scale encapsulation of curcumin at concentrations ranging from micromolar to millimolar and a high structural integrity under physiological conditions. While encapsulation of curcumin using linked γ-CD dimers strongly suppresses its degradation, cellular viability and uptake assays combined with gene expression and fluorescence microscopy reveal that linked γ-CD dimers alone do not produce measurable toxicity on viability or gene expression. The two linked γ-CD dimers are both effective means to deliver curcumin into cells, resulting in inhibition of cellular proliferation, while initiating changes in gene expression similar to that exhibited by curcumin alone. Furthermore, the ICs appears to protect curcumin from binding to BSA, which may result in a more efficient intracellular delivery via cell membranes.

A novel drug nanocarrier system was formulated by encapsulating curcumin and paclitaxel in poly(β-cyclodextrin triazine) (PCDT) for the therapy of four cancer models; ovarian, lung, prostate, and breast cancer (Boztas, 2013). After complexation of the curcumin with a concentration of 16% (w/v) of PCDT, solubility increased up to 35.5 mg/L, approximately 3500 times higher than the original value, which is about 0.01 mg/L. The stability constant resulted 1126 M^{-1}, which is higher than the stability constant of β-CD (134 M^{-1}) and HP-β-CD (424 M^{-1}). Cell viability and colony formation assays revealed enhanced curcumin cytotoxicity upon complexation. Annexin V apoptotic studies showed that the PCDT complexation improved curcumin induced apoptosis in human ovarian cancer cell lines A2780 and SKOV-3, human-nonsmall cell lung carcinoma cell line H1299, and human prostate cancer line DU-145. The bioactivity of combining curcumin and paclitaxel was also investigated. A synergism was found between curcumin and paclitaxel, particularly when complexed with PCDT on A2780, SKOV-3, and H1299 cancer cell lines.

The common phytoestrogens found in soybean are isoflavones and coumestrol. *Coumestrol* (Figure 7) has been reported to have 30–100 times the estrogenic activity of the isoflavones. Coumestrol is a coumarin-like compound with a close structural relationship to stilbestrol as well as to the natural estrogen, estradiol. It has been shown to have receptor dependent anti-estrogenic activity, and its consumation has been associated with a number of beneficial effects such as reduced risk of breast cancer, prostate cancer, heart disease, and a decrease in menopausal symptoms. For the properties above described, Coum finds a wide range of application in the pharmaceutical field. However, its employment is limited by its low water solubility and slow dissolution rate. A possibility to overcome these limitations, and, hence, to provide improved therapeutic options for patients, is the complexation of the free drug with β-CD and HP-β-CD (Cannavà, 2008).

Phase-solubility measurements were performed to evaluate in solution the complexation of the two cyclodextrins, i.e. β-CD and HP-β-CD. The stoichiometry of the complexes was assumed to be 1:1 and stability constants of the coumestrol/β-CD and coumestrol/ HP-β-CD complexes were Kc: 827 M^{-1} and Kc: 1125 M^{-1} respectively. The aqueous solubility of coumestrol has been substantially augmented (about 80%) via complexation with two cyclodextrins.

Hypericin (Figure 7) is a natural polycyclic quinone from *Hypericum perforatum*, commonly known as St John's wort. This natural product is an effective antidepressant and anxiolytic, an antiretroviral agent and a potent photosensitizer suggested for use in photodynamic therapy of cancer. However, Hypericin is very lipophilic and water-insoluble, which makes intravenous injection problematic and restrains its medical applications.

A water-soluble inclusion complex of hypericin with β-CD polymer (CDP) was achieved by supramolecular interactions between hyperforin and CDP. CDP was synthesized by cross-linking β-CD with epichlorohydrin. Compared with hypericin, the water-solubility of hypericin-CDP was greatly enhanced (1.864 gL^{-1}) because of the water-soluble CDP host. The mole ratio of β-CD unit in CDP to hypericin was determined as 2:1. At 25 °C, the dissociated constant of hypericin-CDP was measured as 1.47×10^{-7} mol L^{-1} by UV–vis spectroscopy. In the formation of inclusion complexes, CDP could overcome the β-CD drawbacks – such as the poor water-solubility and the restriction of single cavity size, indicating it was able to use as a universal solubilizer for pharmaceutical application (Zhang, 2013).

Mangiferin (Figure 7) is a naturally occurring polyphenol in several fruits, one being *Mangifera indica* L. (common name: mango), one of the most popular tropical fruit-bearing trees in the world. The interest in mangiferin stems from its wide range of biological actions, for instance, gastroprotective, analgesic, antibacteria) together with cytoprotective (Carvalho, 2009; Duang, 2011). The therapeutic potential of mangiferin has been investigated in the prevention and treatment of periodontitis. A wide spectra of these properties have been attributed to its antioxidant properties. However, the molecules has poor bioavailability, due to its low solubility in water. The complex wiyh β-CD increased its antioxidant activity. Lipid peroxidation assays showed that the complex has a protective effect on the membrane that is as effective as the positive control (Trolox) (da Rocha Ferreira, 2013).

Oleuropein (OLE) (Figure 7) is a major constituent of the olive leaf and green olives (*Olea europaea*). It is an intensely bitter glycoside, is present mainly in green olives and its structure is specified as being that of a heterosidic ester of elenolic acid and dihydroxyethanol. It has antioxidant and therapeutic properties and is valued as a medicinal component. The molecule is a complex phenolic compound constituted by a heterosidic ester of the elenolic acid and dihydroxy ethanol.

However, as oleuropein is an antioxidant, it is easily decomposed when exposed to light, humidity and heat, and one of the possibilities of stabilization of this molecule is by molecular inclusion with cyclodextrins. The IC with natural cyclodextrins were studied (Barão, 2014). From absorbance values data and by appropriate mathematical modeling, the resulting complexation constants are: for OLE/α-CD Kc: 1352.4 M^{-1} and for OLE/β-CD Kc 1827.9 M^{-1}. These results show that OLE has a greater affinity for β-CD than for α-CD and given the relatively high constants, OLE/β-CD complexes have potential for giving longer shelf lives for OLE in in food and pharmaceutical applications. The best preparation was obtained with complexation of oleuropein and β-CD at 1:2 molar ratio.

Efmorfopoulou reported a 1:1 complex for β-CD. The IC produced more haze when complexed with β-CD, significantly less when reacted with α-CD, while complexation with γ-CD gave intermediate values. In all cases, it appears that the amount of the guest molecules complexed with the three CDs has a linear relationship with the time of reaction, reaching a plateau in 1 hour in the case of α- and β-CD and in 2 hours in the case of γ-CD. The guest molecule has much room to move inside the wider cavity of γ-CD and consequently it interacts less with the walls of the cavity. It should also be pointed out that the cavity of γ-CD is so wide that it can accommodate many water molecules, and that their properties resemble water molecules in the bulk of the solvent. As a result, the driving force of complex formation in the case of γ-CD is reduced and consequently the phenomenon of complexation is slower compared to that of the two other CDs (Efmorfopoulou, 2004).

2.3.3. Stylbenes

Resveratrol (Figure 7), a polyphenol, has been found in a wide variety of about 70 plant species, including grapes, mulberries, and peanuts, which could protect against a range of illnesses and diseases, e.g., cardiovascular diseases, cancer, viral attacks, neurodegenerative diseases, and ageing (Delams, 2005; Olas, 2005). However, due to its low water solubility, resveratrol cannot be used for oral administration. This stilbene displays a hydrophobic behaviour, and is also extremely affected by exposure to oxygen, light, and oxidative enzymes, reducing its bioactivity. This limitation could be overcome by the formation of inclusion complexes of the drug with cyclodextrins to protect resveratrol and to increase its solubility, stability and bioactivity was applied in several studies (Li, 2010; Lu, 2009; López-Nicolás, 2010).

The interactions of three native CDs and M-β-CD with resveratrol in aqueous system and the influence of CDs on the thermal stability of the resveratrol was reported by Li et al. (Li, 2010). The results indicated that all of the complexes formed in the aqueous solutions are in 1:1 stoichiometry. The IC formation was favoured by the temperature rise.The binding processes of α-, β- and M-β- cyclodextrin with the guest are mainly driven by enthalpy, while that of γ-cyclodextrin with the drug is driven by both enthalpy and entropy. The encapsulation of γ-CD and resveratrol was the most stable owing to the better fit between the phenolic compound and the CD cavity, since γ-CD had the biggest cavity.

A similar work used the native α-CD and β-CD and 2 derivatives (HP-β-CD and DM-β-CD) to increase solubility and stability. It was observed that the IC with native CD was only capable of complexing with part of the resveratrol molecule and that the HP-β-CD offered a cavity with a better fit to the bioactive molecule (Kumpugdee-Vollrath, 2012).

The encapsulation ameliorates also anti-oxidant and anti-carcinogenic properties of resveratrol. For instance, Lu and co-workers used β-CD and HP-β-CD as resveratrol carrier agents and described the betterment of the scavenging capacity of the IC (Lu, 2009), the inhibition of the lipid peroxidation activity and the cytotoxicity to cancer cells without harming the healthy ones (Lu, 2011). The results of these studies supported the notion that the CD derivative forms a stronger IC with this stilbene (Kc β-CD 1815 M^{-1} or 4317 M^{-1}) as reported by Lucas-Abellán (2007) and Kc HP-β-CD 6778 M^{-1}, related to the easier access of the resveratrol to the HP-β-CD cavity due to enlargement of the cavity opening and the despairing of the intramolecular hydrogen bond network (Lu, 2011 and 2009). Additionally, the employment of HP-β-CD for resveratrol encapsulation also increases its photostability. The host position of the 'guest' molecule inside the CDs did increase the resistance to the degradation by UV radiation without affecting its antioxidant properties. Therefore, the resveratrol/HP-β-CD ICs represent a powerful candidate for the protection of skin against oxidative stressing episodes (Sapino, 2008).

The solubility and antioxidant capacity of resveratrol was increased also by the complexation with G_2-β-CD. Besides the solubilisation, CD was capable of protecting the phenolic compound from rapid oxidation, by entrapping it inside its cavitiy, with a Kc of 5130 M^{-1} (Lucas-Abellán, 2007).

The presence of HP-β-CD can modified the monomer/aggregate equilibrium which is dependent on the protonation state of (E)-resveratrol (López-Nicolás, 2010). The CD reduced the aggregation of the (E)-resveratrol molecules, producing individual molecules of the solute and preventing side effects due to aggregation phenomena. At pH values higher than the pK_{a1} of (E)-resveratrol, the coexistence of different protonated/deprotonated forms of this antioxidant does not permit the fluorimetric determination of the complexation constant of (E)-resveratrol with HP-β-CD. However at physiological pH, this problem was resolved, with a K_C value of 14490 M^{-1} and the stoichiometry of the complexation process was 1:1 (Lu, 2009; López-Nicolás, 2010).

Pinosylvin (Figure 7) is a naturally occurring stilbene present in the wood pulp of pine and eucalyptus trees and is present in tea oils and herbal remedies. Several healthy benefits have been attributed to pinosylvin in recent years. Among its pharmacological properties are a wide range of biological activities, including antimicrobial, antifungal, anticancer, antiinflammatory, antioxidative, and antibacterial properties (Roupe, 2006; Lee, 2005). In recent years, some papers have been published about the antimicrobial properties of a new type of stilbenoid, pinosylvin, for use in the packaging industry.

Despite these beneficial properties, several disadvantages of stilbenoids related to their poor solubility in water, their facility to be oxidized by different agents, or their tendency to be photodegraded have meant that pinosylvin has not been used in active food-packaging films. Steady-state fluorescence was used to demonstrate that natural (α-, β-, and γ-CD) and modified (HP-β-CD, methyl-β-CD, and ethyl-β-CD) CDs are able to complex pinosylvin following a 1:1 stoichiometry (Lopez-Nicolas, 2009). However, substantial differences in the strength of the complexation exist between the CDs tested. Although among natural CDs the interaction of pinosylvin with β-CD was more efficient than with α- and γ-CD, the results show that the complexation constants were higher for all of the modified CDs than for natural CDs, the highest constant being that determined for HP-β-CD-pinosylvin complexes (12112 M^{-1}).

The values were higher at pH values lower than the pKa value of pinosylvin shows that the complexes between HP-β-CD and the protonated form of pinosylvin were more stable than the interaction with the deprotonated forms of this antifungal and antimicrobial compound.

2.4. Sesquiterpenes

Artemisinin (Qinghaosu) (Figure 9) is a sesquiterpene lactone endoperoxide isolated from the Chinese medicinal herb qinghao (*Artemisia annua* L.).

In 1979, it was shown to be an effective antimalarial against chloroquine-resistant strains of P. falciparum. This compound and its derivatives, such as artemether, dihydroartemisinin, arteether, and artesunate, are effective against both chloroquine- resistant and chloroquine-sensitive strains of P. falciparum, as well as against cerebral malaria (Dhingra, 2000). In addition, the anticancer activity of artemisinin has also been reported (Oh, 2009). However, artemisinin's poor solubility in both oil and water, and hydrolytic instability of the lactone function, have led scientists to improve its pharmaceutical aspects.

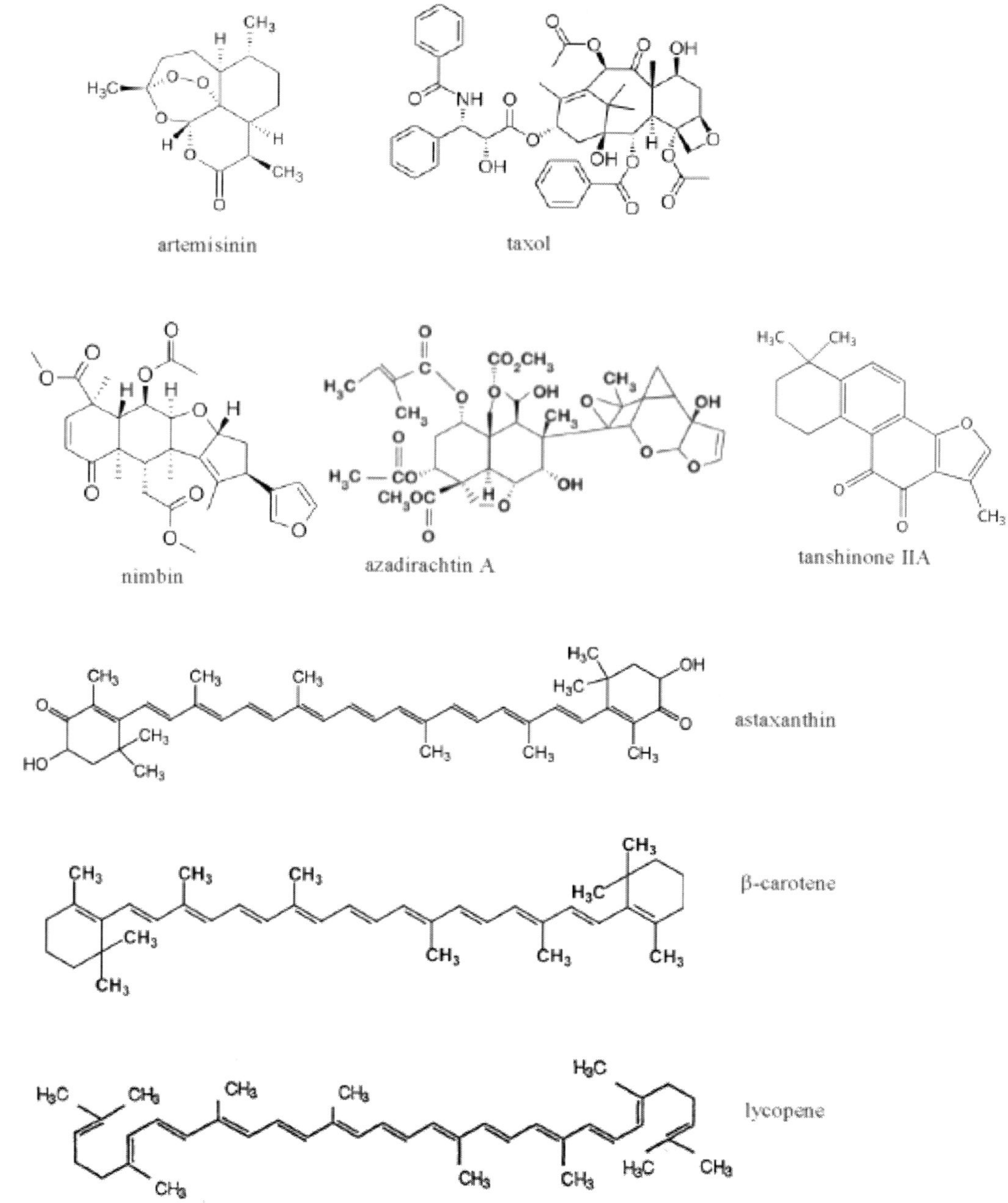

Figure 9. Chemical structures of terpenes and carotenoids.

The inclusion complexation of artemisinin with natural cyclodextrins, namely α-, β-, and γ-CDs with the aim of improving its solubility and dissolution rate were investigated (Wong, 2003 Kakran, 2011). Solubility diagrams indicated that the ICs occurred at a molar ratio of

1:1, and showed a remarkable increase in artemisinin solubility. The complexation capability of CDs with artemisinin increased in the order of α- <γ- <β-CDs and could be ascribed to the structural compatibility between the molecular size of artemisinin and the diameter of the CD cavities. Dissolution profiles of the three complexes demonstrated an increased rate and extent of dissolution compared with those of their respective physical mixtures and a commercial preparation. In solid-state analysis, the γ-CD was capable of complexing the highest percentage of artemisinin, followed by β- and α-CyDs. The respective estimated percentage of sesquiterpene complexed by the CDs were 85%, 40%, and 12%. The Kc artemisinin/β-CD resulted 1355.66 M^{-1} at 24 °C.

The artemisinin complexes with β-CD and γ-CD were shown to display faster rate and higher extent of dissolution compared to the normal form of the drug. The saturation solubility of artemisinin in the complexed form was found to increase approximately two-fold. Both cyclodextrin complexes of artemisinin exhibited a higher rate and extent of artemisinin absorption compared to the reference preparation, in a clinical study. In addition, the absorption of artemisinin appeared to be poor and negligible when the preparations started to arrive in the colon, which could be attributed to poor dissolution of artemisinin in the semi-solid contents of the lower part of the gastrointestinal tract (Wong, 2001; Wong, 2003; Marconi, 2004).

Other cyclodextrins was examined as solubilizers, in particular HP-β-CD, SBE7-β-CD, DM-β-CD, 2,3,6-partially methylated-β-CD (PM-β-CD) as modified CDs, and G1- β-CD, and G2-β-CD as branched CDs (Usuda, 2000). All the solubility curves of artemisinin with CDs can be classified as type AL (1:1 complexes). The constants for artemisinin-β-CD derivative complexes increased in the order of G2-β-CD ~ G1-β-CD ~ PM-β-CD ~ β-CD < HP-β-CD < SBE7-β-CD < DM-β-CD.

Illapakurthy reported a solubility of 4.5 mg/mL for artemisinin in the HP-β-C complex. The stability constants is 475, M^{-1}, respectively (Illapakurthy, 2003).

2.5. Other Terpenes

The neem tree, *Azadirachta indica* A. Juss, is a tropical plant that has long been used in agriculture and medicine. According to folklore, the bark, leaves, and fruit have been used as medicines for the treatment of a variety of human ailments, particularly against diseases of bacterial and fungal origin. Many studies have demonstrated that its seed contains abundant limonoids and simple terpenoids that are responsible for its biological activity. *Azadirachtin A* (Figure 9) is the most important active principle due to its various effects on insects, and has gained considerable attentions as a potential nontoxic, biodegradable, and natural pesticide. Furthermore, azadirachtin also has great application in herbal medicine/healthcare products, especially for major skin diseases, anti-malarial, anti-tuberculosis, anti-worms, anti-clotting, blood detoxifier, anti-viral, anti-periodontitic, anti-bacterial, anti-fungal (Liu, 2005). However, the utility of azadirachtin as biopesticide or herbal medicine is greatly limited by its low water solubility and instability.

The inclusion complexation behavior of azadirachtin with several cyclodextrins and their methylated derivatives has been investigated. The water solubility of azadirachtin was obviously increased after resulting inclusion complex with cyclodextrins. Typically, β-CD

(Kc: 238 M^{-1}), DM-β-CD, permethyl-β-cyclodextrin (TM-β–CD), and HP-β-CD are found to be able to solubilize azadirachtin to high levels up to 2.7, 1.3, 3.5, and 1.6 mg/mL, respectively (Liu, 2005).

Other limonoid with the same activities is *nimbin* (Figure 9). It is widely used in herbal medicine/healthcare products such as anti-inflammatories, antipyretics, antifungals, antihistamines, antiseptics, and antioxidants (Ghimeray, 2009). The molecule has low solubility and stability. Its complexes with β-CD and its derivatives: HP-β-CD, DM-β-CD, and TM-β-CD were reported (Yang, 2010). The 1:1 inclusion stoichiometry was observed in all complexes. The binding constants for the complexation of nimbin with β-CD, HP-β-CD, DM-β-CD, and TM-β-CD were in the following order: HP-β-CD > β-CD > DM-β-CD > TM-β-CD, and they were calculated to be 1279, 1370, 390, 351 M^{-1}, respectively. Considering the structural features of the host and guest, we deduced that the hydrogen bond between the hydroxyl arms of HPβCD or the hydrogen atoms of β-CD and the oxygen atoms of nimbin might strengthen the host-guest association. The results show that the water solubility of the nimbin, compared with that of native nimbin (ca. 50 μg/mL), was remarkably increased to approximately 4.7, 3.8, 1.3, and 2.4 mg/mL by the solubilizing effects of β-CD, HP-β-CD, DM-β-CD, and TM-β-CD, respectively (Yang, 2010).

Danshen, the root of the Chinese herbal plant-*Salvia miltiorrhiza Bunge*, is classified as "Blood invigorating" in traditional Chinese medicine. The chief bioactive ingredients of *Danshen* are the phenanthrofurane quinone derivatives, particularly known as *tanshinones* (Figure 9). The pharmacological tests revealed that tanshinones can dilate coronary arteries, increase coronary flow, and protect the myocardium against ischemia. In addition, tanshinones have attracted particular attention because they exhibit significant antibacterial, antioxidant, antitumor, anti-inflammatory, and anti-platelet aggregation activities (Shi, 200). Among them, the most abundant, effective and structurally representative bioactive constituent in the fraction is tanshinone IIA (TS IIA) (Figure 10), which is selected as the marker component for the quality control of containing *Danshen* preparations in the Chinese Pharmacopoeia. However, its oral bioavailability is very low, and the plasma concentration is well below 10 ng ml^{-1}. This is probably due to TS IIA's poor solubility in water and insufficient dissolution rate.

The effect of HP-β-CD on the aqueous solubility, dissolution rate, and intestinal permeability of TS IIA was investigated (Wang, 2007). The corresponding inclusion complex of TS IIA/HP-β-CD at the molar ratio of 1:1 was obtained by coevaporation, with Kc of 631.4 M^{-1}. The solubility of complexed TS IIA in water at 37 °C was 17 times greater than that for the uncomplexed drug. The dissolution rate of TS IIA was significantly increased by the complexation with HP-β-CD, due to its solubilizing activity. The permeability rates of TS IIA across the intestinal epithelial membrane were enhanced by the formation of IC with HP-β-CD about 5.2, 5.8 and 4.8 times of the uncomplexed TS IIA in duodenum, jejunum and ileum, respectively. It was revealed that the absorption rate-limiting step of TS IIA might be the dissolution process. The present results indicate the potential use of HP-β-CD to improve the gastrointestinal tract absorption of TS IIA (Wang, 2007).

Taxol (Figure 9) is a diterpenoid pseudoalkaloid, isolated from the bark of various Taxus species such as *Taxus baccata I., Taxus brevifolia, Taxus cuspidate*. It is an anticancer agent used in breast, ovarian, lung, head and neck cancers. Taxol shows a hydrophobic chemical structure, characterized by very low water solubility. Due to its high hydrophobicity and the fact that the administration of taxol presents considerable difficulties.

Interaction of taxol with α-CD and carboxymethyl-β-cyclodextrin (CM-β-CD) was was reported (Cserhati, 1995, 1997).

Taxol was also complexed with β-CD, DM-β-CD and TM-β-CD. The Kc of the complexes were 337, 406 and 355 M^{-1} at 25°C, respectively (Alcaro, 2002). Experimental findings are in agreement with molecular modeling analysis, which showed different interaction taxol-CDs as a function of macrocyle methylation. The complexation of taxol within the CD cavity preserved its antitumoral activity, with β-CD having the greater effect.

A solid inclusion 1:1 complex between Taxol and HP-β-CD was prepared by an evaporation method (Agüeros, 2009). The apparent stability constant, resulted 607±41M^{-1}.

RAME-β-CD, DM-β-CD, SBE-β-CD and the different HP-β-CDs were investigated for their taxol solubilising effect (Bouquet, 2007). The highest amount of taxol in PBS was recovered using methylated-β-cyclodextrin. SBE-β-CD and HP-β-CD had a very low affinity for taxol and induced only a small increase of solubility. The degree of substitution of HP-β-CD had a minor effect on taxol inclusion. These results confirmed previous literature reports: methylated-β-cyclodextrins provided a better enhancement of the aqueous solubility of paclitaxel compared to HP-β-CD (Sharma, 1995; Szente, 1999). HP-β-CD, HE-β-CD and DM-β-CD increased taxol solubility 2 x 10^3 fold or more and did not alter the cytostatic properties of taxol *in vitro*. The quantity of drug solubilized increased with the CD concentration, but precipitation upon dilution occurred with some CDs or stoichiometries (Szente, 1999). No precipitation occurred upon dilution with RAME-β-CD. taxol/RAME-β-CD inclusion complexes were stable for at least 6 months when stored at 4 °C.

Oligo(ethylenediamino) bridged bis(β-cyclodextrin)s possessing bridge chains in different length has been investigated in order to improve the water solubility of taxol (Figure 8c), (Liu, 2004). The results obtained show that bis(β-cyclodextrin)s 1 and 2 are able to solubilize paclitaxel to high levels up to 2 and 0.9 mg/mL, respectively. The inclusion complexation stechiometry was 1:2 for 1 and and 1: 1 for 2 CD. Compound 2 showed a relatively weak association with taxol. The Kc of stability of 1/taxol comples was 2.04 x 10^8 M^{-1}. Furthermore, 1/taxol complex increased the antitumor activity, and the thermal stability of parent paclitaxel (Liu, 2004).

2.6. Carotenoids

Carotenoids represent a large family of natural pigments widely distributed in microorganisms, algae, and higher plants. β-Carotene, lycopene and some xanthophylls, e.g. zeaxanthin and lutein, exert antioxidant influences in lipid phases by quenching singlet oxygen or free radicals and are therefore a focus of scientific interest.

Carotenoids show several beneficial functions to human beings, besides these pigments are largely used as colorant in foods, drinks, cosmetics and animal feed, mainly poultry and fish. In addition, the food and cosmetic industries are seeking to substitute the artificial colorants by natural ones. In recent times, the interest on lycopene has increased, due to the association between consumption of lycopene-rich foods, such as tomato and its products, and high levels of lycopene in plasma with a decreased risk of prostate cancer development (Venkateswaran, 2004).

The properties, actions and functions of carotenoids, including lycopene, are closely related to their chemical structure. The polyenic chain is responsible both for visible light absorption and for high molecule reactivity. Most carotenoids in the nature are found in the *trans* due to its higher stability compared to the *cis* isomer.

However, during the processing and storage, carotenoids can easily rearrange in different geometric isomers and also be oxidized, with the consequent decrease or loss of the colorant and biological properties. The most applied alternative to increase stability of carotenoids and to allow their incorporation in hydrophilic environment is the microencapsulation technique, which provides a physical barrier protecting the pigment.

Astaxanthin (Figure 9) is a carotenoid pigment found in certain marine animals and plants such as fish, shrimp and algae. It can be obtained from synthetic and natural sources, e.g. microalgae, yeast, shrimp, crab or lobster by-products. The most important commercial application of astaxanthin is in the aquaculture industry where it is used in the formulation of feed for farmed salmon to provide the typical muscle colour, which is widely accepted by consumers throughout the world. In human nutrition, astaxanthin has been gaining widespread popularity as a dietary supplement due to its powerful antioxidant properties. Because astaxanthin is a highly unsaturated molecule, it can easily be damaged by heat or light, which can cause the loss of biological properties. Furthermore, the poor aqueous solubility of astaxanthin limits its use.

Chen reported the inclusion complex of astaxanthin with □-CD (Chen, 2007). The water solubility of the IC was <0.5 mg/ml, which is better than that of astaxanthin. The complex was 1:4 for astaxanthin/β-CD, with the formation of large aggregates observed in the aqueous solution of the inclusion complex. Furthermore, the stability of the inclusion complex against temperature and light was greatly enhanced compared to that of astaxanthin.

Mele et al. (Mele, 1998) reported inclusion of trans-β-carotene (Figure 9) into β- and γ-CD (molar ratio 1:4) by solid mixture. Light scattering and NMR experiments provide evidence for the formation of large aggregates, like micelles, from β-carotene complexes with β- and γ-cyclodextrin in water.

Polyakov et al. (Polyakov, 2004) prepared complexes as solid mixture using β-carotene and three β-carotene synthetic derivatives and β-CD at molar ratios of 1:1 and 1:2, respectively. Canthaxanthin, β-carotene, and lutein complexation with β-CD was also carried out by the solid mixture method at a molar ratio of 1:10, respectively (Lacrajan, 2001). Me-β-cyclodextrin showed solubilizing, stabilizing effects for β-carotene, lycopene, lutein, and zeaxanthin (Pfitzner, 2000).

Lycopene (Figure 9) is the main carotenoid in tomatoes and tomato products and occurs also naturally in some other fruits and vegetables, guava, watermelon and pink. Among the common dietary carotenoids, lycopene has the highest singlet oxygen capacity in vitro. However, the extremely high lipophilicity of lycopene resulting in its extremely low aqueous solubility and its sensitivity to air and light are significant barriers to its oral formulation and bioavailability.

Mele (Mele, 2002) reported the complex formation using lycopene and α- and β-CD with a molar ratio of 1:4, also confirmed by Nunes (Nunes, 2007).

Multiangle light scattering, ionspray/electrospray ionization mass spectrometry and tandem MS were used to characterize IC (Mele, 2002). The true 1:1 α-CD/lycopene inclusion complex has been observed by IS/ESIMS and confirmed by tandem MS. Large aggregates of particles, on the nanometer-size scale, are present in water, with meaningful differences in the shape of the α-CD/lycopene aggregates with respect to β-CD/lycopene analogues.

In the Nunes' study, the complex lycopene-β-CD was formed using the liquid method, resulting in a translucid pink solution. At a molar ratio of 1:4, about 50% of the added lycopene remained undissolved, whereas when a molar ratio of 1:1 was used, there was no formation of the complex, being all lycopene undissolved and the resulted solution stayed transparent. Matioli and Rodriguez-Amaya (Amaya, 2003) prepared the complexes lycopene-CDs in liquid form, using a molar ratio of 1:50 to 1:200.

An optimized kneading method for the preparation of lycopene–cyclodextrin binary systems was developed leading to solubilization of lycopene in water and in 5% (w/v) dextrose solution. Solubilization ability of CDs increased in the order β-CD< Me-β-CD < HP-β-CD, when water was used to collect the paste after the kneading process. Solubilization affinity of CDs increased in the order Me-β-CD < β-CD < HP-β-CD in presence of dextrose. The presence of dextrose favored the formation of lycopene binary systems with HP-β-CD and β-CD, as it was assured by the increased concentration of lycopene in these solutions, and revealed a dramatic decrease of lycopene concentration in solutions with Me-β-CD. In the presence of cyclodextrins, lycopene concentration in water was 8.0, 27.1 and 16.0 mg/ml for β-CD, HP-β-CD and Me-β-CD, respectively. In 5% (w/v) aqueous dextrose solutions the corresponding values were 16.0, 48.0 and 4.0 mg/ml, respectively. Storage stability characteristics of the binary systems were studied at 4 °C in solution and at -20 °C in the lyophilized products. At 4 °C, storage stability of lycopene–cyclodextrin binary systems in water or 5% (w/v) aqueous dextrose solutions, was limited ($t_{1/2}$ = 1–4 days). Presence of dextrose improved the stability of lycopene in aqueous solution of b-cyclodextrin while it did not practically affect the stability of lycopene in HP-β-CD and Me-β-CD aqueous solutions.

2.7. Other Compounds

Kavalactones represent the active constituents of kava–kava (*Piper methysticum* G. Forster), endowed with sedative and anaesthetic properties. Kavalactones are polar constituents, but poorly soluble in water with a low bioavailability. In this study, the formation of inclusion complexes of one of the most representative kavalactone isolated from kava-kava extract, *(S)-7,8-dihydrokavain* (DHK) (Figure 7), with β-CD was investigated mainly by spectroscopic methods (Bergonzi 2010).

In particular, ^{1}H-NMR diffusion measurements (DOSY) demonstrated that in the presence of β-CD the translational diffusion of kavalactone is sizably slowed down (2.5×10^{-10} m^2/s) with respect to the free drug (4.4×10^{-10} m^2/s) according to the inclusion of DHK in the cavity of β-CD. ROESY experiments confirmed the inclusion of DHK in the hydrophobic pocket of β-CD through the primary hydroxyl rim, being the most relevant interactions between the H3- of β-CD and the ortho protons on the phenyl ring of the DHK, and between H5- of β-CD and the meta/para protons of DHK phenyl ring. The inclusion complexation, with phenyl ring of DHK in the CD cavity and lactone moiety outside of CD, was also confirmed by the induced circular dichroism effects. The solubility of DHK was increased twice in the range of 2-4 mM of β-CD.

10-Undecyn-1-ol is a chemical substance poorly soluble in water present in the leaf extract of neem tree or *Azadirachta indica*. The plant have potential use in toiletries, pharmaceuticals, the manufacture of agricultural implements and furniture, cattle and poultry feeds, nitrification of soils for various agricultural crops, and pest control. It is well-known for its pesticidal properties and has been used for centuries for medicinal and pestmanagement purposes (Liu, 2005; Kaushik, 2000). Results of storage tests mostly indicate that the leaf powder, the seed oil, and all kinds of extracts do indeed have a negative effect on seed-eating insects. Neem leaf extract showed also antifungal activity by spore germination inhibition against *Fusarium oxysporum* and *Colletotrichum lindermuthianum*. The activity was correlated to the high concentration of 10-undecyn-1-ol.

To ameliorate the solubility of this compounds, the inclusion complexation behavior between 10-undecyn-1-ol and cyclodextrin derivatives, namely, RM-β-CD and HP-β-CD, was studied (Neoh, 2008).

The aqueous solubility of 10-undecyn-1-ol was greatly improved through complexation with the CD derivatives. RM-β-CD is comparatively more efficient in solubilizing 10-undecyn-1-ol with an apparent stability constant outstripping that of HP-β-CD by about an order of magnitude. The 1:1 apparent stability constants of these inclusion complexes were calculated from the phase solubility diagrams. The computed values were 10.7 and 1.03 mM^{-1} for RM-β-CD and HP-β-CD, respectively. Comparative *in vitro* evaluations of the growth inhibition effects of inclusion complex solutions toward *Rosellinia necatrix*, a phytopathogenic fungus, were performed. Appreciable improvements of the antifungal activity of 10-undecyn-1-ol through the addition of CD derivatives were observed visually. The improvement was evaluated in terms of area covered by the mycelia of *Rosellinia necatrix* and their growth rate. RM-β-CD was proven to be more effective compared to HP-β-CD with regard to the reduction of both fungal mycelium-covered area and growth rate constant, presumably owing to greater solubility enhancement by RM-β-CD and thus the bioavailability of 10-undecyn-1-ol.

3. Extracts

3.1. *Allium cepa* L.

The work of Ding et al. (Ding, 2013) aims to evaluate the encapsulation of a flavonoid-rich *Allium cepa* L. var. agrogatum Don extract (ACADFE) in β-CD by analyzing the percutaneous penetration *in vitro* and *in vivo*, leading to an explanation of the physical mechanism during transdermal transport.

In China and other Asian countries, many herbs and their extracts are traditionally used in cooking flavonoid-rich foods (e.g., Allium, soybean, Houttuynia, pumpkin flowers, Zanthoxylum); these extracts are used not only as flavoring and coloring agents but also for promoting beneficial health effects due to their antioxidant activities and antimicrobial effects. ACADFE contains quercetin and its glycosides, such as quercetin-3,4-O-β-D-glucopyranoside (199.3 ± 27.8 mg g^{-1}), quercetin-4-O-β-D-glucopyranoside (181.7 ± 25.4 mg g^{-1}), and quercetin (157.9 ± 23.1 mg g^{-1}). A 95% ethanol extract from ACAD, was encapsulated in β-CD by coprecipitation and subsequent freeze-drying. The optimal inclusion compound containing extract in β-CD was prepared in a 1:3 molar ratio (based on quercetin's Mr). It was concluded that the inclusion complex could improve the aqueous solubility and bioavailability of extract. The inclusion complex exhibited significant enhancement of percutaneous absorption both *in vitro* and *in vivo*. Confocal scanning laser microscopy shows that penetration of the extract-β-CD inclusion complex proceeds across skin via both follicular and transcellular routes.

3.2. Green Coffee Bean Extract

Budryn purified aqueous green coffee bean extract to increase the concentration of chlorogenic acids (CHAs) with a simultaneous decrease of caffeine quantity (Budryn, 2014). In addition, the authors prepared inclusion complexes of both crude and purified extracts with β-CD. The obtained concentration of chlorogenic acids in a complex with β-CD amounted to 19 g/100 g. The formation of inclusion complexes was confirmed with the use of ESI-MS/MS and differential scanning calorimetry (DSC) techniques. The complexation efficiency increased as a result of extract purification and also of increase from 1:1 to 1:2 molar ratio of β-CD to CHAs from green coffee extracts during complexation. However β-CD complex has benefits like its sweet taste and low digestive absorption (in relation to β-CD), thus it can partly replace saccharose, simultaneously decreasing the calorific value of a product. It can be assumed that chlorogenic acids after breakdown of the complexes by digestive enzymes become fully bioavailable, but this should be confirmed in similar studies on chlorogenic acids. The analysis of antioxidant capacity of inclusion complexes with β-CD revealed that crude and purified extracts showed about 2-fold higher antioxidant capacity than the obtained complex. Better protection against oxidation of chlorogenic acids in inclusion complexes with β-CD causes their higher stability and lack of darkening and bitter tast, while they remain highly bioavailable and antioxidant active antioxidant active.

When using chlorogenic acids as food additives a problem of their interaction with amino acids, peptides and proteins may occur. This may cause a retardation of their absorption and a reduction of antioxidant activity, especially in foods submitted to heating. The presence of CHAs in food products may additionally result in a bitter taste. These compounds may furthermore undergo oxidative damage (both enzymatic and non-enzymatic), which may lead to a darkening of products. Because of these factors, it may be beneficial to encapsulate chlorogenic acids by forming inclusion complexes with β-CD and to add them in this form to foods and food supplements.

3.3. *Hypericum perforatum* L.

St. John's wort refers to *Hypericum perforatum* L., a perennial herb of Hypericeae family native to Europe, West Asia and North Africa. The herb has been used in folk medicine as healing and anti-inflammatory agent since antiquity. Nowadays, *Hypericum perforatum* L. is widely used as an antidepressant remedy in Europe and the United States, effective in mild to moderate depression. The antidepressant activity is attributed to phloroglucinols (hyperforin, adhyperforin) located in the unripe fruit, and naphthodianthrones (hypericin, pseudohypericin, protohypericin, protopseudohypericin), found in flowers and leafs of the plant. Theherb contains also phenolic acids and a broad range of flavonoids like hyperoside, rutin, quercetin, quercitrin, isoquercitrin, miquelianin, astilbin, I3,II8-biapigenin and amentoflavone Purified methanolic extracts of *Hypericum perforatum* from Northern Greece were very rich in flavonoids (epicatechin, catechin and quercetin glucosides) *Hypericum perforatum* extract exerted significant antioxidant activity. The encapsulation of extract in β-CD was tested to improves the thermal stability of nutraceutical antioxidants present in St John's wort extract (Kalogeropoulos, 2010). The IC was prepared by mixing 1:4 mass ratios of its components in aqueous media and subsequent freeze-drying. The encapsulation was verified by DSC and NMR studies, and encapsulation efficiencies resulted 27.5, 30.0 and 35.0% for catechin, epicatechin and quercetin respectively. The IC remained intact at temperatures where the free *Hypericum perforatum* extract was oxidized complex could serve as a flavonoids-rich food supplement or a novel additive to enhance the antioxidant capacity of fresh or thermally processed food.

3.4. *Melissa Officinalis* (Lemon Balm) Extract

β-CD was also used with the same aim in an inclusion complex with *Melissa Officinalis* (lemon balm) extract (Mourtzinos, 2011). DSC studies confirmed the IC formation and the experiments under oxidative conditions indicated that the complex of lemon balm extracts with CD was protected against oxidation, since they remained intact at temperatures where the free lemon balm extract was oxidized. Encapsulation of lemon balm's active compounds makes the active compound more soluble in water than the free molecule. Moreover the dissolution of β-CD in the extraction solvent increased the antioxidant activity of the obtained extract.

3.5. Milk Thistle

Milk thistle contains a mixture of flavonoidic molecules, named silymarin, that has high antiradical and antioxidant activity on skin cells. Indeed, the extract is noted as able to decrease cellular peroxidation and protect the skin from photoinduced tumors (Nichols, 2010; Singh, 2009).

Further studies already showed that topical application of silymarin suppresses intracellular production of hydrogen peroxide and nitric oxide and reduces depletion of catalase activity in UVB irradiated mouse skin. For these properties Milk thistle was chosen in order to obtain a potential anti-inflammatory and antioxidant topic formulation useful as skin protectant from toxic factors as UV rays, smoke, smog able to induce peroxidation processes and consequent skin damage. Properties of Milk thistle are moreover added to the nutritive and restorative qualities of several excipients used for emulsions.

However penetration capability of a such lypophilic and complex mixture can be not easily achieved, and appear necessary to found a good carrier able to permits the reaction of the flavonoidic molecules with the oxidation factors commonly diffused in the skin.

Spada studied the effect of HP-β-CD on the transport of dry extract through the stratum corneum by exploiting the penetration enhancement activity of CD (Spada, 2013).

The binary complex was composed (1:4 w/w) by Milk thistle and HP-β-CD, correspondent to 1% (w/w) in sylimarine. The complex weas introduced in two different emulsion and its effect in *in vitro* and *ex vivo* tests was performed. From the results, cyclodextrins appear good excipients for sylimarine complexation and administration; in particular, work demonstrated that flavonoids degradation can be efficiently avoided thanks to the phytocomplex formation. The apparent decrement of *in vitro* antioxidant activity could be compensated by the possibility to have slow flavonoids release regulated by the disruption of the binary complex; this aspect can open the possibility to fill the gap about in vivo studies of antioxidant activity. Despite of *in vitro* and *ex vivo* results, cyclodextrins act *in vivo* as valid penetration enhancers for topical administration respect to reference formulations.

3.6. Olive Leaf Extract

The leaves of the olive tree *Olea europaea*, a member of the family Oleaceae, have been widely used in folk medicine in regions around Mediterranean Sea and the islands therein.

The main constituents of olive leaves are secoiridoids like oleuropein, ligostroside, dimethyloleuropein and oleoside. Olive leaves also contain flavonoids (apigenin, kaempferol, luteolin) as well as phenolic compounds (caffeic acid, tyrosol, hydroxytytrosol).

Oleuropein, the main constituent of olive leaf extract, is a complex phenol present in large quantities in olive tree leaves and in low quantities in olive oil and is responsible for the bitter taste and pungent aroma of olive oil. *In vitro* studies have demonstrated that oleuropein acts as an antitumor compound, inhibits platelet-activating factor activity, enhances nitric oxide production by mouse macrophages, and decreases inflammatory mediator production (Visioli, 1998). Oleuropein can be hydrolyzed to hydroxytyrosol, elenolic acid, oleuropein aglycone, and glucose.

The molecular encapsulation of olive leaf extract into β-CD could result in increased aqueous solubility or increased partitioning in the oil/water system, improved protection against oxidation during storage, and, possibly, a better bioavailability. In addition to the above, the encapsulated olive leaf extract could result in a product that can be used readily as a food ingredient, in a better way than olive leaf extract alone, due to its solid form and its possible increased stability against heat and oxidation.

The complex was prepared by using at a mole ratio 1:2 (calculations were based on the molecular weight of oleuropein) olive leaf extract/β-CD (Mourtzinos, 2007). Studies under oxidative conditions indicated that the complex of olive leaf extract with β-CD was protected against oxidation, since it remained intact at temperatures where the free olive leaf extract was oxidized. Phase solubility studies afforded AL type diagrams, then an 1:1 complex stoichiometry, that involves deep insertion of the dihydroxyphenethyl moiety of oleuropein inside the cavity from its secondary side. A moderate binding constant (~300 M^{-1}), and an increase of the aqueous solubility by ~50% were evidenced.

3.7. Red Bell Pepper Pigments

Among the Solanaceae family, the sweet bell pepper (*Capsicum annuum*) is used prominently in the culinary world due to its strong flavour and intense colour. Aside from its sensory properties, sweet bell pepper is also a good source of bioactive compounds, such as the carotenoids. These substances can reduce the risk of developing degenerative disease, inhibit carcinogenic activity and reduce (50% reduced rate) posterior subcapsular cataracts in subjects with higher plasma lutein concentration (Nishino, 2002). These effects have sparked interest in the food industry to add such compounds to functional foods, allowing them to be more bioavailable than their natural sources.

Capsanthin and capsorubin are the pigments responsible for the red coloration of the capsicum species; they are synthesised during maturation. Capsanthin may represent as much as 50% of the total carotenoids in red bell pepper and has been associated with prevention of colon carcinogenesis (Narisawa, 2000).

Some natural pigments, such as carotenoids, exhibit extremely high lipophilicity, low aqueous solubility and sensitivity to air and light that may impair their employment as food additives.

Gomes et al. (Gomes, 2014) prepared inclusion complexes between red bell pepper pigments and β-CD using two different procedures (i.e., magnetic stirring and ultrasonic homogenisation), to characterize the prepared inclusion complexes and to evaluate the colour stability of a selected complex added to yogurt.

The main pigments of the extract were found to be capsanthin, capsorubin, β-cryptoxanthin, β-carotene and the isomers 13- and 9-cis-β-carotene. The mass ratio of extract to β-CD was 1:4 (w/w). The obtained data showed that ultrasonic homogenisation resulted in better yield and inclusion efficiency (about 55%) compared to magnetic stirring.

The DLS analysis indicated that the inclusion complex obtained through the homogenizing ultrasonic procedure had a mean particle diameter of 562 nm, while the result for the physical mixture was 603 nm and for the complex formed by magnetic stirring was 596 nm. Although there was no significant difference among the results, they suggest a reduced particle size for the complex produced by ultrasonic homogenisation, maybe due to the interaction between the apolar carotenoids and the cyclodextrin hydrophobic cavity.

The complex prepared by ultrasonic homogenisation had the lowest polydispersity index in relation to all the other samples, indicating a more uniform particle size distribution due to the ultrasonic procedure.

Yogurt is a food with high nutritional quality due to the variety and amounts of nutrients present. Therefore, the artificial colours that are added to yogurt aim to improve sensory characteristics, making the product more appealing. The extract of red pepper when added to yogurt, can simulate the colour of commercial yogurt of papaya flavour and others, if used in mixture, but due to its low water solubility, there is difficulty in dispersing the extract in the product.

Therefore, in an attempt to improve stability and solubility of the carotenoids from red bell pepper, the IC with β-CD was added in order to increase colour stability. The yogurt with the added complex showed slower variations for the colour compared to yogurt with added extract, indicating a higher protection of the colour during storage by the IC.

4. ESSENTIAL OILS

Essential oils are complex blends of a variety of volatile molecules such as terpenoids, phenol-derived aromatic components, and aliphatic components having a strong interest in pharmaceutical, sanitary, cosmetic, agricultural, and food industries. Since the middle ages, essential oils have been widely used for bactericidal, virucidal, fungicidal, antiparasitical, insecticidal, and other medicinal properties such as analgesic, sedative, anti-inflammatory, spasmolytic, and locally anaesthetic remedies.

EOs are volatile, limpid, and rarely coloured liquids, lipid soluble and soluble in organic solvents with a generally lower density than that of water.They can be synthesized by all plant organs, that is,buds, flowers, leaves, stems, twigs, seeds, fruits, roots,wood, or bark and are stored in secretory cells, cavities, canals, epidermic cells, or glandular trichomes. Constituents are lipophilic and highly volatile secondary plant metabolites, reaching a mass below a molecular weight of 300, that can be physically separated from other plant components or membranous tissue (Sell, 2010).

The majority of the publications is concerning the encapsulation of essential oils with β-CD and its derivatives: RM-β-CD, HP-β-CD, and low ME-β-CD.

Thymol (Figure 10) is a monoterpene present in Lamiaceae plants, specially oreganos and thymes. Cinnamaldehyde (3-phenyl-2-propenal) represents 65–75% of the cinnamon EO. Thymol and cinnamaldehyde are frequently used as flavours, but they are also becoming increasingly important as naturally occurring antimicrobial, antioxidant, and antiseptic agents.

As natural and artificial flavours they are very sensitive to the effects of light, oxygen, humidity, and high temperatures. The study of Hill et al. (Hill, 2013) aimed to elucidate the physicochemical characteristics of essential oils and β-CD (EO-β-CD) inclusion complexes and their resulting antimicrobial activity. *Cinnamon bark extract, transcinnamaldehyde, clove bud extract, eugenol*, and a 2:1 (transcinnamaldehyde: eugenol) mixture were microencapsulated by the freeze-drying method. EO-β-CD complexes were characterized for particle size, morphology, polydispersity index (PI), entrapment efficiency, and phase solubility. All particles showed a spherical shape, smooth surface, no significant differences in size distribution, and strong tendency to agglomerate. The entrapment efficiencies ranged from 41.7 to 84.7%, where pure compounds were higher ($P < 0.05$) than extracts. The oils and their β-CD complexes were analyzed for their antimicrobial activity against *Salmonella enterica* serovar *Typhimurium* LT2 and *Listeria innocua*. All the samples effectively inhibited bacterial growth within the concentration range tested, except free eugenol.

The EO-β-CD complexes were able to inhibit both bacterial strains at lower concentrations than free oils, likely due to their increased water solubility which determined an increased contact between pathogens and essential oils. The cinnamon bark and clove bud olis/β-CD complexes were the most powerful antimicrobials, despite showing the lowest entrapment efficiencies amongst the oils. The results indicate that such EO inclusion complexes could be useful antimicrobial delivery systems with a broad spectrum of application in food systems where Gram-positive and -negative bacteria could present a risk (Hill, 2013).

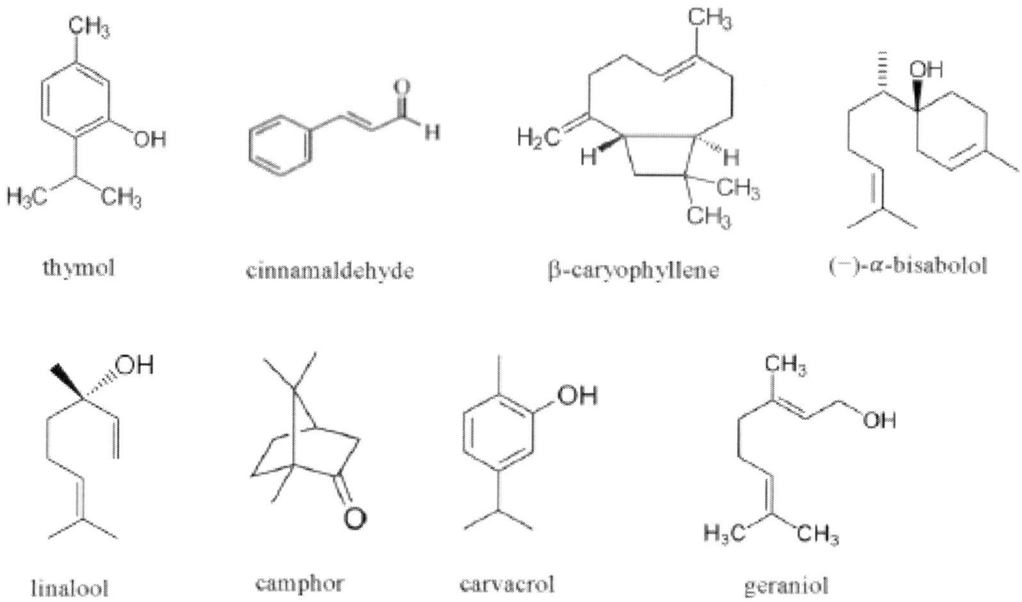

Figure 10. Constituents of essential oils.

The IC of *thymol and cinnamaldehyde* (Figure 10) and β-CD was investigated (Ponce Cevallos, 2010) in order to study the influence of water adsorption by CDs and their complexes on the release of encapsulated compounds. The results showed that β-CD encapsulates efficiently both of them, in a 1:1 molar ratio. The ICs were obtained upon mixing the components in aqueous media and subsequent freeze-drying, as confirmed by differential scanning calorimetry. The samples were stored at constant relative humidity, from 22% to 97%, at 25 °C. The release of encapsulated compounds was determined following the melting enthalpy of each guest. Water sorption isotherms for β-CD and the complexes showed constant and low water sorption at RH < 80%; then the uptake of water increased abruptly. The amount of sorbed water at each RH was smaller for the complexes than for β-CD. The guest molecules displaced water molecules from inside the cavity of *β*-CD. No thymol or cinnamaldehyde release was detected at RH < 84%, and it increased abruptly from 84% RH, coincidentally with the abrupt increase of absorbed water.

Water sorption significantly affects β-CD complexes stability, which is thus governed by the shape of the water sorption isotherm. The stability studies showed that the inclusion complexes thymol-β-CD and cinnamaldehyde-β-CD remain stable up to 75% RH during long storage times. In fact, the guests released from the β-CD complexes were detectable in the region of the water adsorption isotherm at which a sharp increase of water content occurred (84% RH). These results show the relevance of selecting appropriated storage conditions for hydrophobic flavours encapsulated in *β*-CD or for predicting the shelf life of functional products formulated with nanoencapsulated compounds (Ponce Cevallos, 2010).

□-*Caryophyllene* (BCP) (Figure 10) is a bicyclic sesquiterpene compound existing in the essential oil of many plants, e.g. *Eugenia caryophyllata* (Fam. Myrtaceae), *Piper nigrum* L. (Fam. Piperaceae), and *Zingiber nimmonii* (J. Graham) Dalzell (Fam. Zinziberaceae). Due to its woody and spicy odour, BCP has been commonly used as a fragrance and flavouring agent since the 1930s. In recent years, this compound has attracted considerable attention because of its extensive biological activities including antimicrobial, anticarcinogenic, anti-inflammatory, anti-oxidant, anxiolytic- like and local anaesthetic effects (Di Sotto, 2010; Galdino, 2012). However, this compound is volatile, poorly water-soluble, and sensitive to light, oxygen, humidity and high temperature □-CD inclusion complex increased dissolution and water-solubility of β-Caryophyllene.

Liu and coworkers investigated and compared the oral bioavailability and the pharmacokinetics of free BCP and BCP/β-CD IC after a single oral dose of 50 mg/kg on rats (Liu, 2013).

BCP was rapidly released from inclusion complex and the *in vivo* data showed that BCP/β-CD IC displayed earlier Tmax, higher Cmax and the AUC0-12 h showed approximately 2.6 times higher increase than those of free BCP. The β-CD has significantly increased the oral bioavailability of the drug in rats than free BCP (Liu, 2013).

The essential oil of *Chamomilla recutita* (L.) Rauschert, syn. *Matricaria recutita* L., contains up to 50% (−)-*α-bisabolol* (Figure 10) which contributes to the anti-inflammatory properties of camomile oil. Bisabolol is a very lipophilic substance, with a tendency to oxidise decreasing anti-inflammatory activity ca. 50%. (−)-α-Bisabolol was found to form an inclusion complex with β-CD in solution as well as in the solid state. To investigate molecular associations of β-CD with pure (−)-α-bisabolol or (−)-α-bisabolol as a component of camomile EO. Waleczek et al. undertook phase solubility studies (Waleczek, 2003). The complex constant was 273 M^{-1} for the pure (−)-α-bisabolol and 304 M^{-1} for (−)-α-bisabolol as a constituent of the EO. The intrinsic solubility of pure (−)-α-bisabolol (4.85×10^{-4} M) and (−)-α-bisabolol as a component of the EO (1.82×10^{-4} M) differ significantly. Computer simulation proved an inclusion complex having a stoichiometric composition of 2:1 (β-CD : drug) (Waleczek, 2003).

Garlic (*Allium sativum* L.) is a widely distributed plant and is used throughout the world not only as a spice and a food, but also as a folk medicine, and many of the beneficial health-related biological effects have been attributed to its characteristic organosulphur compounds (Rybak, 2004). Steam distillation is widely used to extract and condense the volatile organosulphur compounds in garlic, and the final oily product is called *garlic oil* (GO). The compounds of GO mainly are diallyl disulphide, diallyl trisulphide, allyl propyl disulphide, a small quantity of disulphide, and probably diallyl polysulphide (Pranoto, 2005). GO is recognised to be more potent than aqueous extracts of garlic and exhibits a wide range of pharmacological properties including antimicrobial, antidiabetic, antimutagenic, and anticarcinogenic effects (Agarwal, 1996). However, the application of GO is limited due to its volatility, strong odour, insolubility in water, and low physicochemical stability.

The characterisation of IC of GO/β-CD was investigated (Wang, 2011). The calculated apparent stability constant of IC was 1141 M^{-1}, and the water solubility of GO was significantly improved. Furthermore, the release rate of GO from the inclusion complex was controlled. The results of this study clearly demonstrated that GO could be efficiently complexed with β-CD to form an inclusion complex by the coprecipitation method in a molar ratio of 1:1. The aqueous solubility and stability of GO were significantly increased by inclusion in β-CD (Wang, 2011).

Isothiocyanates (ITCs) are hydrolysis products of sulphur-containing compounds called glucosinolates, which occur naturally in cruciferous vegetables, such as broccoli and cabbage.

Mechanical disruption of cruciferous plant tissues releases ITCs, due to the hydrolysis reaction catalysed by myrosinase bound to the cell wall and possesses antimicrobial activities (Delaquis, 1995). ITCs, in particular allyl isothiocyanate (AITC), have been extensively studied for their antibacterial effect for food. Inclusion complexation reactions between isothiocyanates (ITCs), namely, allyl isothiocyanate (AITC) and phenyl isothiocyanate (PITC), and RM-β-CD were reported (Neoh, 2012).

The apparent activation energy of IC dissociation suggested a reduction of volatility and physical stabilisation by inclusion complexation. RM-β-CD demonstrated a strong solubilising effect on the poorly water-soluble AITC and PITC in the aqueous phase. RM-β-CD was more effective in solubilising PITC though the AITC/RM-β-CD complex had higher apparent stability constants. Despite the fact that a greater amount of AITC was solubilised in the aqueous phase at any given concentration of RM-β-CD, the solid state AITC/RM-β-CD complex showed an inclusion ratio remarkably lower than that of the PITC/RM-β-CD complex. Both of the ITCs may form inclusion complexes with RM-β-CD at guest to host ratios of 1:1 and 1:2 in the aqueous phase (Neoh, 2012).

The inclusion interactions of cyclodextrins (CDs) and β-CD polymers with *linalool* and *camphor* (Figure 10) in *Lavandula angustifolia* EO were investigated in order to prepare novel controlled release systems for the delivery of essential oil used as ambient odours (Ciobanu, 2012).

The complexation behaviour and the retention capacity of α-CD, β-CD, γ-CD, HP-β-CD, randomly methylated-β-cyclodextrin (RAMEB), a low methylated-β-cyclodextrin (CRYSMEB), and crosslinked β-CD polymers for linalool and camphor, two major components of Lavandula angustifolia EO, were studied.The complexation and the retention capacity of CDs and CD polymerswere investigated under solid support or in aqueous media by static headspace gas chromatography. The release profile of aroma from solid support was studied by multiple headspace extraction (MHE). The retention capacity of the CD derivatives was measured in static experiments.

The stability constants with monomeric CD derivatives were determined for standard compounds and for the compounds in essential oil. The RAMEB showed the higher Kc both for the standard compounds (833 M^{-1} for linalool and 1194 M^{-1} for camphor) and for the compounds in the in *Lavandula angustifolia* essential oil (1074 M^{-1} for linalool and 2963 M^{-1} for camphor).All studied CDs and CD polymers reduce the volatility of the aroma compounds and stable 1:1 inclusion complexes are formed. β-CD is themost versatile CDs for the two guests, leading to greater formation constant and retention ability in aqueous phase (Ciobanu, 2012).

Carvacrol (Figure 10) is a phenolic monoterpene constituent of essential oils produced by numerous aromatic plants and spices such as black cumin (*Nigella sativa* L.), marjoram (*Origanum majorana* L.), oregano (*Origanum vulgare* L.), and thyme (*Thymus vulgaris* L.). Many biological effects have been described for carvacrol, such as pronounced antioxidant effect in vitro, free radical scavenger, and antilipidperoxidative agent. Carvacrol is Generally Recognized as Safe (GRAS) for consumption, and it is approved by the US Food and Drug Administration for food use and has been included by the Council of Europe in the list of chemical flavorings that may be added to foodstuffs. However, there are challenges to use carvacrol as antimicrobial and antioxidant in food products: (i) it has an extremely low flavor threshold and can drastically change the sensory properties of the food, (ii) it is highly insoluble in water due to its lipophilic nature and may have limited contact with pathogens in high moisture content foods; and (iii) it is oxidized, decomposed, or evaporated when exposed to the air, light, or heat Inclusion of carvacrol in cyclodextrins (CD) is one method to overcome these problems.

IC with β-CD were prepared using kneading (encapsulation efficacy, EE 83.8%) and freeze drying (EE 91.3%) methods. Differential thermograms and phase solubility study indicated formation of 1:1 stoichiometry inclusion complex, with a K of 155 M^{-1}. Both types of complex inhibited S. Typhimurium and E. coli K12 at a lower concentration than free carvacrol indicating that encapsulation can enhance the mechanism of antimicrobial action and decrease the concentration of antimicrobial compound needed for inhibition. Although both inclusion complexes have shown lower antioxidant activity than free carvacrol, they retained this property over time, similarly free carvacrol, when exposed to light (Santos, 2015).

Guimarães tested the effect of IC carvacrol/β-CD on the nociception induced by tumor cells (Sarcoma 180) in rodents. The complex reduced hyperalgesia, as well as spontaneous and palpation-induced nociception (Guimarães, 2015).

The higher interaction was evidenced in the case of carvacrol/HP-β-CD 1:1 IC. The entrapment efficiency resuted 78.09 in the kneading formulation and 83.74 for freeze drying method. K was 299 M^{-1}. The complex increased the amntimicrobial activity of carvacrol, but also in this case antioxidant activity was lower than free comound (Kamimura, 2014).

Geraniol (Figure 10), a natural component of plant essential oils, having roselike odor and taste is generally used as a fragrance/flavor in food industry to treat infectious diseases and/or preserve the food; therefore, geraniol presents several remarkable properties such as insect repellent, antimicrobial, antioxidant, anti-inflammatory and anticancer. However, sustainability of geraniol is a major challenge because of its volatile nature. The ICs with three types of native CD (α-CD, β-CD and γ-CD were reported (Kajaci, 2014). γ-CD has shown higher complexation efficiency with geraniol, yielding 1:1 molar ratio (geraniol:γ-CD) of solid CD-IC crystals. γ-CD has the largest cavity it is more flexible to accommodate conformers of geraniol with different sizes. The molar ratio of geraniol:CD was 0.78:1, 0.9:1 and 1:1 for geraniol/α-CD-IC, geraniol/β-CD-IC and geraniol/γ-CD-IC, respectively.

CONCLUSION

Plants are a source of a wide range of molecules capable of improve healing: fibre, vitamins, phytosterols, and further sulphur-containing compounds, carotenoids, organic acid anions and polyphenolics.

The use of bioactive molecules from plants has gained a substantial interest during the last decade for food, cosmetic and pharmaceutical applications. Concerning the pharmaceutical applications of these plant-derived bioactive molecules, the current problems are related to the protection of their properties from environmental factors, with their solubility in water and biofluids, and their bioavailability. In order to have biological activity, molecules need to be capable of reaching the action site in the right dose without losing integrity and be able to cross the lipophilic membrane. Hence plants bioactive compounds have restricted application in pharmaceutical field since they have limited water solubility, poor bioavailability, and can be easily modified by environmental factors such as temperature, pH and light. Therefore, in order to preserve bioactive molecules structural integrity, they need to be protected by a finishing formulation with the capacity to deliver them to the physiological targets without losing any bioactivity. The excellent biocompatibility and unique inclusion capability as well as powerful functionalization capacity of cyclodextrins and their derivatives make them especially attractive for engineering novel functional materials for biomedical applications. Consequently, there is an increasing interest to fabricate supramolecular systems for drug and natural compounds.

The hydrophobic cavity of CDs provides them inclusion capacity with a variety of compounds ranging from small molecules, ions, proteins and oligonucleotides. The biomedical applications of CDs are extremely attractive due to their low toxicity and low immunogenicity. CDs have been extensively utilized to form inclusion complexes with natural compounds through host–guest interactions in pharmaceutical science and technologies. The related applications include increasing solubility and stability, enhancing absorption, masking odors and tastes, controlling drug release profiles, alleviating local and systemic toxicity, and improving drug permeability across biological barriers. CDs have a relatively unique capacity of improving solubility of bioactive agents in aqueous systems; protect them from elevated temperatures, pH values, light or the moisture-induced degradations phenomena, which serve to increase their bioavailability. Furthermore, the use of semisynthetic derivatives of cyclodextrins can improve the physicochemical properties of bioactive molecules. Based on the studies described in the literature, the derivatives of β-CD are more appropriate for the improvement of solubility and stability, being the HP-β-CD the most used. However, because natural molecules have a great molecular diversity, it is not possible to generalise the molecular mechanisms responsible of the host-guest interaction, while the efficiency of the encapsulation depends, essentially, on the size of the CD cavity. A fundamental aspect evidenced by the reported studies is that the complexation with CDs improves the physical properties mantaining or increasing the activity of the included botanical substances.

REFERENCES

Agarwal, KC. Therapeutic actions of garlic constituents, Medicinal Research Reviews, 1996, 16, 111–124.

Agüerosa, M; Ruiz-Gatóna, L; Vauthierb, C; Bouchemalb, K; Espuelasa, S; Ponchelb, G; Irache, JM. Combined hydroxypropyl-β-cyclodextrin and poly(anhydride) nanoparticles improve the oral permeability of paclitaxel. *European Journal of Pharmaceutical Sciences*, 38 (2009) 405–413.

Al-Awwadi, NA; Araiz, C; Bornet, A; Delbosc, S; Cristol, JP; Linck, N; Azay, J; Teissedre, PL; Cros, G. Extracts enriched in different polyphenolic families normalize increased cardiac NADPH oxidase expression while having differential effects on insulin resistance, hypertension, and cardiac hypertrophy in high-fructose-fed rats. *Journal of Agricultural and Food Chemistry*, 2005, 53(1), 151–157.

Alcaro, S; Ventura, CA; Paolino, D; Battaglia, D; Ortuso, F; Cattel, L; Puglisi, G; Fresta, M. Preparation, Characterization, Molecular Modeling and In Vitro Activity of Paclitaxel–

Cyclodextrin Complexes. *Bioorganic & Medicinal Chemistry Letters*, 2002, 12, 1637–1641.

Almeida, AP; Cortes, SF; Ferreira, AJ; Lemos, VS. Increase on the coronary flow induced by dioclein in isolated rat heart. *Life Sciences,* 2002, 70, 1121–1128.

Anselmi, C; Centini, M; Maggiore, M; Gaggelli, N; Andreassi, M; Buonocore, A. Non-covalent inclusion of ferulic acid with alpha- cyclodextrin improves photo-stability and delivery: NMR and modeling studies. *Journal of Pharmaceutical and Biomedical Analysis*, 2008, 46(4), 645–652.

Aqil, F; Munagala, R; Jeyabalan, J; Vadhanam, MV. Bioavailability ofphytochemicals and its enhancement by drug delivery systems. *Cancer Letters*, 2013, 334(1), 133–141.

Aridogan, B; Baydaf, H; Kaya, S; Demirci, M; Ozbaw, D; Mumcu, E. Antimi-crobial activity and chemical composition of some essential oils. *Archives of Pharmacal Research*, 2002, 25(6), 860–864.

Azmir, J; Zaidul, ISM; Rahman, MM; Sharif, KM; Mohamed, A; Sahena, F; Jahurul, MHA; Ghafoor, K; Norulaini, NAN; Omar, AKM. Techniques for extraction of bioactive compounds from plant materials: A review. *Journal of Food Engineering*, 117 (2013) 426–436.

Bais, HP; Walker, TS; Kennan, AJ; Stermitz, FR; Vivanco, JM. Structure-dependent phytotoxicity of catechins and other flavonoids: Flavonoid conversions by cell-free protein extracts of Centaurea maculosa (spotted knapweed) roots. *Journal of Agricultural and Food Chemistry*, 2003, 51, 897–901.

Barão, CE; Paiva-Martins, F; Zanin, G; De Moraes, FF. Determination of the inclusion complex constant between oleuropein and cyclodextrins by complexation theory. *Journal of Inclusion Phenomena and Macrocyclic Chemistry*, 2014, 78:465–470.

Belščak-Cvitanovič, A. Encapsulation of polyphenolic antioxidants from medicinal plant extracts in alginate–chitosan system enhanced with ascorbic acid by electrostatic extrusion. *Food Research International*, 2011, 44, 1094–1101.

Benguo, L; Wei, L; Jian, Z; Yang, L; Xiaoai, Z; Guizhao, L. Physicochemical characterisation of the supramolecular structure of luteolin/cyclodextrin inclusion complex. *Food Chemistry*, 2013, 141, 900–906.

Bergonzi, MC; Bilia, AR; Di Bari, L; Mazzi, G; Vincieri, FF. Studies on the interactions between some flavonols and cyclodextrins. *Bioorganic and Medicinal Chemistry Letters*, 2007, 17, 5744–5748.

Bo, Y; Li-Juan, Y; Jun, L; Yong, C; Yu, L. Binding behaviors of scutellarin with α-, β-, γ-cyclodextrins and their derivatives. *Journal of Inclusion Phenomena and Macrocyclic Chemistry*, 2009, 64, 149–155.

Bouquet, W; Ceelen, W; Fritzinger, B; Pattyn, P; Peeters, M; Remon, JP; Vervaet, C. Paclitaxel/b-cyclodextrin complexes for hyperthermic peritoneal perfusion—formulation and stability. *European Journal of Pharmaceutics and Biopharmaceutics*, 2007, 66, 391–397.

Boztas, AB; Karakuzu, O; Galante, G; Ugur, Z; Kocabas, F; Altuntas, CZ; Yazaydin, AO. Synergistic Interaction of Paclitaxel and Curcumin with Cyclodextrin Polymer Complexation in Human Cancer Cells *Molecules* , 2013, 10, 2676–2683.

Budryn, G; Nebesny, E; Pałecz, B; Rachwał-Rosiak, D; Hodurek, P; Miśkiewicz, K; Oracz, J; Żyżelewicz, D. Inclusion complexes of β-cyclodextrin with chlorogenic acids (CHAs) from crude and purified aqueous extracts of green Robusta coffee beans (Coffea canephora L.). *Food Research International*, 61 (2014) 202–213.

Cal, K.; Centkowska, KUse of cyclodextrins in topical formulations: Practical aspects. *European Journal of Pharmaceutics and Biopharmaceutics*, 2008, 68(3),467–478.

Calabrò, ML; Tommasini, S; Donato, P; Raneri, D; Stancanelli, R; Ficarra, P. Effects of alpha- and beta-cyclodextrin complexationon the physico-chemical properties and antioxidant activity of some 3-hydroxyflavones. *Journal of Pharmaceutical and Biomedical Analysis*, 2004, 35(2), 365–377.

Calabrò, ML; Tommasini, S; Donato, P; Stancanelli, R; Raneri, D; Catania, S. The rutin/beta-cyclodextrin interactions in fully aqueous solution: Spectroscopic studies and biological assays. *Journal of Pharmaceutical and Biomedical Analysis,* 2005, 36(5), 1019–1027.

Cannavà, C; Crupi, V; Ficarra, P; Guardo, M; Majolino, D; Mazzaglia, A; Stancanelli, R; Venuti, V. Physico-chemical characterization of an amphiphilic cyclodextrin/genistein complex. *Journal of Pharmaceutical and Biomedical Analysis*, 2010, 51, 1064–1068.

Cannavà, C; Crupi, V; Ficarra, P; Guardo, M; Majolino, D; Stancanelli, R; Venuti, V. Physicochemical characterization of coumestrol/b-cyclodextrins inclusion complexes by UV–vis and FTIR-ATR spectroscopies. *Vibrational Spectroscopy*, 2008, 48 172–178.

Carlotti, ME; Sapino, S; Ugazio, E; Caron, G. On the complexation of quercetin with methyl-β-cyclodextrin: Photostability and antioxidant studies. *Journal of Inclusion Phenomena and Macrocyclic Chemistry*, 2010, 70(1/2), 81–90.

Carvalho, RR; Pellizzon, CH; Justulin, L; Felisbino, SL; Vilegas, W; Bruni, F. Effect of mangiferin on the development of periodontal disease: involvement of lipoxin A4, antichemotaxic action in leukocyte rolling. *Chemico-Biological Interactions*, 2009, 179(2-3), 344-350.

Casolaro, M; Anselmi, C; Picciocchi, G. The protonation thermodynamics of ferulic acid/γ-cyclodextrin inclusion compounds. *Thermochimica Acta*, 2005, 425(1/2), 143–147.

Celik, SE; Ozyürek, M; Tufan, AN; Güclü, K; Apak, R. Spec-troscopic study and antioxidant properties of the inclusion complexes ofrosmarinic acid with natural and derivative cyclodextrins. *SpectrochimicaActa Part A: Molecular and Biomolecular Spectroscopy*, 2011, 78(5), 1615–1624.

Chakraborty, S; Basu, S; Lahiri, A; Basak, S. Inclusion of chrysin in β-cyclodextrin nanocavity and its effect on antioxidant potential of chrysin: A spectroscopic and molecular modeling approach. *Journal of Molecular Structure*, 2010, 977(1–3), 180–188.

Chandra, A; Nair, MG; Iezzoni, AF. Isolation and stabilization of anthocyanins from tart cherries (Prunus cerasus L.). *Journal of Agricultural and Food Chemistry*, 1993, 41(7), 1062–1065.

Chen, X; Chen, R; Guo, Z; Li, C; Li, P. The preparation and stability of the inclusion complex of astaxanthin with β-cyclodextrin. *Food Chemistry*, 2007, 101, 1580–1584.

Chunli, Y; Zhilong, X; Xiaohui, L; Ce, H. Molecular modeling study of β-cyclodextrin complexes with (+)-catechin and (-)-epicatechin. *Journal of Molecular Graphics and Modelling*, 2007, 26, 420–428.

Ciobanu, A; Mallard, I; Landy, D; Brabie, G; Nistor, D; Fourmentin, S. Inclusion interactions of cyclodextrins and crosslinked cyclodextrin polymers with linalool and camphor in Lavandula angustifolia essential oil. *Carbohydrate Polymers*, 2012, 87(3), 1963–1970.

Clifford, MN. Anthocyanins – Nature, occurrence and dietary burden. *Journal of the Science of Food and Agriculture*, 2000, 80(7), 1063–1072.

Crupi, V; Ficarra, R; Guardo, M; Majolino, D; Stancanelli, R; Venuti V. UV–vis and FTIR–ATR spectroscopic techniques to study the inclusion complexes of genistein with β-cyclodextrins. *Journal of Pharmaceutical and Biomedical Analysis*, 2007, 44 110–117.

Cserhati, T. Interaction of some anticancer drugs with carboxymethyl-β-cyclodextrin. International Journal of Pharmaceutics, 1995, 124, 205-211.

Cserhati, T; Forgacs, E; Hollo, J. Interaction of taxol and other anticancer drugs with β-cyclodextrin. *Journal of Pharmaceutical & Biomedical Analysis,* 1995, 13(4/5), 533-541.

Cueva, C; Moreno-Arribas, MV; Martín-Alvarez, PJ; Bills, G; Vicente, MF; Basilio, A. Antimicrobial activity of phenolic acids against commensal, probiotic and pathogenic bacteria. *Research in Microbiology*, 2010, 161(5), 372–382.

da Rocha Ferreira, F; Valentim, IB; Catarí Ramones, EL; Salles Trevisan, MT; Olea-Azar, C; Perez-Cruz, F; Caxico de Abreu, F; Fonseca Goulart. MO. Antioxidant activity of the mangiferin inclusion complex with β-cyclodextrin. *LWT - Food Science and Technology,* 2013, 51 129-134.

da Rosa, CG; Borges, CD; Zambiazi, RC; Nunes, MR; Benvenutti, EV; da Luz, SR; D'Avila, RF; Rutz, JK. Microencapsulation of gallic acid in chitosan, β-cyclodextrin and xanthan. *Industrial Crops and Products*, 2013, 46 138–146

Dandawate, PR; Vyas, A; Ahmad, A; Banerjee, S; Deshpande, J; Swamy, KV. Inclusion complex of novel curcumin analogue CDF and β-cyclodextrin (1:2) and its enhanced in vivo anticancer activty against pancreatic cancer. *Pharmaceutical Research*, 2012, 29(7), 1775–1786.

Dangles, O; Wigand, MC; Brouillard, R. Anthocyanin anti-copigment effect. *Phytochemistry*, 1992, 31(11), 3811–3812.

Daruházi, AE; Kiss, T; Vecsernyés, M; Szente, L; Szoke, E; Lemberkovics, E. Investigation of transport of genistein, daidzein andtheir inclusion complexes prepared with different cyclodextrins on Caco-2 cell line. *Journal of Pharmaceutical and Biomedical Analysis*, 2013, 84C, 112–116.

Delaquis, PJ; Mazza, G. Antimicrobial properties of isothiocyanates in food preservation. *Food Technology*, 1995, 49(11), 73–84.

Delmas, D; Jannin, B; Latruffe, N. Resveratrol: preventing properties against vascular alterations and ageing, *Molecular Nutrition & Food Research*, 2005, 49, 377–395.

Delporte, C; Backhouse, N; Erazo, S; Negrete, R; Vidal, P; Silva, X. Analgesic-antiinflammatory properties of Proustia pyrifolia. *Journal of Ethnopharmacology*, 2005, 99, 119–124.

Dhingra V; Rao KV; Narasu ML. Current status of artemisinin and its derivatives as antimalarial drugs. *Life Sciences,* 2000, 66, 279-300.

Di Sotto, A; Mazzanti, G; Carbone, F; Hrelia, P; Maffei, F. Inhibition by betacaryophyllene of ethyl methanesulfonate-induced clastogenicity in cultured human lymphocytes. *Mutation Research*, 2010, 699, 23–28.

Divakar, S. Structure of a β-cyclodextrin-vanillin inclusion complex. *Journal of Agricultural and Food Chemistry*, 1990, 38, 940–944.

Divakar, S; Maheswaran, M. Structural studies on inclusion compounds of β-cyclodextrin with some substituted phenols. *Journal of Inclusion Phenomena and Macrocyclic*

Chemistry, 1997, 27, 113–126.

Dodziuk, H. *Cyclodextrins and their complexes*. KGaA, Weinheim: Wiley-VCH Verlag GmbH & Co.; 2006, ISBN 3-527-31280-3.

Duang, XY; Wang, Q; Zhou, XD; Huang, DM. Mangiferin: a possible strategy for periodontal disease to therapy. *Medical Hypotheses*, 2011, 76, 486-488.

Duchêne, D; Wouessidjewe, D; Ponchel, G. Cyclodextrins and carrier systems. *J. Control. Release*, 1999, 62, 263–268.

Efmorfopoulou, E.; Rodis, P. Complexation of oleuropein and trans-cinnamic acid with cyclodextrins. *Chemistry of Natursl Compound,* 2004, 40, 362–366.

Fang, Z; Bhandari, B. Encapsulation of polyphenols – A review. *Trends in Food Science & Technology*, 2010, 21(10), 510–523.

Fernandes, A; Ivanova, G; Natércia, F; Mateus, BN; Ramos, MJ; Rangel, M; de Freitas, V. Structural characterization of inclusion complexes between cyanidin-3-O-glucoside and β-cyclodextrin. *Carbohydrate Polymers*, 2014, 102, 269–277.

Ficarra, R; Tommasini, S; Raneri, D; Calabrò, M; Di Bella, M; Rustichelli, C. Study of flavonoids/β-cyclodextrins inclusion complexes by NMR, FT-IR, DSC, X-ray investigation. *Journal of Pharmaceutical and Biomedical Analysis*, 2002, 29 (6), 1005–1014.

Folch-Cano, C; Guerrero, J; Speisky, H; Jullian, C; Olea-Azar, C. NMR and molecular fluorescence spectroscopic study of the structure and thermodynamic parameters of EGCG/β-cyclodextrin inclusion complexes with potential antioxidant activity. *Journal of Inclusion Phenomena and Macrocyclic Chemistry*, 2013, 78, 287-298.

Galdino, PM; Nascimento, MV; Florentino, IF; Lino, RC; Fajemiroye, JO; Chaibub, BA; de Paula, JR; de Lima, TC; Costa, EA. The anxiolytic-like effect of an essential oil derived from Spiranthera odoratissima A. St. Hil. leaves and their major component, β-caryophyllene, in male mice. *Progress in Neuro-psychopharmacology Biological Psychiatry*, 2012, 38, 276–284.

Gao, Y; Gu, ZL; Jiang, XG; Guo, CY. A new progress on the baicalein pharmacological study. *Medicine and Materia Medica Research*, 2010, 21, 1765–1767.

Ge, QH; Zhou, Z; Zhi, XJ; Ma, LL. Pharmacokinetics and absolute bioavailability of breviscapine in Beagle dogs. *Chinese Journal of Pharmaceutics*, 2003, 34, 618–620.

Ghanghoria, R; Kesharwani, P; Agashe, HB; Jain, NK. Transdermal delivery of cyclodextrin-solubilized curcumin. *Drug Delivery and Translational Research*, 2012, 3(3), 272–285.

Ghimeray, AK; Jin, CW; Ghimire, BK; Cho, DH. Antioxidant activity and quantitative estimation of azadirachtin and nimbin in Azadirachta Indica A. Juss grown in foothills of Nepal. *African Journal of Biotechnology*, 2009, 8, 3084–3091.

Gomes, MML; Petito, N; Gonçalves Costa, V; Falcão, DQ; Kátia G. de Lima Araújo. Inclusion complexes of red bell pepper pigments with β-cyclodextrin: Preparation, characterisation and application as natural colorant in yogurt. *Food Chemistry*, 2014, 148 428–436.

Górnas, P; Neunert, G; Baczynski, K; Polewski, K. Beta-cyclodextrin complexes with chlorogenic and caffeic acids from coffee brew: Spectro-scopic, thermodynamic and molecular modelling study. *Food Chemistry*, 2009, 114(1), 190–196.

Gou, J; Zou, Y; Ahn, J. Enhancement of antioxidant and antimicrobial activities of Dianthus superbus, Polygonum aviculare, Sophora flavescens, and Lygodium japonicum by pressure-assisted water extraction. *Food Science and Biotechnology*, 2011, 20(1), 283–287.

Guimaraes, AG; Oliveira, MA; dos Santos Alves, R; dos Passos Menezes, P; Russo Serafini, M; de Souza Araujo, AA; Pereira Bezerra, D; Jose Quintans Junior, J. Encapsulation of carvacrol, a monoterpene present in the essential oil of oregano, with β-cyclodextrin, improves the pharmacological response on cancer pain experimental protocols. *Chemico-Biological Interactions*, 227 (2015) 69–76.

Haidong, L; Fang, Y; Zhihong, T; Changle, R. Study on preparation of β-cyclodextrin encapsulation tea extract. *International Journal of Biological Macromolecules*, 2011; 49(4), 561–566.

Haiyun, D., Jianbin, C., Shuang, Z. G., & Jinhao, P. (2003). Preparation and spectral investigation on inclusion complex of b-cyclodextrin with rutin. Spectrochimica Acta Part A: Molecular and Biomolecular Spectroscopy, 59(14), 3421–3429.

Harada, T; Giorgio, L; Harris, TJ; Pham, DT; Ngo, HT; Need, EF; Coventry, BJ; Lincoln, SF; Easton, CJ; Buchanan, G; Kee, TW. Diamide Linked γ-Cyclodextrin Dimers as Molecular-Scale Delivery Systems for the Medicinal Pigment Curcumin to Prostate. *Cancer Cells Molecule Pharmaceutics* 2013, 10, 4481−4490.

Harada, T; Pham, DT; Leung, MHM; Ngo, HT; Lincoln, SF; Easton, CJ; Kee, TW. Cooperative Binding and Stabilization of the Medicinal Pigment Curcumin by Diamide Linked γ-Cyclodextrin Dimers: A Spectroscopic Characterization. *Journal of Physical Chemistry B* 2011, 115, 1268–1274.

Harborne, JB; Williams, CA. Advances in Flavonoid Research Since 1992. *Phytochemistry*, 2000, 55, 481-504.

Heim, KE; Tagliaferro, AR; Bobilya, DJ. Flavonoid Antioxidants: Chemistry, Metabolism and Structure-Activity Relationships. *The Journal of Nutritional Biochemistry*, 2002, 13, 572-584.

Hill, LE; Gomes, C; Taylor, TM. Characterization of beta-cyclodextrin inclusion complexes containing essential oils (trans-cinnamaldehyde, eugenol, cinnamon bark, and clove bud extracts) for antimicrobial delivery applications. *Food Science and Technology*, 2013, 51, 86–93.

Ho, HH; Chang, CS; Ho, WC; Liao, SY; Wue, CH; Wanga, CJ. Anti-metastasis effects of gallic acid on gastric cancer cells involves inhibition of NF-kB activity and downregulation of PI3 K/AKT/small GTPase signals. *Food Chemistry and Toxicology*, 2010, 48, 2508–2516.

Illapakurthy, AC; Sabnis, YA; Avery, BA; Avery, MA; Wyandt, CM. Interaction of artemisinin and its related compounds with hydroxypropyl-β-cyclodextrin in solution state: Experimental and molecular-modeling studies. *Journal of Pharmaceutical Sciences*, 2003, 92, 649–655.

Iovine, B; Iannella, ML; Gasparri, F; Monfrecola, G; Bevilacqua, MA. Synergic effect of genistein and daidzein on UVB-induced DNA damage: an effective photoprotective combination. *Journal of Biomedical and Biotechnology*, 2011, 692846.

Jayachitra, J; Nalini, N. Effect of naringenin (citrus flavanone) on lipid profile in ethanol-induced toxicity in rats. *Journal of Food Biochemistry*, 2012, 36, 502–511.

Joshi, MPH; Kapoor, VP; Pushpangadan, P; Chaurasia, L. Antifungal activity of wogonin. *Phytotherapy Research*, 17 (2003) 1215–1216.

Ju, W.Z., Chun, J.H., Tan, R.X., Xiong, N.N.: Study on metabolites of scutellarin in gastrointestinal tract by UPLC-MS/MS method. *Chinese Journal of Clinical and Pharmacological Therapy*, 2006, 11, 292–295.

Jullian, C; Brossard, V; Gonzalez, I; Alfaro, M; Olea-Azar, C. Cyclodextrins-kaempferol inclusion complexes: Spectroscopic and reactivity studies. *Journal of Solution Chemistry*, 2011, 40(4), 727–739.

Jullian, C; Miranda, S; Zapata-Torres, G; Mendiza´bal, F; Olea-Azar C. Studies of inclusion complexes of natural and modified cyclodextrin with (+)catechin by NMR and molecular modeling. *Bioorganic & Medicinal Chemistry,* 2007, 15, 3217–3224.

Jullian, C; Moyano, L; Yañez, C; Olea-Azar, C. Complexation ofquercetin with three kinds of cyclodextrins: An antioxidant study. *Spectrochimica Acta Part A: Molecular and Biomolecular Spectroscopy*, 2007, 67(1), 230–234.

Kakran, M; Sahoo, NG; Li, L; Judeh, Z. Dissolution Enhancement of Artemisinin with β-Cyclodextrin. *Chemical Pharmacuetical Bullettin,*2011, 59(5), 646—652.

Kalogeropoulos, N; Yannakopoulou, K; Gioxari, A; Chiou, A; Makris, DP. Polyphenol characterization and encapsulation in β-cyclodextrin of a flavonoid-rich Hypericum perforatum (St John's wort) extract. *LWT - Food Science and Technology,* 2010, 43 882–889.

Kamimura, JA; Santos, EH; Hill, LE; Carmen L. Gomes. Antimicrobial and antioxidant activities of carvacrol microencapsulated in hydroxypropyl-beta-cyclodextrin. LWT - Food Science and Technology, 2014, 57 701e709.

Karathanos, VT; Mourtzinos, I; Yannakopoulou, K; Andrikopoulos, NK. Study of the solubility, antioxidant activity and structure of inclusion complex of vanillin with β-cyclodextrin. *Food Chemistry*, 2007, 101 652–658.

Kaushik, N; Vir, S. Variations in fatty acid composition of neem seeds collected from the Rajasthan state of India. *Biochemical Society Transaction,* 2000, 28, 880–882.

Kayaci, F; Sen, HS; Durgun, E; Uyar, T. Functional electrospun polymeric nanofibers incorporating geraniol–cyclodextrin inclusion complexes: High thermal stability and enhanced durability of geraniol. *Food Research International*, 2014, 62, 424–431.

Kayaci, F; Uyar, T. Solid inclusion complexes of vanillin with cyclodextrins: Formation, characterization and their high temperature stability. *Journal of Agricultural and Food Chemistry*, 2011, 59(21), 11772–11778.

Kerry, N; Abbey, M. The isoflavone genistein inhibits copper and peroxyl radical mediated low density lipoprotein oxidation in vitro, *Atherosclerosis*, 1998, 140 341–347.

Kim, DH; Jeon, SJ; Son, KH; Jung, JW; Lee, S; Yoon, BH. The ameliorating effect of oroxylin A on scopolamine-induced memory impairment in mice. Neurobiology Learning Memory, 2007, 87, 536–546.

Kim, H; Choi, J; Jung, S. Inclusion complexes of modified cyclodextrins with some flavonols. *Journal of Inclusion Phenomena and Macrocyclic Chemistry*, 2009, 64(1/2), 43–47.

Kim, H; Kim, H; Jung, S. Aqueous solubility enhancement of some flavones by complexation with cyclodextrins. *Bulletin of the Korean Chemical Society*, 2008, 29(3), 590–594.

Krishnaswamy, K; Orsat, V; Thangavel, K. Synthesis and characterization of nanoencapsulated catechin by molecular inclusion with β-cyclodextrin. *Journal of Food Engineering*, 2012, 111(2), 255–264.

Kumpugdee-Vollrath, M. Solid state characterization of trans resveratrol complexes with different cyclodextrins. *Journal of the Asian*, 2012, 1(2), 125–136.

Lacrajan, I; Dichl, HA; Socaciu, C; Engelke, M; Zorn-Kruppa, M. Carotenoid incorporation into natural membranes from artificial carriers: liposomes and β-cyclodextrins. *Chemistry and Physics of. Lipids,* 2001, 112, 1-10.

Lee, SK; Lee, HJ; Min, HY; Park, EJ; Lee, KM; Ahn, YH; Cho, YJ; Pyee, JH. Antibacterial and antifungal activity of pinosylvin, a constituent of pine. *Fitoterapia*, 2005, 76, 258–260.

Lewis, CE; Walker, JRL; Lancaster, JE. Effect of polysaccharides on thecolour of anthocyanins. *Food Chemistry,* 1995, 54(3), 315–319.

Li, H; Xu, X; Liu, M; Sun, D; Li, L. Microcalorimetric and spectrographic studies on host–guest interactions of α-, β-, (γ and M-β-cyclodextrin with resveratrol. *Thermochimica Acta*, 2010, 510(1/2), 168–172.

Ling, W; Xuehu, J; Weijuan, X; Chenrui, L. Complexation of tanshinone IIA with 2-hydroxypropyl-β-cyclodextrin: Effect on aqueous solubility, dissolution rate, and intestinal absorption behavior in rats. *International Journal of Pharmaceutics,* 2007, 341, 58–67.

Liu, B; Li, W; An Nguyen, T; Zhao J. Empirical, thermodynamic and quantum-chemical investigations of inclusion complexation between flavanones and (2-hydroxypropyl)-cyclodextrins. *Food Chemistry*, 2012, 134, 926–932.

Liu, B; Zhao, J; Liu, Y; Zhu, X; Zeng J. Physiochemical Properties of the Inclusion Complex of Puerarin and Glucosyl-β-Cyclodextrin. *Journal of Agricultural and Food Chemistry*, 2012, 60, 12501−12507.

Liu, B; Zhu, X; Zeng, J; Zhao, J. Preparation and physicochemical characterization of the supramolecular inclusion complex of naringin dihydrochalcone and hydroxypropyl-β-cyclodextrin. *Food Research International*, 2013, 54, 691–696.

Liu, H; Yang, G; Tang, Y; Cao, D; Qi, T; Qi, Y; Fan, G. Physicochemical characterization and pharmacokinetics evaluation of β-caryophyllene/β-cyclodextrin inclusion complex. *International Journal of Pharmaceutics*, 2013, 450, 304–310.

Liu, H; Yanga, G; Tanga, Y. Physicochemical characterization and pharmacokinetics evaluation of *β*-caryophyllene/*β*-cyclodextrin inclusion complex. *International Journal of Pharmaceutics*, 2013, 450, 304–310.

Liu, Y; Chen, GS; Chen, Y; Cao, DX; Geb, ZQ; Yuan, YJ. Inclusion complexes of paclitaxel and oligo(ethylenediamino) bridged bis (β-cyclodextrin)s: solubilization and antitumor activity. *Bioorganic & Medicinal Chemistry*, 2004, 12, 5767–5775.

Liu, Y; Chen, GS; Chena, Y; Lin, J. Inclusion complexes of azadirachtin with native and methylated cyclodextrins: solubilization and binding ability. *Bioorganic & Medicinal Chemistry*, 2005, 13 4037–4042.

López-Nicolás, JM; García-Carmona, F. Effect of hydroxypropyl-β-cyclodextrin on the aggregation of (E)-resveratrol in different protonation states of the guest molecule. *Food Chemistry*, 2010, 118, 648–655.

López-Nicolás, JM; Rodriguez-Bonilla, P; Garcia-Carmona, F. Complexation of Pinosylvin, an Analogue of Resveratrol with High Antifungal and Antimicrobial Activity, by Different Types of Cyclodextrins. *Journal of Agricultural and Food Chemistry*, 2009, 57, 10175–10180.

López-Tobar, E; Blanch, G.P; Ruiz del Castillo, ML; Sanchez-Cortes, S. Encapsulation and isomerization of curcumin with cyclodextrins character- ized by electronic and vibrational spectroscopy. *Vibrational Spectroscopy*, 2012, 62, 292–298.

Lu, Z; Chen, R; Fu, R; Xiong, J; Hu, Y. Cytotoxicity and inhibition of lipid peroxidation activity of resveratrol/cyclodextrin inclusion complexes. *Journal of Inclusion Phenomena and Macrocyclic Chemistry*, 2011, 73(1–4), 313-320.

Lu, Z; Cheng, B; Hu, Y; Zhang, Y; Zou, G. Complexation of resveratrol with cyclodextrins: Solubility and antioxidant activity. *Food Chemistry*, 2009, 113(1), 17–20.

Lucas-Abellán, C; Fortea, I; López-Nicolás, JM; Núnez-Delicado, E. Cyclodextrins as resveratrol carrier system. Food Chemistry, 2007, 104(1), 39–44.

Ma, SC; Du, J; But, PP; Antiviral Chinese medicinal herbs against respiratory syncytial virus. *Journal of Ethnopharmacology*, 2000, 79, 205-211.

Ma, SX; Chen, W; Yang, XD; Zhang, N; Wang, SJ; Liu, L. Alpinetin/hydroxypropyl-b-cyclodextrin host-guest system: preparation, characterization, inclusion mode, solubilization and stability. *Journal of Pharmaceutical and Biomedical Analysis*, 2012, 67/68, 193–200.

Manach, C; Williamson, G; Morand, C; Scalbert, A; Rémésy, C. Bioavailability and Bioefficacy of Polyphenols In Humans. I. Review of 97 Bioavailability Studies. *American Journal of Clinical Nutrition*, 2005, 81, 230-242.

Mangolim, CS; Moriwaki, C; Nogueira, AC; Sato, F; Baesso, ML; Neto, AM; Matioli, G; Curcumin–β-cyclodextrin inclusion complex: Stability, solubility, characterisation by FT-IR, FT-Raman, X-ray diffraction and photoacoustic spectroscopy, and food application. *Food Chemistry*, 2014, 153, 361–370.

Marconi, G; Monti, S; Manoli, F; Degli Esposti, A; Mayer, B. A circular dichroism and structural study of the inclusion complex artemisinin-β-cyclodextrin. *Chemical Physics Letters*, 2004, 15, 566-571.

Marques, HMC. A review on cyclodextrin encapsulation of essential oils and volatiles. *Flavour and Fragrance Journal*, 2010, 25(5), 313–326.

Martínez-Alonso, A. Encapsulation and solubilization of the antioxidants gallic acid and ethyl, propyl and butyl gallate with β-cyclodextrin, *Journal of Molecular Liquid*, 2015, http://dx.doi.org/10.1016/j. molliq.2014. 12.016.

Matsuda, H; Arima, H. Cyclodextrins in transdermal andrectal delivery. *Advanced Drug Delivery Reviews*, 1999, 36(1), 81–99.

Mele, A; Mendichi, R; Selva, A. Non-covalent associations of cyclomaltooligosaccharides (cyclodextrins) with trans-β-carotene in water: evidence for the formation of large aggregates by light scattering and NMR spectroscopy. *Carbohydrate Research*, 1998, 310, 261-267.

Mele, A; Mendichi, R; Selva, A; Molnar, P; Toth, G. Non-covalent associations of cyclomaltooligosaccharides (cyclodextrins) with carotenoids in water. A study on the α- and β-cyclodextrin/Ψ,Ψ-carotene (lycopene) systems by light scattering, ionspray ionization and tandem mass spectrometry. *Carbohydrate Research*, 2002, 337, 1129–1136.

Mercader-Ros, MT; Lucas-Abellán, C; Fortea, MI; Gabaldón, JA; Nú˜nez-Delicado, E. Effect of HP- b-cyclodextrins complexation on the antioxidant activity of flavonols. *Food Chemistry*, 2010. 118(3), 769–773.

Mercader-Ros, MT; Lucas-Abellán, C; Gabaldón, JA; Fortea, MI; Martínez-Cachá, A; Nú˜nez-Delicado, E. Kaempferol complexation in cyclodextrins at basic pH. *Journal of Agricultural and Food Chemistry*, 2010, 58(8), 4675–4680.

Miyake, K; Arima, H; Hirayama, F; Yamamoto, M; Horikawa, T; Sumiyoshi, H. Improvement of solubility and oral bioavailability of rutin by complexation with 2-hydroxypropyl-beta-cyclodextrin. *Pharmaceutical Development and Technology*, 2000, 5(3), 399–407.

Mohan, PRK; Sreelakshmi, G; Muraleedharan, CV; Joseph, R. Water soluble complexes of curcumin with cyclodextrins: Characterization by FT-Raman spectroscopy. *Vibrational Spectroscopy*, 2012, 62, 77–84.

Monti, D; Tampucci, S; Chetoni, P; Burgalassi, S; Saino, V; Centini, M. Permeation and distribution of ferulic acid and its α-cyclodextrin complex from different formulations in hairless rat skin. *AAPS PharmSciTech*, 2011, 12(2), 514–520.

Mourtzinos, I; Makris, DP; Yannakopoulou, K; Kalogeropoulos, N; Michali, I; Karathanos, VT. Thermal stability of anthocyanin extract of Hibiscussabdariffa L. in the presence of β-cyclodextrin. *Journal of Agricultural and Food Chemistry*, 2008, 56(21), 10303–10310.

Mourtzinos, I; Salta, F; Yannakopoulou, K; Chiou, A; Karathanos, VT. Encapsulation of olive leaf extract in β-cyclodextrin. *Journal of Agricultural and Food Chemistry*, 2007, 55, 8088−8094.

Mourtzinos, L; E. Papadakis, S; Igoumenidis, P; Karathanos, VT. Encapsulation of *Melissa Officinalis* leaf's active compounds in β-cyclodextrin and modified starch. *Procedia Food Science*, 2011, 1, 1679 – 1685.

Munin, A; Edwards-Lévy, F. Encapsulation of natural poly-phenolic compounds: A review. *Pharmaceutics*, 2011, 3(4), 793–829.

Narisawa, T; Fukaura, Y; Hasebe, M; Nomura, S; Oshima, S; Inakuma, T. Prevention of N-methylnitrosourea-induced colon carcinogenesis in rats by oxygenated carotenoids capsanthin and capsanthin-rich paprika juice. Proceeding of the Society for Experimental Biology and Medicine, 2000, 224, 116–122.

Neoh, TL; Tanimoto, T; Ikefuji, S; Yoshii, H; Futura, T. Improvement of Antifungal Activity of 10 Undecyn 1 ol by Inclusion Complexation with Cyclodextrin Derivatives. *Journal of Agricultural and Food Chemistry*, 2008, 56, 3699–3705.

Neoh, TL; Yamamoto, C; Ikefuji, S; Furuta, T; Yoshii, H. Heat stability of allyl isothiocyanate and phenyl isothiocyanate complexed with randomly methylated *β*-cyclodextrin. *Food Chemistry*, 2012, 131(4), 1123–1131.

Nguyen, TA; Liu, B; Zhao, J; Thomas, DS; Hook, JM. An investigation into the supramolecular structure, solubility, stability and antioxidant activity of rutin/cyclodextrin inclusion complex. *Food Chemistry*, 2013, 136(1), 186–192.

Nichols, JA; Katiyar, SK. Skin photoprotection by natural polyphenols:Anti-inflammatory, antioxidant and DNA repair mechanisms. *Archive of Dermatological Research*, 2010, 302, 71–83.

Nishino, H; Murakoshi, M; Ii, T; Takemura, M; Kuchide, M; Kanazawa, M. Carotenoids in cancer chemoprevention. *Cancer and Metastasis Reviews*, 2002, 21, 257–264.

Nunes, IL; Mercadante, AZ. Encapsulation of lycopene using spraydrying and molecular inclusion processes. *Brazilian Archives of Biology and Technology*, 2007, 50, 893–900.

Oh, S; Kim BJ; Singh, NP; Lai, H; Sasaki, T. Synthesis and anti-cancer activity of covalent conjugates of artemisinin and a transferrin-receptor targeting peptide. *Cancer Letters,* 2009, 274, 33-39.

Olas, B: Wachowicz, Resveratrol, B. A phenolic antioxidant with effects on blood platelet functions, Platelets 16 (2005) 251–260. D. Delmas, B. Jannin, N. Latruffe, Resveratrol: preventing properties against vascular alterations and ageing. *Molecular Nutrition & Food Research*, 2005, 49, 377–395.

Ong, CS; Zhou, J; Ong, CN; Shen, HM. Luteolin induces G1 arrest in human nasopharyngeal carcinoma cells via the Akt-GSK-3b-Cyclin D1 pathway. *Cancer Letters*, (2010). 298, 167–175.

Ou, S; Kwok, KC. Ferulic acid: Pharmaceutical functions, preparation and application. *Journal of the Science of Food and Agriculture*, 2004, 84, 1261–1269.

Patro, NM; Sultana, A; Terao, K; Nakata, D; Jo, A; Urano, A. Comparison and correlation of in vitro, in vivo and in silico evaluations of alpha, beta and gamma cyclodextrin complexes of curcumin. *Journal of Inclusion Phenomena and Macrocyclic Chemistry*, 2014, 78, 471–483.

Paul M. Dewick Medicinal Natural Products: A Biosynthetic Approach, 2009, Wiley-Blackwell (an imprint of John Wiley & Sons Ltd) ASIN: B0092J8KL6.

Perrone, D; Farah, A; Donangelo, CM; de Paulis, T; Martin, PR. Comprehensive analysis of major and minor chlorogenic acids and lactones in economically relevant Brazilian coffee cultivars. *Food Chemistry*, 2008, 106, 859–867.

Pescitelli, G; Bilia, AR; Bergonzi, MC; Vincieri, FF; Di Bari, L. Cyclodextrins as carriers for kavalactones in aqueous media: Spectroscopic characterization of (S)-7,8-dihydrokavain and □-cyclodextrin inclusion complex. *Journal of Pharmaceutical and Biomedical Analysis,* 2010, 52, 479–483.

Pfitzner, I; Francz, PI; Biesalski, HK. Carotenoid:methyl-□-cyclodextrin formulations: an improved method for supplementation of cultured cells. *Biochimica et Biophysica Acta*, 2000, 1474, 163-168.

Piber, M; Koehler, P. Identification of dehydro-ferulic acid-tyrosine in rye and wheat: Evidence for a covalent cross-link between arabinoxylans and proteins. *Journal of Agricultural and Food Chemistry*, 2005, 53, 5276–5284.

Pinent, M; Blay, M; Blade, M; Salvado, M; Arola, L; Ardevol, A. Grape seed derived procyanidins have an antihyperglycemic effect in streptozotocininduced diabetic rats and insulinomimetic activity in insulin-sensitive cell lines. *Endocrinology*, 2004, 145(11), 4985-4990.

Pinho, E; Grootveld, M; Soares, G; Henriques M. Cyclodextrins as encapsulation agents for plant bioactive compounds. *Carbohydrate Polymers*, 2014, 101 121– 135.

Polyakov, NE; Leshina, TV; Konovalova, T.A; Hand, EO; Kispert, LD. Inclusion complexes of carotenoids with cyclodextrins: 1H NMR, EPR, and optical studies. *Free Radical Biological Medicine*, 2004, 36, 872-880.

Ponce Cevallos, PA; Buera, MP; Elizalde, BE. Encapsulation of cinnamon and thyme essential oils components (cinnamaldehyde and thymol) in β-cyclodextrin: effect of interactions with water on complex stability. *Journal of Food Engineering*, 2010, 99, 70–75.

Pranoto, Y; Salokhe, VM; Rakshit, SK. Physical and antibacterial properties of alginate-based edible film incorporated with garlic oil. *Food Research International*, 2005, 38, 267–272.

Qiu, N; Chenga, X; Wang, G; Wang, W; Wen, J; Zhang, Y; Song, H; Ma, L; Wei, Y; Peng, A; Chen, L. Inclusion complex of barbigerone with hydroxypropyl-□-cyclodextrin: Preparation and in vitro evaluation. *Carbohydrate Polymers*, 2014, 101 623– 630.

Qiuna, Z; Xiaohui, W; Wei, D; Guixin, C; Zhengtao, W. Preparation and characterization of inclusion complexes formed between baicalein and cyclodextrins. *Carbohydrate Polymers*, 2013, 95, 733– 739.

Quaglia, F; Ostacolo, L; Mazzaglia, A; Villari, V; Zaccaria, D; Sciortino, MT. The intracellular effects of non-ionic amphiphilic cyclodextrin nanoparticles in the delivery of anticancer drugs. *Biomaterials*, 2009, 30 374–382.

Rahman, S; Cao, S; Steadman, KJ; Wei, M; Parekh, HS. Native and □-cyclodextrin-enclosed curcumin: entrapment within liposomes and their in vitro cytotoxicity in lung and colon cancer. *Drug Delivery*, 2012, 19(7), 346–353.

Rezende, BA; Cortes, SF; De Sousa, FB; Lula, IS; Schmitt, M; Sinisterra, RD; Lemos, VS. Complexation with □-cyclodextrin confers oral activity on the flavonoid dioclein. *International Journal of Pharmaceutics*, 2009, 367 133–139.

Rocks, N; Bekaert, S; Coia, I; Paulissen, G; Gueders, M; Evrard, B. Curcumin-cyclodextrin complexes potentiate gemcitabine effects in an orthotopic mouse model of lung cancer. *British Journal of Cancer*, 2012, 107(7), 1083–1092.

Ross, JA; Kasum, CM. Dietary Flavonoids: Bioavailability, Metabolic Effects, and Safety. *Annual Review of Nutrition*, 2002 22, 19-34.

Roupe, KA; Remsberg, CM; Yañez, JA; Davies, NM. Pharmacometrics of stilbenes: seguing towards the clinic. *Current Clinical Pharmacology*, 2006, 1, 81–101.

Rubistein M.H. Pharmaceutical Technology, drug stability. Chichester, England: Ellis Horwood Limited. 1989. Ch.1, p. 1-23.

Rybak, ME; Calvey, E; Harnly, JM. Quantitative determination of allicin in garlic: supercritical fluid extraction and standard addition of alliin. *Journal of Agricultural and Food Chemistry*, 2004, 52, 682–687.

Sabarinathan, D; Vanisree, AJ. Naringenin, a flavanone alters the tumorigenic features of C6 glioma cells. *Biomedicine & Preventive Nutrition*, 2011, 1, 19–24.

Sánchez-Maldonado, AF; Schieber, A; Gänzle, MG. Structure–functionrelationships of the antibacterial activity of phenolic acids and their metabolismby lactic acid bacteria. *Journal of Applied Microbiology*, 2011, 111(5), 1176–1184.

Santos, EH; Kamimura, JA; Hill, LE; Gomes, CL. Characterization of carvacrol beta-cyclodextrin inclusion complexes as delivery systems for antibacterial and antioxidant applications. *LWT - Food Science and Technology*, 2015, 60, 583e592.

Sapino, S; Carlotti, ME; Caron, G; Ugazio, E; Cavalli, R. In silico design, pho- tostability and biological properties of the complex resveratrol/hydroxypropyl-□-cyclodextrin. *Journal of Inclusion Phenomena and Macrocyclic Chemistry*, 2008, 63(1/2), 171–180.

Seelinger, G; Merfort, I; Schempp, CM. Anti-oxidant, anti-inflammatory and anti-allergic activities of luteolin. *Planta Medica*, (2008). 74, 1667–1677.

Sell, C. Chemistry of essential oils, in *Handbook of Essential Oils. Science, Technology, and Applications*, K. H. Başer and G. Buchbauer, Eds., pp. 121–150, CRC Press, Boca Raton, Fla, USA, 2010.

Seridi, L; Boufelfel, A. Wogonin hosted @ β-cyclodextrin: Structural, electronic and nuclear studies. *Journal of Molecular Liquids*, 2013, 188, 13–21.

Sharma, US; Balasubramanian, SV; Straubinger, RM. Pharmaceutical and physical properties of paclitaxel (Taxol) complexes with cyclodextrins, *Journal of Pharmaceutical Sciences*, 1995, 84 1223–1230.

Shen, YC; Chiou, W; Chou, YCh; Chen, ChF. Mechanisms in mediating the anti-inflammatory effects of baicalin and baicalein in human leukocytes. *European Journal of Pharmacology*, 2003, 465, 171–181.

Shi, ZH; He, JT; Chang, WB. Micelle-mediated extraction of tanshinones from *Salvia miltiorrhiza Bunge* with analysis by high-performance liquid chromatography. *Talanta*, 2004, 2, 401–407.

Shiko, AN; Pozharitskaya, ON; Miroshnyk, I; Mirzab, S; Urakova, IN; Hirsjavi, S. Nanodispersions of taxifolin: Impact of solid-state properties on dissolution behavior. *International Journal of Pharmaceutics*, 2009, 377, 148–152.

Shuang, S; Pan, J; Guo, S; Cai, M; Liu, C. Fluorescence study on the inclusion complexes of rutin with □-cyclodextrin, hydroxypropyl-□-cyclodextrin and □-cyclodextrin. *Analytical Letters*, 1997, 30(12), 2261–2270.

Shulman, M; Cohen, M; Soto-Gutierrez, A; Yagi, H; Wang, H; Gold-wasser, J. Enhancement of naringenin bioavailability by complexation with hydroxypropoyl-□-cyclodextrin. *PloSone*, 2011, 6(4), e18033.

Singh, R; Tonnesen, HH; Vogensen, SB; Loftsson, T; Másson, M. Studies of curcumin and curcuminoids. XXXVI. The stoichiometry and complexation constants of cyclodextrin complexes as determined by the phase-solubility method and UV–Vis titration. *Journal of Inclusion Phenomena and Macrocyclic Chemistry*, 2010, 66, 335–348.

Singh, RP; Agarwal, R. Cosmeceuticals and silibinin. *Clinical Dermatology*, 2009, 27, 479–484.

Spada, G; Gavini, E; Cossu, M; Rassu, G. Carta, A; Giunchedi, P. Evaluation of the effect of hydroxypropyl-□-cyclodextrin on topical administration of milk thistle extract. *Carbohydrate Polymers*, 2013, 92, 40– 47.

Sri, KV; Kondaiah, A; Ratna, JV; Annapurna, A. Preparation and characterization of quercetin and rutin cyclodextrin inclusioncomplexes. *Drug Development and Industrial Pharmacy*, 2007, 33(3), 245–253.

Stahl, W; van den Berg, H; Arthur, J; Bast, A; Dainty, J; Faulks, RM; Gärtner, C; Haenen, G; Hollman, P; Holst, B; Kelly, F.J; Polidori, M.C; Rice-Evans, C; Southon, S; van Vliet, T; Viña-Ribes, J; Williamson, G; Astley, SB. Bioavailability and Metabolism. *Molecular Aspects of Medicine*, 2002, 23, 39-100.

Stancanelli, R; Guardo, M; Cannavà, C; Guglielmo, G; Ficarra, P; Villari, V; Micali, N; Mazzaglia, A. Amphiphilic cyclodextrins as nanocarriers of genistein: A spectroscopic investigation pointing out the structural properties of the host/drug complex system. *Journal of Pharmaceutical Sciences*, 2010, 99, 3141–3149.

Stancanelli, R; Mazzaglia, M; Tommasini, S; Calabro, ML; Villari, V; Guardo, M; Ficarra, P; Ficarra, R. The enhancement of isoflavones water solubility by complexation with modified cyclodextrins: a spectroscopic investigation with implications in the pharmaceutical analysis, *Journal of Pharmaceutical and Biomedical Analysis*, 2007, 44, 980-984.

Străzĭsar, M; Andreňsek, S; Smidovnik, A. Effect of □-cyclodextrin on antioxidant activity of coumaric acids. *Food Chemistry*, 2008, 110(3), 636–642.

Szente, L; Vikmon, M; Szema´n, J; Otta, K. Methyl-□-cyclodextrin/paclitaxel aqueous solutions: a tool to test Cremophor-free paclitaxel. *STP Pharmaceutical Sciences*, 1999, 9 243–247.

Tamura, H; Takada, M. The Color Stability and Antioxidative Activ-ity of an Anthocyanin and □-Cyclodextrin Complex. *Functional Foods for Disease Prevention I*, 1998, 701, 157–171.

Tang, B; Ma, L; Wang, H; Zhang, G. Study on the supramolecular interaction of curcumin and beta-cyclodextrin by spectrophotometry and its analytical application. *Journal of Agricultural and Food Chemistry*, 2002, 50(6), 1355–1361.

Tommasini, S; Calabrò, M; Stancanelli, R; Donato, P; Costa, C; Catania, S. The inclusion complexes of hesperetin and its 7-rhamnoglucoside with (2-hydroxypropyl)-beta-cyclodextrin. *Journal of Pharmaceutical and BiomedicalAnalysis*, 2005, 39(3/4), 572–580).

Tomren, MA; Másson, M; Loftsson, T; Tønnesen, HH. Studies on curcumin and curcuminoids XXXI. Symmetric and asymmetric curcuminoids: Stability, activity and complexation with cyclodextrin. *International Journal of Pharmaceutics*, 2007, 338(1/2), 27–34.

Tønnesen, HH; Másson, M; Loftsson, T. Studies of curcumin and curcuminoids. XXVII. Cyclodextrin complexation: solubility, chemical and photochemical stability. *International Journal of Pharmaceutics*, 2002, 244(1/2), 127–135.

Usuda, M; Endo, T; Nagase, H; Tomono, K; Ueda, H. Interaction of antimalarial agent artemisinin with cyclodextrins. *Drug Development and Indutial Pharmacy,* 2000, 26(6), 613-9.

Venkateswaran, V; Fleshner, NE; Sugar, LM; Klotz, LH. Antioxidants block prostate cancer in lady transgenic mice. *Cancer Research*, 2004, 64, 5891-5896.

Vertzoni, M; Kartezini, T; Reppas, C; Archontaki, H; Valsami, G. Solubilization and quantification of lycopene in aqueous media in the form of cyclodextrin binary systems. *International Journal of Pharmaceutics*, 2006, 309, 115–122.

Vianna, HR; Cortes, SF; Ferreira, AJ; Capetinni, LSA; Schimitt, M; Almeida, AP; Massensini, AR; Lemos, VS. Antiarrhythmogenic and antioxidant effect of the flavonoid dioclein in a model of cardiac ischemia/reperfusion. *Planta Medica*, 2006, 72, 300–303.

Visioli, F; Bellosta, S; Galli, C. Oleuropein, the bitter principles of olives, enhances nitric oxide production by mouse macrophages. *Life Sciences,* 1998, 62, 541–546.

Waleczek, KJ; Cabral Marques, HM; Hempel, B; Schmidt, PC. Phase solubility studies of pure (-)-α-bisabolol and camomile essential oil with β-cyclodextrin. *European Journal of Pharmaceutics and Biopharmaceutics,* 2003, 55, 247–251.

Wang, J; Cao, Y; Sun, B; Wang, C. Characterisation of inclusion complex of trans-ferulic acid and hydroxypropyl-□-cyclodextrin. *Food Chemistry*, 2011, 124(3), 1069–1075.

Wang, J; Cao, Y; Sun, B; Wang, C. Physicochemical and release characterization of garlic oil-β- cyclodextrin inclusion complexes. *Food Chemistry*, 2011, 127(4), 1680–1685.

Wong, JW; Yuen, KH. Inclusion Complexation of Artemisinin with □-, □-, and □-Cyclodextrins. *Drug Delivery and Industrial Pharmacy*, 2003, 29, 1035-1044.

Wong, W; Yuen K.H. Improved oral bioavailability of artemisinin through inclusion complexation with β- and γ-cyclodextrins. *International Journal of Pharmaceutics*, 2001, 227, 177-185.

Wu, ZD; Guo, Q; Yang, X; Zhang, B. Encapsulation of a Flavonoid-rich Allium cepa L. var. agrogatum Don Extract in β-Cyclodextrin for Transdermal Drug Delivery. *Journal of Agricultural and Food Chemistry*, 2013, 61, 4914–4920.

Xu, Q; Wu, FG; Cao, JS; Chen, T; Jiang, JY; Saiki, I; Koda, A. Astilbin selectively induces dysfunction of liver-infiltrating cells – novel protection from liver damage. *European Journal of Pharmacology*, 1999, 377, 93–100.

Yamada, T; Komiya, T; Akaki, M. Formation of an inclusion complex of anthocyanin with cyclodextrin. *Agricultural and Biological Chemistry*, 1980, 44(6), 1411–1413.

Yamamoto, N; Kawabata, K; Sawada, K; Ueda, M; Fukuda, I; Kawasaki, K; Murakami, A; Ashida, H. Cardamonin stimulates glucose uptake through translocation of glucose transporter-4 in L6 myotubes, *Phytotherapy Research*, 2011, 25, 1218–1224.

Yang, LJ; Chen, W; Ma, SX; Gao, YT; Huang, R; Yan, SJ. Host–guest system of taxifolin and native cyclodextrin or its derivative: Preparation, characterization, inclusion mode, and solubilization. *Carbohydrate Polymers, 2011*; 85(3), 629–637.

Yang, LJ; Ma, SX; Zhou, SY; Chen, W; Yuan, MW; Yin, YQ. Preparation and characterization of inclusion complexes of naringenin with β -cyclodextrin or its derivative. *Carbohydrate Polymers*, 2013, 98(1), 861–869.

Yang, LJ; Yang, B; Chen, W; Huang, Yan, RS; Lin, J. Host-Guest System of Nimbin and β-Cyclodextrin or Its Derivatives: Preparation, Characterization, Inclusion Mode, and Solubilization. *Journal of Agricultural and Food Chemistry*, 2010, 58, 8545–8552.

Yao, LH; Jiang, YM; Shi, J; Tomás-Barberán, F.A.; Datta, N; Singanusong, R; Chen, SS. Flavonoids in Food and Their Health Benefits. *Plant Foods for Human Nutrition*, 2004, 59, 113-122.

Yatsu, FKJ; Koester, LS; Lula, I; Passos, JJ; Sinisterra, R; Bassani,VL. Multiple complexation of cyclodextrin with soy isoflavones present in an enriched fraction. Carbohydrate Polymers, 2013, 98(1), 726–735.

Yenesew, A; Derese, S; Midiwo, JO; Oketch, Rabah, HA; Lisgarten, J. Anti-plasmodial activities and X ray crystal structures of rotenoids from Millettiausaramensis subspecies usaramensis. *Phytochemistry*, 2003, 64, 773–779.

Yu, Z; Cui, M; Yan, C; Song, F; Liu, Z; Liu, S. Investigation of heptakis (2,6-di-O-methyl)-beta-cyclodextrin inclusion complexes with flavonoid glycosides by electrospray ionization mass spectrometry. *Rapid Communications in Mass Spectrometry: RCM*, 2007, 21(5), 683–690.

Zhang, C; Zhang, L; Yu, H; Bao, J; Sun, Z; Lu, R. Curcumin inhibits invasion and metastasis in K1 papillary thyroid cancer cells. *Food Chemistry*, 2013, 139, 1021–1028.

Zhang, G; Zhao, N; Hu, X; Tian, J. Interaction of alpinetin with bovine serum albumin: probing of the mechanism and binding site by spectroscopic methods, *Spectrochimica Acta A*, 2010, 76, 410–417.

Zhang, M; Li, J; Zhang, L; Chao, J. Preparation and spectral investiga- tion of inclusion complex of caffeic acid with hydroxypropyl-beta-cyclodextrin. *Spectrochimica Acta. Part A: Molecular and Biomolecular Spectroscopy*, 2009, 71(5), 1891–1895

Zhang, QF; Nie, HC; Shangguang, XC; Yin, ZP; Zheng, GD; Chen, JG. Aqueous Solubility and Stability Enhancement of Astilbin through Complexation with Cyclodextrins. *Journal of Agricultural and Food Chemistry*, 2013, 61, 151–156.

Zhang, W; Gong, X; Caia, Y; Zhang, C; Yub, X; Fan, J; Diao, G. Investigation of water-soluble inclusion complex of hypericin with β-cyclodextrin polymer. *Carbohydrate Polymers*, 2013, 95, 366– 370.

Zhao, M; Wang, H; Yang, B; Tao, H. Identification of cyclodextrin inclusion complex of chlorogenic acid and its antimicrobial activity. *Food Chemistry*, 2010; 120(4), 1138–1142.

Zheng, Y; Dong, LN; Liu, M; Chen, A; Feng, S; Wang, B; Sun, D. Effect of pH on the Complexation of Kaempferol-4′-glucoside with three β-Cyclodextrin Derivatives: Isothermal Titration Calorimetry and Spectroscopy Study. *Journal of Agricultural and Food Chemistry*, 2014, 62, 244–250.

In: Cyclodextrins
Editor: Francis G. Ramirez

ISBN: 978-1-63482-788-1

Chapter 5

EFFECT OF INCLUSION COMPLEXATION WITH CYCLODEXTRINS ON THE PHOTODEGRADATION PROFILE OF DRUGS

Snezana Agatonovic-Kustrin[1,*], David W. Morton[1] and Kailash C. Petkar[2]
[1]The School of Pharmacy and Applied Science, La Trobe Institute of Molecular Sciences, La Trobe University, Bendigo, Victoria, Australia
[2]Mil Laboratories Pvt. Ltd., Vadodara, Gujarat, India

ABSTRACT

Most drugs are hydrophobic in nature, and while this is good for their pharmacological activity, it often affects their absorption into the systemic circulation due to poor aqueous solubility. One of the common approaches for the improvement of drug solubility is by complexation with cyclodextrins (CDs). Moreover, it can also make possible the preparation of injectable aqueous formulations of poorly soluble drugs. It has been proven that the presence of CD can significantly increase the aqueous solubility of some drugs, and in addition be a good absorption enhancer. However, cyclodextrins are also known to affect the drug stability, either positively or negatively. The rate of drug photodegradation and distribution of photodegradants may be altered in the presence of CD [1-3]. Therefore, CDs have the potential to control chemical and photochemical reactions. For example, it has been shown that the presence of randomly methylated beta-cyclodextrin (RM-β-CD) slightly decreases the photostability of midazolam. The decreased stability was accompanied by the appearance of a new degradation product, an intermediate, 6-chloro-2-methyl-4-(2-fluorophenyl)-quinazoline and the dimer. In general, photosensitivity reactions have occurred in patients to whom light sensitive drugs have been administered, and have also been associated with the formation of toxic photoproducts. Such drugs decompose to form radical intermediates and highly reactive products, which react with the tissue cells resulting in adverse effects, making the detection and identification of photodegradants important. Thus, there is a need to

* Corresponding author: The School of Pharmacy and Applied Science, La Trobe Institute of Molecular Sciences, La Trobe University, PO Box 199, Bendigo, Victoria, Australia, 3552. Email: snezana@gmail.com.

determine the effect of cyclodextrins on the photodegradation profile of drugs when they are used in drug formulations.

INTRODUCTION

The photodegradation/photodecomposition of drug molecules is important to consider in drug development and drug formulation. Photodecomposition of a drug may result in a loss of potency, which reduces its bioavailability and may give rise to adverse photosensitivity (phototoxic and photoallergic reactions) in patients [4]. CDs have the ability to improve the solubility of drugs, but stability may be either enhanced or reduced when they complex with CD.

CDs are cyclic oligosaccharides consisting of α-1,4-linked D-glucopyranose units [5]. The glucose units are all in the classical C1 chair conformation [6]. This, in combination with the lack of free rotation around the glycosidic bonds results in the characteristic conical shape [6] of the CD molecule. CDs, comprising only of 6-8 glucopyranoside units, can be topologically represented as toroids with the larger and the smaller openings of the toroid exposing to the solvent secondary and primary hydroxyl groups respectively. The secondary hydroxyl groups on the C-2 and C-3 atoms of the glucose units are located on the wider edge of the torus, while the primary hydroxyls on C-6 form the narrower side (Figure 1). Because of this arrangement, the interior of the toroid is not hydrophobic, but substantially less hydrophilic than the aqueous environment making it suitable to host other hydrophobic molecules, while the hydrophilic nature of the exterior contributes to the water solubility of CDs.

The most common naturally occurring CDs are the α-, β- and γ-CDs, with six, seven or eight glucopyranose units, respectively. CDs containing nine, ten, eleven, twelve and thirteen glucopyranose (designated as δ-, ε-, ζ-, η-and θ-CD) have been reported, but their complexing abilities are negligible [5, 7]. β-CD has a distinctly low water solubility (1.85 % w/v) when compared to α-CD (14.5 % w/v) and γ-CD (23.2 % w/v), due to formation of intramolecular hydrogen bonds between the secondary hydroxyl groups (HO-3·HO-2) of adjacent glucose units in the β-CD torus. This hydrogen bonding increases molecular rigidity, making it more inclined to crystallization [8], which prevents hydration of the CD molecule resulting in poor aqueous solubility. In α-CD, a ring curvature increases the hydrogen bond lengths and consequently weakens the bonds to such an extent, that some bond breaking occurs. As a result, α-CD is flexible in solution and more easily hydrated [9]. γ-CD has a non-coplanar, flexible structure and is therefore the most soluble of the three [10].

The CD cavity is lined by two rings of C-H groups (H-3 and H-5) and a ring of glycosidic ethereal oxygen atoms. The non-bonding electron pairs of the glycosidic oxygen bridges are directed towards the inside of the cavity, producing a high electron density and Lewis base character. The C-H groups comprising H-1, H-2 and H-3 and the polar sugar hydroxy groups are located on the exterior of the molecule [10-12]. As a result, the interior of the cavity is lipophilic, while the external faces of the CDs are hydrophilic (Figure 1). Due to this amphipathic property, CDs can form soluble, reversible inclusion complexes with poorly water-soluble compounds. They are commonly used to improve the solubility of poorly water-soluble drugs, by a process of inclusion complexation of the drug into the CD cavity. This cavity size is the major factor in determining which CD should be used for complexation with a particular drug. α-CDs, having a smaller cavity (sixglucopyranose units) can incorporate low molecular weight compounds or compounds with aliphatic side chains. γ-CDs, on the other hand have eight-glucopyranose units, resulting in a substantially larger cavity so that the CD hydrophobic charges cannot effectively interact with many molecules to facilitate complexation. γ-CDs will therefore accommodate larger molecules such as macrocyclic and steroidal molecules. The cavity diameter of β-CDs (seven glucopyranose units) is therefore able to accommodate most drug compounds and will complex aromatic and heterocyclic molecules. Hence, β-CD is most the most commonly used CD complexing agent in drug formulations [13].

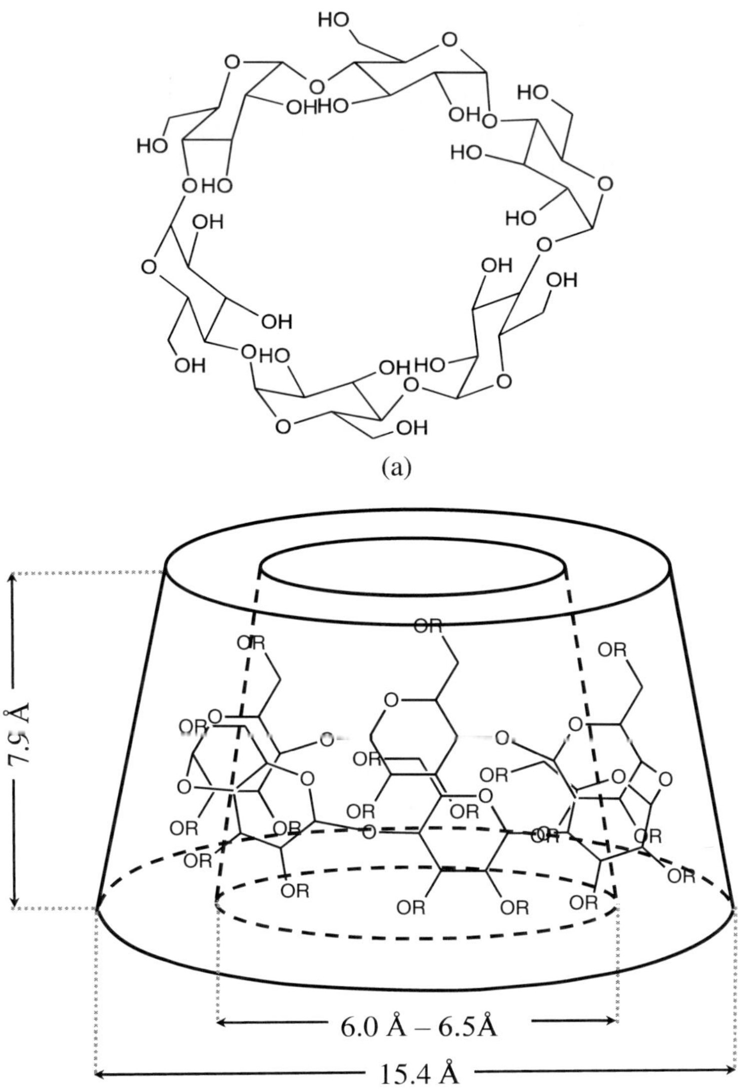

(a)

(b)

Figure 1. Chemical structure of a-cyclodextrins (a) and schematic representation (b) of the cyclodextrin molecule.

Cyclodextrin Derivatives

The poor aqueous solubility of β-CD, its tendency to crystallize in aqueous solution, and the toxicity effects observed after parenteral administration, means that it is generally not suitable for use in drug formulations [14]. To overcome these limitations, a number of modified β-CDs derivatives with improved physicochemical properties and reduced parenteral toxicity have been developed. Since intramoleculer hydrogen bonding decreases aqueous solubility, substitution of any of the hydrogen bond forming hydroxyl groups, dramatically increases water solubility [10, 15]. Permethylation of β-CD also increases the height of the torus, reduces the cavity diameter at the primary side and enlarges the central cavity. When two-thirds of the hydroxyl groups are selectively methylated, as in case of 2,6-*O*-dimethyl-β-CD (DM-β-CD), water solubility is increased to 57 % w/v at 25°C. Complete alkylation results in heptakis (2,3,6-tri-*O*-methyl)-β-CD (TM-β-CD) which has a reduced water solubility of 31 % w/v [14, 16]. The solubility of the methylated derivatives decreases with increasing temperature [15].

Hydroxyalkylated Cyclodextrins

Condensation of β-CD with propylene oxide in alkaline solution produces highly soluble 2-hydroxypropyl-β-CD (2HP-β-CD), with a water solubility exceeding 60% w/v [17]. The increased water solubility and favourable toxicological profile of 2HP-β-CD [18, 19], has prompted many pharmaceutical companies to investigate its usefulness as a drug carrier in parenteral and oral formulations [20, 21]. The number of 2HP-β-CD derivatives generated, if random substitution is assumed, is roughly 130,000, and considering that the 2-hydroxypropyl groups introduce optically active centres, the total number of geometric and optical isomers is even greater [22]. The secondary hydroxyl groups (OH-2 and OH-3) on the CD molecule are slightly more acidic than the primary hydroxyl groups (OH-6). Therefore, alkylation of the least sterically crowded functionality (OH-6), will be favoured in strong basic solutions, while alkylation of OH-2, the most acidic and hindered hydroxyl groups, is favoured in weakly basic solutions [22]. The chemical alkylation of CDs is therefore not totally random and some degree of regioselectivity is possible.

Requirements for Inclusion Complexation

The most important requirement for the formation of a stable complex is that the guest molecule must fit entirely, or at least partially, into the CD cavity. Complex formation is possible with fairly bulky molecules, provided that the guest molecule contains side chains or groups which, based on their molecular dimensions, are amenable to inclusion. In these instances, partial inclusion complexes are formed [9]. The stability of an inclusion complex also depends on the polarity of the guest molecule. In general, hydrophobic molecules or residues show greater affinity for the CD cavity than hydrophilic ones, while highly hydrophilic molecules complex very weakly or not at all [23].

MECHANISM OF INCLUSION COMPLEXATION

The structure of the inclusion complex in solution may differ appreciably from the solid state. In solution, the guest molecule fits wholly or partially into the central cavity of the CD host molecule and the whole complex is surrounded and solvated by a multilayer hydrate shell. In the crystalline state, however, the guest molecule may fit into a void space or interstice in the crystal lattice and not necessarily into the central cavity of the CD torus [11]. No covalent bonds are formed or broken during drug - CD complex formation. When they dissolve in water, the complexes rapidly dissociate with free drug molecules in equilibrium with drug molecules bound within the CD cavity [22].

The formation of a CD inclusion complex in aqueous solution occurs via the following steps [10, 24].

1. The guest and CD molecules approach each other;
2. Water is released from the cavity into the solution ($+\Delta H$);
3. Complex forms due to interaction (Van der Waals interactions and/or hydrogen bonding) of the guest molecule or its substituent groups with groups on the rim or inside the empty CD cavity;
4. Rehydration of the exposed part of the guest molecule after complex formation.

Although release of water molecules from the CD cavity is probably an important driving force for complex formation, other forces may also be involved. These include van der Waals interactions (permanent induced-dipole-dipole interactions and London dispersion forces), hydrogen bonding between the polar functional groups of guest molecules and the hydroxyl groups of CDs, hydrophobic interactions, release of strain energy in the ring frame system of the CD and changes in solvent surface tension [23].

PHARMACEUTICAL APPLICATIONS OF CYCLODEXTRINS

The huge costs involved in developing new pharmacologically active chemical entities has prompted pharmaceutical companies to focus on designing more effective formulations of old drugs. CD technology provides formulators the opportunity to overcome negative aspects of the physicochemical properties of valuable drug molecules and potentially improve their pharmacological, pharmacokinetic and/or adverse effect profiles.

The most common pharmaceutical application of CDs is to increase the aqueous solubility of poorly soluble drugs and to increase their bioavailability and stability [10, 25]. In addition, drug-CD complexes can be used to reduce gastrointestinal drug irritation, convert liquid drugs into microcrystalline or amorphous powder, and prevent drug-drug and drug-excipient interactions. The extent to which CDs increase drug solubility cannot be accurately predicted, although three empirically-based trends have been observed:

1. The lower the aqueous solubility of a drug, the greater improvement in its solubility when it complexes with CD [26];
2. The hydrophobicity of the CD used will affect its solubilising capacity. For instance, the most hydrophobic derivative (e.g. DM-β-CD) is generally found to have the highest solubilising potential [8, 27]; and
3. CD derivatives with a lower degree of substitution are better solubilizers than the same derivatives with a higher degree of substitution [22].

The CD complex of a low water soluble drug is usually more soluble than the free drug itself. The drug-CD complexalso has improved wettability and dissolves more rapidly. Complexation may also produce a marked reduction in crystal size or a decrease in crystallinity leading to a more rapidly dissolving amorphous state [28]. Therefore, the increased dissolution rate is generally the result of improved wettability, decreased crystallinity, increased surface area, and the increased solubility of the drug-CD complex [10] resulting in more rapid uptake into the blood stream and consequently leading to an increase in bioavailability [29].

Enhanced bioavailability is not only limited to CD complexes administered orally, but may also be observed for dermally, rectally and parenterally administered complex formulations [28]. The advantage of incorporating a CD into an intravenous formulation is that higher than normal concentrations of a poorly water-soluble drug in the formulation are possible resulting in higher drug plasma concentrations when they are administered to patients.

DRUG STABILITY

CD inclusion complexation may influence drug stability and depending on the inclusion mode of the drug in the CD cavity, the degradation rate may be decreased, increased or remain unaffected [11]. Also note that the stereochemistry and/or stoichiometry of the drug-CD interaction may affect the stabilizing /destabilizing effects of complexation. Degradation reactions that are observed include hydrolysis, oxidation, stereochemical transformations, racemization and other forms of isomerization, polymerization, thermal decomposition, photodegradation and even enzymatic degradation [10, 15, 28].

For indomethacin CD complexes (Figure 2), hydrolysis is accelerated in the presence of *α*-CD, but retarded by β-CD. The larger cavity of β-CD may be able to include more of the indomethacin molecule or its reactive group than *α*-CD resulting in a lower rate of degradation. Where indomethacin is only partly included into the smaller *α*-CD cavity and the reactive centre is sterically fixed in close proximity to the hydroxyl groups of the CD, a catalytic effect is observed [28] resulting in an increased degradation rate.

Tranilast (an anti-allergic drug) (Figure 3), forms a 2:1 (guest:host) complex with *γ*-CD at low CD concentrations which results in an increase in the rate of the tranilast dimerization degradation reaction. However, a significant reduction in the rate of dimerization is observed when 1:1 and 1:2 complexes form at higher CD concentrations [30]. It is likely that the reduction in the rate of dimerization is due to the complete encapsulation of tranilast within the CD cavity in the 1:1 and 1:2 complexes, while only partial inclusion occurs in the 2:1 complex.

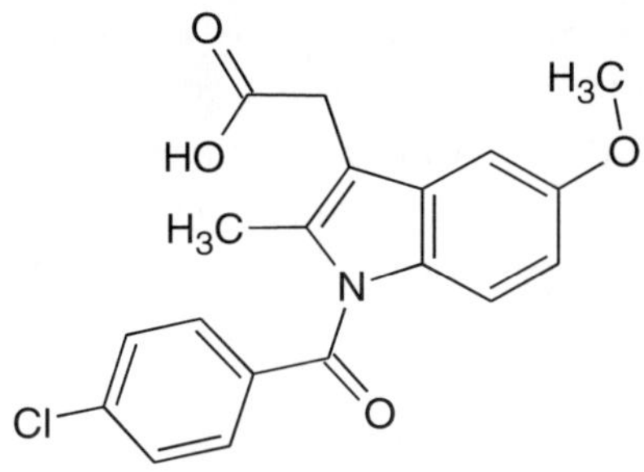

Figure 2. Chemical structure of indomethacin.

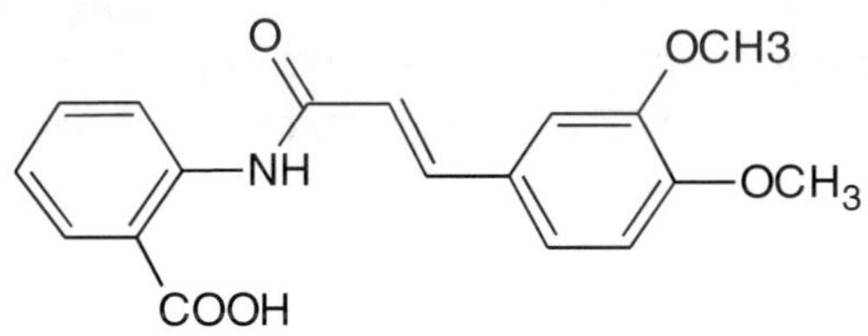

Figure 3. Chemical structure of tranilast.

CD complexes of steroids have been used to enhance solubility, improve dissolution rates, increase drug product stability, and improve chromatographic selectivity. While many applications of steroid-CD complexes have been reported, very little is known about the nature of these complexes. It is possible for either the steroidal A/B ring or C/D ring to be included into the CD cavity. The structural features of a steroidal molecule that lead to stabilization when it complexes with CD are not well understood. Torricelli et al. reported an improvement in the stability of kneaded mixtures of 6-methylenandrosta-1,4-diene-3,17-dione (a steroidal molecule) with β-CD [31]. However, there was no observed improvement in stability for co-ground mixtures of the two components. This means thatit is the inclusion of the steroid molecule into the CD cavity that results in the decreased degradation rate, rather than just the presence of β-CD in the mixture. Work by Andersen and Bundgaard highlight the importance of the size of the cyclodextrin cavity on complex stability. They studied the effect of CD complexation on the base catalyzed rearrangement of betamethasone 17-valerate [32] and found that the rate of degradation increased when β-CD was used, but decreased when eitherTM-β-CD or γ-CD were used.

Cyclodextrin Inclusion Complexation and Drug Photostability

The lipophilic CD central cavity is lined with skeletal carbons and ethereal oxygens of the glucose residues and is similar in polarity to an aqueous ethanolic solution [33]. The cavity provides a lipophilic microenvironment into which suitably sized drug molecules may enter. No covalent bonds are formed or broken during drug-CD complex formation, so these complexes readily dissociate in aqueous solution. For some drugs, complexation with CD results in a change to their physical and chemical properties. CDs have the potential to control chemical and photochemical reactions. This is due to the microplanarity of the host cavity and limited molecular mobility of the guest molecule, due to steric constraints. The nature of the lowest excited states of the guest molecule, deactivation pathways and the fate of the reaction may be modulated by the CD microenvironment. Thus, structural changes in drug molecules that occur when they complex with CD may also accelerate drug degradation [34]. The rate of secondary reactions may also be influenced and the chemical evolution of reaction intermediates controlled [3].

Because CDs are mainly used to increase the solubility of poorly soluble drugs, most investigations have focused on the photostability of drugs in solution [35] so there is a limited amount of published information on the effect of drug-CD photostability in the solid state [36]. Regardless of whether a drug is in a solution or solid form, an improvement in the photostability of a drug resulting from complexation with CD depends on the particular CD used. For example, the photostability of emetine and cephaeline in aqueous solution was improved by complexation with γ-CD and DM-β-CD, but decreased slightly when they were complexed with β-CD [37].

The effects of CD inclusion complexation on drug photostability is similar to its impact on general drug stability, in that it may increase, decrease, or have no effect on the rate of photodegradation of the drug and may even change the nature of the photodegradants [38]. The photodegradation kinetics and degradation pathways of a drug molecule are different within the CD cavity when compared to the external aqueous environment, because the interior of the cavity constitutes an isolated environment where the included species is usually present as a single molecule. The photochemistry is therefore generally restricted to intramolecular events, except in cases where there is more than one molecule present within the cavity [11].

Aqueous dispersions of metronidazolebenzoate (Figure 4a) turn yellow within 2-3 days of exposure to daylight due to photochemical surface oxidation of the crystals. However, suspensions of the β-CD inclusion complex remain colourless (and stable) for more than 3 months [39]. However, complexation of chlorpromazine (Figure 4b) with DM-β-CD accelerates the rate of dechlorination to promazine, while complexation of phenothiazine (Figure 4c) with DM-β-CD unexpectedly inhibits the oxidation and polymerization degradation reaction.

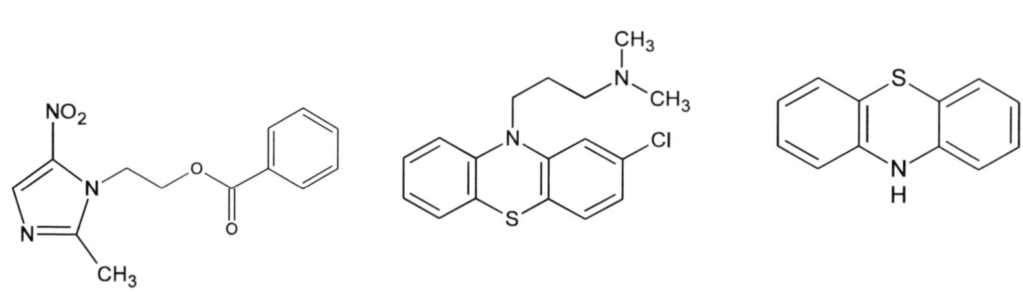

Figure 4. Chemical structures of: (a) metronidazole benzoate, (b) chlorpromazine, and (c) phenothiazine.

The rate of clomipramine hydrochloride photodegradation in solution was decreased bycomplexation with β-CD under both aerobic and anaerobic conditions [40]. For aqueous solutions of emetine and cephaeline, photostability was improved when they formed complexes with both γ-CD and DM-β-CD, but slightly decreased when they formed complexes withβ-CD [37]. Sofalcone (2'-carboxymethoxy-4,4"-bis (3- methyl -2-butenyloxy) chalcone) (CBC) (Figure 5) is an oral gastrointestinal drug [41] and has apreventive effect against various types of gastric ulcers. It is a synthetic derivative of sophoradin [42], a natural product isolated from the root of the Chinese medical plant *Sophoratonkinensis* (also known as *Sophorasubprostrata*). However, the cinnamoyl moiety undergoes (*E*) - (*Z*) photoisomerization resulting in the loss of pharmacological activity. If it forms a 1:1 complex with α-CD, the inclusion of CBC into the small cavity severely hinders rotation of the double bond of the cinnamoyl group, resulting in a decrease in the rate of isomerization. However, in a 1:2 complex with γ-CD, the rate of photoisomerization of CBC is significantly higher. Similar increases in degradation rates have been reported for CBC in organic solvents (e.g. methanol and ethanol). It was therefore proposed that since the inclusion process introduces CBC into a less hydrated environment, namely the cavity of the γ-CD dimer, it would be expected that the rate of photoisomerization would increase [2]. Photostability of the shikon in enantiomer, the main pharmacologically active component of a wound healing preparation used in the East, is increased two-fold in the presence of 15-20 % w/v HP-β-CD due to itsreactive side chain being present in the CD cavity [43].

The rate of photodegradation of vitamin D_3 and protriptyline (Figure 6) are reduced bycomplexation with DM-β-CD in aqueous solution [16, 44, 45]. Irradiation of protriptyline in oxygenated solutions results in the formation of an epoxide and various hydroxylated products. Epoxide formation is suppressed by the addition of DM-β-CD, but the formation of hydroxypropyl- and desmethyl-photodegradants increases. Under anaerobic conditions, degradation of protriptyline also results in the formation of a cyclobutyl dimer degradant. In the presence of DM-β-CD, this dimerization reaction is effectively suppressed. This selectivity is attributed to steric shielding of the reactive site, brought about by partial inclusion of protriptyline in the hydrophobic CD cavity [45].

For benzaldehyde, the rate of photo-oxidation is reduced significantly in the solid-state by α-CD, and to a lesser extent by β-CD and γ-CD [46], while the poor photostability of molsidomine could not be improved by complexation with α-CD [47]. In addition, γ-CD did not provide better photoprotection of molsidomine than lactose, mannitol or sodium chloride in the same weight ratios [48]. The rate of photodegradation of clofibrate decreased in solid 1:1 complexes with β-CD and γ-CD [49] and was subsequently confirmed by Tomono et al. [36].

A number of studies have tried to improve the solid-state photostability of nifedipine using CD complexation. One of the first attempts to enhance itsphotostability was reported by Tomono et al. [36] in 1988 (Figure 7). While Mielcarek et al. [50] observed a 5-fold improvement in solid-state nifedipine photostability for a kneaded β-CD 1:1 complex, others noted only marginal improvements when α-CD, β-CD, γ-CD, DM-β-CD, TM-β-CD, and β-CD polymer inclusion complexes were used [36]. Complexation of nifedipine with β-CD, γ-CD, HP-β-CD and RM-β-CD did not to prevent significant photodegradation in aqueous solution [36]. In the solid-state, the extent of photodegradation of 1:1 α-CD, β-CD and γ-CD spray-dried complexes were comparable to a 1:1 nifedipine: HP-β-CD physical mixture, suggesting that there was minimal photoprotection provided to nifedipine by these CDs [47]. All four CDs were ineffective in stabilizing solid-state nifedipine and appeared to promote photodegradation. Nifedipine, alone or in the presence of DM-β-CD, underwent 25-30 % degradation after 6 hours exposure to light, but this increased to between 36 % and 44 % in the presence of β-CD, TM-β-CD, and the β-CD polymer [47].

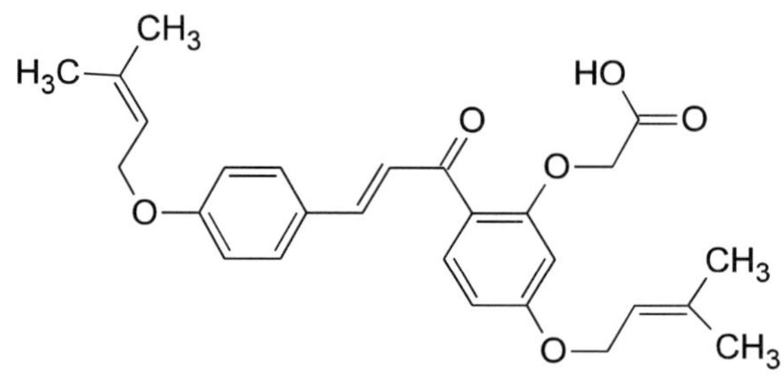

Figure 5. Chemical structure of sofalcone.

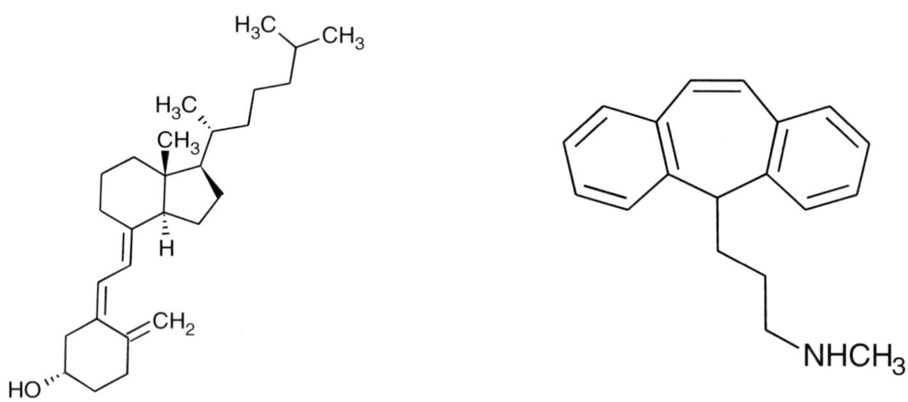

Figure 6. Chemical structures of vitamin D3 and protriptyline.

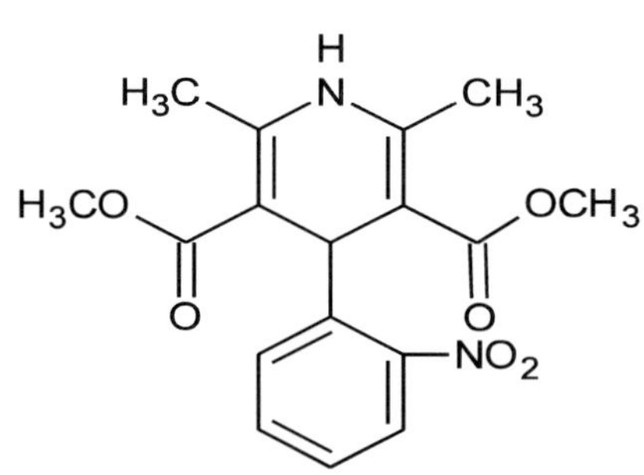

Figure 7. Chemical structure of nifedipine.

Quantitative structure activity relationship (QSAR) studies indicate that the major driving forces for nifedipine complexation, in addition to CD concentration, are hydrophobicity and van der Waals interactions (3D solubility parameters, hydrophilic surface area and differential connectivity index) [29].

In the solid state, furosemide (Figure 8) degrades by less than 2 % after 2 hours exposure to light in the presence of DM-β-CD and TM-β-CD, but this increases to approximately 12 % and 20 % for β-CD and the β-CD polymer, respectively. Although photostability was increased in most cases, the extent of protection varied considerably.

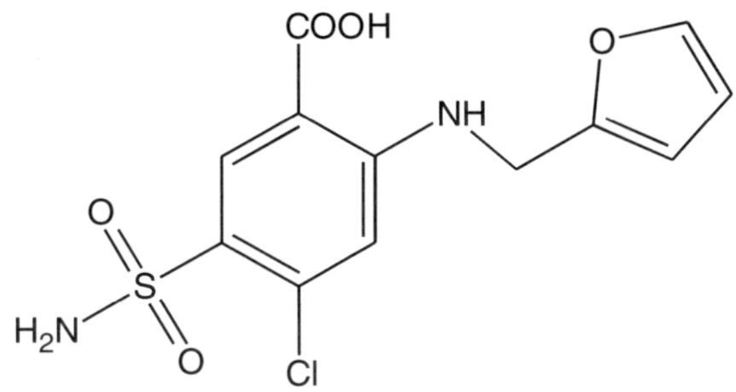

Figure 8. Chemical structure of furosemide.

For midazolam (Figure 9), the rate of photodegradation and the nature and distribution of photodegradants may be altered dueto complexation with CDs [1-3]. It is a potent anxiolytic drug with sedative properties and is susceptible to light degradation. Due to its short half-life, myorelaxant, and amnestic properties, midazolamis are commonly used as a preoperative anesthetic agent, especially for pediatric patients [51, 52]. Intranasal administration of midazolam has been extensively studied in epileptic patients and found to be an effective, safe and inexpensive means of rapidly achieving seizure control [53-56]. However, in order to deliver therapeutic midazolam doses, nasal preparations with a midazolam concentration, exceeding the aqueous solubility of midazolam are required to be developed. Midazolam's solubility is increased under acidic conditions (pH < 4) due to ionization and ring opening, where midazolam is reversibly converted to the corresponding benzophenone, an open ring form with a highly ionizable primary amino group, resulting in the increased water solubility. However, intranasal administration of acidic midazolam solutions would result in severe irritation and swelling in the nasal cavity. If the solubility of midazolam is improved by complexation with RM-β-CD, it undergoes a highly sequence-selective photoreaction pathway. RM-β-CD alters the photobehaviour of a midazolam molecule, by changing the ground state distribution of reactive and non-reactive conformers ("conformational control") resulting in selectivity [57], and thus the reaction can be caused to proceed along one of the competing pathways [58].

A variety of non-covalent forces (i.e. van der Waals forces, hydrophobic interactions and other interactions) are involved in the formation of a stable complex with RM-β-CD that should protect the midazolam molecule against the attack of other reactive molecules and thus increase its chemical stability [15]. CDs are known to accelerate or slow down various reactions. For example, when an ester group of a guest molecule is fixed close to the catalytic site of CD (hydroxyl group of the sugar), the ester will be hydrolyzed faster. The rate of hydrolysis will decrease when the ester is inside the CD cavity. The molecular size of midazolam, calculated with Molecular Modeling pro 5.10 [6] is estimated to be 14.67 Å, in molecular length (x), width (y) 11.59 Å, and molecular depth (z) 4.27 Å. On the other hand, the β-CD diameter inside cavity is 6.0-6.5 Å, while its outside diameter is 15.4 Å and its height is 7.9 Å [35, 36], which corresponds approximately to the size of an aromatic ring. The calculated size of an aromatic ring is 6.18 Å in length, 6.82 Å in width and 3.54 Å in depth. Therefore, the β-CD cavity can accommodate slightly more than one aromatic ring.

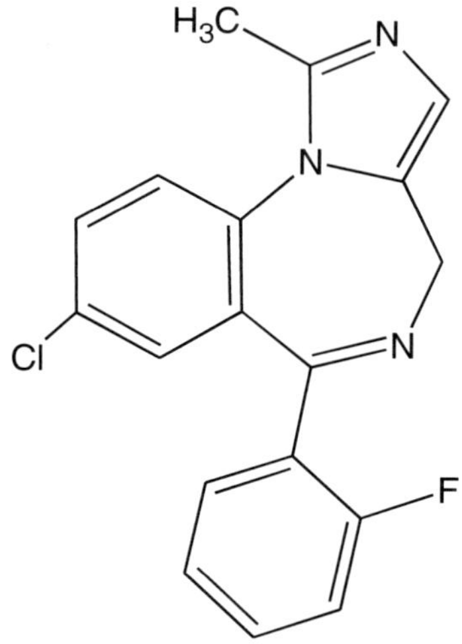

Figure 9. Chemical structure of midazolam.

NMR studies have confirmed the midazolam-β-CD complex formation by the shift in the peaks observed for both the guest molecule and the CD molecule [59]. Proton and ^{13}C-NMR have provided an idea how the guest substrate is positioned in the CD cavity. The shielding of the β-CD cavity protons and associated shift changes in the signal for the midazolam protons in the mixtures of CD and midazolam are attributed to the aromatic ring penetrating into the CD cavity, thereby confirming the formation of the inclusion complex. The structure of the midazolam-β-CD complex was established using two dimensional NMR (ROESY) spectral data [60]. The signals for the protons belonging to the aromatic ring containing fluorine, exhibit strong cross correlation peaks with CD cavity protons. Taking into account the 1:1 stoichiometry of the complex it was concluded that the fluorine containing ring penetrates the β-CD cavity resulting in the formation of 1:1 complex. The signals for the H-8 of the chlorine containing aromatic ring (next to the chlorine substituent) also exhibited a strong interaction with the protons inside the CD cavity. However, H-9 and H-10 did not show any cross peaks with the CD cavity protons (Figure 10). Thus, the possibility of another 1:1 complex involving penetration of the chlorine containing ring was ruled out because the interaction of the protons from the ring containing chlorine with the cavity protons was not evident. We can conclude that the fluorine containing aromatic ring and the chlorine substituent and H-8 from the other aromatic ring are inside the RM- β-CD cavity. Additionally, the fact that midazolam enters the RM-β-CD by the wider rim and there are some additional interactions due to the methoxy groups as compared to the β-CD-midazolam inclusion complex is consistent with the findings of Mathiron et al. who proposed the following structure for the 2,6-di-O-methyl-β-CD (DIMEB)-midazolam complex (Figure 10) [11].

The "cage effect" of inclusion of molecules in the CD cavity cause radical pairs (1) to undergo condensation reaction to the molecule (2) before being able to diffuse into the surrounding medium (Figure 11). This occurs to a lesser extent in the solution outside of the CD cavity due to more rapid diffusion [61].

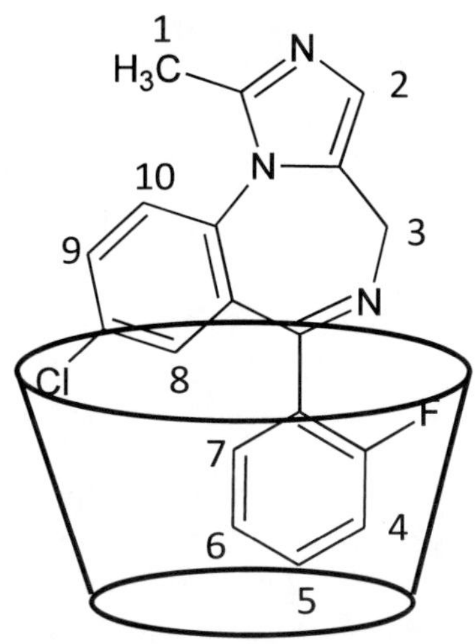

Figure 10. Proposed model for the midazolam RM-β-CD inclusion complex.

Another interesting example is the natural food dye curcumin (Figure 12), which could be used as a pharmaceutical excipient. However, the use of curcumin is limited due to its low water solubility in acidic or neutral conditions, high decomposition rate in alkaline media, and photodegradation in organic solvents. In an attempt to study stabilizing effect of CDs on curcumin, Tønnesen et al. investigated the effect of ionic and non-ionic CDs on the solubilization and the stability of curcumin in aqueous solution. They reported strong improvement in the hydrolytic stability of curcumin under alkaline conditions when complexed with CDs. However, increased photodecomposition rate was reported under alkaline conditions when compared to a curcumin solution in organic solvents. The cavity size, the charge and bulkiness of the CD side-chains, influences the stability constant for complexation and the degradation rate of the curcumin molecule [62].

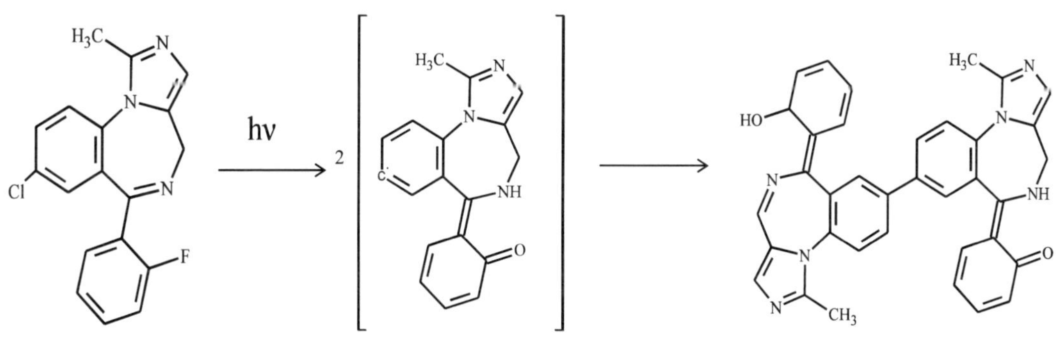

(1) (2)

Figure 11. Condensation of the midazolam photodegradant radical inside the CD cavity.

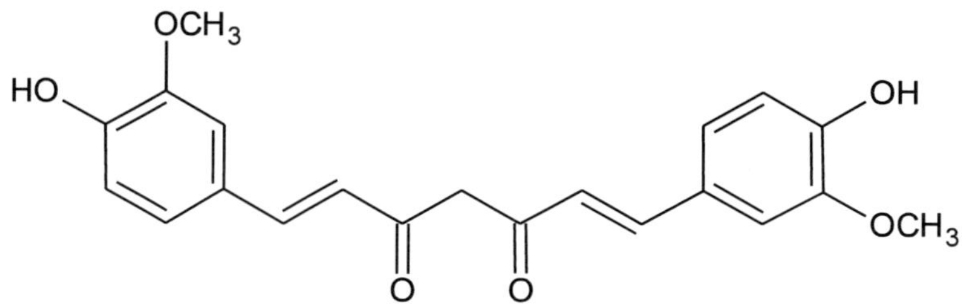

Figure 12. Chemical structure of curcumin.

Piroxicam (4-hydroxy-2-methyl-*N*-pyrido-2-yl-2*H*-1,2-benzothiazine-3-carboxamide-1,1-dioxide) (Figure 13) is a non-steroidal anti-inflammatory drug with analgesic, antipyretic and anti-inflammatory activity. However, its low aqueous solubility and relatively high rate of photodegradation limits its use. Photodegradation of piroxicam results in the formation of the less potent nonacidic ether carbonate ampiroxicam with the degradation rate dependent on the wavelength, intensity and time of exposure to light [63]. Complex formation with β-CD does not have a significant effect on the rate of photodegradation, possibly due to the site of photodegradation on the piroxicam molecule being outside of the CD cavity [64]. Also, the rate of photodegradation follows a non-linear relationship indicating the reaction scheme of photodegradation is more complex than a simple bimolecular encounter, with secondary photodecomposition and autocatalytic processes compounding the kinetic complexity.

More recent work has shown the effect of 1:2complexationofpiroxicamwith 2HP-β-CD resulted in 2-methyl-1,2-benzisothiazol-3(2*H*)-one-1,1-dioxide being the most common (60%) photodegradation product of piroxicam [63, 65]. Piroxicam was stable under the influence of daylight up to 30 days. Thereafter, the complex with CD began to decompose as evidenced by the change of position, intensity and number of bands in its FTIR spectrum. In contrast to earlier reports, they reported that the photostability of 1:2 piroxicam 2HP-β-CDcomplex increased, when compared to uncomplexed piroxicam during one month of exposure.

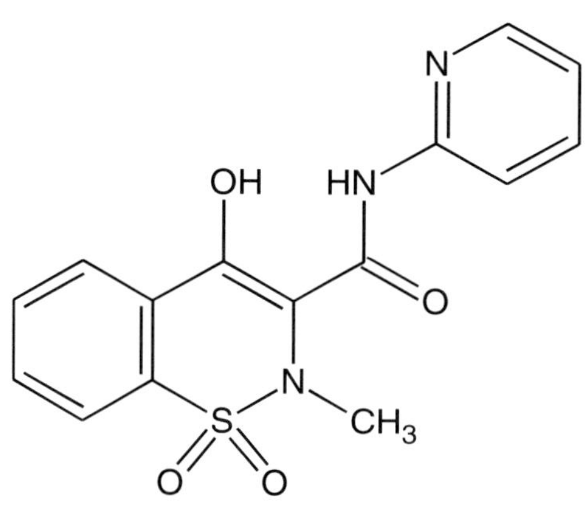

Figure 13. Chemical structure of piroxicam.

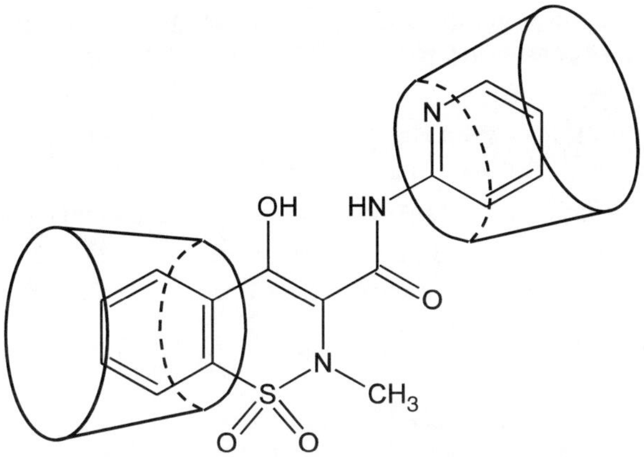

B. D. Glass, M. E. Brown, S. Daya, et al., Influence of cyclodextrins on the photostability of selected drug molecules in solution and the solid-state, International Journal of Photoenergy, vol. 3, no. 4, pp. 205-211, 2001. doi:10.1155/S1110662X01000277.

Figure 14. Proposed model for the inclusion of piroxicam into β-CD.

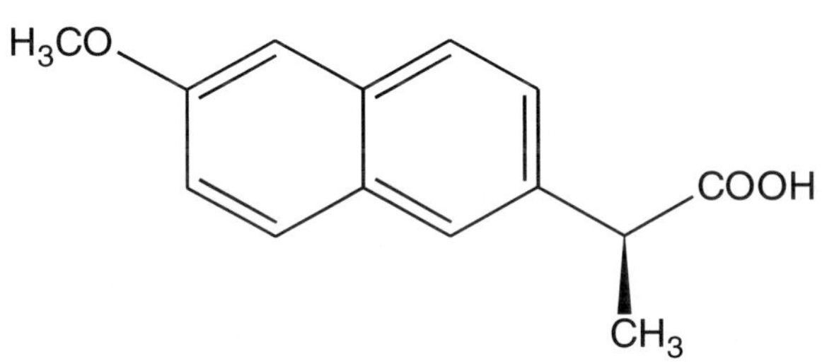

Figure 15. Chemical structure of naproxen.

Another example of a photosensitive nonsteroidal anti-inflammatory drug is naproxen ((+)-(S)-2-(6-methoxynaphthalen-2-yl) propanoic acid) (Figure 15). It is photosensitive low aqueous solubility compound, which can be degraded upon exposure to light giving rise to pharmacologically inactive and unsafe degradants. In addition, the photodegradants were found to be more hydrophobic inducing increased toxicity as compared to naproxen. Photodegradation of naproxen produces two ketoanologues of naproxen, namely 4-isobutylacetophenone, 1-(6-methoxy-2-naphthyl) ethanol and 2-acetyl-6-methoxynaphthalene belonging to the group of aromatic ketones, whose presence increases the rate of photodegradation upon exposure to light. Several structural studies indicated that the inclusion complex of β -CD increases photostability of naproxen, as compared to the uncomplexed naproxen [66].

Conclusion

Although there is long list of drug molecules with enhanced photostability resulting from CD complexation (e.g. midazolam, chlorpromazine, clomipramine, vitamin D3, nifedipine, clofibrate, pyridoxine, retinol acetate etc.), the presence of CDs may either decelerate (diclofenac) or show no effect (e.g. piroxicam, molsidomine) on the rate of photodegradation of certain other drug molecules, and may even change the nature of the photodegradantsas in the case of midazolam [38]. The differences in the photostability of drugs when complexed with CDs may be attributed to the mode of inclusion complexation which helps to understand whether the site of degradation is encapsulated or not into the cavity of the CD molecule. The photodegradation chemistry of a drug molecule is different when encapsulated because the interior of the CD cavity constitutes an isolated environment where the included species is usually present as a single molecule. The photochemistry is therefore generally restricted to intramolecular events, except in cases of multiple occupancy of the CD cavity [11].

REFERENCES

[1] Mielcarek J. Photochemical stability of the inclusion complexes formed by modified 1,4-dihydropyridine derivatives with beta-cyclodextrin. *Journal of Biomedial and Pharmaceutical Analysis.* 1997;15(6):681-6.

[2] Tadanobu U., Keiko I., Fumitoshi H., Kaneto U. Stoichiometry-dependent changes of solubility and photoreactivity of an antiulcer agent, 2′-carboxymethoxy-4,4′-bis(3-methyl-2-butenyloxy) chalcone, in cyclodextrin inclusion complexes. *European Journal of Pharmaceutical Sciences.* 1993;1(2):81-7.

[3] Bortolus P., Grabner G., Kohler G., Monti S. Photochemistry of cyclodextrin host-guest complexes. *Coordin Chem Rev.* 1993;125(1-2): 261-8.

[4] Damiani E., Tursilli R., Casolari A., Astolfi P., Greci L., Scalia S. Effect of complexation with randomly methylated β-cyclodextrin on the aqueous solubility, photostability and antioxidant activity of an indolinonic nitroxide radical. *Free Radical Research.* 2005;39(1):41-9.

[5] Freudenberg K., Cramer F., Plieninger H., A.-G. K., *inventorsVerfahren zur Herstellung von Einschlusverbindungen physiologisch wirksamer organischer Verbindungen.* Germany 1953.

[6] Szejtli J. Introduction and General Overview of Cyclodextrin Chemistry. *Chemical Reviews.* 1998;98(5):1743-54.

[7] Loftsson T., Friðriksdóttir H., Guðmundsdóttir K. T. The effect of water-soluble polymers on aqueous solubility of drugs. *Int J Pharm.* 1996; 127(2):293-6.

[8] Szejtli J. Medicinal Applications of Cyclodextrins. *Medicinal Research Reviews.* 1994;14(3):353-86.

[9] Cabral-Marques H. M. Structure and properties of cyclodextrins. Inclusion complex formation. *Rev Port Farm.* 1994; 44:77-84.

[10] Bekers O., Uijtendaal E. V., Beijnen J. H., Bult A., Underberg W. J. M. Cyclodextrins in the pharmaceutical field. *Drug Development and Industrial Pharmacy.* 1991;17(11):1503-49.

[11] Jones S. P., Grant D. J. W., Hadgraft J., Parr G. D. Cyclodextrins in the pharmaceutical sciences. Part I: Preparation, structure and properties of cyclodextrins and cyclodextrin inclusion compounds. *Acta Pharmaceutica Technologica.* 1984;30(3):213-23.

[12] Myles A. M. C., Barlow D. J., France G., Lawrence M. J. Analysis and modelling of the structures of β-cyclodextrin complexes. *Biochim Biophys Acta.* 1994;1199(1):27-36.

[13] Tiwari G., Tiwari R., Rai A. K. Cyclodextrins in delivery systems: Applications. *J Pharm Bioall Sci.* 2010;2(2):72-9.

[14] Uekama K. Pharmaceutical applications of methylated cyclodextrins. *Pharm Int.* 1985;6:61-5.

[15] Loftsson T., Brewster M. E. Pharmaceutical applications of cyclodextrins. 1. Drug solubilization and stabilization. *J Pharm Sci.* 1996;85(10):1017-25.

[16] Szejtli J. Dimethyl-β-cyclodextrin as parenteral drug carrier. *Journal of Inclusion Phenomena.* 1983;1(2):135-50.

[17] Pitha J., Pitha J. Amorphous water-soluble derivatives of cyclodextrins: Nontoxic dissolution enhancing excipients. *J Pharm Sci.* 2006;74:987-90.

[18] Gould S., Scott R. C. 2-Hydroxypropyl-beta-cyclodextrin (HP-beta-CD): a toxicology review. *Food Chem Toxicol.* 2005;43(10):1451-9.

[19] Uekama K., Hirayama F., Irie T. Cyclodextrin Drug Carrier Systems. *Chem Rev* 1998;98:2045-76.

[20] Strattan C. E. 2-Hydroxypropyl-□-cyclodextrin, part II: safety and manufacturing issues. *Pharm Technol.* 1992;16(2):52-8.

[21] Loftsson T., Duchene D. Cyclodextrins and their pharmaceutical applications. *Int J Pharm.* 2007;329:1-11.

[22] Loftsson T., Brewster M. E. Pharmaceutical applications of cyclodextrins 1:drug solubilization and stabilization. *Journal of pharmaceutical sciences.* 1996;85(10): 1017-25.

[23] Li S., Purdy W. C. Cyclodextrins and their applications in analytical chemistry. *Chemical Reviews.* 1992;92(6):1457-70.

[24] Cabral-Marques H. M. Applications of cyclodextrins. Thermodynamic aspects of cyclodextrin complexes. *Rev Port Farm.* 1994;44:85-96.

[25] Rajewski R. A., Stella V. J. Pharmaceutical applications of cyclodextrins. 2. *In vivo* drug delivery. *Journal of pharmaceutical sciences.* 1996;85(11):1142-69.

[26] Lach J. L., Cohen J. Interaction of pharmaceuticals with Schardinger dextrins. II. Interaction with selected compounds. *J Pharm Sci.* 1963;52: 137-42.

[27] Yoshida A., Arima H., Uekama K., Pitha J. Pharmaceutical evaluation of hydroxyalkyl ethers of □-cyclodextrins. *Int J Pharm.* 1988;46:217-22.

[28] Cabral Marques H. M. Applications of cyclodextrins. Thermodynamic aspects of cyclodextrin complexes. *Rev Port Farm.* 1994;44:85-96.

[29] Agatonovic-Kustrin S., Morton D. W., Worthington M. S., Glass B. D. Modeling the effect of selected cyclodextrins on nifedipine solubility. *Current drug discovery technologies.* 2011;8(2):146-54.

[30] Loftsson T., Brewster M. E. Pharmaceutical applications of cyclodextrins. 1. Drug solubilization and stabilization. *Journal of pharmaceutical sciences.* 1996;85(10): 1017-25.

[31] Torricelli C., Martini A., Muggetti L., Eli M., De Ponti R. Stability studies on steroidal drug/β-cyclodextrin kneaded systems. International *Journal of Pharmaceutics.* 1991 9/20/;75(2-3):147-53.

[32] Andersen F. M., Bundgaard H. The influence of cyclodextrin complexation on the stability of betamethasone-17-valerate. *International Journal of Pharmaceutics.* 1984; 20(1-2):155-62.

[33] Frömming K. H., Szejtli J. *Cyclodextrins in pharmacy.* Topics in inclusion science: Dordrecht: Kluwer Academic Publishers; 1994.

[34] Sortino S., Giuffrida S., De Guldi G., Chillemi R., Petralia S., Marconi G. et al., The photochemistry of flutamide and its inclusion complex with beta-cyclodextrin. Dramatic effect of the microenvironment on the nature and on the efficiency of the photodegradation pathways. *Photochemistry and photobiology.* 2001;73(1):6-13.

[35] Hidaka S., Tokumura T., Tomono K., Suzuki T., Ueda H., Nagai T. et al., Effect of beta-cyclodextrin on the degradation rate of amoxicillin in acidic solution. Yakugaku zasshi: *Journal of the Pharmaceutical Society of Japan.* 2010;130(6):889-93.

[36] Tomono K., Gotoh H., Okamura M., Ueda H., Saitoh T., Nagai T. Effect of □-cyclodextrin and its derivatives on the photostability of photosensitive drugs. *Yakuzaigaku.* 1988;48(4):322-5.

[37] Teshima D., Otsubo K., Higuchi S., Hirayama F., Uekama K., Aoyama T. Effects of cyclodextrins on degradations of emetine and cephaeline in aqueous solution. *Chem Pharm Bull.* 1989;37(6):1591-4.

[38] Agatonovic-Kustrin S., Glass B. D., Michael E., Brown M. E., Lebete M. L. For and Against the use of Methylated Cyclodextrins in the Development of Midazolam Solution for Nasal Application. *Austin J Pharmacol Ther.* 2014;2(2):3-4.

[39] Andersen F. M., Bundgaard H. Inclusion complexation of metronidazole benzoate with β-cyclodextrin and its depression of anhydrate-hydrate transition in aqueous suspensions. *International Journal of Pharmaceutics.* 1984;19(2):189-97.

[40] Hoshino T., Irie T., Hirayama F., Uekama K., Yamasaki M. Alternation in Photochemical Reaction of Clomipramine by Inclusion Complexations of β- and Dimethyl-β-cyclodextrins. Yakugaku zasshi: *Journal of the Pharmaceutical Society of Japan.* 1989;109(2):107-12.

[41] Zayachkivska O. S., Konturek S. J., Drozdowicz D., Brzozowski T., Gzhegotsky M. R. Influence of plant-originated gastroproteciive and antiulcer substances on gastric mucosal repair. *Fiziol Zh.* 2004;50(6): 118-27.

[42] Konturek S. J., Mrzozowski T., Drozdowicz D., Pawlik W., Sendur R. Gastroprotective and ulcer healing effects of solon, a synthetic flavonoid derivative of sophoradin. *Hepato-gastroenterology.* 1987;34(4):164-70.

[43] Chen C.-Y., Chen F.-A., Wu A.-B., Hsu H.-C., Kang J.-J., Cheng H.-W. Effect of hydroxypropyl-β-cyclodextrin on the solubility, photostability and *in vitro* permeability of alkannin/shikonin enantiomers. *International Journal of Pharmaceutics.* 1996;141(1-2):171-8.

[44] Ou D., Ueda H., Nagase H., Endo T., Nagai T. Some Pharmaceutical Properties of 2,3,6-partially Methylated-β- cyclodextrin and its Solubilizing and Stabilizing Abilities. *Drug Dev Ind Pharm.* 1994;20 (12):2005-16.

[45] Hoshino T., Hirayama F., Uekama K., Yamasaki M. Reduction of protriptyline-photosensitized hemolysis by β-cyclodextrin complexations. *International Journal of Pharmaceutics.* 1989;50(1):45-52.

[46] Uekama K., Narisawa S., Hirayama F., Otagiri M., Kawano K., Ohtani T. et al., Improvement of thermal and photochemical stability of benzaldehyde by cyclodextrin complexation. *International Journal of Pharmaceutics.* 1983;13(3):253-61.

[47] Thoma K. *Photodecomposition and stabilization of compounds in dosage forms.* In: Tønnesen HH, editor. The Photostability of Drugs and Drug Formulations. London: Taylor and Francis; 1996. p. 111-40.

[48] Szejtli J. *Cyclodextrins and Their Inclusion Complexes* (Cyclodextrine und ihre Einschlußkomplexe). Budapest: Verlag der Ungarischen Akademie der Wissenschaften. Akadémiai Kiadó; 1982.

[49] Uekama K., Oh K., Otagiri M., Seo H., Tsuruoka M. Improvement of some pharmaceutical properties of clofibrate by cyclodextrin complexation. *Pharm Acta Helv.* 1983;58(12):338-42.

[50] Mielcarek J., Sadaj A. Inclusion compounds of nifedipine and other 1,4-dihydropyridine derivatives with ß-cyclodextrins II. Photochemical stability of the inclusion compounds formed by nifedipine with ß-cyclodextrins. *Acta Pol Pharm.* 1994;51:21-4.

[51] Wermeling D. P., Record K. A., Kelly T. H., Archer S. M., Clinch T., Rudy A. C. Pharmacokinetics and pharmacodynamics of a new intranasal midazolam formulation in healthy volunteers. *Anesthesia and Analgesia.* 2006;103(2):344-9.

[52] Products TEAftEoM. *Assessment of the paediatric needs*: Anaesthesiology. 2006.

[53] Björkman S., Rigemar G., Idvall J. Pharmacokinetics of midazolam given as an intranasal spray to adult surgical patients. *British Journal of Anaesthesia.* 1997;79(5): 575-80.

[54] Burstein A. H., Modica R., Hatton M., Forrest A., Gengo F. M. Pharmacokinetics and pharmacodynamics of midazolam after intranasal administration. *Journal of Clinical Pharmacology.* 1997;37(8):711-8.

[55] Wolfe T. R., Macfarlane T. C. Intranasal midazolam therapy for pediatric status epilepticus. *Am J Emerg Med.* 2006;24(3):343-6.

[56] Mahmoudian T., Zadeh M. M. Comparison of intranasal midazolam with intravenous diazepam for treating acute seizures in children. *Epilepsy Behav.* 2004;5(2):253-5.

[57] Rao B. N., Turro N. J., Ramamurthy V. Modification of chemical reactivity via inclusion complex formation: photochemistry of dibenzyl ketones and benzyl phenylacetates. *The Journal of Organic Chemistry.* 1986;51(4):460-4.

[58] Reddy G. D., Ramamurthy V. Modification of photochemical reactivity by cyclodextrin complexation: alteration of photochemical behavior via restriction of translational and rotational motions. Alkyldeoxybenzoins. *The Journal of Organic Chemistry.* 1987;52 (25): 5521-8.

[59] Hedges A. R. Industrial Applications of Cyclodextrins. *Chem Rev.* 1998;98(5):2035-44.

[60] Ali S. M., Upadhyay S. K. Complexation study of midazolam hydrochloride with β-cyclodextrin: NMR spectroscopic study in solution. *Magnetic Resonance in Chemistry.* 2008;46(7):676-9.

[61] Reddy G. D., Usha G., Ramanathan K. V., Ramamurthy V. Modification of photochemistry by cyclodextrin complexation. Competitive Norrish type I and type II reactions of benzoin alkyl ethers. *The Journal of Organic Chemistry.* 1986;51(16): 3085-93.

[62] Tønnesen H. H., Másson M., Loftsson T. Studies of curcumin and curcuminoids. XXVII. Cyclodextrin complexation: solubility, chemical and photochemical stability. *International Journal of Pharmaceutics.* 2002;244(1-2):127-35.

[63] Glass B. D., Brown M. E., Daya S., Worthington M. S., Drummond P., Antunes E. et al., Influence of cyclodextrins on the photostability of selected drug molecules in solution and the solid-state. *International Journal of Photoenergy.* 2001;3(4):205-11.

[64] Fronza G., Mele A., Redenti E., Ventura P. Proton nuclear magnetic resonance spectroscopy studies of the inclusion complex of piroxicam with beta-cyclodextrin. *J Pharm Sci.* 1992;81(12):1162-5.

[65] Nikolić V., Ilić-Stojanović S., Nikolić L., Cakić M., Zdravković A., Kapor A. J. et al., Photostability of piroxicam in the inclusion complex with 2-hydroxypropyl-β-cyclodextrin. Hemijska Industrija. 2014;68(1): 107-16.

[66] Ilić-Stojanović S., D. N. V., B. N. L., S. Z. A., J. KA., M. P. et al., The improved photostability of naproxen in the inclusion complex with 2-hydroxypropyl-β-cyclodextrin. *Hemijska Industrija* 2015;68(1):107-17.

In: Cyclodextrins
Editor: Francis G. Ramirez
ISBN: 978-1-63482-788-1

Chapter 6

Cyclodextrin Inclusion of Nutraceuticals, from the Bench to Your Table

Ana Bárbara Pereira and Susana S. Braga
Department of Chemistry, QOPNA – Organic Chemistry, Natural and Agro-Food Products Research Unit, University of Aveiro, Aveiro, Portugal

Abstract

The most interesting property of cyclodextrins (CDs) is, undoubtedly, their ability to form inclusion complexes (ICs) through the encapsulation into their central cavity of guest molecules of suitable size, geometry and hydrophobicity. These guest molecules may be apolar, low solubility drugs, volatile compounds warranting stabilization or light-sensitive compounds.

The present chapter deals with applications of cyclodextrins in stabilizing or solubilizing ingredients of pharmaceutical relevance used in the food industry, which, for the combination of these two functions, are commonly called nutraceuticals. This emerging field is gaining relevance and acceptance amidst the consumers, and the use of cyclodextrins as carriers for antioxidant compounds will help meet the growing market demands. Furthermore, the cyclodextrins themselves are also revealing intrinsic nutraceutical properties, as is the case of incorporating α-CD in starchy meals to lower post-prandial glycemia and of the delayed release of glucose by γ-CD. The benefits and challenges of this new field of application for the native cyclodextrins, from the physico-chemical to the formulation standpoints, as well as a thorough revision of the regulatory status, are presented and discussed.

Introduction

The term "nutraceutical", which results from the agglutination of the words *nutrition* and *pharmaceutical*, was coined in 1989 by Stephen DeFelice. [1] Nowadays, it is commonly used in marketing, regardless of the fact that an exact regulatory definition of nutraceutical is still lacking. [2] Indeed, the definition of nutraceutical, referring to "a food or part of a food that provides medical or health benefits, including the prevention and/or treatment of a disease" [1] is somewhat broad and it englobes the concept of functional food, being thus partially overlapped with this category of foods. The term “functional food” refers to any foodstuff possessing an additional function as a result of being fortified with health-promoting ingredients (such as probiotics, vitamins and others). The concept of functional foods was created in Japan and it applies exclusively to foodstuff with a scientifically proven benefit on human health, which is then approved by the Japanese Ministry of Health to receive the FOSHU classification (Food for Specified Health Uses). [3] This way, the main characteristic of functional foods is that there is already a clear scientific evidence of their beneficial action on the human health.

The compounds englobed in the definition of nutraceutical comprise a wide variety of molecules, distributed over several classes according to their different chemical natures: [4]

- terpenoids/isoprenoids, where carotenoids and saponins are included;
- phenolic compounds, including coumarins, isoflavones and flavonols;
- lipidic compounds such as polyunsaturated fatty acids (PUFA) and lecithin;
- aminoacid or peptidic compounds such as folates and choline;
- carbohydrates and derivatives, such as ascorbic acid and oligossacharides.

The strong hydrophobicity associated with most of the terpenoids, phenols or lipids implies low aqueous solubility and thus these will have low bioavailability from ingested foods. Additionally, the low solubilities raise issues when trying to formulate foods supplemented with these compounds, and the use of solubilizing agents is often the solution.

Cyclodextrins (CDs) are naturally occurring cyclic oligosaccharides of 1,4-linked α-D-glucose, formed by the bacterial degradation of starch [5] and able to solubilize hydrophilic compounds by including them into their ring cavities. For this reason, cyclodextrins are considered molecular capsules. The most abundant and affordable native cyclodextrins occur with six to eight glucose units and receive the names of α-CD, β-CD and γ-CD, respectively. Chemical modification of these CDs affords a variety of derivatives with different physico-chemical and biological properties. Of the existing derivatives, now more than 1500, [6] the most relevant for human use is 2-hydroxypropyl-β-cyclodextrin (HPβCD), which combines excellent aqueous solubility and solubilizing properties with a good tolerability *in vivo*, being an FDA-approved inactive ingredient. [7]

Native Cyclodextrins As Nutraceuticals

Origins and Uses

The utilization of native cyclodextrins in food applications began when these materials became commercially available from production sites in Japan and Hungary. [8] In Japan, from the late 1970s to the early 1980s, four companies established processes for the production of native CDs on an industrial scale using only water as a solvent and introduced these products to supply the markets. [9] The native CDs produced in this way are recognized in Japan as natural products and sold without restrictions except for considerations of purity. It was thus not surprising to find that, over a short period of four years (from 1984 to 1987), as much as 88 new patents on food applications of CDs were registered in Japan, and that the food products are the most significant application of CDs in this country, accounting for 90% of the market of cyclodextrins. [9]

In their early applications, CDs were thought to be components with a relative biological inertness and used as preserving agents in a variety of spices, flavors, tea, vinegar and seeds, or as emulsification enhancers in whipped creams, mayonnaises and dressings. Only two decades later, as the introduction of CDs into the food markets of Europe and the United states began to be pondered, did more studies on the biological fate of these compounds become available, and data on the alterations to biochemical parameters of human subjects (blood analysis) which result of the ingestion of CDs was also released. This collection of results, coming not only from academic research work but also from the companies trying to develop new food products and to patent new applications for CDs, has ultimately led to the recognition of the nutraceutical properties which are intrinsic to α-CD and, in some extent, also observed with γ-CD.

Studies with animal models have shown that the native CDs are practically not absorbed when ingested (table 1), because of their bulky and hydrophilic nature. Any absorption is by passive diffusion. The parent α-CD and β-CD are resistant to stomach acid and to salivary and pancreatic enzyme digestion, whereas γ-CD is digested partly by amylases in the large intestine. [10] Another study has revealed that the three native CDs are able to inhibit the action of pancreatic α-amylase (obtained from pigs) on the substrates amylose and maltoheptaose, and that on both substrates the inhibitory action is α-cyclodextrin < β-cyclodextrin < γ-cyclodextrin. [11]

Table 1. The absorption of native cyclodextrins following their oral administration

Cyclodextrin	Species	Dose (mg/kg)	% absorbed	Reference
α-CD	rat	200	1	[12]
β-CD	rat	500	0.6	[13]
γ-CD	rat	1000	0.02	[14]

The regulatory status of native CDs in foods differs between Europe, the USA and the southern countries of Australia and New Zealand (where they was last approved). Native CDs are allowed to be commercialized in the USA under the GRAS status (FDA list of food additives that are 'generally recognized as safe').

In Australia and New Zealand, only α- and γ-CDs are allowed to be sold, under the classification of Novel Foods. In Europe α- and γ-CDs are allowed to be used without restriction, while the use of β-CD in food is limited to a maximum daily dose (0-5 mg/kg). The full details on the regulatory status for each cyclodextrin are presented as an introduction to their corresponding subsections below.

α-Cyclodextrin – Nutraceutical Properties and Uses in Foodstuff

Alpha-cyclodextrin (α-CD) is approved by the joint WHO/FAO expert committee on food additives (JEFCA) as a food additive since 2002 [15] and as a dietary fiber since 2006, [16] in both uses having an acceptable daily intake (ADI) "not specified"; this means that it may be used in food in unrestricted quantities. In the United States it has been awarded the GRAS status in 2004 [17] and it was approved as a New Food by FSAZN (Food Standards of Australia and New Zealand) in 2003 [18] and in Europe in 2008. [19]

α-Cyclodextrin As a Nutraceutical Supplement

Alpha-cyclodextrin is sold for alimentary uses either as a bulk solid for manufacturers (CAS no.: 10016-20-3) or as tablets ready for consumer intake. The tablets available in the market contain 1g of α-CD each and are sold under the brand names "Calorease", from SFI (Soho Flordis International) and "Alpha-Fibe", from BioForm Essentials. The recommendation is to take two tablets at the main meals (max 6g *per* day), which are claimed to reduce the absorption of fat and assist in weight loss. These beneficial actions of α-CD are associated with its ability to form inclusion complexes with the lipidic molecules it encounters along the enteric trajectory, particularly the saturated triglycerides. [20] Even though the implications of this fat-complexing effect inside the gut are still under investigation and remain far from being fully understood, there are already some preliminary results revealing promising benefits. The most immediate outcome is the reduction in the problems associated with the passage of free fat through the gastrointestinal tract (steatorrhea and bowel incontinence). [21] There is also an improvement of the blood lipidic profile. Recent studies with healthy humans adults have demonstrated that the presence of α-CD in a meal affords reduction of the acute postprandial blood triglyceride levels, [22] of the levels of total and LDL cholesterol (LDL = low-density lipoprotein), and of the body weight. [21] Rats on a diet with 21% milk fat and 2.1% of α-CD showed reduced seric proatherogenic lipoproteins and trans-fatty acids. [23]

The most relevant and well-documented property of α-CD, however, is its ability to lower the glycemic index of starchy foods. [24] This effect is thought to derive from α-CD's strong inhibitory action on pancreatic α-amylase [11] and it was first demonstrated in a double-blind, randomised, cross-over design study involving 10 human subjects being fed with boiled white rice (50 g of digestible carbohydrate). Significant, dose-dependant reductions in the post-prandial glycemic responses were observed when 5 and 10 g of α-CD were added. [25]

α-Cyclodextrin As a Dietary Fiber in Food

Recent food formulation trends propose the incorporation of α-CD into bakery products and water-based beverages as a source of fiber. The claimed advantages for the bakery industry are associated with the elimination of the reducing sugars, present in the conventional fibers, which react with the flour proteins to form Maillard compounds and cause browning of the food; additionally, some Maillard compounds have unpleasant odours. The use of α-CD, which is a non-reducing sugar, is presented as an alternative way to convey fiber into these foods while circumventing the reactions of Maillard.

In beverages, the good aqueous solubility of α-CD and the crystallinity of the resulting solutions permit its easy incorporation without causing changes in the appearance of the beverage. [26]

A curios publication which has been recently made available as an e-book is a compilation of ten recipes to prepare food using α-CD. [27] Indeed, α-CD seems to be slowly but firmly making its way into our everyday lives.

Side Effects of α-Cyclodextrin

A few words of caution are required as a final note to this section, as one must always bear in mind that the ingestion of α-CD is not free of side effects. α-CD is not digested by the digestive enzymes of the intestinal tract to any relevant extent, and only a very small fraction is absorbed in the small intestine (roughly 1% of the administered dose), as denoted by the observation of traces of ^{14}C-marked α-CD (0.34%) in the urine of test animals (Winstar rats). [12] Upon reaching the large intestine, *c.a* 99% of the ingested α-CD is thus available for fermentation by the enteric flora. In these rats, is has been shown to undergo ring opening by microbial enzymes to form linear malto-oligosaccharides which are further hydrolyzed and fermented by pathways similar to those of resistant starch. [12, 28] As a consequence of these fermentation processes, bloating, flatulence, nausea and diarrhea were observed both in rats and beagle dogs subject to the regular consumption of α-CD (4-13 weeks). [29] These symptoms were also reported by roughly 30% of the human subjects who participated in the previously mentioned study on boiled rice with α-CD. [25]

β-Cyclodextrin – Properties and Uses in Foodstuff

The cyclodextrin beta (β-CD) is unique among the three most abundant native CDs for its set of distinct physico-chemical properties. In particular, β-CD has a much lower aqueous solubility than the other two native CDs, which results from the intramolecular hydrogen bonds formed between the secondary hydroxyl groups of one glucose units and the adjacent units. [30] These intramolecular hydrogen bonds are homodromic, forming a closed, monodirectional circle which contributes to stabilize the macrocyclic conformation of β-CD. [30] However, they also increase the rigidity of the molecule and lead to reduced interactions with the surrounding water molecules, and, ultimately, to the aforementioned low solubility.

Table 2. FDA [32] and GSFA [33] provisions for β-cyclodextrin

Food category	Max level in food (wt.%)	Reference
Chewing gum	2	[32, 33]
Compressed candies	2	[32]
Breakfast cereal and baked goods prepared from dry mixes	2	[32]
Pre-cooked pastas and noodles and like products	0.1	[33]
Processed cheese products	0.1	[32]
Gelatins and puddings	0.1	[32]
Flavored coffee and tea	0.1	[32]
Dry mixtures for beverages	0.1	[32]
Water-based flavoured drinks, including "sport," "energy," or "electrolyte" drinks and particulated drinks	0.05	[33]
Snacks - potato, cereal, flour or starch based (from roots, tubers, pulses and legumes)	0.05	[32, 33]
Dry mixtures for soups	0.02	[32]

It is thus not surprising to find that the biological behavior of β-CD is also different from that of α- and γ-CDs. As a result of this, β-CD is the only native cyclodextrin with consumption restrictions (except for Japan, as explained in the introduction of this section). In

Australia and New Zealand its use in foodstuff is not approved and in Europe and the United States, [31] food-grade β-CD is available as a bulk powder (CAS No.: 7585-39-9) but it has an ADI limited to 0-5mg per kg of body weight. [32] In order to determine how this value translates into the maximum amount of β-CD allowed in a food product, one has to consider the average weight of the adult consumer (usually estimated at 70 kg), the number of servings per day and also the possible intake of β-CD from other dietary sources. Both the FDA's GRAS notice of acceptance and the Codex General Standard for Food Additives (GSFA) [33] have established some limitations for the amount of β-CD which is safe to incorporate into foods; these are compiled in the table 2.

One may observe from this table that, as expected, a product with a small serving size, such as a chewing gum, is allowed a higher maximum amount of β-CD than foods with larger serving size, such as a water-based beverage (usually 200-330ml per serving) or a starch-based snack (50-150 g per serving). [34]

β-Cyclodextrin As a Functional Ingredient in Food

Similarly to α-CD, β-CD is considered a soluble dietary fiber, easily incorporated into water-based beverages. In beverages, β-CD often brings taste masking action as an additional benefit, since it may include undesired flavours. An example is the reduction of the bitterness of ginseng drinks. β-CD, either alone or in combination with γ-CD, was able to reduce the bitter flavor of ginseng by roughly one half, when incorporated into a ginseng-based energy drink formulation. [35]

β-CD may be also used as a partial fat substitute. Its effect on the flavor-release profiles of non-fat flavored milks and yogurts was analyzed by headspace gas chromatography to reveal that these products mimic quite well their corresponding higher-fat-content counterparts; a follow-up sensory evaluation study was carried out with a number of yogurts and salad dressings. The panel found that the addition of β-CD to the nonfat yogurt did not provide it with the flavor and texture matching those of the fat-containing yogurts, but when evaluating salad dressings (Italian- and ranch-style) they found that flavor profiles of nonfat dressings were indeed more similar to the corresponding full-fat products. [36]

β-Cyclodextrin As a Technological Agent for Modified Foods

The restrictions to the consumption of β-CD make it an interesting molecule for technological applications in which it interacts with food to include and remove undesirable components with subsequent removal of the inclusion complex. In particular, the ability of β-CD to include and sequester cholesterol is applied to generate a variety of free- or low-cholesterol foods: milk, [37, 38] butter, [38-40] and even fresh eggs (by feeding the β-CD to the chickens). [41]

β-CD was also investigated as a retardant of the stalling of bread. Significant impact was observed for both the crust of bread containing β-CD, which exhibited less changes of hardness, cohesiveness and springiness. The retrogradation, by crystallization of the linear amylase, is significantly lowered (especially in the crumb), as a result of interaction of amylase and lipids with β-CD. [42]

γ-Cyclodextrin – Nutraceutical Properties and Uses in Foodstuff

γ-CD, with eight units of glucose, has the larger and most flexible ring among the native CDs, being thus susceptible to ring opening and degradation by the action of the amylases in the human gastrointestinal tract. [43] This way, while α- and β-CDs are indigestible *in vivo*, γ-CD is considered a slowly digestible carbohydrate. [44] For this reason, γ-CD was the first native CD to be approved for uses in foodstuff in countries outside of Japan. In 1999, the JEFCA approved γ-CD as a food additive with no restrictions to the quantities used ("not specified" ADI), [45] followed by approval as GRAS in the USA in the year 2000. [46] Furthermore, γ-CD is a New Food in Australia and New Zealand (since 2004) [47] and in Europe (since 2012), [48] being commercialized as a bulk powder (CAS No.: 17465-86-0).

γ-Cyclodextrin Uses in Food

γ-CD is commonly incorporated into the dough of baked products, such as breads, rolls, muffins and bagels and labeled under the category of preservative. An example is HealthyLife's whole wheat bread (by Lewis Bakeries Inc.). [49]

γ-Cyclodextrin As a Nutraceutical

The delayed release of glucose from γ-CD makes it an almost ideal source of glucose, with low impact on blood sugar and insulin levels. In fact, the intake of a drink containing γ-CD (25g per each serving of 240 ml) was shown to effectively lower post-prandial glycemia and insulinemia when compared to an equivalent maltodextrin-containing drink, in a study conducted on 32 healthy adults with a double-masked, randomized, crossover design. [50] The drink containing γ-CD was patented, [51] as well as a nutritional product (containing γ-CD in a fraction of 5 to 100 wt.% of the total carbohydrates), designed for women with gestational diabetes and claimed to reduce the postprandial insulin secretion, normalize blood glucose levels and prevent night-time hypoglycemia. [52]

Digestibility and Gastrointestinal Tolerability of γ-Cyclodextrin

The full digestibility of γ-CD in the small intestine was demonstrated on dogs canulated at the ileum, [53] and on rats. [14] First examinations into its biological fate in humans were carried out *in vitro*, showing that γ-CD is digested by salivary and pancreatic human amylases. [43] A follow up study with 24 volunteers and a double-blind, placebo-controlled, cross-over, randomized design has revealed that the ingestion of 8g of γ-CD incorporated into 100g of yogurt was well tolerated and did not cause any gastrointestinal discomfort. [54] Another set of clinical studies used even higher amounts – 25g of γ-CD dissolved into a drink of 240 ml – and, again, no significant side effects reported were recorded in the 24 hours after γ-CD ingestion. [50,51] These results show that γ-CD has a good gastrointestinal tolerability.

Molecular Encapsulation with Cyclodextrins

Cyclodextrins are used to form inclusion complexes with a wide variety of nutraceutical compounds and also with some complex, multi-component products featuring nutraceutical properties, namely plant essential oils and other plant liquid extracts. The use of cyclodextrins brings many advantages: increased solubility, making liquid substances easier to formulate by turning them into solids, taste masking and stabilization of volatile aromas.

Carotenoids

Carotenoinds are organic pigments available only from biosynthesis in higher plants and microorganisms (algae, fungi, and bacteria). [55] There are more than over 600 compounds in the carotenoid family, among which the best known are (i) the provitamin-A compounds α-carotene, β-carotene and cryptoxanthin, which are converted *in vivo* to retinal, and (ii) the antioxidants lutein, lycopene, astaxanthin, and zeaxanthin. The carotenoids are divided into two classes, one comprising only unsaturated hydrocarbons, called carotenes, and another one with hydroxylated derivatives which are called xantophyls.

The formulation of carotenoids into foodstuff has two major difficulties to overcome, one being their very low aqueous solubility due to the fact that they are predominantly hydrophobic compounds occurring mostly in the all-*trans* configuration. In addition, their strong antioxidant properties imply that they are very likely to suffer oxidation when incorporated into complex food matrices. Inclusion into CDs is thus a good solution to circumvent these two formulation issues. The Table 3 presents the summary of the carotenoid inclusion complexes reported in the literature, and the benefits of inclusion.

Table 3. Inclusion of carotenoids with various cyclodextrins

Carotenoid	Cyclodextrin	H:G[a]	Benefits of inclusion	Ref.
Xanthophylls				
Astaxanthin	β-CD	4:1	Increased solubility and stability to heat and light	[56]
Astaxanthin	HP-β-CD	42:1	Increased stability to oxygen and light and much higher solubility	[57]
Canthaxantin	HP-β-CD	1:1	Increased solubility, storage stability and antioxidant action	[58]
Lutein	γ-CD	n.s.	Improved bioavailability	[59]
Carotenes				
Licopene	α-CD or HP-β-CD	n.s.	Increased solubility, stability to heat and light and bioavailability	[60]
Licopene	γ-CD	1:1.	Increased solubility and stability to light	[61]
β-carotene	α-CD	36:1	Increased solubility, flavor masking	[62]
β-carotene	β-CD	n.s.	Increased solubility, isolation of the *trans*-form	[63]
Bixin	α-CD	n.s.	Increased solubility and stability to air and light	[64]

[a] H:G represents the host (CD) to guest (carotenoid) stoichiometry reported for each inclusion complex.
n.s. denotes complexes with unspecified stoichiometry of inclusion.
Note: the solubility mentioned in this table is always referring to water as a solvent.

Cyclodextrins are also able to include, solubilize and stabilize complex substances containing mixtures of several carotenoids and other components, such as those obtained from the pepper plants. Recent reports describe the inclusion by β-CD of the oleoresin of habanero chilli (*Capsicum chinense*), containing several carotenoids and the capsaicinoids capsaicin and dihydrocapsaicin; [65] and of the extract of sweet bell pepper (*Capsicum annuum*) containing capsanthin in amounts up to 50% of the total carotenoids. [66] The inclusion product of bell pepper extract with β-CD was tested as a colorant for yogurt, having demonstrated higher stability of the color indices than those of yogurt colored with crude red bell pepper extract. [66]

Essential Oils and Their Main Active Components

Plant essential oils are the naturally occurring volatile oils obtained by steam distillation or cold pressing (expression), having the characteristic aroma of the plant part from which they were derived. This subsection is dedicated to the cyclodextrin inclusion of essential oils comprised mainly of monoterpenes and hydroxyterpenoids, which constitute a class of nutraceuticals with a very broad and promising set of properties. Asides from the fragrances conveyed by these oils, many have antioxidant or antimicrobial actions, thus being suited for food preservation. [67] A few essential oils also display anti-inflammatory, immunomodulating, anti-spasmodic, carminative, venotonic or endorphin-releasing actions.

The average molecular dimensions of the main components of essential oils, which are, as mentioned, monoterpene derivatives, implies that the cyclodextrin with strongest affinity to form inclusion complexes with them is β-CD. Three examples of this adequate host-to-guest fitting are the β-CD inclusion complexes with the main components of three essential oils: β-CD·thymol, β-CD·carvacrol and β-CD·eugenol. These were isolated in the crystalline form and their structures collected by X-ray diffraction, see the Figure 1.

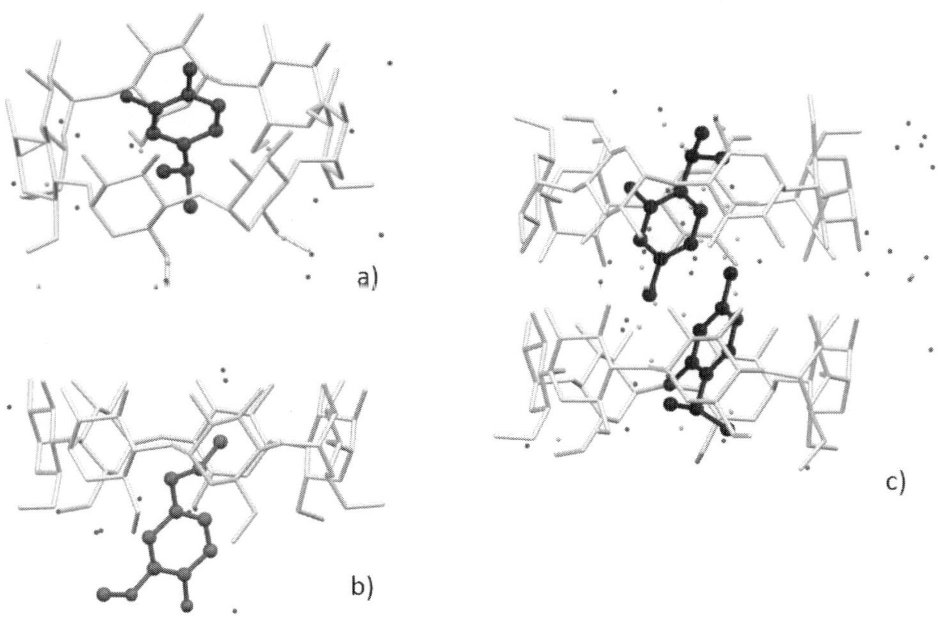

Figure 1. Inclusion geometry of the complexes a) β-CD·carvacrol, b) β-CD·eugenol and β-CD·thymol. The CD rings are drawn in light grey and the guests are highlighted by darker tones and the ball-and-stick representation; free atoms surrounding the inclusion complexes represent water molecules. Hydrogen atoms are omitted for clarity. Redrawn from the atomic coordinates collected by Bethanis et al., [68b] using the software Mercury 3.1 (Copyright CCDC 2001-2012).

Even though β-CD is the most adequate host for the inclusion of essential oils, it was possible to find in the available literature a few punctual reports of inclusion with other hosts. The table 4 presents a compilation of the inclusion complexes of essential oils or their pure terpenoid main active components; focus was given to the inclusion complexes with proven or potential applications in food.

Table 4. Inclusion of plant essential oils and their main components with various cyclodextrins

Guest	Cyclodextrin	H:G[a]	Benefits of inclusion	Ref.
Terpenoids				
Thymol	β-CD	1:1	Increased shelf-life at relative humidity ≥ 75%	[68,69]
Carvacrol	α-CD	n.s.	Increased solubility and antimicrobial action	[70]
Carvacrol	β-CD	1:1	Increased solubility and antimicrobial action	[68, 70]
Carvacrol	HP-β-CD	n.s.	Increased solubility, stability to light and antimicrobial action	[70, 71]
Eugenol	β-CD	2:3	Increased solubility and antimicrobial action	[68, 70]
Eugenol	α-CD or HP-β-CD	n.s.	Increased solubility and antimicrobial action	[70]
Eugenol	γ-CD	n.s.	Turns guest into a dry powder	[72]
Linanol	α-CD, β-CD or HP-β-CD	n.s.	Increased solubility and antimicrobial action	[70]
Cynnamaldehyde	β-CD	1:1	Increased shelf-life at relative humidity ≥ 75% and enhanced antimicrobial action	[69, 73]
Cinnamic acid	α-CD	5:3	Increased solubility	[74]
Cinnamic acid	β-CD	3:2	Increased solubility	[74]
p-cymene	β-CD	1:1	Improved analgesic and anti-inflammatory actions	[75]
D-limonene	α-CD	n.s.	Reduced volatility	[72]
D-limonene	β-CD or γ-CD	n.s.	Transforms D-limonene into a dry powder	[72,76]
α-(−)-bisabolol	β-CD	2:1	Increased stability to heat	[77]
D-carvone	β-CD	1:1	Potential new powder for potato sprout inhibition	[78]
Geraniol	β-CD	1:1	High thermal stability, sustained release and increased anti-convulsant action	[79]
Estragole	α-CD, β-CD or HP-β-CD	n.s.	Increased stability to light and stronger antioxidant action	[80]
Citronellol	α-CD	5:2	Increased solubility	[74]
Citronellol	β-CD	4:3	Increased solubility	[74]
Anethole	α-CD, β-CD or HP-β-CD		Increased solubility and photostability	[81]
Essential oils				
Citrus × limon	β-CD	n.s.	Increased stability to light	[82]
Citrus × sinensis	β-CD	n.s.	Increased stability to light	[82]
Humulus lupulus	β-CD	n.s.	Increased stability to light	[82]

Guest	Cyclodextrin	H:G[a]	Benefits of inclusion	Ref.
Essential oils				
Chamomila recutita	γ-CD	n.s.	Slower deterioration of chamazulene	[82]
Citrus sphaerocarpa	β-CD	n.s.	Makes kabosu essential oil water soluble and easy to disperse uniformly into food	[83]
Angelica sinensis	β-CD	n.s.	Increased stability	[84]
Cymbopogon spp	β-CD	n.s.	Turns oil into a crystalline solid and masks flavor	[85]
Thymus vulgaris	β-CD	1:1	Sustained antifungal action	[86, 89]
Lippia graveolens	β-CD	1:1	Improved antioxidant action	[87]
Mentha × piperita	β-CD	1:1	Controlled release of the aroma components	[88]
Carum carvi	β-CD	n.s.	—	[89]
Coriandrum sativum	β-CD	n.s.	—	[88]
Foeniculum vulgare	β-CD	n.s.	Increased solubility and stability	[89, 90]
Anethum graveolens	β-CD	n.s.	—	[89]
Majorana hortensis	β-CD	n.s.	—	[91]
Lavandula angustifolia	β-CD	n.s.	—	[91]
Ocimum basilicum	α-CD, β-CD or HP-β-CD	n.s.	Increased stability to high-temperature oxidation and improved antioxidant action	[80, 91, 92]
Artemisia dracunculus	α-CD, β-CD or HP-β-CD	n.s.	Increased stability to light and improved antioxidant action	[80]
Salvia sclarea	β-CD or HP-β-CD	1:1	Increased stability	[93, 94]
Lippia gracilis	HP-β-CD	1:1	Turns oil into an easy to handle powder	[95]
Cinnamomum verum	β-CD	1:1	Taste masking effect, increased solubility and antimicrobial action	[96, 97]
Eugenia caryophyllata	β-CD	1:1	Taste masking effect, increased solubility and antimicrobial action	[97]
Satureja montana	β-CD	1:1	Increased solubility, antioxidant, antibacterial and antifungal actions	[98]

[a] H:G represents the host (CD) to guest (oil or terpenoid guest) molar stoichiometry.
n.s. denotes complexes with unspecified stoichiometry of inclusion.
Note: the solubility mentioned in this table is always referring to water as a solvent.

The most common application of the CD inclusion complexes with plant essential oils is in food packaging, taking advantage of their antimicrobial properties. [99] Inclusion complexes of carvacrol [100] and cinnamomum oil, [101] which combine a good antimicrobial action with a mild, pleasant aroma, are particularly appreciated for the incorporation into packaging films and fibers. The β-CD inclusion complex of cinnamaldehyde was also recently applied in the preparation of an antimicrobial edible film for coating pieces of freshly-cut melon. The film is deposited onto the melon chunks using a layer-by-layer technique and allows these pieces of fruit to stand at 4ºC with an increased shelf-life, lasting roughly one week longer than the untreated ones.. [102]

Inclusion complexes of essential oils present the additional advantage of being less toxic than the synthetic pesticides commonly used in farming. The product of the inclusion of Chenopodium (wormseed) oil with β-CD is patented in China for the prevention and treatment of fungal infections of strawberries (grey mold) and tomatoes (anthrachnose). [103]

Flavonoids

Flavonoids are one of the most important classes of nutraceuticals found in plants and the largest class of polyphenols, comprising more than 2000 known compounds. They feature antioxidant and metal chelating properties and many of them are being investigated for anti-inflammatory, antimicrobial, antitumoral and tumor preventive actions.

The flavonoids share the common structure of diphenylpropanes (C6-C3-C6), consisting of two aromatic rings linked through three carbons. The six major subclasses of flavonoids include the flavones (e.g., apigenin, luteolin), flavonols (e.g., quercetin, myricetin), flavanones (e.g., naringenin and hesperidin), flavanols, also called catechins (e.g., epicatechin and gallocatechin), flavylium salts, also called anthocyanidins (cyanidin and pelargonidin), and isoflavones (e.g., genistein and daidezin). [104,105]

The low aqueous solubility and enteric absorption of most flavonols means that cyclodextrin inclusion is a suitable method for the amelioration of their physico-chemical and biological properties. The known inclusion complexes of flavonols with CDs are listed in the table 5.

A relatively large number of these complexes are made with chemically modified cyclodextrins, including the FDA-approved (2-hydroxypropyl)-β-CD (HP-β-CD) and a methoxylated derivative, heptakis 2,6-*bis*-O-methyl-β-CD (DIMEB), which, to the best of our knowledge, is only approved in the EU and the USA for topical use, being part of ocular and nasal formulations; foodgrade DIMEB is available from China retailers, but translated documents on its local regulatory status were not found in our research.

Table 5. Inclusion of flavonoids with various cyclodextrins

Flavonoid	Cyclodextrin	H:G[a]	Benefits of inclusion	Ref.
Anthocyanidin				
Cyanidin	β-CD	c.e.	The product preserves color and flavor of freeze-dryed strawberries and other fruits	[106]
Pelargonidin	β-CD, DIMEB or HP-β-CD	n.s.	Significant retardation of the discoloration of pelargonidin in aqueous solution at pH 3.0	[107]
Flavanols				
(+) Catechin	β-CD	1:1	—	[108,109]
(+) Catechin	DIMEB or HP-β-CD	1:1	—	[109]
(−)-Epicatechin	β-CD	1:1	Sustained release, retained antioxidant activity	[108,110]
(+)-Gallocatechin Gallate	β-CD	1:2	—	[111]
(−)-Epicatechin Gallate	β-CD	1:1	—	[111]
(−)-Epigallocatechin Gallate	γ-CD	1:1	Increased solubility and antioxidant activity	[113]

Flavonoid	Cyclodextrin	H:G[a]	Benefits of inclusion	Ref.
Flavanols				
(–)-Epigallocatechin Gallate	γ-CD	1:1	Increased solubility and antioxidant activity	[113]
Flavanones				
Naringenin	β-CD	1:1	Increased solubility and thermal stability	[114, 115, 116]
Naringenin	HP-β-CD	c.e.	Bioavailability increased up to 12 times	[117]
Naringenin	DIMEB	1:1	Increased solubility and thermal stability	[115]
Hesperetin	β-CD	1:1	Increased solubility	[116]
Hesperetin	HP-β-CD	1:1	Increased solubility and antioxidant potential	[118]
Flavones				
Chrysin	β-CD	1:1	Increased solubility and antioxidant potential	[119]
Luteolin	β-CD, DIMEB or HP-β-CD	1:1	Increased antioxidant potential	[120]
Morin	β-CD, DIMEB or HP-β-CD	1:1	Increased solubility and antioxidant potential	[121]
Diosmin	β-CD	1:1	Increased solubility and in vitro bioactivity	[122]
Flavonols				
Quercetin	β-CD or HP-β-CD	1:1	Increased solubility and antioxidant activity	[123,124]
Quercetin	γ-CD	5:3	—	[125]
Galangin	β-CD, DIMEB or HP-β-CD	1:1	Increased solubility	[126]
Resveratrol	β-CD or HP-β-CD	1:1	Increased solubility while retaining the antioxidant activity	[127]
Myrcetin	HP-β-CD	n.s	Increased antioxidant activity and stability to free-radical rapid oxidation	[124]
Coumaric acids	β-CD	1:1	Increased antioxidant activity	[128]
Rutin	β-CD, HP-α-CD or HP-β-CD	1:1	Increased stability to light and heat and improved antioxidant activity	[129]
Kaempferol	β-CD or HP-β-CD	1:1	Increased antioxidant action and stability to oxidation	[124,130]
Isoflavones				
Daidzein	β-CD, γ-CD or HP-β-CD	1:1	Increased solubility and intestinal membrane permeability (tested on Caco-2 cells)	[131]
Genistein	β-CD, γ-CD or HP-β-CD	1:1	Increased solubility, intestinal membrane permeability (tested on Caco-2 cells) and anti-inflammatory action	

[a] H:G represents the host (CD) to guest (carotenoid) stoichiometry reported for each inclusion complex.

n.s. denotes complexes with unspecified stoichiometry of inclusion; c.e. denotes reports in which a large excess of cyclodextrin was employed.

Note: the solubility mentioned in this table is always referring to water as a solvent.

Inclusion complexes or dispersion products of flavonoids and CDs are commonly found in foods. For instance, combinations of β-CD and/or γ-CD with catechin appear in the composition of many pre-prepared beverages, such as instant teas [133] and refreshments. [134]

Sulfur-Containing Phytochemicals

There is a wide variety of naturally occurring organosulfur compounds, but those with greater relevance as nutraceuticals due to their antioxidant, antimicrobial, antitumoral anti-inflammatory and cardioprotective actions are obtained mostly from garlic and from mustard seed oils. The benefits and applications of the inclusion of these phytochemicals are described in the following subsections.

Another organosulfur compound worthy of mention is α-lipoic acid, even though its biosynthesis is not exclusive to plants, occurring also in the animal kingdom. It is thus present in almost every category of foods, but in very low amounts, thus being often taken as a supplement. In the body, α-lipoic acid helps regenerate other antioxidants such as vitamin C, vitamin E, coenzyme Q10 and glutathione. The recycling action on these vital antioxidants brings benefits in energy replenishment, for both diabetic and healthy individuals. The stabilization of α-lipoic acid (ALA) as an ingredient for supplements is achieved by inclusion into α-CD. The ALA·α-CD complex, traded as CAVAMAX® W6 ALA, is one of the rare examples of inclusion complexes commercially available as raw materials.

Garlic Oil and Its Organosulfur Compounds

Garlic extract contains alliin (*S*-Allyl-L-cysteine sulfoxide), allicin (diallyl thiosulfinate) and diallyl sulphide as the main ingredients, having also some amounts of ajoene, allitridin (diallyl trisulfide) and diallyl tetrasulfide.

The inclusion complexes of different cyclodextrins isolated with a well-known stoichiometry are presented in the table 6; note that TRIMEB (heptakis *tris*-O-methyl-β-CD) is not suitable for human use and thus its inclusion complex with ajoene is of merely academic interest, given that it is the only one within this class of natural compounds having structural data for cyclodextrin inclusion complex, determined using single-crystal X-ray diffraction (see the Figure 2 at the end of this subsection).

Table 6. Inclusion of garlic oil or the compounds occurring naturally in garlic with various cyclodextrins

Guest	Cyclodextrin	H:G[a]	Benefits of inclusion	Ref.
Allicin	β-CD	1:1	Prolonged shelf-life and antimicrobial action	[135]
Diallyl trisulfide	HP-β-CD	1:1	Increased aqueous solubility	[136]
Ajoene	TRIMEB	1:1	Not applicable	[137]
Whole garlic oil	β-CD	1:1	Significant increase in aqueous solubility	[138]

[a] H:G represents the host (CD) to guest (garlic compound) molar stoichiometry.

The most significant advantages of including the garlic compounds or garlic oil into cyclodextrins comprise the obvious increase in aqueous solubility and the masking of their strong odor. The formulations, either foods or supplements, which contain these inclusion complexes are thus much better accepted by the consumers and are of strong economic interest. This is demonstrated by the high number of patents on deodorizing garlic (or its components) by means of CD inclusion. We have found 23 patents on cyclodextrin inclusion complexes (sometimes also called clathrates) with garlic or its components and about one half of these (12 patents) claimed the elimination of odor as an advantage of the invention.

The majority of applications for associations of garlic and cyclodextrins are found within the supplements market, followed by their incorporation into the composition of sauces [139] and seasonings. [140] Recently, powdered egg yolk containing garlic and CDs was patented, [141] with the claim of better stability and absence of garlic odor. In Asian countries, incorporation of garlic into beverages seems to be well accepted. An herbal tea containing extracts of flowers (rose, peony, chrysanthemum and honeysuckle), black garlic extract and β-CD is patented as a nutritional beverage [142] and a garlic coffee drink was also invented, claiming to "maintain the heavy coffee drinker's health". [143]

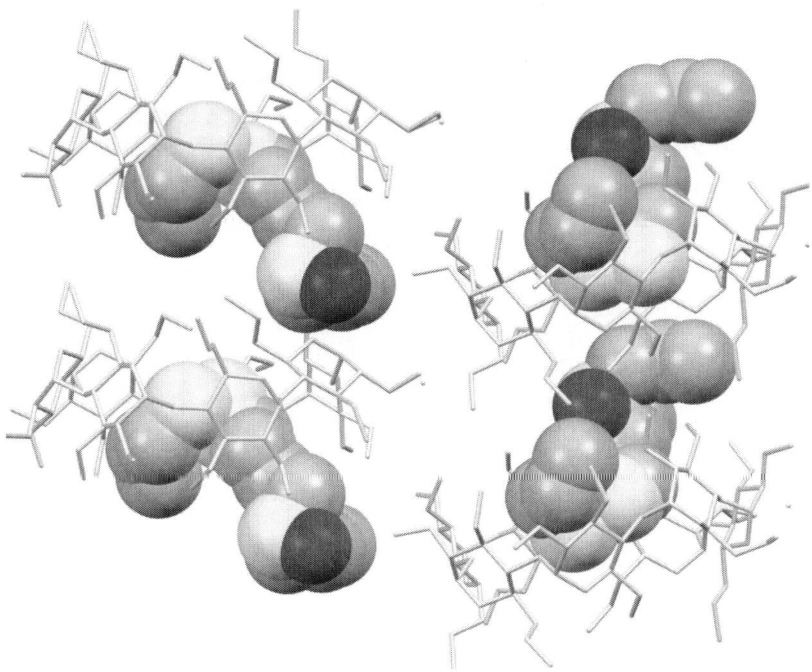

Figure 2. Inclusion geometry and packing of the inclusion complex TRIMEB·*E*-ajoene. The TRIMEB units, represented in light grey as sticks, stack in the head-to-tail fashion to form infinite channels; note that these channels run in antiparallel directions: on the right side, TRIMEB molecules have the "head" (the wider rim) facing upwards and the channel on the left has the TRIMEB "heads" turned downwards. The *E*-ajoene molecules are curled inside the TRIMEB cavities, which is better evidenced by using the spacefill style. This image was redrawn from the atomic coordinates of the complexes collected by Smith et al., [137] using the software Mercury 3.1 (Copyright CCDC 2001-2012). Hydrogen atoms were omitted for clarity.

Isothiocyanates from Mustards, Wasabi and Other Brassicaceae

Mustard (*Brassica juncea*), radishes and wasabi (the Japanese horseradish, comprising several plant species) are spices with a pungent flavor that is attributed to the presence of volatile isothiocyanates. The most relevant and abundant of these compounds is allyl isothiocyanate ($H_2C{=}CHCH_2N{=}C{=}S$), but other compounds have been found in these plants of the Brassicaceae family, namely 6-methylthiohexyl isothiocyanate, 7-methylthioheptyl isothiocyanate and 8-methylthiooctyl isothiocyanate, which also contribute to the pungent flavor of wasabi, and the less known allylcarbinyl isothipcyanate ($H_2C{=}CH_2CH_2N{=}C{=}S$), once identified in the oil of rape (or colza – *Brassica napus*). [144]

Allyl isothiocyanate (AITC) presents interesting nutraceutical properties, being a strong inhibitor of microbial growth and having also some antitumoral effects. Nevertheless, its volatility, high reactivity and easy hydrolysis make it difficult to formulate without the help of a suitable carrier and thus its inclusion into cyclodextrins has been the object of numerous studies and patents. Allyl isothiocyanate interacts both with α-CD and β-CD, but being essentially an alkenyl tiocyanate it has a narrow elongated shape and the inclusion complex is much stronger with the smaller α-CD cavity; this stronger affinity results in a marked reduction of AITC hydrolysis when the complex α-CD·AITC is in an aqueous solution. [145] α-CD also forms complexes with butyl isothiocyanate, isolated in the form of solid crystals. [146]

The inclusion of AITC into CDs affords products with sustained release [147] and helps to reduce its strong flavor, thus allowing it to be incorporated, in very small quantities (ppm range) into different foods to bring an antimicrobial action without causing much changes in their organoleptic properties. This application is patented for bread, confectioneries, rice cake, dairy foods, marine pasty foods, fruit-processed foods, seasonings, noodles and canned foods. [148] Another application is in preventing or delaying the stalling of fresh fruits and vegetables. [149] Cyclodextrin inclusion complexes of wasabi extract (containing roughly 99% AITC) are also incorporated into foods in larger amounts aiming at flavor potentiation and stabilization, for instance, in salad dressings. [150]

CD·AITC complexes can be found in food packages [151,152] and transparent foodwraps, [152,153] to which they convey antimicrobial and preservative properties. They are also incorporated into dry powders, [154] granules [155] or tablets [156] that release AITC gradually and afford an anti-septic environment to confined spaces such as refrigerators or shoe closets, [154] and to plastic boxes for food storage. [153]

Poly Unsaturated Fatty Acids (PUFAs)

The poly unsaturated fatty acids with relevant nutraceutical properties are found within the class of omega-3 fatty acids, characterized by the presence of an unsaturated bond at the third carbon atom from the end of the carbon chain. Three of the omega-3 fatty acids are deemed essential fatty acids, because they are needed for human biosynthetic pathways but not endogenously produced, thus having to be acquired from dietary sources. These are α-linolenic acid (ALA), found in plant oils and particularly abundant in chia, flaxseed, canola and walnut oils, eicosapentaenoic acid (EPA) and docosahexaenoic acid (DHA), both commonly found in oils of fatty fish such as salmon, herring, mackerel and sardine which acquire it from the phytoplankton they eat. It is not surprising, then, that these omega-3 fatty acids have a fish-like smell accompanied by a relatively bitter flavor, which need to be strongly reduced or even eliminated in order to make them acceptable for incorporation into fortified foods or nutritional supplements; such action is often achieved by cyclodextrin incorporation. Furthermore, the CDs help in protecting the double bonds of these fatty acids, which are quite susceptible to oxidation, and delay their otherwise quick and facile rancidity.

The formation of inclusion complexes of EPA and DHA with the native cyclodextrins or HP-β-CD was shown to strongly reduce their odor and taste; [157] furthermore, EPA complexes with α-CD and β-CD are quite stable to oxidation. [158] The association of DHA with γ-CD and a short-chain starch affords its complete odor and taste suppression [159] and allows its incorporation into tablets, claimed to taste "delicious". [160]

Cyclodextrins also contribute to the improvement of some of the medicinal properties of omega-3 fatty acids. A randomly methylated derivative of β-CD name RAMEB was shown to optimize the anti-inflammatory action of ALA, DHA and, most significantly, of EPA (by reducing the expression of CD1a and the secretion of IL-6). [161] The effect of omega-3 fatty acids in lowering seric triglyceride levels is also well-known, [162] and the reduction in seric triglycerides is slightly more pronounced upon intake of equivalent amounts of γ-CD complexes with EPA/DHA. [163] Associations of DHA and cyclodextrins are found in a variety of health-promoting foods and drinks, [164] fortified red date juices to enhance immunity [165] or help memory and brain development, [166] and even in an ice-cream powder, designed for the brain development of adolescents and claimed to have an "excellent taste". [167]

CONCLUSION

This boochapter compiles information on the myriad of cyclodextrin inclusion complexes with nutraceuticals. The large number of academic studies and patented products, methods and applications for nutraceuticals with CDs are a strong evidence that the cyclodextrins are no longer substances of the exclusive use of chemistry researchers and pharmacists but are firmly expanding their grounds to the alimentary and agricultural fields, making their way into our kitchens and featuring on the labels of our pre-prepared drinks and meals.

If in Japan the presence of cyclodextrins in foods is well- known and accepted for three decades, having made possible the preparation of popular products like green-tea flavored ice-cream and wasabi air fresheners for refrigerator and lunchboxes, in the United States its introduction has occurred later on and, in Europe, the approval of the labeling of CDs as new foods is even more recent (2012 for γ-CD). Many new products have appeared in the latest years, from promising new inclusion complexes sold as raw materials – e.g. the complexes γ-CD·curcumin and γ-CD·Coenzyme Q10 (tradenames Cavamax® W8 Curcumin and Cavamax® W8 CoQ10) which can already be found in some dietary supplements – to sophisticated fortified foods such as, for instance, "cactus healthcare bread", [168] a product that uses the valuable taste-masking properties of cyclodextrins to conceal the bitterness of *aloe vera*.

It is important to remember, however, that cyclodextrins are still widely unknown for the majority of the average consumer. This way, building the required confidence in these new ingredients will imply learning more about them. Both industry and academics need to continue their research efforts in order to achieve and a comprehensive understanding of the benefits and risks of the new nutraceutical·CD complexes.

ACKNOWLEDGMENTS

The authors acknowledge Fundação para a Ciência e a Tecnologia (Portugal), the European Union, QREN, European Fund for Regional Development (FEDER), through the programme COMPETE, for general funding to the QOPNA research unit (project PEst C-QUI/UI0062/2013; FCOMP-01-0124-FEDER-037296).

REFERENCES

[1] Brower V, Nutraceuticals: poised for a healthy slice of the healthcare market? *Nat Biotechnol* 1998, 16:728–731.

[2] Zeisel SH, Regulation of "Nutraceuticals." *Science* 1999, 285:185–186.

[3] Ministry of Health, Labor and Welfare Government of Japan, "FOSHU", http://www.mhlw.go.jp/english/topics/foodsafety/fhc/02.html [last accessed on 5 of January, 2015].

[4] Wildman R, Kelley M, In *Handbook of nutraceuticals and functional foods*; Wildman R, Ed, CRC Press: *Boca Raton,* 2001; Chapter 1:1-12.

[5] Villiers A, Sur la fermentation de la fécule par l'action du ferment butyrique. *Compt Rend Acad Sci* 1891, 112:536–538.

[6] Nitalikar MM, Sakarkar DM, Jain PV, The Cyclodextrins: A Review. *J Curr Pharm Res* 2012, 10 (1):01–06.

[7] FDA database of Inactive Ingredients for Approved Drug Products, http:// www.accessdata.fda.gov/scripts/cder/iig/index.cfm [Accessed on 27/08/2014]

[8] Shahidi F, Pegg R, Encapsulation and Controlled Release of Food Ingredients and Bioactives, In *Handbook of food preservation*; Shafiur Rahman M, Ed, *CRC Press: Boca Raton,* 2007 (2nd edition); Chapter 1:517–519.

[9] Hashimoto H, Aplication of Cyclodextrins to Foods, Toiletries and Other Products in Japan, In *Advances in Inclusion Science, Vol. 5: Proceedings of the Fourth International Symposium on Cyclodextrins*, Huber O and Szejtli J, Eds, Kluwer Academic Publishers, 1988, 533–543.

[10] Stella VJ, He QR, Cyclodextrins, *Toxicol Pathol* 2008, 36:30–42.

[11] Koukiekolo R, Desseaux V, Moreau Y, Marchis-Mouren G, Santimone M, Mechanism of porcine pancreatic α-amylase: Inhibition of amylose and maltopentaose hydrolysis by α-, β- and γ-cyclodextrins, *Eur J Biochem* 2001, 268:841–848.

[12] Van Ommen B, De Bie AT, Bär A, Disposition of ^{14}C-α-cyclodextrin in germ-free and conventional rats, *Regul Toxicol Pharmacol* 2004, 39(Suppl.1):57–66.

[13] Kubota Y, Fukuda M, Muroguchi M, Koizumi K, Absorption, distribution and excretion of β-cyclodextrin and glucosyl-β-cyclodextrin in rats. *Biol Pharm Bull* 1996, 19:1068–1072.

[14] De Bie AT, Van Ommen B, Bär A, Disposition of γ-cyclodextrin in germ-free and conventional rats. *Regul Toxicol Pharmacol* 1998, 27:150–158.

[15] World Health Organization, Safety evaluation of certain food additives and contaminants, *WHO Food Additives Series:* 48 (α-cyclodextrin), 2002.

[16] World Health Organization, Safety evaluation of certain food additives (α-cyclodextrin). 63rd meeting of the Joint FAO/WHO Expert Committee on Food Additives (JECFA). *WHO Food Additives Series* 2006, 54, 3–15.

[17] US Food and Drug Administration, Agency Response Letter Gras notice GRN No. 155, CFSAN (Center for Food Safety and Applied Nutrition)/Office of Food Additive Safety, 2004.

[18] Food Standards Australia and New Zealand, Final Assessment Report, Application A494. *Alpha-cyclodextrin as a novel food*, 2004.

[19] European Commission, 2008/413/EC: Commission Decision of 26 May 2008 authorising the placing on the market of alpha-cyclodextrin as a novel food ingredient under Regulation (EC) No 258/97 of the European Parliament and of the Council, Official Journal of the European Union 2008, 51:12–15. http://eur-lex.europa.eu/legal-content/EN/TXT/?uri= uriserv:OJ.L_.2008.146.01.0012.01.ENG

[20] Gallaher DD, Gallaher CM, Plank DW, Alpha-Cyclodextrin fecal excretion of saturated fats, *FASEB J* 2007, 21:A730.

[21] Comerford KB, Artiss JD, Jen KLC, Karakas SE, The beneficial effects of α-cyclodextrin on blood lipids and weigh loss in healthy humans, *Obesity J* 2011, 19:1200–1204.

[22] Jarosz PA, Fletcher E, Elserafy E, Artiss JD, Jen KLC, The effect of α-cyclodextrin on postprandial lipid and glycemic responses to a fat-containing meal, *Metabolis* 2013, 62:1443–1447.

[23] Wagner EM, Jen KLC, Artiss JD, Remaley AT, Dietary α-cyclodextrin lowers low-density lipoprotein cholesterol and alters plasma fatty acid profile in low-density lipoprotein receptor knockout mice on a high-fat diet, *Metabolis* 2008, 7:1046–1051.

[24] Antlsperger G, Reuscher H, Schmid G, Method for reducing the glycemic index for food, Patent EP1447014-A1, Wacker-Chemie GmbH, 2004; b) Schmid G, Reuscher H, Antlsperger G, Mixing cyclodextrins with food; antidiabetic agents; metabolism disease; obesity; hypotensive agents, Patent US20040161526 A1, Wacker-Chemie GmbH, 2004.

[25] Buckley JD, Thorp AA, Murphy KJ, Howe PRC, Dose-Dependent Inhibition of the Post-Prandial Glycaemic Response to a Standard Carbohydrate Meal following Incorporation of Alpha-Cyclodextrin, *Ann Nutr Metab* 2006, 50, 108–114.

[26] Reuscher H, Flexible Building Blocks for Healthy Eating, *Wacker World Wide Corporate Magazine* 2004, 2.4:23–25.

[27] Humphries J, in 10 Ways to Use Cyclodextrin (Recipe Book), Humphries J, Ed, MicJames, 2014.

[28] Andersen GH, Robbins FM, Domingues FJ, Moores RG, Long CL, The utilization of Schardinger dextrins by the rat, *Toxicol Appl Pharmacol* 1963, 5:257–266.

[29] Lina BAR, Bär A, Subchronic oral toxicity studies with alpha-cyclodextrin in rats, *Regulat Toxicol Pharmacol* 2004, 39:14–26; b) Lina BAR, Bär A, Subchronic (13-week) oral toxicity studies with alpha-cyclodextrin in dogs, *Regulat Toxicol Pharmacol* 2004, 39:27–33.

[30] Alfred Karpfen A, Liedl E, Snor W, Weiss-Greiler P, Viernstein H, Wolschann P, Homodromic hydrogen bonds in low-energy conformations of single molecule cyclodextrins, *J Incl Phenom Macro* 2007, 57:35–38.

[31] US Food and Drug Administration, Agency Response Letter Gras notice GRN No. 74, CFSAN/Office of Food Additive Safety, 2001. http://www.fda.gov/Food/Ingredients PackagingLabeling/GRAS/NoticeInventory/ucm154182.htm

[32] World Health Organization, Safety evaluation of certain food additives. Joint FAO/WHO Expert Committee on Food Additives (JECFA). WHO Food Additives Series 1993, 32:173–193. b) World Health Organization, Safety evaluation of certain food additives (β-cyclodextrin). Joint FAO/WHO Expert Committee on Food Additives (JECFA). *WHO Food Additives Series* 1996, 35: 257–268.

[33] Codex General Standard for Food Additives (GSFA). Updated up to the 37th Session of the Codex Alimentarius Commission, 2014. http://www.codexalimentarius.net /gsfaonline/additives/details.html?id=289

[34] Acosta E, Regulatory aspects of nutrient delivery systems, in *Delivery and controlled release of bioactives in foods and nutraceuticals*, Garti N, Ed, Woodhead Publishing Ltd Abbington Hall, Cambridge, UK, 2008, Chapter 17:429–449.

[35] Tamamoto LC, Schmidt SJ, Lee SY, *J Food Sci* 2010, 75: S341–S347.

[36] Reineccius TA, Reineccius GA, Peppard TL, Potential for β-cyclodextrin as partial fat replacer in low-fat foods, *J Food Sci* 2004, 69: FCT334–FCT341.

[37] Lee DK, Ahn J, Kwak HS, Cholesterol removal from homogenized milk with beta-cyclodextrin, *J Dairy Sci* 1999, 82: 2327–2330.

[38] Alonso L, Cuesta P, Fontecha J, Juarez M, Gilliland SE, Use of beta-cyclodextrin to decrease the level of cholesterol in milk fat, *J Dairy Sci* 2009, 92: 863–869.

[39] Dias HMAM, Berbicz F, Pedrochi F, Baesso ML, G Matioli, Butter cholesterol removal using different complexation methods with beta-cyclodextrin, and the contribution of photoacoustic spectroscopy to the evaluation of the complex, *Food Res Int* 2010, 43: 1104–1110.

[40] Kim JJ, Jung TH, Ahn J, Kwak HS, Properties of cholesterol-reduced butter made with beta-cyclodextrin and added evening primrose oil and phytosterols, *J Dairy Sci* 2006, 89:4503–4510.

[41] Park BS, Kang HK, Jang A, Influence of Feeding β-Cyclodextrin to Laying Hens on the Egg Production and Cholesterol Content of Egg Yolk, Asian Australas *J Anim Sci* 2005, 18:835–840.

[42] Tian YQ, Li Y, Jin ZY, Xu XM, Wang JP, Jiao AQ, Yu B, Talba T, β-Cyclodextrin (β-CD): A new approach in bread staling, *Thermochim Acta* 2009, 489:22–26.

[43] Marshal JJ, Miwa I, Kinetic difference between hydrolyses of γ-cyclodextrin by human salivary and pancreatic α-amylases, Biochim Biophys Acta 1981, 661:142–147; b) Harangi J, Beke G, Mótyán JA, The digestable parent cyclodextrin, *J Incl Phenom Macro* 2012, 73: 335–339.

[44] Tovar J, Rascón A, Starch Hydrolysis products with Physiological Activity in Humans, in *Food Oligosaccharides: production, analysis and bioactivity*, Moreno FJ and Sanz ML, Eds, *ITF Press/Wiley*, NY, 2014, chapter 7:107–117.

[45] World Health Organization, Safety evaluation of certain food additives (γ-cyclodextrin). 51st meeting of the Joint FAO/WHO Expert Committee on Food Additives (JECFA). WHO Food Additives Series 1999, 42:937. b) World Health

Organization, Safety evaluation of certain food additives (γ-cyclodextrin). 53rd meeting of the Joint FAO/WHO Expert Committee on Food Additives (JECFA). *WHO Food Additives Series* 2000, 44: 969.

[46] US Food and Drug Administration, Agency Response Letter Gras notice GRN No. 46, CFSAN/Office of Food Additive Safety, 2000. http://www.fda.gov/ucm/groups/fdagov-public/@fdagov-foods-gen /documents/document/ucm261675.pdf

[47] Food Standards Australia and New Zealand, Final Assessment Report, Application A438. Gamma-cyclodextrin as a novel food, 2003.

[48] European Comission, 2012/288/EU: Commission Implementing Decision of 1 June 2012, authorising the placing on the market of Gamma-Cyclodextrin as a novel food ingredient under Regulation (EC) No 258/97 of the European Parliament and of the Council. Official Journal of the European Comission 2012, 55: 41–42. http://eur-lex.europa.eu/legal-content/EN/TXT/?uri=uriserv:OJ.L_.2012.144.01.00 41.01.ENG

[49] http://www.healthylifebread.com/pages/original/100-whole-wheat-bread.php [accessed on 2015-03-08]

[50] Asp ML, Hertzler SR, Chow JM, Wolf BW, gamma-cyclodextrin lowers post-prandial glycemia and insulinemia without carbohydrate malabsorption in healthy adults, *J Am Coll Nutr* 2006, 25:49–55.

[51] Lai C, Chow JM, Wolf BW, Producing a blunted glycemic response and/or reducing post-prandial insulin secretion by administering nutritional product comprising gamma cyclodextrin, Patent US2005215523-A1, Lai C (LAIC-Individual), 2005.

[52] Lai C, Chow JM, Wolf BW, Method of providing insulin dependent individual with nutrition while reducing postprandial insulin secretion, and normalizing blood glucose involves preparing nutritional product comprising gamma-cyclodextrin; and administering the product, Patent US2008254100-A1, *Abbot Lab* (ABBO-C), 2008.

[53] Spears JK, Karr-Lillienthal LK, Grieshop CM, Flickinger EA, Wolf BW, Fahey GC, Pullulans and gamma-cyclodextrin affect apparent digestibility and metabolism in healthy adult ileal canulated dogs, *J Nutr* 2005, 135:1946–1952.

[54] Koutsou GA, Storey DM, Bär A, Gastrointestinal tolerance of γ-cyclodextrin in humans, *Food Addit Contam* 1999, 16:313–317.

[55] Boussiba S, Fan L, Vonshak A, Enhancement and determination of astaxanthin accumulation in green alga Haematococcus pluvialis, 1992, *Meth Enzymol,* 213:386–391; b) Castenmiller JJ, West CE, Bioavailability and bioconversion of carotenoids, *Annu Rev Nutr* 1998, 18:19–38.

[56] Chen X, Cheng R, Guo Z, Li C, Li P, The preparation and stability of the inclusion complex of astaxanthin with β-cyclodextrin, *Food Chem* 2007, 101:1580–1584; b) Kim S, Cho E, Yoo J, Cho E, Choi SJ, Son SM, Lee JM, In MJ, Kim DC, Kim JH, Chae HJ, β-CD-mediated Encapsulation Enhanced Stability and Solubility of Astaxanthin, *J Korean Soc Appl Bi* 2010, 53:559–565.

[57] Yuan C, Jin Z, Xu X, Zhuang H, Shen W, Preparation and stability of the inclusion complex of astaxanthin with hydroxypropyl-βcyclodextrin, Food Chem, 2008, 109:264–268; b) Yuan C, Du L, Jin Z, Xu X, Storage stability and antioxidant activity of complex of astaxanthin with hydroxypropyl-β-cyclodextrin, *Carbohyd Polym,* 2013, 91:385–389.

[58] Gharibzahedi SMT, Razavi SH, Mousavi M, Characterizing the natural canthaxanthin/2-hydroxypropyl-beta-cyclodextrin inclusion complex, *Carbohyd Polym* 2014, 101: 1147–1153.

[59] Madhavi DL, Kagan DI, Cyclodextrin or carotenoid complex, useful for e.g. human ingestion, and in nutritional supplement, comprises freeze-dried gamma-cyclodextrin/lutein complex, Patent US7446101-B1, Bioactives LLC (BIOA-Non-standard) 2008.

[60] Leuenberger B, Stoller H, New inclusion compounds of apocarotinal or lycopene in cyclodextrin - are soluble in polar solvents, Patent EP501267-A, Hoffmann La Roche & Co Ag F (HOFF-C), 1992.

[61] Kagome KK, A water-soluble lycopene prepn. - contains lycopene included in gamma-cyclodextrin and glutene and/or ascorbic acid, Japasese Patent JP8259829-A, Kagome KK (KAGO-C), 1996.

[62] Meiji Milk Prod, New beta-carotene-cyclodextrin inclusion cpds. - useful as antioxidants oxidant, colouring materials, pro-vitamin-A etc, Japanese Patent JP4244059-A, Meiji Milk Prod Co Ltd (MEIP-C) 1992. b) Meiji Milk Prod, Colourless beta-carotene prepn. - comprises adding oily beta-carotene to alpha-cyclodextrin soln., adjusting inclusion cpd. by stirring, adding dye to reduce chroma, Japanese Patent JP6025156-A, Meiji Milk Prod Co Ltd (MEIP-C) 1994.

[63] Tozzi R, Mulinacci N, Storlikken K, Pasquali I, Vincieri FF, Bettini R, Supercritical Extraction of Carotenoids from Rosa canina L. Hips and their Formulation with β-Cyclodextrin, *AAPS Pharm Sci Tech* 2008, 9:693–700.

[64] Lyng SM, Passos M, Fontana JD, Bixin and alpha-cyclodextrin inclusion complex and stability tests, *Process Biochem* 2005, 40:865–872.

[65] Domínguez-Cañedo IL, Beristain-Guevara CI, Díaz-Sobac R, Vázquez-Luna A, Degradación de carotenoides y capsaicina en el complejo de inclusión molecular de oleorresina de chile habanero (Capsicum chinense) con β-ciclodextrina, *CyTA – Journal of Food*, 13:151–158.

[66] Gomes LMM, Petito N, Costa VG, Falcão DQ, Araújo KGL, Inclusion complexes of red bell pepper pigments with β-cyclodextrin: Preparation, characterisation and application as natural colourants in yogurt, *Food Chem* 2014, 148: 428–436.

[67] Tongnuanchan P, Benjakul S, Essential Oils: Extraction, Bioactivities, and Their Uses for Food Preservation, *J Food Sci* 2014, 79:R1231–R1249.

[68] Locci E, Lai SM, Piras A, Marongiu B, Lai A, C-13-CPMAS and H-1-NMR study of the inclusion complexes of beta-cyclodextrin with carvacrol, thymol, and eugenol prepared in supercritical carbon dioxide, *Chem Biodiv* 2004, 1: 1354–1366. b) Bethanis K, Tzamalis P, Tsorteki F, Kokkinou A, Christoforides E, Mentzafos D, Structural study of the inclusion compounds of thymol, carvacrol and eugenol in beta-cyclodextrin by X-ray crystallography, *J Incl Phenom Macro* 2013, 77: 163–173. c) Fan Z, Guo MJ, Dong B, Jing ZL, Diao CH, Chen X, β-cyclodextrin 2-isopropyl-5-methylphenol clathrate hydrate, Huaxue Xuebao (= Acta Chim Sinica) 2010, 68:798.

[69] Cevallos PAP, MP Buera, Elizalde BE, Encapsulation of cinnamon and thyme essential oils components (cinnamaldehyde and thymol) in beta-cyclodextrin: Effect of interactions with water on complex stability, *J Food Eng* 2010, 99:70–75.

[70] Liang H, Yuan QP, Vriesekoop F, Lv F, Effects of cyclodextrins on the antimicrobial activity of plant-derived essential oil compounds, *Food Chem* 2012, 135:1020–1027.

[71] Kamimura JA, Santos EH, Hill LE, Gomes CL, Antimicrobial and antioxidant activities of carvacrol microencapsulated in hydroxypropyl-beta-cyclodextrin, *LWT Food Sci Tech* 2014, 57:701–709.

[72] Reineccius TA, Reineccius GA, Peppard TL, Encapsulation of Flavors using Cyclodextrins: Comparison of Flavor Retention in Alpha, Beta, and Gamma Types, *J Food Sci* 2002, 67:3271–3279.

[73] Chun JY, Jo YJ, Bjrapha P, Choi MJ, Min SG, Antimicrobial Effect of alpha- or beta-Cyclodextrin Complexes with Trans-Cinnamaldehyde Against Staphylococcus aureus and Escherichia coli, *Drying Tech* 2015, 33:377–383.

[74] Samperio C, Boyer R, Eigel WN, Holland KW, McKinney JS, O'Keefe SF, Smith R, Marcy JE, Enhancement of Plant Essential Oils' Aqueous Solubility and Stability Using Alpha and Beta Cyclodextrin, *J Agr Food Chem* 2010, 58: 12950–12956.

[75] Quintans JDS, Menezes PP, Santos MRV, Bonjardim LR, Almeida JRGS, Gelain DP, Araujo AAD, Quintans LJ, Improvement of p-cymene antinociceptive and anti-inflammatory effects by inclusion in beta-cyclodextrin, *Phytomed* 2013, 20:436–440.

[76] Furuta T, Yoshii H, Kobayashi T, Nishitarumi T, Yasunishi A, Powdery encapsulation of D-limonene by kneading with mixed powders of beta-cyclodextrin and maltodextrin at low water-content, *Biosci Biotech Bioch* 1994, 58: 847–850.

[77] Waleczek KJ, Marques HMC, Hempel B, Schmidt PC, Phase solubility studies of pure (-)-alpha-bisabolol and chamomile essential oil with beta-cyclodextrin, *Eur J Pharm Biopharm* 2003, 55: 247–251.

[78] Silva MCE, Galhano CIC, Da Silva AMGM, A new sprout inhibitor of potato tuber based on carvone/beta-cyclodextrin inclusion compound, *J Incl Phenom Macro* 2007, 57:121–124.

[79] Menezes PP, Serafini MR, Santana BV, Nunes RS, Quintans LJ, Silva GF, Medeiros IA, Marchioro M, Fraga BP, Santos MRV, Araujo AAS, Solid-state β-cyclodextrin complexes containing geraniol, *Termochim Acta* 2012, 548:45–50. b) Kayaci F, Sen HS, Durgun E, Uyar T, Functional electrospun polymeric nanofibers incorporating geraniol-cyclodextrin inclusion complexes: High thermal stability and enhanced durability of geraniol, *Food Res Internat* 2014, 62:424–431; c) Lins LCR, Santos IMA, de Melo MS, Menezes PD, Araujo AAS, Nunes RD, Dos Santos MRV, de Medeiros IA, Ribeiro EAN, dos Santos JR, Marchioro M, The anticonvulsant effect of geraniol and inclusion complex geraniol: beta-cyclodextrin, B Latinoam Caribe Pl 2014, 13:557–565.

[80] Kfoury M, Auezova L, Ruellan S, Greige-Gerges H, Fourmentin S, Complexation of estragole as pure compound and as main component of basil and tarragon essential oils with cyclodextrins, *Carbohyd Polym* 2015, 118:156–164.

[81] Kfoury M, Auezova L, Greige-Gerges H, Ruellan S, Fourmentin S, Cyclodextrin, an efficient tool for trans-anethole encapsulation: Chromatographic, spectroscopic, thermal and structural studies, *Food Chem* 2014, 164:454–461.

[82] Thoss M, Schwabe L, Fromming KH, Storage and photostability of cyclodextrin inclusion-compounds of lemon, orange, hop, and chamomile oil, *Farmazie* 1994, 49:252–257.

[83] Tamai K, Maehara M, Method for preparing aqueous solution mix of essential oil useful as foodstuff raw material for producing e.g. kabosu ice cream, involves

dispersing kabosu essential oil in beta-cyclodextrin using ethyl alcohol as adjuvant, *Japanese Patent* JP2004267186-A, Oita Ken (Oita prefecture), 2004.

[84] Zhou CX, Zou JK, Zhao Y, Chen YZ, [Application of beta-cyclodextrin inclusion technique in new dosage form of angelica sinensis essential oil], Zhongguo zhongyao zazhi = *China journal of Chinese materia medica* 2002, 27:832–834.

[85] Lemongrass food KK, Preparation of lemon grass material used in drinks - by extracting lemon grass oil and occluding extract with beta cyclodextrin, Japanese Patent JP61139360-A, Lemongrass food KK (LEMO-non standard), 1986.

[86] Del Toro-Sanchez C, Ayala-Zavala JF, Machi L, Santacruz H, Villegas-Ochoa MA, Alvarez-Parrilla E, Gonzalez-Aguilar GA, Controlled release of antifungal volatiles of thyme essential oil from beta-cyclodextrin capsules, *J Incl Phenom Macro* 2010, 67:431–441.

[87] Arana-Sanchez A, Estarron-Espinosa M, Obledo-Vazquez EN, Padilla-Camberos E, Silva-Vazquez R, Lugo-Cervantes E, Antimicrobial and antioxidant activities of Mexican oregano essential oils (Lippia graveolens H. B. K.) with different composition when microencapsulated in beta-cyclodextrin, *Lett Appl Microbiol* 2010, 50:585–590.

[88] Ciobanu A, Mallard I, Landy D, Brabie G, Nistor D, Fourmentin S, Retention of aroma compounds from Mentha piperita essential oil by cyclodextrins and crosslinked cyclodextrin polymers, *Food Chem* 2013, 138:291–297.

[89] Hădărugă NG, Hădărugă DI, Lupea AX, Paunescu V, Tatu C, Bioactive nanoparticles - 7. Essential oil from Apiaceae and Pinaceae family plants/beta-cyclodextrin supramolecular system, *Rev Chimie* 2005, 56: 876–882.

[90] Xu W, Fennel essential oil-mixed cyclodextrin clathrate compound, comprises mixture of alpha cyclodextrin and beta cyclodextrin, Chinese Patent CN103271957-A, Jiangsu Fengyuan Biotech Co Ltd (JIAN-Non-standard), 2013.

[91] Hădărugă NG, Hădărugă DI, Rivis I, Paunescu V, Costescu CI, Lupea AX, Bioactive nanoparticles - Essential oils from Lamiaceae family plants/beta-cyclodextrin - Supramolecular systems, *Rev Chimie* 2007, 58: 909–914.

[92] Hădărugă NG, Hădărugă DI, Costescu CI, David I, Gruia AT, Thermal and oxidative stability of the *Ocimum basilicum* L. essential oil/β-cyclodextrin supramolecular system, *Belstein J Org Chem* 2014, 10: 2809–2820.

[93] Hădărugă NG, Hădărugă DI, Resiga D, Pode V, Dumbrava D, Lupea AX, Obtaining and characterization of sage (*Salvia sclarea* L.) essential OR/beta-cyclodextrin supramolecular system, *Rev Chimie* 2007, 58: 566–573.

[94] Tian XN, Jiang ZT, Li R, Inclusion interactions and molecular microcapsule of *Salvia sclarea* L. essential oil with β-cyclodextrin derivatives, *Eur Food Res Technol* 2008, 227:1001–1007.

[95] Marreto RN, Almeida EECV, Alves PB, Niculau ES, Nunes RS, Matos CRS, Araujo AAS, Thermal analysis and gas chromatography coupled mass spectrometry analyses of hydroxypropyl-beta-cyclodextrin inclusion complex containing *Lippia gracilis* essential oil, *Termochim Acta* 2008, 475: 53–58.

[96] Petrović GM, Stojanović GS, Radulović NS, Encapsulation of cinnamon oil in β-cyclodextrin, *J Med Plants Res* 2010, 4(14):1382–1390.

[97] Hill LE, Gomes C, Taylor TM, Characterization of beta-cyclodextrin inclusion complexes containing essential oils (trans-cinnamaldehyde, eugenol, cinnamon bark,

and clove bud extracts) for antimicrobial delivery applications, *LWT Food Sci Tech* 2013, 51:86–93.

[98] Haloci E, Toska V, Shkreli R, Goci E, Vertuani S, Manfredini S, Encapsulation of Satureja montana essential oil in beta-cyclodextrin, *J Incl Phenom Macro* 2014, 80: 147–153.

[99] Deng J, Hao X, Li W, Tan X, Active packaging film comprises essential oil and beta-cyclodextrin inclusion compound, *Chinese Patent* CN102585412-A, *Deng J* (DENG-Individual), 2012.

[100] Higueras L, Lopez-Carballo G, Hernandez-Munoz P, Catala R, Gavara R, Antimicrobial packaging of chicken fillets based on the release of carvacrol from chitosan/cyclodextrin films, *Int J Food Micro* 2014, 188:53–59.

[101] Lou W, Wu H, Zong M, Zhu D, Wen P, Preparation of antibacterial agent nanofiber composite film used for food package material, by preparing polyvinyl alcohol solution, adding beta-cyclodextrin and cinnamon essential oil, adding into injector, and vacuum drying, Chinese Patent CN103266424-A, Univ South China Technology (UYSC-C), 2013.

[102] Moreira SP, de Carvalho WM, Alexandrino AC, de Paula HCB, Rodrigues MCP, de Figueiredo RW, Maia GA, de Figueiredo EMAT, Brasil IM, Freshness retention of minimally processed melon using different packages and multilayered edible coating containing microencapsulated essential oil, *Int J Food Sci Tech* 2014, 49: 2192–2203.

[103] Chen L, Xiao R, Ning W, Zhi Y, Zhou S, Microcapsule used for preventing and treating strawberry grey mold disease and tomato anthracnose, comprises Chenopodium essential oil and beta-cyclodextrin, Xinyang Agric & Forestry College (XINY-Non-standard), 2014.

[104] Tapas AR, Sakarkar DM, Kakde RB, Flavonoids as Nutraceuticals*: A Review, Trop J Pharm Res* 2008, 7:1089–1099.

[105] Lin JK, Weng MS, Flavonoids as Nutraceuticals, in *The Science of Flavonoids*, Grotewold E, Ed, *Springer Science,* USA 2006, Chapter 8, pp 213–238.

[106] Wu C, Strawberry freeze-drying process includes compounding water solution of beta-cyclodextrin, mixing cyanidin and fresh strawberry inside some vacuum container and vacuum pumping, Chinese Patent CN1672534-A, Wu C (WUCC individual) 2006.

[107] Matsui Y, Sunouchi A, Yamamoto T, Inhibitory Effect of Cyclodextrins on the Discoloration Reaction of an Anthocyanidin, Pelargonidin Chloride, in Acidic Media, *J Incl Phenom Mol* 1998, 32:56–67.

[108] Ishizu T, Kintsu K, Yamamoto H, NMR Study of the Solution Structures of the Inclusion Complexes of β-Cyclodextrin with (+)-Catechin and (-)-Epicatechin, *J Phys Chem B* 1999, 103:8992–8997.

[109] Jullian C, Miranda S, Zapata-Torres G, Mendizábal F, Olea-Azar C. Studies of inclusion complexes of natural and modified cyclodextrin with (+) catechin by NMR and molecular modeling. *Bioorg Med Chem* 2007:15(9):3217–24.

[110] da Rosa GC, Borges CD, Zambiazi RC, Rutz JK, da Luz SR, Krumreich FD, Benvenutti EV, Nunes MR, Encapsulation of the phenolic compounds of the blackberry (Rubus fruticosus), *LTW Food Sci Tech* 2014, 58:527–533.

[111] Ishizu T, Kajitani S, Tsutsumi H, Yamamoto H, Harano K, Diastereomeric difference of inclusion modes between (-)-epicatechin gallate, (-)-epigallocatechin gallate and

(+)-gallocatechin gallate, with beta-cyclodextrin in aqueous solvent, *Magn Reson Chem* 2008, 46(5):448–456.

[112] Ishizu T, Hirata C, Yamamoto H, Harano K, Structure and intramolecular flexibility of beta-cyclodextrin complex with (-)-epigallocatechin gallate in aqueous solvent, *Magn Reson Chem* 2006, 44:776–783.

[113] Ikeda T, Morohoshi T, Nakayama M, Yui K, The Inclusion Complex Formation of Epigallocatechin Gallate in γ-Cyclodextrin and Its Effect on the Antioxidant Activity, *Trans Mat Res Soc Japan* 2013, 38: 681–685.

[114] Ficarra R, Tommasini S, Rustichelli C, Raneri D, Calabrò ML, Gamberini MC, Ficarra P, Di Bella MR, Study of flavonoids/β-cyclodextrins inclusion complexes by NMR, FT-IR, DSC, X-ray investigation, *J Pharm Biomed Anal* 2004, 29: 1005–1014.

[115] Yang L-J, Ma S-X, Zhou S-Y, Chen W, Yuan M-W, Yin Y-Q, Yang XD, Preparation and characterization of inclusion complexes of naringenin with β-cyclodextrin or its derivative. *Carbohydr Polym* 2013, 98:861–869.

[116] Tommasini S, Ranieri D, Ficarra R, Calabrò ML, Stancanelli R, Ficarra P, Improvement in solubility and dissolution rate of flavonoids by complexation with beta-cyclodextrin, *J Pharm Biomed Anal* 2004, 29: 379–387.

[117] Shulman M, Cohen M, Soto-Gutierrez A, Yagi H, Wang HY, Goldwasser J, Lee-Parsons CW, Benny-Ratsaby O, Yarmush ML, Nahmias Y, Enhancement of Naringenin Bioavailability by Complexation with Hydroxypropoyl-beta-Cyclodextrin, *Plos One* 2011, 6: e18033.

[118] Tommasini S, Calabrò ML, Stancanelli R, Donato P, Costa C, Catania S, et al. The inclusion complexes of hesperetin and its 7-rhamnoglucoside with (2-hydroxypropyl)-beta-cyclodextrin. *J Pharm Biomed Anal* 2005, 39:572–580.

[119] Chakraborty S, Basu S, Lahiri A, Basak S, Inclusion of chrysin in β-cyclodextrin nanocavity and its effect on antioxidant potential of chrysin: A spectroscopic and molecular modeling approach. *J Mol Struct* 2010, 977:180–188.

[120] Jullian C, Cifuentes C, Alfaro M, Miranda S, Barriga G, Olea-Azar C, Spectroscopic characterization of the inclusion complexes of luteolin with native and derivatized beta-cyclodextrin, *Bioorg Med Chem* 2010, 18: 5025–5031.

[121] Jullian C, Orosteguis T, Pérez-Cruz F, Sánchez P, Mendizabal F, Olea-Azar C, Complexation of morin with three kinds of cyclodextrin. A thermodynamic and reactivity study. *Spectrochim Acta A Mol Biomol Spectrosc* 2008, 71:269–275.

[122] Anwer MK, Jamil S, Ansari MJ, Al-Shdefat R, Ali BE, Ganaie MA, Abdel-Kader MS, Shakeel F, Water soluble binary and ternary complexes of diosmin with β-cyclodextrin: Spectroscopic characterization, release studies and anti-oxidant activity. *J Mol Liq* 2014, 199:35–41.

[123] Pralhad T, Rajendrakumar K, Study of freeze-dried quercetin-cyclodextrin binary systems by DSC, FT-IR, X-ray diffraction and SEM analysis, *J Pharm Biomed Anal* 2004, 34:333–339.

[124] Mercader-Ros MT, Lucas-Abellán C, Fortea MI, Gabaldón JA, Núñez-Delicado E, Effect of HP-β-cyclodextrins complexation on the antioxidant activity of flavonols. *Food Chem* 2010, 118:769–773.

[125] Koontz JL, Marcy JE, O'Keefe SF, Duncan SE, Cyclodextrin Inclusion Complex Formation and Solid-State Characterization of the Natural Antioxidants α-Tocopherol and Quercetin, *J Agric Food Chem* 2009, 57:1162–1171.

[126] Jullian C, Improvement of galangin solubility using native and derivative cyclodextrins. An UV-VIS and NMR study, *J Chil Chem Soc* 2009, 54:201–203.

[127] Lu Z, Cheng B, Hu Y, Zhang Y, Zou G, Complexation of resveratrol with cyclodextrins: Solubility and antioxidant activity, *Food Chem* 2009, 113:17–20.

[128] Stražišar M, Andrenšek S, Šmidovnik A, Effect of β-cyclodextrin on antioxidant activity of coumaric acids. *Food Chem* 2008, 110:636–642.

[129] Nguyen TA, Liu B, Zhao J, Thomas DS, Hook JM, An investigation into the supramolecular structure, solubility, stability and antioxidant activity of rutin/cyclodextrin inclusion complex, *Food Chem* 2013, 136:186–192.

[130] Mercader-Ros MT, Lucas-Abellán C, Gabaldón JA, Fortea MI, Martínez-Cachá A, Núñez-Delicado E, Kaempferol complexation in cyclodextrins at basic pH, *J Agric Food Chem* 2010, 58:4675–4680.

[131] Stancanelli R, Mazzaglia A, Tommasini S, Calabrò ML, Villari V, Guardo M, Ficarra P, Ficarra R, The enhancement of isoflavones water solubility by complexation with modified cyclodextrins: A spectroscopic investigation with implications in the pharmaceutical analysis, J Pharm Biomed Anal 2007, 44: 980–984. Daruházi AE, Kiss T, Vecsernyés M, Szente L, Szőke E, Lemberkovics E, Investigation of transport of genistein, daidzein and their inclusion complexes prepared with different cyclodextrins on Caco-2 cell line, *J Pharm Biomed Anal* 2013, 84:112–116.

[132] Crupi V, Ficarra R, Guardo M, Majolino D, Stancanelli R, Venuti V, UV–vis and FTIR-ATR spectroscopic techniques to study the inclusion complexes of genistein with β-cyclodextrins, J Pharm Biomed Anal 2007, 44:110–117. b) Danciu C, Soica C, Csanyi E, Ambrus R, Feflea S, Peev CDehelean C, Changes in the anti-inflammatory activity of soy isoflavonoid genistein versus genistein incorporated in two types of cyclodextrin derivatives, *Chem Cent J* 2012, 6:58.

[133] Instant green, black or oolong tea powder drink for use as health drink, contains non-polymer catechin, preset amount of oxalic acid component, and cyclodextrin as main ingredients, Japanese Patent JP2004159505-A, Kao Corp (KAOS-C), 2004. b) Takatsu H, Otsuksa K, Manufacture of container-packed tea beverage involves adding cyclodextrin and glucan to preparation liquid of concentrate or purified substance of tea extract, heat sterilizing, and adjusting non-polymer catechin content, Japanese Patent JP2006174748-A, Kao Corp (KAOS-C), 2006. c) Takatsu H, Otsuksa K, Container-packed tea beverage as green tea drink, contains preset amount of non-polymeric catechin blended with gamma-cyclodextrin and beta-cyclodextrin in specific weight ratio, Japanese Patent JP2006180711-A, Kao Corp (KAOS-C), 2006. d) Murao T, Aomi M, Powder instant tea for producing tea beverage, contains preset amount of cyclodextrin and thaumatin with respect to non-polymer catechin contained in tea extract powder as main ingredients, Japanese Patent JP2006136244-A, House Shokuhin Kogyo KK, 2006. e) Furuichi N, Itaya E, Takahashi H, Container-packed green tea drink contains non-polymer catechin, theanine, and glutamic acid (salt) in preset mass ratio, Japanese Patent JP2008011834-A, Kao Corp, 2008. f) Fukuda M, Takahashi H, Container-packed drink such as non-tea-type drink, non-fermented-tea drink and half-fermented-tea drink, contains preset amount of non-polymer catechin, beta-cyclodextrin and gamma-cyclodextrin, Japanese Patent JP2007319075-A, Kao Corp, 2008.

[134] Ichitani M, Takihara T, Tsuru C, Container packed drink contains theaflavin, non-polymer catechin, and ascorbic acid and/or cyclodextrin, Japanese Patent JP2010094084-A, Itoen KK, 2010. b) Takatsu H, Container-packed drink contains predetermined amounts of non-polymer catechin, starch, and alpha-cyclodextrin or gamma-cyclodextrin, Japanese Patent JP2010200642-A, Kao Corp, 2010.

[135] Nikolic V, Stankovic M, Kapor A, Nikolic L, Cvetkovic D, Stamenković J, Allylthiosulfinate: beta-cyclodextrin inclusion complex: preparation, characterization and microbiological activity, *Pharmazie* 2004, 59(11):845–848.

[136] Bai YX, Yu B, Xu XM, Jin ZY, Tian YQ, Lu L, Comparison of encapsulation properties of major garlic oil components by hydroxypropyl beta-cyclodextrin, *Eur Food Res Technol* 2010, 231:519–524.

[137] Smith VJ, Caira MR, Hunter R, Bourne SA, Characterization of (E)- and (Z)-ajoene inclusion complexes with α-, β-, γ- and permethylated- β-cyclodextrin using PXRD, single-crystal X-ray diffraction and thermal analysis, *Acta Cryst* 2004, A60:s217.

[138] Wang J, Cao YP, Sun BG, Wang CT, Physicochemical and release characterisation of garlic oil-beta-cyclodextrin inclusion complexes, *Food Chem* 2011, 127:1680–1685.

[139] Liu Y, Tomato chili sauce includes edible salt, xanthan gum, modified starch, garlic powder, hot pepper powder, tomato full powder, white granulated sugar, cyclodextrin, plant oil, potassium sorbate, Chinese blister beetle, and ginger powder, Chinese Patent CN102885285-A, Tianjin Jianfeng Natural Prod R&D Co, 2013. b) Liu Y, Xu Y, Celery chilli sauce comprises salt, xanthan gum, modified starch, garlic powder, chilli, whole celery powder, white granulated sugar, cyclodextrin, vegetable oil, potassium sorbate, ginger powder, monosodium glutamate and water, Chinese Patent CN102166009-A, Tianjin Jianfeng Natural Prod R&D Co, 2011. c) Miao G, Yuan J, Liang S, Yao Z, Wen K, Feng L, Welsh onion ketchup preparation as foodstuffs, comprises common fennel, fresh garlic, ginger, ketchup, flavor essence, ethyl maltol, white sugar, potassium sorbate, welsh onion, soybean oil, salt, and white sesame seed, Chinese Patent CN101124975-A, Beijing Keweihua Food Eng Technology Co, 2008.

[140] Shen G, Sun T, He S, Xu M, Shallot-garlic-ginger compound paste seasoning and preparing method, Chinese Patent CN1849923-A, *Zhejiang Prov Aagric Ssi Acad*, 2007.

[141] Garlic egg yolk powder useful as health food and food additive, comprises predetermined amount of cyclodextrin, Japanese Patent JP2003047426-A, Sun Protect KK, 2003.

[142] Pan B, Zhang R, Low sugar black garlic beverage includes black garlic extract, honeysuckle extract, peony flower extract, rose extract, Chrysanthemum extract, maca extract, honey, white vinegar, maltodextrin, xylitol, malic acid and beta-cyclodextrin, Chinese Patent CN103960729-A , Shandong Jinbaoluo Food & Beverage Co (SHAN-Non-standard), 2014.

[143] Do DH, Woo JS, Method for producing garlic-coffee containing garlic components, Do DH (Individual), 2003.

[144] Ettlinger MG, Hodgkins JE, The mustard oil of rape seed, allylcarbinyl isothiocyanate, and synthetic isomers, *J Am Chem Soc* 1955, 77:1831–1836.

[145] Yoshio O, Takatani K, Kawakishi S, Kinetic and thermodynamic analysis of the cyclodextrin-allyl isothiocyanate inclusion complex in an aqueous solution, *Biosci biotechnol biochem* 2000, 64:190–193.

[146] Sicard-Roselli C, Perly B, Le Bas G, The respective benefits of X-ray crystallography and NMR for the structural determination of the inclusion complex between butyl-isothiocyanate and alpha-cyclodextrin, *J Incl Phenom Macro* 2001, 39:333–337.

[147] Li XH, Jin ZY, Wang J, Complexation of allyl isothiocyanate by alpha- and beta-cyclodextrin and its controlled release characteristics, *Food Chem* 2007, 103:461–466

[148] Food contg. cyclodextrin-allyl isothiocyanate mixt. as preserver - prepd. by mechanical mixing and adding to food, reducing stimulative smell and volatilisation rate, Japanese Patent JP5176733-A, *Green Cross Corp & Nipon Shokuhin Kako KK,* 1993.

[149] Agent used for preserving the freshness of vegetables - comprises isothiocyanate cpd. diluted with lipophilic solvent e.g. cyclodextrin, Japanese Patent JP2135048-A, QP Corp, 1990. b) Guan W, Ma J, Yan R, Zhang N, Antistaling agent used in preserving fruits and vegetables, e.g. apple, grape, banana, potato, onion and garlic, comprises mustard essential oil, beta-cyclodextrin and stearic acid, Chinese patent CN101711533-A, *Nat Farming Prod Preservation Technology*, 2010.

[150] Nishimoto S, Iwase A, Hasegawa H, Dressing type seasoning composition containing encapsulated flavour of wasabi, European Patent EP0761104 A1, Riken Vitamin Co Ltd, 1997.

[151] Antibacterial base material for packaging material for perishable foods - contg. cyclodextrin including iso:thiocyanate cpd or terpene cpd., formed on surface of base material, Japanese Patent JP6286775-A, Sekisui Jushi KK, 1994.

[152] Oikawa T, Fukushima Y, Ishii K, Takesue M, MUTO Y, Fujii K, Fukaya S, Antimicrobial laminate for bags for foodstuffs for gradual release - comprises substrate impervious to steam and pervious film superposed through adhesive layer contg. allyl isothiocyanate cyclodextrin cpd., Patent AU9514811-A / JP8080597-A / US6150004-A, Kyodo Printing Co Ltd, 1996.

[153] Antibacterial film and its prepn - comprises synthetic resin film contg cyclodextrin cpd, isothiocyanate cpd and/or terpene cpd, Japanese Patent JP6191562-A, Sekisui Jushi KK, Ensuiko Sugar Refining Co Ltd & Yakushin KK, 1994. b) Antibiotic wrapping material for foodstuff - contains cyclodextrin inclusion cpd. contg. isothiocyanate, Japanese Patent JP7046973-A, Sekisui Jushi KK, 1995. c) Antibacterial self-adhesive tape for lunch boxes, etc. - has base sheet and self adhesive layer both opt contg anti-fungus agent consisting of iso:thio:cyanuric acid ester clathrated with cyclodextrin, Japanese Patent JP6158002-A, Sekisui Plastics Co Ltd, 1999.

[154] Antimicrobial agent for refrigerator and shoe cupboard - consists of cyclodextrin clathrates having wasabi essential oil, adsorbed on dextrin adsorbent, Japanese Patent JP11180812-A, Sekisui Chem Ind Co Ltd & Sekisui Jushi KK, 1999.

[155] Antibacterial material for foods, drugs and feeds - contains oxygen absorber, inclusion cpd. contg. volatile organic antibacterial agent in cyclodextrin and amorphous calcium phosphate, Japanese Patent JP8176204-A, Sekisui Plastics Co Ltd, 1996.

[156] Ishimatsu N, Deoxygenating agent in which capacity controlling bacterial activity is improved - comprises adding allyl isothiocyanate to deoxygenating agent, Japanese Patent JP4179464-A, Ishimatsu N (Individual), 1992.

[157] Bruzzese T, Mozzi G, Compsn. for dietetic or therapeutic use - comprising complex of cyclodextrin and long chain polyunsaturated fatty acid or deriv., European Patent EP470452-A2, Staroil Ltd, 1992.

[158] Yoshii H, Furuta T, Yasunishi A, Linko YY, Linko P, Oxidation stability of eicosapentaenoic and docosahexaenoic acid included in cyclodextrins, *J Incl Phen Macro* 1996, 25:217–220.

[159] Matsunaga K, Shimizu Y, Eliminating odor or bitter taste of specific component e.g. docosahexaenoic acid in composition, involves mixing composition with low molecular weight polysaccharide and gamma-cyclodextrin, and processing mixture, Japanese Patent JP2010202603-A, Bizen Kasei KK, 2010.

[160] Jian X, Tablet comprises whole milk powder, glucose, maltodextrin, cyclodextrin, sucrose powder, cream powder, citric acid, magnesium stearate, Bifidobacterium, arachidonic acid powder, docosahexaenoic acid powder, and fresh milk essence, Chinese Patent CN103651785-A, Wuhan Shutai Technology Co Ltd, 2014.

[161] Rajnavölgyi E, Laczik R, Kun V, Szente L, Fenyvesi E, Effects of RAMEA-complexed polyunsaturated fatty acids on the response of human dendritic cells to inflammatory signals, *Belstein J org Chem* 2014, 10: 3152–3160.

[162] Harris WS, n-3 fatty acids and serum lipoproteins: Human studies, *Am J Clin Nutr* 1997, 65(Suppl):S1645–S1654.

[163] Kobayashi K, Hamazaki K, Fujioka S, Terao K, Yamamoto J, Kobayashi S, The effect of n-3 PUFA/gamma-cyclodextrin complex on serum lipids in healthy volunteers - a randomized, placebo-controlled, double-blind trial, *Asia Pac J Clin Nutr* 2007, 16:429–434.

[164] Tsuji H, Takahashi C, Seto A, Sakai M, Drinks and foods for mothers and babies - contain fat and oil for high accumulation of docosa-hexa:enoic acid Japan Patent JP9075000-A, Nisshin Oil Mills Ltd, 1997. b) Lee HJ, Ryu MS, Composition of health drink for infant and its preparation, Patent KR2001000741-A, Choi S (Individual) & Kwon L (Individual), 2001. c) Lee HJ, Ryu MS, Composition of health drink for pregnant woman and preparation method thereof, Patent KR2001000743-A, Choi S (Individual) & Kwon L (Individual), 2002.

[165] Ma Q, Walnut red dates concentrated slurry comprises walnut, jujube, docosahexaenoic acid, medlar, Chinese yam, hawthorn, oligoisomaltose, honey fructose syrup, stabilizer, sodium carboxymethyl cellulose, xanthan gum, and beta-cyclodextrin, Chinese Patent CN102511881-A, Shanxi Hanbo Food CO Ltd, 2012.

[166] Ma Q, Concentrated red date pulp useful e.g. for improving memory, growth and development and promoting brain development comprises tuckahoe, spina date seed, phospholipid, docosahexaenoic acid, calcium, iron, zinc and concentrated red date juice, Chinese Patent CN102511874-A, Shanxi Hanbo Food Co Ltd, 2012.

[167] Chen E, Ice cream powder for improving cerebral cortex and retina, comprises beta-cyclodextrin, non-dairy creamer, whole milk powder, whey powder, docosahexaenoic acid, food flavor, stabilizer, emulsifier and vitamin E, Chinese Patent CN101214002-A, Sweet Style Dietary Culture Media Beijing Co Ltd, 2008.

[168] Jing F, Cactus healthcare bread useful for e.g. decreasing blood pressure and reducing blood sugar, comprises flour, eggs, milk, cactus, white sugar, salt, vegetable oil, yeast, algae powder, cyclodextrin, spice and water, Chinese Patent CN101953390-A, 2011.

In: Cyclodextrins
Editor: Francis G. Ramirez

ISBN: 978-1-63482-788-1

Chapter 7

THE IMPORTANCE OF INCLUSION COMPLEXES WITH CYCLODEXTRINS IN PHARMACY

***Vesna D. Nikolic*[1], *Agnes J. Kapor*[2], *Ljubisa B. Nikolic*[1], *Ivan M. Savic*[1] *and Ivana M. Savic-Gajic*[1*]**

[1]Faculty of Technology, University of Nis, Leskovac, Serbia
[2]Department of Physics, Faculty of Sciences, University of Novi Sad, Novi Sad, Serbia

ABSTRACT

Cyclodextrins due to their hydrophobic cavity and hydrophilic surface can form the "host-guest" inclusion complexes with various compounds. The central hydrophobic cavity allows that the cyclodextrins have the role of the "host" molecules. The preparation of inclusion complexes leads to the changes in the physical and chemical properties of "guest" molecules. In this way it is possible to protect the compounds sensitive to the effect of light, heat and oxygen from the air. Also, the aqueous solubility of poorly soluble substances can be increased. Because of these properties and negligible toxicity, the cyclodextrins have found a broad application in the pharmaceutical industry. Today, the drugs based on the cyclodextrins represent about 40% of the total available drugs on the market. Their use in the pharmaceutical formulations is justified, because they improve the aqueous solubility of the hydrophobic drugs without the addition of the organic solvents, surfactants or lipids, increase the dissolution rates, oral bioavailability and stability. The cyclodextrin inclusion complexes reduce the gastrointestinal and ocular irritations, eliminate the unpleasant odors or tastes and prevent the drug-drug or drug-excipients interactions. The controlled release is also one of the important properties of the cyclodextrins which justifies their use in order to improve a drug delivery from a pharmaceutical formulation. Also, the cyclodextrins are used in the preparation of deodorants, lotions, diapers, paper towels, etc. Their role is to form the complexes with smelly substances in order to reduce their volatility.

---

* Corresponding author: Ivana M. Savic-Gajic, Ph.D., Faculty of Technology, University of Nis, Bulevar oslobodjenja 124, 16000 Leskovac, Serbia, Tel.: +381 16 247 203; Fax: +381 16 242 859, E-mail: vana.savic@yahoo.com.

Keywords: Inclusion complexes, drugs, stability, carrier, drug delivery

INTRODUCTION

The improvement of the quality, efficiency and safety of the drugs and medicinal preparations has become a major topic in pharmaceutical industry. For these reasons, the pharmaceutical industry is required a constant innovation in terms of manufacturing processes and technology, and use of better and more contemporary excipients during the manufacturing process. Also, information technology can play an important role to prevent adverse drug events [1]. The bioavailability, which is in direct correlation with the absorption of a drug substance can be changed to the reduced solubility and permeability of the drug substance. Given that about 70% of the drug substances of newer generation are poorly soluble in water, this area of pharmaceutical technology has a great importance for further research. Based on the solubility and permeability, all drug substances are classified into four groups by Biopharmaceutical System of Classification (BSK) [2]. The soluble substances are dissolved in 250 ml of water at pH 1.0 to 7.5 and a temperature of 37 ± 0.5 °C in an amount that represents its maximum therapeutic dose. If it does not meet these criteria, it is considered as poor soluble substance. In terms of permeability, it is very important to analyze the effective permeability through a membrane of human jejunum. The highly permeable substances are absorbed in the percentage greater than or equal to 90% of the applied amount of drug, if there is no data about its instability in the gastrointestinal tract. Otherwise, it is considered as a low permeable [3].

A particular object of the study are compounds II and IV groups of BSK. Their poor water solubility remains a major obstacle in the development and clinical application of the pharmaceutical formulations containing this substance. The substances with these properties have limited digestive absorption after release from a pharmaceutical dosage form, i.e. low bioavailability and poor pharmacokinetics. Accordingly, the numerous studies have been directed towards improving solubility behaviour of the drugs. The different carriers and methods should be applied in order to overcome these problems. One of the most promising strategies to improve the oral bioavailability of poorly water soluble drugs is decreasing a drug particle size (solid dispersions) [4]. In this way drug wettability may be significantly improved. Several nanonization techniques are applied, which include drug nanocrystals [5], nanoemulsions [6] and polymeric micelles [7]. The salt formation is also the most common and effective method of increasing solubility of the acidic and basic drugs [8]. Prodrug strategies are based on the chemical derivatization which utilizes esterification of a drug hydroxyl, amine or carboxyl group with a moiety (progroup) designed to introduce an ionizable function or reduce intermolecular interactions responsible for low solubility [9]. Other various techniques such as the solvent deposition, supercritical fluid process, use of surfactants and complexation have been applied. It is considered that the complexation technique is more precisely technique for improving the stability and dissolution of poorly water soluble drugs.

INCLUSION COMPLEXES

Inclusion complexes represents a special group known as "non-classical" complexes which are formed under the effect of mechanical factors. The components of inclusion complexes are the molecules of host and guest. The host molecules are the compounds which in their crystal lattice or their aggregates have the sufficiently large cavities to incorporate the other components (the guest molecules).

The preparation of the inclusion complexes depends on the size and shape of the guest molecules, as well as the dimensions of cavities formed from the aggregates of the host molecules. In addition to weak Van der Waals attractive forces, hydrogen bonds and polar interactions, there are no the chemical interactions either interatomic or molecular type. Depending on the nature of the cavity of host molecule, there are a cage and canal inclusion complexes [10]. The advantages that provide these inclusion complexes in the pharmaceutical industry are:

- The increase or decrease of the solubility and the dissolution rate of the complex in relation to the active substance depending on the nature of the guest and host molecules;
- A better bioavailability of the drug;
- The change of reactivity of the active substance after the protection of the some functional groups;
- extended the life of the drug (improved physical and chemical stability);
- the reduction of the contraindications;
- the correction of odor and flavour.

The active substances that include in the cavity of host molecule have the following characteristics:

- they are poorly soluble in water and their bioavailability is low;
- the achievement time of the effective level in the blood is too long;
- they can not be used for the preparation of injections due to poor solubility,
- they are suitable for the preparation of drug in the forms of tablet or powder;
- they are chemically and physically unstable without complexation;
- they have more polymorphic crystalline forms;
- they have an unpleasant odor, bitter or irritating flavour;
- they have not a homogeneous content;
- they are incompatible with the excipients;
- they have a higher number of contraindications in relation to the inclusion complexes;
- they have a high toxicity in relation to the complexes.

The various methods can be used for the preparation of inclusion complexes.

THE PREPARATION OF LIQUID INCLUSION COMPLEXES

The preparation of complexes in the solution are performed by addition of increasing amount of guest molecule in the aqueous solution of the host molecule. The prepared suspensions are equilibrated at the defined temperature about 7 days, and then filtered or centrifuged in order to form a clear solution of guest-host complex.

THE PREPARATION OF SOLID INCLUSION COMPLEXES

The solid complexes can be prepared by evaporation, lyophilization or spray-drying of the aqueous solutions of inclusion complexes. The co-precipitation, neutralization, kneading and imprinting methods can be also used for the preparation of inclusion complexes. According to the kneading method, the component that includes is added to the mixture of water and host component. The mixture is homogenized, so as to obtain a thick paste, which is then dried.

Co-precipitation is also one of methods for the preparation of solid inclusion complexes using the organic solvents, which have the role as the agent for precipitation. Due to the ability of the solvent to participate in the complexation, this method is much less applied.

The solid complexes which ionize are prepared using the neutralization method. According to this method, the guest components are dissolved in the acidic (if it is the base component) or in the alkaline (if it is the acidic component) aqueous solutions of the host component. The solubility of the guest component is then reduced by adjustment of pH values, what affects the precipitation of the complex from the solution, i.e. the non-ionized active substance is formed.

The solid inclusion complexes can be prepared and using imprinting techniques. This method is based on the preparation of physical mixture of guest and host components, which is then heated in a closed vessel at higher temperature (up to 90 °C).

Generally, all complexes including and inclusion complexes are characterized by the stability constant (K_S) or the dissociation constant (K_d), which are the measure of physico-chemical changes of the component which is included [11]. It is especially important to determine the stability constant for the inclusion complexes that are used in the drug formulations as a measure of their disintegration after dissolution in the gastric and intestinal juices, when a drug is taken orally. The complex [H-G] in the solution is in equilibrium with the host (H) and guest (G) molecules. If the value of stability constant of complex is low, it can be expected a very rapid resorption of the drug. For very stable complexes, the equilibrium is shifted towards the formation of complexes. Their stability constant is high so that the resorption of active component is very low. The drug resorption can be regulated by adjustment of the conditions for decreasing or increasing the stability constant of the complexes. The scheme of pharmacokinetic behavior of the inclusion complex in the formulation after oral administration is presented in Figure 1, where k_r is the rate constant of dissolution of inclusion complex, k_d is the rate constant of dissociation of inclusion complex in the solution, k_a is the rate constant of association, i.e. the formation of complex in the solution, k_{abs} is the rate constant of absorption of guest molecule, i.e. drug.

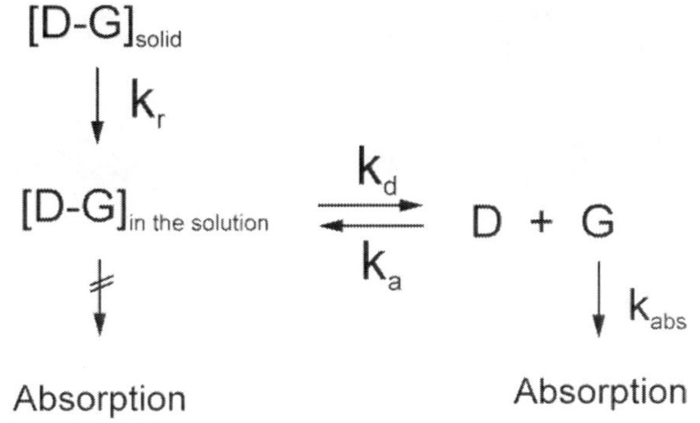

Figure 1. Pharmacokinetics of the inclusion complex in the formulation after oral administration.

CYCLODEXTRINS

Although a number of techniques have been suggested for enhancement of drug activity or the reduction of side effects, cyclodextrins (CDs) have emerged as a productive approach. CDs are oligosaccharides formed by six to eight glucopyranose units, bound via α-1,4-glycosidic bonds [12]. The arrangement of monomers in the CD molecules is such that it can be visualized as a truncated cone with a hydrophilic exterior (Figure 2). Molecular masses of CD molecule are in the range of 1000-2000 Da. These molecules had a large number of hydrogen donors and acceptors.

XRD analysis has shown that the glucoside residues in the CDs have a thermodynamically stable conformation of chair, because all substituents are in the equatorial position. As a consequence of this conformation, all secondary hydroxyl groups (OH-2 and OH-3) are located on one side of cylindrical molecule, while all primary hydroxyl groups (OH-6) are on the other side of the torus molecule. This position of hydroxyl groups on both sides of the cavity shows that its broader side is surrounded with a secondary hydroxyl groups, i.e. that the external side of the CD is more hydrophilic and allows the dissolution in the aqueous solution. The inner side of the cavity is more hydrophobic, because of the free electron pairs of oxygen with a glucosyl oxygen bridge are located inside the cavity. This cavity can act as a host for various lipophilic molecules.

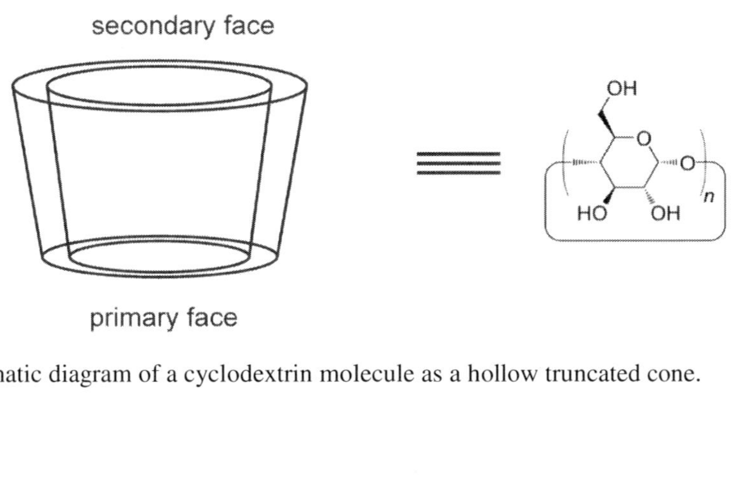

Figure 2. Schematic diagram of a cyclodextrin molecule as a hollow truncated cone.

Commercially, CDs are produced from starch by enzymatic digestion. Cyclodextrin glycosyltransferase produces a mixture of CDs which consisting of 6, 7 and 8 glucose units (α-, β- and γ-CD, respectively) Figure 3 [13,14]. These three CDs are crystalline, homogeneous and nonhygroscopic substances.

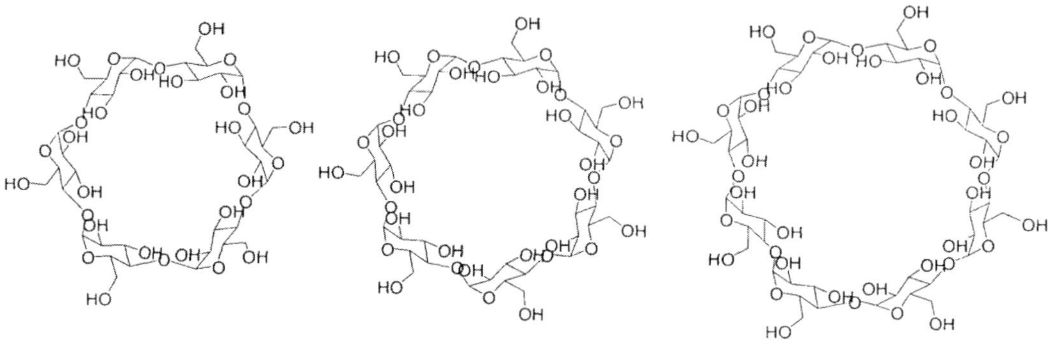

Figure 3. The structure of α-, β- and γ-cyclodextrin.

Due to the presence of different number of glucose units in the structure, there are the slight differences in the conformational structure, flexibility and size of the ring in terms of diameter. The difference in the solubility was also noticed, i.e. the higher water solubility have γ and α–CDs in compared to β-CD [15]. These differences are the result of higher exposure of hydroxyl groups to hydrogen bonding in the aqueous mediums. Thus, CDs are useful molecular chelating agents, which have a cage-like supramolecular structure. They carry out the chemical reactions that involve intramolecular interactions where covalent bonds are not formed between interacting molecules, ions or radicals. Most of them belong to the "host–guest" type.

CDs' Derivatives

In addition to the native CDs, the CD derivates are increasingly used in the pharmaceutical industry. The derivatives can be prepared by the chemical and enzymatic reactions, and can be classified according to the nature of substituents in the following way: alkylated, hydroxylalkylated, esterified, sugar and polymerized derivatives of CDs. The main reason of the modification is the change of their undesirable physico-chemical properties, the improvement of inclusion capacity of the native CDs, as well as the reduction of their parenteral toxicity [16]. The commonly used derivatives are hydroxylated derivatives of β-CD and γ-CD. (2-Hydroxypropyl)-β-cyclodextrin (HP-β-CD) is the most used modified CD, and its popularity can be attributed to the size of cavity. This compound can change the physico-chemical properties of lipophilic compounds, enable their easier diffusion through the biological membranes and to increase their aqueous solubility and stability. Also, HP-β-CD can increase the antimicrobial effectiveness of chemical agents by increasing their release rate. The randomly methylated β-cyclodextrin (RM-β-CD), sulfobutylether β-cyclodextrin

(SBE-β-CD), monochlorotriazinyl β-cyclodextrin (MCT-β-CD), heptakis-β-cyclodextrin and the so called branched CDs, such as maltosyl-β-cyclodextrin (G2-β-CD) are also some of the CD derivatives which have found a practical application.

Pharmacokinetic studies have established that the CDs are poorly absorbed through the biological membranes, while the natural α- and β-CDs cannot hydrolyze by human salivary and pancreatic amylases. Unlike γ-CD, the natural CDs and their derivatives are used in the topical and oral dosage forms. β-CD cannot be used in the parenteral formulations because of its nephrotoxicity. γ-CD forms the visible aggregates in aqueous solutions, so it is not a suitable for parenteral formulations [17]. Extensive toxicological studies for HP-β-CD and SBE-β-CD have shown that these compounds can be found in marketed parenteral formulations at relatively high concentrations. Lipophilic CD derivatives, such as methylated CDs, are to some extent absorbed from the gastrointestinal tract into the systemic circulation and have been shown to be toxic after parenteral administration. Presently, an oral administration of methylated β-CD is limited due to its potential toxicity.

Cyclodextrin Inclusion Complexes and Their Application

Because of this unique physicochemical property, CD and its derivates can be found on the list of substances used for forming inclusion complexes with a very broad range of solid, liquid and gaseous hydrophobic drugs resulting in the improvement of their aqueous solubility without the use of organic solvents, surfactants or lipids. The only one guest molecule can be incorporated in the CDs cavities with diameter of 0.5-0.9 nm. In the crystalline state, the individual helically wrapped CD molecules are arranged one above the other and form a long channel in which can be putted the molecules of organic substances, what represents the so-called molecular encapsulation [10]. In order to achieve an adequate absorption of the drug, the drug should be released in its dissolved form from the formulation. Reducing the hydrophobicity of drugs by CD complexation improves their absorption in human orgnisam, i.e. dissolution rates and bioavailability. In order to prepare economically drug/CD complexes with desirable properties, it is necessary to have data about the process variables which can influence the inclusion complex formation. Type of CD, process temperature and method of preparation are some of the key process variables. It is considered that the inclusion complexation can be more successful if the cavity size of CD is suitable to accommodate a drug molecule of particular size and if CD and the drug have the opposite charge. It was confirmed that the ionic forms of drugs are the weaker complex forming agents in compared with their nonionic forms. In most of the cases, the increase in system temperature leads to the reduction of drug/CD interaction forces, such as Van der Waals and hydrophobic forces. Also, the apparent stability constant of the inclusion complexes are reduced. The method can influence the drug/CD complexation. The effectiveness of a method depends on the nature of the drug and CD used in the preparation [18].

In addition to the solubilization improvement, CD complexation is of great importance for improvement of the chemical, physical and thermal stability of drugs. In this way there is an enhancement of their shelf-life and reduction of the concentrations of the agent required to achieve a biological effect.

Also, the CDs serves to improve the undesired side effects of drugs. Several oral pharmaceutical products have the unpleasant and bitter-taste active components. This is one of several important formulation problems that can cause low patient compliance, especially for pediatric patients. This problem has been the subject of numerous research studies, given that any pharmaceutical formulation with a pleasing taste is desirable due to better compliance and therapeutic value for the patient and more business and profits for the company. In order to overcome the aforementioned problems, the taste-masking techniques, such as the physical barrier coatings, chemical modification and sensory masking have been developed. Among chemical modification, the inclusion complexation of drugs with CDs can provide the effective bitter taste-masking effects without complicated formulation and/or delayed dissolution of drugs [19]. This method is most suitable only for low dose drugs. For instance, the strong bitter taste of carbetapentane citrate syrup was reduced to approximately 50% by preparing a 1:1 complex with CD [20]. Famotidine, an H_2-histamine receptor antagonist, is used for the treatment of ulcers like peptic ulcer, a condition where the stomach produces too much acid [21]. In this case, the commonly used dosage forms are the conventional solid dosage form such as tablets and capsules. However, some of the patients may have swallowing problems of these dosage form. One way of overcoming this problem is the use of orally disintegrating dosage forms that disintegrate in the saliva and are swallowed without water [22]. However, in order to use these rapid dissolving tablets, a bitter taste of famotidine was firstly masked. The various techniques are used for the masking of unpleasant taste such as flavors and sweeteners, ion exchange resins, spray drying, coating with water insoluble polymers, and freeze drying process [23]. Verma et al. [24] have used HP-β-CD as an approach for taste masking of famotidine. This complex was prepared using lyophilisation technique. The obtained results of the physical characterization of the inclusion complex showed that the complex between famotidine and HP-β-CD was successfully formed. The encapsulation efficiency was found to be 84-97%. Also, the reduction of bitterness of antihistaminic drugs in the solutions was achieved through the formation of inclusion complexes with CD [25]. The masking bitter taste of active compounds is implemented in order to the preparation of oral and injectable formulations. Wei et al. [26] tryed to mask the bitter taste of dental lidocaine HCl injection using HP-β-CD. Actually, the inclusion complex was prepared by the solution method in 1:1 molar ratio and characterized for stability and for taste using an Alpha MOS ASTREE electronic tongue. The characterization studies have confirmed that the injectable formulations were stable up to 6 months at 4, 25 and 40 °C. The taste evaluation study indicated that HP-β-CD significantly improved the bitter taste of lidocaine.

Undesirable side effect of drugs such as irritation represents a limiting factor for the use of such substances in the pharmacy. Encapsulation of drug substances with these properties within a CD cavity can alleviate their side effects. The prepared CD inclusion complexes reduce the local concentration of the free drug below the irritation threshold. As the complex gradually dissociates and the free drug is released, it gets absorbed into the body and its local free concentration always remains below the levels that might be irritating to the mucosa. The numerous examples of application of CD for alleviation of the drugs side effects can be found in the literature. For instance, topical tretinoin currently is used to treat photoaged skin, but due to the undesirable side effects such as erythema, irritation, and scaling its application is reduced [27]. The use of CD complex overcomes the drawbacks of tretinoin. One of the most commonly used topical agents for the treatment of psoriasis is calcipotriol. It was noticed that this active substance in the pure form leads to skin irritation in patients. In order to improve the solubility and reduce the undesirable side effect of calcipotriol, an inclusion complex of calcipotriol with CD was prepared [28]. This prepared complex was used for the preparation of a new topical drug delivery system. In this case, the inclusion complex was formulated in an emulgel base. The dissolution profiles of calcipotriol complex from emulgel formulations indicate that the drug release was significantly increased with the emulgel formulations compared to the commercial cream product. Recently, the sustained release aqueous eye drops of dexamethasone based on CD nanogels were designed [29]. This formulation consisted of an aqueous solution of dexamethasone in HP-β-CD medium containing CD nanogels. After optimization of the content, the prepared formulation was tested *in vivo*. Based on the obtained results, it can be conclude that the dexamethasone nanogel eye drops were well tolerated with no macroscopic signs of irritation, redness or other toxic effects. Also, through complexation with HP-β-CD, the solubility and dissolution rate of cantharidin were improved significantly, and the irritation of musous was relieved [30].

The substances that are oils/liquids at room temperature are difficult to handle and formulate into stable solid dosage forms. The complexation with CDs may convert such substances into the microcrystalline or amorphous powders which can be easily handled and formulated into the solid dosage forms by the conventional production processes and equipment. Also, the incompatibility with each other or with other inactive ingredients present in a formulation was prevented by the complexation with CDs. A better stability of the formulation was achieved by physically separating the components.

Cyclodextrins As Delivery Efficiency Enhancers

Apart from their inherent properties, CDs enhance drug delivery through biomembranes by increasing the drug availability at the membrane surface. Two additional features of CDs which improve the drug bioavailability are: their membrane absorption enhancing properties and their ability to stabilize biomolecules in physiological media by shielding them from non-specific interactions [31]. CDs interaction with biological membranes results in the release of certain membrane components cholesterol or phospholipids and consequently, their destabilization and permeabilization. This property of CDs and their derivatives was used in almost all drug delivery systems like oral, sublingual, buccal, ocular, nasal, transdermal, rectal, pulmonary, parenteral, colon-specific delivery, brain drug delivery etc. The efficiency of gene delivery systems can be also improved by the addition of CD derivatives as formulation excipients [31]. Today, it intensively works on commercial application of CDs in the novel drug delivery systems such as nanoparticles, liposomes, microspheres and microcapsules.

Cyclodextrin in Cosmetics

CDs have been used in the cosmetic preparations due to variety of reasons. It is capable to interact with some representative molecules responsible to cause body malodor, such as carboxylic acids, thiols and steroids, present in sweat and body secretions, forming inclusion complexes. For this reason, β-CD can be used for the preparation of care formulations, such as deodorants [32]. Also, the β-CD microspheres were used for the production of rose perfume [33]. CDs have a significant role as an active ingredient in antiacne treatments, improve the stability of hair dyes and reduce the odor in mercaptan systems. CDs may be also used as emulsifiers in the creams or lotions. Particularly β-CD is interesting for the preparation of lotions for mosquito repellent [34]. β-CD can be utilized as photostabilizer additive for enhancement of the photostability of sunscreen agents' molecules (e.g. UV absorber) [35]. In this way it is possible to improve the effectiveness of sunscreen formulations and reduce the skin irritation. The satisfactory penetration properties of CDs, which are most likely due to increased aqueous solubility of the drugs and hence the thermodynamic driving force for permeation across the skin, have provided the application of CDs at the transdermal drug delivery [36].

EFFECT OF CYCLODEXTRIN ON SOLUBILITY AND STABILITY OF SOME DRUGS

There are the numerous examples of molecular inclusion complexes of CDs and their derivatives especially with the various types of drugs including corticosteroids, prostaglandins, antibacterials, as well as NSAIDs.

Amlodipine besylate belongs to the group of dihydropyridines and has the possibility to block the calcium channels [37]. This property provides to apply in the treatment of hypertension, most often the only successful drug for the regulation of blood pressure with the majority of patients. However, the application of amlodipine besylate in the pure state is limited, given that in its structure has a 1,4 dihydropyridine ring which is subject to the photodegradation, i.e. it is sensitive to the effect of light. From these reasons, amlodipine besylate is a photosensitive substance. The obtained photodegradation products (pyridine derivative) cause the loss of pharmacological activity and also exhibits some toxicity, and is a potential cause of cancer. In order to protect amlodipine besylate from photodegradation, attention is being directed towards the production of supramolecular structures. Kapor et al. [38] was described the procedure for the preparation of inclusion complexes of amlodipine besylate with β-CD and HP-β-CD. These molecular inclusion complexes are prepared by the coprecipitation method and characterized by the application of spectroscopic methods FTIR, ^{1}H-NMR, DSC and XRD. The prepared complexes have proven to be more stable on the effect of daylight with a degradation level much lower than in the solid amlodipine besylate. Thus, the new supramolecular systems based on amlodipine besylate could be of interest for the production of antihypertensive formulations.

Nifedipine also belongs to the class of calcium-channel blockers and it is broadly used for the treatment of cardiovascular disease [39]. The various oral formulations of nifedipine, such as capsule and an extended-release (long-acting) tablet have been listed on the market. However, this drug has the several disadvantages, including its low oral bioavailability, due to its poor water solubility. Given that nifedipine is a dihydropyridine derivative, it is extremely photolabile substance. Nifedipine is decomposed to the nitrosopyridine derivative after the effect of daylight, i.e. to the oxidation product of nitropyridine derivative after the effect of UV light. In order to overcome these limitations, the various approaches such as solid dispersions, nanocrystal and micronization of nifedipine, complexation with different types of CDs were applied. After studying the photochemical stability of nifedipine, it can be concluded that the free nifedipin is degraded four to five times faster than the β-CD:nifedipin inclusion complex in the solid state under the natural daylight. Based on the experimental results, Nikolic et al. [40] found that the photosensitivity of nifedipin in the inclusion complex is lower in the solid state in compared with the solution. Glass et al. [41] have considered that the photodegradation of nifedipine could be reduced as a result of intermolecular hydrogen bonding between HP-β-CD and the 1,4 protons of the dihydropyridine ring. Unlike them, Muller and Albers [42] claimed that the photostabilization of nifedipine in the solution could not be achieved by the complexation with HP-β-CD.

Naproxen belongs to class II of the Biopharmaceutics Classification System (BCS) and is characterized by a low bioavailability and low dissolution rate. These factors are a limiting step for its absorption and biological availability. The sudies of numerous researchers were directed towards the improvement of naproxen solubility. Bolten et al. [43] investigated the influence of different stabilizers (Tween and PVP) in aqueous solutions on the solubility of naproxen. Liquid and solid self-microemulsifying drug delivery systems [44], microemulsion [45], solid dispersions [46] are also promising approaches for solubility enhancement of naproxen. The preparation of inclusion complex between β-CD and naproxen was also investigated [47]. In addition to poor solubility, naproxen has still one disadvantage and that is the high photosensitivity. Actually, the photodegradation leads to the formation of pharmacologically inactive products. The molecular inclusion complex based on naproxen and HP-β-CD was prepared in order to enhance the photostability of naproxen [48]. The complex was prepared by the chemical co-precipitation method in the solid state and in the molar ratio of 1:1. Stojanovic et al. [48] analyzed the photosensitive groups of naproxen in a pure and complexed form using the FTIR and UV-VIS methods. FTIR analysis showed that the degradation started 15 days after the exposure of naproxen to daylight, while the inclusion complex of naproxen:HP-β-CD was a photostable for 30 days.

Sulfanilamide is the synthetic drug which is used in the treatment of microbial infections [49,50]. It is known that sulfanilamide is a poor soluble in water and a sensitive to the photo-oxidation due to the presence of aniline in its structure [51]. Because of these unfavorable physico-chemical properties, their application in therapy as the pharmaceutical active substances is limited [2]. In order to improve the solubility and photostability of synthesized sulfanilamide, Tacic et al. [52] were prepared the inclusion complexes with β-CD and HP-β-CD by the co-precipitation method. Using the phase solubility techniques, it is confirmed that the solubility of sulfanilamide was improved by addition of CD, i.e. in the presence of HP-βCD. Based on the photostability studies, the inclusion complex based on HP-β-CD has proven to be a more photostable in compared with the β-CD complex and sulfanilamide. Thus, sulfanilamide with these improved properties can be used for the preparation of oral formulations.

Usnic acid (UA), a phenolic compound, is an optically active derivative of dibenzofurane. In the pure state is a yellow crystalline substance. It is known because it has a broad spectrum of biological activities: antibacterial, antifungal, antimycobacterial [53], antiparasitic [54], and antineoplastic activities [55], anti-inflammatory and analgesic. Having in mind that UA has a pronounced antibacterial activity, it was found the application in the production of pharmaceutical (septolete-lingualete, tooth paste) and cosmetic preparations (preservative in the hydrated creams, inhibitor of Gram-positive microorganisms in deodorant sprays). However, due to its low solubility in the aqueous solutions and hepatotoxicity (about 0.06 mg/cm^3), the application of UA in the pharmaceutical and cosmetic industry is limited. In order to overcome these disadvantage, the micellar solubilisation with surfactants, cosolvents, encapsulation of UA into nanocapsules and sodium salt (Na-usninate) were applied. Among these techniques it can be found and complexation of UA with CDs [56]. Based on the experimental data it was confirmed that the solubility of UA in water was increased about 5 and 70 times after complexation of UA with β-CD and HP-β-CD, respectively. The studies indicated that both complexes had a higher antimicrobial activity in relation to UA. The prepared supramolecular systems with these properties could be of interest for the development of pharmaceutical formulations with satisfactory bioavailability, as well as a broad range of the effects.

Piroxicam, a non-steroidal anti-inflammatory drug with analgesic, antipyretic and anti-inflammatory effects, shows polymorphism and can exist in several conformations [57] and prototrophic shapes. The second disadvantage of piroxicam is its low solubility in the aqueous media. These are limiting factors for the application of piroxicam in the pharmaceutical industry. The complexation of piroxicam with CDs is one of way for solving these problems. For the preparation of inclusion complex, the organic solvents are commonly used as media [58] given that piroxicam has a poor water solubility. However, due to toxicities and the high concentration of residues in the final inclusion complex, such approach of the preparation of complex are changed by other techniques. The use of supercritical carbon dioxide [59] has provided the preparation of inclusion complex without the use of ammonia and any solvent. Nikolic et al. [60] used the water medium for the preparation of complex, but due to poor solubility, the prepared solution was then subjected to ultrasound under the following conditions: temperature 30 °C, ultrasound wave frequency of 40 kHz and power of 150 W, the total nominal power of 3×50 W and exposure time was 1 h. Preclinical and clinical studies of piroxicam-β-CD inclusion complex showed that this complex has a more rapid onset of action after oral administration and improved GI tolerability, because of minimization of the piroxicam gastric effects. Also, the photostability of piroxicam was significantly improved after the preparation of piroxicam:HP-β-CD inclusion complex. Thanks to these properties, the commercial application of piroxicam inclusion complex is longer than 25 years. The solid dosage forms such as tablets, effervescent tablets and powders are currently available.

Allicin is one of the active principles of fresh aqueous extract of garlic [61]. In the pure state is a transparent to slightly yellow oil with typical smell of fresh garlic (*Allium sativum* L.). Allicin has an antibacterial activity against a broad range of Gram-negative and Gram-positive bacteria [62], hypoglycemic activity, anti-inflammatory [63] and antioxidant properties, and reduce elevated blood pressure [64]. Based on the results from the recent studies, allicin can be used for treating diseases such as cancer and diabetes. However, from the technological point of view, there is a problem of the allicin stability. Depending on the conditions, it can be transformed to other products that may be pharmacologically active. Also, allicin loses the microbiological activity during the storage. These are the limiting factors for the application of allicin in the pharmaceutical industry. After complexation, the physical and chemical properties of allicin could be changed, which, in turn, alters the pharmacokinetic properties of this drug. The formation of a new supramolecular compound allicin:β-CD inclusion complex shows a greater stability and microbiological activity compared to free allicin. In this way, the unpleasant smell of allicin is masked.

Iron is an essential element in human body [65]. It is needed for a number of highly complex processes that continuously take place on a molecular level and that are indispensable to human life. Iron-deficiency anemia is the most common type of anemia, which occurs when human organism does not have enough of the mineral iron. The intravenous iron is used for treatment of iron deficit in organism.

This way of application is commonly used in the patients who do not absorb iron well in the gastrointestinal tract, with severe iron deficiency or chronic blood loss, or patients who cannot tolerate oral iron. For parenteral formulations, iron is used as iron dextran, iron sucrose and ferric gluconate. The large doses of iron can be administered at one time when using iron dextran. The serious side effects other than allergic reactions are rare for intravenous iron. However, the oral formulations, tablets and liquid, are more often use for treatment of iron deficit due to a more suitable way of drug application. These formulations as a pharmaceutical active substance contain the organic and inorganic compounds of iron, ferrous sulfate, fumarate, or gluconate. After oral drug administration, iron is absorbed in the small intestine (duodenum and first part of the jejunum). However, the possible side effects of iron tablets are abdominal discomfort, nausea, vomiting, diarrhea, constipation, and dark stools.

Today, the formulations based on iron(II) fumarate are increasingly used for treatment of iron deficit in organism, because of its good properties such as good adsorption affinity in the body, low toxicity and minimal side effects. Based on the analysis it has confirmed that iron(II) fumarate was a poor solubility in water (about 0.14 g/100 cm^3) and therefore exhibit increased side effects while used in a longer period of time. Kapor et al. [66] tried to improve a poor water solubility of the synthesized iron(II) fumarate and to reduce its side effects by the preparation of inclusion complexes with β-CD and HP-β-CD. The complexes were prepared using a co-precipitation method in the molar ratio of 1:1. The best solubility was achieved after complexation with HP-β-CD (2.71 mg/cm^3), and slightly lower with β-CD (2.05 mg/cm^3).

CONCLUSION

There is a growing trend towards more complex and sophisticated pharmaceutical products with patients expecting improved product performance and acceptability. CDs are useful solubilizing excipients with an ever increasing list of beneficial properties and functionalities. Thus, CDs are used in the pharmaceutical applications for numerous purposes, including improving the solubility, bioavailability, masking of bitter taste, reduce volatility of drugs, and convert liquid substance into solid compounds. So, CDs are able to form water-soluble inclusion complexes with many poorly soluble lipophilic drugs. Today, intensively studying the application of CDs in novel drug delivery systems such as nanoparticles, liposomes, microspheares and targeted drug delivery and they become commercially available in the future. Another interesting application of CDs is their incorporation in the food packaging materials as antiseptic or conserving agents. Actually, the great importance of CDs is due to to their biocompatibility, multi-functional characteristics and low toxicities in the animals and humans.

ACKNOWLEDGMENTS

The financial support was provided from the Ministry of Education, Science and Technological Development of the Republic of Serbia under the project TRp-34012.

REFERENCES

[1] M. Lainer, E. Mann and A. Sönnichsen, *Int. J. Qual.Health C.,* p. mzt043, 2013.

[2] M..E. Aulton, Pharmaceutics: The science of dosage form design, Edinburgh: Churchill Livingstone, 2002.

[3] G.L. Amidon, H. Lennernas, V.P. Shah and J.R. Crison. *Pharm. Res.*, Vol. 12, p. 413, 1995.

[4] S.K. Das, S. Roy, Y. Kalimuthu, J. Khanam and A. Nanda, *Int. J. Pharmacol. Pharm. Technol.,* Vol. 1, p. 37, 2012.

[5] V.B. Junyaprasert and B. Morakul, *Asian J. Pharm. Sci.,* Vol. 10, p. 13, 2015.

[6] S. Kotta, A.W. Khan, K. Pramod, S.H. Ansari, R.K. Sharma and J. Ali, *Expert Opin. Drug Del.*, Vol. 9, p. 585, 2012.

[7] Y. Lu and K. Park, *Int. J. Pharmaceut.*, Vol. 453, p. 198, 2013.

[8] D. Hasa, D. Voinovich, B. Perissutti, M. Grassi, A. Bonifacio, V. Sergo and S. Invernizzi, *Pharm. Res.*, Vol. 28, p. 1870, 2011.

[9] N. Xi, Y. Zhang, Z. Wang, T. Lin and Q. Wang, *Cancer Res.*, Vol. 74, p. 2532, 2014.

[10] M.Lj. Mihailovic, Osnovi teorijske organske hemije i stereohemije, Gradjevinska knjiga, Beograd, 1990.

[11] M.M. Meier, M.T.B. Luiz, P.J. Farmer, B. Szpoganicz, *J. Incl. Phenom. Macro.*, Vol.40, p.291, 2001.

[12] R. Breslow, Cyclodextrins. Molecular Encapsulation: Organic Reactions in Constrained System*s*, p. 43, 2012.

[13] G. Schmid, *Tibtech*, Vol. 7, p. 244, 1989.

[14] G. Schmid, Preparation and application of γ-cyclodextrin. In: Duchêne D, ed. New trends in cyclodextrins and derivatives, Editions de Santé, Paris, p. 25, 1991.

[15] R.G. Winkler, S. Fioravanti, G. Ciccotti, C. Margheritis and M. Villa, *J. Comput. Aided Mol. Des.*, Vol. 14, p. 659, 2000.

[16] E. Pinho, M. Grootveld, G. Soares and M. Henriques, *Carbohyd. Polym.*, Vol. 101, p. 121, 2014.

[17] L. Szente, J. Szejtli and G. Kis, *J. Pharm. Sci.*, Vol. 87, p. 778, 1998.

[18] K.P.R. Chowdary and B.N. Nalluri, *Drug Dev. Ind. Pharm.*, Vol. 26, p. 1217, 2000.

[19] 19 [x1] H. Arima, T. Higashi and K. Motoyama, *Therapeutic delivery*, Vol. 3, p. 633, 2012.

[20] H. Arima, T. Kondo, T. Irie, F. Hirayama, K. Uekama, T. Miyaji and Y. Inoue, *Yakugaku Zasshi*, Vol. 112, p. 65, 1992.

[21] Xu. Jianchen, Li. Bovet and K. Zhao, *Int. J. Pharm.*, Vol. 359, p. 63, 2008.

[22] J.E. Aguilar-Díaz, E.G. Montoya, *Eur. J. Pharm. Biopharm.*, Vol. 80, p. 638, 2012.

[23] J.B. Naik, S.M.Rashid, *Lat. Am. J. Pharm.*, Vol. 29, p. 667, 2010.

[24] U. Verma, J.B. Naik and V.J. Mokale, Preparation and Characterisation of the Inclusion Complex of Famotidine with (2-Hydroxy Propyl)–B-Cyclodextrin & Pvp K-30: Effects on Solubility and Bitter Taste Mask. Int. Conf. on Advances in Chemical Engineering and Technology, ICACE TKMCE ‘p. 63, 2014.
[25] N. Ono, Y. Miyamoto, T. Ishiguro, K. Motoyama, F. Hirayama, D. Iohara and K. Uekama, *J. Pharm. Sci.*, Vol. 100, p. 1935, 2011.
[26] Y. Wei, M.P. Nedley, S.B. Bhaduri, X. Bredzinski and S.H. Boddu, *AAPS Pharm. Sci. Tech.*, p. 1, 2014.
[27] T. Miura, A. Takada and M. Ooe, *Aesthe. Plast. Surg.*, Vol. 36, p. 971, 2012.
[28] U. Badilli, G. Amasya, T. Şen and N. Tarimci, *J. Incl. Phenom. Macro.*, Vol. 78, p. 249, 2014.
[29] M.D. Moya-Ortega, T.F. Alves, C. Alvarez-Lorenzo, A. Concheiro, E. Stefánsson, M. Thorsteinsdóttir and T. Loftsson, *Int. J. Pharm.*, Vol. 441, p. 507, 2013.
[30] L.N. An, Y.J. Dang, C.H. Hu and C.Y. Zhu, *C. H. M.*, Vol. 4, p. 224, 2012.
[31] C.O. Mellet, J.M.G. Fernández and J.M. Benito, *Chem. Soc. Rev.*, Vol. 40, p. 1586, 2011.
[32] Lopedota, A. Cutrignelli, V. Laquintana, M. Franco, D. Donelli, L. Ragni and N. Denora, *Int. J. Cosmetic Sci.,* 2015, *in press.*
[33] X.X., Su, Y.N. Yang, X.S. Geng and Z.J. Li, *Fine Chemicals*, Vol. 6, p. 011, 2012.
[34] S. Songkro, N. Hayook, J. Jaisawang, D. Maneenuan, T. Chuchome and N. Kaewnopparat, *J. Incl. Phenom. Macro.*, Vol. 72, p. 339, 2012.
[35] N.A. Al-Rawashdeh, K.S. Al-Sadeh and M.B. Al-Bitar, *J. Spectrosc.*, Vol. 2013, p. 1, 2013.
[36] M. Cirri, M. Bragagni, N. Mennini and P. Mura, *Eur. J. Pharm. Biopharm.*, Vol. 80, p. 46, 2012.
[37] N.Y. Choi, H. Choi, H.H. Park, E.H. Lee, H.J. Yu, K.Y. Lee and S.H. Koh, *Brain Res.*, Vol. 1551, p. 1, 2014.
[38] Kapor, V. Nikolić, L. Nikolić, M. Stanković, M. Cakić, L. Stanojević and D. Ilić, *Cent. Eur. J. Chem.*, Vol. 8, p. 834, 2010.
[39] C.D. Furberg, B.M. Psaty and J.V. Meyer, *Circulation*, Vol. 92, p. 1326, 1995.
[40] V. Nikolić, D. Ilić, Lj. Nikolić, M. Stanković, M. Cakić, Lj. Stanojević and M. Popsavin, *Cent. Eur. J. Chem.*, Vol. 8, p. 744, 2010.
[41] B.D. Glass, M.E. Brown, S. Daya, M.S. Worthington, P. Drummond, E. Antunes and D. Maharaj, *Int. J. Photoenergy*, Vol. 3, p. 205, 2001.
[42] B.W. Muller and E. Albers, *Int. J. Pharm.* Vol. 79, p. 273, 1992.
[43] D. Bolten, R. Lietzow and M. Türk, *Chem. Eng. Technol.,* Vol. 36, p. 426, 2013.
[44] K. Cerpnjak, A. Zvonar, F. Vrecer and M. Gašperlin, *Drug Dev. Ind. Pharm.*, p. 1, 2014.
[45] E. Moghimipour, A. Salimi and S. Eftekhari, *Adv. Pharm. Bull.*, Vol. 3, p. 63, 2013.
[46] V. Nagabandi, A.K. Chandragiri, S. Thota and P. Katakam, *J. Pharm. Sci. Res.*, Vol. 6, 2014.
[47] H.E. Grandelli, B. Stickle, A. Whittington and E. Kiran, *J. Incl. Phenom. Macro.*, Vol. 77, 269, 2013.
[48] S.S. Ilić-Stojanović, V. Nikolić , Lj. Nikolić , A. Zdravković, A. Kapor, M. Popsavin and S. Petrović, *Hem. Ind.*, in press.

[49] G.L. Perlovich, N.N. Strakhova, V.P. Kazachenko, T.V. Volkova, V.V. Tkachev, K.J. Schaper and O.A. Raevsky, *Int. J. Pharm.*, Vol. 349, p. 300, 2008.
[50] C. Garnero, V. Aiassa and M. Longhi, *J. Pharm. Biomed.*, Vol. 63, p. 74, 2012.
[51] Albini and E. Fasani, Rationalizing the photochemistry of drugs. In: Tønnesen, H.H. (ed.) Photostability of Drugs and Drug Formulations, Chapter 4. CRC Press, Boca Raton, 2004.
[52] Tačić, I. Savić, V. Nikolić, I. Savić, S. Ilić-Stojanović, D. Ilić and A. Kapor, *J. Incl. Phenom. Macro.*, Vol. 80, p. 113, 2014.
[53] Ranković, M. Mišić, S. Sukdolak, *World J. Microbiol. Biotechnol.*, Vol. 24, p. 1239, 2008.
[54] H. Cetin, O. Tufan-Cetin, A.O. Türk, et al., *J. Parasitol. Res.*, Vol. 102, p. 1277, 2008.
[55] M. Bačkorová, M. Bačkor, J. Mikeš, R. Jendželovský and P. Fedoročko, *Toxicol. Vitro.*, Vol. 25, p. 37, 2011.
[56] V. Nikolić, M. Stanković, Lj. Nikolić, G. Nikolić, S. Ilić-Stojanović, M. Popsavin and T. Kundaković, *J. Incl. Phenom. Macro.*,Vol. 76, p. 173, 2013.
[57] A.R. Sheth, S. Bates, F.X. Muller and D.J. Grant, *Cryst. Growth Des.*, Vol. 4, p. 1091, 2004.
[58] S.Y. Lin and Y.H. Kao, *Int. J. Pharm.*, Vol. 56, p. 249, 1989.
[59] T. van Hees, G. Piel, B. Evrard, X. Otte, L. Thunus and L. Delattre, *Pharm. Res.*, Vol. 16, p. 1864, 1999.
[60] V.D. Nikolić, S.S. Ilić-Stojanović, Lj.B. Nikolić, M.D. Cakić, A.S. Zdravković, A.J. Kapor and M.M. Popsavin, *Chem. Ind.*, Vol. 68, p. 107, 2014.
[61] V. Nikolic, M. Stankovic, Lj. Nikolic, D. Skala, *Chem. Ind.*, Vol. 55, p. 120, 2001.
[62] F.J. Moghadam, T. Navidifar and M. Amin, *Int. J. Enteri.c Pathog.*, Vol. 2, p. e16749, 2014.
[63] S. Bose, B. Laha and S. Banerjee, *J. Adv. Pharm. Educ. Res.,* Vol. 3, 2013.
[64] J.Y.Y. Chan, A.C.Y. Yuen, R.Y.K. Chan and S.W. Chan, *Phytother. Res.*, Vol. 27, p. 637, 2013.
[65] B.K. Fuqua, C.D. Vulpe and G.J. Anderson, *J. Trace Elem. Med. Bio.*, Vol. 26, p. 115, 2012.
[66] A.J. Kapor, Lj.B. Nikolić, V.D. Nikolić, M.Z. Stanković, M.D. Cakić, D.P. Ilić and I. Ristić, *Adv. Technol.*, Vol. 1, p. 7, 2012.

In: Cyclodextrins
Editor: Francis G. Ramirez
ISBN: 978-1-63482-788-1

Chapter 8

β-CYCLODEXTRIN DIMERIC INCLUSION COMPLEXES: A REVIEW OF SOME CRYSTAL STRUCTURE ASPECTS

Dimitris Mentzafos and Kostas Bethanis*
Agricultural University of Athens, Department of Biotechnology,
Physics Laboratory, Athens, Greece

ABSTRACT

β-Cyclodextrins (β-CD), consisting of seven α-(1→4)-linked D-glucopyranose units, have the tendency to form dimers by linking two molecules through hydrogen bonds relating their O3*n*-H hydroxyl groups. Inside the formed dimeric cavity, a host environment is created where a plethora of guest molecules, disposing an appropriate shape and conformation, can be included forming complexes both in solution and in solid state. The guest molecules are held inside the cavity not by covalent bonds but by weak interactions, like hydrogen bonds, van der Waals interactions, hydrophobic effects, etc. The formed inclusion compounds have wide applications, as the encapsulation of a molecule inside the CD cavity changes beneficially its physicochemical properties enhancing its water solubility and protecting it from hydrolysis, oxidation, photolysis, volatility and chemical reactivity. Moreover, due to the weak non-covalent interactions between the host and the guest molecules, these inclusion complexes are not fixed or permanent but rather in a dynamic equilibrium with their free components in solution, thus offering a slow release of the guests and sustaining their action. The β-CD complexes has been investigated by many methods like NMR, FTIR, UV-visible spectroscopy, spectrofluorometric studies, thermal studies, molecular dynamics and finally by X-ray crystallography.

In this chapter, the main aspects of the crystal structures of the dimeric β-CD inclusion complexes reported so far are reviewed. During the last four decades several crystals of dimeric β-CD complexes have been formed and analyzed by X-ray crystallography, disposing many common features like: the crystallographic space group (the great majority of them being P1, C2, $P2_1$ or $C222_1$), the crystal packing mode (the

* E-mail: kbeth@aua.gr.

majority of them being Channel, Screw Channel, Intermediate, or Chess Board), isostructurality, polymorphism etc. The structural studies of these inclusion complexes have also revealed in detail the inclusion geometry, the guest's conformation, the encapsulation of water molecules inside the dimeric interface in the cases of Channel or Screw Channel packing modes usually being hydrogen bonded either with the guest or the host molecules, host-guest CH-π interactions, π-π interactions between the guests disposing phenyl or naphthyl moieties etc.

Keywords: Dimeric *β*-Cyclodextrin inclusion complexes, crystal structure, molecular packing, isomorphism, polymorphism, π-π interactions

1. Introduction

Cyclodextrins (CDs), called also Cycloamyloses or Cyclomaltoses, are cyclic oligomers obtained by enzymatic degradation of starch by the action of some *Bacillus* as *Bacillus macerans*, *Bacillus megaterium* or *Bacillus stereothermophilus* producing the appropriate enzyme [1]. Therefore, they are considered as natural products. The most common and thoroughly studied CDs are composed of six, seven or eight *α*-(1→4)-linked D-glucose residues, called *α*-, *β*- or *γ*-CD respectively and they dispose the shape of an empty truncated cone (Figure 1). CDs having less than six glucose residues cannot be formed due to steric reasons [2]. On the contrary, CDs containing as many as thirty-one glucose units are identified [3] but only the *α*-, *β*- and *γ*-CDs have been studied in details. The hydrophobic cavity of the CDs provides a microenvironment having the ability to include the hydrophobic moieties of guest molecules [4] whereas the attached hydrophilic groups (primary O6*n*-H at the narrower rim of the CDs and secondary O2*n*-H and O3*n*-H at their wider rim) form a hydrophilic periphery.

Inside the cavity a plethora of molecules, having appropriate shape and conformation, can be included and CD inclusion complexes are formed both in solution and in solid state. The driving forces of the formation of these inclusion compounds are: hydrophobic effects inducing the apolar group of the guest molecule to enter inside the cavity and release the high enthalpy entrapped water molecules, steric effects due to the shape, size and polarity of guests, electrostatic and charge-transfer interactions, hydrogen bonds, van der Waals and other type interactions between the host and guest molecules etc. [5, 2]. The extent of the contribution of each factor depends on the nature and size of the guest, the host and the solvent molecules.

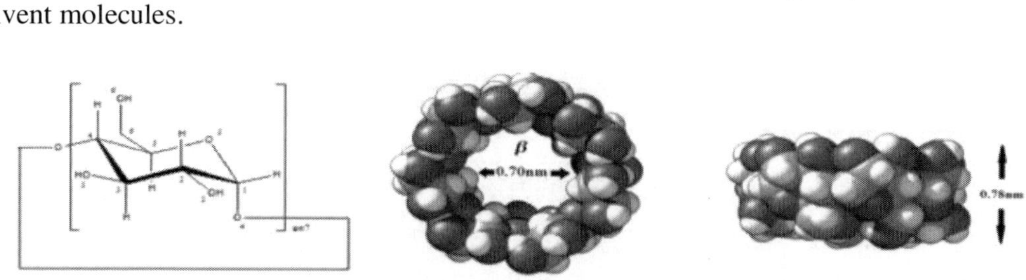

(a) (b)

Figure 1. (**a**) A schematic representation of the chemical structure the *β*-CD (**b**) The 3D shape and the dimensions of the *β*-CD.

This inclusion has advantages over starch, presenting similar inclusion properties, since the CDs do not provide a cultural medium for microorganisms [6]. CDs are readily soluble in water effecting a significant increase in the solubility of the unsolved or poorly water soluble guest molecules of their inclusion compounds. Moreover, the encapsulation of a molecule inside the CD cavity protects it from hydrolysis, oxidation, photolysis, volatility, solution and chemical reactivity, changing beneficially its physicochemical properties [2]. Therefore, CDs constitute an important subject of interest in biology, physical chemistry, synthetic chemistry, computational chemistry, supramolecular chemistry, crystallography etc. [7] and they have been widely used in pharmaceutics, cosmetics, food industry and agriculture [8-10].

Among the various types of CDs, *β*-CD (Figure 1) has the highest yield during the enzymatic degradation of the starch, due to its low solubility, and consequently it has the lowest price in commerce. The abnormally low solubility of *β*-CD may be explained by the formation of aggregates which interact with the surrounding net of water molecules [11]. On account of its low price, the formation of *β*-CD inclusion complexes attracts the highest interest for commercial applications and therefore it has been studied the most. An important characteristic of the *β*-CD inclusion complexes is their tendency to form dimers as two host *β*-CD molecules are connected in a head-to-head mode by intermolecular hydrogen bonds between their O3*n*-H hydroxyl groups (Figure 2) producing a more stable conformation [12].

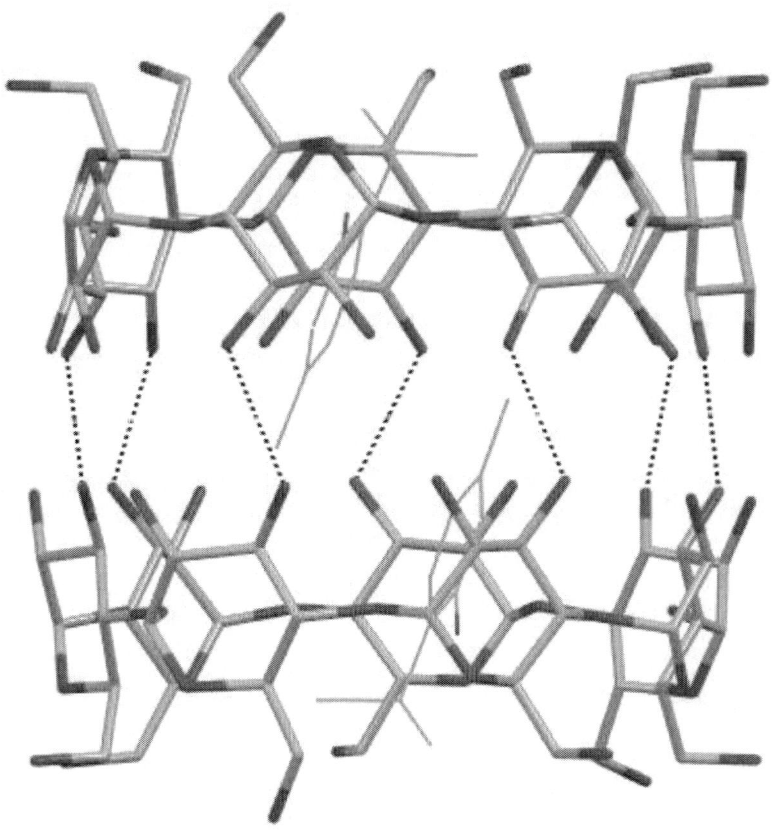

Figure 2. Crystal structure of Thymol/*β*-CD dimeric inclusion complex showing the hydrogen bonds between the secondary O3*n*-H hydroxyl groups of the hosts forming a head-to-head dimer (H-atoms and water molecules are omitted for clarity).

The techniques used for the analysis of the CD complexes are the UV, IR, Raman, NMR, mass spectroscopy, thin-layer chromatography, thermo-gravimetry, differential scanning calorimetry and XRD. X-Ray analysis not only confirms the formation of inclusion complexes but also reveals the 3D structure of theses complexes by determining all their atomic coordinates and thus offers an analytical view of the host-guest-solvent conformation and interactions. During the last decades the structure of numerous β-CD complexes has been determined by X-Ray diffraction mainly in single crystals and in a few cases [13-16] in powder samples.

2. Conformation of the CD Macromolecules

Owing to their biological origin, only dextrorotatory enantiomers of CDs are formed. Their glucopyranoside residues have the $^4C_1$ chair conformation. The primary hydroxyl groups, O6*n*-H, can freely rotate and their majority has the *gauche-gauche* orientation, pointing outwards the CD cavity. Some of them have the *gauche-trans* orientation pointing inwards the cavity, blocking it partially. The *trans-gauche* orientation, unfavorable energetically, is extremely rare and have been noticed only recently in some substituted CDs [17, 18]. The primary hydroxyl groups are frequently disordered over two or even three sites. The major site has, usually, the *gauche-gauche* orientation and the minor site the *gauche-trans* orientation. Contrary, to the ability of the primary hydroxyl groups to rotate, the secondary hydroxyl groups, O3*n*-H and O2*n*-H, are held relatively firm, as the hydroxyl groups O3*n*-H and the O2(*n*+1)-H of the adjacent glycoside residues are hydrogen bonded. The O2*n*-H and the O3*n*-H hydroxyl groups point inwards and outwards the cavity respectively [6].

3. The Isostructurality of the β-CD Inclusion Compounds Crystal Structures

Two or more crystal structures are said to be isostructural if they crystallize in the same space group with similar unit cell dimensions and similar arrangement of the molecules in the asymmetric unit resulting to similar atomic fractional coordinates. The isostructurality is a feature observed very frequently to crystals of the native or substituted β-CD inclusion compounds [13]. The CD complexes may be classified in several groups of isostructure complexes, crystallizing according to given space groups and packing modes.

The isostructurality of the CD complexes is of great importance for the elucidation of a crystal structure. A widely used method for determing an initial model of a CD crystal structure is the isostructure replacement, e.g. the use of the fractional atomic coordinates of the host molecule of a known isostructure CD complex as an initial model for the unknown structure. If the two crystal structures are isostructural the use of this initial model results a satisfactory initial R factor. A fast method used for the verification of the isostructurality of two or more crystal structures is the inspection of their PXRD patterns and the detection of possible similarities between them [13]. In the cases where no isostructural complex can be found the method known as the molecular replacement is used to obtain an initial model.

4. The Packing Modes of the Dimeric β-CD Complexes

The dimeric cavity arising by the head-to-head coupling of two β-CD host molecules hydrogen bonded via their O3*n*-H hydroxyl groups (Fig. 2) has more than twice the volume of the monomeric cavity and therefore is able to include bulky molecules as guests [19]. Moreover, in many cases, small molecules have been found hosted inside β-CD dimeric cavities along with entrapped ethanol or water solvent molecules [20].

At the beginning of the '90s, an analysis of the 34 crystal structures of dimeric β-CD complexes that had been reported at this time [21], showed that they crystallize in four space groups *P*1, *C*2, $P2_1$ and $C222_1$ and that their molecular packing can be classified into four isostructural classes (Table 1): Channel (*CH*), Intermediate (*IM*; frequently characterized as "cage type" by many authors), Screw Channel (*SC*) and Chess Board (*CB*) (Figure 3) [22]. The β-CD dimeric complexes accommodated along two crystal axes form C-centered two-dimensional "invariant layers" stacking along the third crystal axis defining the various packing modes. In the cases where the structures crystallize in the $P2_1$ or $C222_1$ space groups this third crystal axis is the twofold screw axis. According to the former classification, the dimers crystallizing in the *C*2, $P2_1$ and $C222_1$ space groups are always stacked in the *CH*, *SC* and *CB* packing modes respectively and those crystallizing in the *P*1 space group stack either according to the *CH* packing mode, if their cell dimensions are all about 15.5 Å, or to the *IM* packing mode in the case that one of the cell dimensions is more than 17 Å the two others being also about 15.5 Å.

Nowadays, after the publication of many new crystal structures of dimeric β-CD complexes, it is obvious that there is no univocal correspondence between the packing modes and the space groups along with the unit cell dimensions [23]. The structures belonging to the same packing mode are not inevitably isostructural. The cell dimensions, the space groups and the packing modes of some β-CD dimeric complexes published after the former classification are reported in Table 2.

4.1. The *CH* Packing Mode

Two β-CD dimeric complexes crystallizing in space group $P2_1$ stacking according to *CH* and not to *SC* packing mode have been recently reported: the MCPA/β-CD [24] and geraniol/β-CD [25].

Same results have been arisen during the elucidation of the (benzoato.OH-OU_2)/β-CD complex [26] crystallizing in two different forms I and II. Form II crystallizes in space group *C*2 and the dimers stack according to the *CH* packing mode as usual. On the contrary Form I crystallizes in space group $P2_1$ but the molecular packing is also *CH* and not *SC*. In these three structures the channel is developed along the *a*-axis. The cell dimensions are similar to those of the other *CH* structures, but the lengths of the *a* and *c* axes are interchanged. Therefore, the complexes stacked according to the *CH* packing mode may crystallize in three space groups: *C*2, *P*1 and $P2_1$. The lateral shifting of the adjacent layers ranges from 1.14 Å [31] to 3.1 Å [22].

4.2. *SC* and *CB* Packing Modes

The β-CD dimeric complexes including paroxetine [27] or adamantanone [28] crystallize in Space Group $P2_1$ and it was expected that the dimers would stack according to the *SC* packing mode forming a Screw Channel along the twofold screw *b*-axis. However, the distances of the projections of the centers of two adjacent dimers on the *ac* plane are much greater from those observed in the other *SC* crystal modes (Table 1). These distances have been found to be 8.13 Å in paroxetine/β-CD complex and 8.80 Å in adamantanone/β-CD complex. Hence, their packing modes should be characterized rather as *CB*, though their space group is not the $C222_1$ (mean value of the distances of the projections being 8.5 Å) but the $P2_1$.

4.3. Tetrad Packing Mode (*TT*)

A new packing mode was observed in two β-CD dimeric structures crystallizing in space group *C*2: the 4,7-dimethyl-coumarin/β-CD [29] and 4CPA/β-CD [30].

Table 1. Packing modes and cell dimensions of the dimeric β-CD complexes according to Mentzafos *et al.* [22]

Space Group	packing mode	cell dimensions *a* (Å), *b*(Å), *c*(Å) α (°), β (°), γ (°)	lateral shifting (Å)
*C*2	*CH*	19.3, 24.6, 15.8 109.5	2.7
*P*1	*CH*	15.6, 15.56, 15.83 102.1, 102.1, 103.6	3.1
*P*1	*IM*	18.0, 15.41, 15.44 103.0, 112, 101	6.0
$P2_1$	*SC*	15.3, 32.5, 15.4 102.1	2.6
$C222_1$	*CB*	19.18, 24.4, 32.7	8.7

This packing mode was designated as *Tetrad* [29]. The lattice parameters of these two crystal structures differ slightly from those of the structures crystallizing in the space group *C*2 stacking according to the *CH* packing mode. The crystal packing consists of layers stacking along the *c*-axis, a common feature of all the complexes crystallizing in the space group *C*2. The relative shifting of two adjacent C-layers observed in the cases of the aforementioned crystal structures is 6.87 Å and 7.00 Å respectively. These distances are significantly greater than those observed in the *CH* packing mode. The mean planes of the heptagons constituted by the glycosidic O4*n* atoms form an angle of 9.54 Å with the *ab* crystal plane, similar to the one observed in the structures of *CH* packing mode. However, their orientation towards the *ab* crystal plane is opposite to that observed in the *CH* structures (Figure 4). The above described different characteristics and the relatively high value of the shift observed between the successive dimers (greater than the value of 6.0Å observed in the *IM* packing mode) were led to the introduction of the *TT* as a new packing mode. However, only these two dimeric *β*-CD structures have been found to stack according to the *TT* packing mode until now.

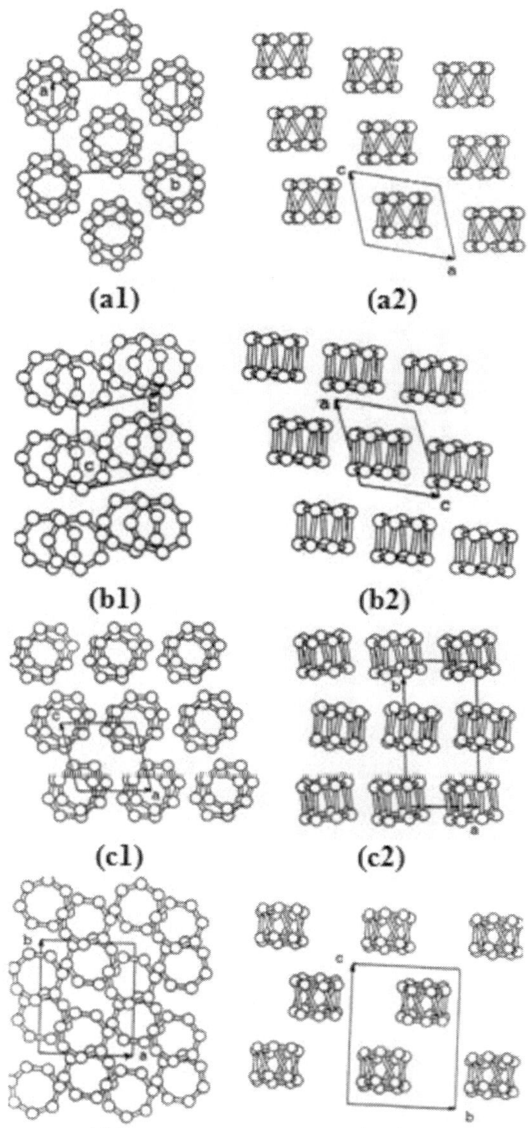

Figure 3. Projections of *β*-CD dimers for the (a1 and a2) CH (b1 and b2) IM (c1 and c2) SC and (d1 and d2) CB packing mode. Each *β*-CD molecule is presented by its O(4) heptagons.

Table 2. *β*-CD dimeric complexes crystal structures classified in old and new packing modes

guest	cell dimensions *a* (Å), *b*(Å), *c*(Å) *α* (°), *β* (°), *γ* (°)	space group	packing mode	Ref.
MCPA	15.860, 24.275, 19.334 108.80	$P2_1$	***CH***	24
geraniol	15.584, 24.289, 18.682 110.91	$P2_1$	***CH***	25
benzoato.OH-OU_2	16.143, 24.81, 19.11 92.03	$P2_1$	***CH***	26
paroxetine	15.2262, 34.4771, 15.6739 104.320	$P2_1$	***CB***	27
adamantanone	15.428, 32.545, 15.437 103.56	$P2_1$	***CB***	28
4,7-dimethyl-coumrin	19.514, 24.024, 16.414 104.49	*C*2	***TT***	29
4CPA	18.998, 24.820, 16.662 105.29	*C*2	***TT***	30
p-hydroxybenzal-dehyde	15.262, 15.728, 16.350 92.67, 96.97, 103.31	*P*1	ideal channel	31
benzoic acid	15.210, 15.678, 15.687 89.13, 74.64, 76.40	*P*1	*IM-TT*	32
neohesperidin-dihydrochalcone	15.125, 30.523, 41.332	$C222_1$	*SC*	33
cobaloxime	15.165, 19.032, 29.252	$P2_12_12_1$	helical	34
1,4-bis(imidazol-l-ylmethyl)/benzene	19.101, 24.032, 32.580 93.763	*C*2	*SC*	35
1,2-bis(4-aminiphe-nyl)/ethane	19.319, 24.19, 33.315 103.924	*C*2	*SC*	23
2,2'-dipyridine	16.427, 24.100, 33.457	*C*2	*SC*	36
$(barbital)_4$	34.341, 15.529, 15.568 103.82	*P*1	*IM*	37
(*S*)-(-)-methyl-*p*-tolysulfoxide	15.495, 65.04, 15.471 102.60	$P2_1$	*SC*	38
cyclizine	15.246, 65.075, 15.609 102.62	$P2_1$	*SC*	39
ferrocene	19.378, 24.094, 32.776 91.963	*C*2	*CH*	40
4-pyridinealdazine	15.067, 14.478, 23.36 105.10	*P*1	*IM*	42
benzoic-acid	19.413, 24.306, 32.975 104.41	*C*2	*CH*	45

4.4. Other Packing Modes

Two dimeric complexes, *p*-hydoxybenzaldehyde/β-CD [31] and benzoic acid/β-CD [32] crystallize in space group *P*1.

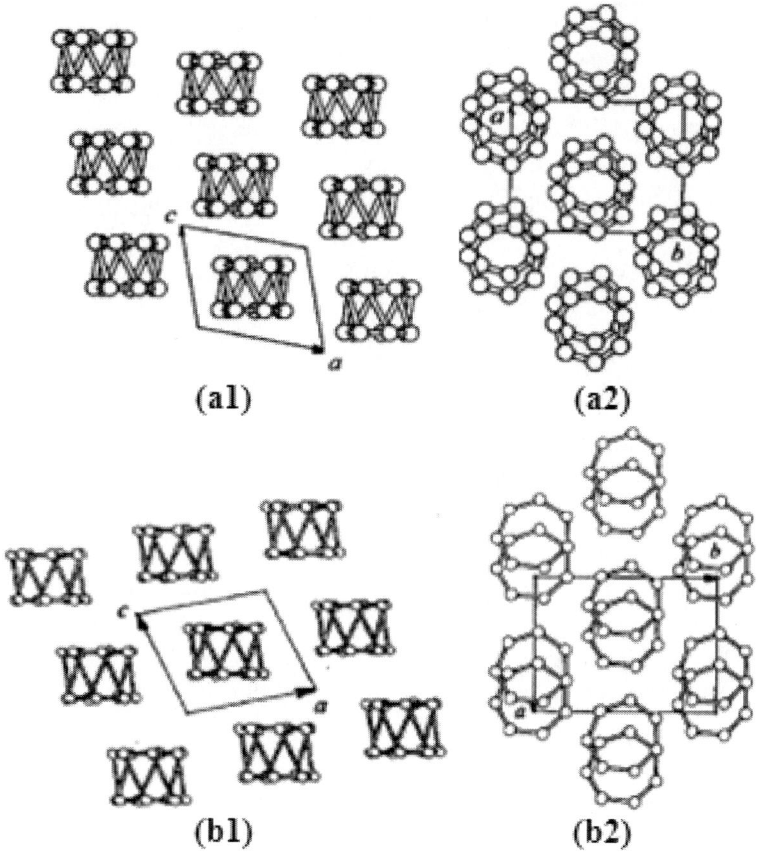

Figure 4. Projection of the β-CD dimers crystallizing according to the (a1 and a2) CH and (b1 and b2) TT packing mode onto the ab plane and along the b axis. Each β-CD molecule is represented by its O4n heptagon.

Their cell dimensions are similar to those of the dimeric β-CD structures crystallizing according to *CH* (Space Group *P*1) or *IM* packing modes but their unit-cell angles are different. Any attempt to reduce these unit cells to those observed in the cases of previously reported packing modes was failed. Layers of *p*-hydroxybenzaldeyde/β-CD dimeric complex stack along the *c*-axis and the two subsequent layers shift by only 1.14 Å. This is less than the 2.7 Å observed in *CH* packing modes. The mean planes of the glucosidic O4*n* atoms of the host dimer form angles of 11.1° and 9.8° with the *ab* plane. The lateral shift of the two adjacent layers of the benzoic acid/β-CD complex, stacking along the *c*-axis, is 6.49 Å. This shifting is about the average of the one observed in *IM* and *TT* packing modes (mean values 6.0 Å and 6.93 Å). The mean planes of their glucosidic O4*n* atoms form angles of 2.6° and 2.8° with the *ab* plane. Therefore, the *p*-hydroxybenzaldeyde/β-CD complex tends to form an ideal channel whereas the packing mode of the benzoic acid/β-CD complex is somewhat between *IM* and *TT*. As a consequence, these two structures are not isostructural to each other or to any other and have unique packing modes, which have never been observed before.

The neohesperidin-dihydrochalcone/β-CD complex [33] crystallizes in the $C222_1$ space group but its cell dimensions are quite different (Table 2) compared with those of the other $C222_1$ structures reported so far stacking according to the *CB* packing mode. In this structure the dimers stack also along the *c*-axis and the mean planes of the glucosidic O4*n* form angles of 7.77° with the *ab* plane. However, the packing mode is not *CB*. Instead, the dimers form a "screw column" and the distance between the adjacent projections of the centers of the O4*n* heptagons on the *ab* plane is about 4.4 Å which is almost the half of the 8.7 Å observed in the other structures crystallizing in $C222_1$. The hydrophobic isovanilin moiety of the guest molecule is included inside the hydrophobic dimeric cavity, while the rest of the guest molecule is accommodated at the interface of the dimers of the column.

The cobaloxime/β-CD dimeric complex [34] crystallizes in the orthorhombic space group $P2_12_12_1$. It consists of two β-CD molecules arranged in a head-to-head mode, giving the impression of a dimeric structure. But the O3*n*-H hydroxyl groups of the two hosts are not hydrogen bonded and therefore no dimers are formed. The structure may be characterized as pseudodimer. The formation of a real dimer is inhibited by the accommodation of two bulky cobaloxime molecules at the interface of the head-to-head arranged host molecules. The complexes are located along the two-fold screw *c*-axis and the successive complexes are arranged perpendicularly to each other. Thus, the molecular packing has the form of a helical arrangement.

The 1,4-bis(imidazol-l-ylmethyl)/benzene/β-CD complex [35] crystallizes in *C*2 space group having different cell dimensions from the majority of the crystal structures of the complexes crystallizing in this space group. Therefore, its molecular packing is not *CH* and resembles rather to *CB* with the approximate seven-fold molecular axis of the formed dimers oriented along the *c*-axis.

4.5. Special Cases of Dimeric β-CD Inclusion Complexes

The 1,2-bis(4-aminophenyl)ethane/β-CD complex [23] crystallizes in the *C*2 space group. The *c*-axis of its unit cell is equal to 33.315 Å having the double length of that observed in the other crystal structures of dimeric complexes crystallizing also in *C*2. This is due to the presence of two different β-CD dimeric complexes in the unit cell including guests with different conformations. Its crystal mode resembles the *SC* mode, but the screw channel is developed along the *c*-axis and the adjacent dimers are crystallographicaly independent. The angles of the approximately sevenfold axes of the two dimers with the *c*-axis are 23° and 3°, respectively, and their lateral displacement is 2.45 Å. Another similar case is the benzoic-acid/β-CD dimer [36] which crystallizes in four forms. One of them, form II, crystallizes in space group *C*2 and the length of the *c*-axis is about the double of that observed in the crystal structures of the complexes crystallizing in space group *C*2. This is due to the two different dimeric complexes presented in the unit cell and accommodated along the *c*-axis forming a *SC* packing mode. Same characteristics have been observed in the 2,2'-dipyridine/β-CD complex [37] crystallizing in space group *C*2 and stacking also according to the *SC* packing mode.

The (barbital)$_4$/β-CD complex [38], crystallizing in $P1$ space group, has an a-axis equal to 34.341 Å, the b and c axes being about 15.5 Å. The two different dimers included inside the unit cell form a tetrameric channel, and the tetramers stack according to the *IM* packing mode.

In the asymmetric unit of the (S)-(-)-methyl-p-tolysulfoxide/β-CD complex [39], crystallizing in space group $P2_1$, there are two dimers, forming also tetrameric channels. The length of the b-axis is 65.04 Å, approximately double the mean value of the c-axis of the β-CD dimeric structures crystallizing in space group $C222_1$. However, the complexes stack according to the *SC* instead of the *CB* packing mode. The cyclizine/β-CD complex [40], having similar cell dimensions and space group with the (S)-(-)-methyl-p-tolysulfoxide/β-CD complex [39], seems to be isostructural to it. Its crystal packing consists also of tetramers having a unique host:guest stoichiometry 4:3, due to the different stoichiometry of the two dimers of the asymmetric unit as one dimer includes one guest molecule while the other includes two guests.

The ferrocene/β-CD complex [41] crystallizes in space group $C2$ with cell dimensions that differ significantly from any other similar crystal structure reported so far. Its particularity is a result of a host:guest supramolecular arrangement described as follows: Two dimers are arranged successively with their approximately seven-fold molecular axes of the β-CD molecules along the c-axis. Inside each dimeric cavity two guest molecules are encapsulated whereas a fifth ferrocene molecule is localized at the interdimeric space. The orientation of this ferrocene is perpendicular to the orientation of the ferrocene guest molecules included inside the dimeric cavities. Therefore, the complexes form tetramers arranged along the c-axis and a Channel which is interrupted by the fifth guest molecule is developed.

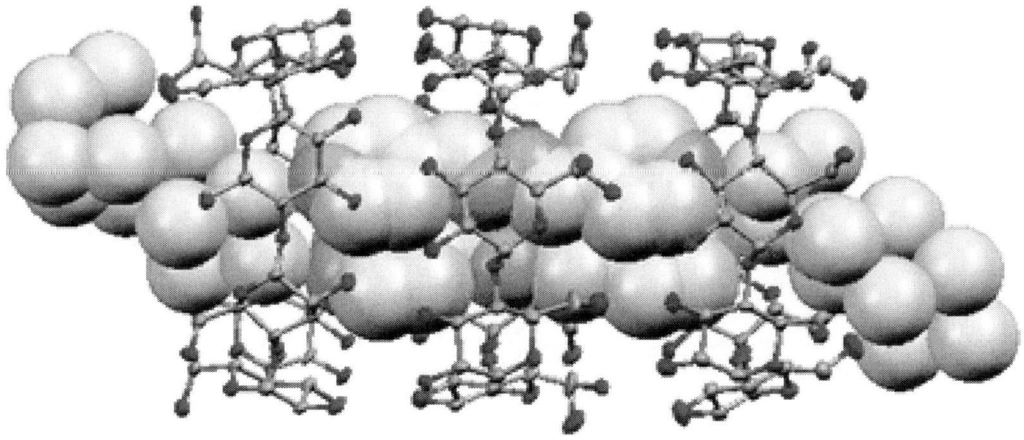

Figure 5. The 4-pyridinealdazine/β-CD trimer. Three host molecules are arranged in a head-to-head-to-tail mode forming channels inside of which two guests are included.

The crystal of the 4-pyridinealdazine/β-CD complex [42] crystallizes in space group $P1$. The length of the unit cell c-axis is about 6 Å greater than that observed in all the other cases of complexes crystallizing in this space group and stacking according to the *IM* packing mode. This structure is quite unique as the asymmetric unit consists of three host molecules.

Two host molecules are arranged in a head-to-head mode forming the usual β-CD dimer and the third host molecule is arranged in a head-to-tail mode with a host molecule of the dimer. This arrangement results to the formation of a trimer (Figure 5). The formed trimers host two guest molecules with their approximated molecular axes about parallel to the approximate seven-fold molecular axis of the β-CD molecules. Some water molecules are located in the free space between the trimers. The molecular packing is a Channel developed along the *c*-axis.

4.6. Conclusion

By reviewing the crystal structures of the dimeric β-CD complexes reported so far we conclude that they crystallize according to four space groups namely $C2$, $P1$, $P2_1$ and $C222_1$ and the majority of their packing modes are *CH*, *IM*, *SC* and *CB* (Table 2). Note that one β-CD complex crystallizes in space group $P2_12_12_1$ its packing mode being a helix but this complex forms pseudodimers because although the two hosts are arranged in a head-to-head mode, not all their secondary hydroxyl groups O3*n*-H are hydrogen bonded. Two complexes crystallizing in space group $C2$ stack according to the *TT* packing mode. One dimeric complex, crystallizing in $P1$ space group, stacks according to a combination of *IM* and *TT* packing modes while the dimers of an other dimeric complex, also crystallizing in $P1$ space group, form almost an ideal channel. Two different dimers hosting guests of different conformation have been found in the unit cell of the crystal structure in two cases of dimeric complexes. These dimers stack according to the *SC* packing mode though they crystallize in space group $C2$. In three cases the complexes form tetramers and in another case the inclusion complex consists of β-CD trimers. One complex was found with the unique 4:3 host:guest stoichiometry. One complex having two dimers in its asymmetric unit contains one guest molecule between two successive dimers. As the study of the crystal structures of the β-CD complexes continues, it is possible that some new packing modes may be revealed later.

5. The Polymorphism of the β-CD Complexes

Although at a given temperature and pressure, only one arrangement of a given substance in the solid state is in principle thermodynamically stable, it is possible to appear more than one crystal structures of the same substance. This ability of a solid material to exist in more than one crystalline phase, having different molecular arrangements or conformations, is called polymorphism [43]. The polymorphism is possible when the lattice energies (in general being about a few kJ/mol) of these different crystal structures are similar. Therefore, slight variations in preparative conditions of a crystal may lead the system to different polymorphs [44]. This can happen even for crystals from the same batch and thus if an individual crystal is selected for X-ray diffraction and structure determination it is not necessarily representative of the structure of the bulk material [45].

Sometimes polymorph structures can be derived using different methods or conditions of crystallization of the same substance (vide infra). The polymorphism of a structure may be ought to conformational differences of the substance or to different packing arrangement [46].

The polymorphism is considered as a generally rare phenomenon for CD complexes [15]. However, in the recent years at least five polymorph crystal structures of β-CD dimeric complexes were reported. The cases in which the inclusion compounds have same hosts and guests but different stoichiometry or a different number of water molecules are designated not as polymorphs but as pseudopolymorphs [47, 48] a term that, at any rate, is controversial [49]. The following complexes are referred in the literature as polymorphs or pseudopolymorphs:

n-propanol/β-CD [50, 51]: Two forms of dimeric crystals of *n*-propanol/β-CD, CI and CII, have been prepared under different conditions of crystallization. CI has been formed by vapor diffusion of *n*-propanol into a 1:1 *n*-propanol:water solution of β-CD. CII crystallizes from a 3:2 *n*-propanol:water solution of β-CD by slow cooling.

Although both polymorphs crystallize in the same Space Group $P1$, the dimers of CI stack according to the *IM* packing mode and those of CII according to the *CH* packing mode. CII crystals are metastable changing into CI after several hours.

ethanol/β-CD [52, 53, 54, 20]: Two ethanol/β-CD complexes are elucidated by X-ray [52, 20] while two others are investigated by neutron crystal structure analysis [53, 54]. Three of the aforementioned studied complexes were found crystallize in space group $P2_1$ forming monomers and stacking as herringbones. The forth complex forms dimers [20], crystallizing in space group $C2$ stacking according to *CH* packing mode.

benzoato.OH-U(VI)O₂/β-CD [26]: Two dimeric polymorphs, CI and CII, have been prepared by slow evaporation at 303 K from a benzoato.*Cl*-U(IV)O_2:β-CD 1:0.5 aqueous solution. The host:guest stoichiometry of CI is 2:1 and this of CII 2:2. CI and CII crystallize in space groups $P2_1$ and $C2$ respectively.

methylparaben/β-CD [27]: Two dimeric inclusion complexes, CI and CII, have been prepared using different preparative methods. In the CI crystal, taken from a water solution of the host and guest molecules at 280 K, the complex found crystallizing in space group $P1$ and the formed dimers stacking according to the *IM* packing mode. The CII crystal formed in a same mother solution but at room temperature and the complex found crystallizing in space group $C2$ stacking according to *CH* packing mode. Just as the *n*-propanol/β-CD complex the CI polymorph, stacking also according to *IM* crystal packing mode, is more stable than the CII.

benzoic acid/β-CD [36]: This dimeric complex crystallizes in four forms I, II, III and IV. As it has already been mentioned in paragraph 4.5, the form II crystallizes in space group $C2$. Form I crystallizes in space group $P1$ having all the unit cell dimensions slightly greater than 15 Å and dimers stacking according to *CH* packing mode. Form III crystallizes also in space group $P1$ with a and b axes lengths slightly greater than 15Å and c-axis about 18 Å. The dimers stack according to *IM* packing mode. The form IV, like Form II, crystallizes in space group $C2$ with cell dimensions of about a = 19 Å, b = 24 Å and c = 15 Å. These values are quite similar to the majority of the cell dimensions observed in complexes crystallizing in space group $C2$. Thus, the dimers stack according to *CH* packing mode. Crystals of the Forms II and IV have been created after 1.5 year storage of the batch at 298 K and they are stable. Crystals of the Forms I and III have been collected from the batch after a storage of only 7 days also at 298 K.

5.1. Conclusion

Only five polymorphic crystals of *β*-CD complexes have been published till now. Therefore, the polymorphism of this kind of crystals can indeed be considered as a rare phenomenon [15].

The different forms of the polymorphic crystals correspond to structures of different stacking modes and/or space groups whereas in a case an inclusion complex is found crystallizing both as a monomer and as a dimer.

The conditions of crystallization of the polymorphs are in many cases different as regards the temperature, the crystal incubation time and the mother liquor.

Finally, in one case a metastable crystal structure which changes into a more stable form has been reported.

6. Water Molecules Encapsulated Inside the Dimeric Cavity or the Formed Channel

During the process of the complexation the guest molecules replace the water molecules of high enthalpy that are entrapped inside the hydrophobic cavity of the host [55]. However, in the cases of dimeric complexes it is possible to find encapsulated water molecules inside the formed large dimeric cavity, especially at the region of the interface between the head-to-head arranged host CDs. This region is rich of hydroxyl groups and despite the formation of a hydrogen-bond net between them is not as hydrophobic as the internal cavity of a monomer. The possibility of encapsulation of water molecules inside the dimeric cavity depends on the mode of accommodation of the guest molecules inside the cavity and their conformation. If the guest molecule included inside the cavity disposes hydrophilic groups, it can be hydrogen bonded with water molecules also entrapped inside the cavity. Moreover, it is possible to find water molecules in the free space of the channels formed by the complexes crystallizing in *CH* or *SC* packing modes though this area is also hydrophobic as some of the hydroxyls of the primary rims of the dimer are hydrogen bonded with the hydroxyls of the primary rim of the vicinal dimer. A search in the literature has given the following cases of complexes with water molecules entrapped inside the dimeric cavity or the interstices between the dimers:

p-Ethylaniline/β-CD [56]: The complex crystallizes in space group $P2_1$ and the accommodation of two guest molecules inside the dimeric cavity is such that it leaves enough space for the encapsulation also of two water molecules, hydrogen bonded with the amine groups of the guest molecules.

1-adamantanemethanol/β-CD [57]: The complex crystallizes in space group $C222_1$. A water molecule is entrapped in the interface of the dimeric cavity. The water molecule is found disordered occupying three sites. Although hydrogen bonds can be formed between the occupied sites, no hydrogen bond can exist between them and the guest (the hydrophilic moieties of the guests are located outside of the dimeric cavity) or the host molecules.

3,3-dimethylbutylamine/β-CD [58]: The complex crystallizes in space group *C*2. One water molecule is entrapped inside the dimeric cavity hydrogen bonded with the amino groups of both guests.

benzocaine/β-CD [59]: The complex crystallizes in space group *C*2 and the dimers stack according to the *CH* packing mode. One water molecule, entrapped at the interface of the dimeric cavity, is hydrogen bonded with the ester groups of both guests accommodated inside the cavity. Another water molecule, located in the free space between the vicinal dimers of the Channel, is hydrogen bonded with the amine groups protruding from the primary hydroxyl groups region of the dimer.

p-bromoacetaninilide/β-CD [60]: The complex crystallizes in space group *P*1 stacking according to *CH* packing mode. Two guests are encapsulated inside the dimeric cavity and their acetyl residues, disordered over two sites, are located in the dimeric interface. Their amino and C=O groups are hydrogen bonded with five water molecules encapsulated inside the dimeric cavity. These water molecules are also hydrogen bonded to each other.

trans-cinnamic acid/β-CD [61]: The complex crystallizes in space group *C*2 stacking according to *CH* packing mode. Two water molecules are located in the free space of the channel bridging via hydrogen bonds the carboxyl groups of the vicinal guest molecules, protruding from the cavities (Figure 6).

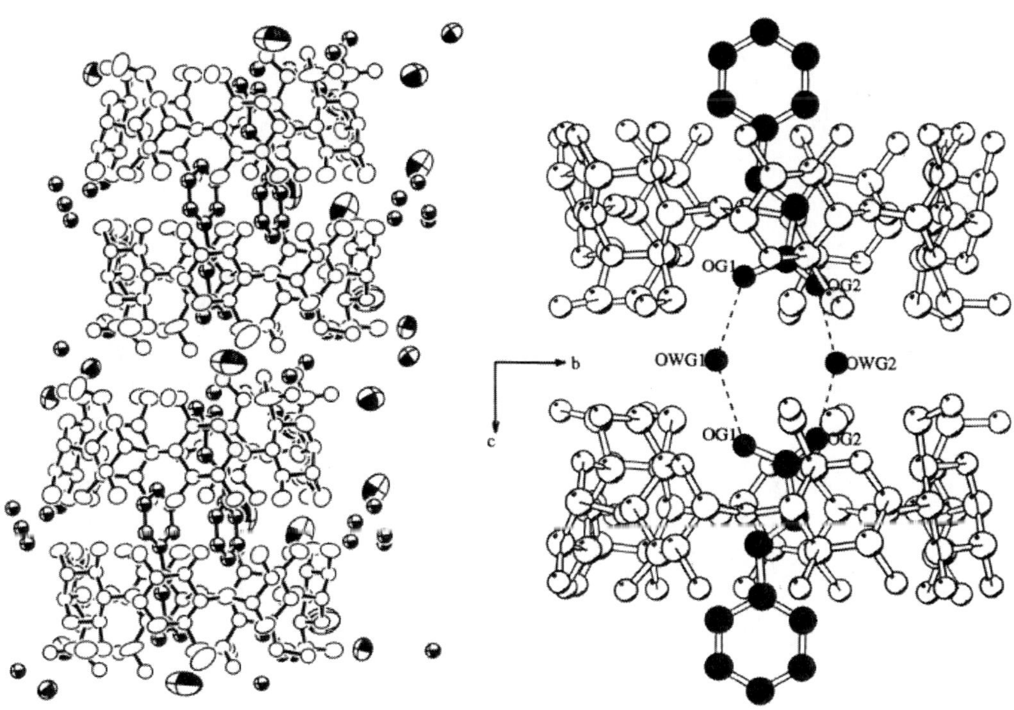

(a) (b)

Figure 6. (**a**) The formed channel by head-to-head dimers of *trans*-cinnamic acid/β-CD complex. (**b**) OWG1 and OWG2 water molecules located in the free space of the channel bridging via hydrogen bonds the carboxyl groups of the vicinal guest molecules.

1,12-dodecanedioic acid/β-CD [62]: The complex crystallizes in space group *P*1 stacking according to the *IM* packing mode. Several water molecules are accommodated in the free space between the complexes; two of them are hydrogen bonded with the carboxyl groups of the guests of the vicinal complexes forming bridges linking them.

ethanol/β-CD [20, 32, 62]: The complex crystallizes in two polymorphic forms I and II. The complexes of the form I stack according to the *CH* packing mode [20]. One ethanol and two water molecules are entrapped inside the dimeric interface. The water molecules are hydrogen bonded between them and with the ethanol molecule. The complexes of the form II stack also according to the *CH* packing mode [32]. Three water molecules are entrapped inside the dimeric cavity. One disordered ethanol molecule entrapped also inside the dimeric cavity is hydrogen bonded with one water molecule. Besides, two water molecules are hydrogen bonded with the secondary hydroxyl groups and two others with the primary hydroxyl groups of the host molecules.

triclosan/β-CD [63]: The packing mode of the *β*-CD dimeric complex is *CH*. Five water molecules are enclosed inside the dimeric cavity at the interface of the hosts' head-to-head dimer arranged on a plane perpendicular to the approximate sevenfold molecular axis of the *β*-CD.

benzoic acid/β-CD [32]: The complexes are stacked according to the *CH* packing mode. One ethanol and two water molecules are located inside the dimeric cavity laying at the interface of the head-to-head formed dimer. One of the water molecules is hydrogen bonded with the ethanol molecule. The carboxyl group of the guests protrudes from the primary rims of the dimer towards the interdimeric space of the channel hydrogen bonded with water molecules.

MCPA/β-CD [24]: The guest molecule is disordered over three sites. One water molecule is entrapped inside the cavity located in the interface of the dimer and hydrogen bonded with the chlorine atom of the guest (Figure 7).

2,4-dichlorophenoxyacetic acid/β-CD [64]: The dimeric inclusion complex crystallizes in space group *P*1 with the dimers accommodated along the *a* crystal axis and forming Channels. The host:guest stoichiometry is 2:1 and the guest molecule is disordered over three sites having different orientations inside the dimeric cavity. Up to four water molecules have also been found encapsulated, occupying five possible sites at the dimeric interface and hydrogen bonded between them and with the Chlorine atom of the guest (Figures 8a and b).

aspirin-salicylic acid [65]: Inside the dimeric cavity two aspirin and one salicylic acid molecules are located. The two aspirin molecules are accommodated inside the two host hydrophobic cavities with their carboxyl groups protrude from the primary hydroxyl groups' rim. The salicylic guest molecule, which is the hydrolyzed product of the aspirin, is sandwiched at the interface of the dimer.Although the dimers of the complex stack according to the *IM* packing mode the carboxyl groups of the guests are hydrogen bonded with water molecules located in the space between the vicinal dimers.

6.1. Conclusion

Water molecules can be entrapped inside the dimeric cavity, especially at the region of the dimeric interface, if there is enough space for them to be encapsulated. They are usually hydrogen bonded to the hydrophilic groups of the guest molecules, like carboxyl or amino groups, or to the secondary hydroxyl groups of the hosts which form the dimer. Water molecules have also been found located at the relatively hydrophobic inter-dimeric space of the channels formed in some cases by the dimeric complexes. These water molecules are hydrogen bonded to the hydroxyls of the primary rims of the dimers or to the hydrophilic groups of the guest molecules protruding from the dimeric cavity.

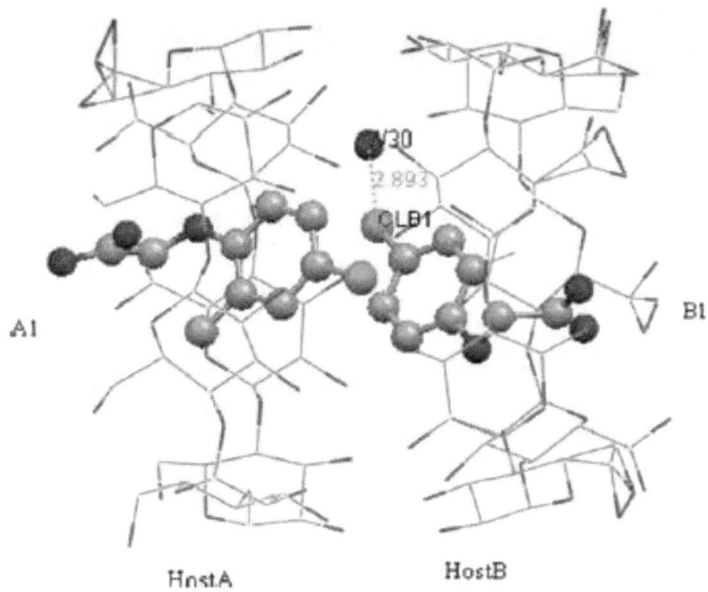

Figure 7. Crystal structure of the MCPA/β-CD complex. The entrapped water molecule, W30, is located in the dimer's interface, hydrogen bonded with the Cl atom of the guest MCPA.

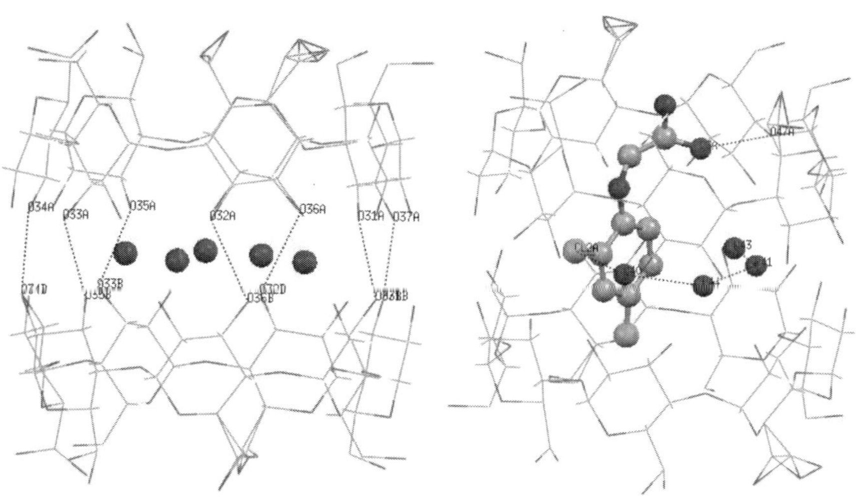

(a) (b)

Figure 8. 2,4-dichlorophenoxyacetic acid/β-CD complex (**a**) The five sites found occupied by water molecules are located in the intra-dimeric β-CD interface (guest molecule is omitted for clarity) (**b**) The guest molecule occupying the one possible site while its Cl atom is hydrogen bonded with a water molecule and three more water molecules are laying at the dimeric interface, connected via hydrogen bonds with each other.

7. Guest Molecules Sandwiched at the Interface of the Head-to-Head Cyclodextrin Dimer

Guest molecules disposing a planar moiety such as a phenyl or a naphthyl ring can be accommodated in the interstitial space between the cyclodextrins of a head-to-head dimer in a sandwich mode. In these cases the planar moiety of the guest is found lying almost parallel to the mean plane of the glucosidic O4*n* atoms of the host molecules of the dimer.

aspirin and salicylic acid/β-CD [65]: One aspirin molecule is included inside the hydrophobic cavity of each host molecule of the dimer and one salicylic acid molecule, the hydrolyzed product of the aspirin, is accommodated in the dimeric interface in a sandwich mode, disordered over three sites.

methyl-p-tolylsulfoxide/β-CD [38]: Three guest molecules are encapsulated inside the dimeric cavity of the complex the host:guest complex stoichiometry being therefore 2:3. One guest is accommodated inside each hydrophobic cavity of the hosts forming the dimer whereas the third guest molecule is sandwiched highly disordered in the dimeric interface.

3,5-dimethylbenzoic acid/β-CD [66]: The guest molecule is disordered over six sites. In two of them, the carboxyl group of the guest is oriented towards the primary hydroxyl groups region of the hosts of the dimer whereas in two others it is oriented towards the secondary hydroxyl groups region. In the last two disordered sites (with occupation factors of 0.44 and 0.41), the guest is accommodated in the dimeric interface its carboxyl group being hydrogen bonded with the carboxyl group of another entrapped guest molecule.

n-octanol-pyrene/β-CD and *cyclohexanol-pyrene/β-CD* [67]: The complexes crystallize in space group *C*2 stacking according to the *CH* molecular packing. In both structures the pyrene molecule lays in the interstitial space of the dimer nearly perpendicular to the approximate seven-fold molecular axis of the hosts. One *n*-octanol molecule protrudes from the primary hydroxyls rim towards the vicinal dimeric cavity of the Channel while two cyclohexanol molecules are accommodated inside the hydrophobic host cavities of the dimer.

1,10-phenanthroline/β-CD [68]: The host:guest stoichiometry is 2:3. The hydrophobic cavities of the hosts forming the dimer are filled with two guest molecules accommodated inside them whereas the third guest is located in the interstitial space its phenanthroline plane forming an angle of 89.3° with the approximate seven-fold molecular axis of the host.

bithiophen/β-CD [69, 70]: The host:guest stoichiometry is 2:3. Two guest molecules are located inside the host cavities and the third one is accommodated in the dimeric interface with its mean plane about perpendicular to the approximate seven-fold axes of the hosts.

8-nitro-quinoline/β-CD [71]: The host:guest stoichiometry is 2:3. Two guest molecules are located in the cavities of the hosts and the third one lies sandwiched in the dimeric interface. Van der Waals and hydrophobic interactions hold this guest inside the dimeric cavity.

o-aminobenzoic acid/β-CD [72]: The host:guest stoichiometry is 2:3, two guest molecules are accommodated inside the two host molecules' cavities and the third, disordered over two sites with occupancy factors 0.72 and 0.28, is located in the interstitial space. This guest is no hydrogen bonded but Van der Waals and hydrophobic interactions hold it inside the dimeric cavity.

N-methylanthranilin acid/β-CD [73]: The complex crystallizes in space group *P*1 and its packing mode is *IM*. The host:guest stoichiometry is 2:1 as only one guest molecule is included inside the dimeric cavity accommodated in the dimeric interface being stabilized only by hydrophobic interactions.

4-hydroxycoumarin/β-CD [74]: Three guest molecules are accommodated inside the dimeric cavity. Therefore, the host:guest stoichiometry is 2:3 and one of the guests is located in the dimeric interface accommodated parallel to the mean plane of the glycosidic O4*n* atoms.

α-naphthylacetic acid/β-CD [75]: The dimer crystallizes in space group *P*1 and its molecular packing is *IM*. Two ethanol molecules and one *α*-naphthylacetic acid are included inside the dimeric cavity. Each ethanol molecule is located inside the monomeric hosts' cavity whereas the *α*-naphthylacetic acid is accommodated in the dimeric interface its naphthyl plane being parallel to the O4*n* mean plane and its carboxyl group hydrogen bonded with a secondary hydroxyl group of the host (Figure 9).

eugenol/β-CD [76]: Three eugenol molecules are encapsulated inside the dimeric cavity the host:guest stoichiometry being, therefore, 2:3. Two of them are located inside the host cavities and their hydrophilic moieties protrude from the primary hydroxyl groups' rim. The third guest molecule is accommodated in the interstitial space of the dimer disordered over two sites (Figure 10).

7.1. Conclusion

The majority of the complexes having a guest molecule sandwiched in the interstitial space of the dimeric complex dispose a phenyl group. Fewer complexes dispose a naphthyl group. Other molecules with planar moieties located also in the interstitial space are the pyrene, the phenanthroline and the bithiophen.

The majority of the complexes is characterized by a host:guest stoichiometry equal to 2:3 with the two guest molecules located inside the hydrophobic cavities of the hosts forming the dimer whereas the third guest is located in the dimeric interface. In the cases of 2:1 host:guest stoichiometry the sole guest molecule encapsulated inside the dimeric cavity is sandwiched in the dimer's interface with its planar moiety parallel to the O4n mean planes of the hosts.

The dimeric complexes with a disordered sandwiched guest molecule usually crystallize in space group *C*2 and two complexes linked by a twofold crystal axis form the dimer.

8. π-π INTERACTIONS BETWEEN THE GUEST MOLECULES

The π-π interactions and the hydrogen bonds are two important secondary electrostatic interactions for the formation of the supramolecular materials [77]. A π-system is composed by a σ-framework sandwiched between the two negative charged π-electron clouds.

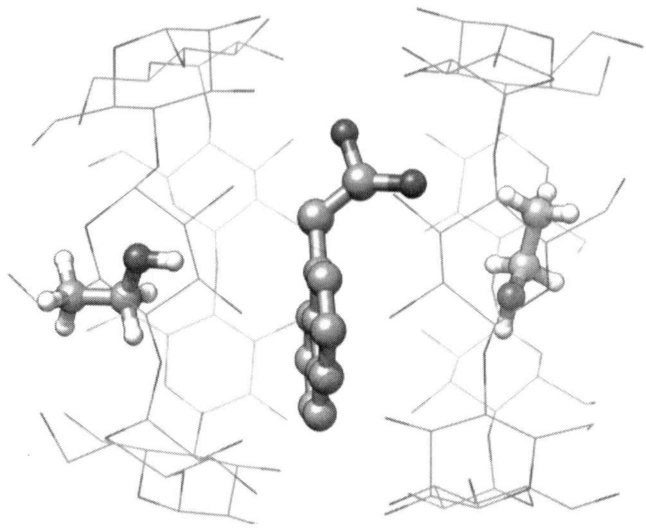

Figure 9. The sandwich-like inclusion mode of the α-naphthylacetic acid in β-CD dimer. The hydrophobic cavities of the β-CD monomers are filled with ethanol molecules.

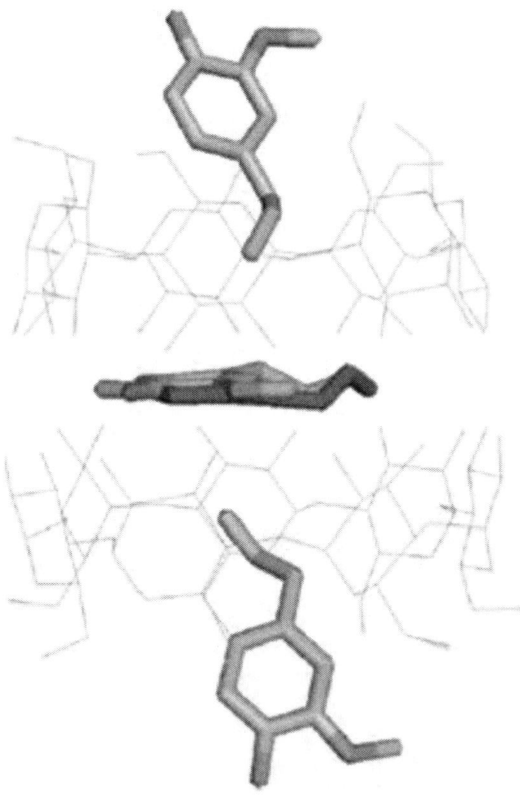

Figure 10. The eugenol/ β-CD 3/2 dimeric inclusion complex.

The distance between two attractive π-systems ranges from 3.4Å to 4.0Å whereas for shorter distances repulsions between the π-electron clouds of the vicinal π-systems are observed [78]. The perfect face-to-face stacking of these systems is extremely rare as repulsive forces are exerted between them [79]. Attractive forces are exerted when one π-system is rotated by up to 90° or it is offset relative to a vicinal one. Many stacking modes have been observed such as: edge, point-to-face, T-shape, offset-slipped facial arrangements [80].

In the β-CD dimeric inclusion complexes the π-π stacking of the encapsulated guest molecules depends mainly on the steric constraints imposed by the formed hosting cavity. π-π interactions have been observed between couples of guest molecules containing phenyl or naphthyl π-systems. Molecules containing anthracene, tetracene or pentacene moieties, although they can be π-π stacked, they cannot be encapsulated as a couple inside the cavity of the β-CD dimers due to their size and thus no such cases have been reported.

A search in the literature shows that π-π interactions are observed between guest molecules that can be hosted as couples inside the dimeric β-CD cavity and contain phenyl or naphthyl rings.

In *2,2-bis(4-aminophenyl)ethane/β-CD* complex [23], two dimers are formed each containing two guest molecules. Between the guests of the two dimers π…π interactions exist their mean distances being 3.69 Å and 3.53Å.

In *p-Ethylaniline/β-CD* complex [56], π…π interactions are observed between the two guest molecules hosted inside the dimeric cavity.

In *4,4'-dipyridine/β-CD* complex [81], two dipyridine molecules are entirely included inside the dimeric host cavity. The pyridine ring of the one guest is stacked with that of the other by π…π interaction. The separation of the centroids of their rings is 3.634 Å and the angle between their planes is 42.2°.

In *thymol/β-CD* complex [76], the phenyl moieties of the two guests included inside the dimer are associated by π…π interactions in a mode of offset-slipped facial arrangement (Figure 11).

The *benzoic acid/β-CD* complex [42] crystallizes in two Forms I and II. A π…π interaction exists between the phenyl groups guest molecules of the II form their interplanar distance being 3.56 Å.

The asymmetric unit of the *p-hydroxybenzaldehyde/β-CD* complex crystal structure [31] contains two pairs of guest molecules, one encapsulated inside the dimeric cavity and another located in the interdimeric space. The phenyl rings of each pair are nearly parallel to each other but they are offset and they are stabilized by π…π interactions.

The guest molecules of the *p-bromoacetarilide/β-CD* complex [60] adopt an approximately planar conformation with a π…π interaction between them.

A face-to-face π…π interaction appears between the phenyl rings of the guest molecules of the *4-hydroxybiphenyl/β-CD* complex [82]. The phenyl rings are not quite parallel to each other and a distance of 0.37 Å between their cendroids offsets is observed.

In the cases of the *p-nitrobenzoic acid/β-CD* [83], *p-nitroacetanilide/β-CD complex* [84] and *neohesperidin dihydrochalcone/β-CD* [33] dimeric complexes, π…π interaction between the phenyl moieties of the two guests hosted inside the dimer is observed. In the case of the *p-nitroaniline/β-CD* complex [85], the phenyl rings of the two guests included inside the dimeric cavity are placed at an average interplanar distance of 3.40 Å forming an angle of 8.52°. It is evident that a π…π interaction stabilizes the structure.

In the case of the *bisphenol/β-CD* complex [86] the two guest molecules, each one possessing two phenyl groups named as 'a' and 'b', are encapsulated inside the dimeric cavity. A π…π interaction is observed between the 'a' phenyl group of the one guest and the 'b' phenyl group of the other. The phenyl rings are not parallel and the distance between their centroids is 3.947 Å.

A π…π interaction is also observed between the phenyl moieties of the two guests of the *p-amino-p'-nitrobipheny/β-CD* dimeric complex [87].

In the case of the *p-nitrophenoxyacetic acid/β-CD* complex [88], a π…π interaction is observed between the naphthyl rings of the two guests. The distance of the cendroids of the planar moieties is 4.56 Å and they form an angle of 26.5°.

The phenyl groups of the two guests of the *ferbinac/β-CD* complex [89] are face-to-face π…π stacked, the angle between them being 13.9°.

In the case of the *cis-dioxovanadium(V)/β-CD* complex [90], only the biphenyl side chains of both guests are located inside the hydrophobic cavity and there is a π…π interaction between them. The distance between the two phenyl rings is 3.63 Å. Only one paroxetine molecule is encapsulated inside the dimeric cavity of the *paroxetine/β-CD* complex [27], having the conformation of a hairpin and stabilized by an intermolecular π…π interaction between the phenyl rings.

Finally, a π…π interaction between phenyl rings of the guests is also observed in the case of the *S-flurbiprofen/β-CD* complex [91].

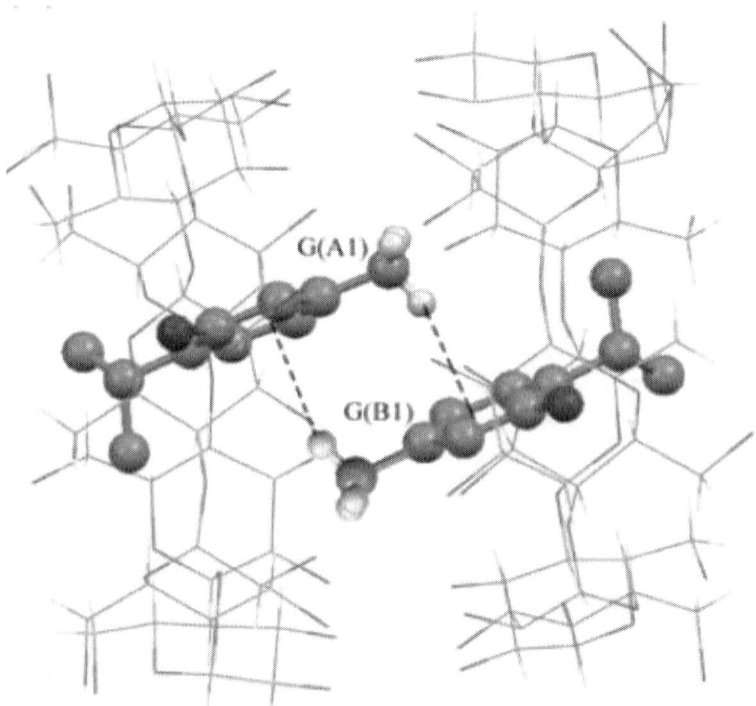

Figure 11. The thymol/β-CD 2/2 dimeric inclusion complex. Mutual CH/π interactions between the thymol guest molecules. For clarity, only the H-atoms of the –CH3 group of the guests are drawn.

REFERENCES

[1] Saenger, W. Angew. Cyclodextrin inclusion compounds in research and industry. *Chem. Int. Ed. Engl.,* 19, 344-362 (1980).

[2] Crini, G. and Morcellet, M. Synthesis and application of adsorbents containing cyclodextrins. *J. Sep. Sci.,* 25, 789-813 (2002).

[3] Larsen, K. L. Large cyclodextrins. *J. Incl. Phenom. and Macrocyclic Chem.,* 43, 1-13 (2002).

[4] Del Valle, E. M. M. Cyclodextrins and their uses. *Process. Biochem.*, 39, 1033-1046 (2004).

[5] Liu, L. and Guo, Q.-X. The driving forces in the inclusion complexation of cyclodextrins. *J. Incl. Phenom. and Macrocyclic Chem.*, 42, 1-14 (2002).

[6] Wenz, G. Cyclodextrins as building blocs for supramolecular structures and functional units. *Angew. Chem. Int. Ed. Engl.*, 33, 803-822 (1994).

[7] Szejtli, J. Introduction and general overview of cyclodextrin chemistry. *Chem. Rev., 98,* 1743-1753 (1998).

[8] Uekama, K., Hirayama, F. and Irie, T. Cyclodextrin drug carrier systems. *Chem. Rev.*, 98, 2045-2076 (1998).

[9] Buschmann, H.-J., Knittel, D. and Schollmeyer, E. New textile applications of cyclodextrins. *J. Incl. Phen. and Macr. Chem.*, 40, 169-172 (2001).

[10] Irie, T. and Uekama, K. Pharmaceutical applications of cyclodextrins. III. Toxicological issues and safety evaluation. *J. Pharmac. Scien.*, 86, 2, 147-162 (1997).

[11] Coleman, A., Nicolis, I., Keller, N. and Balbiez, J. P. Aggregation of cyclodextrins: an explanation of the abnormal solubility of β-Cyclodextrin. *J. Incl. Phen. and Mol. Rec. Chem.*, 13, 139-143 (1992).

[12] Bonnet, P., Jaime, C. and Morin-Allory, L. Structure and Thermodynamics of α-, β- and γ-Cyclodextrin Dimers. Molecular Dynamics Studies of the Solvent Effect and Free Binding Energies. *J. Org. Chem.*, 67, 8602-8609 (2002).

[13] Caira, M. On the isostructurality of cyclodextrin inclusion complexes and its practical utility. *Rev. Roumaine Chim.*, 46, 4, 371-386 (2001).

[14] Pop, M. M., Goubitz, K., Borodi, G., Bogdan, M., De Ridder, D. J. A., Peschar, R. and Schenk, H. Crystal structure of the inclusion complex of β-Cyclodextrin with mefenamic acid from high-resolution powder-diffraction data in combination with molecular mechanics calculation. *Acta Cryst. B58*, 1036-1043 (2002).

[15] Borodi, G., Bratu, I, Dragam, F., Peschar, R., Helmholdt, R. and Hernaanz, A. Spectroscopic investigations and crystal structure from synchrotron powder data of the inclusion complex of β-Cyclodextrin with atenol. *Spectroch. Acta Part A70*, 1041-1048 (2008).

[16] Guo, P., Su, Y., Cheng, Q., Pan, Q. and Li, H. Crystal structure of the β-Cyclodextrin-*p*-aminobenzoic acid inclusion complex from powder diffraction data. *Carb. Res.*, 346, 986-990 (2011).

[17] Anibarro, M., Gessler, K., Uson, I., Sheldrick, G. M. Harata, K., Uekama, K., Hirayama, F. Abe, Y. and Saenger, W. Effect of peracylation of β-Cyclodextrin on the molecular structure and the formation of inclusion complexes: an X-ray study. *J. Amer. Chem. Soc.,* 123, 11854-11862 (2001).

[18] Tsorteki, F., Bethanis, K. and Mentzafos, D. Structure of the inclusion complexes of heptakis(2,3,6-tri-*O*-methyl)-*β*-Cyclodextrin with indole-3-butyric acid and 2,4-dichlorophenoxyacetic acid. *Carbohydr. Res.*, 339, 233-240 (2004).

[19] Harata, K. Structural aspects of stereodifferentiation in solid state. *Chem. Rev.,* 98, 1803-1827 (1998).

[20] Aree, T. and Chaichit, N. A new crystal form of *β*-Cyclodextrin-ethanol inclusion complex: channel-type structure without long guest molecules. *Carb. Res.,* 338, 1581-1589 (2003).

[21] Allen, F. H., Kennard, O. and Taylor, R. Systematic analysis of structural data as a research technique in organic chemistry. *Acc. Chem. Res.*, 16, 146-153 (1983).

[22] D. Mentzafos, I. M. Mavridis, G. Le Bas and G. Tsoucaris. Structure of the 4-*tert*-butylbenzyl alcohol-*β*-Cyclodextrin complex. Common features in the geometry of *β*-Cyclodextrin dimeric complexes. *Acta Cryst. B47,* 746-757 (1991).

[23] Giastas, P., Yannakopoulou, K. and Mavridis, I. M. Molecular structures of the inclusion complexes *β*-cyclodextrin-1,2-bis(4-aminophenyl)ethane and *β*-cyclodextrin-4,4'-diaminobiphnyl; packing of dimeric *β*-cyclodextrin inclusion complexes. *Acta Cryst B59*, 287-299 (2003).

[24] Triantafyllopoulou, V. Mentzafos, D., Tsorteki, F. and Bethanis, K. Inclusion compounds of plant growth regulators in Cyclodextrins. Study of the inclusion compound of MCPA in *β*-Cyclodextrin by X-Ray crystallography. *J. Chem. Crystallogr.*, 42, 238-244 (2012).

[25] Bethanis, K., Bouclaki, V.-S., Christophorides, H., Kokkinou, A., Tsorteki, F. and Mentzafos D. Volatile substances encapsulated in cyclodextrins. Part III. Study of the inclusion compounds of geraniol with *β*-cyclodextrin, and heptakis(2,3,6-tri-*O*-methyl)-*β*-cyclodextrinby X-ray crystallography. In preparation.

[26] Navaza, A. Iroulart, M. G. and Navaza, J. A monomeric uranyl hydroxide system obtained by inclusion in the *β*-Cyclodextrin cavity. *J. Coord. Chem.,* 51, 153-168 (2000).

[27] Caira, M., De Vries, E., Nassimbeni, L. and Jacewicz, V. Inclusion of the andidepressant paroxetin in *β*-cyclodextrin. *J. Incl. Phen. and Macroc. Chemistry,* 46, 37-42 (2003).

[28] Sanchez-Ruiz, X., Alvarez-Larena, A., Jaime, C., Piniella, J., Redondo, J., Virgili, A., Sanchez-Ferrando, F., Germain, G. and Baert, F. Molecular and crystal structure of the 1:1 complex of adamantanone with *β*-cyclodextrin. *Supramol. Chem.*, 10, 219-223 (1999).

[29] Brett, T. J., Alexander, J. M. and Stezowski, J. J. Chemical insight from crystallographic disorder-structural studies of supramolecular photochemical systems. PART 3. The β-cyclodextrin-7-hydroxy-4-methylcoumarin inclusion complex: direct observation of photodimerization by X-ray crystallography. *J. Chem. Soc. Perkin Trans., 2*, 1095-1103 (2000).

[30] Tsorteki, F., Bethanis, K., Pinotsis, N., Giastas P. and Mentzafos D. Inclusion compounds of plant growth regulators in cyclodextrins. V. 4-chlorophenoxyacetic acid encapsulated in *β*-cyclodextrin and heptakis(2,3,6-tri-*O*-methyl)-*β*- cyclodextrin. *Acta Cryst. B61,* 207-217 (2005).

[31] Braga, S., Aree, T., Imamura, K., Vertut, P., Boal-Palheiros, I., Saenger, W. and Teixeira-Dias, J. Structure of the β-cyclodextrin-p-hydroxybenzaldehyde inclusion complex in aqueous solution and in the crystalline state. *J. Incl. Phen. and Mol. Chem.*, 43, 115-125 (2002).

[32] Aree, T. and Chaichit, N. Crystal structure of β-cyclodextrin-benzoic acid inclusion complex. *Carb. Res.*, 338, 439-446 (2003).

[33] Malpezzi, L., Fronza, G., Fugati, C., Mele, A. and Bruckner, S. Crystal architecture and conformational properties of the inclusion complex, neohesperidin dihydrochalcone-cyclomaltoheptaose (β-cyclodextrin), by X-ray diffraction. *Carb. Res.,* 339, 2117-2125 (2004).

[34] Zhang, X., Li, Y.-Z., Yan, H., Hu, J. and Chen, H.-L. A new β-CD inclusion complex with a seven-membered ring Co-CH_2-CH_2-CH_2-CH_2O-N bridged organocobaloxime. *In. Chem. Comm.*, 9, 907-910 (2006).

[35] Shen, X.-J., Chen, H.-L., Yu, Fei, Zgang, Y.-C., Yang, X.-H. and Li, Y.-Z. A new type of [2] and [3] pseudorotaxane composed of β-cyclodextrin and bisimidazolyl compounds. *Tetr. Letters*, 45, 6813-6817 (2004).

[36] Aree, T., Chaichit, N. and Engkakul, C. Polymorphism in β-Cyclodextrin-benzoic acid inclusion complex: a kinetically controlled crystal growth according to the Ostwald's rule. *Carb. Res.,* 343, 2451-2458 (2008).

[37] Liu, Y., Zhao, Y.-L., Zhang, H.-Y., Yang, E.-C. and Guan, X.-D. Binding ability and assembly behaviour of β-cyclodextrin complexes with 2,2'-dipyridine and 4.4'-dipyridine. *J. Org. Chem.*, 69, 3383-3390 (2004).

[38] Nakanishi, I., Arai, Fujiwara, T. and Tomita, K. X-ray structural studies on two forms of β-cyclodextrin-barbital complexes. *J. Incl. Phen.*, 2, 689-699 (1984).

[39] Fujiwara, T., Tomita, K. I., Marseigne, I. and Vicens, J. X-ray structural studies of some β-cyclodextrin inclusion complexes with stereoisomeric methyl-p-tolylsulfoxides, protective effect during irradiation in solid state. *Mol. Cryst. Liq. Cryst. Inc. Nonlin. Opt.,* 156, 393-404 (1988).

[40] Caira, M., Dodds, D. and Nassimbenni, L. R. Diverse modes of guest inclusion in a β-cyclodextrin: X-ray structural and thermal characterization of a 4:3 β-cyclodextrin-cyclizine complex. *Supramolecular Chem.*, 13, 61-70 (2001).

[41] Liu,Y., Zhong, R.-Q., Zhang, H.-Y. and Song, H.-B. A unique tetramer of 4:5 β-cyclodextrin-ferrocene in the solid state. *Chem. Comm.*, 2211-2213 (2005).

[42] Chatziefthimiou, S., Yannakopoulou, K. and Mavridis, I. M. β-Cyclodextrin trimers enclosing an unusual organization of the guest: The inclusion complex of β-cyclodextrin/4-pyridinealdazine. *Cryst. Eng. Comm.,* 9, 976-979 (2007).

[43] Bernstein, J., Davey, R. and Henck J.-O. Concomitant polymorphs. *Angew. Chem. Int. Ed.,* 38, 3440-3461 (1999).

[44] Caira, M. Crystalline polymorphism of organic compounds. *Top. Cur. Chem.*, 198, 163-208 (1998).

[45] Aakeröy, C. B. and Seddon, K. R. The hydrogen bond and crystal engineering. *Chem. Soc. Rev.*, 397-407 (1993).

[46] Kalman, A. Fabian, L. Argay, G., Bernath, G. and Gyarmati, A. Dipole-inclusion polymorphs of *trans*-2-hydroxycycloheptanecarboxylic acid with virtually the same unit cell. *J. Am. Chem. Soc.,* 125, 34-35 (2003).

[47] Stezowski, J., Parker, W., Hilgenkamp, S. and Gdaniec, M. Pseudopolymorphism in Tetradeca-2,6-O-methyl-*β*-cyclodextrin: The crystal structures for two new hydrates-Conformation variability in the alcylated *β*-cyclodextrin molecule. *J. Am. Chem. Soc.*, 123, 3919-3926 (2001).

[48] Ibragimov, B., Beketov, M., Weber, E., Seidel, J. Sumarna, O., Makhkamov, K. and Köhnke, K. Formation and polymorphic phase transitions of crystalline inclusion compounds between 2,2'-bis(9-hydroxy-9-fluorenyl)biphenyl host and acetone guest. *J. Phys. Org. Chem.*, 14, 697-703 (2001).

[49] Bernstein, J. ... And another comment on pseudopolymorphim. *Cryst. Growth and Des.*, 5, 5, 1661-1662 (2005).

[50] Stezowski, J., Jogun, K., Eckle, E. and Bartels, K. Dimeric *β*-cyclodextrin complexes may mimic membrane diffusion transport. *Nature*, 274, 617-619 (1978).

[51] Jogun, K. H. and Stezowski, J. J. Metastable crystals of *β*-cyclodextrin complexes and the membrane diffusion model. *Nature,* 278, 667-668 (1979).

[52] Tokuoka, R., Abe, M., Fujiwara, T., Tomita, K. and Saenger, W. Crystal structure of a *β*-cyclodextrin.ethanol.octahydrate. *Chem. Lett.*, 491-494 (1980).

[53] Steiner, T., Mason, S. and Saenger, W. Cooperative O.-H. O hydrogen bonds in *β*-cyclodextrin-ethanol-octahydrate at 15 K; A neutron diffraction study. *J. Am. Chem. Soc.*, 112, 6184-6190 (1990).

[54] Steiner, T., Mason, S. and Saenger, W. Disordered guest and water molecules. Three-center and Flip-Flop O.-H...O hydrogen bonds in crystalline *β*-cyclodextrin ethanol octahydrate at T = 295 K; A neutron and X-ray diffraction study. *J. Am. Chem. Soc.*, 113, 5676-5687 (1991).

[55] Giordano, F., Novak, C. and Ramon Moyano, J. Thermal analysis of cyclodextrins and their inclusion compounds. *Thermochimica Acta,* 380, 123-151 (2001).

[56] Tokuoka, R., Fujiwara, T. and Tomita, K.-I. Structure of the *β*-Cyclodextin (*β*-CD) Inclusion Complex with *p*-Ethylaniline (PEA). *Acta Cryst. B37,* 1158-1160 (1981).

[57] Hamilton, J. Structure of the inclusion complexes of cyclomaltoheotaose (cycloheptaamylose): crystal structure of the 1-adamantanemethanol adduct. *Carb. Res.,* 142, 21-37 (1985).

[58] Mavridis, I. M., Hadjoudis, E. and Tsoucaris, G. The crystal structure of the inclusion complex of cyclomaltoheptaose (β-cyclodextrin) with 3,3-dimethylbutylamine. *Carb. Res.,* 220, 11-21 (1991).

[59] Hamilton, J. and Sabesan, M. Structures of inclusion complexes of cyclomaltoheptaose (cycloheptaamylose): crystal structure of the benzocaine adduct. *Carb. Res.,* 102, 31-46 (1991).

[60] Caira, M. and Dodds, D. Inclusion of nonopiate analgetic drugs in cyclodextrins. I. X-ray structure of a 1:1 *β*-cyclodextrin-*p*-bromoacetanilide complex. *J. Incl. Phen.and Macroc. Chem.,* 34, 19-29 (1999).

[61] Kokkinou, A., Makedonopoulou, S. and Mentzafos, D. The crystal structure of the 1:1 complex of β-cyclodextrin with *trans*-cinnamic acid. *Carb. Res.*, 328, 135-140 (2000).

[62] Makedonopoulou, S. and Mavridou, I. M. Structure of the inclusion complex of *β*-cyclodextrin with 1,12-dodecanedioic acid using synchrotron radiation data; a detailed dimeric *β*-cyclodextrin structure. *Acta Cryst. B56,* 322-331 (2000).

[63] Paulidou, A., Maffeo, D., Yannakopoulou, K. and Mavridis, I. M. Crystal structure of the inclusion complex of the antibacterial agent triclosan with cyclomaltoheptaose and NMR study of its molecular encapsulation in positively and negatively charged cyclomaltoheptaose derivatives. *Carb. Res.*, 343, 2634-2640 (2008).

[64] Triantafyllopoulou, V., Tsorteki, F, Kokkinou, A., Mentzafos, D. and Bethanis, K. Study of the 2,4-diclorophenoxyacetic acid/*β*-cyclodextrin and 2,4,5-triclorophenoxyacetic acid/heptakis(2,3,6-tri-*O*-methyl)-*β*-cyclodextrin inclusion complexes by X-ray crystallography and molecular dynamics. In preparation.

[65] Nashioka, F., Nakanishi, I., Fujiwara, T. and Tomita, K. The crystal and molecular structure of the *β*-cyclodextrin inclusion complex with aspirin and salicylic acid. *J. Incl. Phen.*, 2, 701-714 (1984).

[66] Rontoyianni, A. and Mavridis, I. M. The crystal structure of the inclusion complex of cyclomaltoheptaose (*β*-cyclodextrin) with 3,5-dimethylbenzoic acid. *J. Incl. Phen. and Mol. Rec. Chem.*, 18, 211-227 (1994).

[67] Udachin, K. and Ripmeester, J. A novel mode of inclusion for pyrene in *β*-cyclodextrin compounds: the crystal structures of *β*-cyclodextrin with cyclohexanol and pyrene, and with *n*-octanol and pyrene. *J. Am. Chem. Soc.*, 120, 1080-1081 (1998).

[68] Liu, Y., Chen, G.-S., Zhang, H.-Y., Song, H.-B. and Ding, F. Interaction between β-cyclodextrin and 1,10-phenanthroline: uncommon 2:3 inclusion complex in the solid state. *Carb. Res.*, 339, 1649-1654 (2004).

[69] Takashima, Y., Oizumi, Y., Sakamoto, K., Miyauchi, M., Kamitiri, S. and Harada, A. Crystal structure of the complex of *β*-cyclodextrin with bithiophene and their oxidation polymerization in water. *Macromolecules,* 37, 3962-3964 (2004).

[70] Takashima, Y., Sakamoto, K., Oizumi, Y., Yamaguchi, H., Kamitori, S. and Harada, A. Complex formation of cyclodextrins with various thiophenes and their polymerisation in water: preparation of poly-pseudo-rotaxanes containing poly(thiophene)s. *J. Incl. Phen. and Macr. Chem.*, DOI 10.1007/s10847-006-9059-4 (2006).

[71] Fan, Z., Diao, C.-H., Yu, M., Jing, Z.-L., Chen, X. and Deng, Q.-L. An investigation of the inclusion complex of *β*-cyclodextrin with 8-nitro-quinoline in the solid state. *Supr. Molec. Chem.*, 18 (1), 7-11 (2006).

[72] Zhang, Y., Yu, S. and Bao, F. Crystal structure of cyclomaltoheptaose (*β*-cyclodextrin) complexes with *p*-aminobenzoic acid and *o*-aminobenzoic acid. *Carb. Res.*, 343, 2504-2508 (2008).

[73] Lien, N. and Telford, J. An investigation of the inclusion complex of cyclomaltoheptaose (β-cyclodextrin) with N-methylanthranilic acid in the solid state. *Carb. Res.,* 344, 2606-2608 (2009).

[74] Wang, E., Guangying, C. and Changri, H. Crystal structures of *β*-cyclodextrin inclusion complexes with 7-hydroxycoumarin and 4-hydroxycoumarin and substituent effects on inclusion geometry. *Chin. J. Chem.,* 29, 617-622 (2011).

[75] Wang, E., Chen, G.-Y. and Han, C. Crystal structures of a novel sandwich inclusion complex of *β*-cyclodextrin with *α*-naphthylacetic acid. *Chem. Res. Chinese Universities*, 27(5), 730-733 (2011).

[76] Bethanis, K., Tzamalis, P., Tsorteki, F., Kokkinou, A., Christoforides, E. and Mentzafos, D. Structural study of the inclusion compounds of thymol, carvacrol and eugenol in *β*-cyclodextrin by X-ray crystallography. *J. Incl. Phenom. Macrocycl. Chem.*, 77, 163-173 (2013).

[77] Hooben, F., Jonkheijm, P., Meijer, E. and Schenning, A. About supramolecular assemblies of π-conjugated systems. *Chem. Rev.,* 195, 1491-1546 (2005).

[78] Hunter C. and Sanders K. The nature of π-π interactions. *J. Am. Chem. Soc.,* 112, 5525-5534, (1990).

[79] S.-Q. Bai and Z.-Y. Fu. Pi stacking interaction in two TPT derivatives and their photoluminescence characters. *Chinese J. Struct, Chem.*, 25 (11), 1433-1438 (2009).

[80] Pickering, A. Seeber,G. Long. D. L. and Cronin, L. The importance of π-π, π-CH and N-CH interactions in the crystal packing of Schiff derivatives of *cis*, *cis*- and *cis*, *trans*-1,3,5-thiaminocyclohexane. *Cryst. Eng. Comm.,* 7, 504-510 (2005).

[81] Liu, Y., Zhao, Y.-L., Zhang, H.-Y. and Song, H.-B. Polymeric rotaxane constructed from the inclusion complex of 4,4'-dipyridine by coordination with Nickel(II) ions. *Angew. Chem. Int. Ed.*, 42, 3260-3263 (2003).

[82] Wang E. J., Yan, Z. and Cai, J. Crystal structure of a cyclomaltoheptaose-4-hydroxybiphenyl inclusion complex. *Carb. Res.,* 342, 1530-1534 (2007).

[83] Fan, Z., Diao, C.-H., Guo, M.-J., Du, R.-J., Song, Y.-F., Jing, Z. L. and Yu, M. An investigation of the inclusion complex of β-cyclodextrin with p-nitrobenzoic acid in solid state. *Carb. Res.*, 342, 2500-2503 (2007).

[84] Harding, M., MacLennsan, J. and Paton, M. Structure of the complex cycloheptaamylose-*p*-nitroacetanilide. *Nature,* 274, 621-623 (1978).

[85] Rashid, A., Kirschbaum, K. and Shoemaker, R. Origin of a second harmonic generation in a supramolecular inclusion complex between *p*-nitroaniline and β-cyclodextrin. *J. Mol. Struct.*, 785, 1-6 (2006).

[86] Yang, Z.-X., Chen, Y. and Liu, Y. Inclusion complexes of bipshenol A with cyclomaltoheptaose (β-cyclodextrin): solubilization and structure. *Carb. Res.,* 343, 2439-2442 (2008).

[87] Brett, T., Liu, S., Coppens, P. and Stezowski, J. The 20 K structure of *p*-amino-*p'*-nitrobiphenyl in the non-constraining environment of its β-cyclodextrin inclusion complex. *Chem. Commun.*, 551-552 (1999).

[88] Guo, M.-J., Diao, C.-H., Jing, Z. L., Dong, B., Fan, Z. and Wang, M. The structure of inclusion complex of *β*-cyclodextrin with *p*-nitrophenoxyacetic acid in solution and the solid state. *J. Incl. Phenom. Macrocycl. Chem.*, 67, 393-398 (2010).

[89] Wang, E., Chen, G. and Liu, H. Crystal structure of *β*-cyclodextrin-ferbinac inclusion complex. *Chin. Jour. Chem.,* 27, 2097-2101 (2009).

[90] Lippold, I., Görls, H. and Plass, W. New aspects for modeling supramolecular interactions in vanadium haloperoxidases: β-cyclodextrin inclusion compounds of *cis*-dioxovanadium(V) complexes. *Eur. J. Inorg. Chem.*, 1487-1491 (2007).

[91] Harata, K., Uekama, K., Otagiri, M. and Hirayama, F. Crystal structures of cyclodextrin complexes with chiral molecules. *J. Incl. Phen.*, 2, 583-594 (1984).

In: Cyclodextrins
Editor: Francis G. Ramirez

ISBN: 978-1-63482-788-1

Chapter 9

SALT EFFECTS ON THE INCLUSION COMPLEX FORMATION OF CYCLODEXTRINS WITH SOME BIOLOGICALLY ACTIVE AROMATIC CARBOXYLIC ACIDS

Irina V. Terekhova*
G. A. Krestov Institute of Solution Chemistry of
Russian Academy of Sciences, Ivanovo, Russian Federation

ABSTRACT

The growing interest in cyclodextrins is determined by their successful use as encapsulating materials and based on the ability of these natural cyclic oligosaccharides to form inclusion complexes. Application of inclusion compounds in medicine, cosmetology and food industry implies that they are affected by physiological medium of the living organisms, content of which is rather complicate. Inorganic salts are the main components of these biological liquids. As it is well known, salts are not always inert in the reaction medium, and equilibria occurring in saline solution are sensitive to the nature and concentration of the electrolytes. Therefore, the aim of this work was to reveal the influence of biologically active inorganic ions on complex formation of cyclodextrins with some aromatic carboxylic acids possessing pharmacological and biological activity.

Complex formation of α- and β-cyclodextrins with benzoic, nicotinic, p-aminobenzoic and m-aminobenzoic acids in water and saline solutions (KCl, KBr, KH_2PO_4 and K_2SO_4) was studied. Salt effects were discussed taking into account the possibility of nonspecific and specific interactions of the inorganic ions with cyclodextrins and acids. Possibility of ternary complex formation CD/acid/ion was also considered. Dependence of the manifestation of salt effects on the nature and properties of inorganic ions, the acid structure and size of cyclodextrin cavity was analyzed.

* E-mail: ivt@isc-ras.ru.

Numerous practical applications of cyclodextrins (CDs) are determined by their unique structure and ability to form inclusion complexes. As it is well known, CDs are the cyclic oligosaccharides obtained by the enzymatic disintegration of starch and composed of glucose units. Molecular structure of CDs is characterized by hydrophilic external surface and relatively hydrophobic internal cavity of varying size depending on the number of glucose units (6-8) in the ring. Different guest molecules can penetrate into CD cavity in accordance with the principle of geometric complementarity. Retention of the guest molecules inside the macrocyclic cavity is realized via noncovalent interactions. Complex formation of CDs with a wide variety of guest molecules in solution and solid state is intensively studied by means of different experimental techniques.

It has been demonstrated that inclusion complex formation can dramatically change the physicochemical properties of the guest compounds (e.g. solubility, stability, activity, etc.) [1, 2]. Due to this propensity CDs received increasing popularity in pharmaceutical science and drug design, where they are widely used as drug delivery systems and nontoxic encapsulating materials [3, 4].

To date, detailed information on structure of the complexes, thermodynamics and binding forces of complex formation of CDs with a number of pharmaceutically and biologically active substances has been obtained. To the best of our knowledge, complex formation in these systems has been investigated mainly in water or buffer solutions with physiological pH values. Unfortunately, the works devoted to study the behavior of pharmaceutically important inclusion compounds of CD in more complicated solvents modeling the internal environment of a living organism are very limited [5-10]. However, the knowledge of the influence of the internal environment is of practical importance since it allows to predict the behavior and activity of CD/drug inclusion complexes during their practical administration.

This work reveals the effects of biologically relevant inorganic ions on complex formation of CDs with aromatic carboxylic acids, some of which are used as vitamins and medicines. Nicotinic acid (NA) is well known as vitamin B_3 and pellagra-preventing factor. Benzoic acid (BA) is used for the treatment of fungal skin diseases as well as analgesic and antiseptic. p-Aminobenzoic acid (pABA) is a drug against fibrotic skin disorders and sunscreen agent. Only m-aminobenzoic acid (mABA) has no biological activity and it is considered herein as structural isomer of pABA.

Previously, complex formation of cyclodextrins with BA, NA, pABA and mABA in water has been studied in detail [11, 12]. Present research was undertaken to examine the influence of inorganic salts on the complexation process. For this purpose, investigation of binding affinity of α-CD and β-CD to aromatic carboxylic acids in 0.2M solutions of KCl, KH_2PO_4, KBr and K_2SO_4 was carried out using ^{1}H NMR spectroscopy. It should be noted that the inorganic salts under consideration have a common cation (K^+) and different anions. Specific interactions of these ions with CDs and acids can take place and affect the equilibrium of the inclusion complex formation. Therefore, possibility of these interactions should be taken into account.

INTERACTIONS OF INORGANIC IONS WITH CYCLODEXTRINS

In the literature there are some works [13-28], in which the interactions of inorganic ions with CDs have been described. Complexation in this case is theoretically possible since the CD molecule is polar and ion-dipole interactions can occur [13]. However, there is selectivity in the binding of CDs with ions. As follows from the CD structure, macrocyclic cavity mainly consists of carbon atoms which are positively polarized. As a result, the penetration of anions in the macrocyclic cavity is favorable, whereas the inclusion of cations becomes unfavorable [14].

Interactions of CDs with cations have been studied by computational methods [15-17]. In particular, on the basis of molecular modeling results it has been shown that among alkali metal cations only Na^+ can form complex with β-CD in the gas phase [15]. However, coordination of inorganic cations with the outer surface of CD is not excluded. In this case, alkali metal cations are able to bind with the oxygen atoms of the hydroxyl groups surrounding CD cavity [16, 17]. Thus, binding of K^+ with CDs is unlikely.

In contrast to the cations, inorganic anions can be inserted into CD cavity. It was established that anions are bound to CDs by van der Waals, hydrophobic and ion-dipole interactions [13, 14, 18-20]. Binding mode of CDs with inorganic anions was determined using ^1H NMR spectroscopy [18, 19]. It has been shown that the interactions of Cl^-, Br^-, I^-, SCN^-, ClO_4^- and NO_3^- with α-, β- and γ-cyclodextrins are accompanied by downfield shift of H-5 protons located in the macrocyclic cavity. This fact proves that anions are placed inside the CD cavity at the narrow rim of the molecule. On the contrary, anions $H_2PO_4^-$, HCO_3^- and SO_4^{2-} did not induce significant chemical shift change of CD protons and do not form complexes with CDs.

Table 1. Stability constants of CD/anion complexes

Complex	K, M^{-1}	Ref.
α-CD/Br^-	1.6 ± 0.1	[18]
	0.87	[21]
	3.5	[22]
	1	[23]
	2.3 ± 0.8	[24]
α-CD/Cl^-	n/c	[25]
	<<1	[23]
	3 (NaCl)	[26]
	5 (KCl)	[26]
α-CD/SO_4^{2-}	1	[23]
β-CD/Br^-	6.5	[25]
	0.81	[27]
β-CD/Cl^-	0.46	[27]
	2.56	[25]
	1.3 ± 0.4	[14]
β-CD/SO_4^{2-}	< 0	[14]
	< 0	[28]

Stability constants of CD complexes with anions summarized in Table 1 have been determined by different experimental techniques. Analysis of K values reported in Table 1 shows that the CD complexes with anions are weak, and in some cases complexation is absent. Thermodynamic parameters of complex formation were derived only for binding of α-CD with Br^- (Table 2). Complexation is characterized by the negative enthalpy and entropy changes, i.e., complexes are only enthalpy stabilized.

Table 2. Thermodynamics of α-CD complex formation with Br^- in water

$\Delta_c G$, kJ/mol	$\Delta_c H$, kJ/mol	$T\Delta_c S$, kJ/mol	Ref.
-1.2	-15.7 ± 1.0	-14.5 ± 0.9	[18]
-2.4	-6 ± 1	-3.6 ± 1.5	[24]

Stability of the CD/anion complexes is determined by the structure and physicochemical properties of the anions, such as the radius, charge density distribution, hydration energy and others. In some papers [18, 24] devoted to calorimetric and ^{1}H NMR study of CD binding with a number of inorganic salts it has been demonstrated that the larger the radius of the anion, the higher the complex stability. Small anions are not retained in the CD cavity, so less stable complexes are formed in this case. For instance, inclusion of more bulky and more hydrophobic ClO_4^- into β-CD cavity is preferable in comparison with Cl^- [13]. Spencer et al. [24] revealed a quantitative relationship, according to which the ratio of the anion radius (r) to the α-CD cavity radius (r_{cav}) determines the thermodynamic parameters of complex formation. Higher stability constants and more negative $\Delta_c H$ and $\Delta_c S$ values are observed when the ratio r/r_{cav} is close to unity.

When considering the complexation of CDs with anions the effects of counter ions should be taken into account. Electrolyte exists in solution as an ion pair. It was proposed that when the anion is included in the CD cavity, the cation follows it due to the electrostatic interactions. Sanemasa et al. [26] investigated the formation of inclusion complexes of α-CD with salts having the same anion and different cations (e.g., KCl and NaCl; $LiClO_4$, $KClO_4$ and $NaClO_4$). It has been observed that the effect of the cations as counter ions manifested in the following order $K^+ < Na^+ < Li^+$. Thus, influence of K^+ on the binding is less pronounced.

Inspection of the data presented in Table 1 allows to note that results obtained for binding of CDs with Cl^- are contradictory. Rohrbach et al. [25] as well as Hersey et al. [23] demonstrated an absence of complex formation between CD and Cl^-. On the contrary, $K > 1$ has been reported by Sanemasa et al. [26]. It should be mentioned that weak complexes with $\lg K \leq 1$ is quite difficult to detect, because the accuracy of the method used, the experimental conditions (temperature, solvent properties, concentration range, etc.) and data processing may significantly affect the results. Therefore, to give more insight in binding of CDs with inorganic anions we investigated these systems using ^{1}H NMR spectroscopy under the same experimental conditions in which the further study of complexation of CDs with aromatic carboxylic acids in salt solutions was performed.

To this purpose, ^{1}H NMR spectra of α-CD and β-CD were recorded in deuterated water in the presence of variable amounts of KCl, KBr, KH_2PO_4 and K_2SO_4. The ^{1}H NMR spectrum of CDs consists of the signals from external protons H-1, H-2, H-4 and H-6 located outside of the macrocyclic cavity, and the internal protons H-3 and H-5 placed inside the cavity at the wide and narrow rims of CD molecules, respectively Assignment of the signals in the ^1H NMR spectrum has been done on the basis of available in literature data [29]. Figure 1 shows dependences of the experimentally obtained chemical shift changes ($\Delta\delta$) of CD protons on the salt concentration. As can be seen from Fig. 1, addition of an excess amount of salt causes significant changes of chemical shifts of all CD protons. Probably, high values of $\Delta\delta$ are caused by the electric field generated by electrolytes. With the exception of KBr, binding isotherms are linear, indicating that the interactions of CDs with Cl^-, $H_2PO_4^-$ and SO_4^{2-} are weak. However, the attention should be drawn to the following fact. Chemical shift changes of H-5 protons of α-CD and β-CD are slightly higher in the presence of KCl compared to KH_2PO_4 and K_2SO_4. This result indicates that Cl^- penetrates into the CD cavity but not retained therein.

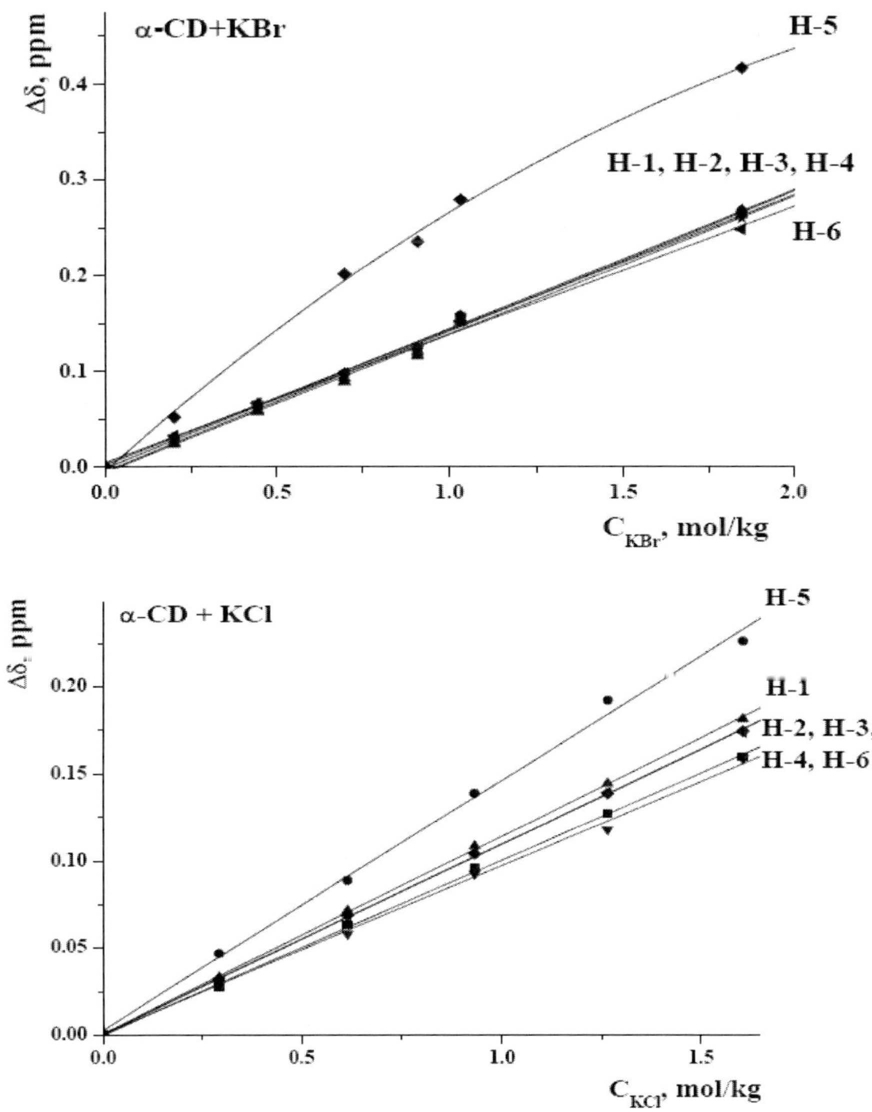

Figure 1. (Continued).

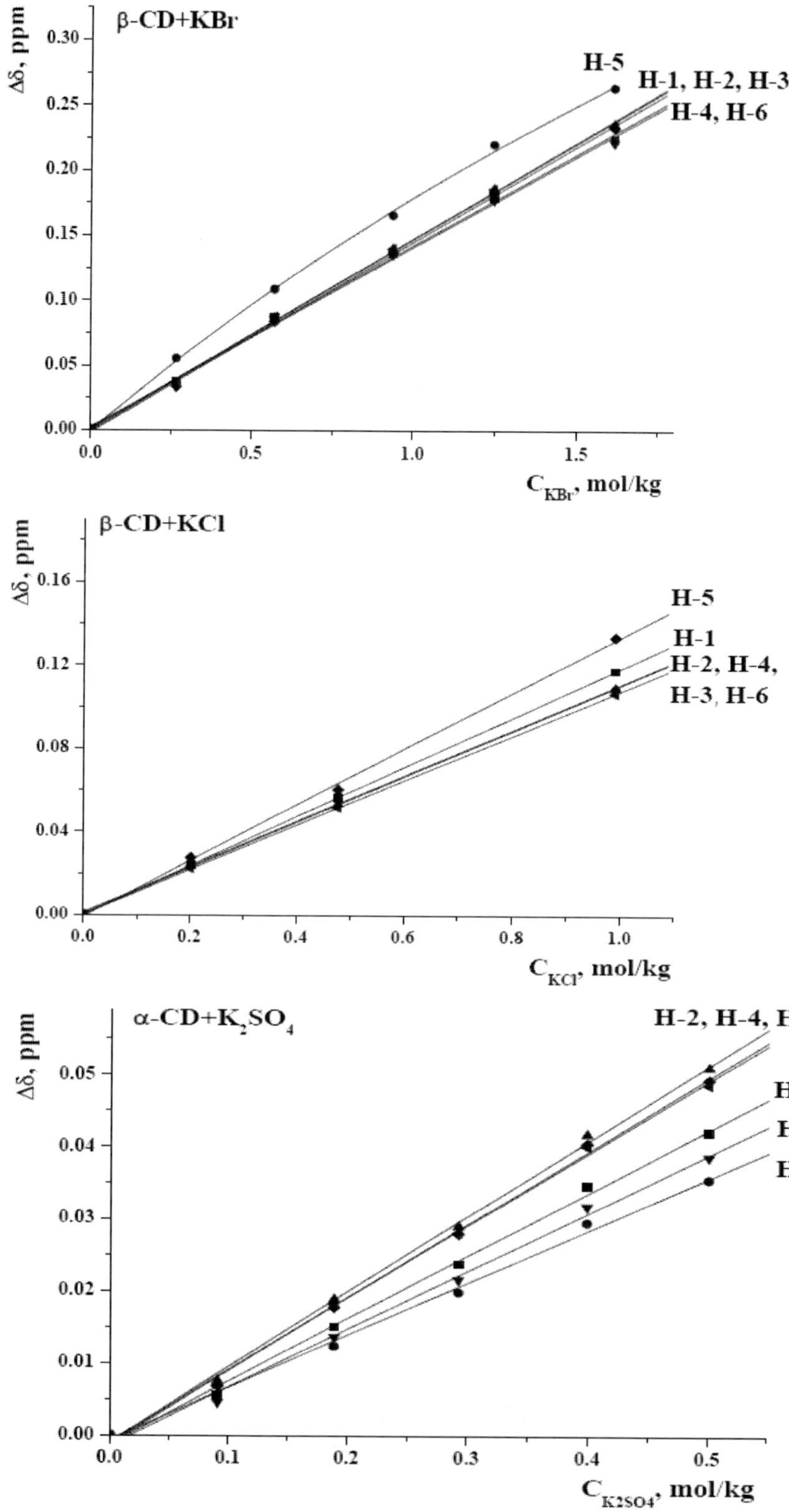

Figure 1. (Continued).

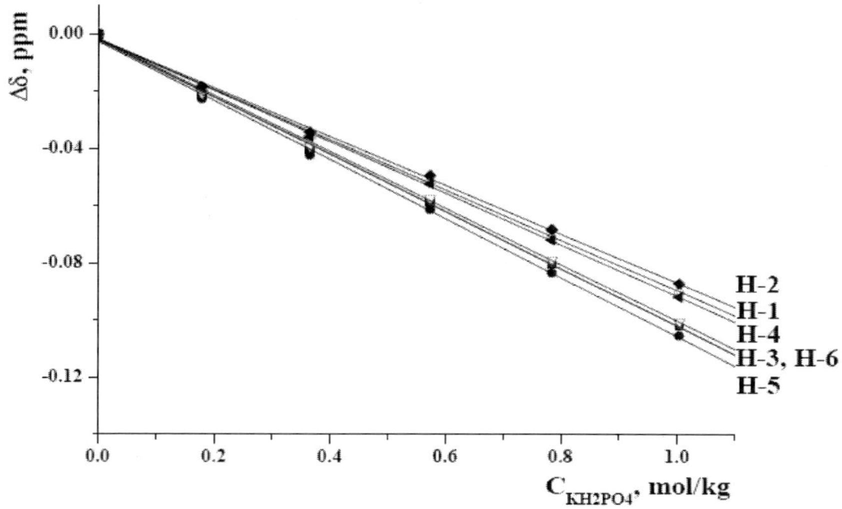

Figure 1. Dependences of chemical shift changes of CD protons on salt concentration (T=298.15 K).

Deviation of the concentration dependences from the linearity, which was observed only for H-5 CD protons in the presence of KBr (Fig. 1), is caused by the insertion of Br^- into macrocyclic cavity. In accordance with ^1H NMR results, Br^- is located near the narrow rim of the CD molecule. This is in agreement with the binding mode proposed in [18]. Stability constants of α-CD/Br^- and β-CD/Br^- complexes calculated on the basis of the concentration dependences shown in Fig. 1 are equal to 0.24±0.05 and 0.17±0.05 kg/mol, respectively. As one can see, complexes of CDs with Br^- are weak. The values of *K* are very close to *K* obtained by Gelb et al. [21] and Lima et al. [27] and presented in Table 1.

The revealed difference in the binding affinity of CDs to the inorganic salts under study is caused by the properties of the anions. Ionic radius and enthalpy of hydration of the considered anions are given in Table 3. As follows from Table 3, Br^- and Cl^- are less hydrated in comparison with $H_2PO_4^-$ and SO_4^{2-} and, therefore, their ability to complex formation with CDs is higher. On the other hand, Br^- is more bulky compared to Cl^-. Consequently, retention of Br^- inside the macrocyclic cavity via van der Waals forces is more favorable. Increase in the macrocyclic cavity size results in the weakening van der Wals interactions and decreasing of β-CD affinity to Br^-.

Table 3. Ionic radius and enthalpy of hydration of the anions [30]

Ion	*r*, nm	$\Delta_{hydr}H$, kJ/mol
Cl^-	0.181	-367
Br^-	0.196	-336
$H_2PO_4^-$	0.200	-522
SO_4^{2-}	0.230	-1138

Thus, it was shown that among the considered ions only Br^- is capable to inclusion complex formation with α- and β-cyclodextrins in aqueous solution. Complexation in this case takes place in accordance with the principles of geometric and energetic complementarity.

Complex Formation of α- and β-Cyclodextrins with Aromatic Carboxylic Acids in the Presence of Inorganic Salts

Complex formation of α-CD and β-CD with BA, NA, pABA and mABA in water and 0.2M solutions of KCl, KBr, KH_2PO_4 and K_2SO_4 was studied by ^1H NMR. Previously, binding mode of CD complexation with these acids in water has been determined [11, 12, 31]. It was shown that insertion of the aromatic carboxylic acids into smaller cavity of α-CD is shallow. In this case, carboxylic group is located near H-5 protons, while a part of the aromatic ring is placed near H-3 protons (Fig. 2). Signals of CD protons of H-3 and H-5 were shifted upfield and downfield, respectively (Table 4). Deeper inclusion of the acids into β-CD is accompanied by the positioning of the whole aromatic ring inside the macrocyclic cavity (Fig. 2). This binding mode is characterized by the upfield shift of H-3 and H-5 protons (Table 4).

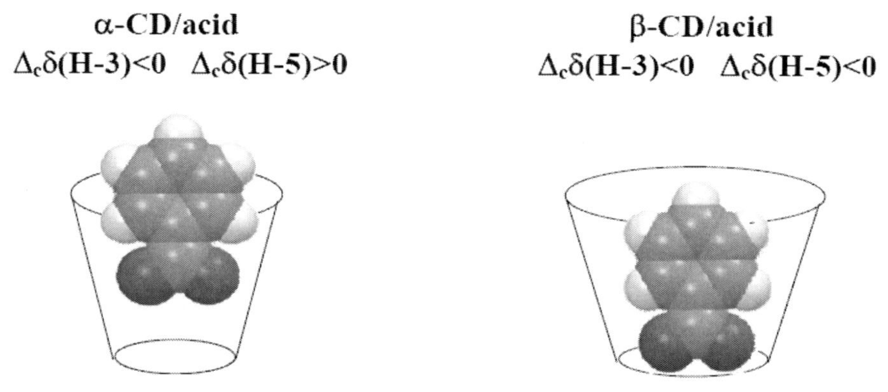

Figure 2. Proposed structures of CD complexes with aromatic carboxylic acids in water.

It was found that binding mode of both CDs with the acids is not changed in the presence of the salts under study. Table 4 lists chemical shift changes of all CD protons induced by 100% complex formation with the acids in water and 0.2M saline solutions. As one can see, there is no significant difference between $\Delta_c\delta$ values. Thus, inorganic salts have no influence on the binding mode.

Stability constants of the CD/acid inclusion complexes were calculated from the binding isotherms obtained by means of ^1H NMR (Fig. 3). Values of K are given in Table 5. Comparison of the binding constants shows a decrease of complex stability in salt solutions. To explain this fact, specific and nonspecific salt effects should be taken into account.

Table 4. Chemical shift changes of CD protons induced by complexation with the acids in water and saline solutions (0.2M) at 298.15 K

Complex	Exp. conditions	$\Delta_c\delta$, ppm H-1	H-2	H-3	H-4	H-5	H-6
α-CD/BA	H_2O	-0.07	-0.08	-0.37	-0.02	0.13	~0
	KBr	-0.06	-0.07	-0.39	-0.02	0.13	-0.03
	KCl	-0.06	-0.06	-0.39	-0.02	0.10	-0.03
	K_2SO_4	-0.06	-0.06	-0.38	-0.03	0.14	-0.03
	KH_2PO_4	-0.06	-0.07	-0.38	-0.03	0.13	-0.03
α-CD/pABA	H_2O	-0.06	-0.06	-0.31	-0.02	0.13	-0.02
	KBr	-0.05	-0.06	-0.30	-0.01	0.07	-0.05
	KCl	-0.06	-0.06	-0.30	-0.02	0.12	-0.03
	K_2SO_4	-0.06	-0.06	-0.32	-0.02	0.13	-0.04
	KH_2PO_4	-0.06	-0.06	-0.31	-0.02	0.13	-0.03
α-CD/mABA	H_2O	-0.08	-0.08	-0.31	-0.04	0.16	-0.05
	KBr	-0.07	-0.08	-0.31	-0.06	0.16	-0.06
	KCl	-0.07	-0.08	-0.31	-0.05	0.15	-0.07
	K_2SO_4	-0.07	-0.07	-0.32	-0.04	0.19	-0.03
	KH_2PO_4	-0.05	-0.08	-0.32	-0.04	0.14	-0.06
α-CD/NA	H_2O	-0.08	-0.07	-0.46	-0.02	0.10	-0.04
	KBr	-0.08	-0.08	-0.49	-0.05	0.05	-0.09
	KCl	-0.07	-0.07	-0.47	-0.02	0.05	-0.04
	K_2SO_4	-0.09	-0.09	-0.48	-0.04	0.06	-0.05
	KH_2PO_4	-0.09	-0.09	-0.47	-0.02	0.06	-0.04
β-CD/BA	H_2O	-0.05	-0.05	-0.18	-0.03	-0.24	-0.05
	KBr	-0.05	-0.04	-0.18	-0.03	-0.23	-0.04
	KCl	-0.05	-0.05	-0.17	-0.03	-0.23	-0.04
	K_2SO_4	-0.05	-0.05	-0.18	-0.03	-0.23	-0.05
	KH_2PO_4	-0.05	-0.05	-0.16	-0.03	-0.21	-0.05
β-CD/pABA	H_2O	-0.04	-0.04	-0.18	-0.02	-0.23	-0.06
	KBr	-0.04	-0.03	-0.15	-0.03	-0.23	-0.06
	KCl	-0.04	-0.03	-0.15	-0.03	-0.23	-0.06
	K_2SO_4	-0.04	-0.03	-0.15	-0.03	-0.23	-0.06
	KH_2PO_4	-0.04	-0.03	-0.15	-0.02	-0.23	-0.06
β-CD/mABA	H_2O	-0.07	-0.06	-0.18	-0.07	-0.20	-0.07
	KBr	-0.06	-0.05	-0.19	-0.04	-0.26	-0.07
	KCl	-0.06	-0.05	-0.19	-0.04	-0.21	-0.06
	K_2SO_4	-0.06	-0.06	-0.19	-0.04	-0.21	-0.07
	KH_2PO_4	-0.06	-0.06	-0.19	-0.05	-0.21	-0.07
β-CD/NA	H_2O	-0.08	-0.08	-0.25	-0.08	-0.26	-0.06
	KBr	-0.08	-0.08	-0.29	-0.07	-0.31	-0.08
	KCl	-0.09	-0.09	-0.29	-0.08	-0.29	-0.08
	K_2SO_4	-0.09	-0.09	-0.30	-0.08	-0.31	-0.09
	KH_2PO_4	-0.08	-0.08	-0.27	-0.07	-0.28	-0.07

Table 5. Stability constants of CD/acid complexes in water and salt solutions at 298.15 K

Complex	Exp. conditions	*K*, kg/mol
α-CD/NA	H_2O	27.0 ± 0.4
	0.2M KCl	26.1 ± 0.6
	0.6M KCl	23.1 ± 0.5
	0.2M KBr	21.1 ± 0.7
	0.2M KH_2PO_4	23.8 ± 0.3
	0.2M K_2SO_4	23.6 ± 0.2
α-CD/pABA	H_2O	1259 ± 42
	0.2M KCl	1230 ± 70
	0.2M KBr	1058 ± 29
	0.2M KH_2PO_4	1163 ± 63
	0.2M K_2SO_4	1131 ± 95
α-CD/mABA	H_2O	59 ± 1
	0.2M KCl	55 ± 2
	0.6M KCl	53 ± 1
	0.2M KBr	47 ± 3
	0.2M KH_2PO_4	54 ± 1
	0.2M K_2SO_4	53 ± 1
α-CD/BA	H_2O	897 ± 20
	0.2M KCl	862 ± 20
	0.6M KCl	827 ± 15
	0.2M KBr	773 ± 19
	0.2M KH_2PO_4	599 ± 18
	0.2M K_2SO_4	817 ± 18
β-CD/NA	H_2O	8.0 ± 0.5
	0.2M KCl	7.0 ± 0.1
	0.2 KBr	5.3 ± 0.3
	0.2M KH_2PO_4	7.3 ± 0.5
	0.2M K_2SO_4	5.7 ± 0.3
β-CD/pABA	H_2O	721 ± 24
	0.2M KCl	731 ± 33
	0.2M KBr	532 ± 35
	0.2M KH_2PO_4	703 ± 22
	0.2M K_2SO_4	743 ± 26
β-CD/mABA	H_2O	61 ± 1
	0.2M KCl	54 ± 3
	0.2MKBr	54 ± 2
	0.2M KH_2PO_4	58 ± 1
	0.2M K_2SO_4	53 ± 2
β-CD/BA	H_2O	390 ± 25
	0.2M KCl	403 ± 26
	0.2M KBr	331 ± 22
	0.2M KH_2PO_4	636 ± 17
	0.2M K_2SO_4	372 ± 28

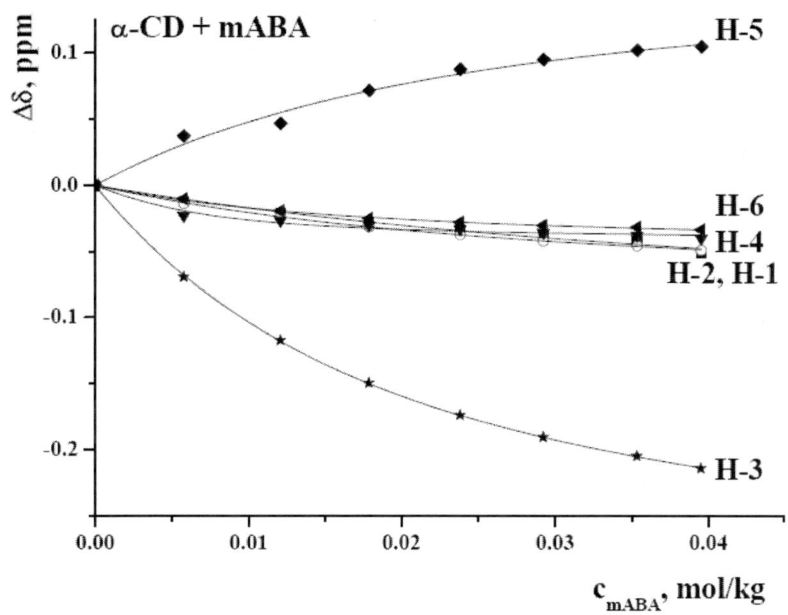

Figure 3. Dependences of α-CD (0.005 mol/kg) chemical shift changes on mABA concentration (T=298.15 K).

As it is well known, manifestation of nonspecific salt effects is reflected in the change of the ionic strength of the solution. From this point of view, the influence of divalent anions (SO_4^{2-}) creating stronger electric field should be more pronounced compared to monovalent anions (Cl^-, Br^- and $H_2PO_4^-$) of the same concentration. In this case, more significant change of the binding constants in the presence of 0.2M K_2SO_4 should be expected. To confirm this assumption the dependences of lg*K* on the ionic strength (*I*) were plotted (Fig. 4). Additional points corresponding to the inclusion complex formation in 0.6M KCl solution were also used on these dependences with the purpose to show the nonspecific action of K_2SO_4. KCl is indifferent electrolyte since it does not interact with CDs and acids. As it was mentioned above, K^+ cations do not bind to CDs [15-17], whereas attraction of K^+ to –COO^- group of the acid and formation of contact ion pairs can be assumed. There are some publications in the literature [32-35], in accordance with which the affinity of K^+ to carboxylate anion is very low. In particular, ion-specific effects of K^+ towards the carboxylic groups of isolated amino acids and short peptides were quantified by means of molecular dynamics simulations and conductivity measurements [32]. The relative interaction strength of the monovalent cations with acetate and formate anions was examined by Uejio et al. [33]. Fedorov et al. [34] studied the effect of potassium cations on α-poly-L-glutamate conformation in aqueous solution. It has been demonstrated in all these works that attraction of K^+ ions to the carboxylates is very weak. Recently, we also showed that potassium ions have much less potency to make direct contact pairs with the carboxylic group of BA and NA [35]. Thus, one can say that strong affinity of K^+ to carboxylic group of BA, NA, pABA and mABA is practically absent.

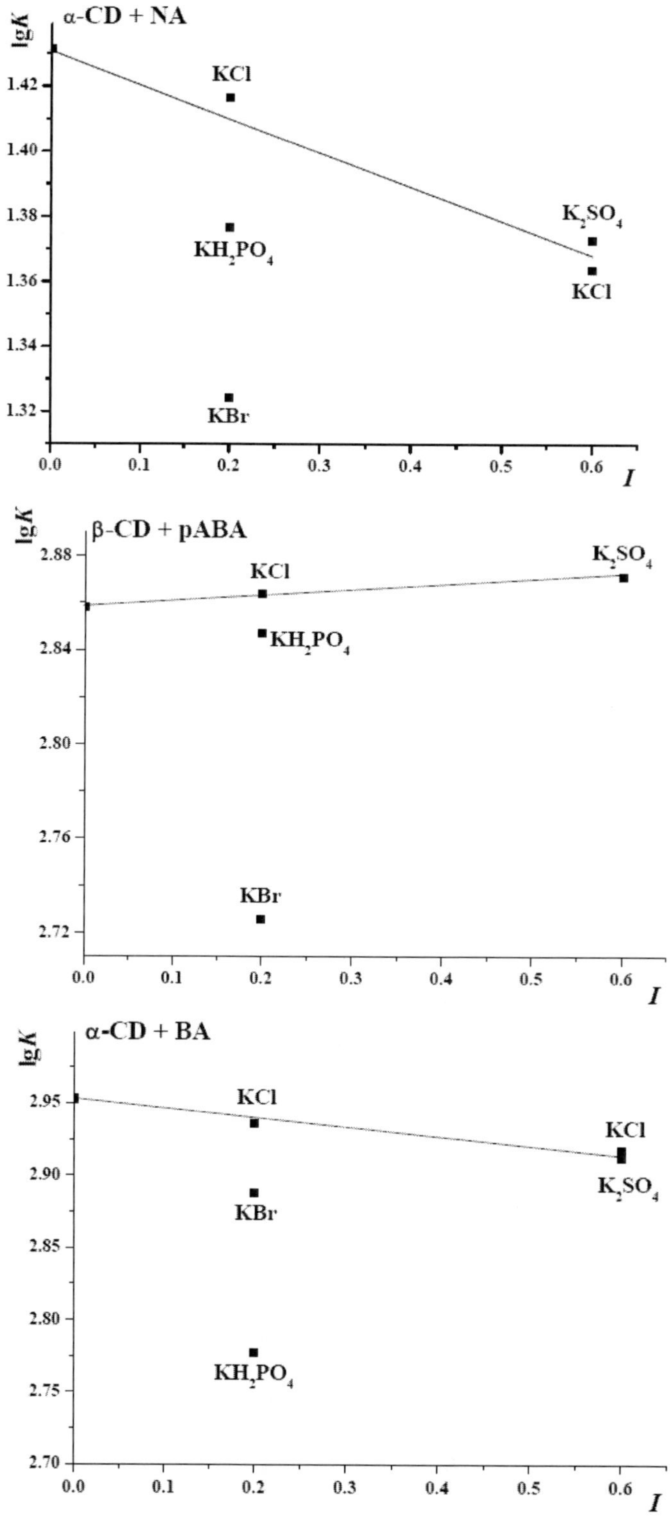

Figure 4. lg*K* of CD inclusion complexes with acids *versus* the ionic strength (T=298.15 K).

It is not difficult to see from Fig. 4 that values of lg*K* decrease linearly with increasing *I*. The points corresponding to 0.6M KCl and 0.2M K_2SO_4, which create the same ionic strength, agree. Thus, the obtained results testified that KCl and K_2SO_4 display nonspecific action on the inclusion complex formation of CDs with aromatic carboxylic acids. On the other hand, points corresponding to the systems with KBr and KH_2PO_4 do not belong to these dependences. Consequently, effects of KBr and KH_2PO_4 are specific due to propensity of these anions to interact with CDs or/and acids. The presence of these salts in solution results in more noticeable change of the binding constants (Table 5).

It was established using ^1H NMR that among anions under study only Br^- penetrates the cavity of α-CD and β-CD. As the result, there are two equilibria which take place in solution containing CD, acid and KBr:

$$CD + acid = CD \cdot acid \quad (1)$$

$$CD + Br^- = CD \cdot Br^- \quad (2)$$

Processes (1) and (2) are competing and their coexistence leads to a weakening of the binding of α-CD and β-CD with the acids. As can be seen from Table 5, stability constants are lower in the presence of KBr. Competition between inorganic anion and guest molecule for the CD cavity was also observed in the binding of 3-chloroperbenzoic acid with α-CD in solution of $NaClO_4$ [36] as well as in complex formation of β-CD and HP-β-CD with bile acids in the presence of NaJ и NaSCN [5]. In all these systems insertion of inorganic anions into macrocyclic cavity was unfavorable for inclusion complex formation of CDs with the acids.

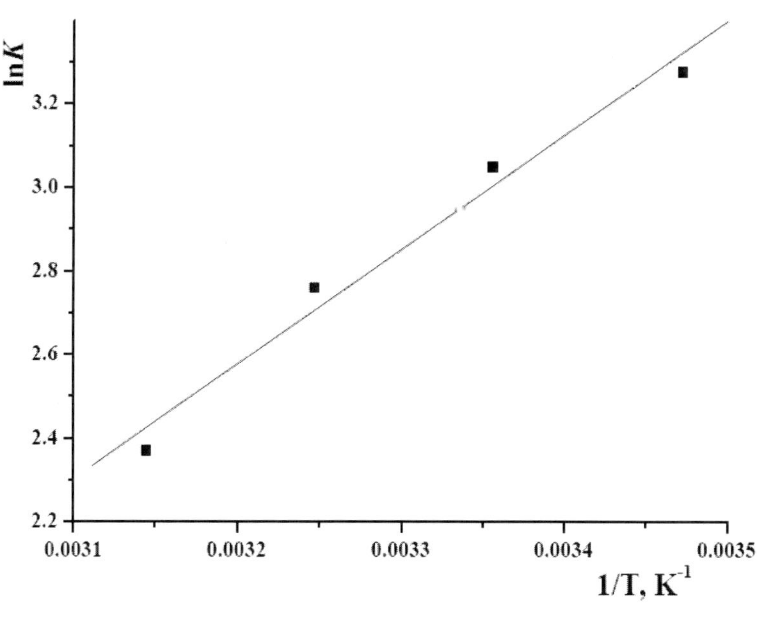

Figure 5. ln*K* *versus* 1/T (example is given for complex formation of α-CD witn NA in KBr solution).

Let us consider and compare the enthalpy and entropy changes characterizing the inclusion complex formation of CDs with aromatic carboxylic acids in water and aqueous solutions of KCl and KBr. These thermodynamic parameters were obtained by performing the ^{1}H NMR experiments at different temperatures (293.15 – 318.15 K). Values of $\Delta_c H$ and $\Delta_c S$ calculated from the temperature dependence of ln$K$ (Fig. 5) are summarized in Table 6. Reliability of the reported data is evident from the comparison of $\Delta_c H$ and $\Delta_c S$ derived using van´t Hoff approach and direct calorimetric measurements [11, 12]. One can see a good agreement between these values.

Table 6. Thermodynamic parameters of complex formation of CDs with aromatic carboxylic acids in water and salt solutions (0.2M) at 298.15 K

Complex	$\Delta_c G$	$\Delta_c H$	$T\Delta_c S$
	kJ/mol		
α-CD/NA (H_2O)	-8.2 ± 1.2	-25.2 ± 0.3	-17.0 ± 1.2
	-8.7*	-26.4 ± 0.3*	-17.7*
α-CD/NA (KCl)	-8.4 ± 1.4	-24.0 ± 0.3	-15.6 ± 1.4
α-CD/NA (KBr)	-7.6 ± 0.3	-23.0 ± 1.1	-15.4 ± 1.0
α-CD/mABA (H_2O)	-10.2 ± 0.7	-20.9 ± 0.7	-10.7 ± 1.0
α-CD/mABA (KCl)	-10.3 ± 0.7	-22.5 ± 0.9	-12.2 ± 1.1
α-CD/pABA (H_2O)	-17.7 ± 0.3	-39.0 ± 2.0	-21.3 ± 2.0
	-18.3*	-41.0 ± 0.4*	-22.7*
α-CD/pABA (KBr)	-17.3 ± 1.1	-36.2 ± 1.1	-18.9 ± 1.5
β-CD/pABA (H_2O)	-16.3 ± 1.6	-19.3 ± 0.6	-3.0 ± 1.7
β-CD/pABA (KCl)	-16.3 ± 0.7	-18.2 ± 1.7	-1.9 ± 2.1
β-CD/pABA (KBr)	-15.6 ± 1.0	-15.3 ± 0.9	0.3 ± 1.4

* - from calorimetric experiments [11, 12]

As follows from Table 6, binding of CDs with the acids in water is accompanied by the negative enthalpy and entropy changes. Influence of KCl on the thermodynamics of complex formation is negligible, while the effect of KBr is more pronounced. The enthalpy and entropy changes of complex formation were slightly increased in the presence of KBr. In this case, dehydration is the main factor determining the increase of thermodynamic parameters. Dehydration of CDs during complex formation with acids in the presence of KBr is more intensive in comparison with complexation in pure water, since CD interacts with both Br^- and acid. Thus, enhancement of $\Delta_c H$ is the result of the increased positive contribution from dehydration of the species participating in the inclusion complex formation. The important role of solvent reorganization is confirmed by the increased entropy changes observed for complexation in KBr solutions (Table 6). Occupation of CD cavity by acid and Br^- results in the release of larger amount of water molecules from the macrocyclic cavity to the bulk solvent. Moreover, solvent reorganization occurs due to the partial disruption of the hydration shells of CD, acid and Br^- during complex formation. All these processes give rise to the entropy gain.

Analysis of the obtained results (Table 5) shows that manifestation of the Br^- effect is not dependent on the CD cavity size. Comparison of K values (Table 5) showed that decrease in K caused by Br^- was approximately the same for α-CD and β-CD complexes. This is due to the identical affinity of α-CD and β-CD to Br^- anions. On the other hand, the acid structure influences the manifestation of Br^- effect. In particular, the observed decrease in the binding constants in KBr solution was more pronounced for complexes of CDs with NA and mABA (≥20%), while it does not exceed 15% for complexes with BA and pABA (Table 5). The revealed difference is caused by different ionization state of the acids in aqueous solution.

Considering in this work BA, NA, pABA and mABA are weak aromatic carboxylic acids. It is well known from the literature that BA and pABA exist in aqueous solution as neutral molecules [37, 38], whereas NA and mABA are zwitterions [39, 40]. Bipolar nature of NA and mABA implies that carboxylic group is dissociated and amino group is protonated in aqueous solution. In this case, electrostatic attraction of inorganic ions to the ionized groups of NA and mABA can be expected. Therefore, possibility of these interactions should be examined. To this end, some additional experiments were carried out. To reveal the possibility of attraction of Br^- and Cl^- to the positively charged groups of NA ($-NH^+$) and mABA ($-NH_3^+$), the UV-spectra of NA and mABA were recorded in the presence of the excess amounts of KBr and KCl. Absorbance change without displacement of the absorption maxima was observed under salt addition. Fig. 6 depicts the concentration dependences of the absorbance. In all these systems, absorbance is changed linearly with an increase of salt concentration. The linearity of the obtained dependences points out the weak interactions, which are not accompanied by complex formation.

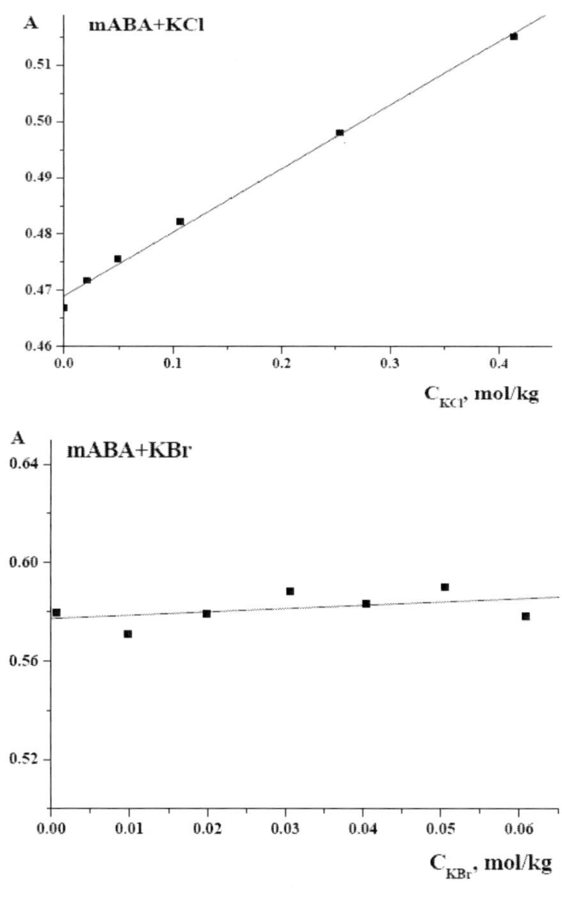

Figure 6. (Continued).

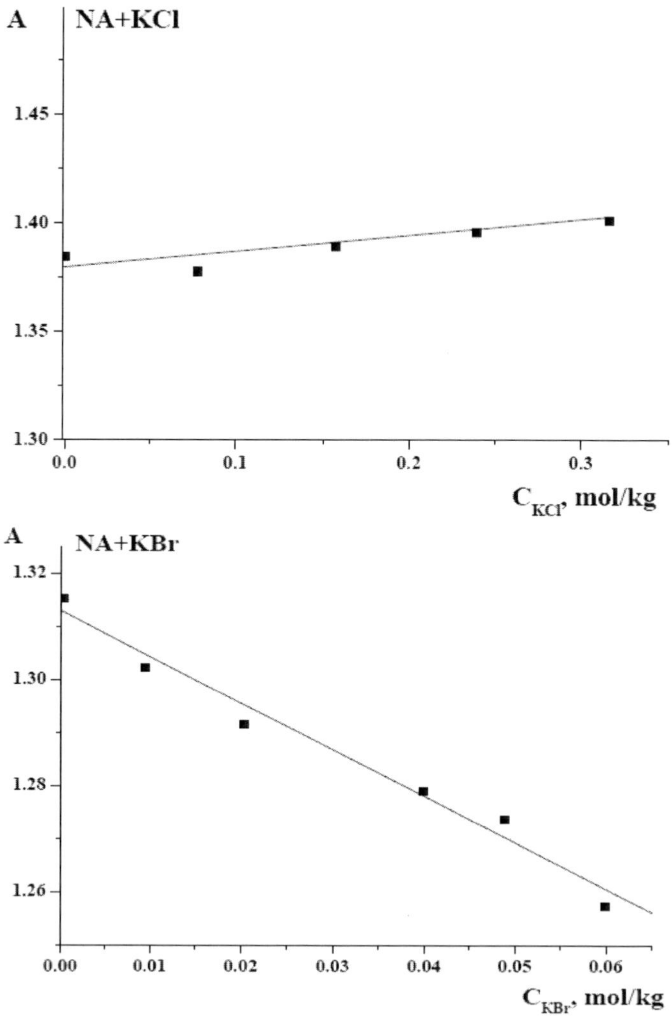

Figure 6. Dependences of the absorbance of aqueous solutions of mABA (0.0007 mol/kg) and NA (0.0003 mol/kg) *versus* salt concentration.

Besides the complex formation, contact ionic pairs between inorganic anion and zwitterionic acid can be formed. We have attempted to study this process by means of solubility method. For this purpose, solubility of NA and mABA was determined in water and solutions of KBr, KCl, KH_2PO_4 and K_2SO_4. It was demonstrated that all these salts act as salting-in agents. More probably, salting-in action is caused by the formation of ionic pairs between inorganic anions and zwitterions of the acids. The following trends of the solubilizing action of the salts were found

for NA $SO_4^{2-} \approx H_2PO_4^- > Cl^- > Br^-$
for mABA $SO_4^{2-} > H_2PO_4^- > Br^- > Cl^-$

In both cases, salting-in effect of Br^- as well as Cl^- can be explained by the attraction of these anions to the zwitterions. Thus, effect of Br^- in complex formation of CDs with NA and mABA is due to inclusion of Br^- into macrocyclic cavity and ion pairing of Br^- with the zwitterions.

Compared to Br^- and Cl^-, anions SO_4^{2-} and $H_2PO_4^-$ display more pronounced salting-in effects towards the NA and mABA. Divalent anions SO_4^{2-} have a high charge density distribution and, therefore, their stronger attraction to the zwitterions is evident. The behavior of $H_2PO_4^-$ is ambiguous. On the one hand, $H_2PO_4^-$ can form ionic pairs or H-bonds with NA and mABA. On the other hand, salting-in effect of $H_2PO_4^-$ may be due to its possible participation in the acid-base equilibria. Influence of $H_2PO_4^-$ on the distribution of different ionized forms of NA and mABA was demonstrated by means of pH measuring of the solutions. In particular, pH of the aqueous solution of NA (0.01M) was increased from 3.6 to 4.0 under addition of 0.2M KH_2PO_4. The same situation was observed for the aqueous solution of mABA (0.04 M). In this case, pH was changed from 3.9 to 4.1 in the presence of 0.2M KH_2PO_4. Thus, KH_2PO_4 shifts the equilibrium *zwitterion* ↔ *anion* to the right. Anions of NA and mABA are strongly hydrated compared to zwitterions [41] and, therefore, they have higher solubility in water and lower affinity to CDs. It has been shown in several works [11, 12] that binding constants of CDs with anionic guests are significantly lower than K of the CD complexes with the neutral or zwitterionic guests.

It is interesting to note that variation of pH induced by KH_2PO_4 is not a sole reason of K changing (Table 5). It is evident from Fig. 7. One of the dependences presented in Fig. 7 illustrates the enhancement of experimentally obtained pH values of the aqueous solution of mABA with increasing KH_2PO_4 concentration. The second dependence shows the decrease of theoretically calculated pH of the same system. Calculations were done on the basis of pK_a of mABA and $H_2PO_4^-$ assuming the absence of interaction between the acid and inorganic anion. The observed significant difference between these dependences can be due to possible interaction between zwitterionic acids and $H_2PO_4^-$ ions. Thus, shifting of the acid-base equilibria as well as the interactions occurring between $H_2PO_4^-$ and zwitterions are the key factors responsible for the observed decrease of complex formation of CDs with the NA and mABA in solution containing KH_2PO_4 (Table 5). More pronounced decreasing K in the presence of KH_2PO_4 compared to KCl points out the specific effects of $H_2PO_4^-$. Competition between $H_2PO_4^-$ and CD for binding with zwitterionic NA and mABA takes place and shifts the equilibrium of inclusion complex formation in the opposite direction.

Inspection of data given in Table 5 reveals the singular effect of KH_2PO_4 in complex formation of CDs with BA. Presence of KH_2PO_4 in the reaction medium results in dramatic decrease of stability constant of α-CD/BA complexes, whereas an unexpected increase of K was observed for β-CD/BA complexes. However, these effects are significantly weaker in complexation of CDs with pABA, which likewise BA exists in aqueous solution as neutral molecule. It seems that $H_2PO_4^-$ anions possess affinity to the carboxylic group of BA and selectively interact with it. Probably, this interaction is not so strong and realized through the hydrogen bonding. There were some attempts to confirm this assumption.

Solubility of BA and pABA was determined in the presence of KH_2PO_4 as well as KCl, KBr and K_2SO_4. Solubility diagrams are shown in Fig. 8. As can be seen, KH_2PO_4 displays salting-in effect, while the salting-out effect was revealed for all other salts under study. Salting-out action of KCl, KBr and K_2SO_4 towards BA [42-44] and pABA [45] has been previously demonstrated.

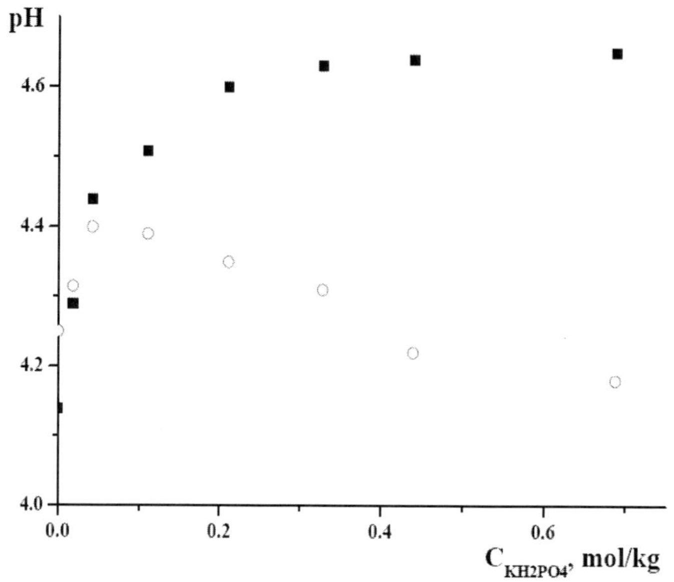

Figure 7. Experimental (!) and calculated (.) pH of mABA solutions *versus* KH_2PO_4 concentration.

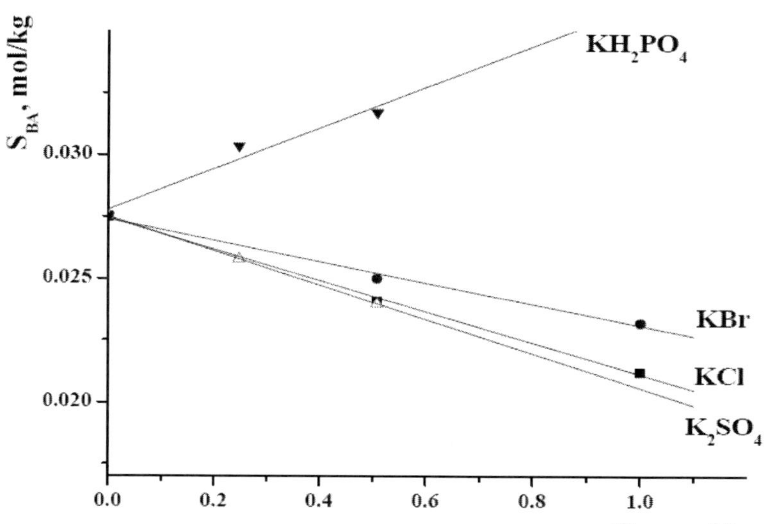

Figure 8. Solubility of BA in saline solutions.

Salting-in effect of KH_2PO_4 can be explained by taking into account the ability of $H_2PO_4^-$ ions to change the pH of the reaction medium and participate in the hydrogen bonding. For example, pH of BA solution ($2\cdot10^{-4}$ mol/kg) was increased from 4.0 to 4.3 by adding KH_2PO_4 (0.1M), while the further increase of salt concentration up to 0.7M has no a noticeable effect on the pH. Thus, addition on KH_2PO_4 to the solution of BA or pABA results in a partial dissociation of the acids and appearance of the anionic forms. In this connection, a decrease of stability constants of α-CD/BA, α-CD/pABA and β-CD/pABA complexes in KH_2PO_4 solution (Table 5) is due to the weaker binding affinity of CDs to the ionized guests. For instance, significant difference in the CD binding with benzoic acid and benzoate anion has been obtained in our recent publications [35, 46] as well as in the literature [31]. Assuming that influence of KH_2PO_4 on the acid-base equilibria is a sole factor affecting the inclusion complex formation one should expect the similarity of KH_2PO_4 effects in complex formation of CDs with BA and pABA. However, it is not true (Table 5), and more pronounced influence of KH_2PO_4 was observed for complexation with BA. In this case, formation of H-bonds between BA and $H_2PO_4^-$ was proposed.

As it was shown above, influence of KH_2PO_4 on CD binding with BA is rather complicate and, therefore, it should be considered in detail. When comparing the stability constants of α-CD/BA and β-CD/BA complexes in water and KH_2PO_4 solution (Table 5), one can found a noticeable difference. In particular, KH_2PO_4 causes a decrease in the stability of α-CD/BA complexes, whereas the opposite influence of this salt was revealed for β-CD/BA complexes. This phenomenon is due to different binding modes of BA with α-CD and β-CD.

It has been stated earlier on the basis of ^{1}H NMR data [31, 46] that shallow insertion of BA into α-CD cavity is accompanied by the location of a part of benzene ring with carboxylic group inside the cavity. In accordance with this binding mode, anions $H_2PO_4^-$ are able to interact only with BA molecules existing in the bulk solvent. As a result, two competing processes occur in solution

$$BA + \alpha\text{-}CD = BA\cdot\alpha\text{-}CD \quad (3)$$

$$BA + H_2PO_4^- = BA\ldots H_2PO_4^- \quad (4)$$

Mutual influence of reactions (3) and (4) is reflected in the decreasing stability constant of α-CD/BA complexes.

The opposite effect of KH_2PO_4 was demonstrated in the binding of BA with β-CD. For this system, increasing complex stability was obtained. Generally, an increase in the stability constants due to the addition of salt has been explained in the literature [8, 47, 48] by ternary complex formation. In this case, ion acts as link between the guest and host and interacts with them via electrostatic forces and/or H-bonding. There are some examples of ternary complexes: complexes of β-CD with substituted benzothiazoles and ClO_4^- [48], complexes of β-CD with tryptophan and carbonate ion [8] and CD/pentamineruthenium(II) complexes with carbonate or phosphate [47]. Thus, relying on the published data it is possible to predict the formation of β-CD/BA/$H_2PO_4^-$ complexes.

According to ^{1}H NMR data [46], BA deeply penetrates into β-CD cavity. It means that BA benzene ring is completely inside the cavity and carboxylic group is located outside at the narrow rim of β-CD molecule. In this case, formation of H-bonds between BA carboxylic group and β-CD hydroxyls can take place. Anions $H_2PO_4^-$ ingressed in the solution are also able to participate in the hydrogen bonding [47-49]. Consequently, the formation of ternary complex is possible.

To confirm β-CD/BA/$H_2PO_4^-$ comlexation, some additional ^{1}H NMR experiments were carried out. To this end, binding of methylated β-CD (M-β-CD) with BA was studied. Partial substitution of the outer OH-groups of β-CD by $-CH_3$ groups results in weakening ability of M-β-CD to form H-bonds compared to native β-CD. Therefore, formation of the ternary complex M-β-CD/BA/$H_2PO_4^-$ expected to be less probable.

^{1}H NMR spectra of BA were recorded in the presence of variable concentrations of M-β-CD in deuterated water and aqueous solution of 0.2M KH_2PO_4 at 298.15 K Binding isotherms shown in Fig. 9 were used for calculation of the stability constants of M-β-CD complexes with BA. Values of *K* were equal to 457±12 and 370±12 kg/mol for D_2O and solution of KH_2PO_4, respectively. As can be seen from the obtained results, stability of the complexes was reduced in saline solution. This result illustrates the absence of ternary complex formation. Furthermore, comparison of the dependences depicted in Fig. 9 allows to notice the difference in the behavior of $\Delta\delta$ corresponding to H-1′ and H-5′ BA protons located near the carboxylic group. For complex formation in pure water, values of $\Delta\delta$ for H-1′ and H-5′ protons are in the error limits (< 0.005 ppm), i.e. carboxylic group does not participate in the binding with M-β-CD. On the contrary, $\Delta\delta$ increases up to 0.05 ppm in KH_2PO_4 solution. It is known that chemical shift is the average value, which reflects the contributions from different states of the compound (e.g. free and complexed). Taking into account that interaction of BA carboxylic group with M-β-CD is absent, the observed increase of $\Delta\delta$(H-1′, H-5′) in KH_2PO_4 solution (Fig. 9) can be explained by the interaction of –COOH with $H_2PO_4^-$ anion. This interaction being in competition with the inclusion complex formation results in decrease of stability of M-β-CD/BA complex in KH_2PO_4 solution.

In conclusion, thermodynamic parameters of complex formation of α- and β-cyclodextrins with benzoic, nicotinic, p-aminobenzoic and m-animobenzoic acids in water and solutions of biologically active inorganic salts (KCl, KBr, KH_2PO_4 and K_2SO_4) were obtained and analyzed. It was demonstrated that observed salt effects are mainly caused by inorganic anions. The revealed influence of Cl^- and SO_4^{2-} on the inclusion complex formation is nonspecific and insignificant. It is only determined by the change of ionic strength of the solution. On the contrary, specific action of Br^- and $H_2PO_4^-$ was found as more pronounced. Competing interactions such as CD/Br^- and acid...$H_2PO_4^-$ shift the equilibrium of inclusion complex formation in the opposite direction and considerably weaken the stability of the inclusion complexes. Formation of ternary complexes was detected only for binding of benzoic acid with β-cyclodextrin in the presence of KH_2PO_4.

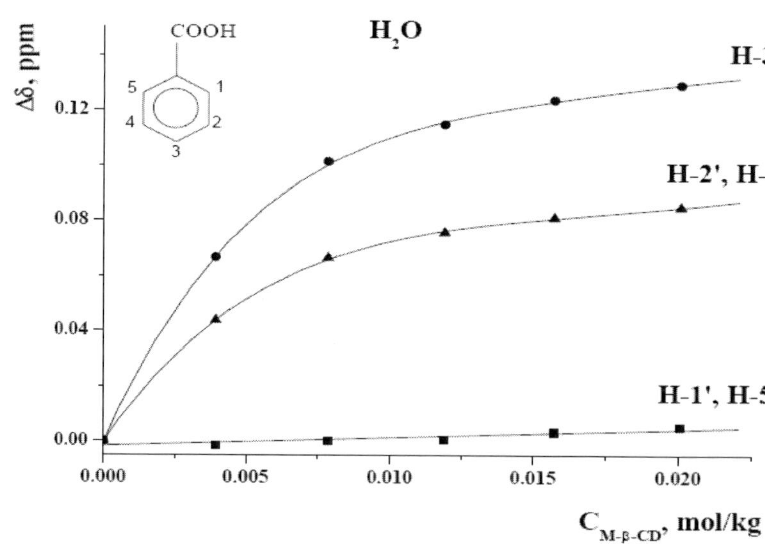

Figure 9. (Continued).

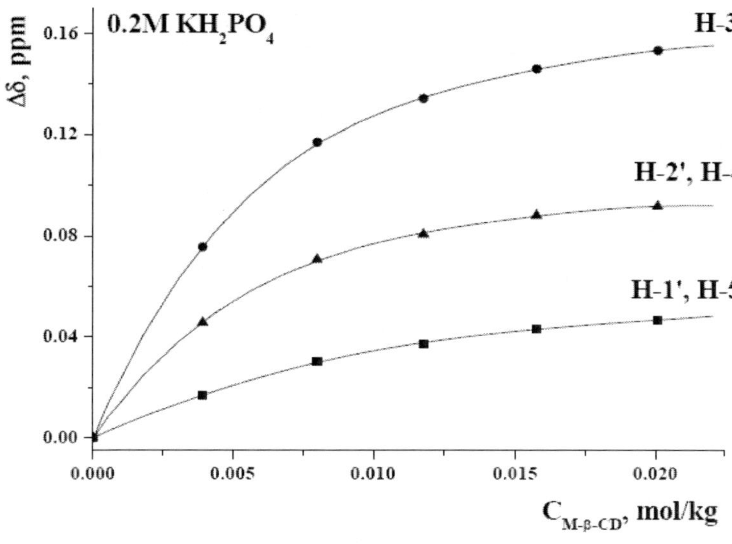

Figure 9. Dependences of chemical shift changes of BA (0.005 mol/kg) protons on M-β-CD concentration in water and solution of 0.2M KH_2PO_4 (T=298.15K).

This work was supported by Russian Foundation for Basic Research (grant №12-03-97516-r-center-a). We thank "The upper Volga region centre of physicochemical research" for scientific equipment provided.

REFERENCES

[1] Cyclodextrins and their complexes // *Ed. H. Dodziuk. Wiley-VCH.* 2006.

[2] Del Valle E.M.M. Cyclodextrins and their uses: a review // Process *Biochem.* – 2004. – V. 39. – № 9. – P. 1033-1046.

[3] Uekama K. Recent aspects of pharmaceutical application of cyclodextrins // *J. Incl. Phenom. Macrocycl. Chem.* – 2002. – V. 44. – № 1-4. – P. 3-7.

[4] Chaudhary V.B., Patel J.K. Cyclodextrin inclusion complex to enhance solubility of poorly water soluble drugs: a review // *IJPSR*. – 2013. – V. 4. – № 1. – P. 68-76.

[5] Holm R., Schönbeck C., Somprasirt P., Westh P., Mu H. A study of salt effects on the complexation between β-cyclodextrins and bile salts based on the Hofmeister series // *J. Incl. Phenom. Macrocycl. Chem.* – 2014. – V. 80. – №3-4. – P. 243-251.

[6] Junquera E., Aicart E. Potentiometric study of the encapsulation of ketoprophen by hydroxypropyl-*β*-cyclodextrin. Temperature, solvent, and salt effects // *J. Phys. Chem. B*. – 1997. – V. 101. – №36. – P. 7163-7171.

[7] Zhu X., Tan W., Zhou F., Li Z., Duan L. The effect of phosphate buffer solutions on uniconazole complexation with hydroxypropyl-β-cyclodextrin and methyl-β-cyclodextrin // *J. Incl. Phenom. Macrocycl. Chem.* – 2012. – V. 73. – P. 193-198.

[8] Song L.X., Bai L. Old drugs, new tricks: the effect of molecule-ion interactions on the precipitation-dissolution equilibrium of lithium carbonate in aqueous solution and on

the chiral recognition of cyclodextrins to D-,L-tryptophan // *J. Phys. Chem. B*. – 2009. – V. 113. – №34. – P. 11724-11731

[9] Brandariz I., Iglesias-Martínez E. Complexation constants of procaine with β-cyclodextrin in NaCl media // *J.* Chem. Eng. Data. – 2009. – V. 54. – №7. – P. 2103-2105.

[10] Goyenechea N., Sánchez M., Vélaz I., Martín C., Martínez-Ohárriz C., González-Gaitano G. Inclusion complexes of nabumetone with *β*-cyclodextrins: thermodynamics and molecular modelling studies. Influence of sodium perchlorate // *Luminescence*. – 2001. – V. 16. – №2. – P. 117-127.

[11] Terekhova I.V. Selective interactions of native and modified cyclodextrins with some B-vitamins in aqueous solution // Ch. 4 in monograph "Solution Chemistry Research Progress". Ed. D.V. Bostrelli. *Nova Science Publ. Inc*. 2007. P.93.

[12] Terekhova I.V. Inclusion complex formation of cyclodextrins with aminobenzoic acids in aqueous solution // Ch. 8 in Macrocyclic Chemistry: New Research Developments. Eds. D.W. Fitzpatrick and H.J. Ulrich. *Nova Sc. Publ*. 2010. P.263.

[13] Fu Y., Liu L., Guo Q.-X. A Theoretical study on the inclusion complexation of cyclodextrins with inorganic cations and anions // *J. Incl. Phenom. Macrocycl. Chem*. – 2002. – V. 43. – P. 223-229.

[14] Yi Z., Zhao C., Huang Z., Chen H., Yu J. Investigation of buffer-cyclodextrin systems // *Phys. Chem. Chem. Phys*. – 1999. – V. 1. – P. 441-444.

[15] Reale S., Teixidó E., De Angelis F. Study of alkali metal cations binding selectivity of beta-cyclodextrin by ESI-MS // *Ann. Chim*. – 2005. – V. 95. – №6. – P. 375-381.

[16] Gámez F., Hurtado P., Hortal AR, Martínez-Haya B., Berden G., Oomens J. Cations in a molecular funnel: vibrational spectroscopy of isolated cyclodextrin complexes with alkali metals // *ChemPhysChem*. – 2013. – V. 14. – P. 400-407.

[17] Stachowicz A., Styrcz A., Korchowiec J., Modaressi A., Rogalski M. DFT studies of cation binding by β-cyclodextrin // *Theor. Chem. Acc*. – 2011. – V. 130. – P. 939-953.

[18] Matsui Y., Ono M., Tokunaga S. NMR spectroscopy of cyclodextrin-inorganic anion systems // *Bull. Chem. Jpn*. – 1997. – V. 70. – P. 535-541.

[19] Taraszewska J., Wójcik J. Complexation of inorganic anions by β-cyclodextrin studied by polarography and ^{1}H NMR // *Supramolec. Chem*. – 1993. – V. 2. – P. 337-343.

[20] Mochida K., Kagita K., Matsui Y., Date Y. Effects of inorganic salts on the dissociation of a complex of β-cyclodextrin with an azo dye in an aqueous solution // *Bull. Chem. Soc Jpn*. – 1973. – V. 46. – № 12. – P. 3703-3707

[21] Gelb, R.I., Schwartz, L.M., Radeos, M., & Laufer, D.A. Cycloamylose complexation of inorganic anions // *J. Phys. Chem*. – 1983. – V. 87. – P. 3349-3354.

[22] Wojclk J.F., Rohrbach R.P. Small anion binding to cycloamylose. Equilibrium constants // *J. Phys. Chem*. – 1975. – V. 79. – № 27. – P. 2251-2253.

[23] Hersey A., Robinson B.H., Kelly H.C. Mechanism of inclusion-compound formation for binding of organic dyes, ions and surfactants to *α*-cyclodextrin studied by kinetic methods based on competition experiments // *J. Chem. Soc. Faraday Trans.1*. – 1986. – V. 82. – №5. – P. 1271-1287.

[24] Spencer J.N., He Q., Ke X., Wu Zh., Fetter E. Complexation of inorganic anions and small organic molecules with alpha-cyclodextrin in water // *J. Solution Chem*. – 1998. – V. 27. – № 11. – P. 1009-1019.

[25] Rohrbach R.P., Rodriguez L.J., Eyring E.M., Wojclk J.F. An equilibrium and kinetic investigation of salt-cycloamylose complexes // *J. Phys. Chem.* – 1977. – V. 81. – № 10. – P. 944-948.

[26] Sanemasa I., Fujiki M., Deguchi T. A new method for determining cyclodextrin complex formation constants with electrolytes in aqueous medium // *Bull. Chem. Soc.* – 1988. – V. 61. – P. 2663-2665.

[27] Lima S., Goodfellow B.J., Teixeira-Dias J.J.C. How inorganic anions affect the inclusion of hexanoic and decanoic acid in β-cyclodextrin // J. Phys. Chem. A. – 2004. – V. 108. – №46. – P. 10044-10049.

[28] Buvári A., Barcza L. Complex formation of inorganic salts with β-cyclodextrin // *J. Incl. Phenom. Macrocycl. Chem.* – 1989. – V. 7. – P. 379-389.

[29] Schneider H.-J., Hacket F., Rüdiger V. NMR studies of cyclodextrins and cyclodextrin complexes // *Chem. Rev.* – 1998. – V. 98. – P. 1755-1785.

[30] Marcus, Y. Ion Properties. / in Ionic interactions in natural and synthetic macromolecules // Ed. A. Ciferri & A. Perico. *John Wiley & Sons, Inc.* 2012. Chapter 1. pp. 3-33.

[31] Simova S., Schneider H-J. NMR analyses of cyclodextrin complexes with substituted benzoic acids and benzoate anions // *J. Chem. Soc. Perkin Trans.* 2. – 2000. – № 10. – P. 1717-1722.

[32] Vrbka L., Vondrásěk J., Jagoda-Cwiklik B., Vácha R., Jungwirth P. Quantification and rationalization of the higher affinity of sodium over potassium to protein surfaces // *PNAS.* –2006. – V. 103. – № 42. – P. 15440-15444.

[33] Uejio J.S., Schwartz C.P., Duffin A.M., Drisdell W.S., Cohen R.C., Saykally R.J. Characterization of selective binding of alkali cations with carboxylate by x-ray absorption spectroscopy of liquid microjets // *PNAS.* – 2008. – V. 105. – № 19. – P. 6809-6812.

[34] Fedorov M.V., Goodman J.M., Schumm S. To switch or not to switch: the effects of potassium and sodium ions on α-poly-l-glutamate conformations in aqueous solutions // *J. Am. Chem. Soc.* – 2009. – V. 131. – № 31. – P. 10854-10856.

[35] Terekhova I.V., Romanova A.O., Kumeev R.S., Fedorov M.V. Selective Na^+/K^+ effects on the formation of α cyclodextrin complexes with aromatic carboxylic acids: competition for the guest // *J. Phys. Chem. B.* – 2010. – V. 114. – №39. – P. 12607-12613.

[36] Deary M.E., Mousa S.M., Davies D.M. Effect of kosmotrope and chaotrope anions on rate and equilibria processes for the α-cyclodextrin catalysed reaction of 3-chloroperbenzoic acid with iodide // *J. Incl. Phenom. Macrocycl. Chem.* – 2014. – V. 78. №1-4. – P. 127-136.

[37] Laufer D.A., Gelb R.I., Schwartz L.M. Carbon-13 NMR determination of acid-base tautomerization equilibriums // *J. Org. Chem.* – 1984. – V. 49. – № 4. – P. 691-696.

[38] Venkateswaran K., Stadlbauer J.M., Wu Zh., Gillis H.A., Walker D.C. Effect of pH on muonium reactions in aqueous solutions of *p*-aminobenzoic acid // *J. Phys. Chem.* – 1996. – V. 100. – № 9. – P. 3569-3572.

[39] Cohn E.J., McMeekin T.L., Edsall J.T., Blanchard M.H. Studies in the physical chemistry of amino acids, peptides and related substances. I. The apparent molar

volume and electrostriction of the solvent // *J. Am. Chem. Soc.* – 1934. – V. 56. – №4. – P. 784-794.

[40] García B., Ibeas S., Leal J.M. Zwitterionic pyridinecarboxylic acids // *J. Phys. Org. Chem.* – 1996. – V. 9. – № 9. – P. 593-597.

[41] Romanova A.O., Chibunova E.S., Kumeev R.S., Fedorov M.V., Terekhova I.V. α-Cyclodextrin/aminobenzoic acid binding in salt solutions at different pH: dependence on guest structure // *Intern. J. Biol. Macromol.* – 2013. – V. 57. – P. 255-258.

[42] Kim M., Kim J., Hyun S. Solubility of organic acids in various methanol and salt concentrations: the implication on organic acid sorption in a solvent system // *Chemosphere.* – 2012. – V. 89. – №3. – P. 262-268.

[43] Chase E.F., Kilpatrick M. The classical dissociation constant of benzoic acid and the activity coefficient of molecular benzoic acid in potassium chloride solutions // *J. Am. Chem. Soc.* – 1931. – V. 53. – P. 2589-2597.

[44] Goeller G.M., Osol A. The salting-out of molecular benzoic acid in aqueous salt solutions at 35° // *J. Am. Chem. Soc.* – 1937. – V. 59. – P. 2132-2134.

[45] Sugunan S., Thomas B. Thermodynamics of transfer of aminobenzoic acids from water to aqueous salt solutions // *J. Chem. Eng. Data.* – 1995. – V. 40. №2. – P. 415-416.

[46] Terekhova I.V. Comparative thermodynamic study on complex formation of native and hydroxypropylated cyclodextrins with benzoic acid // *Thermochim. Acta.* – 2011. – V. 526. – №1-2. – P. 118-121.

[47] Johnson M.D., Bernard J.G. Hydrogen bonding effects on the cyclodextrin encapsulation of transition metal complexes: "molecular snaps" // *Chem. Commun.* – 1996. – №2. – P. 185-186.

[48] Dey J., Roberts E.L., Warner I.M. Effect of sodium perchlorate on the binding of 2-(4'-aminophenyl)- and 2-(4'-(*N,N'*-dimethylamino)phenyl) benzothiazole with β-cyclodextrin in aqueous solution // *J. Phys. Chem. A.* – 1998. – V. 102. – №1. – P. 301-305.

[49] Liu L., Wang M. Competitive recognition between 1: 2 ternary complex of zwitterion and H_3PO_4 towards 3D cucurbit[8]uril frameworks // *J. Incl. Phenom. Macrocycl. Chem.* – 2015. – V. 81. – P. 509-516.

In: Cyclodextrins
Editor: Francis G. Ramirez
ISBN: 978-1-63482-788-1

Chapter 10

SULFOBUTYLETHER-CYCLODEXTRINS: STRUCTURE, DEGREE OF SUBSTITUTION AND FUNCTIONAL PERFORMANCE

István Puskás, Erzsébet Varga, Kata Tuza, Julianna Szemán, Éva Fenyvesi, Tamás Sohajda and Lajos Szente
CycloLab Cyclodextrin Research and Development Laboratory Ltd., Budapest, Hungary

ABSTRACT

Intermolecular interactions between guest molecules and various sulfobutyl ether cyclodextrins (SBECD) differing in degree of substitution and number of anhydroglucose units were studied. For pharmaceutical purposes a special grade of SBECD has been authorized as excipient (i.e. complying US Pharmacopoeia 37) which is a highly specified product of strictly determined chemical properties. The aim of the study was to reveal how the functional properties would change by alteration from the pharmacopoeial specifications. For the characterization of the composition of this group of substituted cyclodextrins, three independent methods were compared. By NMR limited but undistorted structural information may be obtained, while capillary electrophoresis is able to provide the compositional profile, too. A HPLC method is presented by which information on the compositional profile as well as the quantity of typical impurities may be obtained. Several drug molecules were tested with a sulfobutyl ether cyclodextrin library using capillary electrophoresis. Based on the collected data it has been demonstrated that multiple factors may play significant role in the binding process ensuring some sort of selectivity amongst the members of the cyclodextrin library used. Besides the dimensions of the cavities, the steric effect of the relatively bulky sulfobutyl ether moieties as well as the charge born on the terminal of the substituent chains influence the strength of interaction. These data may serve as examples how the great variety of sulfobutylether cyclodextrins may be exploited in specific fields of application.

1. INTRODUCTION

1.1. Significance of Degree of Substitution in Relation to Cyclodextrin Derivatives

A major issue associated with the use of cyclodextrins (CD) is that the chemical term of *cyclodextrin* became a hypernym of a vast variety of substances. At the dawn of the history of these compounds, the word *cyclodextrin* became simply a non-eponymous term for the three Schardinger dextrins. The chemical modification of the Schardinger dextrins has sown the seed of semantic misconception associated with the exact nomenclature and structural information of CD derivatives persistent for the last decades. Present chapter is intended to illustrate the importance of the detailed characterization of thecomposition of substituted CDs by the example of sulfobutylether cyclodextrins. Besides, some insight into the applicability of the two less used sulfobutylether derivatized α-CD and γ-CD is provided. The effect of the degree of substitution (DS) on various properties is represented. In present chapter under DS the average number of substituents per cyclodextrin ring is understood.

In general CD substitution reactions may be carried out resulting in:

A. A single compound, wherein the DS is a known integer number. Such material is Sugammadex [1], or 6-monotosyl-, monoazido- and monoamino-β-CD [2,3], 6-fluoresceinylthioureido-β-CD [4], per-6-halogeno [5, 6], sulfo [7] and amino [8, 9] derivatives of α-, β- and γ-CD, as well as heptakis(2,3,6-tri-*O*-methyl)-β-CD [10]. However, over- and undersubstituted species might be always present as impurities. It is important to note that a cyclodextrin compound simply characterized in that its DS equals 1 does not necessarily mean that it is a single compound: Stella et al. prepared and characterized three monosubstituted sulfobutylether β-cyclodextrin derivatives which were identified as the products modified at the 2-, 3-, and 6-hydroxyl groups of a glucose unit, respectively [11].
B. A composition of substituted cyclodextrins, wherein the number and the position of substituents may vary molecule by molecule resulting in a combinatoric (random) mixture of substances. The degree of substitution can be expressed as an average value only. The distribution profile can be characterized by harmonicity (i.e. close to Gaussian distribution) as well as wideness [12]. Such examples are sulfobutylether cyclodextrins and 2-hydroxypropyl-β-cyclodextrin of low DS. An even more complicated case occurs when the substituents themselves are not identical in the sample, which can be different on purpose (see the sophisticated syntheses of Sollogoub at al. [13]). It is known that upon synthesizing high DS 2-hydroxypropyl-β-cyclodextrin, propylene oxide can react with the hydroxyl group of the hydroxypropyl substituents forming oligomeric and even longer chain polypropylene glycol side chains [14]. Consequently in addition to their number and position, also the distribution profile of different substituents is characteristic to the chemical structure of these multicomponent mixtures.
C. A composition of substituted cyclodextrins, wherein a selected single isomer component of particular importance is standardized. DIMEB-50 - for example - is a mixture of differently methylated cyclodextrin species, but the quality of the

substance depends on the content of a single isomer 2,6-di-O-methyl-β-cyclodextrin (which is ~50% for DIMEB-50). The nomenclature and characterization of such methylated cyclodextrin compositions or single isomers have been recently reviewed [15].

In addition of the above, using different derivatization strategies, various oligomeric and polymeric CD-structures may be produced, but the enumeration of these materials is beyond the scope of present chapter.

As these examples show, the plant-to-plant, laboratory-to-laboratory and even batch-to-batch variation between such compositions can be remarkable. Reliable analytical techniques to characterize the compositional profile of these derivatives are essential for the successful application of these compounds, especially in the field of pharmaceutical use.

1.2. Cyclodextrin Derivatization for the Purpose of Pharmaceutical Use

Intravenous application of parent cyclodextrins is highly limited due to their nephrotoxicity. From the list of already approved drug formulations only a special injectable product can be mentioned which contains a parent cyclodextrin [16]. The α-CD solubilized Prostaglandin E1 (Prostavasin®, Viridal®, Alprostadil®, Edex®, Caverject® produced by Ono and Schwarz, respectively) was granted marketing authorization due to the relatively low α-cyclodextrin exposure.

Currently two cyclodextrin derivatives are available showing both effective inclusion complexation potential and outstanding *in vivo* parenteral safety for biomedical uses:

a) polyanionic variably substituted sulfobutyl ether of β-cyclodextrin (Betadex Sulfobutyl Ether Sodium, SBEBCD or more commonly SBECD). SBEBCD has been developed for pharmaceutical use by Cydex (brand name: Captisol®). The substance has become generic worldwide in 2011. The chemical formula of a representative SBEBCD (DS 6) species is shown in Figure 1.
b) (2-hydroxypropyl) beta cyclodextrin (HPBCD, Hydroxypropyl betadex) developed for pharmaceutical use by Janssen Pharmaceuticals, Inc. / Johnson & Johnson [17]. Unlike SBEBCD, HPBCD is a non-charged derivative. Janssen's patent no longer inhibits the worldwide use of HPBCD in pharmaceutical applications, but in the USA the patent is still valid until 2019.

In order to highlight the special properties of SBEBCD it is worth comparing these two injectable derivatives.

Both CD derivatives have undergone thorough toxicological safety studies. HPBCD has a monograph in both United States Pharmacopoeia (USP) and European Pharmacopoeia (EP), while SBEBCD is listed only in USP (the EP monograph is currently in preparation). It is noteworthy that while the pharmacopoeial specification for average DS permits the wide range of 2.8-10.5 in the case of HPBCD, SBEBCD must meet the strict requirement of 6.2-6.9. Unlike HPBCD, SBEBCD is not only a solubilizing but a definitive osmotic agent, too

[18]. Their pharmaceutically and physico-chemically relevant properties are listed in Tables 1-3.

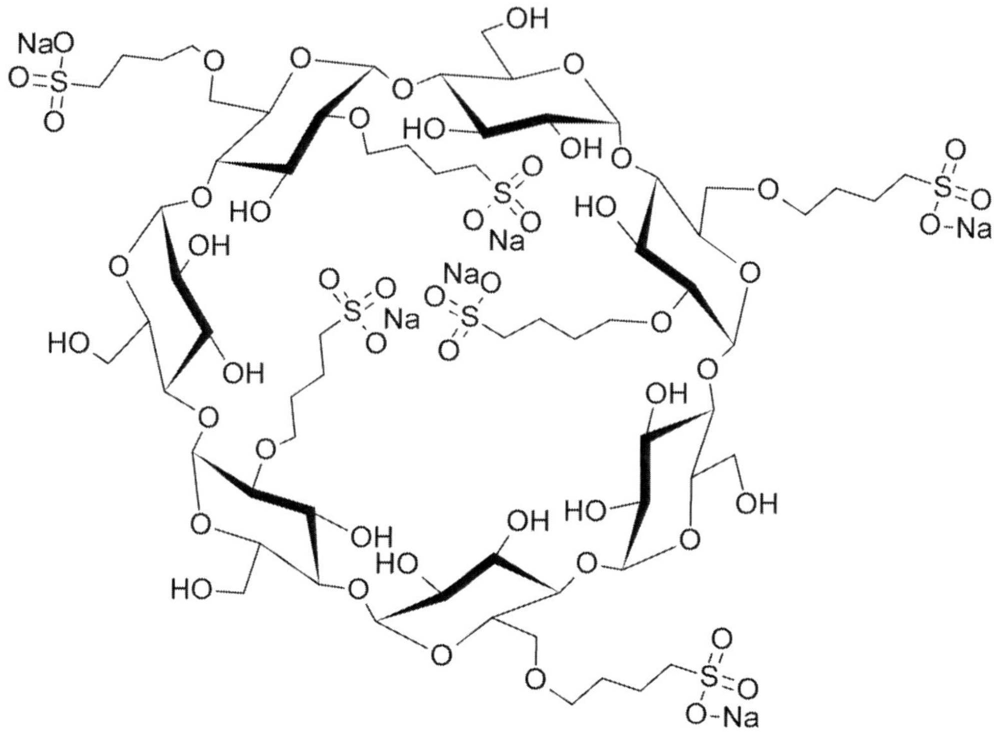

Figure 1. Chemical formula of a representative species of SBEBCD (DS 6).

Table 1. Comparison of relevant physico-chemical properties of HPBCD and SBEBCD

	HPBCD	SBEBCD
Crystallinity /amorphousness	Amorphous	Amorphous
Color	white	white
Solubility in water	> 1200 mg/ml	> 1200 mg/ml
Administration route (typical examples)*	Oral, intravenous, intramuscular, ophthalmic, nasal and suppository	Oral, intravenous, intramuscular, subcutaneous, ophthalmic, nasal and inhalation
Acceptable DS according to pharmacopoeiae	2.8-10.5	6.2-6.9
Charge	Neutral	Polyanionic (Na^+ salt)
Suitability for taste masking	Suitable	Not suitable due to salty taste

*The administration routes of marketed and investigational products are listed.

Both HPBCD and SBEBCD may interact with neutral drugs to facilitate solubility and chemical stability, but SBEBCD owing to its polyanionic nature binds especially well to cationic, nitrogen containing compounds [19, 20].

Table 2. Osmolality of aqueous HPBCD and SBEBCD solutions

Concentration (w/w%)	Osmolality (mOsm/kg)	
	HPBCD (DS=4.6)	SBEBCD (DS=6.5)
5	37	135
10	81	286
15	129	501
20	196	785

Table 3. Viscosity of aqueous HPBCD and SBEBCD solutions at 25°C

Concentration (w/w%)	Viscosity (cP)	
	HPBCD (DS=4.6)	SBEBCD (DS=6.5)
5	1.16	1.17
10	1.34	1.37
15	1.60	1.68
20	2.00	2.10
25	2.60	2.85
30	3.38	4.09

The pharmacokinetic properties, tissue distribution, and cellular effects of some representative CDs, including SBEBCD and HPBCD, have been reviewed by Stella and He [21]. In this review numerous drug complexes of these CDs have been shown to rapidly dissociate followed by parenteral drug administration. It has been demonstrated that such compositions do not have tissue-irritating effects after intramuscular dosing, besides that the complexation may result in improved oral bioavailability of poorly water-soluble drugs. Human experience with CD derivatives indicates that SBEBCD is well tolerated in humans and has no adverse effects on the kidneys or other organs following either oral or intravenous administration.

1.3. Synthesis of Sulfoalkylated Cyclodextrins

The preparation of sulfoalkylated cyclodextrins (SAE-CDs) was first described by Parmerter [22] in 1969 mainly as flocculant and paper industry auxiliary materials. It is known from Parameter's work that the reaction usually requires four compounds: a natural cyclodextrin, solvent, base and alkylating agent as shown in Scheme 1. As mentioned earlier, the resulting material is a multicomposite mixture of randomly substituted cyclodextrins. The distribution profile (i.e. average DS as well as the DS distribution) greatly depends on the production parameters, therefore needs to be carefully optimized. Well after Parmerter's pioneering work when the use of CDs became industrially feasible, based on the experiences of numerous analogous CD derivatizing reactions and later studies, the need of elaboration of a consistent industrial manufacturing process arose. This task required the controlling of the parameters in the alkylation technology which included the optimization of the characteristics of the solvent, the quantity of the used base as well as the alkylating agent, the temperature of the reaction, the reaction time, the purification steps and the water removal (by lyophilization on the laboratory scale and spray drying on the industrial scale) [23].

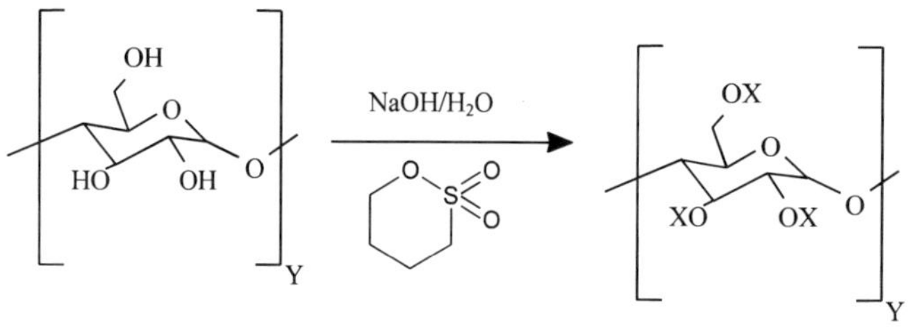

Scheme 1. Synthesis of sulfobutylether cyclodextrins.

SBEBCD is far the most promising candidate for interaction studies with compounds abundant in nitrogen containing moieties, since besides molecular host-guest inclusion, electrostatic interactions may contribute to successful solubilizing. To illustrate this statement some drugs successfully marketed in SBEBCD complex form are shown in Table 4.

The first patent application claiming that sulfoalkylether cyclodextrins are suitable for pharmaceutical use was submitted by Stella and Rajewski [24]. Herein it is underlined that these derivatives possess low cell membrane disrupting effect.

The critical feature of having minimized amount of residual (unreacted) cyclodextrin is also emphasized. The same authors issued another patent application claiming low unreacted cyclodextrin containing compositions [25].

In pharmaceutical use, highly specific quality parameters apply (average DS, substitution pattern, residual β-CD and butane sultone content, content of hydrolysis products originating from butane sultone) besides compliance to the specifications of general pharmacopoeial chapters (e.g. appearance; solution clarity, pH as well as water, endotoxin and heavy metal content) [26].

There are two hydrolysis products originating from butane sultone. The first is 4-hydroxybutane-1 sulfonic acid (HBSA) as denoted in USP: please note that in fact its sodium salt forms. The second is bis(4-sulfobutyl)ether disodium (DBSA). The hydrolysis products may be removed for example by dialysis/diafiltration or ion chromatography.

Later on different intellectual property protection strategies were utilized to gain exclusivity to some extent on the sulfoalkylated cyclodextrin market. One of Antle's patent application [27] directs to the use of SAE-CD of low phosphate content having low UV absorbance. Due to the spectral properties of SAE-CD, significant UV absorption in the specified ranges (245-270 nm as well as 320-350 nm) may indicate the presence of unwanted contaminants detrimental for pharmaceutical use. Follow-up patent applications of the same inventor specifies the product and process of the same features characterized in that the substance contains low amounts of chloride ion [28, 29]. Sklavounos et al. filed a patent application [30] claiming a process wherein the alkylation reaction is partially conducted in a relatively mild alkalic milieu (i.e. pH 8-11) until the residual unreacted cyclodextrin falls below a critical value. Then additional hydroxide is required to be added to complete the reaction. Cydex Pharmaceuticals, Inc. (owned by Ligand Pharmaceuticals, Inc.), holds several closely related patents [31, 32 33, 34] specifying an SAE-CD composition consisting of agglomerated particles having a pre-defined bulk density (0.38-0.66 g/cm^3) having favorable compression characteristics for tablet pressing. The product is isolated by a 3 chamber fluidized bed spray drier wherein the fine particles are recycled. Iványi et al. (CycloLab Ltd.) were granted a patent for their invention disclosing the use of additional ethanol followed by the alkylation reaction facilitating the elimination of residual carcinogenic butane sultone [35]. Even though the above variations of the production processes as well as products are patent protected, SBEBCD as a substance has already become generic for pharmaceutical use. Since 2011, an USP-compliant generic SBEBCD (Dexolve®, CycloLab Ltd.) has been available besides the originator's excipient.

1.4. Sulfobutylether Cyclodextrins: Variability of DS and Performance

The major scope of present chapter is the demonstration of the applicability of different sulfobutylether cyclodextrins in the light of different cavity sizes and average DS. In the 1990's SBEBCD of different average DS were concurrently available and used in early phase drug formulation and toxicology studies. Some publications dealt with the use of SBEBCD of lower DS [36, 37, 38, 39]. Qu et al. synthesized sulfobutyl ether-β-cyclodextrin of DS 2.5: 1H NMR indicated that the primary hydroxyl group was mainly subject to the substitution. MS spectra showed that no more than one substitution occurred on a single glucose unit [40]. Comparative studies were demonstrated in the early patent of Stella et al. [24] wherein the association constants (*K*) of SBE7-β-CD, SBE4-β-CD and SBE1-β-CD (i.e. SBEBCD of DS=7; 4 and 1, respectively) with different substances were enumerated. In one case the DS versus *K* correlation had a minimum type of curve (for progesterone), while digoxin and phenytoin showed a descending correlation (SBEBCD of low DS binds the strongest) and testosterone showed an ascending correlation. Thompson [41] compared (amongst other CD derivatives) to what extent sulfobutylether cyclodextrin (SAE-CD) derivatives such as SBE7-β-CD, SBE4-β-CD and SBE1-β-CD show hemolytic behavior in concentrations typically used to solubilize pharmaceutical formulations. The hemolytical activity was found in the order of SBE7-β-CD < SBE4-β-CD << SBE1-β-CD << β-CD. From the data it may be derived that the use of SBE7-β-CD was considered the safest.

Mosher et al. demonstrated the applicability of SBEBCD of different average degrees of substitution (4, 7 and 11) for solubilizing amiodarone resulting in a liquid formulation which is dilutable without the risk of drug precipitation [42]. SBEBCD of average DS of 7 was found the preferred composition.

The effect of DS on the applicability of different CD derivatives in capillary electrophoresis for chiral separations in pharmaceutical analysis (SBEBCD included) was reviewed by de Boer et al. [43]. Herein it is emphasized that the use of commercially available CDs having a defined DS may lead to a better modeling, optimization, reproducibility, and to a more rugged separation system.

From the year 2000 the use of SBEBCD of DS 7 is overwhelming, since SBE7-β-CD was selected for the clinical studies and later this material was specified in the USP. It must be noted herein that the denotation of DS 7 is somewhat misleading, since the average DS of the USP compliant substance is between 6.2-6.9. The denotation reflects that the required distribution profile may be attained easier if the DS is closer to 7 (see later Section 4.1).

Table 4. Drugs (also) marketed in SBEBCD containing formulations

	Drug	Trade name
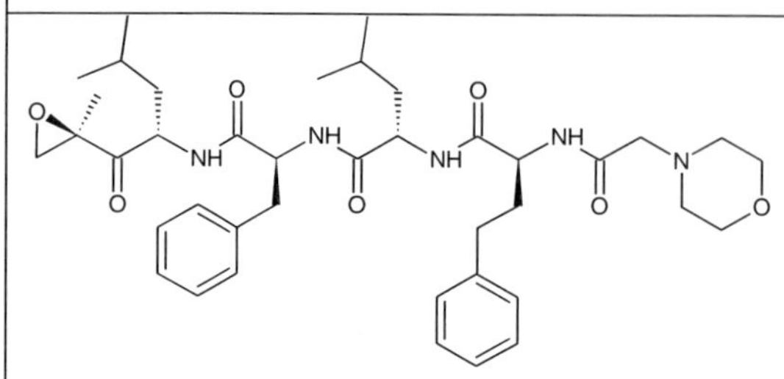	Carfilzomib	Kyprolis®
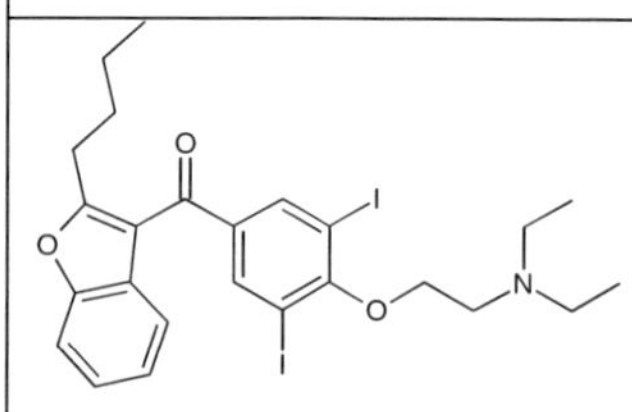	Amiodarone	Nexterone®
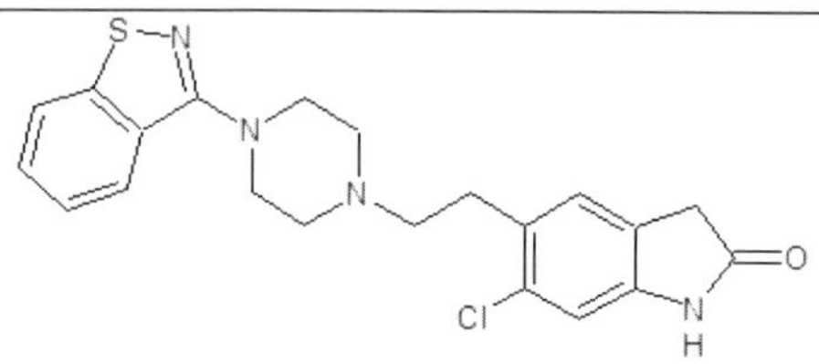	Ziprasidone	Geodon®
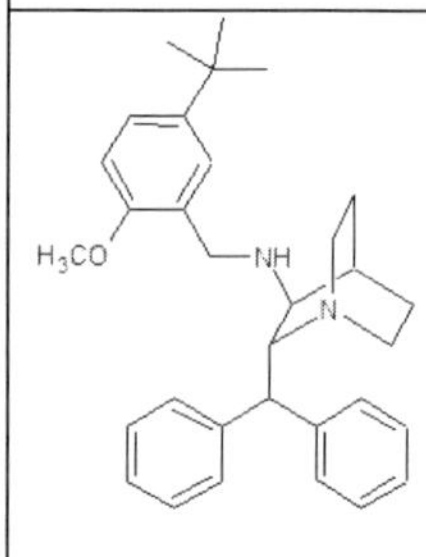	Maropitant	Cerenia®
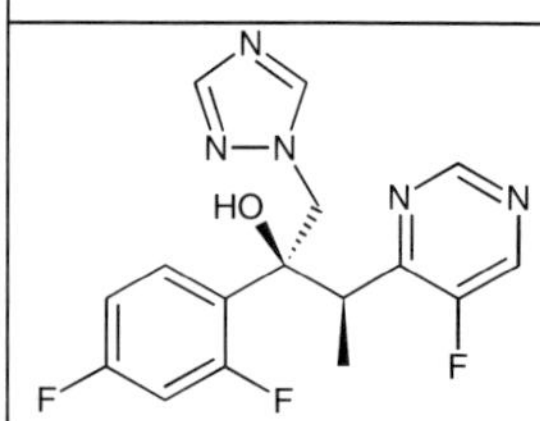	Voriconazole	Vfend®
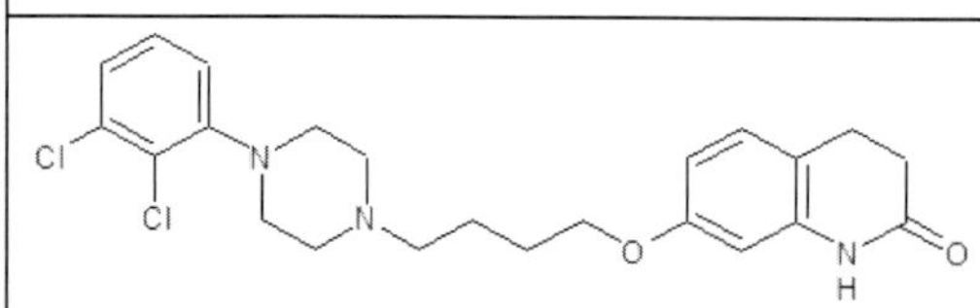	Aripiprazole	Abilify®

Zia et al. attempted to find correlation between the DS and binding of molecules to SBEBCDs with varying degrees of sulfobutyl ether substitution. Complexation constants of molecules to SBEBCDs were calculated as a function of temperature, enabling the estimation of thermodynamic parameters, including the enthalpy and entropy of binding. Binding constants of various molecules to SBEBCDs did not show a uniform trend to total degree of sulfobutylether substitution. However, a distinct pattern was observed with the enthalpy and entropy of complexation. The results showed the complexation of substrates to SBEBCDs to be more entropy-favored as the number of sulfobutyl ether groups increased. This favorable entropy was compensated by a less favorable enthalpy of interaction [44].

Skanchy et al. elaborated a separation technique for the enantiomers of the basic drug bidisomide from five closely related known process impurities using novel sulfobutylether derivative mixtures and separated fractions having a specific DS (i.e isolated SBEBCD types having degrees of substitution from one to seven). Fractions having a lower DS provided adequate chiral and achiral selectivity allowing both chiral and achiral purity to be determined in a single run [45].

It may be therefore concluded from all these prior studies that the number of the sulfobutyl ether moieties (presumably owing to a combined effect of steric and electrostatic factors) may play a significant role in the binding process. Strict standardization of the DS profile evidently yields greater trust in the reproducible performance in the pharmaceutical and analytical fields of use. However, the versatile features of the potential variable SBEBCD compositions may not be fully exploited this way. In the following sections the applicability of relatively rarely used sulfobutylether cyclodextrin derivatives (other than Captisol® and Dexolve®) are discussed and compared.

2. Materials

The following sulfobutyl CDs (all of them are sodium salts) used throughout the investigations are the fine chemical products of CycloLab Ltd., Hungary.

- Sulfobutyl-α-cyclodextrin (SBEACD), DS ~ 4,
- Sulfobutyl-β-cyclodextrin (SBEBCD), DS 3.9, DS 4.8, DS 6.5, DS 10.4, respectively
- Sulfobutyl-γ-cyclodextrin (SBEGCD), DS ~ 4.

3. Methods

3.1. Determination of Average Degree of Substitution by Capillary Electrophoresis

As highlighted earlier, SAE-CDs are isomeric mixtures of differently substituted derivatives. Determination of the average DS value was based upon capillary electrophoresis (CE) measurements according to the Betadex Sulfobutyl Ether Sodium USP N.F. monograph. During the test an uncoated fused-silica capillary of 58.5 cm total length and 50 cm effective length with 50 μm inner diameter was used. The measurement was conducted in a 30 mM benzoic acid / 100 mM TRIS buffer (pH 8.3 - 8.7 without adjustment) at 25 °C using a positive applied voltage gradient and indirect detection.

Calculation of DS

After recording the electropherogram, the peak responses for the individual peaks (I to X) were measured. The corrected peak areas were calculated for each peak in the electropherogram:

$$\text{Corrected Peak Area}, A_i = \frac{\text{Peak Area x Effective Capillary Length (cm)}}{\text{Migration Time (min)}}$$

The corrected peak areas were normalized by presenting each as a percentage of the total corrected substitution envelope area (n= highest level of substitution):

$$\text{Normalized Area}, NA_i = \frac{A_i}{\sum_{i=1}^{n} A_i} \times 100$$

The average degree of substitution was determined by the equation below:

$$\text{Average Degree of Substitution}, DS = \frac{\sum_{n=1}^{n} (\text{Level of Substitution for Peak}_i \, x \, NA_i)}{100}$$

Besides placing a requirement for average degree of substitution (DS= 6.2-6.9) the USP monograph also sets acceptance criteria for the peak area % for each of differently substituted SBEBCD fractions.

3.2. Determination of Average Degree of Substitution – Nuclear Magnetic Resonance (NMR) Spectroscopy

Besides CE, the average degree of substitution can also be determined using NMR spectroscopy. As peak distribution pattern cannot be examined with NMR, similar criteria are not included for this test. Herein, the DS determination is shown on the example of SBEBCD.

The SBEBCD sample was dissolved in D_2O at a minimum of 14 mg/mL concentration. A free induction decay (FID) was acquired for 4 sec with at least 8 transients and with a recycle delay of at least 15 sec using a spectral window from at least 0 to 8 ppm, with the solvent peak located at 4.76 ppm at 25°C. The spectrum was zerofilled at least 2 times and the FID was Fourier transformed with no Gaussian line broadening and no more than 0.3 Hz of Lorentzian line broadening (Figure 2).

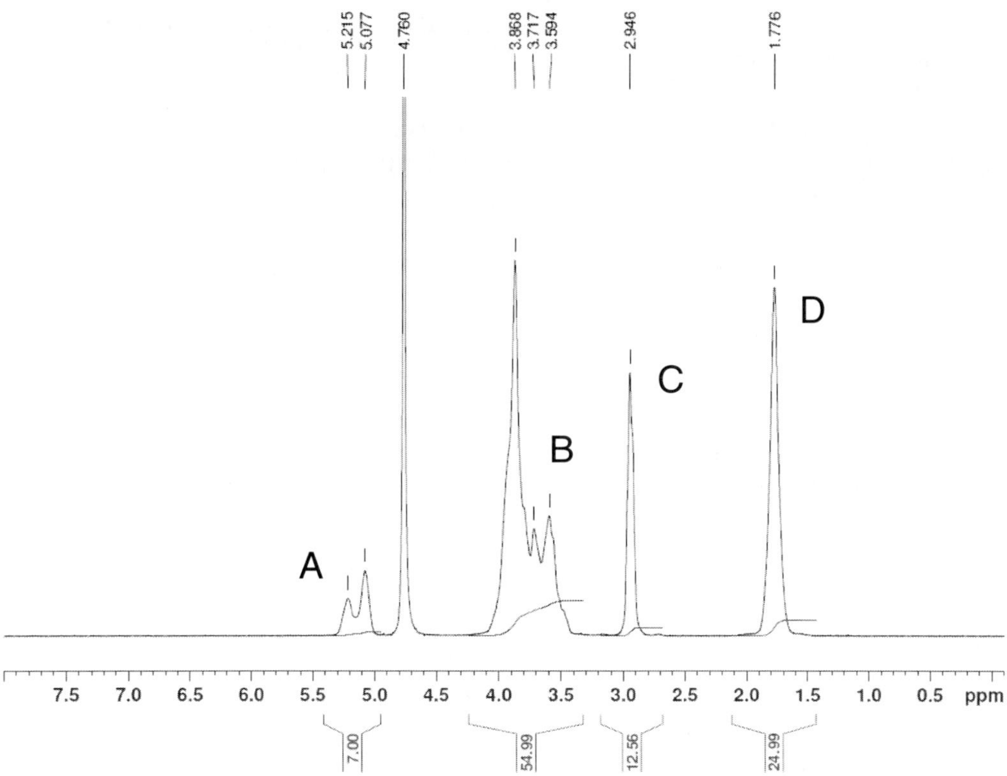

Figure 2. Representative NMR spectrum recorded on a SBEBCD sample of DS 6.6 (as determined by CE)

The peak areas of the resonances were determined for the four regions indicated in Figure 2. As a reference, integral value of peak A, representing the anomeric protons was set to 7.00, since there are 7 such protons in case of β-CD derivatives. DS was calculated from the dataset:

DS = D/4,

in which D is the area arising from four methylene protons shown in bold of the sulfobutylether functional group, -O-CH_2-$\mathbf{CH_2}$-$\mathbf{CH_2}$-CH_2-SO_3^- (region D).

3.4. Phase Solubility Studies

CD solutions of various concentrations were prepared in distilled water. Excess amount of substance to be solubilized (20-30 mg) was added to these solutions (3 ml) and stirred for 24 hours at room temperature. Then the solutions were filtered and the concentrations of the solubilized substances were measured by UV photometry (after proper dilution with 50 v/v% aqueous ethanol).

3.5. Determination of Apparent Interaction Affinity (Stability Constants) by Capillary Electrophoresis

The determination of apparent interaction affinity (stability constants) was carried out as described by Rundlet and Armstrong [46] , by running the analyte first in the background electrolyte (BGE, without CD added), then in BGEs containing increasing concentrations of each CD derivative investigated.

The presence of the CD affects the migration velocity of the guest in case there exists a detectable interaction (*Note: the influence is dependent on the charge of both the CD and the guest molecule as well as the applied voltage and several further factors*). The migration time of the guest is always related to that of the electroosmotic flow (EOF) marker. The EOF represents the migration velocity of neutral species in the system, a flow that affects all the species that are within the capillary.

This can be considered as a benchmark for each experiment. Obviously, according to changes in the BGE (e.g. increase in viscosity or ionic strength due to CD addition), the migration time of the neutral species may change as well. By relating the migration time of the guest to that of the EOF, these effects are compensated during the calculations.

The stability constants can be determined with the x-reciprocal method (among other similar approaches) commonly applied for complex stability determinations with CE [47]. For the determination, all the below detailed steps should be performed for at least three CD concentration levels as well as in a system containing no CD.

The calculation begins with the determination of effective mobilities according to the following:

- As first response factors the migration times of the electroosmotic flow (EOF, a compound serving as a marker) and the analyte are recorded.
- The mobilities (μ) of EOF and analyte are calculated using the formula below:
  $\mu = (l_{eff} \times l_{total})/(t_{mig} \times U)$
  where l_{eff} and l_{total} are the effective and total lengths of the capillary,
  t_{mig} is the migration time in minutes and U is the applied voltage in V.
- The effective mobility is calculated from the subtraction of the two mobility values:
  $\mu_{eff} = \mu_{EOF} - \mu_{Analyte}$

X-reciprocal methods need some further calculations. Mobility differences are calculated first by the subtraction of the averaged μ_{eff} value obtained at 0 mM CD concentration from actual μ_{eff} values at each CD levels (practically: $\mu_{eff\_withCD} - \mu_{eff\_withoutCD}$). For the calculation $(\mu_{eff\_withCD} - \mu_{eff\_withoutCD})/[CD]$ is plotted as a function of $(\mu_{eff\_withCD} - \mu_{eff\_withoutCD})$. The negative of the slope equals to the stability constant (*K*).

The *K* values presented in this chapter were determined in 50 mM NaH_2PO_4 buffer (pH=7.2) at 25 °C. SAECD concentration of 0.1-10 mM was applied.

4. Results

4.1. Discussion on the DS determination by CE

By using the CE method described in USP, not only the pharmaceutical grade SBEBCDs can be characterized. The DS of the CDs used throughout the study were determined by CE. In Figure 3, electropherograms of SBEBCDs of different DS may be compared.

Table 5. Comparison of SBEBCD samples complying the average DS criterion of USP

SBEBCD sodium peaks	Limit range (peak area %) (USP)	Dexolve® Complying USP (DS=6.5)	Fine chemical grade SBEBCD (DS=6.2)	Fine chemical grade SBEBCD (DS=6.9)
I (DS-1)	0-0.3	0.1	0.2	0.1
II (DS-2)	0-0.9	0.6	1.0 (OOS*)	0.3
III (DS-3)	0.5-5.0	2.2	3.2	1.3
IV (DS-4)	2.0-10.0	5.7	8.1	4.0
V (DS-5)	10.0-20.0	12.2	14.9	9.6 (OOS*)
VI (DS-6)	15.0-25.0	19.9	21.6	17.0
VII (DS-7)	20.0-30.0	25.6	24.6	25.5
VIII (DS-8)	10.0-25.0	22.2	18.3	25.3 (OOS*)
IX (DS-9)	2.0-12.0	9.7	7.1	13.6 (OOS*)
X (DS-10)	0-4.0	1.8	0.9	3.2

* OOS: Out of specification (USP NF 37).

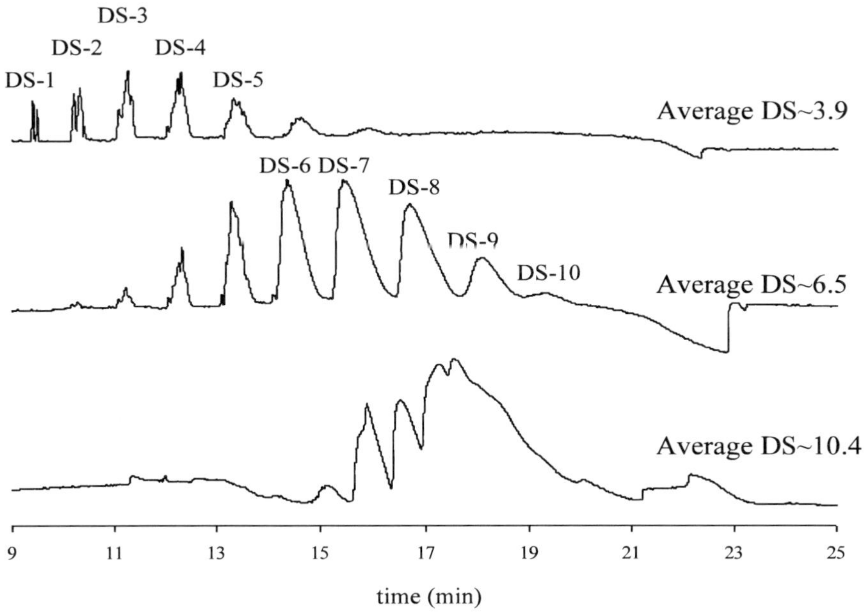

Figure 3. Comparison of electropherograms recorded on SBEBCDs of different DS.

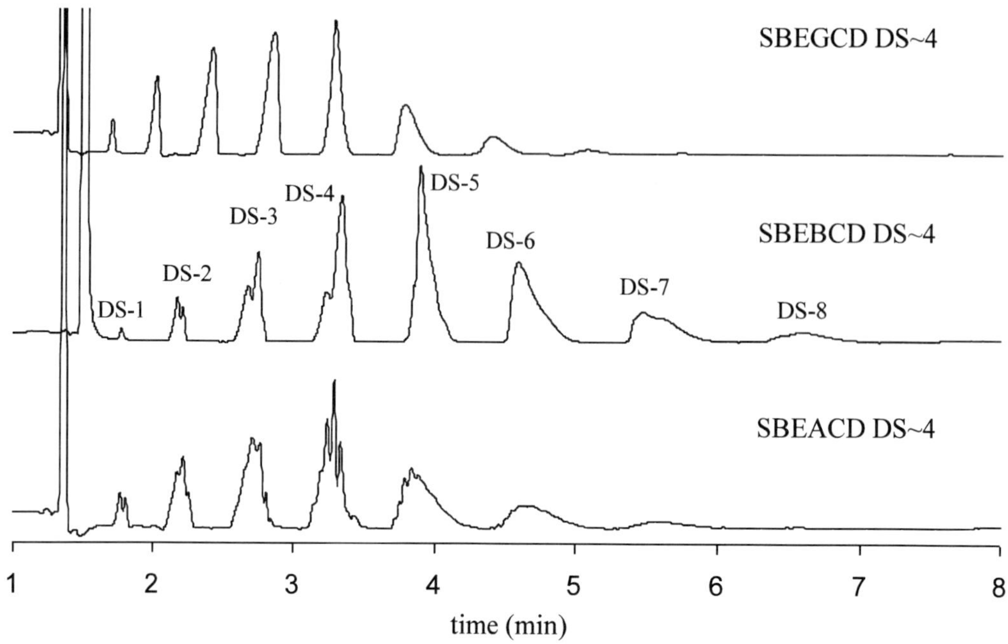

Figure 4. Comparison of electropherograms recorded on sulfobutylether CDs of different cavity sizes.

As Figure 3 shows, the technique is suitable to qualify SBEBCD of approximately DS~8 and below, yet above DS 8, due to significant peak overlapping the correct evaluation is not possible. The quantitative analysis of each peak is provided for theree samples having *average* DS falling in the range of the pharmacopoeial limit (Table 5).

By all means, two of these samples fail to comply the *peak distribution* requirement. As it can be seen, there is a correlation between the average DS and the shape of the Gaussian peak distribution curve of the SBEBCD fractions meaning that an average DS of 6.5-6.6 is ideal to achieve the prescribed distribution, whereas it is difficult to meet with DS values near the limits of the acceptance range. In Figure 4, electropherograms of sulfobutylether CDs of nearly identical DS, but different cavity sizes can be compared.

It may be concluded from the results presented in Figure 4 that all three SAE-CDs of similar DS values and different cavity sizes provide similar fingerprint in the CE analysis. It can also be observed that the resolution of each peak is different for the derivatives: peaks of SBEACD are the most resolved, while the fingerprint of the SBEGCD seems to consist of single peaks.

The suggested underlying reason for this phenomenon is that the apparent interaction strength of the positional isomers in SBEACD is more different towards the buffer component benzoic acid enabling a better resolution of these fractions, whereas the interactions strengths of SBEGCD positional isomers is highly similar towards the benzoic acid providing poor separation in this case.

4.2. Comparison of DS Determination Methods: CE and NMR

The accuracy of the DS determination method was tested by calculation of average DS *via* two independent (both validated) methods: capillary electrophoresis and NMR. As well known from the literature the capillary electrophoretic determination of DS has several disadvantages [23, 48]. The highly substituted fractions of the cyclodextrin composition (migrating last) have different response factors than the lower substituted ones, and thus are differently weighted in the calculation of the DS. This difference leads to different DS values determined using CE and other, independent methods e.g. NMR or elemental analysis. DS value determined by NMR (as shown earlier) is based on the signal areas provided by anomeric glucose and sulfobutyl side-chain protons and therefore bears no such distortion. Table 6 displays the average DS values of five different SBEBCD samples determined with both techniques.

Table 6. Average degrees of substitution determined with CE method according to USP and NMR

Sample	DS with NMR	DS with CE (USP)
Batch 1	6.2	6.7
Batch 2	6.1	6.7
Batch 3	6.1	6.7
Batch 4	6.2	6.6
Batch 5	6.1	6.7

As it can be clearly seen from the data of Table 6, DS determined with CE is approximately 0.4 to 0.6 units higher than DS determined with NMR. These findings confirm the previously reported distortions of the CE method and question its suitability for the determination of real average degree of substitution.

The strengths of the CE method are good resolution, recording a fingerprint eminently suitable for identification, and determination of the DS. Based on the in-house experiences at CycloLab, among weaknesses the poor intermediate precision, moderate pH and BGE composition sensitivity, questionable accuracy and limitation of the characterization of high-DS SAE-CDs can be mentioned.

4.3. Characterization of SBEBCD by HPLC

SBEBCD can be characterized with HPLC methods, as well. Grard et al applied ion pair reversed-phase chromatography with evaporative light scattering detection to separate the component groups based on their degree of substitution [49]. Later they used anion exchange chromatography to obtain the fingerprint of SBEBCD [50].

ChiroQuest Ltd. developed a special HPLC stationary phase (CD-Screen-DAP) for analysis of anionic cyclodextrin derivatives. The separation of SBEBCD components is based on anion exchange and inclusion complexation providing efficient separation of not only the SBEBCD components, but also residual β-CD (BCD) and its synthesis-related impurities,

HBSA, DBSA (Figure 5). The method, however, can not be used for calculation of degree of substitution because of the non-linear characteristics of the ELSD detection method.

Figure 6 shows the comparison of SBEBCD samples having different DS. Comparing these chromatograms to the CE fingerprint in Figure 3 it can be concluded that the HPLC method is better to characterize the qualitative component distribution of SBEBCD than the CE method keeping in mind that the results obtained are not directly comparable due to the non-linear characteristics of the HPLC detector.

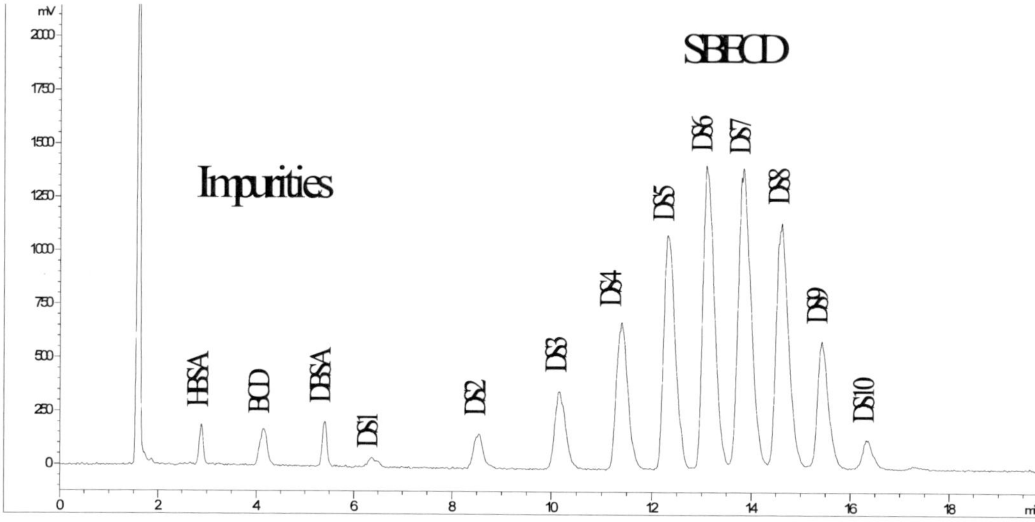

Figure 5. HPLC separation of SBEBCD component groups and impurities using CD-Screen-DAP column. (Triethylamine acetate buffer at pH=5.0 and acetonitrile solvent gradient, ELSD).

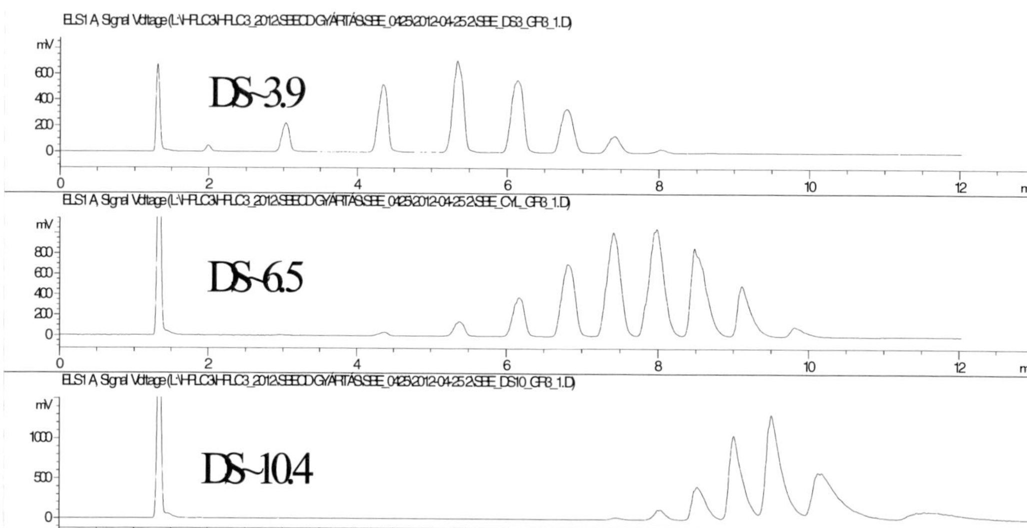

Figure 6. Comparison of SBEBCD component distribution with HPLC method (Column: CD-Screen-DAP, Triethylamine acetate buffer at pH=5.0 and acetonitrile solvent gradient, ELSD).

Table 7. Complex stability constants (M^{-1}) of sulfobutylether-CD complexes of steroid drugs pH=7.2; 25°C

	Ethinyl estradiol	17β-Estradiol	Progesterone	Medroxyprogesterone acetate	Norethisterone	Cyproterone
SBEACD	360 ± 10	330 ± 10	415 ± 10	295 ± 5	500 ± 5	240 ± 5
SBEBCD	**50000** ± 100	**30000** ± 5000	1525 ± 160	NA	8901 ± 1	**1640 ± 5**
SBEGCD	24385 ± 50	NA	**1720** ± 100	**6260** ± 2	**18040** ± 10	1560 ± 5

NA: no data are available because of the irregular shape of the peaks in the electropherogram.

Table 8. Complex stability constants (M^{-1}) of sulfobutylether-CD complexes with drugs of acidic character (or its salt) at pH=7.2; 25°C

	Diclofenac sodium*	Ibuprofen	Naproxen	Ketoprofen	Gemfibrozil	Bezafibrate
SBEACD	100 ± 10	**2200** ± 100	**2640** ± 100	100 ± 10	74 ± 2	150 ± 2
SBEBCD	70 ± 10	1400 ± 100	NA	220 ± 20	**362** ± 2	220 ± 5
SBEGCD	**150** ± 10	55 ± 10	80 ± 10	**290** ± 20	67 ± 3	**590** ± 20

* The very weak basicity of NH-moiety of diclofenac is not relevant relating to complex formation.

NA no data are available because of the irregular shape of the peaks in the electropherogram.

Table 9. Complex stability constants (M^{-1}) of sulfobutylether-CD complexes of basic character at pH=7.2; 25°C

	Nicotine	Carbamazepine	Metoprolol	Verapamil	Caffeine	Sulfamethoxazole
SBEACD	40 ± 1	188 ± 10	480 ± 20	202 ± 2	**237** ± 3	588 ± 2
SBEBCD	**85** ± 10	**547** ± 1	**640** ± 50	**1620** ± 10	87 ± 5	305 ± 5
SBEGCD	**82** ± 2	372 ± 4	95 ± 10	370 ± 10	87 ± 5	**1150** ± 20

Table 10. Complex stability constants (M^{-1}) of sulfobutylether-CD with drugs without peculiar acid/base character under the studied conditions (pH=7.2; 25°C)

	Propofol	Paracetamol
SBEACD	155 ± 2	78 ± 5
SBEBCD	**1288** ± 10	**146** ± 2
SBEGCD	380 ± 5	117 ± 2

4.4. Interaction of Different SAE-CDs with Guest Molecules: Effect of Cavity Size

Twenty drug molecules (which are amongst those pharmacons produced in highest amounts worldwide) were investigated regarding their binding affinity towards sulfobutylated cyclodextrins using CE. This technique is primarily useful for studying charged compounds, but the interaction between uncharged compounds with charged molecules may be also studied by CE.

The sulfobutylated CDs bearing charged substituents are appropriate host molecules for such investigations. Dissolving sulfobutylether CDs in the buffer the migration of the analyte will change in consequence of both the ionic interactions with the anionic sulfobutyl substituents and the hydrophobic interactions with the CD cavity. The dependence on the concentration of the CD gives the possibility to calculate the complex association constants. The higher the change in the migration in the presence of SBEBCDs the higher is the interaction between the host and the guest and the higher association constant is calculated. The effect of the cavity size was studied in the first series of experiments applying sulfobutylether α-, β- and γ-CD derivatives of approximately identical DS (~4). The results are summarized in Tables 7-10. The values obtained for the optimal ring size are printed in bold.

There are only two (ibuprofen and naproxen) among the 20 studied compounds which prefer the small cavity size of SBEACD. The largest cavity size gave the best fit for 7 of the 20 molecules, while the most applied β-CD derivative gave the highest association constants for 10 substances (nicotine forms complexes of low stability with about identical association constants with SBEBCD and SBEGCD).

The highest binding constants were obtained for steroid drugs: ethinyl estradiol/SBEBCD and /SBEGCD as well as 17β-estradiol/SBEBCD followed by the norethisteron/SBEGCD system (Table 7). The association constants of these steroid/SBEBCD complexes are of 10^4 (M^{-1}) order of magnitude.

The association constants of 10^3 (M^{-1}) order of magnitude are still considered to characterize complexes of high stability: norethisterone/SBEBCD and medroxiprogesterone acetate/SBEGCD, naproxen/SBEACD and /SBEBCD, ibuprofen/SBEACD and /SBEBCD, verapamil/SBEBCD, propofol/SBEBCD, progesterone and cyproterone/SBEGCD and /SBEBCD, sulfamethoxazole/SBEGCD systems belong to this group. The other systems showed lower interaction strength.

The stability constants obtained for weak acids (Table 8) as well as drugs of basic character (Table 9) can be compared. The data obtained for paracetamol and propofol (neither of these have acid/base character at physiological pH) are summarized in Table 10.

4.5. Comparison of Interaction Strength of Different Drugs with Charged SBEBCD versus Uncharged HPBCD

A set of drug molecules were selected for model experiments to compare their interaction with CD derivatives of high solubilizing and stabilizing efficiency. Capillary electrophoresis studies were conducted for determination of the binding (association constants). Some of these association constants obtained with SBEBCD were compared with those obtained for the same drugs under identical conditions with HPBCD (these are unpublished data of the authors). Figure 7 clearly shows the benign effect of the presence of the negatively charged substituents on SBEBCD in the case of the drugs with basic character and the opposite for the acidic drugs. The most representative examples are ibuprofen and verapamil. The weak acid ibuprofen prefers HPBCD to SBEBCD, while the association constant of basic verapamil/SBEBCD complex is about 10 times higher than that of verapamil/HPBCD. This specific behavior was confirmed to drugs where high *K* values were obtained. Bezafibrate is the only example amongst the studied drugs, where the opposite trend was found noting that the stability constants were found fairly low. Diclofenac-Na is the sodium salt of an amphoteric compound (diclofenac is both a carboxylic acid and an amine): due to the presence of the diphenilamine moiety it is a very weak base (pK_a = 0.79) [51], therefore at physiological conditions, the carboxylic acid salt feature prevails.

These results show the main advantage of the application of SBEBCD over HPBCD in the complexation of basic drugs mainly those containing nitrogen atom(s).

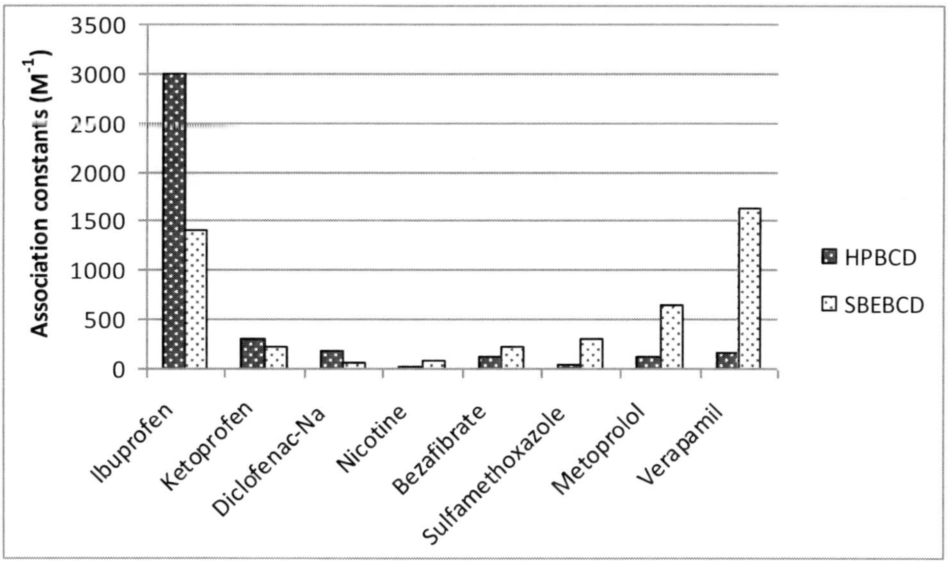

Figure 7. Comparison of HPBCD and SBEBCD in complexing acidic and basic drugs [CycloLab unpublished data].

4.6. Interaction of Different SBEBCD with Guest Molecules: Effect of DS

The effect of degree of substitution on complex stability constants was investigated using 12 test molecules. Many of these molecules prefer the cavity of β-CD (see Tables 7-10). Complex stability constants were determined with CE by calculation using the x-reciprocal method. The data are summarized in Table 11. The stability constants of the most stable complexes are written in bold numbers. The obtained stability constant values were plotted as a function of DS (Figures 8-9).

Table 11. Complex stability constants of some drugs with SBEBCD of different DS (pH=7.2; 25°C)

	Complex stability constants (M^{-1})		
	SBEBCD DS 3.9	SBEBCD DS 6.5	SBEBCD DS 10.4
Gemfibrozil	**362**	175	125
Verapamil	**1620**	750	1030
Sulfamethoxazole	**305**	50	145
Carbamazepine	**547**	500	460
Ketoprofen	**220**	85	85
Hydrocortisone	**2400**	1640	1740
Bezafibrate	220	**640**	65
Paracetamol	146	**260**	195
Voriconazole	314	370	**385**
Diclofenac sodium	70	50	**105**
Metoprolol	640	600	**1120**
Nicotine	85	65	**115**

The compounds were classified into four groups according to the type of trend.

I. Minimum curve (sulfamethoxazole, verapamil, hydrocortisone, metoprolol, nicotine, diclofenac sodium)
II. Maximum curve (bezafibrate, paracetamol)
III. Ascending curve (voriconazole)
IV. Descending curve (gemfibrozil, ketoprofen, carbamazepine)

The selection of the 12 drugs was based on the potential utility and due to the fact that these pharmacons are commonly used. Their structures represent a wide versatility, therefore firm correlation on the found trends and the chemical characteristics should not be made. Nevertheless it is assumed that two major factors influence the affinity of SBEBCDs ranked in increasing DS.

- The bulky sulfobutylether substituents block the cavity of the β-CD exposing a steric hindrance for the binding process (geometric factor).

- By increasing the DS, the charge of the host increases making the SBEBCD molecule more attractive to guest of opposite charge (electrostatic factor).

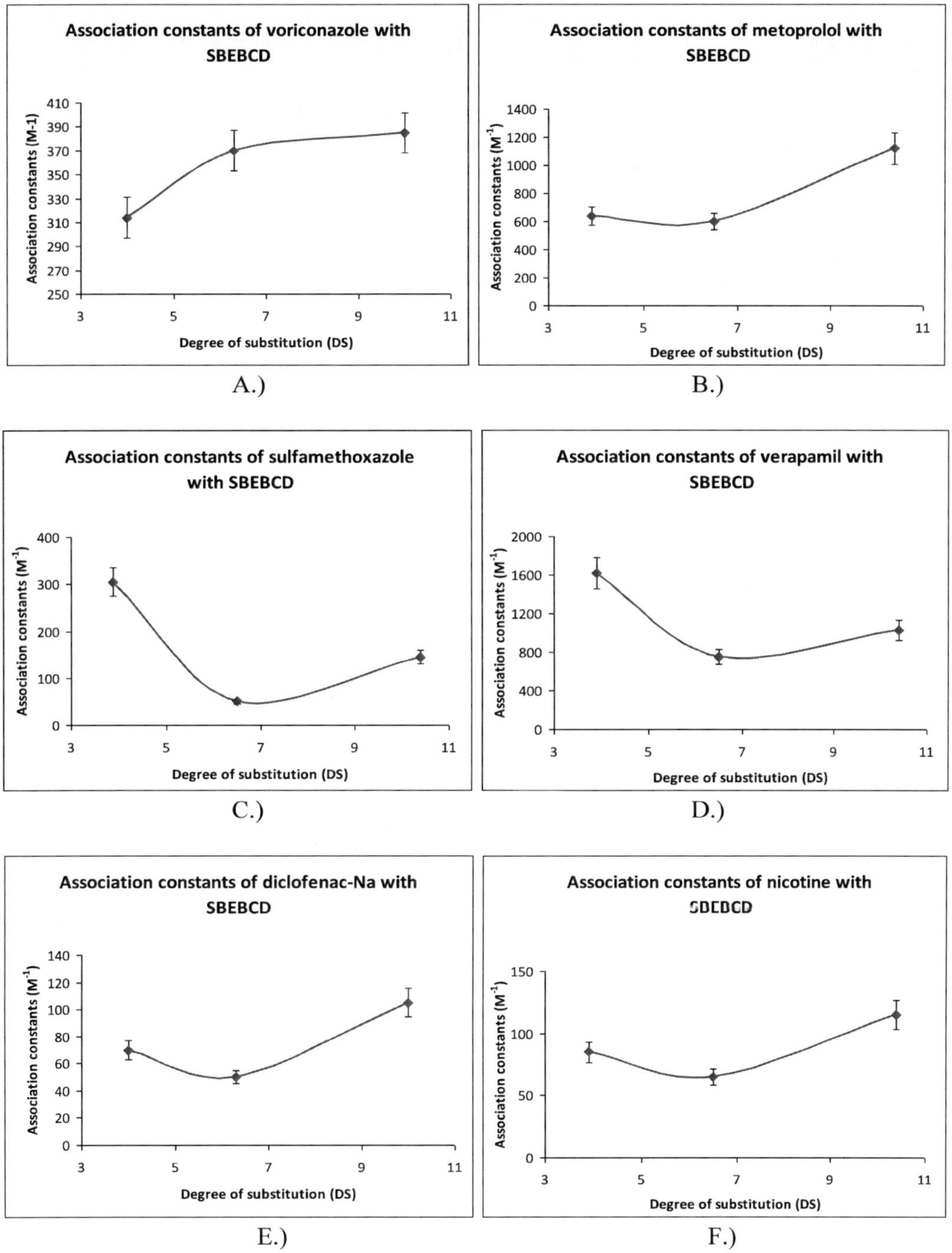

Figure 8. Association constants of different drugs with SBEBCD (at pH=7.2; 25°C) as a function of the DS of the solublizing SBEBCD (ascending or minimum curve types) A.) voriconazole, B.) metoprolol, C.) sulfamethoxazole, D.) verapamil, E.) diclofenac-Na, F.) nicotine.

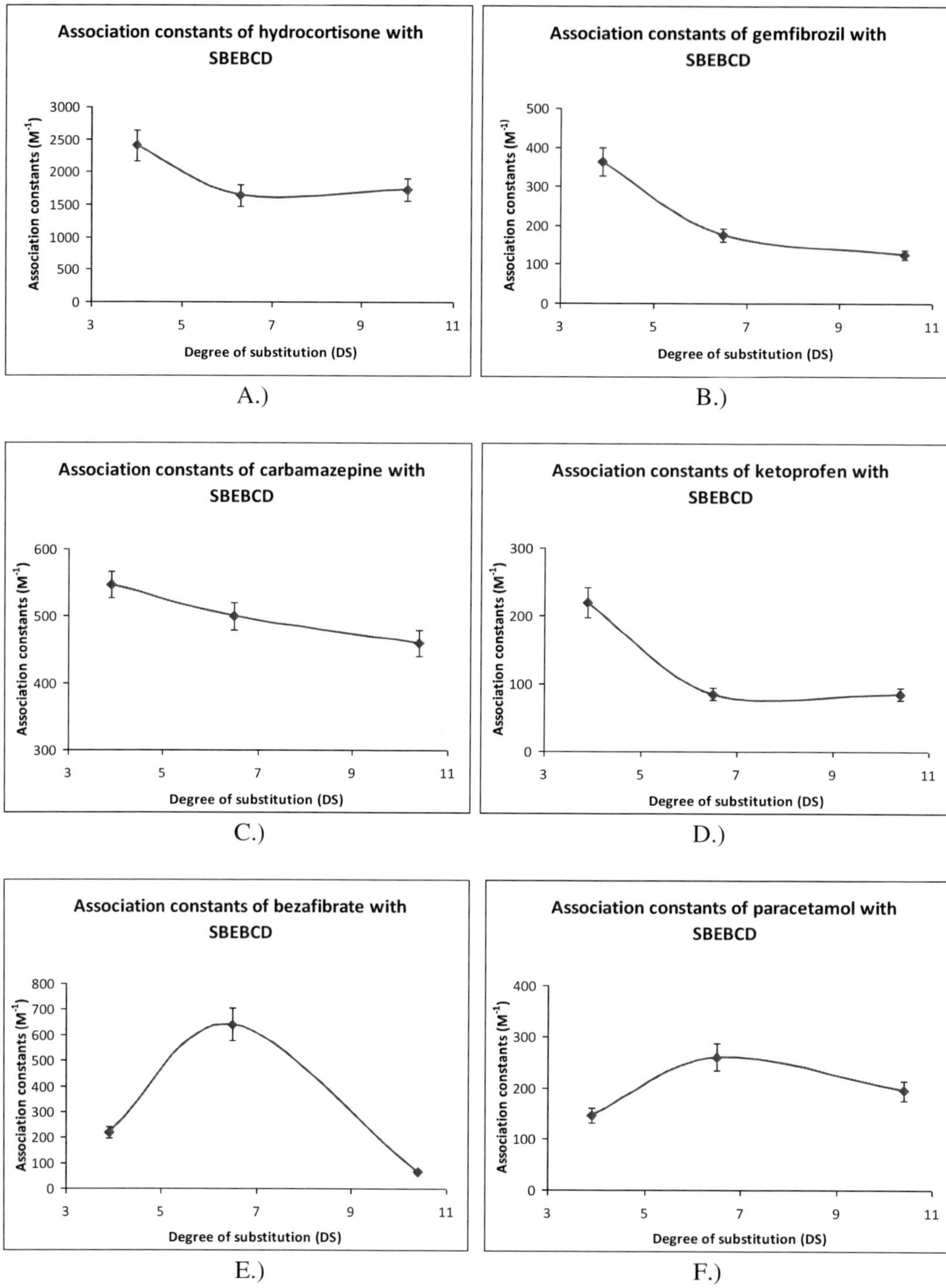

Figure 9. Association constants of different drugs with SBEBCD (at pH=7.2; 25 °C) as a function of the DS of the solublizing SBEBCD (descending or maximum curve types) A.) hydrocortisone, B.) gemfibrozil, C.) carbamazepine, D.) ketoprofen, E.) bezafibrate, F.) paracetamol.

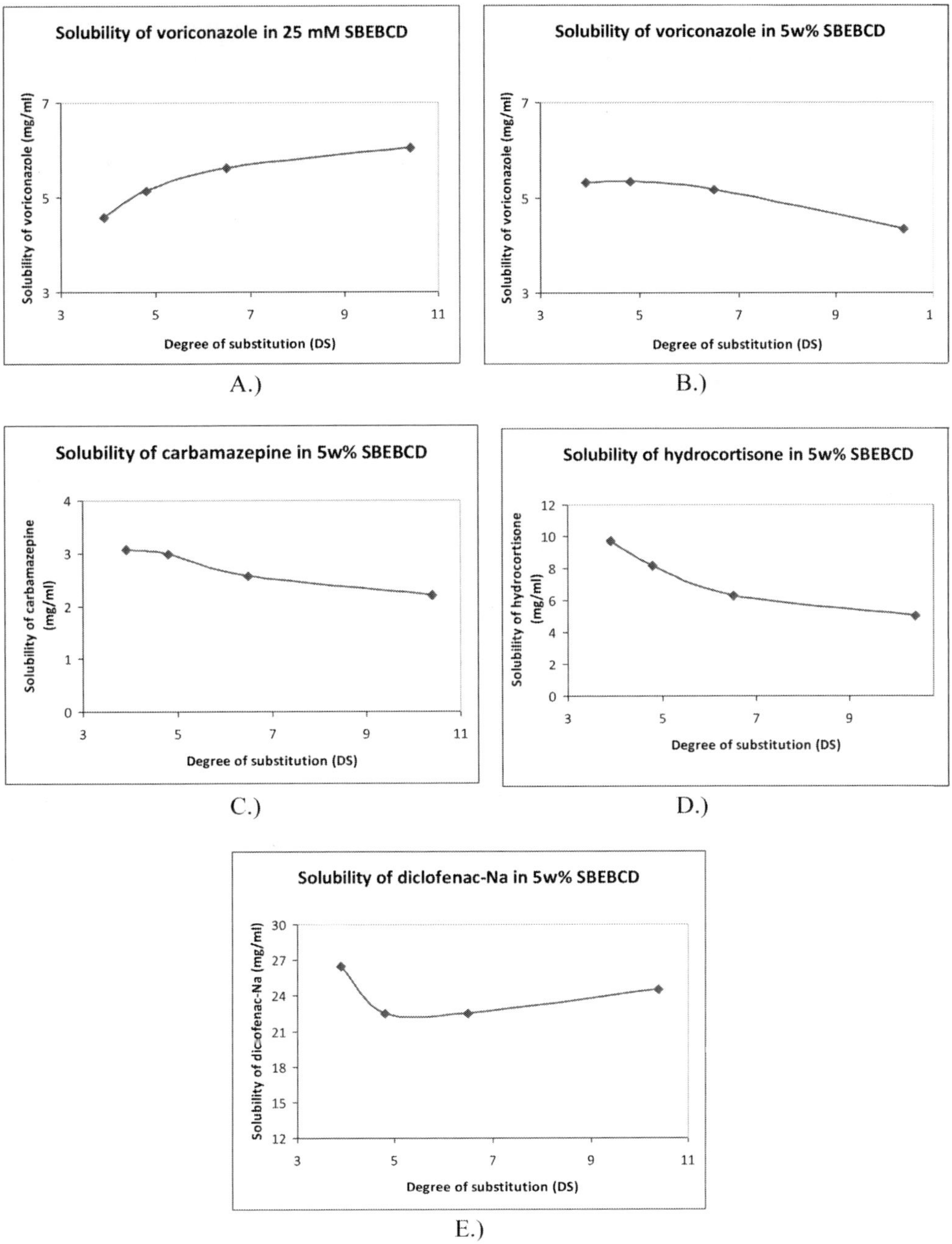

Figure 10: Solubility of different guest molecules as a function of average DS of the solublizing SBEBCD (at a single SBEBCD concentration at 25°C) A.) voriconazole in 25 mM SBEBCD, B.) voriconazole in 5w% SBECD, C.) carbamazepine, D.) hydrocortisone, E.) diclofenac-Na.

In conclusion it may be postulated that in certain cases the association may be primarily driven by geometric factors (the descending trend of carbamazepine may be attributed to the relative bulkiness of the molecule) or primarily by electrostatic factors (the acidic gemfibrozil and ketoprofen show descending trend, while voriconazole especially abundant in basic nitrogen containing moieties show ascending trend). For the minimum and maximum curve type drugs, probably the resultant of both factors influence the interaction.

The interaction of drugs with SBEBCDs having different DS was also studied using the phase solubility method. The effect of DS on the solubility of guest substances was investigated using four test molecules. The usual representation of the phase solubility tests is the expression of dissolved drug amount versus CD concentration (in w% units). This interpretation of the data gives information directly applicable for the pharmaceutical technologists. On the other hand, the complex stability constants characterize the interaction on molar basis. The average molecular weight of SBEBCD highly depends on DS: SBEBCD of DS 3.9 has the average molecular weight of 1752 g/mol, while the same value for SBEBCD of DS 10.4 is 2780 g/mol. The example of voriconazole shows that the way of ranking SBEBCD samples of different DS regarding solubilizing potential might be different if the same molar quantity (Figure 10A) or the same w% is applied (Figure 10B). The voriconazole solubility enhancing effect of high DS SBECD was found the highest if the solutions contained the same molar amount of SBECD (which is very well correlating with the trend found for the association constant determinations). On the other hand, the high average molecular weight SBECD compositions perform the poorest if the same weight of SBEBCD was added to the solubility test samples.

For those drugs which gave descending or minimum type trends in their stability constant versus DS functions (carbamazepine, hydrocortisone, diclofenac-Na), the solubility versus DS curves (expressed in w%) are well in line with the trend of the *K* values (Figures 10C-E). This finding was expected, since in such cases the molecular weight differences between the SBEBCD samples theoretically would not change the trend to its opposite when transforming the data from molar quantities to weight based concentrations.

By all means, from the pharmaceutical technology perspective, the weight / volume (or weight / weight) concentration has still higher relevance than the scientifically sound thermodynamic (molarity based) approach.

CONCLUSION

In present chapter the possible impact of the degree of substitution and number of anhydroglucose units on the applicability of sulfobutylated cyclodextrins were studied in several aspects.

First, the utility of three different analytical techniques (HPLC, capillary electrophoresis and NMR) for studying these cyclodextrin compositions were evaluated and compared. The strength of NMR is that the determined DS value is not distorted since the result is based on the signal areas corresponding to the anomeric glucose and sulfobutyl side-chain protons. On the other hand NMR only provides the average DS value, while by capillary electrophoresis (CE) the distribution profile may be obtained, too. The drawback of using CE is poor intermediate precision, moderate pH and background electrolyte composition sensitivity, questionable accuracy and limitation is the characterization of high-DS SAE-CDs. An alternative HPLC method was elaborated to overcome such disadvantages. By this technique the separation of SBEBCD components is based on anion exchange and inclusion complexation providing efficient separation of not only the SBEBCD components, but also residual beta-cyclodextrin and the synthesis related impurities. The HPLC method, however, can not be used for calculation of degree of substitution because of the non-linear characteristics of the evaporative light scattering detection method.

CE and solubility tests were utilized for the characterization of the host-guest type interactions between drug molecules and various sulfobutyl ether cyclodextrin compositions. Comparative data were collected how SAE-CD compositions differing from the favored, highly specified grade complying US Pharmacopoeia 37 may be used for pharmaceutical purposes. The aim of the study was to establish grounds for future extension of the applicability of sulfobutylated cyclodextrins in the field of e.g. pharmaceuticals, analytical method development and environmental technologies. Based on the collected data it has been demonstrated that multiple factors may play significant role in the binding process ensuring some sort of selectivity amongst the members of the cyclodextrin library used. Besides the dimensions of the cavities, the steric effect of the relatively bulky sulfobutyl ether moieties as well as the charge born on the terminal of the substituent chains influence the strength of interaction.

References

[1] Adam, JM; Bennett DJ; Bom, A; Clark, JK; Feilden, H; Hutchinson, EJ; Palin, R; Prosser, A. Rees, DC; Rosair, GM; Stevenson, D; Tarver, GJ; Zhang MQ. Cyclodextrin-derived host molecules as reversal agents for the neuromuscular blocker rocuronium bromide: synthesis and structure-activity relationships. *J. Med. Chem.*, 2002, 45, 1806-1816.

[2] Jicsinszky, L; Iványi, R. Catalytic transfer hydrogenation of sugar derivatives. *Carbohydr. Polym.*, 2001, 45(2) 139-145.

[3] Tang, W; Ng, SC. Facile synthesis of mono-6-amino-6-deoxy-α-, β-,γ-cyclodextrin hydrochlorides for molecular recognition, chiral separation and drug delivery; *Nature Protocols*, 2008, 3(4) 691-697.

[4] Krauter, I; Herrmann, W; Wenz, G. Self organization of fluorescent molecular necklaces in aqueous solution. *J. Inclusion Phenom. Mol. Recognit. Chem.*, 1996, 25(1-3), 93-96.

[5] Chmurski, K; Defaye, J. An improved synthesis of per(6-deoxyhalo) cyclodextrins using n-halosuccinimides - triphenylphosphine in dimethylformamide. *Proceedings of the Ninth International Symposium on Cyclodextrins*, 1999, 183-186.

[6] Chmurski, K; Defaye, J. An improved synthesis of per(6-deoxyhalo) cyclodextrins using n-halosuccinimides - triphenylphosphine in dimethylformamide. *Supramolecular Chemistry,* 2000, 12(2), 221-224.

[7] Zhu, W; Vigh, G. A family of single-isomer, sulfated γ-cyclodextrin chiral resolving agents for capillary electrophoresis: Octa(6-O-sulfo)-γ-cyclodextrin. *Electrophoresis*, 2003, 24 (1-2) 130–138.

[8] Umezawa, S; Tatsuta, K. Studies of aminosugars. XVIII. Syntheses of amino derivatives of Schardinger α-dextrin and raffinose. *Bulletin of the Chemical Society of Japan*, 1968, 41(2), 464-468.

[9] Ashton, PR; Königer, R; Stoddart, JF.; Amino acid derivatives of β-cyclodextrin. *J. Org. Chem.*, 1996, 61, 903-909.

[10] Lipták, A; Fügedi, P; Szurmai, Z; Imre, J; Nánási, P; Szejtli, J. The chemistry of cyclodextrin derivatives. J. Szejtli (Ed.): Proceedings of the First International Symposium on Cyclodextrins. *Advances in Inclusion Science,* 1982, 1, 275-287.

[11] Luna, EA; Vander Velde, DG; Tait, JR; Thompson, DO; Rajewski, RA; Stella, VJ. Isolation and characterization by NMR spectroscopy of three monosubstituted 4-sulfobutyl ether derivatives of cyclomaltoheptaose (β-cyclodextrin). *Carbohydrate Research,* 1997, 299(3), 111-118.

[12] Luna, EA; Bornancini, ERN; Thompson, DO; Rajewski, RA; Stella VJ. Fractionation and characterization of 4-sulfobutyl ether derivatives of cyclomaltoheptaose (β-cyclodextrin). *Carbohydrate Research,* 1997, 299(3), 103-110.

[13] Sollogoub, M. Site-selective heterofunctionalization of cyclodextrins. Discovery, development and use in catalysis. *SYNLETT,* 2014, 24, 2629-2640.

[14] Frömming, K-H; Szejtli, J. Cyclodextrins in Pharmacy, *Series: Topics in Inclusion Science,* Vol. 5, Dordrecht: Kluwer Academic Publishers, 1994, 24-28.

[15] Fenyvesi, É; Szemán, J; Csabai, K; Malanga, M; Szente, L. Methyl-beta-cyclodextrins: the role of number and types of substituents in solubilizing power. *Journal of Pharmaceutical Sciences*, 2014, 103, 1443-1452.

[16] Fenyvesi, É. Approved pharmaceutical products containing cyclodextrins. *Cyclodextrin News*, 2013, 27(2), 1.

[17] Lindberg, B; Pitha, J. Regioselective substitutions in cyclodextrins. PCT Patent application WO9012035, 1990.

[18] Okimoto, K; Ohike, A; Ibuki, R; Aoki, O; Ohnishi, N; Rajewski, RA; Stella, VJ; Irie, T; Uekama, K. Factors affecting membrane-controlled drug release for an osmotic pump tablet (OPT) utilizing $(SBE)_{7m}$-β-CD as both a solubilizer and osmotic agent. *Journal of Controlled Release*, 1999, 60(2–3), 311–319.

[19] Miyajima, M; Ozeki, T; Stella, VJ. Binding constants for aromatic amino acids and their derivatives with sulfobutyl ether β-cyclodextrin determined using capillary electrophoresis. *Journal of Drug Delivery Science and Technology*, 2004, 14(5), 383-387.

[20] Zia, V; Rajewski, RA; Stella, VJ. Effect of cyclodextrin charge on complexation of neutral and charged substrates: comparison of $(SBE)_{7M}$-beta-CD to HP-beta-CD. *Pharm Res.*, 2001, 18(5), 667-673.

[21] Stella VJ; He Q. Cyclodextrins. Toxicologic Pathology, 2008, 36, 30-42.

[22] Parmerter SM; Allen EE; Hull GA. Cyclodextrins with anionic properties. US patent 3426011, 1969.

[23] Luna, EA; Bornancini, ERN; Tait, RJ; Thompson, DO; Stobaugh, JF; Rajewski, RA; Stella, VJ. Evaluation of the utility of capillary electrophoresis for the analysis of sulfobutyl ether β-cyclodextrin mixtures, *Journal of Pharmaceutical and Biomedical Analysis*, 1996, 15(1), 63-71.

[24] Stella, VJ; Rajewski, RA. Derivatives of cyclodextrins exhibiting enhanced aqueous solubility and the use thereof. PCT patent application WO9111172, 1991.

[25] Stella, VJ; Rajewski, RA. Derivatives of cyclodextrins exhibiting enhanced aqueous solubility and the use thereof. PCT patent application WO9402518, 1994.

[26] Official Monographs for NF 30 (First Supplement). The United States Pharmacopeial Convention, 2012.
[27] Antle, V. Sulfoalkyl ether cyclodextrin compositions. PCT patent application WO2009134347, 2009.
[28] Antle, V. Sulfoalkyl ether cyclodextrin compositions. PCT patent application WO2013130666, 2013.
[29] Antle, V. Alkylated cyclodextrin compositions and processes for preparing and using the same PCT patent application WO2014066274, 2014.
[30] Sklavounos, C; Shah, BK. Process for making a cyclodextrin. US Patent US6153746, 2000.
[31] Pipkin, JD; Mosher, GL; Hecker, DB. Sulfoalkyl ether cyclodextrin compositions and methods of preparation thereof. US patent 7629331, 2009.
[32] Mosher, GL; Pipkin JD; Hecker, DB. Sulfoalkyl ether cyclodextrin compositions and methods of preparation thereof. US patent 8049003, 2011.
[33] Mosher, GL; Pipkin JD; Hecker, DB. Sulfoalkyl ether cyclodextrin compositions and methods of preparation thereof. US patent 8829182, 2014.
[34] Mosher, GL; Pipkin JD; Hecker, DB. Sulfoalkyl ether cyclodextrin compositions and methods of preparation thereof. US patent 8846901, 2014.
[35] Iványi, R; Jicsinszky, L; Szejtli, HG; Vadász, Z; Szente, L. Process for preparing sulphobutylated cyclodextrins of pharmaceutical additive quality. Hungarian patent HU228817, 2013.
[36] Stella VJ; Lee HK; Thompson DO. The effect of SBE4-β-CD on i.m. prednisolone pharmacokinetics and tissue damage in rabbits: Comparison to a co-solvent solution and a water-soluble prodrug. *International Journal of Pharmaceutics*, 1995, 120(2), 197-204.
[37] Järvinen, K; Järvinen, T; Thompson, DO; Stella, VJ. The effect of a modified beta-cyclodextrin, SBE4-beta-CD, on the aqueous stability and ocular absorption of pilocarpine. *Curr. Eye Res.*, 1994, 13, 897-905.
[38] Stella, VJ; Lee, HK; Thompson DO. The effect of SBE4-β-CD on i.v. methylprednisolone pharmacokinetics in rats: Comparison to a co-solvent solution and two water-soluble prodrugs. *International Journal of Pharmaceutics,* 1995, 120(2), 189-195.
[39] Gorecka, BA; Sanzgiri, YD; Bindra, DS; Stella, VJ. Effect of SBE4-β-CD, a sulfobutyl ether β-cyclodextrin, on the stability and solubility of O6-benzylguanine (NSC-637037) in aqueous solutions. *International Journal of Pharmaceutics,* 1995, 125(1), 55-61.
[40] Qu, Q; Tucker, E; Christian, SD. Sulfoalkyl ether β-cyclodextrin derivatives: Synthesis and characterizations. *Journal of inclusion phenomena and macrocyclic chemistry*, 2002, 43(3-4) 213-222.
[41] Thompson, DO. Cyclodextrins-enabling excipients: their present and future use in pharmaceuticals. *Critical reviews in therapeutic drug carrier systems*, 1997, 14(1), 1-104.
[42] Mosher, GL; Johnson, KT; Gayed, AA. US patent application US2003216353, 2003.
[43] de Boer, T; de Zeeuw, RA; de Jong, GJ; Ensing, K. The use of charged cyclodextrins in capillary electrophoresis for chiral separations in pharmaceutical analysis. *Electrophoresis*, 2000, 21, 3220-3239.

[44] Zia, V; Rajewski, RA; Stella, VJ; Thermodynamics of binding of neutral molecules to sulfobutyl ether beta-cyclodextrins (SBE-beta-CDs): the effect of total degree of substitution. *Pharm Res.*, 2000, 17(8), 936-941.

[45] Skanchy, DJ; Xie, GH; Tait, RJ; Luna, E; Demarest, C; Stobaugh, JF. Application of sulfobutylether-beta-cyclodextrin with specific degrees of substitution for the enantioseparation of pharmaceutical mixtures by capillary electrophoresis. *Electrophoresis,* 1999, 20(13), 2638-49.

[46] Rundlett, KL; Armstrong, DW. Examination of the origin, variation, and proper use of expressions for the estimation of association constants by capillary electrophoresis, *J. Chromatogr.* A, 1996, 721, 173-186.

[47] Wallingford, RA; Ewing, AG. Capillary electrophoresis. *Adv Chromatogr.* 1989, 29, 1-76.

[48] Tait, RJ; Skanchy, DJ; Thompson, DP; Chetwyn, NC; Dunshee, DA; Rajewski, RA; Stella, VJ; Stobaugh, JF. Characterization of sulphoalkyl ether derivatives of β-cyclodextrin by capillary electrophoresis with indirect UV detection. *J Pharm Biomed Anal*, 1992, 10(9), 615-622.

[49] Grard, S; Elfakir, C; Dreux, M. Sulfobutyl ether-β-cyclodextrin fingerprint using ion pair reversed-phase chromatography. *Chromatographia,* 1999, 50(11-12), 695-700.

[50] Grard, S; Elfakir, C; Dreux M. Characterization of sulfobutyl ether-beta-cyclodextrins mixtures by anion-exchange chromatography using evaporative light scattering detection. *J Chromatogr A.,* 2000, 897(1-2), 185-93.

[51] Gordon, AJ; Ford, RA. *The chemist's companion: a handbook of practical data, techniques, and references.* Wiley, 1972, 59.

In: Cyclodextrins
Editor: Francis G. Ramirez

ISBN: 978-1-63482-788-1

Chapter 11

DYE INCLUSION SINGLE CRYSTALS: PREPARATIONS, STRUCTURES AND FUNCTIONS

Youichi Tsuchiya*
Institute of Systems, Information Technologies and Nanotechnologies (ISIT), Fukuoka industry-academia Symphonicity (FiaS), Nishi-ku, Fukuoka, Japan

ABSTRACT

The dyes which have the large π system easily associate themselves. To inhibit their π–π stacking, host-guest interaction with cyclodextrins are often employed in water medium. Porphyrins and fullerenes are such typical dyes as guest molecules. Still there are many reports for their host-guest interactions, the structures of their inclusion complexes have been discussed only on the basis of the theoretical calculations and the spectral properties in solution, e.g. NMR and absorption spectra.

It is really surprising that, so far, nobody knows the real complex structure determined from single-crystal analysis. In 2011, we reported the first successful isolation of a dye inclusion single crystal (DISC) between trimethyl-β-cyclodextrin (TMβCD) and tetrapyridyl porphyrin, and analyzed its X-ray crystal structure. It revealed the real complex structure which has the prospective barrel shape bicapped complex.

Furthermore, the X-ray crystal structure provided several newly information in the molecular conformation. Recently, several crystal structures of dye inclusion complexes using cyclodextrin were reported from other groups. The crystallography of dye inclusion complex should invite the new field of functional materials designed by supramolecular chemistry.

* Corresponding author: Youichi Tsuchiya. Institute of Systems, Information Technologies and Nanotechnologies (ISIT), Fukuoka industry-academia Symphonicity (FiaS), 4-1, Kyudai Shinmachi, Nishi-ku, Fukuoka, 819-0388 Japan. E-mail: tsuchiya@isit.or.jp.

INTRODUCTION

In nature, the chromophoric groups typified by the heme play various roles to produce the biological functions, for example: oxygen carrier in blood, light-sensitizer in photosynthesis, electron transporter in respiration and active site in enzyme.

These functions are achieved in the many case by the control of their aggregation mode or the isolation from the external environment. In artificial systems, the accumulation of chromophores is often required to construct the high performance functional systems, e.g. photon-to-electron conversion system. Many researchers constructed the dye accumulated systems through controlling of its aggregation mode [1, 2]. Similarly, inhibiting the aggregation of dyes is one of the large research targets to bring out their functions [3-17].

The inclusion ability of cyclodextrins (CDs) is often employed to inhibit the π–π stacking of dyes which have the large π system [7]. Specially, the host-guest chemistry between CDs and porphyrins has attracted great interest from supramolecular chemists because of their unique structures and functions mimicking the heme proteins [8-17].

The first host-guest complex between CD and porphyrin was reported in 1990 [18]. Manka et al. isolated the complex as a precipitate from the aqueous solution of tetraamino-tetraphenyl porphyrin with 10 equiv. of heptakis (2,6-di-*O*-methyl)-β-cyclodextrin (2,6-DMβCD) by the addition of sodium tetraphenylborate (Figure 1). They proposed that the porphyrin was encapsulated with two CDs as a pseudo-rotaxane from the results of NMR, MS, UV-titration, combustion analysis, and CPK modeling. Since then, many host-guest systems have been reported by using CD and porphyrin system. In these reports, the structures of complexes have been discussed only on the basis of the spectral properties in solution and theoretical calculations. In 2011, we reported the single-crystal structure for the inclusion complex of porphyrin bicapped with CDs [19]. It was the first successful example of a dye inclusion single crystal (DISC) between CDs and porphyrins.

This chapter will describe the preparation techniques of DISCs, the structural details about the dye inclusion complex analyzed by X-ray and the properties of porphyrin DISC.

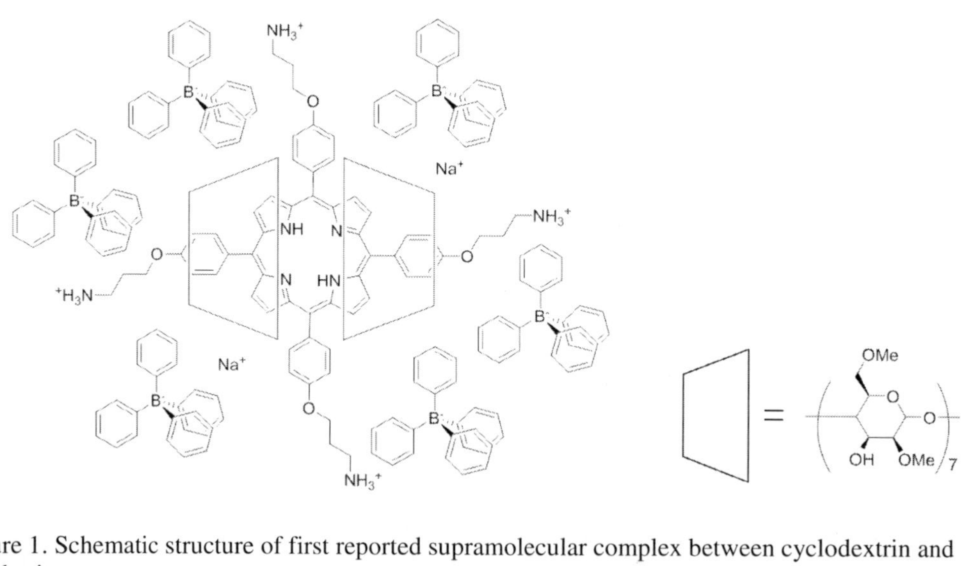

Figure 1. Schematic structure of first reported supramolecular complex between cyclodextrin and porphyrin.

How to Isolate Discs of Cyclodextrin-Porphyrin Complex

To isolate the DISC, it should be necessary to obtain the stable complex solution as a mother liquor. In general, the host-guest complexes are prepared by using CD under the thermodynamically equilibrium system in the aqueous solution. The major driving force of guest inclusion of CDs is hydrophobic interaction. When the cavity size of CD in the aqueous dye solution matches the dye size, the dyes are included into the CD cavity.

Meanwhile, the many water soluble dyes which have the large π surface form small aggregates in water because of their hydrophobicity and π–π interaction.

These two equilibrium systems of inclusion/elimination of CD and association/ dissociation of dye make difficult to prepare DISCs. Fortunately, there is the detail information for the interaction between CDs and porphyrins. Kano et al. reported the details of interaction between the several water soluble tetraaryl porphyrins and the various CDs (Figure 2) [20-23]. These reports give us the three important information. First, the cavity size of CD is important to interact with tetraaryl porphyrin.

β-CD consisting seven glucoses has the highest inclusion ability to *meso*-aryl group on the porphyrin in comparison to α-CD and γ-CD, which are consisting six and eight glucose units respectively. Second, the ring flexibility of CD affects to its binding ability. 2,3-DMβCD or heptakis (2,3,6-tri-*O*-methyl)-β-cyclodextrin (TMβCD) have higher inclusion ability than 2,6-DMβCD or unmodified β-CD. This difference is due to the flexibility of CD ring. The CD rings of β-CD and 2,6-DMβCD are stabilized by the intramolecular hydrogen-bonds between O2 and O3 on the neighboring glucoses. On the other hand, the CD rings of 2,3-DMβCD and TMβCD have the flexibility by the loss of secondary hydroxyl groups. As a result, they can enjoy more strong interaction by the deformation of CD ring, calling induced-fit type interaction, with guest porphyrin than β-CD and 2,6-DMβCD. Third, more hydrophobic *meso*-aryl group of porphyrin has stronger interaction to the CD cavity.

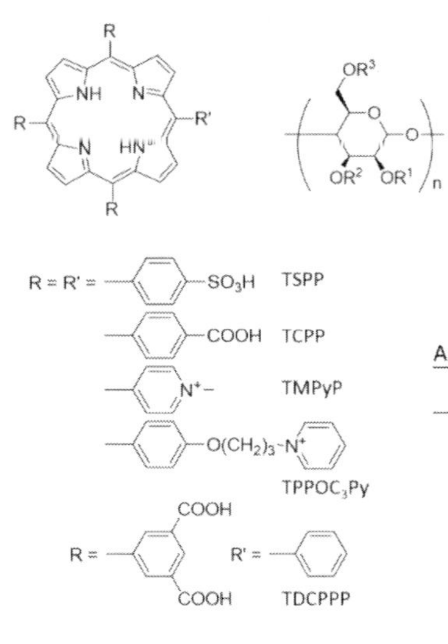

Association constants between TMβCD and porphyrins.[17]

	K_1/M^{-1}	K_2/M^{-1}
TSPP	2.0 x 10^4	5.8x10^4
TCPP	1.7 x 10^4	2.0x10^5
TMPyP	-	-
TPPOC₃Py	2.3 x 10^3	9.4x10^3

(in EG: H_2O = 3: 1)

Association constant between CDs and TDCPPP. [20]

	n	R^1	R^2	R^3	K/M^{-1}
α-CD	6	H	H	H	33
TMαCD	6	CH_3	CH_3	CH_3	-
β-CD	7	H	H	H	1.2 x 10^3
2,6-DMβCD	7	CH_3	H	CH_3	1.2 x 10^4
2,3-DMβCD	7	CH_3	CH_3	H	8.5 x 10^6
TMβCD	7	CH_3	CH_3	CH_3	6.9 x 10^6
γ-CD	8	H	H	H	6.4 x 10^2
TMγCD	8	CH_3	CH_3	CH_3	2.9 x 10^4

Figure 2. Association constant between various cyclodextrins and porphyrins reported in literature [17, 20].

These pieces of information suggest that the flexible β-CD, e.g. TMβCD, and hydrophobicaryl porphyrin should be the best pair to obtain the stable inclusion complex in aqueous solution.

As for a guest porphyrin, 5,10,15,20-tetrapyridin-4-ylporphyrin (TPyP) was selected because of its water solubility. The water solubility of TPyP can be adjusted by solution pH and the compound disperses well in acidic aqueous solution. Basically, there is scarce interaction between *meso*-tetra pyridinium groups and the CD cavity in acidic solution.

This is attributed primarily to the enhanced hydrophilicity of the *meso*-pyridinium groups and secondly to the slightly positively polarized interior cavity of CDs [22, 23].

As a result, protonated TPyP shows only a low affinity for CDs. Whereas in neutral solution, TPyP becomes water-insoluble and the hydrophobic *meso*-pyridyl groups have more chance to be included into the CD cavity. As expected, the acidic TPyP solution (0.1 mM) provided the stable inclusion complex solution by the neutralization in the presence of 4 eq. TMβCD. This complex solution showed the absorption spectrum as an isolated TPyP with the circular dichroism signals [24].

To isolate the DISCs, we have still another problem; the mother liquor for the crystallization must be put into the supersaturation condition. There are several ideas to make the supersaturation state. The most major idea is cooling of the solution. However, no crystals were obtained by cooling of TMβCD·TPyP complex solution. Next idea should be increase of the concentration by vaporization of solvent. However, the crystals of TMβCD and TPyP itself were deposited separately. It is because that the mother liquor containing large excess amount of TMβCD versus TPyP, and there is the elimination of TPyP from host cavity. The poor solvent often employ to the crystallization. Thus, several organic solvents were introduced to the TMβCD·TPyP complex solution. As a result, the guest TPyP quickly precipitated. It is because that the small poor solvent molecules interact to the hydrophobic CD cavity. Therefore, the poor solvent inhibits the interaction between CD and guest.

We noticed that there are some reports which prepare the inclusion crystals of CD at the high temperature [25]. We thus employed the high temperature crystallization to isolate DISCs. A TMβCD·TPyP DISC was obtained from an aqueous solution of TPyP with excess TMβCD at 60 °C. After the incubation for 3 days, the red-brown rhombohedral crystal was isolated (Figure 3a). It may be a little curious that crystallization of the TMβCD·TPyP complex was performed at such high temperature.

In aqueous solution, CDs and their derivatives are strongly solvated with water molecules through hydrogen-bonding interactions. Under heating, it is expected that the intermolecular hydrogen-bonds are cleaved and the hydrophobic nature becomes the stronger effect. In particular, TMβCD behaves as a more hydrophobic host molecule than β-CD and TMβCD complexes in water are more stabilized at higher temperatures. As a result, high temperature condition provides the advantages for crystal growth of DISCs. However, when 2,6-DMβCD was used instead of TMβCD, colorless crystals of 2,6-DMβCD and dark purple micro-crystals of TPyP were obtained separately.

This difference is caused by the flexibility of their CD ring. 2,6-DMβCD have less flexibility of CD ring by the intramolecular hydrogen-bonding between the hydroxyl group on a glucose C3 and the methoxy group on the neighboring glucose C2. Therefore, flexible TMβCD can enjoy the stronger binding with the aryl porphyrin than rigid 2,6-DMβCD through the induced-fit type interaction. Accordingly, TMβCD provide TPyP DISCs, but 2,6-DMβCD does not provide it.

These methods to provide the mother liquor and to isolate DISC can be apply to other porphyrin which has pH-dependent solubility change. TCPP can be solubilized into water by alkaline dissociation of the carboxylic acid groups. In the solution state, there are several preceding studies on the inclusion complex between TMβCD and TCPP [8-10]. The binding constants are K_1=1.7 x $10^4 M^{-1}$ (for a 1:1 complex) and K_2=2.0 x $10^5 M^{-1}$ (for a 1:2 bicapped complex) in aqueous buffer solution at pH 7 [17].

Therefore, the gradual acid neutralization of the dissociated carboxylate groups would further enhance the association constants between TMβCD and TCPP. In fact, A DISC of TMβCD·TCPP complex was obtained from an aqueous solution of TCPP containing an excess amount of TMβCD at 50 °C (Figure 3b). For the highly water-soluble 5,10,15,20-tetrakis(4-sulfonatophenyl)-porphyrin (TSPP), it was difficult to isolate the DISC by using the same procedure. In such a case, a high salt concentration and high temperature may assist to isolation of the inclusion complex. However, the crystal quality of TMβCD·TSPP was not sufficient for X-ray analysis.

When more hydrophobic guests typified with 5,10,15,20-tetraphenyl porphyrin (TPP) are used, it is difficult to prepare the complex solution as a mother liquor. Komatsu et al. provided us a solution for this problem. They reported that the supramolecular complexation of fullerenes C_{60} with γ-CD was achieved by the mechanochemical method [26]. The aqueous solution of γ-CD·C_{60} complex shows the same spectra to monomolecular dispersed C_{60} in organic solvent. Therefore, we apply to prepare the complexation between TMβCD and hydrophobic tetraaryl porphyrins. The microcrystals of TPP and excess TMβCD were finely ground in an agate mortar. After dispersing the ground mixture into water with stirring, the cloudy solution was filtered to obtain a clear solution that contained the inclusion complex. Then, the TMβCD·TPP DISC was obtained by keeping the solution at 50°C (Figure 3c). This method is widely applicable to complex formation between CDs and hydrophobic guests, e.g. metallo porphyrins. The Ni^{II}-coordinated TPP (Ni-TPP) DISC was also obtained by using the same procedure (Figure 3d). It should be the same for the other metal coordinated TPPs.

STRUCTURES OF CYCLODEXTRIN-PORPHYRIN COMPLEX IN CRYSTALLINE STATE

Figure 4a and 4b shows the crystal structures and crystal packing of TPyP, TCPP, TPP and Ni-TPP DISCs [24]. Regardless of the difference in the *meso*-aryl groups, all complex structure consist of one porphyrin capped with two TMβCD molecules.

Figure 3. Photograph of the DISCs between TMβCD and (a) TPyP, (b) TCPP, (c) TPP or (d) Ni-TPP.

In addition, two opposite *meso*-aryl groups of the guest porphyrins penetrate into the TMβCD cavity from the O2, O3 side to the O6 side. This pseudo-rotaxane type bicapped host-guest complex structure is consistent with that proposed, so far, for the solution complexes on the basis of the spectral studies [8-18]. Very interestingly, TMβCD·TCPP DISC has the different crystal system to other three porphyrins, in spite of the same complex structure. DISC of TPyP, TPP, and Ni-TPP complexes commonly adopt triclinic crystalline systems that are remarkably similar to one another, whereas TCPP DISC have a monoclinic crystalline system.

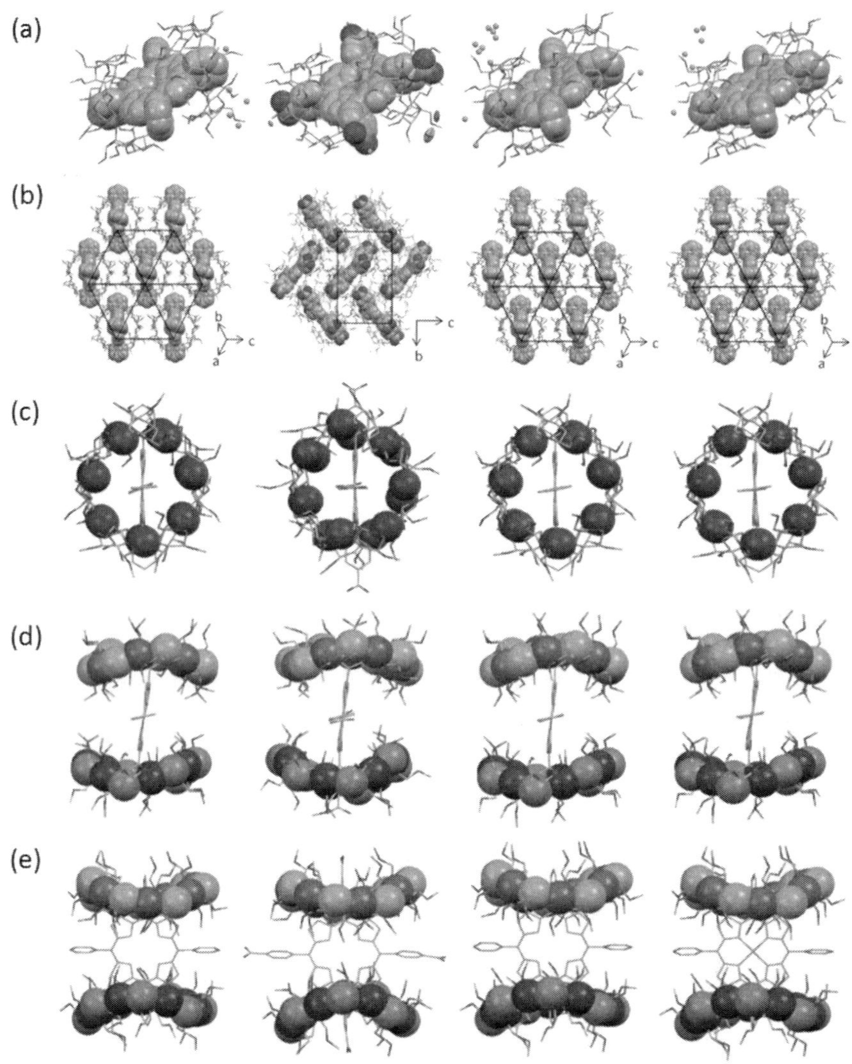

Figure 4. Crystal structures of TMβCD·TPyP, TMβCD·TCPP, TMβCD·TPP, and TMβCD·Ni-TPP DISCs (left–right); (a) unit cell: water molecules of crystallization are illustrated as light blue ellipsoids (occ.=1.0) or balls (occ. <0.5), (b) molecular packing structure: viewed along the (111) or (100) plane, (c) top view with the emphasized oxygen atoms of the glycoside bond, (d) side view with emphasized C1 and C4 atoms, (e) front view with emphasized C1 and C4 atoms. Hydrogen atoms and water molecules of crystallization have been omitted for clarity.

This difference is due to the arrangement of inclusion complexes. In DISC of TPyP, TPP, Ni-TPP, the porphyrins are arranged unidirectionally and maintain a periodical distance, whereas the TCPP DISC adopts an alternate orientation. The unique alternate orientation of TCPP stems from hydrogen-bonding network that extends between the carboxyl groups on TCPP, the water molecule, and the oxygen atoms of methoxy groups on TMβCD.

In fact, six water molecules are found in the crystal structure of TMβCD·TCPP complex. Their respective occupation probability is 1.0, whereas the water molecules found in other three complexes are much lower (< 0.5) than this value. It suggests that the *meso*-substituents of porphyrin should control the arrangement of isolated porphyrins in the DISC state. In addition, the difference of packing structure induces the different crystal shapes (Figure 3) [24].

STRUCTURES OF PORPHYRIN AND CYCLODEXTRIN IN DISC PHASE

Because of the presence of induced-fit type interaction, the CD ring should deform to the ellipsoidal structure along the included porphyrin in the crystal structure of DISC. Contrary to our expectation, the ellipsoidally-deformation of CD rings was not so clear from the top view of TMβCD·TPyP complex structure (Figure 5) [19].

More obvious is finding that the π system of the porphyrin ring is twisted to the left by 7.3°. The twisted π system of the porphyrin ring is a general trend for all free-base porphyrin complexes. Interestingly, twist angle of Ni-TPP is increased up to 9.1°.

However, the significant difference in the crystal packing between TMβCD·TPP and TMβCD·Ni-TPP was not observed, except for the twist angle of porphyrins. It is already known that the π plane of certain metallo porphyrins, such as Ni-TPP, is more distorted than free-base porphyrins [27]. This trend can be further emphasized to left handed twisting in the chiral environment of DISC state.

Moreover, there is a significant structural difference between TMβCD·TCPP and other complex; that is the position of glucose units in TMβCD with respect to the included porphyrin. The position of glucose units in the crystal structure could be assigned by using the position of glucoside oxygen atoms linking the glucose units together (Figure 4c) [24].

In TMβCD·TPyP, TMβCD·TPP and TMβCD·Ni-TPP, the glucose units on upper and lower TMβCDs occupy almost regularly aligned position between the two TMβCDs in the top view. Conversely, their positions in TMβCD·TCPP are not aligned but rather appear to alternate. This difference should be induced by the presence of hydrogen-bonding between the carboxyl groups on TCPP and the oxygen atoms of methoxy groups on TMβCD through the water molecule. However, all free-base porphyrins are equally twisted to the left by 7.3° in their DISCs. These results indicate that the twisted π system of porphyrins is not induced by steric crowding that arises from the positional relationship between the porphyrin and the glucose of TMβCD. Alternate reason which twists the porphyrin π system is found as the very important common structural characteristic of their inclusion complexes. When the C1 and C4 atoms connected to the glucoside oxygen atoms are emphasized, the strains in the glucose ring belt of TMβCDs emerge clearly in the side view and front view of their crystal structures (Figure 4 d,e).

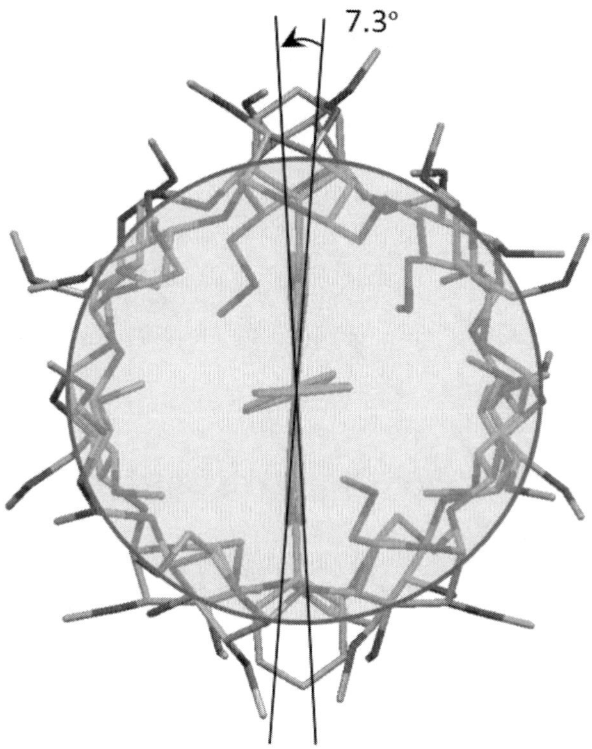

Figure 5. Top view of TMβCD·TPyP complex with circle and lines indicating circular CD ring and porphyrin strain, respectively.

Probably, this type of strain is more advantageous than CD ring strain along the porphyrin to enhance the hydrophobic contact with porphyrins. Kano et al. reported that the bent structure of permethylated CDs that have no OH groups plays an important role in chiral recognition [21]. Therefore, this bent structure of CD ring belt is the essential character of induced-fit type interactions and induce chiral twisting of the included porphyrin π system.

Recently, Watanabe et al. reported an X-ray structure of porphyrin DISC using the permethylated CD dimer connected with a pyridine spacer on O3 position of CDs (Py3OCD) [28]. This CD dimer strongly binds the water soluble Fe(III) coordinated porphyrin (Fe-TPPOH) by both hydrophobic interaction and axial coordination, and forms 1:1 complex (Figure 6). It was found that the general trends discussed above in its crystal structure; that are the twist of porphyrin, aligned glucose position between upper and lower CDs, and strain of CD rings. Interestingly, the pyridine ligand at the linker position coordinate to the ferric center with a slope (tilt angle = 29°), and the hydroxylate ion coordinates as a sixth axial ligand. The plane of the axial pyridine is aligned with the line connecting the nitrogen atom of pyrrole with that of the opposite pyrrole (eclipsed conformation). They discussed the latter orientation of axial ligand to correlate with the similar that in the case of myoglobin. In addition, the pyridine ligand orientation can be explained by the general trend among the porphyrin DISC with TMβCD; that is the aligned position of upper and lower glucose on CD ring. From the odd number of glucose group on a CD ring, the glucoses are absence in the right beside of porphyrin in the top view of crystal structure. Therefore, pyridine spacer should coordinate to the ferric center with a slope.

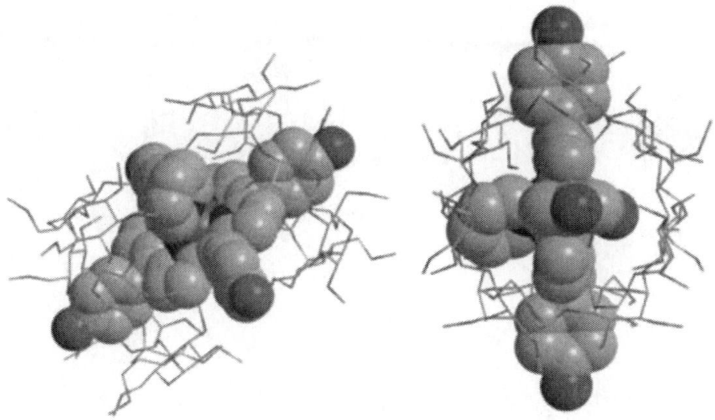

a b

Figure 6. Crystal structure of Py3OCD·Fe-TPPOH inclusion complex (CCDC deposition number, 929036), (a) complex structure, (b) top view along CD cavity [28]. Hydrogen atoms have been omitted for clarity.

Furthermore, to connect the O3 positions between the oppositely aligned glucoses, the pyridine spacer should become the eclipsed conformation to the nitrogen atoms on porphyrin.

Optical Properties of Porphyrin DISC

DISC of TMβCD·TPyP shows unique optical anisotropy (Figure 7) [24]. The oblique extinction was observed under crossed Nicol prisms. The extinction angle used for this experiment was 45°. In one Nicol observation, the extinction of polarized light was observed at 135° with respect to the major axis of the crystal.

This angle lies approximately evenly between the major and minor axes. More interestingly, when the fluorescence was observed through the polarizer, the fluorescence intensity was minimized at around 60° and maximized at around 150° for the major axis of the crystal. These angles can be explained rationally by the neat arrangement of TPyP molecules in the crystal. Figure 8 shows the arrangement of porphyrin molecules viewed along the (001) plane, which is same as the plane indices of crystal in Figure 7.

The angles of the *meso*-axis of TPyP with respect to the *a* axis are similar to the angles of maximum and minimum intensity of the liner polarized fluorescence with respect to the major axis of the crystal. If one *meso*-axis has an angle of 55° with respect to the major axis of the crystal, TPyP is tilted around 56° towards the (001) plane. In contrast, if the other *meso*-axis has an angle of 145° with respect to the major axis of the crystal, TPyP is arranged almost parallel (<2°).

DISC of TPP, which has a crystal structure similar to the DISC of TPyP, showed the same optical properties. This characteristic alignment of porphyrins should be mainly associated with linearly polarized luminescence properties. The DISC of Ni-TPP, which is known to be nonfluorescent, showed the same optical anisotropy.

On the other hand, significant optical anisotropy was not observed for TMβCD·TCPP. This difference is likely to be associated with the difference in the arrangement of porphyrins in the DISC, that is, whether they are unidirectional or not.

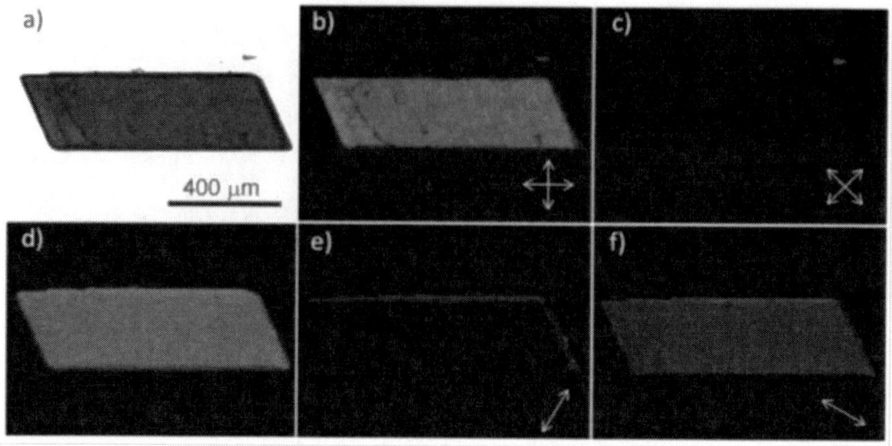

Figure 7. Microscope images of the TMβCD·TPyP DISC; a) bright-field image, b) crossed Nicols polarized image (0–90°), c) crossed Nicols polarized image (45–135°), d) fluorescent image, e) fluorescent image with polarizer (60°), and f) fluorescent image with polarizer (150°).

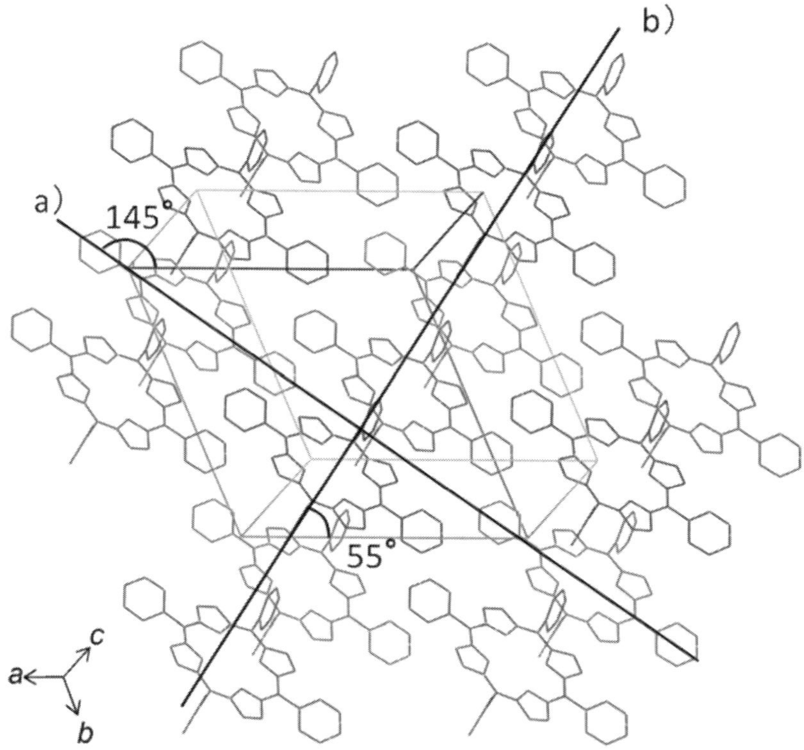

Figure 8. Arrangement of TPyP molecules in the TMβCD·TPyP DISC viewed along the (001) plane. a) *meso*-Axis of TPyP that provides the maximum polarized fluorescence. b) *meso*-Axis of TPyP that provides the minimum polarized fluorescence.

The DISC of fluorescent porphyrin shows the similar fluorescent ability to the aqueous solution of their complex. For example, Figure 9 shows the fluorescence spectra of TMβCD·TPyP in the aqueous solution phase and in the DISC phase.

The fluorescence quantum yields were estimated to be 5.2 and 3.5%, respectively. In the solution phase, the fluorescence intensity ratio of the higher peak (λ_{em} = 704 nm) versus lower peak (λ_{em} = 647 nm) was 0.22. Conversely, the peak ratio in the DISC phase was 0.47 ((λ_{em} = 695 and 650 nm).

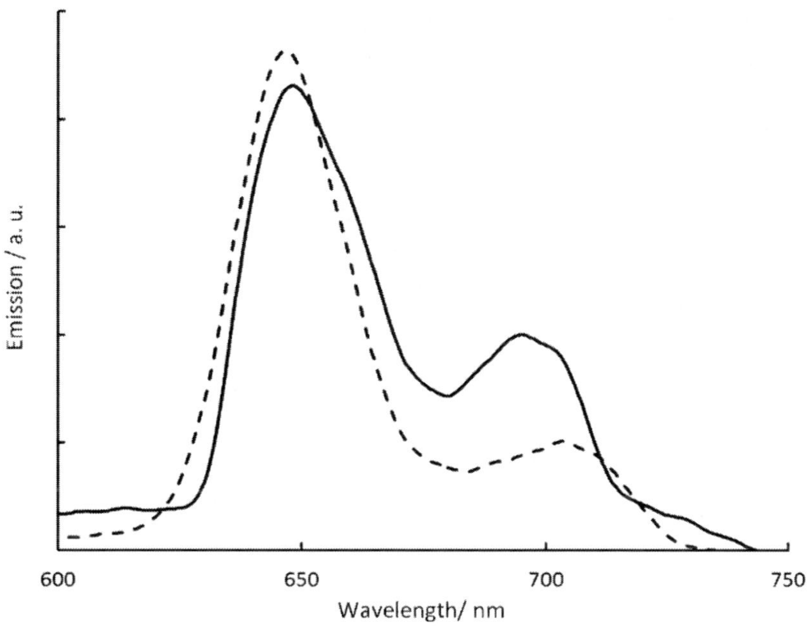

Figure 9. Fluorescence spectra of TMβCD TPyP in aqueous solution (broken line, [2TMβCD·TPyP]=6.6 μM) and DISC; λ_{ex}=510 nm.

One may consider that the change in the fluorescence peak ratio and the shift are induced by the porphyrin structure being fixed in the CD cavity. In addition, the fluorescence lifetime of DISC phase at 650 and 705 nm are 9.9 and 10.3 ns, respectively. Whereas the fluorescence lifetime of aqueous solution phase at 650 and 705 nm are 11.8 and 13.2 ns, respectively [24, 29]. Takagi et al. reported that the energy transfer in porphyrin arranged system can take place very efficiently [30, 31]. In that system, the porphyrins was arranged on the inorganic planer host with the distance about 2.4 nm. In the DISC phase, porphyrins was arranged with the distance between 1.7 and 2.9 nm. Therefore, the shorter fluorescence lifetime of DISC phase than solution phase should be based on the energy transfer between neighboring porphyrins.

Crystal Study in Other Dye Inclusion Crystals

X-Ray crystal structure for the inclusion complex between CD and other dye is still limited to a few reports. Most of them have comparatively small π systems, e.g. bithiophen and pyrene. Takashima et al. reported that the intermolecular hydrogen-bonding affects to the crystalline complex structure between β–CD and bithiophene [25, 32].

Despite of the 1:1 stoichiometry of complex between β–CD and bithiophene [33], their crystal structure shows the 2:3 inclusion complex (Figure 10). One of the bithiophene was located at the center of the dimer cavity on the secondary hydroxyl side of the β–CDs, which are aligned vertically with the cavity axis.

This bithiophene molecule is stabilized by the intermolecular hydrogen-bonds between secondary hydroxyl groups of two β–CDs. Meanwhile, the conformation of 2,6-DMβCD is stabilized by the intramolecular hydrogen-bonds connecting secondary hydroxyl groups at C3 with the neighboring methoxy groups at C2.

Therefore, there is no intermolecular hydrogen-bonding to other 2,6-DMβCD molecules. As a result, the 1:1 complex was found in the crystal which use 2,6-DMβCD instead of β–CD [32]. Similarly, the pyrene molecule is stabilized by the intermolecular hydrogen-bonds in β–CD·pyrene crystal. To enclose the protruding portion of pyrene interacting to an apolar cavity of CD, the 2:1 complex, like β–CD·porphyrin complex, was proposed from the results of their stoichiometry, kinetics and computational study [34-40].

Meanwhile, there are many reports that other molecules, especially alcohols, enhance the solubility and the photochemical properties of aqueous β–CD·pyrene solution [41-50]. It has been postulated that co-inclusion takes place in the CD cavity, yielding very stable complexes. Udachin et al. succeeded to isolate the crystal of β–CD·pyrene complex from aqueous solution containing octanol [51]. The pyrene that crystal couched perpendicularly between the two β–CDs stabilized by the intermolecular hydrogen-bonds, as same as the bithiophene (Figure 11). Recently, Kida et al. reported the arrangement of pyrene in the CD complex from two dimensional NMR and crystal studies using C6 position modified hydrophobic β–CD (heptakis(6-O-triisopropylsilyl)-β-CD, TIPS-β-CD) [52].

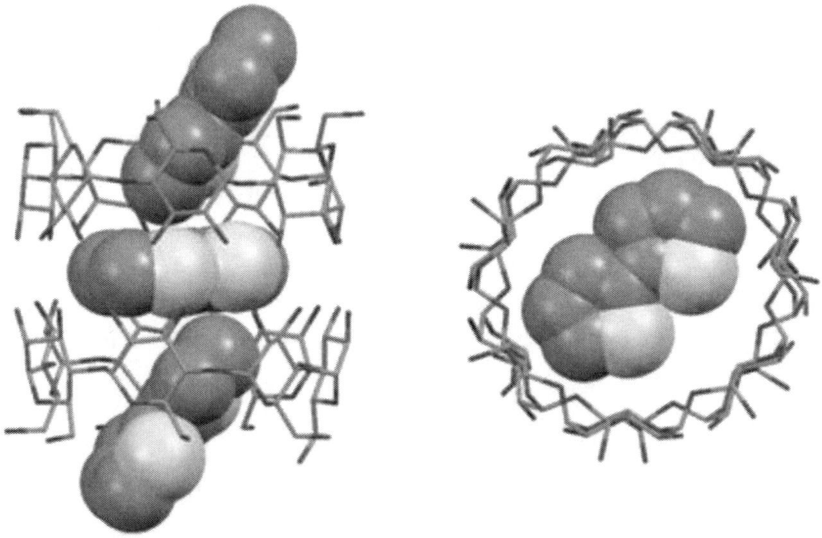

a b

Figure 10. (a) Crystal structure of β–CD bithiophene inclusion complex (CCDC deposition number, 267515) [25]. (b) Orientation of bithiophen perpendicular to cavity axis of β–CDs. Hydrogen atoms have been omitted for clarity.

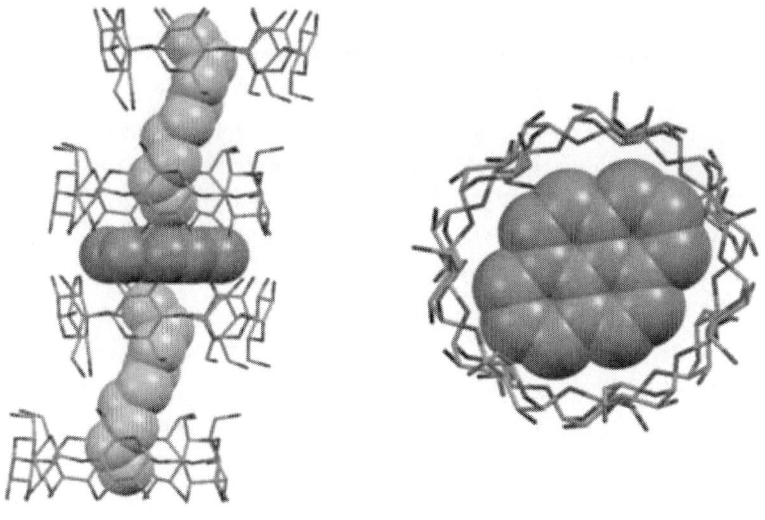

a b

Figure 11. (a) Crystal structure of β–CD octanol pyrene inclusion complex (CSD refcode, PUKPIU) [51]. (b) View of pyrene molecular orientation with respect to β–CD cavity. Hydrogen atoms have been omitted for clarity.

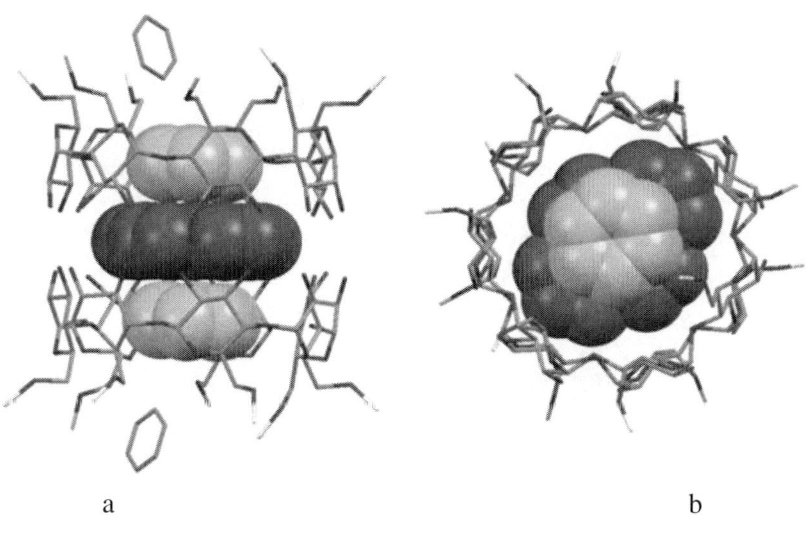

a b

Figure 12. (a) Crystal structure of TIPS-β-CD pyrene inclusion complex crystalized from benzene solution (CCDC deposition number, 823544) [52]. (b) View of pyrene benzen aggregate orientation with respect to β–CD cavity. Hydrogen atoms and isopropyl groups have been omitted for clarity.

From the NMR study, they proposed two different inclusion modes depending on the solvent. In benzene, the pyrene molecule was included with long axis almost perpendicular to the axis of TIPS-β-CD. On the other hand, in cyclohexane, long axis of the incorporated pyrene molecule tilts to the axis of TIPS-β-CD, and thus the pyrene molecule penetrates more deeply into the CD cavity as compared with the case of benzene. They succeeded to isolate the TIPS-β-CD·pyrene crystal from benzene solution.

Interestingly, the pyrene molecule forms a sandwich-type complex with two benzene molecules through π–π interaction and is located at the center of the dimer cavity (Figure 12).

Two additional benzene molecules were also included near the C6 position within the TIPS-β-CD. The location of pyrene in the crystalline state showed reasonable agreement with the proposed one in the benzene solution. These results indicating that the arrangement mode of the small dyes in inclusion crystals should depend on their preparation conditions.

In views of the great interest in the isolation of large π system in water, the complex between γ–CD and fullerene C_{60} has attracted many researchers.

^{1}H and ^{13}C NMR spectroscopy and elemental analysis have shown that the pristine γ–CD·C_{60} complex has a 2:1 stoichiometric ratio [53, 54]. Furthermore, molecular simulation studies have indicated that the 2:1 complex is a reasonable structure and that two γ–CD s can form 16 intermolecular hydrogen-bonds in the fully optimized structure [55-57].

However, single crystals of the γ–CD·C_{60} complex between sufficient for X-ray crystal structure analysis had not been obtained for long time [54, 58, 59].

Ikeda et al. recently reported the crystal structure of γ–CD·C_{60} derivative complex [60]. We can find several interesting points in its crystal structure. First, the complex has 2:1 stoichiometry (Figure 13). Second, the complex forms a pseudo-rotaxane conformation. Third, intermolecular hydrogen-bonds form among the secondary hydroxyl groups of the γ–CD. The average intermolecular distance between two oxygen atoms are within the range 2.766-2.985 Å in 16 places. These values are within the range of the distance between two oxygen atoms in hydrogen-bonding.

Therefore, they concluded that the multipoint hydrogen-bonding interactions contributed to stabilization of the γ–CD·C_{60} derivative complex. In addition, because a substituent of C_{60} derivative did not hinder the formation of the 2:1 complex, the 2:1 complex between γ–CD and pristine C_{60} may have a similar conformation and intermolecular hydrogen-bonding plays an important role in the stability of their complex structure.

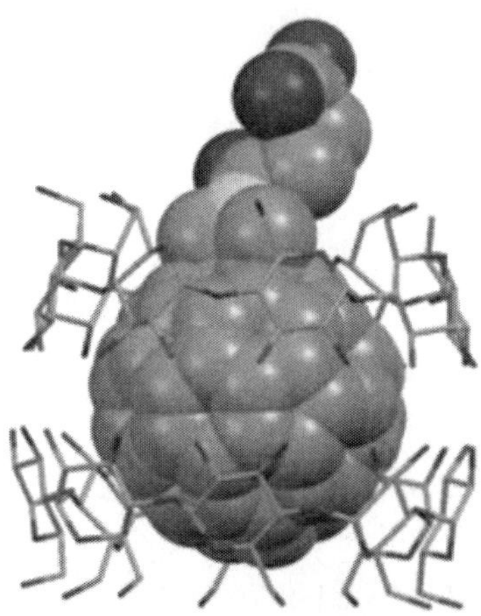

a

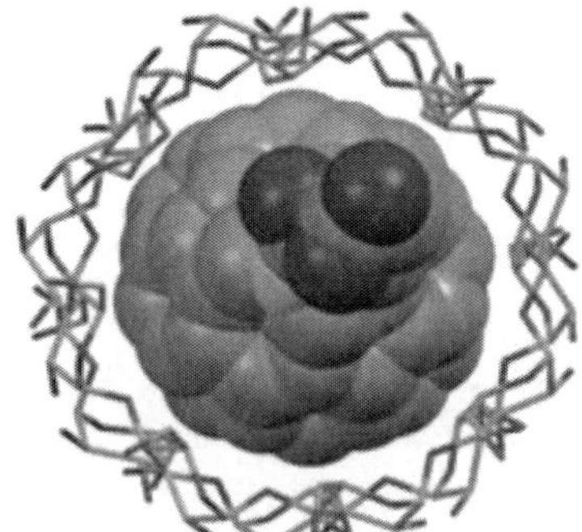

b

Figure 13. Crystal structure of γ–CD·modified C_{60} inclusion complex (CCDC deposition number, 837832) [61]; (a) side view, (b) top view along CD cavity.

Conclusion

Recently, the X-ray structures for various dye inclusion complexes of CDs were reported, not only for the small dyes but also for the large dyes. These structures revealed the detail complex structure, interaction modes between host and guest and the dye arrangement in the inclusion crystal.

These pieces of information are very important to construct the bio-mimic organization systems.

Moreover, the isolated and arranged porphyrin in the DISC phase shows several interesting optical properties. We thus believe that the study of DISCs has important implications to design new functional materials that use the concept of supramolecular chemistry.

Acknowledgments

The author is grateful for useful discussion with Prof. S. Shinkai (ISIT, Kyushu University), Dr. B. Roy (Kyushu University), Prof. K. Kano (Doshisha University), Dr. A. Yamano and Dr. T. Matsumoto (Rigaku Corporation). This work was financially supported by the Ministry of Education, Culture, Sports, Science and Technology (MEXT), Grant-in-Aid for Young Scientists (B) (No. 24750144).

References

[1] Inoue, H., Funyu, S., Shimada, Y., Takagi, S. *Pure Appl. Chem.* 2005, 77, 1019-1033.

[2] Satake, A., Kobuke, Y. *Org. Biomol. Chem.* 2007, 5, 1679-1691.

[3] Huijser, A., J. Savenije, T., Kroeze, J. E., Siebbeles, L. D. A. *J. Phys. Chem. B* 2005, 109, 20166-20173.

[4] Eu, S., Hayashi, S., Umeyama, T., Oguro, A., Kawasaki, M., Kadota, N., Matano, Y., Imahori, H. *J. Phys. Chem. C* 2007, 111, 3528-3537.

[5] Hatano, T., Ikeda, A., Akiyama, T., Yamada, S., Sano, M., Kanekiyo, Y., Shinkai, S. *J. Chem. Soc. Perkin. Trans.* 22000, 909-912.

[6] Ikeda, A., Hatano, T., Shinkai, S., Akiyama, T., Yamada, S. *J. Am. Chem. Soc.* 2001, 123, 4855-4856.

[7] Harata, K. Chapter 7. Crystallographic Study of Cyclodextrins and Their Inclusion Complexes; H. Dodziuk, H.; Ed.; *Cyclodextrins and Their Complexes*; Wiley-VCH, Weinheim, 2006, pp. 147-198.

[8] Zhon, H., Groves, J. T. *J. Porphyrins Phthalocyanines* 2004, 8, 125-140.

[9] Kano, K. *Colloid Polym. Sci.* 2008, 286, 79-84.

[10] Kano, K., Itoh, Y., Kitagishi, H., Hayashi, T., Hirota, S. *J. Am. Chem. Soc.* 2008, 130, 8006-8015.

[11] Ikeda, A., Hatano, T., Konishi, T., Kikuchi, J., Shinkai, S. *Tetrahedron* 2003, 59, 3537-3540.

[12] Mosseri, S., Mialocq, J. C., Perly, B. *J. Phys. Chem.* 1991, 95, 2196-2203.

[13] Jiang, T., Li, M., Lawrence, D. S. *J. Org. Chem.* 1995, 60, 7293-7297.
[14] Venema, F., Rowman, A. E., Nolte, R. J. M. *J. Am. Chem. Soc.*, 1996, 118, 257-258.
[15] Carofiglio, T., Fornasier, R., Lucchini, V., Rosso, C., Tonellato, U. *Tetrahedron Lett.* 1996, 37, 8019-8022.
[16] Hamai, S., Koshiyama, T., *J. Photochem. Photobiol. A.* 1999, 127, 135-141.
[17] Kano, K., Nishiyabu, R., Asada, T., Kuroda, Y. *J. Am. Chem. Soc.*, 2002, 124, 9937-9944.
[18] Manka, J. S., Laurence, D. S. *J. Am. Chem. Soc.*, 1990, 112, 2440-2442.
[19] Tsuchiya, Y., Yamano, A., Shiraki, T., Sada, K., Shinkai, S. *Chem. Lett.* 2011, 40, 99-101.
[20] Kano, K., Nishiyabu, R., Doi, R. *J. Org. Chem.* 2005, 70, 3667-3673.
[21] Kano, K., Hasegawa, H., *J. Am. Chem. Soc.* 2001, 123, 10616-10627.
[22] Kano, K., Tanaka, N., Minamizono, H., Kawakita, Y. *Chem. Lett.* 1996, 925-926.
[23] Mosinger, J., Slavětínská, L., Lang, K., Coufal, P., Kubát, P. *Org. Biomol. Chem.* 2009, 7, 3797-3804.
[24] Tsuchiya, Y., Shiraki, T., Matsumoto, T., Sugikawa, K., Sada, K., Yamano, A., Shinkai, S. *Chem. Eur. J.* 2012, 18, 456-465.
[25] Takashima, Y., Oizumi, Y., Sakamoto, K., Miyauchi, M., Kamitori, S., Harada, A. *Macromol.* 2004, 37, 3962-3964.
[26] Komatsu, K., Fujiwara, K., Murata, Y., Braun, T. *J. Chem. Soc. Perkin. Trans.* 11999, 2963-2966.
[27] Jentzen, W., Unger, E., Song, X.-Z., Jia, S.-L., Turowska-Tyrk, I., Schweitzer-Stenner, R., Dreybrodt, W., Scheidt, W. R., Shelnutt, J. A. *J. Phys. Chem. A* 1997, 101, 5789-5798.
[28] Watanabe, K., Kitagishi, H., Kano, K. *Angew. Chem. Int. Ed.* 2013, 52, 6894-6897.
[29] Cosma, P., Catucci, L., Fini, P., Dentuto, P. L., Agostiano, A., Angelini, N., Scolaro, L. M. *Photochem. Photobiol.* 2006, 82, 563-569.
[30] Takagi, S., Shimada, T., Eguchi, M., Yui, T., Yoshida, H., Tryk, D. A., Inoue, H., *Langumuir* 2002, 18, 2265-2272.
[31] Takagi, S., Eguchi, M., Tryk, D. A., Inoue, H., *J. Photochem. Photobiol. C* 2006, 7, 104-126.
[32] Takashima, Y., Sakamoto, K., Oizumi, Y., Yamaguchi, H., Kamitori, S., Harada, A. *J. Incl. Phenom. Macrocycl. Chem.* 2006, 56, 45-53.
[33] Lagrost, C., Chane-Ching, K. I., Lacroix, J.-C., Aeiyach, S., Jouini, M., Lacaze, P.-C., Tanguy, J. *J. Mater. Chem.* 1999, 9, 2351-2358.
[34] Defeyter, S., Vanstam, J., Boens, N., Deschryver, F. C. *Chem. Phys. Lett.* 1996, 249, 46-52.
[35] Kobayashi, N., Saito, R., Hino, H., Ueno, A., Osa, T. *J. Chem. Soc. Perkin. Trans.* 2 1983, 1031-1035.
[36] Kano, K., Takenoshita, I., Ogawa, T. *Chem. Lett.* 1982, 321-324.
[37] Patonay, G., Shapira, A., Diamond, P., Warner, I. M. *J. Phys. Chem.* 1986, 90, 1963-1966.
[38] Hamai, S. *J. Phys. Chem.* 1989, 93, 2074-2078.
[39] Patonay, G., Fowler, K., Nelson, G., Warner, I. M. *Anal. Chim. Acta* 1988, 207, 251-258.

[40] Xu, W., Demas, J. N., DeGraff, B. A., Whaley, M., *J. Phys. Chem.* 1993, 97, 6546-6554.

[41] Nakajima, A. *Bull. Chem. Soc. Jpn.* 1984, 57, 1143-1144.

[42] Nelson, G., Patonay, G., Warner, I. M. *Talanta* 1989, 36, 199-203.

[43] Hamai, S. *J. Phys. Chem.* 1990, 94, 2595-2600.

[44] Nelson, G., Warner, I. M. *J. Am. Chem. Soc.* 1990, 94, 576-581.

[45] Munoz de la Pena, A., Ndou, T. T., Zung, J. B., Greene, K. L., Live, D. H., Warner, I. M. *J. Am. Chem. Soc.* 1991, 113, 1572-1577.

[46] Elliott, N. B., Ndou, T. T., Warner, I. M. *J. Inclusion Phenom.* 1993, 16, 99.

[47] Schuette, J. M., Will, A. Y., Agbaria, R. A., Warner, I. M. *Appl. Spectrosc.* 1994, 48, 581-586.

[48] Yang, H., Bohne, C. *J. Photochem. Photobiol. A* 1995, 86, 209-217.

[49] Anibogu, V. C., Warner, I. M. *Appl. Spectrosc.* 1996, 50, 995-999.

[50] Yang, H., Bohne, C. *J. Phys. Chem.* 1996, 100, 14533-14539.

[51] Udachin, K. A., Ripmeester, J. A., *J. Am. Chem. Soc.* 1998, 120, 1080-1081.

[52] Kida, T., Iwamoto, T., Fujino, Y., Tohnai, N., Miyata, M., Akashi, M. *Org. Lett.* 2011, 13, 4570-4573.

[53] Sundahl, M., Andersson, T., Nilsson, K., Wennerström, O., Westman, G. *Synth. Met.* 1993, 56, 3252-3257.

[54] Yoshida, Z., Takekuma, H., Takekuma, S., Matsubara, Y. *Angew. Chem. Int. Ed.* 1994, 33, 1597-1599.

[55] Andersson, T., Nilsson, K., Sundahl, M., Westman, G., Wennerström, O. *J. Chem. Soc. Chem. Commun.* 1992, 604-606.

[56] Marconi, G., Mayer, B., Klein, Ch. Th., Köhler, G. *Chem. Phys. Lett.* 1996, 260, 589-594.

[57] Raffaini, G., Ganazzoli, F. *J. Phys. Chem. B* 2010, 114, 7133-7139.

[58] Andersson, T., Westman, G., Wennerström, O., Sundahl, M. *J. Chem. Soc. Perkin. Trans.* 2 1994, 1097-1101.

[59] Andersson, T., Westman, G., Stenhegen, G., Sundahl, M., Wennerström, O. *Tetrahedron Lett.* 1995, 36, 597-600.

[60] Ikeda, A., Aono, R., Maekubo, N., Katao, S., Kikuchi, J., Akiyama, M. *Chem. Commun.* 2011, 47, 12795-12797.

INDEX

B

C

D

E

G

H

L

M

N

O

P

Q

R

T

U

V

W

X

Y

Z

Forms Liberate

Reclaiming the Jurisprudence of Lon L Fuller

KRISTEN RUNDLE

·HART·
PUBLISHING

OXFORD AND PORTLAND, OREGON
2012

Published in the United Kingdom by Hart Publishing Ltd
16C Worcester Place, Oxford, OX1 2JW
Telephone: +44 (0)1865 517530
Fax: +44 (0)1865 510710
E-mail: mail@hartpub.co.uk
Website: http://www.hartpub.co.uk

Published in North America (US and Canada) by
Hart Publishing
c/o International Specialized Book Services
920 NE 58th Avenue, Suite 300
Portland, OR 97213-3786
USA
Tel: +1 503 287 3093 or toll-free: (1) 800 944 6190
Fax: +1 503 280 8832
E-mail: orders@isbs.com
Website: http://www.isbs.com

British Library Cataloguing in Publication Data
Data Available

ISBN: 978-1-84946-104-7

Typeset by Hope Services, Abingdon
Printed and bound in Great Britain by
Lightning Source UK Ltd

MIX
Paper from responsible sources
FSC® C013056

Acknowledgements

Completing this book would not have been possible without the support and inspiration provided by many, only a handful of whom I acknowledge here.

Given that it began its life as a doctoral thesis, the history of this book is inseparable from the superb experience I had as a graduate student at the University of Toronto. Here, I owe my deepest gratitude to David Dyzenhaus, who deserves a medal for his years of patience with me and this project, discharged, then and since, with a generosity far beyond the call of supervisory duty. Mayo Moran's disarming knack for asking precisely the right questions at each point of the supervisory process was also immensely valuable, as was the privilege of being the beneficiary of Ernie Weinrib's guidance and formidable knowledge of Fuller. I am equally indebted to Melissa Williams for welcoming me into her community of fellows at the Centre for Ethics for the 2008–2009 academic year, which proved to be an ideal environment for bringing my doctoral project to a close. Then, at the examination stage, my internal examiner, Jutta Brunnée, and my external examiner, Mark Walters of Queen's Law School, raised numerous questions and criticisms that were instrumental in setting the path for the major revision of my doctorate that is reflected in this book. Last but not least, I will be eternally grateful to the powers that be for selecting such a fabulous group of colleagues with whom to endure the trials of doctoral study. Special honours here go to Zoran Oklopcic, Rayner Thwaites, Emily Hammond, Kim Stanton and Rose Ann MacGillivray, each of whom was not only close at hand at all times, but added laughter and lasting friendship to the mix.

The pages to follow stand as a record of the enormous value gained, over the course of its life as both doctoral project and a book in progress, from opportunities to engage with scholars and resources internationally. Symposia convened by Jeremy Waldron at NYU Law School, and Peter Cane at the College of Law, Australian National University, to commemorate the fortuitously-timed 50th anniversary of the Hart-Fuller debate in 2008 provided a unique chance to be exposed to the thinking of leading legal philosophers on questions of core concern to my project. In the later stages, I also profited from opportunities to present work in progress at Cornell Law School, Birmingham Law School and the Jurisprudence Discussion Group at Oxford University, where I benefited from the illuminating input of Nicola Lacey and Pavlos Eleftheriadis as discussants, and Jeremy Waldron as a generous audience participant. Right near the finish line, a visit to NYU enabled crucial clarifying conversations with Liam Murphy, to whom I am also indebted for his generous gesture of offering comments on the entire manuscript just prior to submission.

In terms of the shape that this book has ultimately taken, my two visits to the archive of Fuller's private papers at Harvard University proved to be critical to the arguments and insights I have been able to offer through it. Sincere thanks thus go to Lesley Schoenfeld from the Special Collections Library at Harvard Law School, whose efficient and friendly assistance enabled my access to key materials while at Harvard and subsequently. Both of these trips to Harvard also saw me benefit from conversations with Ken Winston, who shared both his time as well as his own research into Fuller's archive in a spirit of absolute generosity. My second research trip in September 2010, made possible through the LSE by a STICERD research grant, also saw me profit from conversations with Robert Summers, Stephen Perry, Colleen Murphy and Margaret Martin. This grant equally enabled me to gain from the research assistance of Ryan Rafaty, who proved to be a valuable conversational partner as I muddled my way through a rethinking of my claims.

My appointment to the Department of Law, LSE, in September 2009, offered ideal conditions for bringing this book to completion, as well as for testing my instincts on many of the arguments that I develop in it. The value gained from being thrown into the deep end of an undergraduate jurisprudence course cannot be overstated, nor can the brilliance and sheer fun of my students over the course of 2009–2011. Here special acknowledgement goes to Grace Cheng, Natasha Chan, Ryan Raymond, Olly Wiseman, Shou Jie Eng, Eugene Sng, Joel Keh, Alexandra Avlonitis and Alex Rodin, each of whom made their own distinctive contribution – even if they did not know it – to how I have ultimately set out and defended the ideas contained in the pages to follow. Above all, I have benefited from the spectacularly collegial environment that is the Department of Law at LSE, where senior and junior colleagues alike have each left their helpful fingerprints on the final product of this book. I am especially grateful for the wise and warm mentoring received from Linda Mulcahy, Nicola Lacey and Susan Marks, and for numerous conversations with Anne Barron and Grégoire Webber that helped to clarify and shape my core concerns.

On the personal front, my family in Australia deserves enduring credit for lovingly acceding to my unreasonable request that they not ask me when I expected this book to be finished. In this, as in so many things, they knew that I just had to get it done my way, and that I'd be sure to report when it was all over. Whitney Smith discharged the unenviable task of cohabiting with this project for the great majority of its life span with grace and unswerving faith. His insight into the creative process was a godsend, and for this I am deeply thankful. A small army of women also earned my heartfelt gratitude for their role in satisfying my odd but highly productive fetish for writing in homes other than my own. Here all thanks go to Gina Lewis Watts, Adrienne Christie, Mary Purcell, Susan Marks and Emily Jackson for their generous hospitality, and, above all, to Claire Finnegan for providing the refuge of the sunny room overlooking a garden in Finsbury Park where, against the odds, this book found its way to completion.

Finally, tracing the genesis of the ideas contained in these pages to conversations in 2001, I wish to convey my deep thanks to Rod Macdonald for his inspiration and generosity, then and since. My inner jury might still be out on whether you ought to be forgiven for tricking me into this academic path, but while the deliberations continue, it is my privilege to dedicate this book to you.

Kristen Rundle
London
April 2012

Contents

1

Reclaiming Fuller

Lon L Fuller was something of an outsider within the intellectual climate of mid-twentieth century legal philosophy, which during his time came to be increasingly dominated by the legal positivist jurisprudence of HLA Hart. Today, among contemporary legal philosophers, Fuller still remains mostly known as the natural lawyer who apparently lost the debate about the connection between law and morality to his analytically superior opponent, with the consequence that his contribution to legal philosophy has often been cast in terms that suggest he offers little to enlighten the enduring debates of the discipline.[1] There is also a general impression, in many ways justified, that although rich with insights into the way that law works in practice, Fuller's jurisprudence is a scattered affair, unsystematic in focus and lacking any obvious internal coherence.

Why, then, should we consider taking the time to revisit Fuller, indeed, to reclaim his jurisprudence and situate it more securely on the agenda of twenty-first century legal philosophy? One way to begin answering this question is to invoke the image to which the title of this book refers, and which comes from an untitled and undated working note that can be found among Fuller's private papers that are held at the Harvard Law School Library. The note is comprised of no more than four lines of text, all of which have been crossed out in thick black pen, leaving only two exposed and circled in red: 'forms liberate'.[2]

In one sense there is nothing especially remarkable about this note, in so far as it shares much in common with many other notes from Fuller's archive that are littered with the crossings-out, exclamation marks and scrawled annotations of works in progress. Certainly, on its face it seems to say so little that one might wonder why it has been kept for posterity at all, which, indeed was my own thought when I first saw it. But as I became more immersed in Fuller's jurisprudence, and as my intuitions about how that jurisprudence should be interpreted

[1] The full exchange of the 'Hart-Fuller debate' consists of HLA Hart, 'Positivism and the Separation of Law and Morals' (1958) 71 *Harvard Law Review* 593, and Lon L Fuller, 'Positivism and Fidelity to Law: A Reply to Professor Hart' (1958) 71 *Harvard Law Review* 630. Hart and Fuller's exchange then continues through HLA Hart, *The Concept of Law* (Oxford, Oxford University Press, 1961); Lon L Fuller, *The Morality of Law* (New Haven, Yale University Press, 1964); HLA Hart, 'Lon L. Fuller: *The Morality of Law*' (1965) 78 *Harvard Law Review* 1281, reprinted in HLA Hart, *Essays in Jurisprudence and Philosophy* (Oxford, Clarendon Press, 1983); and Lon L Fuller, 'A Reply to Critics', in the 2nd edition of *The Morality of Law* (New Haven, Yale University Press, 1969).

[2] The Papers of Lon L Fuller, Harvard Law School Library, Box 12, Folder 1 (notes for 'Reply to Critics').

took shape, I began to think that this peculiar little working note might in fact have captured something fundamental, not only about the message of Fuller's legal philosophy, but also, in its frustrated air, about the difficulties that he experienced in bringing that message to expression.

I have thus chosen those words, 'forms liberate', as the frame for the project of this book precisely because of how they invite us to approach Fuller's jurisprudence from an underexplored angle: that is, from the angle of his interest in the connection between the form of law and its relationship to human agency. I hope to persuade the reader that a proper understanding of Fuller's jurisprudence requires that we begin with his enduring interest in the distinctiveness of law's form, and then, from this starting point, witness how he proceeds to interrogate the implications of that form for the character, existence and normativity of law, and, indeed, for the enterprise of legal philosophy itself. The result is as much a criticism of the dominant positivist account of law as it is a conception of law in its own right, and its essence can be stated as follows. For Fuller, there can be no meaningful concept of law that does not include a meaningful limitation of the lawgiver's power in favour of the agency of the legal subject. This is not a moral objective that is imposed on the enterprise of lawgiving from without. It is, rather, simply something that follows from the formal distinctiveness of law as the enterprise of subjecting human conduct to the governance of general rules.

My task in the chapters to follow is to draw out and explain the core ideas, and their interconnection, that give content to this jurisprudential claim. For introductory purposes, however, the basic thrust of Fuller's jurisprudence might be summarised in these terms. Fuller's vision of law begins and never sways from the view that to label something as 'law' is to designate a distinctive mode of governance. Law is a formally recognisable alternative to rule by men, and this difference is made especially clear when we consider the status that is enjoyed by the subjects of a legal as opposed to some other kind of order. To be a legal subject, Fuller insists, is not merely to be a member of 'a subservient populace ready to do what they are told to do',[3] but rather to be a participant in a distinctly constituted social condition in which one is respected as an agent. This respect, and the social condition that speaks to and constitutes it, arises from the particular way that a legal order creates and communicates its norms, namely, through observance of principles of generality, promulgation, clarity, avoidance of contradiction and of impossibility, constancy through time, non-retroactivity, and the requirement that there be congruence between official action and declared rule. These eight principles, Fuller famously argued, constitute law's 'internal morality'.

To understand what Fuller was driving at when he designated law's distinctive mode of creating and communicating norms as a 'morality', it is necessary to call to mind the context within which he articulated this idea. When he began developing his thinking on the constitutive features of law, Fuller did so against the backdrop of the age-old jurisprudential debate between legal positivists and

[3] Undated and untitled document, The Papers of Lon L Fuller, Harvard Law School Library, Box 12, Folder 1 (notes for 'Reply to Critics').

natural lawyers on the question of the connection between law and morality. This means that, when we read Fuller, we need to see that he is working in two, in his view compatible, directions. First, he is seeking to investigate and to explain the distinctiveness of law's form: its attributes, presuppositions, formal features, and so forth. But then, second, he is also attempting to situate that investigation within the extant debates of jurisprudence; most particularly, within debates about the connection between law and morality.

This is the background against which Fuller advanced the idea that law is an intrinsically moral phenomenon, and which he defended along two interconnected lines, one relating to the moral demands of lawgiving, and the other to its moral value from the point of view of the legal subject. As Fuller explained it, when we take seriously the idea that law finds expression through a distinctive form, we come to see that to create and maintain that form requires the adoption of a distinctive ethos, a special understanding of the demands of his role, on the part of a lawgiver. Law is thus intrinsically moral in the sense that it is constitutively dependent on the observance of this ethos. But second, and itself a key part of the demands of this ethos, law is also intrinsically moral for how its form—that of governance of general rules—presupposes the legal subject's status as a responsible agent. Thus, law is also intrinsically moral for how, if it is to function, it must maintain and communicate respect for that status of agency.

It is these morally significant commitments that Fuller sees as distinguishing rule through law from rule by men, and he insisted that they are internal to law because of how they arise from the presuppositions that must be in place in order to support law as a distinctive form of social ordering. This leads in turn to Fuller's explanation of law's normativity, or, at least, his explanation of at least one of the reasons for law's normative force. If a lawgiver fails to observe the requirements of the internal morality of law, the legal subject can justifiably withdraw her fidelity because, as an agent and bearer of dignity, she cannot be expected to comply with the lawgiver's demands in the face of such disrespect for her status.[4] Thus, if the necessary reciprocity between lawgiver and subject that creates and maintains the distinctive attributes of a legal order disappears, so too must law, because the lawgiver has disavowed his commitment to law and is now proceeding through a different mode of ordering.

There have been multiple barriers to understanding Fuller's claims about law's morality in these terms. The primary reason for this is because analyses of Fuller's jurisprudence to this point have tended to be married to a particular, and particularly narrow, understanding of the kinds of connections between law and morality that the debates of jurisprudence ought to be interested in. As these analyses have it, exemplified in HLA Hart's own approach not only to debates about law and morality generally but in response to Fuller specifically, the main or indeed only question of importance is whether the fact of something being law guarantees the moral status of the ends pursued through it. This neat and powerful prism

[4] *Morality of Law* (n 1) 40.

of debate, historically, has been the primary one through which Fuller's contribution to legal philosophy has been read and evaluated.

As I explain in more detail below, it is not my aim in this book to marginalise discussion of what Fuller did have to say on this question of law's connection to the substantive morality of its ends, although it is important at the outset to note that at no point in his writings does Fuller claim any necessary conceptual connection between the two. But I am concerned to ensure that this particular way of understanding debates about law and morality no longer be allowed to obscure the other conversations that are extant in Fuller's jurisprudence, and which largely remain, to this day, under-acknowledged and under-explored. I hope to persuade the reader that the conversation Fuller thought we needed to have in jurisprudence, and which he also insisted was salient to traditional contests between positivism and natural law, was one about the distinctiveness of law's form and the implications of this distinctiveness for lawgiver and legal subject alike.

What, then, has prevented us from interpreting Fuller's contribution in this way, or, indeed, what is the cause of the apparent resistance with which legal philosophers have met Fuller's attempt to initiate a conversation in these terms? Although the strength of the assumption just explained that to debate about the connections between law and morality is to debate about the connections between law and substantive justice cannot be overstated, there have also been other powerful obstacles to both understanding and embracing Fuller's jurisprudence on the terms that he intended for it. One of these is the avowedly idealistic tenor of Fuller's project; how the demands of the internal morality of law clearly set a high bar for a lawgiver, and thus also a high bar for what ought to be designated as a legal system. It is a normatively demanding view of law.

Yet none of this escaped Fuller himself. There is ample evidence in his writings to suggest that he was clearly alive to the conditions that might bear upon whether this standard could be met in the actual practice of an enterprise 'dependent for its success on the energy, insight, intelligence, and conscientiousness of those who conduct it, and fated, because of this dependence, to fall always somewhat short of a full attainment of its goals'.[5] Still, his jurisprudence is uncompromisingly committed to the idea that the only meaningful account of law's nature is one that is capable of absorbing and accounting for these formal demands and the limits on the exercise of power that they impose. In that jurisprudence, the question of whether one is in fact governing through law is always measurable by the extent to which the morally significant demands of lawgiving are met. To abuse law is, at a certain point, to lose law.

But there is much about the way that Fuller gave expression to these claims that has not assisted their ready understanding. Although he did at times offer quite disciplined statements of his position, such as his exegesis of the eight principles of the internal morality of law in chapter two of *The Morality of Law*, the key insights of his jurisprudence are otherwise often scattered throughout his writ-

[5] ibid 145.

ings. To complicate matters further, Fuller could be undeniably sloppy in his choice of language, using terms interchangeably or being inattentive to their potential for contradiction, with the consequence that he often fed the fuel of his critics' interpretations of his jurisprudence much more readily than he guarded his position against them. These are just some of the reasons why reading Fuller can be an exasperating experience for philosophers trained in standards of clarity and rigour, and who expect the same from any work that makes claim to a philosophical engagement.

It is equally true that Fuller never managed or arguably adequately sought to reclaim his project from Hart's much narrower agenda. His 1958 reply to Hart's seminal essay, 'Positivism and the Separation of Law and Morals', might show much effort in this vein,[6] but when Fuller took the task on much more squarely in his 1969 'Reply to Critics' (itself an under-explored resource in analysis of Fuller's thought) the attempt was both too little and too late.[7] Surveying the writings linked to the Hart-Fuller debate over the course of those 11 years, therefore, we are ultimately presented with a strange mix of Fuller developing his distinctive jurisprudential project; Fuller being captive to Hart's agenda; Fuller losing himself and some of his best ideas to the challenge of understanding why Hart and others had dismissed him so harshly; Fuller saying some ridiculously polemical things; Fuller offering some profound insights; and then Fuller attempting to bring it all to a close by seeking to understand what it was about the priorities and methods of his critics' projects that might explain their apparent desire to keep his own jurisprudential agenda at a solid distance from the 'proper' concerns of legal philosophy.

Yet whether or not he made an adequate attempt to defend the terms of his own jurisprudential agenda against Hart's much narrower one, one thing that is clear is that this whole notion that legal philosophy has its 'proper' and 'improper' concerns was a thorn in Fuller's side from the very beginning. As a lawyer untrained in the tools of analytic philosophy, Fuller was well aware of how decisive the possession of a particular analytic skill-set had become to the possibility of being heard in twentieth century jurisprudence. The distress this caused him, and the sense of vulnerability and frustration that it precipitated, is made clear in his private correspondence and working notes. Referring to his exchanges with Hart in a letter to a colleague, Fuller confided that 'I genuinely do not have the philosophic insight to know just how to reproduce my own convictions properly, and if I did, I would not know what language to use to convey those convictions undistorted to my readers'.[8]

But it is important not to focus too much here on the extent to which Fuller was on the back foot. Although keenly aware of the disadvantage he faced in engaging effectively with 'real' philosophers, Fuller was also keenly sensitive to how the vocabulary of analytic philosophy operated to bring certain views about

[6] See chapter three.

[7] See chapter five.

[8] Letter from Fuller to Professor Boris I Bittker, 4 April 1960, The Papers of Lon L Fuller, Harvard Law School Library, Box 14, Folder 1 ('The Forms and Limits of Adjudication').

the nature of law into focus at the same time as it actively marginalised others. The strength of Fuller's awareness of this agenda-setting force of the vocabulary of analytic legal philosophy is evident in how throughout his writings Fuller repeatedly tried to expose what is at stake in conceding to narrow standards of analysis as definitive of the agenda and the methodology of jurisprudence. This is perhaps expressed no better than in the closing words of his 'Reply to Critics', where he appeals to legal philosophers to give up their 'endless debates about definitions' and efforts to build 'conceptual models' in favour of 'an analysis of the social processes that constitute the reality of law'.[9]

The point, then, is that although the project of 'reclaiming' Fuller need not be an exercise in condemning the standards of analytical precision that in many ways have served legal philosophy well, it must nonetheless involve an appeal to the intellectual credibility of a wider view that accepts certain compromises of clarity in favour of a sensitivity to the insights that we might take from practice. Despite the numerous instances of hyperbole that infect some of his claims, Fuller actually never overstated the wider ambitions of his project. For example, he never suggested that his claims were to be accepted as universally applicable, in the style of general jurisprudence. Indeed, if we are to situate Fuller's project, we might say that it is loosely historicised around the phenomenon of modern law; cues we receive, amongst other resources, from the references to seventeenth century English case law that accompany the tale of King Rex in *The Morality of Law*,[10] and his reference in his final 'Reply to Critics' to the responsibilities involved in maintaining the rule of law in a modern state.[11] But what we can have little doubt about is that Fuller held very strong views about what, as participants in the enterprise of jurisprudence, we need to be asking ourselves. Though he might have framed it in various ways, the question itself is consistent: how well do our theoretical understandings of law, and the methods through which we pursue those understandings, serve our complex legal realities?

There is much evidence among recent writings of increasing sympathy for this question and, more generally, for the wider agenda for jurisprudence that Fuller sought. Progress in this direction seems in particular to have been triggered by the renewed attention given to Fuller's thought on the occasion of the 50th anniversary of Hart and Fuller's famous exchange in the 1958 *Harvard Law Review*.[12] This

[9] 'Reply to Critics' (n 1) 242.

[10] *Morality of Law* (n 1) 33.

[11] 'Reply to Critics' (n 1) 234.

[12] See especially the following contributions to the *New York University Law Review* symposium, 'The Hart-Fuller Debate at Fifty': David Dyzenhaus, 'The Grudge Informer Case Revisited' (2008) 83 *New York University Law Review* 1000; Nicola Lacey, 'Philosophy, Political Morality, and History: Explaining the Enduring Resonance of the Hart-Fuller Debate' (2008) 83 *New York University Law Review* 1059; and Jeremy Waldron, 'Positivism and Legality: Hart's Equivocal Response to Fuller' (2008) 83 *New York University Law Review* 1135. See further Peter Cane (ed), *The Hart-Fuller Debate in the Twenty-first Century* (Oxford, Hart Publishing, 2009), and especially, in that volume, Nicola Lacey, 'Out of the "Witches" Cauldron?: Reinterpreting the Context and Reassessing the Significance of the Hart-Fuller Debate'; Martin Krygier, 'The Hart-Fuller Debate, Transitional Societies and the Rule of Law'; and Leslie Green, 'Law as a Means'.

occasion, however, was not an isolated moment of contact with Fuller, but rather built upon an emerging trend that has seen several scholars sympathetic to Fuller's position in his exchanges with Hart exploring those elements of his thought that suggest a link between law and freedom, or the status of the legal subject as an agent, and that attempt to make a case for a conceptual relationship between law and the rule of law.[13] A more nuanced picture has also been acknowledged by some who would ordinarily understand themselves as Fuller's critics, and yet who have recently affirmed aspects of his position.[14]

How these developments might be clarified and strengthened by the reading I supply in this book, and how they might be put into conversation with the dominant contemporary projects of legal philosophy, is a matter that I address in chapters six and seven, as well as in my conclusions in chapter eight. But here, as a way of bringing the concerns of this book into focus, it is helpful to dwell on the very idea of 'affirming' Fuller's position. As foreshadowed above, the common view still seems to be one in which to appraise Fuller's jurisprudence is to evaluate whether he advanced a compelling objection to Hart's positivist position on the connection between law and morality, and, specifically, with respect to the moral quality of legal ends. Thus, to affirm Fuller, this approach likely suggests, is to affirm some kind of connection between law and substantively moral outcomes, as if Fuller remains the modern guardian of this claim in our jurisprudential imagination.

In some contrast to this view—though, as already noted, without discounting fruitful conversations that might be had about Fuller's sense of this kind of connection between law and morality—the idea I wish to advance in this book is that the most meaningful concession that can be made to Fuller is one that sees the form of law as significant conceptually and salient morally. To reclaim this Fuller, we need to focus on his interest in the distinctiveness of law's form and its connection to human agency, and to do this it is helpful to return to the sense of his jurisprudential ambitions that has in the past been articulated especially insightfully, albeit with different aims in view, by two scholars in particular: Kenneth Winston and Jeremy Waldron.

The chapters to follow will reveal the extent of my debt to Winston's effort of compiling the archive of Fuller's private papers in the early 1980s. These archival materials have been essential to piecing together and lending support to my view that Fuller's jurisprudence is a much more coherent undertaking than he has generally been given credit for. But Winston's own analysis of Fuller's jurisprudence,

[13] See, most notably, Jeremy Waldron, 'Why Law? Efficacy, Freedom or Fidelity?' (1994) 13 *Law and Philosophy* 259; Colleen Murphy, 'Lon Fuller and the Moral Value of the Rule of Law' (2005) 24 *Law and Philosophy* 239; Jennifer Nadler, 'Hart, Fuller, and the Connection Between Law and Justice' (2007) 27 *Law and Philosophy* 1; Evan Fox-Decent, 'Is the Rule of Law Really Indifferent to Human Rights?' (2008) 27 *Law and Philosophy* 533. See also Nigel Simmonds, *Law as a Moral Idea* (Oxford, Oxford University Press, 2007). See further chapter six.

[14] Nicola Lacey makes this point in 'Out of the "Witches" Cauldron' (n 12) 4, noting Leslie Green's moves to this effect in 'Positivism and the Inseparability of Law and Morals' (2008) 83 *New York University Law Review* 1035.

read above all through the prism of Fuller's unfinished *eunomics* project, has equally been crucial to laying the foundations for the perspective I offer in this book.[15] As I explain in chapter two, this is because Fuller's eunomics project was an attempt to uncover the forms, limits and different kinds of moral value that attach to different modes of social ordering: a project, even if this has been greatly underrecognised, that shares many of its fundamentals with the jurisprudential position that Fuller developed in the context of his exchanges with Hart. Moreover, Winston has repeatedly emphasised Fuller's interest in agency,[16] and has suggested that the requirements set out in Fuller's idea of the internal morality of law are moral because they constitute what it means for a lawgiver to treat the legal subject with respect as a responsible agent.[17] My own reading of Fuller picks up and develops these intuitions in multiple ways.

Waldron's engagement with Fuller, which is ongoing, has generally been more squarely concerned with Fuller's contribution to questions of mainstream jurisprudence. In recent works, however, Waldron's exploration of that contribution has seen a turn towards the place of the agent within Fuller's jurisprudence, as part of questioning whether legal philosophers ought to be satisfied with the undiscriminating concept of law offered by positivism.[18] But Waldron's intuitions about the right way to approach Fuller's jurisprudence trace back to his 1994 essay, 'Why Law? Efficacy, Freedom or Fidelity?', where he suggests that Fuller's writings on the morality of law might be regarded as the initiation of a research programme that seeks to explain what the connection between legal forms and fidelity to law might actually consist in.[19]

This question of what we might learn from Fuller's investigation of the implications that flow from the form through which law is constituted and expressed has not, in my view, yet been given the attention that it deserves. My aim in this book is to provide at least a starting point for redressing this neglect and this, indeed, is why the 'forms liberate' image provides such a helpful frame for pursuing that endeavour. It points us to where we might look if we are to begin to understand Fuller on his own terms.

I Form and Agency

It should by now be apparent that my aim in this book is not only to advance the idea that the connection between legal form and human agency is a consistent

[15] Eunomics is also emphasised in Robert S Summers, *Lon L. Fuller* (Stanford, Stanford University Press, 1984).

[16] See especially Kenneth I Winston, 'Legislators and Liberty' (1994) 13 *Law and Philosophy* 389, 390, and generally 'Introduction to the Revised Edition' in Kenneth I Winston, *The Principles of Social Order: Selected Essays of Lon L. Fuller*, revised edn (Portland, Hart Publishing, 2001).

[17] Winston, 'Introduction' (n 16) 51.

[18] Jeremy Waldron, 'The Concept and the Rule of Law' (2008) 43 *Georgia Law Review* 1 , 13, 15.

[19] Jeremy Waldron, 'Why Law? Efficacy, Freedom or Fidelity?' (1994) 13 *Law and Philosophy* 259, 276.

theme in Fuller's writings, but also to insist that this theme offers a crucial and yet under-explored resource for both interpreting Fuller's claims in his debate with Hart, and for developing the insights of his jurisprudence more generally.

That said, the ambition of presenting and elaborating the connection between form of law and human agency does not appear to have been the conscious aim of Fuller's forays into the questions of jurisprudence. Instead, it seems to have been something that he uncovered along the way as he explored what it is that defines the form of law and distinguishes it from other modes of social ordering. Fuller was a lawyer, not a philosopher. Thus, in his writings we do not find, as we do in Hobbes or Kant, a trajectory of inquiry that starts with the premise of human agency and moves from this to the necessity of law (even if, when we go back to his 1941 polemic against positivism, *The Law in Quest of Itself*, we see him criticise modern positivists for forgetting what led, on Hobbes' account, to the advent of positive law in the first place: namely, the pursuit of peace and order among irreconcilable, or at least unreconciled, agents).[20]

It is thus helpful to set the stage for the chapters to follow by drawing attention to how consistently Fuller's writings return to questions such as the following, and attempt to flesh out intuitions that might offer answers to them. What is it about the form of law that is distinctive? How does it relate to our understanding of what law is, and of its normativity? Does the form of law do morally significant work in its own right? What is it about the self-understanding of the project of legal philosophy that causes legal philosophers to persist in neglecting such questions?

The most well-known route through which Fuller attempted to answer these questions was, of course, his idea that law contains an 'internal morality' that can be expressed through eight principles of lawgiving: principles of generality, promulgation, clarity, avoiding contradiction between laws, avoiding impossibility, constancy through time, non-retroactivity, and congruence between official action and declared rule. For Fuller, then, to speak of the form of law is to speak of law's distinctive shape, and especially how, as is widely accepted among legal philosophers, law is an enterprise of general rules.

To understand Fuller's conception of form is thus to understand, and indeed to bring within the very idea of form an appreciation of the implications that are generated by this phenomenon of generality. To begin, once a decision is made to proceed with governance through general rules, the other seven principles of Fuller's internal morality of law follow naturally, because, together, it is observance of these requirements that makes governance through general rules not just effective, but possible. Crucially, however, there is more to what constitutes and gives expression to law's distinctive form than these eight principles. Law's formal attributes implicate not only a set of responsibilities, as captured by the eight principles, that burden the lawgiving task, but also rely upon a particular conception of the legal subject as someone who must possess the necessary

[20] Lon L Fuller, *The Law in Quest of Itself* (Chicago, The Foundation Press, 1940) 19–20. See further chapter two.

responsible agency to be able to interact with general rules, and the conditions through which they are communicated, in the first place. For Fuller, then, there is not a form called law that acts upon the legal subject. Rather, properly understood, the form of law includes the legal subject's capacity for agency within it.

As for agency itself, the best way to approach this idea is to begin with Fuller's understanding of what it means to be an agent: a person 'capable of purposive action',[21] in possession of her capacities, and who is to be regarded as an end in herself. An agent, therefore, is more than someone simply capable of responding to a lawgiver's direction, even if that direction is entirely favourable to her. She is instead a bearer of dignity, with a life to live of her own. Further, and crucially for how we might situate the distinctiveness of Fuller's claims about the relationship between legal form and agency vis-à-vis other jurisprudential projects that also keep the status of the legal subject firmly in view,[22] Fuller envisages the legal subject as coming to law in this state, in the sense that law's form not only respects and speaks to her status as such, but presupposes it. As he put it in *The Morality of Law*, the very enterprise of law involves 'a commitment to the view that man is, or can become, a responsible agent, capable of understanding and following rules, and answerable for his defaults'.[23]

I look more closely at this conception of the legal subject as an agent in chapter four. But the point to note for preliminary purposes is how Fuller's movement towards the subject, specifically, to a conception of the legal subject as an agent, links in critical ways to his view that among law's defining attributes is how it mediates the way that power is experienced by those who are subject to it. The very form of law, Fuller argues, instantiates a 'relationship with persons generally' that demands 'more than forbearances',[24] and that is revealed to us once we examine the relationships between participants that the feature of law's expression through general rules necessarily constitutes.

This is why, as I also explain in detail in chapter four, it is crucial not to understand Fuller's eight principles as a simple checklist for the features which law must possess. The principles instead speak to an idea that is broader, and more fundamental, than the sum of eight parts: namely, the distinctive kind of relationship between power and those subject to it that law constitutes, and which is brought into being, regulated and served by the principles of the internal morality of law. Once we appreciate this, we can also appreciate why Fuller sees no contradiction in the idea that there can be permissible departures from his eight criteria.[25] Such departures might be permissible, that is, *if* they operate to serve the particular quality of relationship between lawgiver and subject that Fuller associates with a condition of legality; a relationship in which the lawgiver can pursue his ends efficaciously at the same time as the subject is respected as an agent.

[21] Lon L Fuller, 'Freedom: A Suggested Analysis' (1955) 68 *Harvard Law Review* 1305, 1307.
[22] See my analysis of Dworkin's jurisprudence in chapter seven.
[23] *Morality of Law* (n 1) 162.
[24] ibid 42.
[25] See chapter four, IA 'The story of King Rex'.

It does not follow, therefore, that because Fuller turned his mind so directly to the position of the legal subject and her contribution to the possibility of law that he was necessarily uninterested in the source of lawgiving and the considerations of efficacy that shape the pursuit of governance through law. Fuller looked to the position of the lawgiver and to considerations of efficacy repeatedly, and his tale of King Rex in *The Morality of Law* offers an obvious example. But what is crucial not to lose sight of is how closely Fuller's jurisprudential ambitions were associated with his criticism of the tendency of the legal philosophies of his era, above all legal positivism, to set their theoretical sights exclusively on the office of the lawgiver. For Fuller, the problem with the source-oriented impulse of the dominant school of jurisprudence was not that it was wholly misguided, but simply that it led to an understanding of law that it is fundamentally incomplete: it fails to keep both sides in the picture. Thus, to borrow terminology from Ronald Dworkin, to which I return in chapter seven, Fuller's jurisprudence in fact orients around two 'centres of gravity': morality *and* efficacy.[26] It is not, as the response by Hart that was so decisive to Fuller's dismissal from the inner circle of twentieth century legal philosophy would have it, a matter of one or the other.[27]

II What is Being 'Reclaimed'?

The broad overview of the ambitions of this book just offered will have provided a general sense of what it is I seek to reclaim by revisiting Fuller's jurisprudence through a more direct focus on his interest in the moral and other implications that attach to law's distinctive form. But it is helpful to take the space in this foundational chapter to map the levels of that reclamatory project more closely, because arriving at a position where we can read Fuller on his own terms requires many smaller reclamations in order to be achieved.

To begin, at the widest level, it is necessary to reclaim an idea of 'jurisprudence'—of intuitive wisdom about law—within a field that over the course of Fuller's lifetime came to define the scope of its proper concerns in increasingly narrow terms. During and since Fuller's time, for instance, the project of jurisprudence has often been proclaimed to be one concerned only with truth claims about law universally true, descriptive in its orientation and kept at a safe distance from normative, or political, claims. This is, of course, not true for projects such as those offered by John Finnis[28] or Ronald Dworkin,[29] or indeed for elements of the work of Joseph Raz,[30] and it is increasingly also not true in the work of younger

[26] Ronald Dworkin, 'Hart's Postscript and the Character of Political Philosophy' (2004) 24 *Oxford Journal of Legal Studies* 1, 26.
[27] See chapter four, II 'Hart's review'.
[28] See further my introductory comments in chapter six.
[29] See chapter seven.
[30] See chapter six.

contemporary legal philosophers who now seek to build bridges across what, in Fuller's time, were very sharp divides.[31] Still, my choice to use the term 'jurisprudence' rather than 'legal philosophy' in the title of this book is deliberate: though not intended to suggest that a project such as Fuller's lay outside of the field of legal philosophy, properly understood, it *is* intended to bring the simple wisdom and creative possibilities of the idea of 'jurisprudence' back into focus.

At the next level, this book stands as an attempt to reclaim the Hart-Fuller debate, in the very specific sense of reading that famous exchange from Fuller's perspective. Here, it is not just a matter of clarifying what Fuller actually sought to convey in his claims about law's internal morality, and distancing those claims from interpretations that have served to distort our understanding of his contribution to debates about the connections between law and morality. It is also, indeed more importantly, a matter of demonstrating how many of the questions that Fuller put to Hart in the context of that debate were in fact never adequately answered either by Hart or by positivists since. The attention given to Fuller's position that was generated by the 50th anniversary of the Hart-Fuller debate in 2008 has, however, done much to draw attention to these omissions, and to invite awareness of how Fuller's responses to Hart were shaped by a distinctive agenda; one that some scholars have helpfully framed in terms of a challenge to Hart to explain how positivists understand the connections between law and 'legality', or 'the rule of law'.[32] Reading the Hart-Fuller debate from Fuller's perspective is thus to accept the possibility that his responses to Hart express a rather different set of challenges to the dominant mode of legal philosophy than those which are ordinarily attributed to it.

One such challenge that becomes apparent when we revisit the Hart-Fuller debate with twenty-first century eyes is the robust methodological critique that Fuller raised against Hart's positivism; a challenge that, in content and orientation, mirrors much of what became a much more famous exchange between Hart and Dworkin. In the terminology of how these issues are often now discussed, Fuller's was an unapologetically internal approach to doing legal philosophy, even if he did not explicitly possess or argue for a methodological approach along these lines.[33] In many ways akin to John Rawls's method of 'reflective equilibrium',[34] Fuller's way of doing jurisprudence is to move back and forth between insights gained from a lawyer's wisdom of law in practice, and more abstract theoretical propositions, as indeed we see in his notion that law has an 'internal morality'. As such, it is a methodology that in some moments seeks a descriptive account of what law is that is compatible with the kind of methodology pursued by positivists, while in others turns to combine this with precisely what positivists attempt to avoid or simply reject: namely, a turn towards practice and—with practice in

[31] See further my discussion of Scott Shapiro's recent book, *Legality*, in chapter eight.

[32] See especially David Dyzenhaus, 'Grudge Informer Case' (n 12) and Jeremy Waldron, 'Positivism and Legality' (n 12).

[33] See Frederick Schauer, 'Fuller's Internal Point of View' (1994) 13 *Law and Philosophy* 285.

[34] John Rawls, *A Theory of Justice*, revised edn (Cambridge, Harvard University Press, 1999) 40–46.

view—towards prescription. Fuller's own understanding of this method is captured especially nicely in a letter to a colleague about his essay 'The Forms and Limits of Adjudication', which I would suggest equally illuminates the method we see in his exchanges with Hart:

> Many seem to assume that my definition of adjudication is something with which I started. In fact, this definition resulted from a process almost identical with that illustrated in my horse thief cases. I don't know just what that process is, but I know the clarity of premise comes at the end. This happened over about ten years of brooding over adjudication. So it does not correspond to my own thought processes to say that I derive rules from my definition; my definition seemed to me to give coherence and added meaning to rules already vaguely perceived.[35]

The methodological mix at play here is consistent with the conviction, to which Fuller held unswervingly, that we lose too much in the enterprise of jurisprudence if we adhere uncompromisingly to fixed dichotomies. This, Fuller believed, was the case whether those divisions speak to the strict separation of law and morality, or to the opposition of means and ends, or is and ought, or description and prescription or, indeed, theory and practice.[36] An important part of reading the Hart-Fuller debate from Fuller's perspective, then, is recognising how crucial a rejection of received polarities was to his basic jurisprudential aims. Only then can we see how the contest recorded in the Hart-Fuller exchanges is in fact much less an instance of a traditional debate between positivism and natural law, with the two held out as starkly opposed positions, than it is a more nuanced attempt, on Fuller's part, to challenge Hart to explain whether and in what ways positivists see law is answerable to conditions that go beyond the narrow factual criteria that satisfy a positivist test of legal validity.

It was very clear to Fuller that to pose such a challenge, to bring these conditions into focus, was to move the inquiry of jurisprudence into an atypical space. But this is precisely the space from which Fuller's jurisprudential project arises, and its concerns are not straightforwardly about whether we can ever insulate the concerns of legal philosophy from the concerns of political philosophy, in the sense of whether we can fruitfully insulate the study of law from the reasons why we might come to need, or want, law in the first place. This kind of story is very much present in Fuller's writings, from *The Law in Quest of Itself* onwards. Equally present is an interest in the effect of law on morality, expressed repeatedly in Fuller's writings in terms of how the existence of the stable baselines that are provided by a legal order make a crucial contribution to the very possibility of moral conduct, and so forth.

But the principal and recurrent space from which Fuller's main jurisprudential claims arise is, for him, a distinctly *legal* space. This is especially well captured in a working note where Fuller suggests that what he means by 'morality' when he

[35] Letter from Fuller to Professor Boris I Bittker, 4 April 1960, The Papers of Lon L Fuller, Harvard Law School Library, Box 14, Folder 1 ('The Forms and Limits of Adjudication').

[36] See further chapter two.

criticises Hart and others for what they seem to miss in their way of debating the connections between law and morality is the 'conditions antecedent to law':

> A legal system cannot lift itself into being legal by fiat. Its security and efficacy must rest on opinions formed outside of it which create an attitude of deference towards its human author (say, a royal law-giver), or a constitutional procedure prescribing the rules for enacting valid law. To say that this acceptance is 'moral' means merely that it is antecedent to law.[37]

To reclaim the Hart-Fuller debate and to read it from Fuller's perspective, then, is to begin seeing how he repeatedly asked his critics to explain where consideration of these conditions could be found in their accounts of law. If they were absent, why were they absent? If they were present, why were their implications not teased out more fully as part of the project of doing legal philosophy? Taken together, Fuller's jurisprudential writings reflect a sustained attempt to draw his critics into a conversation about these forms, relationships, commitments and modes of participation within debates about the connections between law and morality, but also, crucially, as subjects of theoretical concern in their own right. Indeed, one of the reasons I rely extensively in this book on archival materials (especially letters and working papers) is not only because they make these priorities clear, but because they bear witness to his frustration at his apparent inability to instantiate meaningful conversation with his peers about them.[38]

Fuller's challenge to positivists to explain where the conditions that make law possible reside within their legal philosophy was equally a challenge to what he perceived to be positivism's highly permissive conception of law, in the sense that it appears to place no meaningful limits on a lawgiver's power. Here, Fuller's concern lay squarely with the two themes that lie at the centre of this book: his interest in the status of the legal subject and her experience of law as opposed to other modes of ordering and, closely related to this, his complaint that positivism offers a fragile basis from which to distinguish law from those other modes of ordering. This, as I will show over the course of chapters three, four and five, is the agenda that unites Fuller's discussion of Nazi law in his 1958 reply to Hart, the tale of King Rex and his exegesis of the eight principles of legality in *The Morality of Law*, as well as his invocation of the distinction between law and managerial direction in his final 'Reply to Critics'.

By the time that Fuller is defending his position in that reply, however, his debate with Hart and others about the claimed moral status of his eight principles of the internal morality of law had taken a very narrow—and for Fuller, very bewildering—turn. It had, in effect, become a debate about whether observance of the internal morality of law made law 'moral' or just simply 'effective'. As Hart famously put it, Fuller gave us no meaningful reason to distinguish his 'internal morality of law'

[37] Untitled and undated document, paginated in hand as p 25, The Papers of Lon L Fuller, Harvard Law School Library, Box 12, Folder 4 ('Encyclopaedia Britannica').

[38] See further chapter five.

from an 'internal morality of poisoning':[39] aided by the principles of efficacy reflected in the so-called internal morality of law, morality enters law only when this instrument is used for moral good. But beyond this, the argument ran, there is no meaningful sense in which those principles ought to be regarded as moral.[40]

This famous retort has had a decisive impact on how the scholarly memory of Fuller's jurisprudence has been shaped. Few, especially among positivists, have challenged it, though here Raz's engagement with Fuller is notable for the much less central place the matter of efficacy occupies, reflecting Raz's appreciation of how, from Fuller's perspective, the Hart-Fuller impasse was never really about this question.[41] But the power of the 'morality versus efficacy' prism through which standard readings of Fuller have largely been filtered cannot be overstated; indeed, this is the reason why diagnosing the foundations as well as the consequences of this standard response to Fuller occupies much of the exegesis that I offer in chapters two to five.

The debate about efficacy, moreover, is arguably also responsible for what some might regard as the lost or at least underdeveloped promise of Fuller's contribution to more traditional debates about law's connection to substantive morality, or other questions relating to law's wider moral project. With a legal positivist as his partner in conversation, Fuller evidently discerned that such a debate was not going to happen, even if he always tried to keep the door to it open in his insistence that the question of what law *does* must be brought into debates about what law *is*. But the point to note is that, mindful of the debate that he was in, and which at Hart's hand became increasingly narrow in its scope, Fuller ultimately posed his questions about law's wider moral project in a manner compatible with the agenda for conversation that had been set by Hart. Whether or not this led to us learning less of Fuller's thinking about law and morality than we might have, had the conversation been convened differently, what it does show is the extent to which Fuller entered his exchange with Hart with a spirit of engagement that, for the most part, he continued to maintain despite the withering criticism that met his efforts.

The point we need to take from this context, and which is important to the task of reclaiming Fuller, is that there is a need to distinguish between the promise, or invitation, that his jurisprudence presents, especially on the question of law's connection to substantive morality, and the primary concerns and content of the claims that he did in fact develop. Though closely related in ways that I will tease out over the chapters to follow, these are, in fact, different projects. As should by now be clear, in this book my energies are directed above all to the latter: to the animating concerns and content of the claims that Fuller did in fact develop, and how these were shaped, positively and negatively, by the context of the debates in which they were articulated.

39 Hart, 'Lon L Fuller' (n 1).
40 See chapter four, II 'Hart's review'.
41 See further chapter six.

It is clear, then, that to engage with Fuller's jurisprudence in any meaningful way, let alone to attempt to reclaim it along under-explored lines, is to accept the towering presence of the agenda for legal philosophy announced by HLA Hart in 1958. That said, it is important not to read anything said in the foregoing paragraphs as suggesting that the Hart-Fuller exchanges were a thorn in Fuller's jurisprudential side, full-stop. The remarkable endurance of that debate is a testament to how the two modes of legal philosophy that found expression within it continue to be drawn into an engagement with each other. It thus remains not only an important prism through which to understand Fuller's jurisprudence, but also the practice of legal philosophy itself.

But within the very specific project of reclaiming Fuller, a caveat does need to be added to this last point. The contest that ostensibly defines the Hart-Fuller debate—whether there is, or should be, a relationship between legal validity and the morality, or otherwise, of the content of law—was a contest that long pre-existed Fuller's intuitions about the questions to which the study of jurisprudence should attend. It is important to acknowledge this explicitly, because the often-sensed awkwardness of Fuller's relationship with the parameters of that debate was in many ways a consequence of how it was not necessarily the ideal forum within which to develop his intuitions about what, for him, was so independently interesting and important, for both theory and practice, about the form of law.

This is why, in the final analysis, the project of reclaiming Fuller must be a project dedicated not merely to clarifying crucial issues of vocabulary, but to reconvening whole conversations. Clearly, words matter. The way that Fuller was and largely continues to be received within the field is a testament to this, with its persistent queries about what exactly he meant by 'morality' or some other term (though, despite this rarely being acknowledged by his critics, Fuller regularly asked the same questions in return, such as in his challenge to Hart in 1958 to explain what precisely it is that positivists seek to exclude when they exclude 'morality' from their accounts of law,[42] or his appeal in the 'Reply' for his critics to explain how exactly they envisage the measure of 'efficacy' being applied to 'the creation and administration of a thing as complex as a whole legal system'[43]).

But though Fuller was clearly mindful of the decisive significance of the words we use to the possibility of fruitful dialogue across philosophical divides, his working notes make clear that the problem was a much bigger one than simply finding the right words: the only way out of the impasse, or simply through it, was to try to change the conversations themselves. Revealing the struggles of a thinker who was trying to find the best way to redeem a debate and initiate a meaningful conversation about his ideas, these notes capture a remarkable diversity of attempts on Fuller's part to frame the impasse between him and his critics through

[42] 'Positivism and Fidelity' (n 1) 635. See further chapter three, IB 'Reorienting the agenda: Fuller's replies'.

[43] 'Reply to Critics' (n 1) 202.

different conversational prisms. Did the impasse, for example, turn upon differences of opinion about whether there is such a thing as 'legal morality'? Or on disputes about 'the structure of purpose'? Or about 'the structure of reciprocity'?[44]

This is why I ultimately close the project of reclaiming Fuller's jurisprudence in my concluding chapter by gesturing to the kinds of conversations that we might now have about the claims, promise or simply the novelty of Fuller's jurisprudence. Indeed, building upon the exposition of Fuller's central claims that I offer in chapters two to five, I hope to show in chapters six and seven that the passage of time has, for the most part, offered little reason to suggest that those conversations ought to be convened in terms greatly different to those through which Fuller himself attempted to convene them. The difference, however, is that he made this attempt at a time when his peers were much less receptive to the invitation than, I will argue, they appear to be now.

III About the Book: Method, Material and Structure

Anyone who has ever attempted to write about Fuller's jurisprudence in the past will likely report that it is a much more difficult undertaking than might be expected. Certainly, others who have entered this territory before me have warned of the traps. Kenneth Winston has counselled that Fuller's eclecticism as a thinker and the wide range of resources that he drew upon pose the danger, in any attempt at reconstruction, of 'making him appear as a more systematic thinker than he actually was'.[45] Peter Teachout has further suggested that any such attempt at systematisation is liable to derail altogether, producing 'the Cinderella syndrome in reverse, where the golden coach gets turned into a pumpkin'.[46]

All of this places the Fuller scholar in the awkward position of not wanting to do too much in a study of his thought, while at the same time maintaining the impulse to not do too little. To confine Fuller to the Hart-Fuller exchanges is definitely too little, but, at the other end of the spectrum, there is minimal value to be achieved in an attempt to draw all of his intellectual interests into one picture. The challenge, therefore, is to not distract attention from the rich array of his intellectual interests and their points of interconnection, while at the same time not allowing this wider view to swallow the coherent jurisprudential vision that inhabits the same space.

[44] Still more of these notes are simply headed with the question, 'what is at stake?', and all can be found in The Papers of Lon L Fuller, Harvard Law School Library, Box 12, Folder 1 (notes for 'Reply to Critics'). See further chapter five.

[45] Kenneth Winston, 'Three Models for the Study of Law', in Willem J Witteveen and Wibren van der Burg (eds), *Rediscovering Fuller: Essays on Implicit Law and Institutional Design* (Amsterdam, Amsterdam University Press, 1999) 69.

[46] Peter Read Teachout, 'The Soul of the Fugue: An Essay on Reading Fuller' (1986) 70 *Minnesota Law Review* 1073, 1088.

To the extent that my aim in this book is mostly to pursue the latter agenda—to illuminate the coherent jurisprudential vision—it is helpful to establish some parameters around what I seek to include and exclude in pursuit of that agenda. At the level of intellectual biography, it is not my intention to retrace Robert Summers' efforts in providing an outline of Fuller's biography,[47] nor Nicola Lacey's valuable recent exploration of the biographical elements of the Hart-Fuller debate.[48] Still, where relevant to the exposition of the ideas under review, a certain level of biographical commentary does creep in at times, for the simple fact that some of the archival materials that I draw upon offer illuminating insight not only into Fuller's ideas but also the personal struggles of the thinker who was trying to bring them to expression.

The place occupied by these archival materials within my project poses questions and challenges in its own right. In this book's previous incarnation as a doctoral thesis, the insights I recovered through a review of Fuller's working papers were largely relegated to the footnotes, offering support for my reading of the major published texts, but rarely attempting to lead that reading. I was cautious in my use of these resources because I did not want to suggest that Fuller only 'got it right' in his embryonically sketched eunomics project, or in the working papers for his 'Reply to Critics', but not in the works that he chose to publish as a representation of his views. Put bluntly, I did not wish to leave the reader with the impression that the only way to make any sense of Lon Fuller's jurisprudence is to immerse oneself in an archive at Harvard and have a eureka moment with an old scrap of paper.

In this book, however, I have opted for a change of course. I embrace these wider resources with much more confidence because, as I hope might also occur for the reader, familiarity with them has played a crucial part in enabling me to see the recurring concerns of Fuller's jurisprudence, not least his consistent attention to the relationship between the form of law and human agency. But to bring such materials to bear on an interpretation of Fuller's jurisprudence naturally raises the question of how much weight they ought to be given. I have undoubtedly exercised a high degree of authorial choice when privileging some of these resources over others and so, the sceptic might rightfully object, it is ultimately just *my* Fuller who is being reclaimed. There is little that can be done to resolve this criticism: any author who undertakes a project like the one represented here will invariably face it. But what should become apparent as my analysis unfolds is that the working papers generally do not introduce anything radically new to our understanding of Fuller, but rather they tend to be supportive, or offer a richer illumination, of the claims advanced in his published writings. This is especially the case for those archival materials which speak to the lawgiver's 'trusteeship' for the fate of persons,[49] how law's formal feature of generality envisages the legal subject, and how the efficacy of law cannot be understood in terms that portray

[47] Summers, *Lon L. Fuller* (n 15).
[48] Lacey, 'Out of the "Witches" Cauldron' (n 12).
[49] See chapter four, IIB 'Efficacy and trusteeship'.

that subject as simply something to be acted upon,[50] and which, in my view, genuinely do illuminate Fuller's position much better than the published writings, standing alone.

The archival materials are also revealing for how they instruct us about the different modes of authorial voice through which Fuller expressed his ideas, and which greatly affected how those ideas were received and interpreted. Comparing the published and unpublished materials, it is hard to avoid the conclusion that Fuller's best articulations of his position were very often *not* expressed in the context of exchanges with his critics where he very evidently often felt intellectually vulnerable: as he commented in a letter to a colleague about the Hart-Fuller debate, 'as this debate now stands I am truly beleaguered, with both the left and the right, a positivist and a moralist, closing in on me'.[51] By contrast, when Fuller understood himself as charged with a more expository task, such as giving a lecture to a general audience, or when he was corresponding with colleagues who he anticipated would be sympathetic to his claims,[52] he often tended to illuminate the basic thrust of his claims, and the interconnections between them, with more clarity and confidence than we see in the much more polemical atmosphere of the Hart-Fuller debate.

Here *Anatomy of the Law*, an example of Fuller writing for the reader of an encyclopaedia rather than of a scholarly debate, is especially striking.[53] In *Anatomy*, Fuller's authorial voice is, for the most part, expository and balanced and, even though the book obviously gives expression to his own jurisprudential viewpoint, the atmosphere of point-scoring that typifies his exchanges with Hart is largely absent. This is especially well demonstrated in the conciliatory note on which the book concludes. Returning to the big picture frame of debates between legal positivism and natural law, Fuller brings *Anatomy* to a close by suggesting that even if the two schools 'bring to expression a fundamental polarity', it is nonetheless the case that 'at particular junctures in human affairs both kinds of counsel can seem pertinent', and so 'it is likely that both kinds will continue to be offered in the future'.[54]

Finally, it is necessary to make some remarks about my own methodology for 'reclaiming' Fuller. Throughout the main expository chapters of the book (chapters two to five), my method is firstly descriptive and secondarily analytical. That is, my first priority is to work closely with Fuller's texts and to provide an account of their main claims, so as to supply an overall account of Fuller's jurisprudence that is true to his own words, and which shows that we need not go beyond his own output (for example, by way of a radical reconstruction) in order to make a case for its coherence.

[50] See chapter five, II 'Generality, efficacy and agency: insights from the archive'.

[51] Letter from Fuller to Professor Walter F Berns, 14 October 1964, The Papers of Lon L Fuller, Harvard Law School Library, Box 11, Folder 17 (correspondence relating to *The Morality of Law*).

[52] Here Fuller's correspondence with Philip Selznick is especially notable.

[53] Lon L Fuller, *Anatomy of the Law* (New York, Frederick A Praeger, 1968).

[54] ibid 119.

Yet this close marriage between my analysis and development of Fuller's jurisprudence and his texts as written is precisely why my efforts are likely to irritate readers on both sides of the jurisprudential divide. Those who are settled in their view that Fuller offered little to enlighten key questions of jurisprudence may well be irritated by my insistence, especially in light of insights recovered from the archival materials, that there is much more to Fuller's contribution than might previously have been thought. Then, from the opposite end of the spectrum, those who look to Fuller's thought as a resource to develop claims about law's wider moral project might conclude that I have not done enough in service of the promise of his writings on this front. Though it is obviously not my aim to alienate either audience or, indeed, those who come to Fuller from still different starting points, my first priority is to be faithful to the texts. The consequence of this is that sometimes I will claim more for those texts than has been claimed before, and at other times, perhaps less.

It is my hope that by the end of this book the reader will have a much richer, and clearer, sense of what Fuller did in fact contribute to the contests of a field in which his contribution was long-sidelined from its declared core concerns. I therefore also hope that the reader will be better placed to assess the place that Fuller's jurisprudence, or even just certain of its insights, might occupy in that field as it stands today. At each point along this trajectory, the reader will likely be confronted with the need to map borderlines and to bring received dichotomies into view at the same time as contesting them. For better or worse, this is something that simply comes with the territory of working with Fuller's thought. Nonetheless, the end goal of this book is to persuade the reader that there is much more to Fuller than the Fuller we know as HLA Hart's respondent in a debate about positivism and natural law. This is why this book stands, in many ways, as an intellectual history of a debate at the same time as it offers a reclamation of Fuller's jurisprudence, on his own terms.

IV Outline of the Chapters

The next four chapters provide an exposition of Fuller's main jurisprudential claims along a primarily chronological course. Chapter two begins this exposition by situating Fuller's thought in the context of the jurisprudential project that he set for himself in 1954: his eunomics theory of 'good order and workable social arrangements'.[55] Though Fuller never completed his work on this theory, nor offered a systematic outline of it, my claim, shared by Kenneth Winston and Robert Summers before me, is that the eunomics project provides important cues

[55] See Lon L Fuller, 'American Legal Philosophy at Mid-Century: A Review of Edwin W. Patterson's *Jurisprudence, Men and Ideas of the Law*' (1954) 6(4) *Journal of Legal Education* 457, 477, and 'Means and Ends', in Kenneth I Winston, *The Principles of Social Order: Selected Essays of Lon L. Fuller*, revised edn (Portland, Hart Publishing, 2001) 62.

for understanding his jurisprudence generally. This is not least because Fuller formulated the project in reaction to the neglect of questions of legal form in contemporary legal philosophy. Chapter two thus surveys Fuller's ambitions for his eunomics project, gleaned from published writings as well as archival material, with a view to illuminating his emerging interest in questions concerning the distinctiveness and animating moral values of different legal forms.

I then move in chapter three to examine Fuller's reply to Hart in the 1958 *Harvard Law Review*. The Hart-Fuller debate is a classic of Anglo-American jurisprudence, above all for how, at Hart's hand, it announced a renewed agenda for positivist legal philosophy. Rarely, however, is consideration given to how this famous exchange might be read from Fuller's perspective. The purpose of chapter three is to offer that reading with a view to showing how Fuller's reply to Hart challenges positivists to answer a range of questions that are not reducible to the terms usually at stake in contests about the connections between law and morality. These include Fuller's challenge to Hart to explain how he understands the relationship between his positivist concept of law and the kinds of conditions that we ordinarily associate with a state of legality, and whether he regards law's form to be distinctive, or, indeed, to relate in any meaningful way to the legal subject's possession and experience of agency. I also consider in chapter three whether Fuller's position in his debate with Hart about Nazi law might yield a standard for legal validity, beyond the rather non-committal gestures he himself makes on that point.

Chapter four then offers a close analysis of Fuller's 1964 text, *The Morality of Law*. To a considerable extent, this famous book stands as evidence of the many ways in which Fuller's contact with Hart enabled him to develop and clarify his own jurisprudential agenda, especially with respect to the enumeration of his eight principles of the 'internal morality of law'. Still, there are many aspects of *The Morality of Law* that have received little attention from Fuller's critics in their assessment of his claims, precisely because they appear to have little to do with the agenda for legal philosophy as it was declared by Hart. In relation to the curiosities and criticisms that have been generated by Fuller's insistence on designating his eight principles as a 'morality', the argument I develop in chapter four is that we need to recognise the two separate but interrelated senses through which Fuller gave content to his 'morality' designation. The first, as I foreshadowed earlier, speaks to the *ethos* of the lawgiver's role: the requirements of rightness associated with the endeavour of creating and maintaining a successful legal order. The second suggests that law is intrinsically moral in the sense that it generates moral *value* for the legal subject, through how it presupposes that subject to be a 'responsible agent, capable of following rules, and answerable for his defaults'.[56]

These two different senses of 'morality' within Fuller's position have often been confused or overlooked by legal philosophers, above all because of how widely the field has accepted Hart's response that the principles of the internal morality of

[56] *Morality of Law* (n 1) 162–63.

law are merely amoral principles that aid law's efficacy. To bring them to light and to develop their implications thus clearly complicates the narrow lines on which Fuller's reply to Hart on the connection between law and morality has most commonly been appraised. But the plot here, and the need to revisit how we understand the debate itself, is made still thicker by how archival materials suggest that Hart's response to Fuller on the morality claim seems closely linked to how, in his review of *The Morality of Law*, Hart omitted any reference to certain key terms that invite us to read Fuller's position in a more nuanced light. Much of chapter four is thus dedicated to tracing why it is that the core of Fuller's jurisprudence came to be understood as reducible to a claim about what makes law efficacious, and what this meant for Fuller in terms of the options available for defending his position. Akin to my foray into the issue of legal validity in chapter three, I also take space in chapter four to consider where Fuller's thinking might have led on the law and morality question had he not been interrupted by Hart's 'efficacy' charge and the need to defend his position against that charge.

Chapter five is then the site in which I explore the intended 'last word' of the Hart-Fuller debate, 'Reply to Critics', that Fuller added to the second edition of *The Morality of Law* in 1969. Guided by the need to diagnose the sources of Hart's interpretation of his claims about the internal morality of law as reducible to the value of efficacy, as well as to respond to the charge itself, the 'Reply' develops our understanding of Fuller's jurisprudence in crucial ways. Above all, this is because the 'Reply' sees Fuller turn his enduring interest in the distinctiveness of the form of law into a critique which contends that the account of law provided by legal positivism aligns more readily with a conception of managerial direction than it does with law. A turn to the working papers for the 'Reply', however, reveals a number of insights that take Fuller's thinking on this distinction to a level deeper than what he ultimately chose to record in the published piece. Most critically, these papers point to the idea that Fuller saw the relationship between law's formal feature of generality and the legal subject's capacity for agency not only to be a defining feature of law, but something that greatly problematised the apparent ease with which his positivist critics had dismissed his claims. Accordingly, my aim in chapter five is to draw out these latent elements of Fuller's position and to assess whether they invite us to read the questions at issue in the Hart-Fuller debate in a still more nuanced light.

I then turn in chapters six and seven to the task of resituating Fuller in the landscape of twenty-first century legal philosophy. I do this by exploring where, or indeed whether, Fuller's interest in the connections between the form of law and human agency might find a place within two projects that are notable not only for the dominant place they have occupied on the agenda of Anglo-American jurisprudence for many decades, but also for how neither has 'signed on' to Fuller's apparent agenda in any meaningful way. I am referring here to the legal positivism of Joseph Raz, and the anti-positivist jurisprudence of Ronald Dworkin.

Raz inherited and has remained largely committed to Hart's instrumental view of law, but he has taken positivism's engagement with Fuller's claims about the moral dimensions of law, and the relationship between the concept of law and the concept of the rule of law, much further than Hart did. In chapter six, therefore, I seek not only to explore this more fruitful engagement, but to ask whether it might be taken a step further in the direction of Raz's work on the nature of legal authority. The argument I develop is that this site within Raz's distinctive contribution to positivist legal philosophy might open up a novel site of engagement with Fuller, but less in the vein of debates about the connections between law and morality (though in no way incompatible with them) than in the direction of a conversation about the relationship between legal form and agency.

Chapter seven then turns to the anti-positivist jurisprudence of Ronald Dworkin, a project which has remained at a distance from Fuller's since its inception. Yet, in his reconceptualisation of the landscape of contemporary jurisprudence as a contest about competing conceptions of the meaning, and value, of the idea of legality, Dworkin has recently arrived in territory that was Fuller's primary turf. This move, however inadvertently on Dworkin's part, thus stands as a direct invitation not only to illuminate the long overlooked similarities between his and Fuller's projects, but to resituate a reclaimed Fuller on the landscape of contemporary legal philosophy as Dworkin has mapped it. My analysis of the connections between Fuller and Dworkin in chapter seven also provides an opportunity to look more closely at Fuller's thinking on adjudication and interpretation as an aspect of Fuller's jurisprudence than I do in the main, expository, chapters of the book.

I then bring the book to a close in chapter eight by gesturing to three closely related conversations that we might develop out of the project of reclaiming and resituating Fuller as I have undertaken it. The first is the conversation that maintains a presence throughout the book and thus which it is helpful to distil and to reflect upon at the point of its close: namely, the conversation about the connections between law and morality. The second, which indeed emerges out of the unique struggles that Fuller had as a participant in the first, is one about instrumentalism; specifically, about where the idea of law as an instrument fits within Fuller's jurisprudence, and how that jurisprudence interrupts the increasingly settled conviction in contemporary legal theory that law is best understood purely in these terms. Finally, and extending upon my analysis in chapter seven of Dworkin's effort to map the landscape of contemporary legal philosophy by reference to the meaning and value of legality, I close the book by responding to Scott Shapiro's recent engagement with Fuller on this question in his book, *Legality*.[57]

The foregoing outline should, I hope, make clear that this book is not a defence of Fuller per se, in the sense that I do not attempt to establish that he was 'right', or that he 'won' the Hart-Fuller debate. My aims are rather more modest and, I hope, ultimately more constructive. As the 'forms liberate' image so nicely captures, my goal has been to draw out how and why Fuller thought the form of law so interesting and important, how he sought to correct its under-explored

[57] Scott Shapiro, *Legality* (Cambridge, The Belknap Press of Harvard University Press, 2011).

status in jurisprudence, how in his writings we can trace an idea about the way that the form of law relates to human agency, and how all of this was both illuminated *and* obscured through his exchanges with Hart and the ways in which those exchanges have since developed.

Those familiar with Fuller's writings will surely agree that he did not necessarily succeed in articulating this jurisprudential vision in a way that might have seen his message better understood and better placed to endure. But an important part of the project of this book is to understand the causes of these failures, and here the perspective of the underdog has much to teach us: few understand better the character of a terrain, its modes of inclusion and exclusion, than those who try and yet don't entirely succeed in securing admission to it. The 'hows' and 'whys' of Fuller's struggles in legal philosophy thus offer important lessons to those writing in the field today about the way we frame conversations, the way we defend those frames, and what we exclude, consciously or unconsciously, in doing so. The message I ultimately hope to convey, therefore, is not that Fuller 'won' the Hart-Fuller debate, but that he hasn't gone away: there is still much to be said about, and much to be developed from, the jurisprudential agenda that he sought to pursue.

2

Before the Debate

If there is ever a site where Fuller's eclectic intellectual interests reveal themselves to be resistant to neat systematisation, it lies in the canon of his work that is, for the most part, *not* framed by the jurisprudential agenda announced by HLA Hart in 1958. Concerned as it is with that canon, this chapter is at risk of being somewhat unruly in its form, scope and aims. But we lose too much if, in the task of reclaiming Fuller's jurisprudence, we do not begin with the projects that he largely initiated of his own accord, as opposed to those which he developed in response to, or that were at least significantly shaped by, agendas set by others.

Foremost among these projects is Fuller's unfinished *eunomics* theory of 'good order and workable social arrangements'. The main argument I wish to elaborate in this chapter is that acquaintance with this project is vital if we are to gain a sense of how we should read, and should always have read, his exchanges with Hart and others on the internal morality of law. If we want to understand Fuller on his own terms, there is much to be learned by starting here, above all because the stated concern of the eunomics inquiry was to explore matters of form that Fuller thought were neglected in mainstream legal philosophy, and because the project was born at least in part out of a desire to challenge the traditional lines of contests between positivism and natural law.

Before turning to eunomics, therefore, it is helpful to begin this chapter with Fuller's early forays into those traditional debates, as this helps us to identify what he thought was so unhelpful, or obstructing of wider inquiry, about the manner in which those debates were typically conducted. A review of these early writings readily reveals the distinctiveness, from the outset, of Fuller's attempts to be a participant in those debates. His criticisms, above all, are methodological: it is the prevailing 'thought styles' of a positivism and realism-dominated field, and the tendency of both (in his view) to close doors of inquiry rather than to open them, that he thinks must be corrected by at least a partial return to the style of natural law thinking. This, certainly, is a key theme of his 1940 book, *The Law in Quest of Itself.* But as Fuller's thinking on these issues developed over the course of the 1940s, this sense of the decisiveness of methodology to the shape and content of the debates of jurisprudence evolves into a growing awareness of how the inflexible dichotomies to which those debates were married—law versus morality, is versus ought, means versus ends, and so forth—kept at bay any consideration of whether the study of law might actually be more fruitfully pursued by bringing these apparent opposites into some kind of interrelation. It becomes clear that he envisages his own jurisprudential contribution as one that must be directed to dissolving, or at the very least rethinking, those received divisions.

This attitude forms the background for the introduction of eunomics. Although Fuller announced the project under that name for the first time in 1954, its key themes had been central to his thinking, under the working title of 'The Principles of Social Order', for much of the preceding decade. While my final view is that Fuller ultimately did not tell us enough about eunomics to allow us to conclude that we ought to read the whole of his jurisprudence through this prism, there is no doubt that eunomics, writ both large and small, is an abundantly instructive starting point for illuminating Fuller's theoretical disposition generally, and the likely content of many of his claims in the Hart-Fuller exchanges specifically.

To approach the eunomics project 'writ large' is to identify and to explore the 'big picture' ambitions and themes that animate Fuller's declared intentions for the project and which we then also see finding expression in his exchanges with Hart: a concern for the possibilities and limits of different legal forms and processes, their ways of engaging the agency of their participants, and the shape of the relationships, demands of integrity, and animating moral values that constitute each of these forms and account for their distinctiveness. Eunomics 'writ small', by contrast, involves an exploration of Fuller's studies of these forms of ordering in their own right: his essays on legislation, adjudication, mediation, contract, and so forth, each examined with a view to understanding how their distinctive designs and modes of participation do the form-constituting, limit-setting, agency-generating, value-nurturing work that they do.

While dividing an analysis of eunomics along these lines is helpful, there is nonetheless a certain futility to attempting to disentangle the 'large' of eunomics from the 'small'. In the same way that Fuller's general methodology moves back and forth between theory and practice, the eunomics project moves back and forth between the models under review and the wider theoretical questions that he sought to explore. This is why, ultimately, it is most fruitful to explore Fuller's ambitions for the eunomics project through the prism of his studies of specific models. Accordingly, the primary focus of my analysis of eunomics 'writ small' is on the essay that we might regard as the most exemplary 'exercise in eunomics' in Fuller's canon: his 'The Forms and Limits of Adjudication'.

Despite its notoriety, and despite being possibly the most reworked and redrafted piece of his entire career (the carbon copies of letters that Fuller wrote to colleagues across the United States, and internationally, soliciting feedback on one or other draft of the essay run to an entire file in his private papers), 'Forms and Limits' remained unpublished at the time of his death.[1] But the very title of 'Forms and Limits' should make clear that there is much to be learned for the story of reclaiming Fuller's jurisprudence from this essay, precisely because Fuller's exploration of the form of adjudication, and the central thesis he elaborates about that form, is guided closely by his wider theoretical intuitions about

[1] The essay, which remained in draft form at the time of Fuller's death, was published posthumously as part of the tribute to Fuller in (1978) 92 *Harvard Law Review* 353.

how the distinctiveness of a given form of social ordering is constituted by the particular modes of participation that it affords to its participants. By bringing these questions and themes into view, 'Forms and Limits' equally gives expression to Fuller's effort to react and respond to the persistent tendency of legal theorists of his era to regard such questions as merely technical or practical matters, and thus not of concern to legal theory proper.

I then turn at the close of the chapter to an analysis of the labels that typically attach to Fuller's jurisprudential contribution, and how we might best situate that contribution among the wide variety of intellectual influences that shaped it. Was Fuller really a 'natural lawyer', or was he a legal theorist in the vein of early American pragmatism? Or was he in fact a sociologist of law or, indeed, a 'legal process' jurist alongside his Harvard Colleagues, Henry Hart and Albert Sacks? Or is the best approach to Fuller to see him as some kind of eclectic and perhaps even defiant mix of them all?

These are important questions to answer within any project that seeks to engage with, let alone to reclaim, Fuller's jurisprudence. By addressing them at the starting point of my own attempts in this vein, my aim is to provide a foundation for recognising, and thus navigating, the scene-setting role these influences and their associated labels play over the course of Fuller's exchanges with Hart and others on the question of the 'internal morality' of law. The main thorn in Fuller's side here was always, of course, the label of 'natural law'. Yet when we examine Fuller's responses to this label—one that he accepted qualifiedly but increasingly grudgingly as he came to see the obstacles it created to meaningful engagement with those who used it—we learn as much about how Fuller himself envisaged the character of his own jurisprudential contribution as we do the apparently default instinct of scholars of the field to divide their stage into two strict camps and to distribute the actors accordingly.

The early writings in particular record Fuller's awareness of what tends to be lost, for the more creative rather than analytical thinker, when they attempt to seek admission to a field that tends to take theories, rather than questions, as its starting point. This worry, and Fuller's sense of its applicability to his way of approaching the theoretical study of law, is nicely captured in a working note for *The Law in Quest of Itself*, where he speaks of his 'deep conviction that a too self-conscious concern with preserving some kind of 'purity' of method is demonstrably incompatible with creative work, and results in the creative work being concealed'.[2] It is thus fitting to close these preliminary comments by noting that many contributors to the field of legal philosophy today, including those who self-identify as working within the strictest understanding of that term, seem much more inclined to build bridges rather than fortresses, to begin with questions rather than theories, when they offer their contribution.[3] This is a matter I return to in my conclusions in chapter eight. But here, at the beginning, it is worth

[2] Working note titled 'Concerning "is" and "ought" in legal science', The Papers of Lon L Fuller, Harvard Law School Library, Box 9, Folder 7 (notes for *The Law in Quest of Itself*).

[3] See, for example, my discussion of Scott Shapiro's recent book, *Legality*, in chapter eight.

moving into Fuller's early writings with an eye to just how much they reveal about his own desire to approach to the field of jurisprudence through the prism of questions rather than received theories, mining those theories only when to do so served his exploration of the questions.

I The Early Fuller: Positivism and Natural Law at Mid-century

Fuller's first book, *The Law in Quest of Itself*, is best described as a polemic against positivist legal philosophy as it stood in 1940, which means with the efforts of John Austin and Hans Kelsen primarily in view.[4] Still, despite its polemical style, and the now mostly outdated targets of its critique, *Law in Quest* remains a valuable resource for contemporary jurisprudence precisely because of the nature of the objections that Fuller raises.

Fuller's ultimate objection to legal positivism in *Law in Quest* lies not with the project of positivism per se, in the sense of its quest for clarity, or strict standards for legal validity, or a desire not to confuse law and morality, and so forth. His objection, rather, is to the intellectual orientation reflected in the efforts of those who have developed the positivist project since Hobbes. As described in *Law in Quest*, Hobbes favoured the idea of an imposed positive law because, given man's innately warring nature, he saw no alternative other than a lawgiving sovereign for the attainment of peace and order. For Hobbes, then, 'there was no uncertainty or ambiguity about the object which he pursued in constructing his theory', which, Fuller suggests, explains why it has been said without paradox that Hobbes 'founded legal positivism on a natural law basis'.[5] Fuller's point, then, is that what is missing in contemporary variants of positivism is Hobbes' purposive sense of why a lawgiver is needed at all.[6] Unlike Hobbes, who kept the reasons for arriving at positive law in view, contemporary positivists move their focus directly to the source of law, without regard to the story, at least in Hobbes' account, that precedes it.

But *Law in Quest* also makes clear that Fuller sees the offences of contemporary positivists as going beyond their loss of the 'why' of law. They are equally at fault on methodological grounds in their tendency to advance a 'disintegrative approach to language and experience' that removes the 'whole view' from our gaze.[7] This grasp of the methodological foundation of the impasse between the two dominant jurisprudential schools, as I foreshadowed in my preliminary remarks above, is something which features regularly among Fuller's early com-

[4] Lon L Fuller, *The Law in Quest of Itself* (Chicago, The Foundation Press, 1940).
[5] ibid 19–20.
[6] ibid 24. See also Lon L Fuller, 'Reason and Fiat in Case Law' (1946) 59 *Harvard Law Review* 376, 388.
[7] ibid 377.

mentaries. It is repeated in Fuller's 1949 class materials, published in temporary form as *The Problems of Jurisprudence*, when he speaks of how 'distortions proceeding from methodological motives' lead us to see 'neat diagrams instead of the complex and moving reality that actually exists', and how, in turn, these motives can be traced to 'the modern tendency to exclude rigidly anything that savours "natural law" from legal thinking'.[8]

The essay 'Reason and Fiat in Case Law', published in 1946, makes a similar point, but here with the strict conceptual dichotomies typical of jurisprudential argument in view. Whether pitting the phenomenon of 'morality' against 'law', or suggesting the necessary opposition of 'reason' and 'fiat', 'means' and 'ends', or 'what is' as opposed to 'what ought to be', Fuller's response to these received prisms of theoretical discussion reveals an intellectual debt to Morris Cohen's 'principle of polarity': the idea that 'notions apparently contradictory form indispensable complements for one another'.[9] An embrace of this principle, Fuller contends in 'Reason and Fiat', is precisely what legal philosophy needs, because it is scarcely to the credit of the field that it achieves harmony within its domain 'if this is accomplished only by barring its frontiers to every vital detail and fruitful idea that might disturb the internal order'.[10]

Fuller's consciousness in these early writings of the decisiveness of methodology to the character as well as content of the debates of jurisprudence is intriguing for positive as well as negative reasons. That is, if this understanding was so firmly held by Fuller in the 1940s, why did he not address it, or at least position it more squarely, in his exchanges with Hart about the internal morality of law? In that context, as I explain further in chapter five, the issue of methodology and its constitutive effect on the apparent impasse between Fuller and his critics does not take a central place until the 'last word' of those exchanges, his 1969 'Reply to Critics'.[11] Similar questions can be raised about how the early writings address the normative capacity of legal philosophy to give 'a profitable and satisfying direction to the application of human energies in the law' in a manner that, again, is not repeated with the same energy or clarity in the Hart-Fuller exchanges.[12] In his rejoinder to Ernest Nagel, for instance, which forms part of an ongoing exchange stemming from his 1956 essay, 'Human Purpose and Natural Law',[13] Fuller not only expresses his concern for the prescriptive ambitions of legal philosophy, but, in a manner not repeated in his more well-known writings, takes this concern towards the question of how to specify the existence conditions for law. Framed by his contest with Nagel about the need for a strict division between 'is' and

[8] Lon L Fuller, *The Problems of Jurisprudence*, temporary edn (Brooklyn, The Foundation Press, 1949) 717.

[9] 'Reason and Fiat' (n 6) 381. Fuller's embrace of Cohen's 'principle of polarity' is discussed at length in Peter Teachout, 'The Soul of the Fugue: An Essay on Reading Fuller' (1986) 70 *Minnesota Law Review* 1073, 1105–10.

[10] 'Reason and Fiat' (n 6) 377, 381–89.

[11] See chapter five.

[12] *Law in Quest* (n 4) 2.

[13] Lon L Fuller, 'Human Purpose and Natural Law' (1956) 53 *Journal of Philosophy* 697.

'ought' in legal philosophy, determining the question of what law *is*, Fuller argues, necessarily implicates a diagnostic inquiry within which a normative question is rightly paramount; that is, the question of whether something which calls itself a legal order 'is missing that target so woefully that it cannot in any meaningful sense be termed a system of law', or that it otherwise has 'so little "value" that it has ceased to "exist"'.[14]

The links between these comments and what Fuller comes to articulate in his response to Hart in the 1958 *Harvard Law Review* shortly afterwards are obvious. This will become clear through my analysis of that exchange in chapter three. But the point to note for now is that, much more than in the works we associate with the Hart-Fuller debate, Fuller's early 'pre-Hart' writings reveal just how necessary he thought it was to clear the decks of the kinds of baggage typically associated with natural law thinking—yet while defending and restoring certain of its insights at the same time—if he was to secure a place for his own distinctive contribution on the agenda of contemporary legal philosophy. As his 1954 essay 'American Legal Philosophy at Mid-Century' makes clear, Fuller traced the rejection by scholars of his era of the natural law tradition to two main causes: the tradition's failure to furnish an enduring and unchanging set of criteria capable of answering the question of what makes law 'law', and its claim that natural law sets itself above positive law and 'counsels a disregard of any enactment that violates its precepts'.[15]

On these points, it is clear from the early writings that Fuller is on the side of the critics: a stance which, despite confusions that often still surround how his claims about the internal morality of law are interpreted, Fuller did not shift from at any point in his exchanges with Hart. But again, it is to the 'pre-Hart' writings that we should look to see this point most clearly. In his 'Rejoinder to Professor Nagel', for instance, Fuller states explicitly that he does not accept any doctrine of natural law which asserts that its demands can be the subject of an authoritative pronouncement, or that there is something called '*the* natural law' capable of concrete application like a written code, or that there is a '"higher law" transcending the concerns of this life against which human enactments must be measured and declared invalid in case of conflict.'[16] Still, in the same context, Fuller actively defends the view that it is not a 'perversion' to hold a position which rejects these commitments but which still lays claim to affinities with natural law thinking; a stance, he suggests, that 'is at least as ancient as Aristotle, in whom I find no trace of the elements I reject'.[17]

It is within this last point that we can uncover the nub of Fuller's 'natural law' position: his interest in 'those principles of social order which will enable men to attain a satisfactory life in common' which, on his reading, was equally an enduring, if secondary, concern of the older natural law theorists.[18] 'Human Purpose

[14] Lon L Fuller, 'A Rejoinder to Professor Nagel' (1958) 3 *Natural Law Forum* 83, 92.

[15] Lon L Fuller, 'American Legal Philosophy at Mid-Century: A Review of Edwin W. Patterson's *Jurisprudence, Men and Ideas of the Law*' (1954) 6(4) *Journal of Legal Education* 457, 467. See also *Problems of Jurisprudence* (n 8) 701–2.

[16] 'Rejoinder to Professor Nagel' (n 14) 84.

[17] ibid.

[18] ibid.

and Natural Law' gives us a glimpse of where Fuller would head with this idea in the years to follow. There he speaks of the 'internal' sense through which we might understand the idea that there are natural laws of social order, and how a form of social order 'makes its own technical demands if it is to be in fact what it purports to be'.[19] Replying to Nagel on this point, the 'Rejoinder to Professor Nagel' sees Fuller specify this idea in terms both more concrete and clearly resonant with the claims that advanced in his 1958 reply to Hart, as well as *The Morality of Law.* No one, Fuller there argues, would regard a legal order to be in existence in a situation where there is 'no discernible correspondence' between the laws issued by a government and the acts of those who purport to enforce them, or where all laws are 'retrospective in effect and no prospective laws are ever enacted'.[20]

These 'pre-Hart' forays into the traditional jurisprudential contest between positivism and natural law speak directly to the foundations of the reply to Hart's law and morality agenda that was shortly to follow. By the same token, these early writings, especially *Law in Quest*, give us a sense of what was likely to go wrong when Fuller entered that territory. As already foreshadowed, a major thorn in Fuller's side here was the tendency of his readers to align his claims more closely with those of traditional natural law thinking than was appropriate to the distinctiveness of those claims. But in the early writings we can also readily identify certain thorns that were largely of Fuller's own making.

As I explained in chapter one, Fuller understood himself to be very much disadvantaged in his forays into traditional jurisprudential debates by his lack of training in conventional philosophy. This disadvantage was already stingingly apparent in the aftermath of the publication of *Law in Quest*, which saw him greatly discouraged by Morris Cohen's scathing review of the book that not only questioned Fuller's grasp on the philosopher's distinction between 'is' and 'ought', but also declared his 'strange definition of positivism and natural law' not only to do 'violence to the historic meaning of these terms', but to hopelessly confuse his claims as well.[21] But these are disadvantages of training or skill-set, as opposed to those of style or intellectual character. The latter was arguably a bigger problem: Fuller held himself out to be alienated by his tendency to be aggressively polemical, even inflammatory, when he attempted to inhabit more traditionally philosophical territory. In some ways, this approach could be fruitful. The very title of *The Law in Quest of Itself*, for instance, is not only provocative, but neatly captures the nub of Fuller's enduring complaint against positivism: how it seeks a concept of law that is insulated from all other moral, social and political phenomena.[22] Yet

[19] 'Human Purpose' (n 13) 704–5.

[20] 'Rejoinder to Professor Nagel' (n 14) 91.

[21] Morris Cohen, 'Should Legal Thought Abandon Clear Distinctions?' (1941) 36 *Illinois Law Review* 239, 244.

[22] In draft notes prepared for a lecture given in 1959, Fuller offers an alternate formulation of this point in terms of positivism's untenable marriage to the idea of the 'self-sufficiency' of legal order: notes for a talk delivered at the University of Vermont, 13 July 1959 (announced title: 'Traditions of Justice Among Western Peoples') p 8 (Fuller's handwritten pagination), The Papers of Lon L Fuller, Harvard Law School Library, Box 11, Folder 8.

in other ways the general tone of *Law in Quest* invited Fuller's early detractors to weigh in against him with respect to style as much as content. Here the collegial but no-nonsense feedback of Thomas Reed Powell is instructive:

> The temper seems to me frequently ugly notwithstanding the Christ-like ethical standards. Woe unto you, Publicans and Sinners. Your mood doesn't seem to be one in which fairness and objectivity are likely to find sustenance. To an extent, therefore, McDougal seems to me to give a fair impression if he would say: *Qui etait dans son Coeur, sinon sur ses levres.* When you indict the undefined, you offer wide opportunities to accept service and plead *volo contendere.*[23]

It might well be, then, that another reader was on to something when, reviewing *Law in Quest*, he suggested that it 'comes as a disappointment chiefly because Professor Fuller's undoubted talents deserve a wider range than is imposed by a battle of the schools'.[24] But enter that battle Fuller did, and the purpose of the foregoing brief foray into some of his early attempts at the same has simply been to signal the style and content of what was to come when he did so in the context of responding to Hart.

II Eunomics: A 'Science or Theory of Good Order and Workable Social Arrangements'

Fuller never explained in his writings from where, or from whom, he borrowed the term *eunomics*, which roughly translates from the Greek (*eu* = good; *nomos* = order) as good order, or good law.[25] That he took the term from Aristotle, who wrote not only of good order but of good law in his *Politics*, seems the most likely explanation. The evidence to suggest this can be found in Fuller's personal copy of Jerome Hall's *Readings in Jurisprudence*, which includes a marginal annotation next to Aristotle's reference to the 'goodness of the laws', a phrase, in the original Greek, that is expressed through the term *eunomia*.[26] Whether Fuller himself read and understood the Greek text of the passage is unclear, but again, there is evidence that he might have done: in his personal copy of his 1949 class materials, *The Problems of Jurisprudence*, the extracts from Aristotle's *Nicomachean Ethics* on 'Justice' are annotated, in hand, in Greek.[27]

[23] Letter to Fuller from Thomas Reed Powell, undated, The Papers of Lon L Fuller, Harvard Law School Library, Box 9, Folder 7 (notes for *The Law in Quest of Itself*).

[24] Hardy C Dillard, 'Book Review: *The Law in Quest of Itself* by Lon L. Fuller' (1941) 27 *Virginia Law Review* 568, 571.

[25] The Oxford English Dictionary definition speaks in terms of 'law-abiding', 'socially well adjusted or ordered', and referable to a political condition of *eunomy*, defined in turn as 'good law well-administered'.

[26] Jerome Hall, *Readings in Jurisprudence* (Indianapolis, Bobbs-Merril Company, 1938).

[27] I am indebted to Kenneth Winston for sharing this intriguing research with me during conversations at the John F Kennedy School of Government at Harvard University on 29 April and 1 May 2008.

Even if we do not know its precise source, what we do know is that Fuller introduced the term eunomics as a framework for a new jurisprudential agenda in a 1954 review of Edwin W Patterson's *Jurisprudence, Men and Ideas of the Law.*[28] That essay, 'American Legal Philosophy at Mid-Century', offers a valuable glimpse of Fuller's preliminary sense of what a jurisprudential project cast in eunomic terms—a 'science, theory, or study of good order and workable social arrangements'[29]—might entail. Ultimately, this glimpse is inseparable from objections advanced in the essay to the dominant preoccupations of legal philosophy in his era, precisely because eunomics is offered as a response to those objections. Foremost among these is what Fuller perceives to be an unqualified acceptance of the 'imperative theory of law' among the legal theorists of his era.[30] Echoing what he later brings to strong expression in his exchanges with Hart, the problem shared by proponents of the imperative theory, Fuller argues, lies in how they characterise law as 'a compliant instrument ready to give sanction to ends that are not, and should not be, imbedded within the legal apparatus of coercion itself'. According to this view, if we are to seek assurance that the ends of law will be good, we must look to education, or adherence to the democratic tradition, or some other extra-legal space to gain that assurance.[31]

For Fuller, what is missing here is any acknowledgment of the reciprocal basis of legal obligation, above all, of how the force which lies behind political authority is 'essentially a moral power' derived from 'the general acceptance of the rules by which the law-making process is conducted'.[32] But for the explicit gesture to matters moral, it might be said that Fuller, in his emphasis on 'acceptance' of 'rules of lawmaking' as foundational to an account of legal obligation, here sounds very much like Hart's claim in his 1958 essay that the 'key to the science of jurisprudence' lies in what it means for a social group to accept the fundamental rules of lawmaking procedure; a thesis then elaborated at length in *The Concept of Law.*[33] But Fuller's main objection in 'American Legal Philosophy' is more fundamental. The very task of legal philosophy, he argues, ought to be something quite different than that suggested by the imperative theory of law. It should instead be about discerning 'those minimum principles that must be accepted in order to make law possible and then to protect the integrity of those principles and to promote a general understanding of them.'[34]

This statement is striking for how it clearly signals the type of intellectual programme that Fuller thought necessary for jurisprudence, and which he ultimately attempted to pursue in his debates with Hart. Other aspects of that programme are equally revealed in the essay, such as in how Fuller suggests the need for an approach to jurisprudence that is grounded in real world facts as much as it is in

[28] 'American Legal Philosophy' (n 15) 477.
[29] ibid 477.
[30] ibid 459.
[31] ibid 463.
[32] ibid 462.
[33] See further chapter three, A 'Hart's claims' and chapter four, IB 'Situating Fuller's claims'.
[34] 'American Legal Philosophy' (n 15) 463.

aspirations: law 'is not a datum', Fuller insists, but 'an achievement that needs ever to be renewed, and that cannot be renewed unless we understand the springs from which its strength derives'.[35]

This is the background against which Fuller introduces his intended theory of eunomics, and by keeping this background in view we are able to see the extent to which the eunomics programme stood as a reaction to Fuller's frustration towards the failure of contemporary legal philosophers to concern themselves in any meaningful way with the 'general principles that will guide choice among the available forms of order'.[36] This ostensibly lawyerly inquiry—which kinds of forms of ordering are suited to which kinds of problems?—is given its jurisprudential significance through how Fuller traces this apparent neglect to 'an exaggerated reaction against the theory of natural law' in an intellectual era which has come to privilege determinate criteria and clarity of analysis.[37] Eunomics, then, is an attempt to reclaim what has there been lost but, crucially, in a highly qualified manner. In its emphasis on 'the means aspect of the means-ends relation',[38] eunomics implies no commitment to 'ultimate ends'.[39] Fuller is therefore clear that his intended project departs substantially from claims more typically associated with the natural law tradition, such as the notion that natural law 'sets itself above positive law and counsels a disregard of any enactment that violates its precepts'.[40] Eunomics, instead, is a natural law project only in so far as it attempts to reclaim the natural law intuition that there are 'compulsions necessarily contained in certain ways of organizing men's relations with one another', such that, by illuminating these compulsions, we might uncover a kind of 'natural law' of social ordering.[41]

We learn little else from 'American Legal Philosophy' about the ambitions of eunomics than what is revealed in these comments. Beyond this, the only other way that Fuller elaborates the possible parameters of the project is through the example of a class exercise (which he suggests reflects a 'problem of eunomics') which poses the question of how the forms of adjudication and contract, in whole or hybrid form, might be brought to the resolution of a boundary dispute between two hypothetical countries. The question of what such an exercise has got to do with legal philosophy is precisely Fuller's point.[42] Why should legal philosophy *not* be concerned with 'two fundamental and pervasive principles of social order', adjudication and contract, the 'principles that underlie' each, and how by studying them 'in this uneasy and somewhat hazardous mixture' we might learn something 'about their social functions generally'?[43]

[35] ibid 467.
[36] ibid 477.
[37] ibid.
[38] ibid.
[39] ibid.
[40] ibid 467.
[41] ibid 473, 476. See also 478–79.
[42] ibid 478.
[43] ibid.

The only other reference to eunomics in Fuller's writings during the period to which 'American Legal Philosophy' belongs can be found in a hand-out that Fuller distributed to his jurisprudence class in 1954–55, which foreshadows an intended new chapter for his class materials, *The Problems of Jurisprudence.* The original, and ultimately unchanged, copy of those class materials concludes with a chapter titled 'The Principles of Order', that sees Fuller set out a general exposition of his own views on the various problems of jurisprudence addressed elsewhere in the volume.[44] The class hand-out, however, suggests that he intended to rewrite this chapter under the title 'The Problems of Jurisprudence: An Essay in Eunomics'. While only skeletal in outline, the hand-out nonetheless contains some instructive suggestions as to the direction that this intended essay might have taken. It speaks, for instance, of Fuller's desire to give closer attention to the means-ends relation, as well as to the relationship between different forms of social organisation and the goals of 'efficacy', 'human satisfaction' and 'human development'. Moreover, in a manner importantly suggestive of the nuances of Fuller's position that have so frequently been overlooked in standard readings of his claims about the internal morality of law, the outline also states, as if to qualify the instrumental associations that so readily accompany talk of 'efficacy', that the envisaged eunomics project is concerned only 'with those forms of "coming together" or association which result in a benefit to all participants'.[45]

There is no further mention of eunomics until Fuller penned the essay, 'Means and Ends', some time in the early 1960s. The essay remained unpublished at the time of Fuller's death in 1978, and appears to have been basically unknown until it was reproduced by Winston as the foundational chapter of *The Principles of Social Order.*[46] Curiously, the early drafts of 'Means and Ends' contain no reference to the term 'eunomics'. It is only in the final draft that Fuller states that he 'once suggested a term for describing the kind of study undertaken in this book', and so the eunomic frame finds its way back in.[47]

Fuller states at the outset of 'Means and Ends' that he intended to develop the essay into a systematic exposition of a 'science, theory, or study of good order and workable social arrangements'. Certainly, there are some significant developments from the sketch presented in 'American Legal Philosophy', most notably, Fuller's clarification of the *kind* of order with which eunomics is concerned: not just any type of order ('the order, say, of a concentration camp') but order that is 'just, fair, workable, effective, and respectful of human dignity'.[48] The purpose of

[44] *Problems of Jurisprudence* (n 8) 693.

[45] See document titled 'Jurisprudence, Supplementary Readings, Chapter VI, 1954–1955, Principles of Eunomics', The Papers of Lon L Fuller, Harvard Law School Library, Box 10, Folder 12 ('The Problems of Jurisprudence').

[46] There are suggestions that Fuller circulated the draft essay to colleagues; see Robert S Summers, 'Professor Fuller's Jurisprudence and America's Dominant Philosophy of Law' (1978) 92(2) *Harvard Law Review* 433, 438, note 34, where Summers also dates the draft essay to 1960.

[47] The Papers of Lon L Fuller, Harvard Law School Library, Box 14, Folder 2 ('Means and Ends').

[48] Lon L Fuller, 'Means and Ends', in Kenneth I Winston, *The Principles of Social Order: Selected Essays of Lon L. Fuller*, revised edn (Portland, Hart Publishing, , 2001) 61.

the eunomic project as stated in 'Means and Ends', therefore, is not simply to examine the principles of social order, but to illuminate the principles of *good* social order.[49] This distinction in turn signals that the phenomena of pathology and health will be key diagnostic tools within the eunomics project, as it seeks to make distinctions between 'sound and unsound social institutions', and to 'stigmatize certain forms of social ordering as perverted and parasitic',[50] while also analysing 'the conditions under which particular forms of social order may be said to approach perfection'.[51]

Above all, however, 'Means and Ends' is an essay about methodology, specifically, about how, in Fuller's view, the basic methodology of 'social philosophy' must be oriented away from its persistent focus on ends in abstraction from means.[52] The exemplar of this methodological offence, it seems, is JS Mill's comment in *Utilitarianism, Liberty, and Representative Government* that rules of action 'must take their whole character and colour from the end to which they are subservient', such that 'when we engage in a pursuit, a clear and precise conception of what we are pursuing would seem to be the first thing we need, instead of the last thing we are to look forward to'.[53] Contesting this approach in favour of one directed to exploring 'the ways open to human beings to arrange their mutual relations so as to achieve their individual and collective ends, whatever those ends may be',[54] the nub of the argument that Fuller puts in reply is that no end in view will remain unaffected, positively or negatively, by the means through which it is pursued. Means and ends mutually transform each other. But we will not see this, he argues, if we remain blinded by five 'modes of thought' that dominate ends-orientation of means-ends analysis: namely, assumptions that the ends served by social institutions can be understood as distinct from each other; that the first task of social philosophy is to arrange human ends hierarchically; that social arrangements are infinitely pliable; that formal structure consists only in ends, and not means; and the very idea that institutional means are merely 'necessary evils'.[55]

It is a methodological critique, but its overriding message is that the vehicles through which ends are pursued, the forms through which social goals are framed and realised, matter in their own right and so deserve our independent theoretical attention. This is not an instrumental understanding of form: such would be antithetical to the whole tenor of 'Means and Ends'.[56] Rather, the point of the eunomic analysis is to encourage us to appreciate that although forms of ordering are instrumentally essential to the pursuit and realisation of desired social ends, there

[49] See ibid 62 (with Fuller's emphasis).
[50] ibid.
[51] ibid.
[52] ibid 63.
[53] ibid.
[54] ibid 64, 69.
[55] ibid 67–72.
[56] See further below, and chapter eight, II 'Instrumentalism'.

are criteria of soundness that ought to accompany our evaluation, and indeed use, of such forms. And, in eunomic terms, a sound institutional arrangement is one which strives to bring about 'a pattern of living that is satisfying and worthy of men's capacities', irrespective of the nature of the end being pursued through it.[57]

A Eunomics 'Writ Large'

It is through Kenneth Winston's work that the idea of eunomics has become associated with Fuller's essays on adjudication, contract, legislation, and other themes relevant to his exploration of questions of institutional design.[58] Until Winston published *The Principles of Social Order*, few scholars had given anything more than a passing reference to the project and, when they had done so, it was usually to indicate no more than that Fuller had announced the need for a line of inquiry cast in such terms. Even then, however, the immediate tendency was not to turn to the offerings of the project itself, but rather to ask what *kind* of project it was. Was eunomics just another name for natural law? Was it really a natural law project at all? How, if at all, did it contribute to debates of legal philosophy about the positive aims of the law?[59]

Given the extent of the efforts by Fuller to finesse the relationship between his intended project and the natural law tradition, so as to avoid precisely this kind of narrow engagement, it is striking how readily these responses default to a standard set of questions drawn from a received understanding of the parameters of jurisprudential debate. One critic, indeed, asked whether the 'so-called "science of eunomics"' was merely 'a fanciful semantic device used to facilitate the return to the study of natural law without directly colliding with the bias of the American intellectual community'.[60] Irrespective of Fuller's ambitions for it, it seems, a natural law project is a natural law project and thus one wholly unsuited to twentieth century scholarly inquiry. Among Winston's achievements in reclaiming Fuller along 'eunomic' lines thus lies in how, without compromising his attention to Fuller's concern to illuminate the moral values that animate a given mode of ordering, he manages to largely avoid a debate about labels and present eunomics instead for what it is: an attempt to uncover the organising principles, features of design and participatory commitments which constitute different models of social ordering, and which make them appropriate for use in a given context:

[57] 'Means and Ends' (n 48) 68–69.

[58] See, eg the essays in Willem Witteveen and Wibren van der Burg (eds), *Rediscovering Fuller: Essays on Implicit Law and Institutional Design* (Amsterdam, Amsterdam University Press, 1999).

[59] See, eg AP d'Entrèves, 'The Case for Natural Law Re-examined' (1956) 1 *Natural Law Forum* 5, 30, 32; Edgar Bodenheimer, 'A Decade of Jurisprudence in the United States of America: 1946–1956' (1958) 3 *Natural Law Forum* 44, 61–62; Joseph P Witherspoon, 'Relation of Philosophy to Jurisprudence' (1958) 3 *Natural Law Forum* 105, 109–13; Charles L Palms, 'The Natural Law Philosophy of Lon L. Fuller' (1965) 11 *Catholic Lawyer* 94, 114; and Julius Stone, *Human Law and Human Justice* (Stanford, Stanford University Press, 1965) 221–26.

[60] See Ernest F Roberts, 'Book Review' (1961) 6 *Villanova Law Review* 109, 113.

> The central task of eunomics is to describe these models in detail and assess the possibilities for their realization. The contextual factors that enhance, or impair, the effectiveness of particular structures determine whether or not an existing social problem can be collectively managed—and how. We want to know where a mechanism works and where it does not—and cannot be made to. Under what conditions will a legal form continue to retain its integrity, and what are the limits beyond which its distinctive aims and capacities are compromised?[61]

This passage is worth dwelling upon, not only for how it fuses the theoretical and practical aims of eunomics, but also for how ideas about 'effectiveness' and 'integrity' are here brought into interrelation. This, indeed, is the nub of eunomics as a theoretical contribution, as it also is of the jurisprudential position that Fuller came to develop in his responses to Hart. As Winston in the quoted passage suggests, the demands of integrity associated with different models of ordering, while clearly instrumental, in the sense of being decisive, to the effectiveness of any such model in its use towards the achievement of a specified end, are for Fuller nonetheless a matter for investigation in their own right. Integrity has its own ways of being constituted and maintained, and these are neither exhausted by, nor wholly referable to, the measure of effectiveness.[62]

This is why it is crucial to accept the instrumentalist orientation of eunomics at the same time as to insist that this orientation is qualified in a distinctly Fullerian way. To develop this point, it is helpful to return to Fuller's idea of form, and how the eunomics inquiry assists us to illuminate the dimensions of that idea further. I sketched this capacious understanding of form in chapter one in the context of outlining the unifying argument of this book that Fuller's jurisprudence is best reclaimed through the prism of his understanding of the connections between the distinctive form of law and human agency. As I explained there, when we read his claims about the internal morality of law through this prism, we come to see that Fuller's jurisprudence is animated by an understanding of form that is inclusive of the legal subject's presupposed status as a responsible agent. There is not a form called law that acts upon the legal subject, but rather, the form of law includes the legal subject's capacity for agency within it.[63]

I revisit this point in the present context because in the eunomics writings, this understanding of form is not something that needs to be teased out: it is foundational to Fuller's whole inquiry. This is made especially clear when we examine the idea of eunomics 'writ small', that is, the insights of Fuller's writings on different models of ordering and the emphasis within these on the distinctive modes for the expression of agency that these models afford to those who participate within them. I return to this point in more detail below.

[61] KI Winston, 'Introduction to the Revised Edition', in *The Principles of Social Order* (n 48) 9.

[62] See further Lon L Fuller, *Anatomy of the Law* (New York, Frederick A Praeger, 1968), especially 57–69.

[63] See further chapter four, IC 'Conception of the person implicit in legality'.

B Eunomics 'Writ Small': The Models

Many essays are associated with Fuller's studies of different modes of ordering, and what I offer here is something far short of a comprehensive engagement with them. Covering adjudication, legislation, custom, contract, mediation, managerial direction, 'principles of human association', and more, each could and indeed has generated a body of scholarship in its own right.[64] Common to all of these studies, however, is a concern for form, its integrity, and the work that forms do, independently of the goals that might, instrumentally, be sought through them.

Fuller's analysis of legislation, or 'enacted law', offers a clear illustration of this orientation. Here, the primary focus is not on the idea that enacted rules are something issued by an authority with an expectation of compliance on the part of those subject to them. Fuller instead directs his analysis to how legislation provides baselines which 'serve to set the limits men must observe in their interactions with one another' so that they might be 'free within those limits to pursue their own goals'.[65] Legislation, accordingly, 'does not tell a man what he should do to accomplish specific ends set by the lawgiver' but rather 'furnishes him with baselines against which to organize life with his fellows'.[66] This is a qualitatively different conception of enacted law than the more traditional view that sees legislation as an 'instrument of social control' that acts upon the citizen.[67] Fuller is equally concerned to illuminate the conditions antecedent, as well as subsequent, to the success of a legislative endeavour, specifically how its efficacy depends on the development of 'stable interactional expectancies' between lawgiver and subject which are in turn maintained by a faith on both sides that they are 'playing the game of law fairly'.[68] Again, then, efficacy becomes a measure inseparable from the presence of certain moral conditions: here, a 'faith' generated by the back and forth between legislator and subject, that constitutes the moral foundations to which both those who are subject and those who enact and administer it are answerable.[69]

Fuller's writings on contract, known far less as exercises in eunomics than as part of a body of doctrinal scholarship for which he was renowned, speak similarly to this idea about how the relationships that constitute the form of a given mode of ordering underscore its moral force.[70] The contract writings return consistently to the question of how the distinctive form of contract generates the authority and legitimacy it has for those who create and sustain it. These writings

[64] See generally the collection of essays on these themes in Winston, *The Principles of Social Order* (n 48).

[65] See especially Lon L Fuller, 'Human Interaction and the Law' (1969) 14 *American Journal of Jurisprudence* 1.

[66] ibid 24.

[67] ibid 20.

[68] ibid 25.

[69] ibid 25.

[70] See Lon L Fuller and William R Perdue, Jr, 'The Reliance Interest in Contract Damages' (1936–37) 46 *Yale Law Journal* 52.

equally revert regularly to insights about how, in order for a contractual arrangement to be effective and binding, it is necessary for each party to recognise the other as bearing an equal entitlement to the exercise of her agency. Thus, read through the prism of eunomics, the message conveyed repeatedly in Fuller's writings on contract is that the distinctive nature of the contractual form—its structure of reciprocity—is not only sustained by, but also generates, the distinctive moral attitudes that make contract itself possible.

This point is made especially clearly in Fuller's famous early essay on contract, 'Consideration and Form', where he speaks of how the formalities of contract shape the quality of the parties' participation within that institution by inducing attitudes appropriate to the creation and maintenance of a contractual relationship.[71] The same is conveyed in a much later piece on contract, where Fuller comments on how contractual negotiations require participants to 'understand why the other makes the demands he does even as he strives to resist or qualify those demands', and to accept the other's 'right to work for a solution that will best serve his own interests'.[72] Contract, in short, 'brings home to man graphically the fact of social cooperation, and forces him to take a conscious part in that process'.[73]

There are numerous other studies that we can associate with these themes of Fuller's eunomics project, many of which are brought together by Winston in *The Principles of Social Order* to offer us, through an assemblage of its possible parts, one picture of what a more developed study in eunomics might have looked like. But in my view we can equally grasp that picture by examining what might be regarded as the most exemplary of Fuller's exercises in eunomics: his well known essay, 'The Forms and Limits of Adjudication'.

The central thesis of 'Forms and Limits' is that adjudication can be distinguished from other forms of ordering by the particular mode of participation that it confers upon the parties to an adjudicated dispute and decision: that of 'presenting proofs and reasoned arguments for a decision in his favor'.[74] Whatever heightens the significance of this participation, Fuller argues, lifts adjudication 'towards its optimum expression', while whatever destroys it destroys 'the integrity of adjudication itself'.[75]

Even in these succinct formulations it is possible to see the key themes of Fuller's jurisprudence that I sketched in chapter one. First, there is the idea that adjudication, as a form of social ordering, is distinctive: it is different, in key ways, from other models of social ordering. Then there is the idea that this distinctiveness is constituted and maintained only if certain demands of formal integrity are met. Above all, this is the demand of maintaining the particular mode through

[71] Lon L Fuller, 'Consideration and Form' (1941) 41 *Columbia Law Review* 799.

[72] Lon L Fuller, 'The Role of Contract in the Ordering Processes of Society Generally', which Winston dates to 1972, reproduced in *The Principles of Social Order* (n 48) 203.

[73] Undated document titled 'Democracy: Philosophic Issues, Natural Law, Last Chapter', The Papers of Lon L Fuller, Harvard Law School Library, Box 16, Folder 5 (notes on 'Law and Democracy').

[74] 'Forms and Limits of Adjudication' (n 1) 364.

[75] ibid.

which the parties to an adjudicated dispute participate in the process from which an adjudicated decision emerges. The resonance between the thesis of 'Forms and Limits' and Fuller's wider jurisprudential agenda is then illuminated still further by how the key actors here, crucially, are not judges (though, as I set out in my analysis of his views on interpretation and the judicial role in chapter seven, Fuller does elsewhere have much to say about the judicial role and its own distinctive demands of integrity) but the litigants.[76] That is, in 'Forms and Limits', the primary agents in view, and to whom Fuller attributes the distinctiveness of the constitutive form of adjudication, are those who come to adjudication so as to resolve a dispute. It is then a small step from acknowledging this distinctive mode of participation to the moral conclusion that forms of ordering that are designed around structures and dynamics of reciprocity in this way manifest a 'certain regard for human dignity' through how they necessarily treat their participants as ends in themselves.[77]

This last point deserves special emphasis for how it highlights the extent to which Fuller's interest in the connections between the form of a given mode of ordering and the opportunities for the expression of agency that it presents cuts across writings that might otherwise be regarded as unrelated. It equally deserves emphasis for how it reveals Fuller's tendency to move from a diagnostic assessment of the features that constitute a given form of ordering to an analysis of the moral values, or demands, that enable those features to be maintained. The structural analysis and the moral analysis are essentially inseparable.

It is thus also helpful to explore the ways in which 'Forms and Limits' brings this moral impulse within Fuller's thinking, specifically that which reflects his concern for how forms of ordering afford opportunities for the expression of agency, into an arguably clearer light than we gain from those of his writings associated with his debates with Hart on the connections between law and morality. Although Fuller's ambitions in 'Forms and Limits' are primarily theoretical, in the sense that the main aim of the essay is to elaborate a thesis about what constitutes the distinctiveness of the form of adjudication, it is also the case that the inquiry of the essay is consciously practical and lawyerly in its orientation.

That is, what Fuller is ultimately doing in 'Forms and Limits' is trying to understand what adjudication is, what it asks of its participants, what it is good for, what kinds of problems it is not well-suited to resolving, and so forth, for the benefit of the theorist and lawyer alike. This grounds the inquiry of 'Forms and Limits', and specifically its interest in the modes of agency afforded by the form of adjudication, in a much less abstract theoretical space than accompanies the exploration of the same connections in Fuller's writings on the internal morality of law. There, framed by the jurisprudential question of whether there is a necessary connection between law and morality, and undoubtedly mindful of the trans-Atlantic setting of the discussion, Fuller largely directs his claims towards a

[76] See chapter seven, IV 'Fuller, Dworkin and interpretation'.
[77] 'Forms and Limits' (n 1) 362.

more abstract philosophical space. In 'Forms and Limits', by contrast, we are privy to a much more tangible sense of the way that specific processes or procedures, such as adjudication, enable respect for the agency of a legal subject to manifest. Processes and procedures make the agent visible and agency itself tangible precisely because they speak directly to her and supply conditions for her interaction with them. Within such a space, therefore, the agent's presently possessed agency finds expression, and, to the extent that the issues that brought her to an adjudicator in the first place are resolved, her prospects for future agency are also addressed.[78]

The point, then, is that in 'Forms and Limits' the conversation is not convened in the vocabulary of form, generality, agency, and so forth, but rather in the language of having a chance to put your case, of arguing for a decision in your favour, of participating in a process with a view to your own ends and, within this, being treated as an end in yourself, yet all the while necessarily respecting another's entitlement to do the same. This is an important contrast between the eunomics writings and the internal morality writings as much as it is an important bridge. That is, Fuller's writings on specific forms or processes of ordering help us to see just how fundamental his interest in the connections between form and agency actually is, at the level of theoretical insight into the nature of law as a distinctive enterprise, as much as to the everyday experience of how legal processes are constituted and enlivened by the ways in which agents participate within them.

The archival resources also strongly suggest that Fuller regarded the inquiry of 'Forms and Limits' as a launching point for a range of other theoretical conversations: if, indeed, he could secure the appropriate conversational partners. Two such attempts in particular, letters to John Rawls, as well as to Hart, written during what seems to be the peak of Fuller's interest in the 'Forms and Limits' project in late 1959, bear noting. The letters are striking, first, for how they clearly situate the conversation that Fuller seeks to initiate about adjudication within the context of his ongoing project of exploring the 'principles' or 'forms' of 'social order' (though here the language of eunomics, introduced five years earlier, is absent).[79] They are equally striking for how Fuller's gestures towards engagement are backed not only by a keen awareness of the concerns of his addressee's current research, but a sense that his own inquiry might benefit from conversations about how those other trajectories of research could inform his own.

The letter to Rawls, while conceding little direct acquaintance with the details of Rawls's work to that point, speaks to what Fuller identifies as a common interest in the idea of reciprocity; an idea, according to Fuller, 'which seems to me to be a fundamental basis of social order' and which is explored at length in 'Forms

[78] See, eg Jeremy Waldron, 'The Rule of Law and the Importance of Procedure', in James E Fleming (ed), *Getting to the Rule of Law* (2011) *Nomos* vol L.

[79] In his letter to Hart, Fuller states explicitly that he plans to include the 'Forms and Limits' essay 'as part of a book, *The Forms of Social Order*': Letter from Fuller to Professor HLA Hart, 19 October 1959, The Papers of Lon L Fuller, Harvard Law School Library, Box 14, Folder 1 ('The Forms and Limits of Adjudication': correspondence).

and Limits'.[80] The letter to Hart is still more interesting for how it attempts to engage Hart in a conversation about a far less obvious common ground; namely, whether Hart's interest in ordinary language philosophy might offer some unexpected cues for theorising about the 'implications' of institutional forms. As Fuller explains it:

> On the problem of legal method implicit in *The Forms and Limits* it occurs to me that there may be some affinity for the study of ordinary linguistic usage. In effect, what is proposed by Austin (I know of no explicit statement by you on this matter) is that the philosopher trace out the complex implications of ordinary language, the theory being that much wisdom will thus be uncovered. How about tracing out the implications of our institutions, such as adjudication? Michael Oakeshott does indeed in his inaugural address suggest studying the 'intimations' of our traditions.[81]

It is not clear from the archival records whether Hart took up this conversation with Fuller at any stage, and, indeed, whether in making the approach Fuller properly understood the character of Hart's interest in JL Austin's philosophy is not the point. The more interesting point here is one about chronology: how Fuller's attempt to instigate this conversation, in late 1959, takes place some time after the conclusion of his exchange with Hart in the *Harvard Law Review*. This strongly suggests that, at least at this juncture, Fuller did not anticipate further exchange with Hart on the 'law and morality' debate, or at least was not letting such interfere with his intention to develop his 'forms of social order' project. None of this would be especially remarkable but for the fact that, come 1961, this 'principles of social order' project seems to fall into a state of neglect, or, at least, no longer to occupy Fuller's thinking as squarely as it did in the years prior. The question worthy of speculation, therefore, is whether the publication of Hart's *The Concept of Law* in 1961 had something to do with this apparent derailment; specifically, Hart's brief but unequivocal assessment in that book of the apparent irrelevance of Fuller's 1958 interventions to the proper ambit of debates about the connections between law and morality.[82] It at least seems viable to surmise that Hart's response to Fuller set the latter off, however reactively, into a new, and more squarely traditional, jurisprudential direction, because matters of 'eunomics' and 'principles of order' do, at that point, seem to fall by the wayside.

But to return to the more general point of what we might learn about the relationship between Fuller's thinking on the forms and limits of different modes of social ordering and the jurisprudential claims that he came to articulate in his later exchanges with Hart and others, the suggestion Fuller makes in his letter to Hart about 'tracing out the implications of our institutions' is, consciously or otherwise, an important gesture towards things to come. This is especially the case

[80] Letter from Fuller to Professor John Rawls, Department of Philosophy, Massachusetts Institute of Technology, 20 September 1961, The Papers of Lon L Fuller, Harvard Law School Library, Box 7, Folder 4 (correspondence).

[81] Letter from Fuller to Professor HLA Hart, 19 October 1959, The Papers of Lon L Fuller, Harvard Law School Library, Box 14, Folder 1 ('The Forms and Limits of Adjudication': correspondence).

[82] See chapter four, IB 'Situating Fuller's claims'.

with respect to how Fuller eventually attempts to resolve the impasse between him and Hart on the question of whether the principles of the internal morality of law are merely reducible to the value of efficacy: that is, by 'tracing out' the conception of the legal subject that is envisaged by law's form, above all by its defining feature of generality.[83] I return to this point in chapter five.

But when Fuller did receive feedback on 'Forms and Limits', he was clearly not always receptive to what came his way. On some key points, he took the criticism on board and pledged to tone down various claims in later drafts: he backpedalled very quickly, for instance, from assertions about the 'essence' of adjudication,[84] and spoke generally of how the essay contained 'some fairly foolish passages'.[85] But the archival resources also reveal Fuller's impatience for those responses to 'Forms and Limits' that he thought missed the subtleties of his analysis,[86] as well as those which disputed his intuition that the limits of adjudication as a model suited to the resolution of certain types of problems (his argument that adjudication is unsuited to the resolution of 'polycentric' disputes being his most well-known claim on this point)[87] was a matter not taken sufficiently seriously by either lawyers or legal scholars. One critic's suggestion that it was 'impossible to conceive of any problem that could not be solved adjudicatively' received the following stinging rebuke:

> If this statement is true, it is only because the concept of possibility is stretched beyond the point where it could have any intelligible meaning in dealing with social problems. By that standard Talleyrand's famous remark that it is possible to do everything with bayonets except to sit on them is patently false. After all, we *can* sit on bayonets.[88]

More importantly for understanding the wider jurisprudential frame of Fuller's inquiry into the forms and limits of different models of social ordering, the archival material also illuminates the extent to which Fuller saw traditional jurisprudential contests between positivism and natural law to be an obstacle, or at least a potential one, to meaningful engagement with the inquiry of 'Forms and Limits'. Here his correspondence with Boris Bittker of Yale Law School is especially telling. Though very much in agreement with his intervener's reservations 'essences', Fuller nonetheless challenges Bittker to consider whether resistance to such ideas might not take its cue from more fundamental divisions within jurisprudential thought:

[83] See further chapter five.

[84] Letter from Fuller to Professor Boris I Bittker, 4 April 1960, The Papers of Lon L Fuller, Harvard Law School Library, Box 14, Folder 1 ('The Forms and Limits of Adjudication').

[85] Letter from Fuller to Philip Selznick, 29 November 1957, The Papers of Lon L Fuller, Harvard Law School Library, Box 7, Folder 6 (correspondence).

[86] See especially letter from Fuller to Professor Robert R Bowie, 13 October 1959, The Papers of Lon L Fuller, Harvard Law School Library, Box 14, Folder 1 ('The Forms and Limits of Adjudication'): 'One criticism . . . you raised was that the paper ignores "power". I would have thought that in one aspect that is precisely what it is about. Social "power" in my view is the product of social forms and the attitudes shaped by those forms'.

[87] 'Forms and Limits' (n 1) 394–404.

[88] Fuller to Professor Robert R Bowie, 13 October 1959, The Papers of Lon L Fuller, Harvard Law School Library, Box 14, Folder 1 ('The Forms and Limits of Adjudication').

> I'm afraid that the notion of an 'essence' of an institution is distasteful because (1) of historical associations, and (2) because people (incorrectly, so far as I am concerned) read supra-human implication into it. I don't think I read into it what you do. Perhaps that word 'natural law' is still sticking in your throat. I notice you say that adherents of natural law very seldom present any conclusions without purporting (speciously) to derive them from natural law premises. Please count me out on that. In my debate with H.L.A. Hart I don't think I presented any conclusion as a deduction from natural law premises, yet nine readers out of ten seem to read that exchange as if I were apologizing for natural law throughout. The result is that I am widely credited by intelligent readers with espousing a doctrine I never proposed, namely, that a law that is evil enough in its intention is not really a law at all. . . .
>
> Like yourself, I find it hopeless to treat every case as *sui generis*. Like yourself I seek order and coherence. If in that quest I overreach myself at times, I hope this will not be charged against the slogan 'natural law'.[89]

This passage provides a fitting segue to the question of how to navigate the different labels which have been attached to Fuller's thought. As I observed in my introductory comments, to address this question prior to heading into the more well-known territory of Fuller's exchanges with Hart on the connections between law and morality helps to bring to light the scene-setting role that labels such as 'natural law' played, not only with respect to how those exchanges were understood—above all, as a contest between 'positivism' and 'natural law'—but equally with respect to how the claims Fuller there advanced were interpreted.

III Navigating the Labels

Ubiquitous as the 'natural law' tag might have become in association with Fuller, it is in fact very difficult to situate his jurisprudence within any one intellectual tradition. As preceding discussion should have made clear, Fuller himself aligned his contribution with the natural law tradition in only a qualified sense, and one that he mapped along two main lines. First, it is clear that Fuller embraced the methodological tenor of natural law inquiry: as he put it in a working note for his 'Reply to Critics', despite its more 'foolish' assertions, the natural law tradition 'does not bar off whole areas from investigation; it closes fewer doors of inquiry'.[90] Second, it is equally apparent that Fuller thought that the idea that there are 'natural laws of social ordering', or as he put it in 'American Legal Philosophy', 'compulsions necessarily contained in certain ways of organizing men's relations with one another' was a preoccupation of the natural law tradition that ought to be revisited by contemporary legal philosophers.[91] Beyond these two commitments, and his

[89] Letter from Fuller to Professor Boris I Bittker, 4 April 1960, The Papers of Lon L Fuller, Harvard Law School Library, Box 14, Folder 1 ('The Forms and Limits of Adjudication').

[90] Undated document titled 'Natural Law', The Papers of Lon L Fuller, Harvard Law School Library, Box 12, Folder 1 (notes for 'Reply to Critics').

[91] 'American Legal Philosophy' (n 15) 476, 478–79.

occasional suggestions to the effect that more traditional natural law claims might in certain instances be defensible[92]—though at no point indicating any affinity between such claims and his own—Fuller vigorously questioned the suitability of the natural law label as a means of capturing the message of his jurisprudence.[93] Indeed, he deliberately coined the term 'eunomics' for his 'forms of social order' project, in a stated effort to avoid 'the confusions invited by the term natural law'.[94]

Yet despite his repeated attempts to situate his contribution in a non-traditional natural law space, these confusions persisted well beyond the reception of eunomics. As I explore further in chapter four, responses to *The Morality of Law* were especially plagued by the attribution, consistent with the dominant traditional view, that Fuller's claims about his principles of the internal morality of law were directed to the question of the moral quality of legal ends.

But the point relevant for present purposes is that another reason why it is problematic to designate Fuller as a 'natural lawyer' is because of how this operates to obscure other important intellectual influences which animate his jurisprudence. Foremost among these is the thinking of the early American pragmatists, especially William James and John Dewey.

The pragmatist strain of Fuller's jurisprudence is most likely lost on many non-American readers of Fuller, but its influence nonetheless finds expression in his writings in a number of different ways.[95] At the most general level, throughout Fuller's writings is the pragmatist's commitment not only to the notion that ideas are answerable to experience, but also to the notion that ideas shape experience. This commitment comes through especially when Fuller addresses the function served by definitions of law which, as he puts it in his 1958 reply to Hart, are in his view not 'mere images of some datum of experience', but 'direction posts for the application of human energies'.[96] The pragmatist current is also evident in how Fuller develops his jurisprudential insights around ideal forms which are offered as models against which to assess any practice that attempts to approximate them and, indeed, to determine when that practice has taken on such a pathological form that it can no longer be regarded as a proper instance of the practice. Moreover, the means-ends analysis which forms the centrepiece of Fuller's articulation of his eunomics project in 'Means and Ends' is taken largely from the thinking of John Dewey, with multiple references to Dewey to be found throughout Fuller's working papers for 'Means and Ends',[97] as well as in correspondence with

[92] See, eg chapter 3 II, 'Reclaiming Fuller through the Nazi law debate'.

[93] See, eg Lon L Fuller, *The Morality of Law* (New Haven, Yale University Press, 1964) 96.

[94] 'American Legal Philosophy' (n 15) 477.

[95] Winston, for instance, quotes Fuller's report of one such encounter in which he was apparently told 'I thought you were an advocate of natural law, but I see you are just an American pragmatist': see Kenneth I Winston, 'Is/Ought Redux: The Pragmatist Context of Lon Fuller's Conception of Law' (1988) 8(3) *Oxford Journal of Legal Studies* 329, 349, note 59. See also Lon L Fuller, 'The Ideal Element in a Definition of Law' (1986) 5 *Law and Philosophy* 89.

[96] Lon L Fuller, 'Positivism and Fidelity to Law: A Reply to Professor Hart' (1958) 71 *Harvard Law Review* 630, 632.

[97] See generally The Papers of Lon L Fuller, Harvard Law School Library, Box 14, Folder 3 ('Means and Ends').

Philip Selznick, who suggests in one letter that Fuller's work might be regarded as bringing 'an elevated pragmatism' to the study of jurisprudence.[98]

But in other ways it is important to exercise caution before interpreting Fuller's jurisprudence through the prism of early American pragmatism. In particular, caution must be exercised against too readily attributing the instrumentalist strains of that tradition, especially as found in Dewey's work, to Fuller's jurisprudential message, as, indeed archival material seems to support. In a working note for 'Means and Ends', Fuller speaks of how American pragmatism is 'capable of many interpretations':

> One meaning, a certain insouciance toward reality; we can shape it; instrumentalism. Another meaning: a respect for the complexity of reality; an abhorrence of deciding questions in the abstract and then applying these solutions to concrete situations that were unforeseen, a dislike of intellectual absentee management . . . Contextualism.[99]

In my view it is very clear that Fuller aligned himself with the second, and not the first, of these two identified meanings.[100] As I will explain further in the chapters to follow, but especially in my conclusions in chapter eight where I look more closely at the idea of instrumentalism and its relationship to the task of reclaiming Fuller's jurisprudence, what is frequently missed in standard readings of Fuller is how he sees the moral and instrumental dimensions of law not as standing in a relationship of polarity, but rather in one of irreducible connection, with this connection arising from the way that the structural features of law interact with its animating moral commitments to together constitute its distinctive form. To the extent, therefore, that we might acknowledge and embrace the pragmatist strains of Fuller's project, it is crucial that these be read in light of these important, and distinctly Fullerian, qualifications.

There are many other influences that bear noting if we are to understand the wider context from which Fuller's jurisprudential claims emerge. *The Morality of Law*, for instance, contains a number of references to Kant, particularly when the point under discussion is how institutional forms do important moral work independently of the specified ends towards which they might be directed.[101] Aristotle often finds a place in the early writings as well as in *The Morality of Law*, and especially in connection with Fuller's own methodological commitment of seeking a Aristotelian-style 'middle way' between the received antinomies of jurisprudential debate.[102] Works by Michael Polanyi, Michael Oakeshott, Georg Simmel and Morris Cohen equally receive frequent citation.

[98] See letter from Philip Selznick to Fuller, 7 July 1965, The Papers of Lon L Fuller, Harvard Law School Library, Box 7, Folder 6 (correspondence).

[99] Undated notes in preparation for 'Means and Ends', The Papers of Lon L Fuller, Harvard Law School Library, Box 14, Folder 3 ('The Principles of Social Order').

[100] See Roderick Macdonald, 'Legislation and Governance' in *Rediscovering Fuller* (n 58) 310.

[101] eg, *Morality of Law* (n 93) 94.

[102] ibid, 5, 19, 64, 94.

There is also, of course, Fuller's association with the 'legal process school' of his Harvard colleagues Henry Hart and Albert Sacks,[103] and Herbert Wechsler,[104] with 'Forms and Limits' being perhaps the most important work associated with Fuller's connection to the project of these theorists. Later writings also suggest a strong case for designating Fuller as a 'sociological jurist', given that he read sociology with increasing enthusiasm as his career progressed, indeed, partly in reaction to the scathing reception that *The Morality of Law* received at the hands of the philosophers.[105] Here again, however, a cautionary note is needed. Even though Fuller saw himself as contributing actively, especially in his late career, to the study of the sociology of law, he was adamantly opposed to the 'immensely popular slogan *Law and* Society', and the associated work of scholars who were 'not concerned to examine legal processes as social processes, but rather seek to examine and appraise the "impact" of something called "law" on something called "society"'.[106] Finally, readers are also likely to detect a pattern of reference in Fuller's writings to examples drawn from the discipline economics; a practice most likely be explained by the fact that Fuller's first degree was in economics.

It is, all in all, an eclectic mix, and Fuller himself never succeeded in reducing this eclecticism into the kind of systematic project or theoretical statement that philosophers tend to prize. Nonetheless, it is important to keep these other influences in view, otherwise so easily eclipsed by the default 'natural law' label, if we are to appreciate just how unconventional a contribution Fuller attempted to make to what appeared, on its face, to merely be a mid-twentieth century instalment of a traditional contest about the connections between law and morality.

IV Conclusion

The overarching purpose of this chapter has been to widen the view of what the task of reclaiming Fuller's jurisprudence entails. There can be little doubt that Fuller sought to be a game-changer within the field of jurisprudence as it stood during his time, but this does not mean that he sought radical reform. His ultimate intention seems instead to have been one of reclamation: that is, to reclaim the wisdom of resources long neglected by the field that he thought might fruitfully illuminate a range of neglected questions, especially those relating to the institutional design of law and legal processes.

[103] Henry M Hart and Albert M Sacks, *The Legal Process: Basic Problems in the Making and Application of Law* (William N Eskridge, Jr and Philip P Frickey (eds), (New York, The Foundation Press, 1994).

[104] Herbert Wechsler, 'Towards Neutral Principles of Constitutional Law' (1959) *Harvard Law Review* 1.

[105] See further Robert S Summers, *Lon L. Fuller* (Stanford, Stanford University Press, 1984) 12–13.

[106] Letter from Fuller to Dr Robert P Corman, 22 April 1971, The Papers of Lon L Fuller, Harvard Law School Library, Box 11, Folder 17.

Eunomics, Fuller's unfinished theory of 'good order and workable social arrangements', was his intended response to this picture: an attempt to move the agenda of legal philosophy away from its preoccupation with source-based imperative theories and their instrumentalist commitments and towards questions relating to the form, limits and opportunities for agency afforded by different modes of social ordering. Throughout the writings associated with the eunomics project, writ large as well as small, Fuller is concerned not only to examine our received forms of social ordering from the perspective of a lawyer's problem-solving skill-set, but also to illuminate and to assess what these forms offer to the agents who participate within them.

This, as I will explain over the course of the chapters to follow, is an enduring theme of Fuller's jurisprudence, and one that comes to the fore especially clearly when he introduces the idea, in his final 'Reply to Critics' in 1969, that the positivist's concept of law is in fact more akin to a model of managerial direction than it is to law. I have reserved my discussion of Fuller's views on managerial direction to chapter five, because although it does make an unlabelled appearance in his 1965 essay 'Irrigation and Tyranny',[107] we learn most about Fuller's thinking on this form of ordering in the 'Reply to Critics'.[108] For present purposes, however, in connection with the themes of eunomics, the point to foreshadow is that Fuller identified a key difference between the constitutive dynamics of forms of ordering such as legislation, contract and adjudication, and those of managerial direction, precisely because the latter does not, by design, respect or nurture the subject as an end in herself.

Finally, it is worth noting the extent to which Fuller's early writings have a distinctly American flavour about them, and not just because of certain associations with the thought of the early American pragmatists, or indeed with the work of Harvard's legal process scholars. The eunomics inquiry, and especially the 'Forms and Limits' project within it, seems largely to have remained an 'at home' endeavour. During the height of its development, over the course of the late 1950s and early 1960s, Fuller's lectures and letters give expression to a very American conversation about institutions and processes, and the under-appreciated moral work that these do, with regular references to the ambitions and achievements of the American constitutional framers (men, he suggests, who were uniquely aware of how 'the moral life of the citizen would to a significant degree be shaped by these new forms')[109] offered by way of illustration of insights lost. It might even be suggested that the central place of the agent within these writings, and with the question of how she participates with different forms of order, reflects a distinctly American embrace of the legal subject, of 'we the people', as the source of law.[110]

[107] Lon L Fuller, 'Irrigation and Tyranny' (1965) 17 *Stanford Law Review* 1021, reproduced as a study of managerial direction in Winston, *The Principles of Social Order* (n 48).

[108] See chapter five.

[109] See 'Legal Philosophy in the United States', p 4, English version manuscript of a publication in French, The Papers of Lon L Fuller, Harvard Law School Library, Box 12, Folder 2. See also *Morality of Law* (n 93) 101–3.

[110] Thanks to Gordon Woodman for suggesting this point.

The Anglo-American and more explicitly philosophical conversation that Fuller sought to initiate when he responded to Hart's positivist defence of the separation of law and morality did much to lift these themes out of an American context and into more abstract discussion about the nature of law. The frame of the 1958 debate is very clearly not about English and American law and legal traditions, but about the big questions of jurisprudence. Yet there are moments in that debate which do seem to see its contestants descend into a trans-Atlantic intellectual contest with attendant moments of cultural criticism. It is thus unlikely that Fuller did not gain some satisfaction, in response to some of Hart's arguments in 'Positivism and the Separation of Law and Morals', from the complaint of his friend, the legal historian Alexander P D'Entrèves, about 'the incurable smugness of our English friends'.[111]

I thus begin my analysis of how we might read that famous exchange between Hart and Fuller with the tensions of this trans-Atlantic theme in mind. Generous and enthusiastic in tone, the letter that Fuller wrote in support of Hart's proposed visit to Harvard clearly reveals his expectation that he and his English colleague would engage in many bridge-building conversations about subjects of common concern. How those conversations ultimately panned out, and how they influenced the reception of Fuller's jurisprudential claims as well as the development of those claims over time, is the subject of the next three chapters.

[111] Letter from Alexander P D'Entrèves to Fuller, 8 April 1958, The Papers of Lon L Fuller, Harvard Law School Library, Box 3, Folder 15 (correspondence relating to the 1958 debate).

3

The 1958 Debate

Few students of Anglo-American jurisprudence in the past half-century have not had some kind of encounter with the exchange between HLA Hart and Lon L Fuller in the 1958 *Harvard Law Review*. The 'Hart-Fuller debate', as it became known, grew out of the occasion of Hart's delivery of the Oliver Wendell Holmes lecture at Harvard Law School during his extended visit to the university in 1957.[1] Fuller, then the Carter Professor of General Jurisprudence at Harvard, had been an academic sponsor of the visit, and his letter of support makes apparent that he anticipated much from Hart's stay in terms of exchange and cross-fertilisation of ideas. As Fuller expresses it in that letter, the preceding three decades had seen the difference in interest and 'general mood and temper' between American and English legal philosophy become so great 'that even the heritage of a common language has at times seemed in danger of being lost in this field'. Hart, however, had been influential in helping to bring jurisprudential thought in England out of an impasse between semantics and conventional philosophy, and so, Fuller hoped, a visit to Harvard might provide an opportunity for him 'to exercise a similar influence in this country, particularly after he has become able to orient his writings towards the current interests of legal philosophy here'.[2]

The story of the Harvard visit and the debate which grew out of the Holmes lecture has been well told in Nicola Lacey's biography of Hart, which reports the bouts of crippling insecurity that Hart wrestled with in the months and weeks leading up to the event, and his palpable relief when the lecture turned out to be a great success.[3] Much less, however, is known about the episode from Fuller's perspective. From Lacey, quoting a former student who was present at the lecture, we learn that Fuller paced agitatedly back and forth at the back of the hall ('like a hungry lion') as he heard Hart's defence of the positivist claim that there is no necessary connection between law and morality unfold.[4] On some reports, Fuller was apparently so agitated that he stormed out of the lecture before Hart had actually finished. Either way, what we do know is that Fuller demanded a right of

[1] The exchange was published as HLA Hart, 'Positivism and the Separation of Law and Morals' (1958) 71 *Harvard Law Review* 593 and Lon L Fuller 'Positivism and Fidelity to Law: A Reply to Professor Hart' (1958) 71 *Harvard Law Review* 630.

[2] Undated notes apparently pertaining to the sponsorship of Hart's visit to Harvard in 1957, The Papers of Lon L Fuller, Harvard Law School Library, Box 3, Folder 14 (correspondence: HLA Hart).

[3] See Nicola Lacey, *A Life of H.L.A. Hart: The Nightmare and the Noble Dream* (Oxford, Oxford University Press, 2004) 197.

[4] ibid.

reply, and this response was published alongside the text of Hart's lecture in the 1958 *Harvard Law Review.*

Lacey's research makes clear that Hart regarded the significance of Fuller's interventions to the development of his own ideas to be negligible. In Hart's private papers, Fuller is either referred to harmlessly as 'a nice New Englander with some quite interesting ideas', or impatiently, indeed scathingly, as the interlocutor who was to put Hart to the inconvenience of having to absorb, and perhaps even respond to his 'piece of logomachy'.[5] Yet a review of the correspondence between the two men reveals that their personal relationship remained at all times candid and collegial, sustained through regular letters and visits from 1957 through to the early 1970s. Certainly, the tone of this private back and forth seems to convincingly override any suggestion that the debaters themselves were anything more than intellectual opponents.

While Fuller's interventions seem barely to have touched Hart's understanding of the ambitions of his own project, the same is hardly true in reverse. It is impossible to overstate the significance of the Hart-Fuller debate to the development of Fuller's jurisprudence. All indications point to how Hart's lecture inflamed a powerful current of creativity in Fuller that was perhaps exceeded only by an equally powerful current of frustration. In many ways, this was a positive thing: Hart's unwelcome agenda provided a foil for Fuller to articulate his jurisprudential instincts in far stronger and perhaps also more philosophical terms than he might have had done if he had dedicated his energies to the development of his eunomics project. The many subjects covered in the debate continued to occupy Fuller's attention in writings from that point until the end of his career.

That is the positive side. On the negative, however, the chapters to follow will starkly reveal the costs to Fuller of having essentially accepted an agenda for debate on Hart's terms, and of having continually attempted, through to the last words of his final 'Reply to Critics' in 1969, to initiate and sustain a conversation that spoke to that agenda. Still more problematic is the enormous extent to which ideas about who won and who lost this or other part of the Hart-Fuller debate have shaped appraisals of the nature and value of Fuller's contribution to legal philosophy more generally. It is this matter, above all, that I hope, if not to remedy, then at least to reorient through the objective of this chapter to read the 1958 Hart-Fuller debate from Fuller's perspective. This objective, however, seeks to carry more than simply an effort to add nuance to a particularly rigid scholarly memory. The Hart-Fuller debate deserves our especially close attention as part of the project of reclaiming Fuller, because in my view Fuller's 1958 reply to Hart, in many ways more than the extended exegesis of his claims that he offers in *The Morality of Law*, can be read as the most instructive record of the core concerns and ambitions of his distinctive jurisprudential agenda. But we are unable to see this if we remain on the well-trodden path that has been taken by most legal philosophers, or at least legal positivists, in the past half century;

[5] ibid 198.

namely, the path that starts and ends with an assessment of how *he* fits with *them*.

The tasks I seek to fulfil in this chapter are broadly three. The first is essentially descriptive: after offering a brief statement of Hart's claims in 'Positivism and the Separation of Law and Morals', as a foundation for understanding Fuller's reply, I sketch the basic architecture and arguments of that reply. None of this, nor my summary of Hart's position, is intended to be a substitute for a careful reading of the debate itself. Indeed, the reader should note that, to avoid unnecessary repetition, I have reserved a detailed account of two sites of Fuller's reply to locations other than this first, essentially descriptive, overview. Fuller's reply to Hart on the matter of interpretation, for instance, forms part of my discussion of his views on interpretation and the judicial role in chapter seven, as these questions are important to the conversation between Fuller and Ronald Dworkin that I convene in that chapter.

I also reserve my account of Fuller's reply to Hart on the matter of Nazi law to the second section of this chapter because, as I explain further below, the centrality of this particular site of exchange between the two scholars to both the popular memory of what the Hart-Fuller debate was about, as well as to a proper understanding of the themes of Fuller's jurisprudence, cannot be overstated. The Nazi law debate thus deserves a detailed analysis in its own right. Bearing these caveats in mind, however, my general aim has been to provide a sufficiently detailed and descriptive account of the sites of contest in the 1958 exchange to enable readers familiar and unfamiliar alike with the debate to understand the basic arc and claims advanced within it.

My decision to give over the central section of this chapter to a close analysis of the Nazi law debate is motivated by my view that this site of exchange illuminates the impasse between Fuller and Hart on the questions that Fuller regards to be central to the project of jurisprudence in sharper terms than any other site of exchange between them. There is, in short, an enormous amount to be learned from the Nazi law debate, not only with respect to insight into Hart and Fuller's respective projects, but also for how we might develop a twenty-first century agenda around the question of the relationship between the concept of law and the concept of legality, or the rule of law, and, indeed, on the question of whether positivist standards for legal validity can satisfactorily explain what it is that makes law law-like.

The third task of this chapter is therefore an exploratory one: to consider whether a jurisprudence convened along Fullerian lines, as I have reclaimed it, is capable of yielding a standard for legal validity; or, to use an alternate vocabulary, for determining the grounds of, or existence conditions for, law. The ideas I advance in this third section are suggestive only: indeed, they are intended to invite further consideration of this generally under-explored question, rather than making any claim to having resolved it. But the absence of any strongly stated argument on the matter of a standard for legal validity in Fuller's 1958 reply to Hart has been an important tool for positivists in their appraisal of the value, or the importance, of his contribution to the core questions of legal philosophy. The task of reclaiming Fuller would thus surely be incomplete without at least some attempt to explore whether this, too, is a site where his jurisprudence deserves to be taken more seriously.

I Mapping the Debate

A Setting the Agenda: Hart's Claims

Hart's goal in 'Positivism and the Separation of Law and Morals' is to elaborate and defend a positivist legal philosophy that is capable of explaining the normativity of law while maintaining the traditional positivist insistence that there is no necessary connection between law and morality, otherwise known as the 'separability thesis'. The six sections of Hart's essay reflect his attempt to address, in turn, six sites of jurisprudential inquiry that raise the question of whether law is in some sense necessarily moral.

The first two sections of Hart's essay speak to the utilitarian roots of legal positivism and its defence of the separability thesis, and amount to a general endorsement of the utilitarian effort in this vein on the basis that to conflate law and morality not only serves to obstruct a clear theoretical analysis of law, but might also discourage moral criticism of law in practice.[6] But Hart is clear that he seeks to depart from the utilitarian version of legal positivism in key ways, most particularly from its adherence to a command conception of law.[7] Gesturing to the seeds of what, in *The Concept of Law*, becomes the thesis that a legal system is comprised of a union of primary and secondary rules, supported by an attitude of acceptance on the part of legal officials, Hart suggests that the real key to the science of jurisprudence lies not in the notion of a command but rather in an acknowledgment of how:

> nothing which legislators do makes law unless they comply with fundamental accepted rules specifying the essential lawmaking procedures. This is true even in a system having a simple unitary constitution like the British. These fundamental accepted rules specifying what the legislature must do to legislate are not commands habitually obeyed, nor can they be expressed as habits of obedience to persons. They lie at the root of a legal system, and what is most missing in the utilitarian scheme is an analysis of what it is for a social group and its officials to accept such rules.[8]

A positivist legal philosophy reinvented along these lines can continue to defend the separation of law and morality, Hart insists, because even if law is born of rules—the 'fundamental accepted rules specifying what the legislature must do

[6] Hart, 'Positivism' (n 1) 598.
[7] ibid 601–2.
[8] ibid 603.

to legislate'—that are not themselves law, it does not follow that these rules are necessarily moral in nature.

This is the foundation from which Hart develops the arguments of his essay, beginning with his turn to the question, presented by the American legal realists, of whether the positivist separability thesis is challenged by the tendency of judges to refer to some notion of what the law ought to be when interpreting a rule whose meaning is in dispute.[9] The substantial space that Hart dedicates to exploring this question is notable not only because it yields his famous 'core and penumbra' model of interpretation, but also for how such is generally consistent with Hart's tendency in the essay to focus and rest his claims on the example of *particular* laws and the challenges they raise for the separability thesis, rather than on the wider questions relating to what constitutes and sustains a legal *system* that we see Fuller seek to emphasise. According to Hart's 'core and penumbra' model of interpretation, law properly so called is comprised only of those core instances where the meaning of law is settled or otherwise clear, and not those penumbral cases of uncertainty where judges must exercise discretion and create new law. Moral considerations might indeed enter the judge's reasoning in this penumbral space, but this need not disturb the positivist separability thesis because, the argument runs, there is no law, as such, in the penumbra. Nor, Hart says, should we think that the separability thesis is disturbed by how considerations of what the law ought to be might affect the judge's reasoning, because such considerations of 'ought' need not implicate a moral judgement or have any other connection to moral standards.[10] Thus, Hart illustrates, by way of example, it would be possible to say that a legal decision is 'as it ought to be' even inside a system that was 'dedicated to the pursuit of the most evil aims'[11] such as was the case in Nazi Germany, where lawmakers were guided above all by considerations of what was needed 'to maintain the state's tyranny effectively'.[12]

The fourth section of Hart's essay sees him turn from the matter of interpretation to the more traditional contest between positivism and natural law on the question of whether the substantive injustice or immorality of a given law ought to go to the question of its validity; a question precipitated by the apparent revival of natural law thinking in the aftermath of Nazism by the German legal philosopher, Gustav Radbruch. As foreshadowed above, I have reserved my account of Hart's response to this question to the second section of this chapter, below. The thrust of Hart's position, however, can be briefly stated. In keeping with the view of his utilitarian predecessors that the existence of law is one thing, and its merit or demerit another, Hart wholly rejects Radbruch's suggestion that the moral quality of the content of law ought to go the question of its status as valid law. As Martin Krygier has neatly captured it, Hart's position on Nazi law ultimately boils

[9] ibid 606.
[10] ibid 613.
[11] ibid 613.
[12] ibid. I return to Hart's claims about interpretation and Fuller's responses to them in chapter seven, 'Fuller, Dworkin, and interpretation'.

down to this: 'Nazis had laws, and they were immoral; not a happy story but a simple one'.[13]

It is only after addressing the question of whether the example of Nazi law challenges the separability of law and morality, conducted as an inquiry into the validity of particular laws that Hart turns in the fifth section of his essay to the question of whether there is some important connection between law and morality in the foundations of a legal *system*. Hart's response to this question, which is essentially repeated in *The Concept of Law* three years later, turns upon two lines of argument. The first speaks to the idea, which Hart readily concedes invites moral inquiry, that every legal system must contain a minimum content of 'natural law' if it is to contribute to the goal of ensuring 'survival in close proximity to our fellows'. Here Hart has in mind the way that a legal system is founded upon rules so fundamental that if it did not have them 'there would be no point in having any other rules at all', such as rules that forbid the free use of violence and which constitute the minimum rights and duties of property.[14] But it is not possible for legal philosophy to say much else beyond this about the foundations of legal order and the gestures that these foundations make to something moral, Hart argues, because 'the purposes men have for living in society are too conflicting and varying to make possible much extension of the argument that some fuller overlap of legal rules and moral standards is "necessary"'.[15] In short, the point is this: there is something of moral significance in the minimum content of a legal system, but it is not significant enough to play a role in philosophical debates about the necessary connections between law and morality.

Hart's second defence of why exploration of the foundations of a legal system need not disturb the positivist separability thesis addresses the implications of law's formal feature of generality. Here, Hart makes a statement, again reiterated in *The Concept of Law*,[16] that becomes important to points that Fuller later seeks to make in his 'Reply to Critics'. As Hart explains it:

> If we attach to a legal system the minimum meaning that it must consist of general rules—general in both the sense that they refer to courses of action, not single actions, and to multiplicities of men, not single individuals—this meaning connotes the principle of treating like cases alike, though the criteria of when cases are alike will be, so far, only the general elements specified in the rules. It is, however, true that *one* essential element of the concept of justice is the principle of treating like cases alike. This is justice in the administration of the law, not justice of the law. So there is, in the very notion of law consisting in general rules, something which prevents us from treating it as if morally it is utterly neutral, without any necessary contact with moral principles.[17]

[13] Martin Krygier, 'The Hart-Fuller Debate, Transitional Societies and the Rule of Law', in Peter Cane (ed), *The Hart-Fuller Debate in the Twenty-First Century* (Oxford, Hart Publishing, 2010) 112.

[14] Hart, 'Positivism' (n 1) 623.

[15] ibid.

[16] HLA Hart, *The Concept of Law* (Oxford, Oxford University Press, 1961) 206.

[17] Hart, 'Positivism' (n 1) 623–24.

This passage sounds very much like a concession to a connection between law and morality that ought to be of interest to legal philosophers. However, Hart goes on to insist that this 'certain overlap' between legal and moral standards ultimately does not compromise the positivist claim that law and morality are necessarily distinct, because:

> a legal system that satisfied these minimum requirements might apply, with the most pedantic impartiality as between persons affected, laws which were hideously oppressive, and might deny to a vast rightless slave population the minimum benefits of protection from violence and theft . . . Only if the rules failed to provide these essential benefits and protection for any one—even for a slave-owning group—would the minimum be unsatisfied and the system sink to the status of a set of meaningless taboos. Of course no one denied those benefits would have any reason to obey except fear and would have every moral reason to revolt.[18]

Having thus dispensed with the question of whether there is a connection salient to debates about law and morality in the very foundations of legal order, Hart returns in the sixth and final site of his essay to the wider questions of the project of legal philosophy with which the essay began. His message, stated briefly, is that a further reason why it is unwise to blur the conceptual distinction between law and morality is because of how such necessarily introduces into debates of legal theory the complex and controversial issues of meta-ethics that belong to the domain of moral rather than legal philosophy. For legal philosophy, then, the agenda must be a narrower one, and this narrower agenda must start by disavowing the natural lawyer's idea that law is a phenomenon to be understood in purposive terms. Instead, the right path forward is to pursue a descriptive theory of law, untainted by evaluative standards, that in this neutrality is capable of serving the goal of theoretical clarity at the same time as it assists us, in practice, 'to see clearly the problems posed by morally bad laws'.[19] Legal positivism, so Hart's message goes, is the only philosophy of law that is able to do both.

Of these six sites of analysis that constitute Hart's defence of the separability thesis, the two most crucial for Fuller's contest with positivism are Hart's treatment of Nazi law and his claims about the absence of any theoretically salient connection between law and morality in the foundations of legal order. I will address the matter of Nazi law at length below, and will explore the debate between Hart and Fuller on the moral significance of law's form as a system of general rules in more detail in chapter five, as this issue became increasingly important to Fuller as he sought to diagnose and navigate the key points of impasse between him and Hart as their exchanges developed.

For present purposes, however, some brief remarks about the general features of Hart's presentation of his defence of positivism on the law and morality question can be made. First, it is helpful to note how consistently Hart's claims rest on the conviction that clarity is the guiding aim of legal philosophy. Equally notable,

[18] ibid 624.

[19] ibid 597.

as I have already emphasised, is how his analysis, overall, orients much more naturally to the example of particular laws as the testing ground for his intuitions. This stands in direct contrast to Fuller, whose reply orients consistently to the broader question of what gives a legal system its character as law—what makes it law-like—and thus which is also implicitly committed to the idea that qualitative concerns ought to be allowed to trump the imperative of clarity in those moments where the very intelligibility of the idea of law is at stake.

B Reorienting the Agenda: Fuller's Replies

(i) The Fidelity Frame

The title of Fuller's *Harvard Law Review* essay has often puzzled his critics. But the title in fact directly addresses the points of criticism that lie at the foundation of that response, and which speak to how, in Fuller's view, the positivist project distorts the aims of legal philosophy in its failure to appreciate that its definitions of what law is 'are not mere images of some datum of experience', but rather 'direction posts for the application of human energies'.[20] As he states it in the opening pages of the essay:

> Law, as something deserving of loyalty, must represent a human achievement; it cannot be a simple fiat of power or a repetitive pattern discernible in the behavior of state officials. The respect we owe to human laws must be something different from the respect we accord to the laws of gravitation. If laws, even bad laws, have a claim to our respect, then law must represent some general direction of human effort that we can understand and describe, and that we can approve in principle even at the moment when it seems to us to miss its mark.[21]

The foundational thesis of Fuller's reply to Hart, then, is that a putative theory of the nature of law must be able to explain not only why law commands our fidelity, but why it is capable of so commanding. How in and what ways might law have a moral claim over us, and how might this relate to the conditions that cause law itself to come into existence? Does, or can, Hart's positivist project answer this question, either explicitly or implicitly?

The 'fidelity frame' that sets up Fuller's reply to Hart thus provides a valuable insight into the general aims of that reply as a whole. With respect to those aims, it is helpful to note the distinctive architectural features of Fuller's essay. For instance, although he does address each of the arguments advanced by Hart, Fuller does so in a different order than that pursued by Hart himself. This is clearly deliberate, designed to problematise Hart's agenda by illuminating the kinds of questions and tensions implicit within his claims when they are viewed from an alternate perspective. Thus, Fuller does not address the question of whether a connection between law and morality is revealed when judges interpret

[20] 'Positivism and Fidelity' (n 1) 631, 632.
[21] ibid 632.

a law whose meaning is in dispute until after he has articulated his arguments about the moral foundations of legal order and the problem of Nazi law: the reverse to the order pursued by Hart. As will be illuminated further in my analysis of the debate on Nazi law, below, this strategy is clearly intended to expose Hart's sense that the problem of Nazi law *was* a problem about the interpretation and status of particular laws, rather than, as Fuller sees it, a problem of the pathology of an entire legal system. The subtitles that Fuller gives to each of the sections of his essay also bear noting as part of a strategy to respond to Hart by approaching his claims from a different perspective. In contrast to Hart's minimalist, numbered sections, Fuller's subtitles frame the problems that he sees as ignored in or problematised by Hart's analysis in evocative and bold terms. An exchange about whether legal positivism can defend the necessary separability of law and morality becomes, under Fuller's pen, a debate about such issues as 'The Definition of Law', 'The Moral Foundations of Legal Order', 'The Morality of Law Itself', and 'The Problem of Restoring Respect for Law and Justice After the Collapse of a Regime that Respected Neither'.

But it is precisely this attempt on Fuller's part to address Hart's agenda at the same time as seeking to reorient it that leads to tensions within his essay that are never fully resolved at any point in the ongoing back and forth between him and Hart on the questions arising from that agenda. Moreover, the fact that all of this is referable to and ultimately grounded in the framework of a debate about the necessary connections between law and morality causes as much trouble for Fuller's attempt to convene a debate as it aids that attempt. There are many occasions where, upon elaborating a given point, Fuller's return to the 'law and morality' framework comes across as somewhat forced. And there are several moments where we gain the sense that the point in question might have been elaborated more naturally within a different frame of reference, such as in a conversation about the conditions of law's existence, or what it means to anoint a given system of social order with the status of law, or how law gains and maintains authority over its subjects. I will highlight some of the more striking examples of this tension in the analysis to follow, as the importance of recognising and, if needed, seeing through and recasting them into a different frame of reference is an important step towards reading Fuller on his own terms, as well as resituating his jurisprudence on the agenda of contemporary legal philosophy. After all, Fuller's apparent failure to raise compelling arguments against Hart's position on the separability thesis was the primary cause of his marginalisation from the inner circle of legal philosophy during his time.

But despite these tensions and obstacles, the argumentative strategy that infuses Fuller's 1958 reply to Hart is very clear. His aim, pursued consistently, is to point out the ways in which Hart's positivism provides an incomplete view of law, and to argue for why it matters to fruitful jurisprudential discussion to bring these missing elements back in. Thus, on each point raised, Fuller implores the reader to widen their gaze beyond Hart's parameters: to ask, for instance, how an effective relationship of communication might be achieved between lawgiver and legal subject, or how a judge understands his responsibility to keep an entire legal system coherent each time he deals with the meaning of a particular law, or to notice how German legal philosophers, legislators and judges alike were charged with the task not just of judging unjust laws, but of transforming a pathological legal order into one worthy of supporting the growth of a new democratic society. All of this grows out of the fidelity to law frame with which Fuller commences his reply.

(ii) Diagnosing the Impasse

Fuller begins his point-by-point engagement with Hart's claims with a diagnosis: the primary objective of positivist legal philosophy is to preserve the integrity of the concept of law.[22] Accordingly, positivists 'have generally sought a precise definition of law, but have not been at pains to state just what it is they mean to exclude by their definitions'.[23] As far as Fuller can see, Hart's particular version of legal positivism remains squarely within this tradition, in so far as it seeks to exclude 'all sorts of extra legal notions about "what ought to be" in service of an apparent concern for how if law is infused with some idea of morality, the morality so infused might be an immoral one'.[24]

Fuller challenges this controversy about 'infusing morality into law' by suggesting that it is linked to Hart's apparent assumption that 'evil aims may have as much coherence and inner logic as good ones'; an assumption that Fuller answers with the assertion that 'coherence and goodness have more affinity than coherence and evil'.[25] This assertion on Fuller's part is intended not as a claim of logic but rather simply as an observation about what, in a later essay, he calls 'the prosaic facts of human life': practice seems to indicate it.[26] Still, Fuller did himself few favours by advancing such an underdeveloped and overstated claim, even if it provides a small window into his enduring intuition that there is something about the form of law, when maintained in an appropriate state, that works against its use for the pursuit of immoral aims. But in the 1958 essay, we are left only with the idea that 'when men are compelled to explain and justify their decisions, the effect will generally be to pull those decisions towards goodness, by whatever standards of ultimate goodness there are'.[27]

These comments provide the background for Fuller's reply to Hart on the question of whether the foundations of a legal order reveal a necessary connection between law and morality. Fuller's diagnosis of Hart's position on this point begins by noting the important common ground between them implicit in Hart's argument that the foundation of a legal system lies not in coercive power, but in certain 'fundamental accepted rules specifying the essential lawmaking proce-

[22] ibid 635.
[23] ibid.
[24] ibid 635, 636.
[25] ibid 636.
[26] Lon L Fuller, 'Reply to Professors Cohen and Dworkin' (1965) 10 *Villanova Law Review* 655, 664.
[27] 'Positivism and Fidelity' (n 1) 637, 643.

dures'.[28] What Fuller evidently saw in this statement by Hart and the emphasis Hart himself placed on it as capturing the 'key' to the science of jurisprudence was the idea that the proper focus of jurisprudence ought to be on the conditions which make lawgiving possible, including, as Hart suggests here, the attitudes of acceptance towards law on the part of both the officials and the 'social group' who together constitute the participants of a legal system.

Fuller thus entreats Hart to conversation on this point by focusing on the phenomenon of acceptance that is at the centre of Hart's claim. According to Fuller, again consistent with the general purpose of his reply to illuminate key sites of incompleteness within Hart's positivism, what is missing in the positivist position is any sense of how the fidelity to law implied by such acceptance 'is something for which we must plan'.[29] A new constitution, for instance, cannot 'lift itself unaided into legality', but rather must rely on a general acceptance from those who will be both its agents and its subjects that it is 'necessary, right, and good'.[30] Fuller's point, in short, is that the moral viewpoint towards law that underscores its acceptance is something that must be planned for and worked towards: this 'sociology of fidelity', as Krygier has astutely called it, does not arise from nowhere.[31] Why, therefore, is there no recognition of this in Hart's analysis?

To my mind, there are few other points in the 1958 exchange that represent such a promising site of conversation between Hart and Fuller on the question of the conditions that make law itself possible than this moment of shared interest in the 'fundamental accepted rules specifying the essential lawmaking procedures'. But much of this promise is left unfulfilled by how Fuller (even if this is wholly explicable by the parameters of the debate itself) appeals to Hart to explain the *nature* of these rules that furnish the framework within which the making of law takes place. If they are not born of law, surely these foundations of legal order must be born of morality, not least because the efficacy which underpins their general acceptance must ultimately rest on a moral perception that they are 'right and necessary'?[32]

This conclusion is, at the very least, simplistic and underdeveloped, even if its aim—to speak to the terms of the debate—is clear enough. But surely the more important point, framed almost out of relevance by those parameters, is that Fuller is here calling upon positivists to explain how it is possible for law to arise, and to gain and maintain its authority over officials and subjects alike, *unless* there is some reason, other than coercion, for those participants to accept that authority. The criticism, therefore, is that if the positivist account side-steps meaningful engagement with this question, it exposes itself as essentially incapable of illuminating some of the most fundamental questions about the nature of law. In doing so, moreover, positivism also substantially disserves the ideal of fidelity to law,

[28] ibid.
[29] ibid.
[30] ibid.
[31] Krygier, 'Transitional Societies' (n 13) 119.
[32] 'Positivism and Fidelity' (n 1) 639.

because it offers little to explain why we go along with law's authority, why we accept it, as we apparently do, at our own initiative.[33]

The dance seen here between the point that Fuller appears to be trying to convey and his need to frame it in terms that speak to a debate about the separability thesis is one that permeates much of the 1958 essay. In this instance, the law and morality prism very arguably functions to channel Fuller's objection into a much more simplistic rejoinder than he might otherwise have advanced, and, in so doing, invites a much narrower conversation than might otherwise have been initiated on the matter of what might explain our attitude of acceptance towards law. A turn to the archival materials is helpful here, in so far as they clearly suggest that Fuller was aware that the vocabulary of 'morality' as he used it in this context was at the very least in need of clarification. An especially instructive working note on this point can be found among the draft papers for what was to become *The Anatomy of Law*, where Fuller offers the following explanation of what he meant by 'morality' when exploring the conditions of law's acceptance in his reply to Hart:

> In whatever form the distinction between law and morality is presented, it is generally assumed to be a characteristic of a legal system that it be 'established', that is, that it have achieved some measure of obedience within the society it purports to regulate. In this sense it is a truism that the foundations of law are 'moral'. A legal system cannot lift itself into being legal by fiat. Its security and efficacy must rest on opinions formed outside of it which create an attitude of deference towards its human author (say, a royal law-giver) or a constitutional procedure prescribing the rules for enacting valid law. To say that this acceptance is 'moral' means merely that it is antecedent to law. The grounds on which it rests need not be elevated; it may proceed from a simple fear of chaos and a conviction that almost any government is better than none.[34]

Fuller's choice to opt for the 'if not law then surely morality' approach to engaging with Hart on the question of the conditions of law's acceptance also sidelined a further conversation that might have been pursued in light of Hart's declaration that the key to jurisprudence lies not only in the phenomenon of rules but the phenomenon of their acceptance on the part of *both* legal officials and the 'social group' (presumably law's subjects) who are governed by law. For anyone interested in the place of the legal subject in the accounts of law offered by different legal philosophies, it is worth noting how this 'social group' disappears in Hart's *The Concept of Law* in favour of a virtually exclusive focus on how the attitudes and practices of legal officials create and sustain the legal order. As I will explain further in chapters four and five, the absence of any meaningful concern for the position of the legal subject and her contribution to the creation and maintenance of legal order is a matter that becomes increasingly important to Fuller's critique of legal positivism, and thus also to the demands that he places on his own jurisprudence to see to her appropriate inclusion. Awareness of the

[33] ibid 642.

[34] Draft notes for 'Anatomy of the Law' at p 25 (Fuller's handwritten pagination), The Papers of Lon L Fuller, Harvard Law School Library, Box 12, Folder 4 ('Encyclopaedia Britannica').

distance between him and Hart on this issue is not, however, entirely lost in the 1958 debate because, as I highlight below, a central element of Fuller's criticism of Hart's treatment of Nazi law relates to how that treatment fails to consider the position in which a debased legal order places its subjects, and thus also the question of whether such debasement has any bearing on the conditions of law's existence.

(iii) Moralities External and Internal

The fourth section of Fuller's essay sees him develop this theme of the conditions of law's acceptance, but here framed in terms of how there is a 'twofold sense in which it is true that law cannot be built on law'.[35] Fuller designates the first of these senses as the 'morality external to law', and explains that this 'external' morality relates to how the authority that law claims for itself 'must be supported by moral attitudes that accord to it the competency it claims'.[36] But Fuller then goes on to explain why this external morality of support for law on the part of its subjects is insufficient to ensure that law itself possesses the attributes that enable it to function successfully. For this aspiration, he argues, we must also respect the 'internal' morality of law.[37]

Here, therefore, we see Fuller reintroduce the idea that he first sketched in the essay, 'Human Purpose and Natural Law',[38] where he argues that the institutional form through which any purposive activity is channelled necessarily implicates certain requirements that can be understood as internal to the activity itself. In his 1958 reply to Hart, however, Fuller explores this idea that law possesses an 'internal' morality by suggesting that the distinction between law and morality, as it is defended by positivists at the level of a whole legal system, might be restated in terms of a distinction between 'order *simpliciter*',[39] and order ('good order') that has moral worth in its correspondence to 'the demands of justice, or morality, or men's notions of what ought to be'.[40]

The point Fuller wants to make is that any such distinction is untenable, because all attempts even at order *simpliciter* must at least be good enough 'to be considered as functioning by some standard or other';[41] a point he extends in a working note by adding that these standards of functionality 'cannot be derived from the concept of order itself'.[42] From this Fuller concludes that the very notion of order itself contains 'what may be called a moral element'.[43]

[35] 'Positivism and Fidelity' (n 1) 645.
[36] ibid.
[37] ibid.
[38] Lon L Fuller, 'Human Purpose and Natural Law' (1956) 53 *Journal of Philosophy* 697, 704–5.
[39] 'Positivism and Fidelity' (n 1) 644.
[40] ibid.
[41] ibid.
[42] Notes for a talk delivered at the University of Vermont, 13 July 1959 (announced title: 'Traditions of Justice Among Western Peoples'), The Papers of Lon L Fuller, Harvard Law School Library, Box 11, Folder 8, p 8 (Fuller's handwritten pagination).
[43] 'Positivism and Fidelity' (n 1) 644.

Again, the 'moral' point here seems awkwardly positioned, and this awkwardness is only partly alleviated by how Fuller then proceeds to introduce the first of what will become several hypothetical tales of hapless or ill-motivated monarchs who attempt, with varying degrees of success, to take on the role of lawgiver. In the 1958 essay, the unnamed monarch, 'whose word is the only law known to his subjects', gets the law game comprehensively wrong on a range of fronts. He issues commands for which he promises rewards for compliance and threatens punishment for disobedience, but then 'habitually punishes loyalty and rewards disobedience',[44] thus failing in his lawgiving by failing to accept 'that minimum self-restraint that will create a meaningful connection between his words and his actions'.[45] He also becomes 'slothful in the phrasing of his commands', such that 'his subjects never have any clear idea what he wants them to do',[46] and so fails again by failing to assume the responsibility of clarity in legal drafting.[47] The examples are brief, but their point is stated strongly. Even if considered merely as order, the monarch's failures show us how law contains its own 'implicit morality' which:

> must be respected if we are to create anything that can be called law, even bad law. Law by itself is powerless to bring this morality into existence. Until our monarch is really ready to face the responsibilities of his position, it will do no good for him to issue still another futile command, this time self-addressed and threatening himself with punishment if he does not mend his ways.[48]

The link between this trajectory of analysis and the criticism he is seeking to advance against Hart is made clear when Fuller suggests that it is precisely this phenomenon of the internal morality of law—of the demands that must be met if order itself is to come into being—that is neglected by positivist legal philosophy. In Hart's case, however, there does seem to be one important exception to this, in the form of his foray into the matter of law's generality and its attendant principle of treating like cases alike. Fuller sees this as a sign of promise, but one that is ultimately and very clearly defeated by Hart's apparent view that the matter of the moral dimensions of law's generality has 'no special relevance to his main enterprise'.[49] As far as Fuller is concerned, therefore, all indications point to the conclusion that law, for Hart, is a mere 'datum projecting itself into human experience and not as an object of human striving'.[50] The point about the moral implications of law's generality is thus left there, with Fuller turning instead to reframe his 'law as datum' versus 'law as striving' distinction in terms of the failure of legal positivists to appreciate how 'the existence of a legal system, even a bad or evil legal system, is always a matter of degree'.[51]

44 ibid.
45 ibid.
46 ibid 645.
47 ibid.
48 ibid.
49 ibid 646.
50 ibid.
51 ibid 646, 647.

This well-known claim goes to the very core of Fuller's jurisprudence, and especially to his concern to offer an account of law that is capable of explaining the phenomenon of legal pathology. I return to this point in more detail below. But for present purposes, in terms of the architecture of his reply to Hart, an important point to notice is how Fuller follows his introduction of this idea with a reference to the example of Nazi law. As he explains it, when we recognise the 'simple fact of everyday legal experience' that the existence of a legal order is something towards which we must strive, it 'becomes impossible to dismiss the problems presented by the Nazi regime with a simple assertion: "Under the Nazis there was law, even if it was bad law"'.[52] Instead, the jurisprudential quandaries presented by Nazi law should be approached by inquiring into 'how much of a legal system survived the general debasement and perversion of all forms of social order that occurred under the Nazi rule, and what moral implications this mutilated system had for the conscientious citizen forced to live under it'.[53] I return to this crucial statement, and what Fuller is inviting us to take from it, in my analysis of Nazi law in section II below.

(iv) The Incompleteness of Positivism

Fuller's essay then turns, consecutively, to two sites of debate that, as noted, I have reserved for elaboration elsewhere: his famous response to Hart's treatment of Nazi law, which I set out immediately below, and his equally well-known response to Hart on whether a connection between law and morality is revealed in how judges decide cases where the meaning of the legal rule is in dispute, which I have chosen to reserve for discussion in chapter seven because it is more continuous with his thinking elsewhere on interpretation and adjudication that I focus upon there than it is with the themes of the 1958 exchange just traversed.

It is therefore notable that Fuller closes his reply to Hart with a discussion, titled 'The Moral and Emotional Foundations of Positivism', which returns to the question of the extent to which an account of law's nature ought to make sense of the phenomenon of fidelity to law. Here, however, Fuller's focus is on Hart's rejection, articulated in the final section of his own essay, of notions of purpose within legal philosophy. Fuller readily acknowledges the source of Hart's worry here: a purposive interpretation of law, carried too far, may pose 'a threat to human freedom and human dignity',[54] as indeed the phenomenon of Nazi law attests to. Nonetheless, Fuller clearly thinks that Hart, in his wholesale rejection of legal philosophies that accommodate the idea that law is a purposive phenomenon, has thrown the baby out with the bathwater. The reason, again, lies with the intelligibility of the idea of fidelity to law. Fuller's point, in short, is that the very idea of fidelity to law, of why law can lay claim to our support, is inseparable from an understanding of its purpose, on the part of all of its participants, lawgivers

[52] ibid 646.
[53] ibid.
[54] ibid 671.

and legal subjects alike, as a distinctive and morally worthwhile enterprise of social ordering.[55] 'Positivism and Fidelity to Law' thus closes with the conclusion that has been sustained by Fuller throughout it: that positivist legal philosophy, even at Hart's reviving hand, is incomplete in crucial ways.

II Reclaiming Fuller through the Nazi Law Debate

A Nazi Law according to Hart

Hart's engagement with the subject of Nazi law represents that site of his defence of legal positivism on the most traditional of contests about the separability of law and morality: the question of whether law that is unjust, or 'evil', in its content ought to be anointed with the status of valid law. As noted earlier, this classic 'natural law versus positivism' debate had been reawakened during Hart's time as a result of the claim of the German legal philosopher, Gustav Radbruch, in response to Nazism, that when a law reaches a certain level of moral iniquity it should be refused the character of law.[56]

As Hart tells the story, prior to Nazism Radbruch had been a committed legal positivist who saw resistance to immoral law as a matter for personal conscience.[57] Upon living through the experience of Nazism, however, Radbruch concluded that the positivist insistence on the separation of law and morality had powerfully contributed to the horrors of the era and especially to the failure of the German legal profession 'to protest against the enormities which they were required to perpetrate in the name of law'.[58] Radbruch's response was to argue for the view that 'the fundamental principles of humanitarian morality were part of the very concept of *Recht* or Legality', such that no positive enactment, however clearly it conformed with the formal criteria of legal validity with a given system, should be regarded as valid law if it contravened basic principles of morality.[59]

According to Hart, the application of the 'Radbruch formula' in post-war Germany as a standard for determining the status of Nazi law required every lawyer and judge to 'denounce statutes that transgressed the fundamental principles not as merely immoral or wrong but as having no legal character'.[60] By way of

[55] ibid 670.

[56] The actual text of what has become known as the 'Radbruch formula' is as follows: 'The positive law, secured by legislation and power, takes precedence even when its content is unjust and fails to benefit the people, unless the conflict between statute and justice reaches such an intolerable degree that the statute, as "flawed law", must yield to justice': Gustav Radbruch, 'Statutory Lawlessness and Supra-Statutory Law (1946)', translated by Bonnie Litschewski Paulson and Stanley Paulson (2006) 26 *Oxford Journal of Legal Studies* 1, 7.

[57] Hart, 'Positivism' (n 1) 616.

[58] ibid 617.

[59] ibid.

[60] ibid.

illustration, Hart refers to a case involving a woman who reported her husband to the Nazi authorities for insulting remarks he had made about Hitler while home on leave from the German army. As Hart recounts the facts, what the husband had said 'was apparently in violation of statutes making it illegal to make statements detrimental to the government of the Third Reich', although the wife was under no legal duty to report him.[61] After the war, the wife was prosecuted in a German court for illegally depriving her husband of his freedom, a crime of the pre-Nazi German criminal law that, the prosecution argued, had remained in force throughout the Nazi years. In her defence the woman pleaded that because her husband's imprisonment was pursuant to a Nazi statute that was lawful at the time, she had not illegally deprived her husband of his liberty and thus had committed no crime. Though this defence succeeded at trial, a court of appeal rejected it on the basis of the Radbruch formula: that is, because the statute under which she acted 'was contrary to the sound conscience and sense of justice of all decent human beings', and so not entitled to the status of law.[62]

This outcome, 'hailed as a triumph of the doctrines of natural law and as signalling the overthrow of positivism', is emphatically condemned by Hart.[63] The postwar German legal system, he argues, had other options available to it for dealing with cases of this nature, such as letting the woman go unpunished, or calling upon the legislature to enact a retrospective statute to declare actions taken under the morally impugned laws to be unlawful.[64] Hart readily accepts that neither of these options is straightforward. Indeed, he sees the choice between them as presenting a moral quandary between letting the woman go unpunished, thus sacrificing the moral value of her punishment, or securing her punishment only through the sacrifice of 'a very precious principle of morality endorsed by most legal systems'.[65] But, according to Hart, to resolve 'grudge informer'[66] problems of this kind through recourse to the Radbruch formula is not only to hide this moral quandary, but to abandon legal principle. The better solution would have been to take the path that best serves clarity: to 'speak plainly' through a retroactive statute and condemn valid laws as too evil to be obeyed,[67] rather than presenting 'the moral criticism of institutions as propositions of a disputable philosophy'.[68]

The message we are left with, then, is that even in the face of Nazi law and the atrocities committed in its name, the positivist position that the validity of law is one thing and its moral merits another remains the best standard for determining legal validity. This, it is suggested, is for two reasons: in its simplicity, the positivist

[61] ibid 618–19.
[62] See generally ibid.
[63] ibid 619.
[64] ibid.
[65] ibid.
[66] I put the term 'grudge informer' in quote marks because Hart himself does not use it. Fuller, however, does, and the case discussed in the Hart-Fuller debate has generally become known to students of jurisprudence as 'the case of the grudge informer'.
[67] ibid 620.
[68] ibid 621.

approach serves the goal of clarity at the same time, ostensibly, as it serves our capacity for moral criticism of law.

B Nazi Law according to Fuller

The title of Fuller's analysis of Nazi law—'The problem of restoring respect for law and justice after the collapse of a regime that respected neither'—is itself instructive of how Fuller sees the jurisprudential quandaries presented by Nazi law as having a distinctly practical bearing.[69] It is with this eye towards the 'truly frightful predicament' that faced those charged with rehabilitating the post-war German legal system that Fuller elaborates his critique of Hart's treatment of Nazi law and its neatly defended conclusion about the superiority of a legal positivist standard for legal validity.

The thrust of Fuller's complaint is immediately made clear: the primary ill that plagues Hart's analysis of Nazi law is his complete disregard of 'the degree to which the Nazis observed what I have called the inner morality of law itself'.[70] No mention is made of the systematic derogations from the internal morality of law that occurred in monstrous form under Hitler, such as how the Nazis took generous advantage of retroactive statutes to cure past legal irregularities, how they attempted to legalise killings by secret statute and how, over the course of their rule, they tolerated and encouraged officials to ignore the letter of the law while terrorising those who contemplated calling such officials to account.[71] Instead, Hart's focus is directed only to the matter of ends: as Fuller puts it curtly, for Hart 'the only difference between Nazi law and, say, English law is that the Nazis used their laws to achieve ends that are odious to an Englishman'.[72] All else seems to lie outside of his assessment of whether Nazi law was law.

These are the foundations from which Fuller defends the decisions of the post-war German courts in the grudge informer cases that Hart criticised so severely. In contrast to Hart's succinct account of the cases and their context, Fuller's analysis begins from a detailed explanation of the two statutes from the Nazi era upon which the accused woman rested her defence. The first of these statutes, he explains, was part of a series of enactments introduced in 1938 that created special wartime criminal offences, including the one at issue in the case which rendered a person liable to be punished by the death penalty if he or she 'publicly' incited a refusal to fulfil the obligations of service in the German army, or otherwise sought 'to injure or destroy the will of the German people or an allied people to assert themselves stalwartly against their enemies'.[73] The second, enacted in 1934, threatened imprisonment for any person who publicly expressed 'spiteful or provocative' statements that disclosed 'a base disposition toward leading person-

[69] 'Positivism and Fidelity' (n 1) 648.
[70] ibid 649–50.
[71] ibid 651, 652.
[72] ibid 650.
[73] ibid.

alities' of the Nazi party.[74] This statute, however, went considerably further than the 1938 enactment, in so far as it stated that malicious utterances not made in public 'shall be treated in the same manner as public utterances when the person making them realized or should have realized that they would reach the public',[75] and conferred on the Minister of Justice a wide discretion to 'determine who shall belong to the class of leading personalities' for the purpose of the statute.[76]

Fuller's analysis of the two statutes develops his idea of the internal morality of law in important ways; indeed, with respect to the second statute, much more substantially than the brief hypothetical of the lawgiving monarch offered earlier in the essay. With respect to the first statute, Fuller's commentary is directed primarily to the perverse principle of interpretation that enabled the relevant Nazi court to determine that the husband's private utterance to his wife was 'public' and thus unlawful, observing that it is 'almost inconceivable that a court of present-day Germany would hold the husband's remarks to his wife . . . to be a violation of the final catch-all provision of this statute'.[77] More, however, about what Fuller has in mind through his idea of the internal morality of law is revealed in how he labels the second statute a 'legislative monstrosity'. The suggestion here seems to be that even though the statute bears the external form of a legislative enactment, there is something distinctly not law-like, at a deeper understanding of law's form, about a statute that hinges upon an uncontrolled administrative discretion and so of its nature represents an invitation to the exercise of arbitrary power.

Although there are no enumerated principles here against which to specifically evaluate the pathologies at issue (for this we must wait for the eight desiderata that Fuller elaborates in *The Morality of Law*) it takes little to extrapolate that Fuller regards the statute in question as pathological for how the demands it makes on the legal subject are entirely unclear, as well as for how it represents a standing risk of retroactivity when the subject discovers, too late, that she has offended a previously undesignated leading personality of the Nazi party. The statute is a legislative monstrosity, Fuller therefore seems to be telling us, because it defies, at some fundamental level, what a legal subject should be entitled to expect from something that makes claim to be law.

All of this is set out by Fuller before any mention is made of the Radbruch formula and Hart's criticisms of it. It is thus clear that Fuller seeks to establish the basis of his own engagement with Hart on the question of Nazi law as an analysis directed to the matter of law's form, before turning to Hart's response to Radbruch and to the actions of the post-war courts more generally. But with respect to the latter, to Fuller's eyes Hart's criticism of the Radbruch formula betrays an internal contradiction. Hart seems to suggest that when courts are faced with an immoral law, and must decide whether or not they will apply it to the case at hand, they are

[74] ibid 654.
[75] ibid.
[76] ibid.
[77] ibid.

in essentially the same position as citizens who might face an immoral law and refuse to obey it on grounds of personal moral objection. That is, Hart seems to be suggesting that judges, too, can simply refuse to apply a law on grounds of personal moral objection. If this is the thrust of Hart's position, Fuller responds, then it hardly resolves a moral dilemma because 'surely moral confusion reaches its height when a court refuses to apply something that it admits to be law'.[78]

Nor was it the case, Fuller continues, that the Radbruch formula was regarded by the post-war German courts, or indeed by Radbruch himself, as a simple fix to the problem of unjust law. Contrary to what Hart seems to suggest, Radbruch clearly appreciated that the application of his formula was problematic and ought to be reserved for only the most extreme cases, so as to 'not conceal from ourselves . . . what frightful dangers for the rule of law can be contained in the notion of "statutory lawlessness" and in refusing the quality of law to duly enacted statutes'.[79] But, Fuller emphasises, there was much more at stake here than questions of legal philosophy and indeed of pursuing a legally principled response to a problematic situation. Radbruch and the German courts were faced with a situation of drastic emergency: not only was the grudge informer problem a pressing one but, if legal institutions were to be rehabilitated in Germany, 'it would not do to allow the people to begin taking the law into their own hands, as might have occurred while the courts were waiting for a statute.'[80]

Thus, by this point in his response to Hart on Nazi law, we are beginning to see how and why Fuller regards the Radbruch formula and its application in post-war Germany as generally defensible: real world problems were at stake, something had to be done, and what was done was carefully policed. He also generally defends, against Hart's censure,[81] Radbruch's suggestion that the culture of legal positivism might have made some contribution to the atrocities committed in the name of law under Nazism: there was surely good cause for Radbruch to seek to explain the fact that the first attacks on established order were 'on ramparts which, if they were manned by anyone, were manned by lawyers and judges'.[82] But, and this is crucial to seeing the distinct differences between Fuller and Radbruch's 'natural law' approaches to the question of the status of Nazi law, while he insists that Radbruch's position is defensible on all counts, Fuller is clear that he personally would not have grounded a solution to the problem of the grudge informer cases by recourse to the Radbruch formula. He would, as the beginning of his analysis foreshadows, have instead evaluated the laws in question by reference to the idea of the internal morality of law:[83]

> To me there is nothing shocking in saying that a dictatorship which clothes itself with a tinsel of legal form can so far depart from the morality of order, from the inner morality

[78] ibid 655.
[79] ibid 655–56.
[80] ibid 655.
[81] Hart, 'Positivism' (n 1) 617–18.
[82] 'Positivism and Fidelity' (n 1) 659.
[83] ibid 659–60.

of law itself, that it ceases to be a legal system. When a system calling itself law is predicated upon a general disregard by judges of the terms of the laws they purport to enforce, when this system habitually cures its legal irregularities, even the grossest, by retroactive statutes, when it has only to resort to forays of terror in the streets, which no one dares challenge, in order to escape even those scant restraints imposed by the pretence of legality,—when all these things have become true of a dictatorship, it is not hard, for me at least, to deny to it the name of law.[84]

This is a crucial passage, and I consider what we are invited to take from it, as a statement about legal validity, in my analysis to follow. But in terms of reading Fuller's jurisprudence through his treatment of Nazi law, it is also important to note how this passage is followed by another that speaks to themes of that jurisprudence that thereafter remain underdeveloped in his exchanges with Hart. There is surely something to be learned, Fuller contends, from how the most serious debasements of legal morality under Hitler occurred in the domain of public law, with no comparable deterioration witnessed in the ordinary branches of private law:[85]

It was in those areas where the ends of law were most odious by ordinary standards of decency that the morality of law itself was most flagrantly disregarded. In other words, where one would have been most tempted to say, 'This is so evil it cannot be a law', one could usually have said instead, 'This thing is the product of a system so oblivious to the morality of law that it is not entitled to be called a law'. I think there is something more than accident here, for the overlapping suggests that legal morality cannot live when it is severed from a striving toward justice and decency.[86]

I return to the invitations expressed in this passage in chapter four, where I examine private correspondence that extends suggestively (and, notably, in the direction of law's relationship to the status of the legal subject as an agent) upon this intuition that there is, in practice, an apparent correlation between the pursuit of oppressive aims and the debasement of the form through which law finds expression. Here, however, in his reply to Hart on Nazi law, Fuller simply invites us to speculate on this connection; that it seems to be 'something more than accident'. Still, even if only a gesture, that moment of speculation is wholly compatible with Fuller's comment earlier in the essay that his 'external' and 'internal' moralities of law stand in a relationship of 'reciprocal influence', such that 'a deterioration of the one will almost inevitably produce a deterioration in the other'.[87]

[84] ibid 660.
[85] ibid 660, 661.
[86] ibid 661.
[87] ibid 645.

C Analysis: Reading Fuller through the Nazi Law Debate

(i) Defending a Distinctive Natural Law Analysis

A great deal of the correspondence in Fuller's archive relating to the 1958 debate speaks only to the Nazi law issue, which, or at least much of the correspondence indicates, is widely understood as a traditional debate between positivism and natural law. Accordingly, much attention is given to the extent to which Fuller's position is distinguishable from a traditional natural law adherence to the idea of a higher law against which the validity of positive law is determined.[88] The persistence of this question from so many of those acquainted with the 1958 debate ('nine readers out of ten', according to one frustrated reply, credited him with 'espousing a doctrine I have never proposed, namely, that a law that is evil enough in its intention is not really law at all')[89] was for Fuller a source of great bewilderment. Indeed, this bewilderment appears to at times have spilled over into infuriation, as is indicated in an ongoing exchange with Robert Summers, then a young professor at Stanford Law School, which reached an especially high pitch. Rebutting the suggestion that his own position amounts to the claim that 'an evil law is a contradiction in terms', Fuller explains the cause of his 'sensitivity to this particular misinterpretation':

> Quite a number of people who have *not* read my exchange with Hart have said to me, 'I understand that you say that the Nazi law was so bad it was not law at all. I don't see how that can be'. I have to answer that it's not quite so simple as that. I then have gone on to point out that the Nazis passed secret laws, that they retroactively approved by 'law' wholesale murder, that they paid no attention to their own laws in judging cases, etc., etc. I then have to explain that I take a distinction between the integrity of a legal system considered just as law, and the morality or immorality of its substantive aims. All this is unfamiliar and complicated. But I have this kind of difficulty usually only with those who have not read my article. When, therefore, someone like yourself who *has* read my article says the same thing, I lose patience.[90]

Even if this frustration on his part seems understandable, it remains the case that Fuller did broadly support Radbruch's approach in his reply to Hart, even if he then suggested that he personally would have resolved the grudge informer problem on different grounds. Fuller was undoubtedly an admirer of Radbruch's; something made clear not only in the essay itself but also in a letter to Radbruch's widow, Lydia, attaching a copy of the *Harvard Law Review* debate, where Fuller

[88] Fuller actually only uses the term 'natural law' once in the entire essay, when he comments on how 'it is chiefly in Roman Catholic writings that the theory of natural law is considered, not simply as a search for those principles that will enable men to live successfully together, but as a quest for something that can be called a "higher law"': ibid 660.

[89] Letter from Fuller to Professor Boris I Bittker, 4 April 1960, The Papers of Lon L Fuller, Harvard Law School Library, Box 14, Folder 1 (correspondence).

[90] Letter from Fuller to Professor Robert S Summers, 15 January 1964, The Papers of Lon L Fuller, Harvard Law School Library, Box 7, Folder 16 (correspondence)

writes that he has always 'had great admiration for your husband, and was privileged to hear him give a few lectures in Heidelberg in the summer of 1929'.[91] And, certainly, the 1958 essay indicates much sympathy towards Radbruch's predicament as someone charged with the task of finding a principled basis on which to clean up the Nazi legal system so that a legal order suited to the aspirations of a new and democratic Germany could take root.

But it also remains the case that Fuller did clearly distinguish his own line of reasoning on the matter of resolving the status of Nazi law from Radbruch's. In addition to his explicit statements to this effect in the essay itself, dissimilarity can also be discerned in how even when Fuller's focus is an individual law, as is the case in his consideration of the statutes relevant to the grudge informer case, his inquiry moves intuitively and quickly to questions concerning the quality of the wider systemic context in which the law in question was enacted, interpreted and applied. Subject to some comments I make below about how the Radbruch formula might be read more broadly, it seems safe to say that a much more focused debate on this question of the validity of Nazi law can be conducted between Radbruch and Hart's positions, than it can between Radbruch and Fuller's, or, indeed, between Hart and Fuller's, because of how Fuller's approach speaks to and engages with a much wider site of evaluation. Consistent with his comment early in the 1958 essay, when framing his reply to Hart in terms of the issue of fidelity to law, even in the Nazi law debate Fuller is concerned above all with the idea that if laws, even bad laws, have a claim to our respect, 'then law must represent some general direction of human effort that we can understand and describe, and that we can approve in principle even at the moment when it seems to us to miss its mark'.[92] It is this big picture, the 'general direction of human effort' that constitutes and maintains a legal order, which is always kept in view.

There is also much to be learned from the 1958 debate on Nazi law about the distance between Hart and Fuller on matters broadly relevant to the issue of methodology in jurisprudence, and which play themselves out in the exchange in at least two ways. The first concerns the question, again raised throughout Fuller's exchanges with Hart, of whether positivism can accommodate a meaningful idea of legal pathology, and if not, why. The second concerns what we might regard as the methodological impulse that underlies Fuller's distinctive turn towards the perspective of the legal subject when assessing the question of whether Nazi law could meaningfully have been regarded as 'law'.

(ii) Positivism and Legal Pathology

The challenge that Fuller presents to Hart to explain whether positivism can accommodate a meaningful idea of legal pathology might not be stated in precisely those terms at any point in the 1958 essay, but there are numerous ways in

[91] Letter from Fuller to Lydia Radbruch, 18 August 1958, The Papers of Lon L Fuller, Harvard Law School Library, Box 14, Folder 1 (correspondence).

[92] 'Positivism and Fidelity' (n 1) 632.

which this challenge pervades the debate about Nazi law. To see this, it is helpful to note how in *The Concept of Law*, to which I return at the beginning of chapter four, Hart elaborates the argument that a system of rules constituted only by primary rules governing acceptable and unacceptable modes of conduct will invariably suffer from problems of stasis as well as clarity unless it is supplemented by secondary rules that enable processes of adjudication and change, and which supply a standard for determining which rules of the system are to be regarded as valid law. It is only the latter—a system comprised of a union of primary and secondary rules—that, on Hart's analysis, is to be properly regarded as a legal system.

But this does not mean, on Hart's analysis, that a system constituted only by primary rules of conduct is pathological: rather, it is simply primitive, unable to serve a society's ends as effectively as can a system comprised of a union of primary rules and secondary rules. Indeed, we see the same basic understanding in the work of Hart's successor, Joseph Raz, on the question of what kind of authority law possesses. As I explain further in chapter six, on Raz's account, a law which possesses merely de facto, as opposed to legitimate, authority is not understood as pathological, but rather simply as a law which has characteristics closer to a command than one that speaks more directly and intelligibly to the legal subject's reasons for accepting it as authoritative.[93]

The point I am seeking to make, therefore, is that there is in fact much more going on in Fuller's remark that Hart assesses the status of Nazi law as if the only difference between Nazi law and English law is that 'the Nazis used law for ends that are odious to the Englishman' than a mere snarky, throwaway line. The objection actually speaks directly, and powerfully, to the question of whether positivism can accommodate a meaningful idea of legal pathology and, if it can't, whether it ought to be regarded as a theoretical position adequate to the task of illuminating the jurisprudential quandaries presented by Nazi law *as* law. Indeed, Hart's omissions in this vein did not go unnoticed by some of Fuller's supporters, whose letters congratulate him on how he attempted to keep the debate focused on 'the immediate implications of law as law',[94] and which suggest that 'on one or two of the more important points you take up [Hart] was, if not silent, at least uncommitted in his article'.[95] But a response to the question implicit throughout Fuller's essay of whether positivism can accommodate a meaningful idea of legal pathology was something that Hart continued to resist as the exchange between the two scholars extended beyond the pages of the 1958 *Harvard Law Review*. Even when Fuller reframes the question as a challenge to Hart to explain whether positivism makes provision for the withdrawal of lawgiving power in the event of its abuse, Hart provides little by way of a meaningful response.[96]

[93] See chapter six, III 'Raz on authority'.

[94] Letter to Fuller from Professor Leo Strauss, 19 March 1958, The Papers of Lon L Fuller, Harvard Law School Library, Box 3, Folder 15 (correspondence relating to the 1958 debate).

[95] Letter from Professor David Derham to Fuller, 28 April 1958, The Papers of Lon L Fuller, Harvard Law School Library, Box 3, Folder 15 (correspondence relating to the 1958 debate).

[96] See chapter four, II 'Hart's review'.

(iii) The Turn to the Subject

Fuller's concern for the phenomenon of legal pathology finds perhaps its most powerful manifestation in the 1958 essay in how Fuller's challenge to Hart on this point sees him turn to the perspective of the legal subject. In direct contrast to Hart's approach, where the question of how Nazi law was experienced by those who were subject to or otherwise participated in it is completely absent, Fuller's analysis sees the legal subject and the question of how law exercises its authority over her brought into full view. As he puts it in the segue that links his account of the moral foundations of legal order to his response to Hart on Nazi law, the really important questions for jurisprudence, in light of the pathologies of Nazi law, are not approached by asserting that '[u]nder the Nazis there was law, even if it was bad law', but rather by asking 'how much of a legal system survived the general debasement and perversion of all forms of social order that occurred under the Nazi rule, and what moral implications this mutilated system had for the conscientious citizen forced to live under it'.[97]

This simple, rhetorical question neatly charts the progression of interconnected ideas that Fuller comes to develop more explicitly over the course of subsequent writings: how the health of the form of law connects to how its normativity might be understood from the subject's perspective, which in turn connects to the question of whether a sufficiently unhealthy legal system, in this formal sense, can meaningfully be understood to be existing at all. In framing his entry into the Nazi law discussion in this way, there is no sense that Fuller is seeking to turn us away from more objective ways in which the phenomenon of law's formal debasement might be witnessed and evaluated. We might still, for instance, take seriously such indicia of pathology as the fact that laws are being 'published' secretly, or are retroactive or impossible to comply with. But what is interesting about Fuller's approach, and what makes it so distinctive, is how he intuitively frames the question of the pathology of Nazi law by reference to how it was experienced by those who were subject to it: whether, to them, it was something capable of commanding the conscientious fidelity that they would ordinarily direct towards law. Or, if we understand Fuller's jurisprudence as gravitating consistently to the question of how law, properly understood, relates to the phenomenon of human agency, the nub of his challenge to Hart here might be phrased in these terms. Can drastically formally debased 'law', generating drastically debased possibilities for the exercise of the subject's agency, be meaningfully regarded as law and capable of establishing and maintaining an attitude of fidelity towards it?

The Hart-Fuller debate is rarely remembered as a contest that pivots around questions of this kind but, when read from Fuller's perspective, my contention is that it should be. Reading Fuller's challenge in this way also helps us to illuminate just how far removed the message of his analysis of Nazi law is from the interpretation of his claims about law's internal morality that comes to define Hart's

[97] 'Positivism and Fidelity' (n 1) 646.

overall response to his jurisprudence; namely, that observance of the internal morality of law merely makes law a more effective instrument.

I return to this issue at length in chapter four, but it is helpful to foreshadow it here, in the context of analysing what we can learn about Fuller's jurisprudence from his discussion of Nazi law in the 1958 debate. Certainly, as I suggest there, some aspects of Fuller's elaboration of the idea that law has an internal morality have strong instrumental overtones, in the sense that one of the messages we receive from his tales of hapless lawmaking monarchs is that observance of the principles of the internal morality of law is essential to producing law that is effective for the lawgiver's purposes. But what is greatly under-explored in discussions of this point, if indeed considered at all, is how an 'efficacy' reading of Fuller's claims about the internal morality of law is possibly reconcilable with his analysis of Nazi law.

In short, there is nothing in his analysis of Nazi law to suggest that Fuller sees the pathology of Nazi legal order in terms of the obstacles to efficacy that it presented for Nazi lawgivers. Quite to the contrary, it is abundantly clear, especially from his concern for the implications that the mutilated Nazi legal system had for 'the conscientious citizen forced to live under it', that, for Fuller, Nazi law was pathological and, depending on the instance in question, possibly undeserving of the title of 'law', because of the confused, incoherent and arbitrary position in which it placed those who were subject to it. Fuller's analysis of Nazi law thus provides not only a concrete example of a problem of the internal morality of law, but also gives us crucial insight into *why* such is a problem; namely, because of its implications for the position of the legal subject and, within this, for the possibility that any such putative law or legal order could meaningfully possess over that subject the kind of authority we associate with law and which distinguishes it from brute power.

(iv) Validity and Existence

It is a small step from these points to the question of how the impasse between Hart and Fuller on Nazi law in the 1958 exchange speaks to the fundamental question of how to understand the conditions of law's existence or, alternately, its validity. The stakes of Fuller's reply to Hart on this question are well captured in a letter from the legal historian Alexander P D'Entrèves, who, though clearly 'entirely on Fuller's side', raises some important criticisms of the latter's treatment of Nazi law. Of these, D'Entrèves reserves his strongest rebuke for Fuller's ultimate choice to align himself with Hart's view that the problems of the grudge informer cases would have been best resolved through the use of a retroactive statute ('Why did you indulge in this after-thought after having so clearly indicated the correct solution on p.660, first paragraph?'—that is, to evaluate the status of Nazi law by reference to considerations pertaining to the internal morality of law).[98] But D'Entrèves' letter, beyond its general support for Fuller's approach

[98] Letter from Alexander P D'Entrèves to Fuller, 8 April 1958, The Papers of Lon L Fuller, Harvard Law School Library, Box 3, Folder 15 (correspondence relating to the 1958 debate).

to Nazi law, is also almost prophetic in how it anticipates what, 11 years later, Fuller suggests in his 'Reply to Critics' is the key point of impasse between him and positivism: namely, what positivists mean when they state that a law 'actually exists'. Do they mean that it is valid by reference to some source-based test or, having regard to its actual acceptance, that it is effective?[99]

Precisely what Hart's position was on the question of what it takes for a legal system to 'exist'—whether, that is, such was exhausted for positivists by the factual criteria revealed in the rule of recognition—remained a pressing one for Fuller from the time of the 1958 exchange through to his final 'Reply to Critics' in 1969. Archival materials indicate and, notably, by reference to the Nazi law debate, that Hart is 'extraordinarily ambiguous on this score'.[100] There is also a sense in these notes that Fuller thought Hart played their ongoing exchange on this point somewhat unfairly, moving too fluidly from the question of law's existence in the face of formal legal pathology to the question of whether observance of the internal morality of law is compatible with the pursuit of iniquitous legal ends. Again, much of the context and significance of this point requires the discussion to follow in chapter four to be illuminated. But the point to note for present purposes is the extent to which the Nazi law debate (or, more accurately, Hart's apparent refusal to meaningfully engage with Fuller's claims in his reply on Nazi law) is 'the elephant in the room' here. As Fuller expresses it in the same working note:

> Outside the Nazi system, he seems to concede that some minimum respect for legality is essential for a legal system . . .
>
> But when he comes to the Nazi system, he has me saying that an iniquitous law is not law. This I never said. I tried to say that the decent German citizen under Hitler might well refuse his obedience to isolated orders on one or two grounds (1) that the law was 'iniquitous'; (2) that the area of legal restraint was so deteriorated that it was no longer law.[101]

As I explain in chapter six, it is not until Joseph Raz's essay, 'The Rule of Law and its Virtue', that it is directly acknowledged by a legal positivist that the question of the conditions of law's existence and the implications of an answer to that question for the positivist defence of the separability thesis was in fact the nub of Fuller's challenge to Hart: that challenge was not about whether observance of the internal morality of law is compatible with law's use, as an instrument, towards the pursuit of iniquitous ends.[102] But the task for now is not to map Hart's engagement with Fuller on this point but rather to ascertain precisely what the claim is with which Fuller challenges the view implicit in his analysis of Nazi law (implicit, that is, because Hart is, at this point, yet to formulate his idea of the rule of recognition) that whatever the officials of a system declare to be valid law is to be regarded as such.

99 ibid.

100 Note titled 'Does a sufficient disregard of legal morality result in a system that is not "law"?', The Papers of Lon L Fuller, Harvard Law School Library, Box 12, Folder 1 (notes for 'Reply to Critics').

101 ibid.

102 See chapter six, II 'Raz on the rule of law'.

We know that Fuller objects to this view, and to the indifference to the formal and qualitative features of a legal system in action that it indicates. But does this mean that Fuller's response to that view offers an alternate standard for legal validity, in its own right? To answer this question, we must start with Fuller's oft-repeated claim that the existence of law is always 'a matter of degree'; that there is a 'continuum' along which law might be said to or to not exist. Fuller readily acknowledged that this was a highly underspecified criteria of analysis; as he put it in a letter to a colleague at Yale, 'a continuum without a definable end is not a continuum at all, but a great gray soup'.[103] But though Fuller willingly conceded that he could not be 'indifferent toward the problem of clarifying what lies at the end' of such a continuum,[104] he was more or less comfortable with how a certain degree of vagueness with respect to the precise point at which the existence or not of law might be determined was inevitable. After all, such is wholly consistent with his equal commitment to the view that law is not a 'datum' but an achievement towards which we must strive.

Still, it remains the case that the way Fuller concludes his debate with Hart on the status of Nazi law with the statement that, on review of the pathologies that serially plagued the Nazi legal order as a whole, 'it is not hard, for me at least, to deny to it the name of law'[105] is both tantalising and infuriatingly underspecified as a response to Hart's position on the same point. The question that needs to be posed, therefore, even if it cannot be unequivocally answered, is this: does Fuller's analysis of Nazi law yield a distinctly Fullerian standard for legal validity, or, at least, the seeds of one?

III Fuller and Legal Validity

Before a conversation about Fuller and legal validity can even be commenced, it is important to add the caveat that the very vocabulary of 'legal validity' sits awkwardly with his jurisprudence. Taking its lineage from Kelsen and carrying through to Hart, the problem with the language of 'legal validity', from a Fullerian standpoint, is that it tends to keep inquiry into the status of law as law hovering around something of a crisis point—a site of 'all or nothing' standards. Little of the subtlety of Fuller's perspective, with its 'continuums' and matters of 'degree', is at home here, nor within the kinds of conversations that such a vocabulary seems to speak to. Yet neither does Fuller's approach sit particularly easily with the more common contemporary vocabulary of determining the 'grounds of law', that is, the question of what determines the content of the law in force, and which materials are brought to bear on this inquiry, and in what ways. This too was not

[103] Letter from Fuller to Professor Boris I Bittker, 4 April 1960, The Papers of Lon L Fuller, Harvard Law School Library, Box 14, Folder 1 ('The Forms and Limits of Adjudication').

[104] ibid.

[105] 'Positivism and Fidelity' (n 1) 660.

Fuller's obvious territory, beyond his general commitment to a purposive approach to interpretation (itself more a position on statutory interpretation than any developed theoretical statement about the grounds of law), and, as I set out in chapter seven, his views about what judges do when charged with the interconnected tasks of resolving the dispute before them, interpreting the relevant law, and maintaining the coherence of the over-arching legal order within which these tasks must be met.[106]

For all the reasons just stated, then, the provision of a clear formula for legal validity was not and was never claimed to be a natural starting point for Fuller's jurisprudence, nor its end-game. Even if he gestures to a validity claim in the Nazi law discussion (to which I will return shortly) Fuller's writings are otherwise concerned with the much broader task of understanding law as a distinctive form, and with rejecting pathological instances of that form as rightly belonging within the meaning of law as a concept. As he put it in a working note some time after the 1958 debate, the focus of jurisprudence needs to be on what can claim to be law in a 'meaningful sense'.[107] It might be said, therefore, that Fuller's approach is more akin to John Finnis' sense of the enterprise of jurisprudence as directed to illuminating the 'central case' of law but, at Fuller's hand, with the question of the health of law's formal features as the arbiter of the boundaries of what ought to be regarded as a central or non-central case.[108] Jurisprudence is about understanding the nature of law, and this requires that we know the difference between the pathological case not worthy of the name, and the case that enables us to see how and why law is different from other modes of ordering.

This is the natural territory of Fuller's claims, and indeed the apparent distance between his jurisprudence and the matter of standards for legal validity is further revealed in how he doesn't speak directly to questions of law's existence or its 'validity' in any significant way again after 1958, other than at the end of the story of Rex in *The Morality of Law* where he also frames the point broadly in terms of how sufficient departures from the internal morality may lead to something that is 'not properly called a legal system at all'.[109] Still, even if Fuller does not see the end-game of jurisprudence to be its capacity to yield a standard for legal validity, it would be wrong to suggest, as we see for Finnis, that he consciously sought to steer clear of such a question. Because when it comes to the crunch, the pathological case of law is, for Fuller, not just a non-central case of law, but not a case of law at all. Moreover, he wants this to matter not just to jurisprudential discussion about the nature of law, but to legal practice as well. So much is clearly indicated in his challenge to Hart to resolve the dilemma, with respect to the intelligibility of her obligation to obey law, of the 'conscientious citizen' forced to live under Nazi law.

[106] See chapter seven, IV 'Fuller, Dworkin and interpretation'.

[107] Draft notes for 'Anatomy of the Law' at p 32 (Fuller's handwritten pagination), The Papers of Lon L Fuller, Harvard Law School Library, Box 12, Folder 4 ('Encyclopaedia Britannica').

[108] John Finnis, *Natural Law and Natural Rights* (Oxford, Clarendon Press, 1980) ch 1.

[109] Lon L Fuller, *The Morality of Law* (New Haven, Yale University Press, 1969) 39.

It might be said, therefore, that Fuller thinks that the Nazi legal system *as a whole* was an 'invalid' legal order, or an order incapable of generating valid law, wholesale. Certainly, the crucial passage where Fuller concludes his analysis by listing a catalogue of pathologies that plagued the Nazi system as a whole, and then remarks that 'it is not hard, for me at least, to deny to it the name of law',[110] seems to invite this interpretation. But in my view it would be premature to accept that invitation, because elsewhere Fuller's position on the validity of Nazi law emerges in more subtle forms. His analysis of the grudge informer cases, for instance, is very much focused on the text and mode of administration of particular Nazi laws, rather than any broad commentary on the system as a whole (even if the laws in question were typical of the kind generated by that system). But the comment that Fuller makes about how a much more severe deterioration of the standards of the internal morality of law tended to be witnessed in areas of Nazi public law than in private law is still more instructive. There is no suggestion here that German private law and private legal acts taken pursuant to it (the settling of contracts, or making or wills, for instance) was a site of 'non-law', or, at least for those who were not the target of Nazi persecution. Rather, the best reading of Fuller on the question of whether Nazi law was 'valid' is one that sometimes answers 'yes', and sometimes 'no'. As I have suggested elsewhere, all depends on the particular law, and the intelligibility of the particular legal relationship, in question.[111]

What I wish to suggest about Fuller's analysis of Nazi law, then, is that he regards his jurisprudence as capable of going to the question of the validity of particular laws, if pushed. Teasing out the precise content of this gesture is, however, complex. First, it is clear that Fuller's starting point is always the idea that, to be properly regarded as such, a legal *system* must maintain a particular standard of formal health, else it be unrecognisable to its subjects as an order of law, as opposed to some other mode. This commitment always remains in the background, even when Fuller is evaluating the characteristics of a particular law, as indeed we see in his analysis of the statutes at issue in the grudge informer cases. There, Fuller makes clear that it is not enough that a given law *look* like a law, as was the case for the statutes in question, which bore the surface characteristics of a legislative enactment. Such a law must also be capable of operating, in practice, in a law-like manner, meaning, in a manner that places the subject in a particular quality of position vis-à-vis the source of power from which the putative law emanates. The idea of law at play here, therefore, is that which I suggested in chapter one is the nub of Fuller's jurisprudence, namely, one that ensures a meaningful limitation on lawgiving power in favour of the agency of the legal subject.

That said, it is also the case that, in his reply to Hart on Nazi law, Fuller never conveys any sense that a positivist, source-based standard for legal validity ought to be rejected outright. Rather, his position belies a presumptive acceptance of a

[110] 'Positivism and Fidelity' (n 1) 660.

[111] Kristen Rundle, 'The Impossibility of an Exterminatory Legality: Law and the Holocaust' (2009) 59 *University of Toronto Law Journal* 65, 84.

positivist test of validity *unless* evidence of formal pathology becomes so flagrant (or, put differently, unless the interaction between the form of law and the agency of the subject becomes so debased) as to disavow law's distinctiveness *as* law. The question, then, is this: is this two-step analysis capable of being translated into a theoretically and practically workable standard for legal validity?

While it is not my purpose here to defend any such standard outright, or to argue that Fuller's jurisprudence clearly yields it, the gesture itself deserves to be taken seriously, as well as situated within extant debates about whether a positivist standard for legal validity is adequate to all cases where the question of the status of law comes up for determination. To this end, it is helpful to return to Fuller's affinities with Radbruch, because there is a good case for interpreting the structure of Fuller's gesture along similar lines to the Radbruch formula. That is, Fuller's gesture might be read as offering a standard for legal validity that is grounded in a positivist, source-based test, and which accepts this test as appropriate to the majority of cases, but which then insists that the formal health of the putative law, rather than its substantive justice, should be the arbiter of how far that acceptance ought to extend. In short, we can accept as valid law that which a source-based factual test for legal validity pronounces to be such, until and unless a point is reached when the legal order from which that law emanates is so plagued by formal pathologies that its output, exemplified by the particular law whose validity stands for determination, should be denied that status.

The real-world possibilities for such a standard might not be as far-fetched as may first appear. For guidance on this, it is helpful to return to the Nazi grudge informer cases, but this time to the research of HO Pappe, which sought to clarify the many confusions of fact that afflict Hart's analysis of those cases in the *Harvard Law Review*.[112] As Pappe explains, the case referred to by Hart was indeed problematic, and was criticised severely at the time, but the basis of its reasoning was quite different to that reported by Hart: the woman's acts were determined to be unlawful above all because she was under no legal duty to inform, and so must have realised that her action in reporting her husband was contrary to the sound conscience and sense of justice of all decent human beings. The point, then, is that the court in that case did not deny the legal validity of the relevant Nazi statutes by application of the Radbruch formula, even if it did acknowledge that the Radbruch formula might apply if the statute in question was one that effected extreme injustice. There were, however, cases with similar facts that were decided at a higher level of the German legal system, but (again contrary to Hart's understanding)[113] in accordance with a different line of reasoning; namely, that in order to be declared as such, law must evidence qualities that go beyond questions of valid legal enactment or justice of content.[114] Legal effect could therefore be removed

[112] HO Pappe, 'On the Validity of Judicial Decisions in the Nazi Era' (1960) 23 *Modern Law Review* 260, 263. Pappe explains that Hart's error flowed from how he evidently relied on a misreport of the case in an earlier article in the *Harvard Law Review*. See further David Dyzenhaus, 'The Grudge Informer Case Revisited' (2008) 83 *New York University Law Review* 1000.

[113] Hart does, however, acknowledge Pappe's research in *The Concept of Law* (n 16) 208, note 1.

[114] Pappe, 'Validity of Judicial Decisions' (n 112) 263–67.

from decisions made under the statutes relied upon by the grudge informers if the legal process within which they were made was so plagued by judicial and administrative abuses, contrary to principles of German legality, as not to deserve recognition as valid law.

Thus, quite apart from how it clarifies the facts, Pappe's research also appears to vindicate the general orientation of Fuller's reasoning in his reply to Hart, which is directed to precisely such considerations of formal health and standards of 'legality'. But this does not mean that any such standard for legal validity that might be developed out of that reasoning is without its obstacles. Certainly, the real-world operationalisation of such a standard would require a much more concrete reference point for determining when the presumptive positivist source-based test ought to run out than such ideas as 'when formal health is grossly debased', or 'when form and agency are no longer in proper interaction' or, only slightly less abstractedly, 'when the principles of the rule of law are not appropriately observed'. One remedy here would likely be to make much more of the matter of procedures than Fuller did or, at least, than he did in his exchanges on questions of mainstream jurisprudence as opposed to in his writings about forms of ordering, especially adjudication, that I discussed in chapter two. As I explained there, procedures offer a much more tangible site on which to tease out the concerns of Fuller's thinking on formal legal pathology and its implications for the assumed status of the legal subject as an agent than do the more abstract standards that animate his jurisprudence. Again, it is not my aim to fully develop these questions here. But it is nonetheless important to note that, at least in one setting, an approach to legal validity that was mindful of the importance of the formal health of law to its intelligibility as law was regarded not only as legally coherent in the face of oppressive laws, but more so than the Radbruch formula.

On this point, however, it is worth pondering at a philosophical level whether any such standard for legal validity that might be developed out of Fuller's jurisprudence would really be so different to Radbruch's at all. That is, it is not entirely clear whether the Radbruch formula, and the primary meaning of 'equality before the law' that it gives to the threshold of 'justice' that qualifies an otherwise standing commitment to a positivist test for legal validity, ultimately speaks as much to an idea of formal justice as it does substantive justice. Radbruch's formula is famously underspecified and, indeed, Fuller himself expressed confusion in his private correspondence about precisely what it amounts to, though noting that whatever its ultimate orientation, Radbruch 'was talking about the Nazi situation, when "laws" were not only directed to evil ends, but drafted, interpreted, applied, and disregarded by government itself in a way that largely destroyed the meaning of law itself'.[115]

It might be said, then, that if Radbruch's formula for legal validity can be summed up by the shorthand 'positivism with a minus sign',[116] so too might a

[115] Letter from Fuller to Professor Robert S Summers, 15 January 1964, The Papers of Lon L Fuller, Harvard Law School Library, Box 7, Folder 16 (correspondence).

[116] See David Dyzenhaus, 'The Juristic Force of Injustice', in David Dyzenhaus and Mayo Moran (eds), *Calling Power to Account* (Toronto, University of Toronto Press, 2005) 280.

Fullerian standard that presumes the applicability of a positivist standard unless and until such ought to be displaced in the face of a sufficiently serious debasement of the relevant law's formal health. But even in the hints we receive from his analysis of Nazi law, it is likely that Fuller's standard would need to be framed in more demanding terms (a 'positivism plus form', for instance) because it is clear that, in his jurisprudence, law's formal health is a matter that should *always* be kept in the picture, and not just raised as a concern in the extreme case.

The problem with such labels, however, or at least from the perspective of reclaiming the insights of Fuller's jurisprudence, lies in how they retain the positivist position as their starting point in a way that might suggest a jurisprudential line in the sand beyond which all other kinds of concerns or questions are optional, rather than core, to inquiry into the nature of law and any relationship between that inquiry and the question of legal validity. Certainly, as already emphasised, there is nothing in Fuller's jurisprudence that disavows the good sense of a positivist standard for legal validity in the great majority of cases. Nor does he ever suggest that the value of legal certainty evidently secured by the positivist formula for legal validity is not a fundamental one, nor, indeed, a core standard against which to test the extent to which a given legal theory speaks meaningfully to the question of what makes law law-like. But it is a giant leap to go from these basic expressions of support for legal certainty and how it is honoured by positivism to the suggestion that all of the implications that follow from the positivist position should be accepted as a necessary part of a standard for legal validity.

This, indeed, is the nub of Fuller's complaint against Hart in his reply on Nazi law. He clearly does seek to distance his own jurisprudential position from the blunter edges of positivism on the questions that he considers central to inquiry in the field; namely, how to explain fidelity to law, how to conceptualise the problem of formal legal pathology, and how to appraise the status of laws that significantly oppress or confuse the expression of agency on the part of those subject to them. It might well be, therefore, that if we are to take his gestures about form, agency, and its implications for legal validity seriously, we should follow Neil MacCormick and suggest that Fuller's contribution to the question of legal validity belongs to a 'post-positivist' space:[117] a space, that is, where there is nothing philosophically aberrant or unworkable about introducing standards of evaluation when standards of fact become incapable of honouring a meaningful idea of law. Such would surely challenge the prevailing positivist standard for legal validity that has been neatly captured in John Gardner's formulation of 'LP*'; namely, that in any legal system whether a given norm is legally valid 'depends on its sources, not its merits (where its merits, in the relevant sense, include the merits of its sources)'.[118] The content of that challenge would simply be to ask whether

[117] Neil MacCormick, *Institutions of Law: An Essay in Legal Theory* (Oxford, Oxford University Press, 2007) 278–79.

[118] John Gardner, 'Legal Positivism: 5½ Myths' (2001) 46 *American Journal of Jurisprudence* 199, 201.

more than Gardner thinks needs to be said about the merits of the source of law in fact does need to be said, from the perspective of law's formal health.[119]

These questions are all larger than the concerns of the present chapter and, indeed, of the present project. But, as I suggested at the outset, to reclaim Fuller's jurisprudence through his core concern for the connections between the form of law and human agency necessarily invites at least their candid exploration. Moreover, it is clearly the case that the validity question is salient to the way that Fuller's contest with positivism returns repeatedly to his intuition that any explanation of law's normativity, framed in his 1958 essay in terms of the question of 'fidelity to law' and the 'external morality' of law's 'acceptance', must link intelligibly to an account of the conditions for law's existence. As Waldron has so nicely captured, Fuller writings point us to the idea that fidelity to law must be at least in part explained by what law *is*, not just what it is used for; an intuition that finds particularly strong expression in the Nazi law debate. So, even if a discussion about the conditions for law's 'existence' might be much more comfortable territory on which to tease out the implications of what Fuller is getting at here, as opposed to contests about standards for 'validity', or for determining the 'grounds of law', there is still much we can learn from a closer appraisal of Fuller's response to Hart on the matter of Nazi law. At the very least, this most famous site of the Hart-Fuller debate, especially when explored with a view to the matter of standards for legal validity, offers a further prism through which to see the ways in which Fuller's escapades with positivism illuminate the boundaries that separate, as well as bridge, his and Hart's projects.

IV Conclusion

The consistent message of the 1958 debate, read from Fuller's perspective, is that Hart's defence of the necessary separability of law and morality, and his account of the nature of law that underscores that defence, simply needs to address more questions than Hart himself seems concerned to acknowledge. Among these questions is that of what it means to anoint law with the status of legality, and whether a deeper appreciation of what is at stake in this question ought to affect our philosophical deliberations about the conditions of law's existence.

Although he does not provide an extended account of what precisely he means by this idea in the 1958 essay, the most crucial arguments that Fuller offers by way of his own response to this question flow from the notion that law contains its own internal morality that is independent from the moral quality of any ends that might be pursued through law. From the errors of his unnamed and ill-willed

[119] Should such a challenge be pursued, it would also necessarily challenge Gardner's view that Fuller was 'not talking about the conditions of legal validity at all, and so is not engaging with (LP*)': ibid 208, note 20.

monarch, we learn that the enterprise of lawgiving may lapse in the face of a flagrant failure in congruence between official action and declared rule, or through a failure to achieve clarity in the articulation of the law. Then, when we turn to his analysis of Nazi law, we also learn that secret laws, ignorance of the letter of the law in favour of arbitrary acts of terror, illogical interpretive practices, and uncontrolled administrative discretion are equally pathological to the internal morality that, he argues, is what lends law its law-like quality.

For clearer insight into *why* these apparent conditions that make law law-like are thought by Fuller to be moral in character, we need to turn to his book, *The Morality of Law*, which is the subject of the next chapter. But even in lieu of that elaboration, we can readily see how Fuller's 1958 essay reveals the coming together of two enduring trajectories of concern within his jurisprudence: his interest in the forms through which law finds expression, and his commitment to a purposive conception of law and the connection he makes between such a conception and the phenomenon of fidelity to law. Moreover, what is striking about the 1958 essay, again consistent with Fuller's jurisprudential approach generally, is how these themes are explored by engaging the perspective of all of law's participants: the lawgiver (the hypothetical monarch), the judge (in the interpretation analysis), and the subject (Nazi law's conscientious citizen).

It seems, then, that there were many reasons why Fuller was justified in thinking that the occasion of Hart's Holmes lecture marked a pivotal moment for a range of pivotal new conversations in jurisprudence; conversations, perhaps, of the kind he had anticipated in his letter of support for Hart's visit to Harvard. For this envisaged promise to have been fulfilled, however, a meaningful rejoinder by Hart to Fuller's *Harvard Law Review* essay was needed. Such, unfortunately, was not forthcoming: Fuller received only a few lines of reply in Hart's seminal work, *The Concept of Law.*

As the next two chapters will make apparent, the exchanges between Fuller and Hart that followed the *Harvard Law Review* essays were ultimately pared down, by Hart, to a much narrower set of questions. Among these, the primary one was that of whether observance of the principles of the internal morality of law is compatible with the pursuit of iniquitous ends, and whether the value of those principles is thus reducible to the morally neutral status of mere efficacy. But I hope that I have shown in this chapter that there were many other directions, and other, arguably richer, questions, towards which the two scholars' subsequent exchanges might have developed. Indeed, from the perspective of those interested in the questions raised by Fuller in 1958, many of which remain, at best, under-explored, the field of legal philosophy is largely yet to recover from the narrowing of focus that Hart's responses to Fuller ultimately precipitated, and which saw these other trajectories of inquiry obscured. My aim in the next two chapters, therefore, is to illuminate the causes of this with a view not only to reclaiming some of Fuller's lost insights, but also to exploring how we might ensure that future conversations do not suffer the same unfortunate limitations.

4

The Morality of Law

The Morality of Law is the product of the 'welcome spur' of an invitation to Fuller to deliver a series of four lectures at Yale Law School in April 1963.[1] The chapters of the book, each having its own fairly distinct sphere of concern, stand as a record of those lectures with apparently few changes. It was Fuller's personal view that the ideas expressed in *The Morality of Law* represented considerable 'progress toward clarity' in his articulation of his position.[2] In some ways this seems a sound assessment. Still, especially in his effort to offer a foundation for the book in the idea that we can distinguish between the morality of 'duty', concerned with the basic rules without which an ordered society is impossible, and the morality of 'aspiration', concerned with the qualities of human potential and striving, Fuller admitted that some of his claims were, at best, a little obscure. Indeed, he registered this dissatisfaction in a letter to Hart, following his receipt of Hart's review of *The Morality of Law*, agreeing with Hart that his analysis of the moralities of duty and aspiration was 'full of open ends', but had he tried to trace out all the relationships implied by it, he 'would never have got past the first lecture'.[3]

The more fruitful place to begin an acquaintance with *The Morality of Law* is Chapter 2: Fuller's famous attempt, through the tale of a hypothetical legislating monarch, King Rex, to explore what it means to fail to make law. Chapter 2 is the right place to start for a number of reasons. First, akin to what we saw in his analysis of Nazi law in the 1958 debate, Chapter 2 sees Fuller head into his claims and analysis on the 'front foot': that is, rather than being on the defensive, he is here squarely in the territory of his core jurisprudential intuitions.

Accordingly, my method for structuring the discussion of the present chapter is as follows. I begin in section I with a detailed exposition of the allegory of King Rex, and the lessons that Fuller seeks to extrapolate from Rex's eight failures to make law. I then turn to Fuller's attempt in Chapter 3 of *The Morality of Law* to situate his claims about law's internal morality within the extant landscape of legal philosophy, including with respect to Hart's *The Concept of Law*, which had been published in the intervening period. I then move from these foundations to the site of analysis in Chapter 4 of the book that has, until relatively recently, been

[1] Lon L Fuller, *The Morality of Law* (New Haven, Yale University Press, 1969) vii.

[2] Letter from Fuller to Walter F Berns, 21 September 1964, The Papers of Lon L Fuller, Harvard Law School Library, Box 11, Folder 17 ('The Morality of Law').

[3] Letter from Fuller to HLA Hart, 3 February 1965, The Papers of Lon L Fuller, Harvard Law School Library, Box 3, Folder 14 (correspondence).

largely neglected in commentaries on Fuller's jurisprudence. But it is here, in defence of his choice to designate his principles of the internal morality of law as a 'morality', that Fuller introduces the idea that implicit in the very possibility of law is a conception of the person as a responsible agent.

My aim in section I of this chapter is thus to show the keys ways in which *The Morality of Law* provides us with crucial insight into the parameters of Fuller's jurisprudence in its own right, but while also positioning itself as a continuation of his debate with Hart. With respect to the latter, Fuller was certainly on the 'back foot': the invitations to conversation that he presented in his 1958 reply to Hart were barely taken up by Hart in *The Concept of Law*, occupying no more than a handful of lines. But it is precisely this bare response that causes Hart's book to loom large over *The Morality of Law*, because Fuller, it seems, simply refused to let Hart close the conversation so dismissively—or, at least, that was his intention.

To a certain extent, his efforts in this vein bore fruit: Hart published a review of *The Morality of Law* in the 1965 *Harvard Law Review*. But this response ultimately turned out to be double-edged for Fuller. Yes, he finally received the engagement from Hart that he had so energetically invited. But that engagement—specifically, Hart's response that the principles of the internal morality of law are merely morally neutral aids to the efficacy of law—basically had the effect of settling the dominant interpretation of Fuller's jurisprudence and its widespread evaluation as an unsuccessful rejoinder to positivism on the matter of the necessary separability of law and morality.

My task in section II, therefore, is to offer a close analysis of Hart's review and its implications for how the scholarly memory of Fuller's contribution to jurisprudence has been constructed and largely maintained to this point. To aid this inquiry, two sites of additional material are drawn upon. First, correspondence between Fuller and Hart is examined, alongside several of Fuller's private working papers, to highlight a problem of misinterpretation, or more accurately, incomplete quotation of his claims on Hart's part, that greatly distressed Fuller in how it suggested that he, too, saw the value of the internal morality of law as reducible to efficacy. Second, gesturing to the response he comes to develop in his final reply to his critics in 1969, I also turn to a little-known essay of Fuller's from 1968 which offers illuminating insight into how he appears to have been thinking through ways to defend his position against that 'efficacy' reading.

I then turn in section III to take this objective of unsettling the scholarly memory of Fuller's contribution a step further. This time, however, I head in the direction of speculation about how Fuller's jurisprudence might have developed but for Hart's reply and the effect it had on narrowing the terms of the debate between them. Here I draw the foundations of my analysis from private correspondence that belongs to the intervening period between the publication of *The Morality of Law* and Hart's review, and which offers tantalising insight into some of the under-explored themes of *The Morality of Law*. My aim in this section, akin to what I pursued in the final section of chapter three in my gestures to a Fullerian standard of legal validity, is to present an analysis that is exploratory and suggestive only. Again, Fuller himself did not provide us with enough to go much further than this, or at least not exclusively in his name. Still, and not least because of how this correspondence engages with the matter of the conception of the legal subject as an agent that Fuller argues in *The Morality of Law* is implicit in the very possibility of a legal order, his brief foray into the idea that slavery is inherently incompatible with law is, at least, an intriguing proposition that invites further thought.

I Mapping *The Morality of Law*

A The Story of King Rex

Fuller's story of the well-intentioned but ultimately hapless monarch who fails in eight ways to make law follows closely from the brief tale of the lawmaking monarch in his 1958 reply to Hart. But in Rex's case, the lessons we are invited to take are spelled out much more carefully and systematically than the more intuitive remarks of the 1958 essay. This is immediately apparent from how Fuller sets up Rex's story in a manner that is importantly suggestive of the arguments that he seeks to convey through the tale.

We are told that Rex, conscious of the fact that the greatest failure of his royal predecessors had been in the field of law, came to the throne with the zeal of a legal reformer and resolved to make his name in history as a great lawgiver.[4] Rex therefore began his reign not just with good intentions, but distinctly legal ones, keen to understand what it takes to excel at the enterprise of lawgiving. Unfortunately, however, his pursuit of this goal was plagued with difficulties from the start. Rex struggled to achieve appropriate generality in his rules and to keep his decisions with respect to those rules consistent.[5] Some of his laws were kept secret while others were applied retroactively.[6] His skills of legislative draftsmanship suffered too greatly from obscurity,[7] his laws perpetuated confusion and then were changed too frequently.[8] Then, in his capacity as an adjudicator, he rendered judicial opinions that bore little relation to the enacted laws on which they were allegedly based.[9] Deeply disillusioned, and facing revolt from his subjects, Rex died a miserable King. His efforts, however, were not entirely futile. Learning from Rex's mistakes, the first act of his successor, Rex II, was to take the powers of government away from the lawyers and to place them in the hands of psychiatrists

[4] *Morality of Law* (n 1) 33–34.
[5] ibid 34.
[6] ibid 35.
[7] ibid 36.
[8] ibid 37.
[9] ibid 38.

and experts in public relations, so that his people might be made happy without rules.[10]

What are we invited to take from this playful tale? By presenting us with an account of legal pathology and failure, of how *not* to create and maintain a legal system, the story of Rex is designed to make us think about what it might take to replace such failure with success. As Fuller enumerates them, the pathologies that rendered Rex's plight so hopeless were (1) a failure to achieve general rules at all; (2) a failure to publicise or make available the rules that citizens are required to observe; (3) the abuse of retroactive legislation; (4) a failure to make rules understandable; (5) the enactment of contradictory rules; (6) rules that require conduct beyond the powers of the affected party; (7) subjecting the rules to too-frequent change; and (8) a failure of congruence between the rules announced and their actual administration.[11] Corresponding to these failures, Fuller suggests, are 'eight kinds of excellence towards which a system of rules may strive', and which together comprise the 'internal morality of law'.[12] A total failure to meet any one of these principles does not simply result in a bad system of law, but in something 'that is not properly called a legal system at all'.[13]

Echoes of this last claim will be recalled from Fuller's 1958 reply to Hart, and I took space in chapter three to explore how such might be read as a claim not only about what constitutes a meaningful idea of law but perhaps also a standard for legal validity. But in Chapter 2 of *The Morality of Law*, the point Fuller seeks to make concerns the connections between observance of the standards of the internal morality of law and the existence of a legal system, and the legal subject's obligation of obedience. As he argues it, 'there can be no rational ground for asserting that a man can have a moral obligation to obey a legal rule that does not exist, or is kept secret from him, or that came into existence only after he had acted, or was unintelligible, or was contradicted by another rule of the same system, or commanded the impossible, or changed every minute'.[14] This release from moral obligation does not arise because it is necessarily impossible to obey a rule that is then disregarded by those charged with its administration, but rather, Fuller argues, because 'at some point obedience becomes futile—as futile, in fact, as casting a vote that will never be counted'.[15]

Fuller's reasoning here is undoubtedly a little obscure. But the argument he is trying to develop becomes clearer when he then proceeds to situate his point within the context of a further claim: that the legal subject's moral obligation to obey law only arises in the first place in response to, or in anticipation of, the lawgiver's corresponding effort to create and maintain a workable legal order within which she might be able to live her life. As Fuller explains it, to secure the legal subject's fidelity to law, a lawgiver must enter into a relationship of 'reciprocity'

[10] ibid.
[11] ibid 39.
[12] ibid 41.
[13] ibid 39.
[14] ibid.
[15] ibid.

with her. But if this bond of reciprocity is finally and completely ruptured, then 'nothing is left on which to ground the citizen's duty' to observe the lawgiver's rules.[16]

The precise content of this bond between lawgiver and subject is not specified, and neither are the kinds of circumstances that might lead to its rupture. But we are told that the respect conveyed by the legal subject to the lawgiver's authority is not only essential for the existence of law, but must not be confused with a 'mere respect for constituted authority': after all, Rex's subjects remained faithful to *him* even while they were not faithful to his law (because, in any event, Rex never actually made any law).[17] The point being made, then, is that fidelity to law is something qualitatively different to deference to authority. Even if wielded by a human person, the possession of lawgiving power specifically is still something to be distinguished from the possession of power generally. Indeed, Fuller suggests that the possession of lawgiving power, discharged through observance of his eight principles of the internal morality of law, concerns 'a relationship with persons generally', and so is something that demands more than mere forbearance on the part of a lawgiver.[18] It instead requires the direction of human energies towards 'a specific kind of achievement': one that appeals 'to a sense of trusteeship and to the pride of the craftsman'.[19]

Fuller moves from this framework of jurisprudential claims to tell us more about the eight principles themselves. He begins with the requirement of law's generality, observing that this 'obvious' and fundamental feature of law in fact receives a 'very inadequate treatment in the literature of jurisprudence'.[20] But Fuller himself also treats generality fairly sparingly in *The Morality of Law*, explaining simply that this requirement stands for the idea that there be general rules of general, rather than particular, application. Generality does, however, later become an important site of reflection for Fuller when he seeks to clarify the impasse between him and his critics that develops with respect to his claims about why law is internally moral.

As for the demand of promulgation, or publicity, Fuller suggests that this is similarly fundamental to a meaningful idea of law, even if it is not possible to generally specify just how much information about the law needs to be conveyed to the citizen, or in what form. But the bottom line is that the law must be readily available to her: she is entitled to know its content, and to criticise it.[21] The desideratum of clarity then stands as recognition of how an obscure or incoherent law can greatly impair the attainment of a condition of legal, as opposed to some other kind of order, whether that obscurity is committed by legislators or judges, though it is notable here that Fuller especially emphasises the responsibility of legislators to make their objectives sufficiently clear so as to minimise the level of

[16] ibid 40.
[17] ibid 40, 41.
[18] ibid 42.
[19] ibid 42, 43.
[20] ibid 46, 48.
[21] ibid 51.

interpretation needed on the part of the courts and administrative tribunals charged with the task of putting the law into effect.[22] The requirement of non-contradiction is portrayed as speaking primarily to the problem of poor draftsmanship, and how this can render the law unable to be followed,[23] while the gist of the requirement of avoiding impossibility is similarly concerned with how a rule that demands what is impossible contradicts 'the basic purpose of a legal order'.[24] Then, the sixth requirement of constancy through time is one that expresses recognition of how too-frequent change in the law can amount to something akin to retroactivity, in so far as citizens are unable to know, or at least are impaired in being able to know, precisely which laws apply to them at a given point in time.[25]

A much more extended treatment is given to the remaining two principles of non-retroactivity and congruence between official action and declared rule. Fuller explains that the reason why non-retroactivity is such a complex demand is because its total prohibition may not necessarily always serve legality. A retroactive statute can, in certain instances, be an important curative measure in repairing the 'various kinds of shipwreck' in which a legal system might find itself.[26] The complexity of the requirement of non-retroactivity thus arises from how and when to know the difference; that is, how to know when its breach represents a tolerable sacrifice of legality, as opposed to when such would be an abuse of the feature of prospectivity that otherwise makes sense of the enterprise of governing through rules.

But it is the eighth principle of congruence between official action and declared rule that Fuller suggests is the most complex of all the desiderata that make up the internal morality of the law. This is because the principle speaks to the relationship between law and its administration; a relationship that may be destroyed, or impaired, in ways ranging from mistaken interpretation, to a lack of insight about what is required to maintain the integrity of a legal system, to simple prejudice.[27] The message, above all, is about how the principle of congruence requires an appreciation, on the part of the relevant legal and administrative actors, of the purposes that legal order itself is intended to fulfil. This point is stated especially succinctly when Fuller explains why, in his view, the 'most subtle element' in the task of maintaining congruence between law and official action lies in 'the problem of interpretation'. As he elaborates it, 'legality' requires that judges and other law-applying officials apply statutory law 'not according to their fancy or with crabbed literalness, but in accordance with principles of interpretation that are appropriate to their position in the whole legal order'.[28]

The eight standards of Fuller's internal morality of law, each ostensibly neatly self-contained and presiding over its own distinct domain, do tend to read like

[22] ibid 64.
[23] ibid 65–70.
[24] ibid 79.
[25] ibid 80.
[26] ibid 53–54.
[27] ibid 81, 82.
[28] ibid 82. See further chapter seven, IV 'Fuller, Dworkin and interpretation'.

something of a checklist for how to create and maintain a legal order. If the eight criteria are met, law exists and is being done the right way. This might seem right—indeed, it has been the dominant reading of Fuller's account for decades. Still, reading Fuller in this narrow way has led to considerable misunderstandings about, or at least unduly narrow interpretations of, his jurisprudence.[29]

Thus, as I foreshadowed in chapter one, the meaning of the internal morality of law in fact needs to be understood in much more capacious and nuanced terms than the 'checklist' reading tends to invite. The key to the correct reading of the internal morality of law lay precisely in what is hinted at in Fuller's suggestion that certain instances of retroactivity might be permissible if they serve the cause of legality. When Fuller speaks of law and its existence, he has in view a particular quality of relationship between the lawgiver and the legal subject, one that is *reflected* in the observance of his eight principles but which is not *exhausted* by them. Certainly, over the course of his writings, Fuller comes to suggest that some of these principles, especially the principles of generality and congruence, occupy a particularly central place within his understanding of what it means to have and to sustain a condition of legality.[30] But if we are to gain a deeper appreciation of the idea of legality of which he speaks, and which the eight principles are designed to instantiate, operationalise and serve, we need to move towards the more capacious understanding of law's 'form' that I sketched in chapter one; that is, a conception of the form of law that is inclusive of the conception of the person as a responsible agent that Fuller argues is implicit in the internal morality of law. Corresponding to this, we need equally to think more deeply about Fuller's repeated references to the relationship of 'reciprocity' that a legal system constitutes, and which signals the equal presence and responsibilities of lawgiver and legal subject alike, as well as his gestures to how the enterprise of lawgiving implicates a 'sense of trusteeship' and a 'relationship with persons'. I turn to each of these points in more detail below.

B Situating Fuller's Claims

Fuller's task in Chapter 3 of *The Morality of Law* is to put his analysis of the internal morality of law into 'its proper relation with prevailing theories of and about law'.[31] He begins with the natural law tradition, and suggests that his model of eight principles might be read as an attempt to discern and articulate the 'natural laws' of a particular kind of human undertaking, namely, 'the enterprise of subjecting human conduct to the governance of rules'.[32] This is a different kind of

[29] There are multiple examples, but an instructive one is the exchange between Edward Rubin and Peter Strauss in (1989) 89 *Columbia Law Review* 369 and 467: see my own analysis of this exchange in Kristen Rundle, 'The Impossibility of an Exterminatory Legality: Law and the Holocaust' (2009) 59 *University of Toronto Law Journal* 65, 108, note 152.

[30] See further chapter five.

[31] *Morality of Law* (n 1) 95.

[32] ibid.

project to traditional natural law theories and their focus on the idea that positive law ought to be evaluated against some kind of higher moral law. As Fuller puts it, if any metaphor of elevation is appropriate, his eight principles should be called 'lower', rather than 'higher', laws, 'like the natural laws of carpentry, or at least those laws respected by a carpenter who wants the house he builds to remain standing and serve the purpose of those who live in it'.[33] The internal morality of law is thus a kind of 'procedural' rather than 'substantive' natural law, concerned with the ways 'a system of rules for governing human conduct must be constructed and administered if it is to be efficacious and at the same time remain what it purports to be'.[34]

As I explained in chapter two, these comments are generally representative of Fuller's very qualified association with the natural law tradition, as the concerns of that tradition are typically understood. At most, Fuller here places himself broadly within the school that, he says, has historically shown some concern for the 'problems of social architecture' involved in creating and maintaining a legal order.[35] But that is about as much as we take from Chapter 3 of *The Morality of Law* with respect to Fuller's sense of his jurisprudence as a natural law project. The chapter is otherwise more squarely concerned with developing the implications of Fuller's claims about the internal morality of law for his debate with legal positivism.

There is undoubtedly a level of mischief intended by the title that Fuller gives to Chapter 3: 'The Concept of Law'. Hart, however, is not singled out immediately; indeed, he is 'passed over' until quite late in the chapter.[36] Fuller instead begins the project of situating his claims by grouping positivists and realists in the same mix of legal philosophies that see the essence of law in 'a pyramidal structure of state power', abstracted from 'the purposive activity necessary to create and maintain a system of legal rules',[37] and thus in contrast to his view of law as 'an activity', and a legal system as 'a product of a sustained purposive effort'.[38]

Fuller immediately identifies several objections that might be advanced against his view of law, especially how to speak of law as an enterprise is to imply that it may be carried on with varying degrees of success, and so also to suggest that the existence of a legal system, and, indeed, sometimes also of particular laws, will always be a matter of degree.[39] But he is strongly resistant to these objections: they simply illuminate the erroneous assumption, even if a dominant one within modern legal philosophy, that 'law is like a piece of inert matter—it is there or not there'.[40]

[33] ibid.
[34] ibid 97.
[35] ibid 35. See further chapter two, III 'Navigating the labels'.
[36] ibid 133.
[37] ibid 110, 106.
[38] ibid 106.
[39] ibid 122.
[40] ibid 123.

This is the background against which Fuller turns to Hart's *The Concept of Law*. The central thesis of *The Concept of Law*, well-known to students of jurisprudence as the backbone of modern positivism, is that legal order is the product of the union of two different types of rules: 'primary rules' that regulate human action by imposing duties, and 'secondary rules' that confer powers to enable the alteration of existing, or the introduction of new, primary rules.[41] The most important of these secondary rules is the 'rule of recognition' which, when supplemented by an 'internal point of view' on the part of legal officials that recognises this rule as establishing the relevant legal system's standard for legal validity, unifies that system by providing an identifiable site of legal authority.[42]

Fuller makes clear at the outset of his engagement that although he regards *The Concept of Law* as a 'contribution to the literature of jurisprudence such as we have not had in a long time', he is in 'virtually complete disagreement' with all aspects of Hart's book.[43] His primary objection relates to how the analysis of *The Concept of Law* proceeds in terms that systematically exclude any consideration of problems of the internal morality of law.[44] The main target of Fuller's criticism here is Hart's rule of recognition, specifically, how Hart's account of this rule seems to suggest that the power to recognise valid laws comes without limitation. As far as Fuller can see, it is as if Hart's account of the very foundations of legal order 'were designed to exclude the notion that there could be any rightful expectation on the part of the citizen that could be violated by the lawgiver'.[45]

These criticisms of the rule of recognition follow closely from the argument Fuller introduced in his 1958 reply to Hart that Hart's version of legal positivism fails to meaningfully accommodate the phenomenon of the abuse of law or, as I put it in chapter three, an idea of legal pathology (a term that Fuller himself here uses).[46] In *The Morality of Law*, however, we see Fuller suggest that this omission on Hart's part not only reflects an apparent disregard for the position of the legal subject and the reciprocities involved in maintaining legal order, but also contradicts Hart's stated aim of moving the theory of legal positivism away from its Austinian association with coercion and sanctions. As he puts it, 'if the rule of recognition means that anything called law by the accredited lawgiver counts as law, the plight of the citizen is in some ways worse than that of the gunman's victim'.[47] That is, if lawgiving power comes without any meaningful limitations in favour of the position of the subject, how is the concept of law that animates Hart's positivism ultimately any different from the idea of law as a command?

It is a pity that Fuller did not extend this objection to the analysis in *The Concept of Law* where Hart explains that the internal point of view towards the rule of

[41] HLA Hart, *The Concept of Law* (Oxford, Oxford University Press, 1961) 81.
[42] *Morality of Law* (n 1) 115, 116.
[43] ibid 131, 133.
[44] ibid 133.
[45] ibid 140.
[46] ibid 157.
[47] ibid 139. See further Lon L Fuller, 'A Reply to Critics', *The Morality of Law* (New Haven, Yale University Press, 1969) 215–16.

recognition need only be held by the officials of the legal system. Such a society 'might be deplorably sheeplike', and the 'sheep might end in the slaughterhouse', Hart says, but so long as the legal officials adopt an internal point of view towards the rules, a legal system will exist and there is no basis for denying it that title.[48] Fuller, however, does not address this passage at any point in *The Morality of Law*, so we can only assume that its basic thrust receives implicit attention in his critique of the potential for abuse of power inherent in Hart's rule of recognition.

The Concept of Law does, however, reappear in a different setting and with different aims in view in Chapter 4 of *The Morality of Law*. There, Fuller begins his discussion of 'The Substantive Aims of Law' by referring to Hart's analysis of the relationship between law and morals, and specifically, within this, his analysis of whether the determination of legal validity is in any way connected to morality. It is this analysis that constitutes the sum total of Hart's engagement with Fuller's reply to him in the 1958 *Harvard Law Review*. As Hart explains it:

> Further aspects of this minimum form of justice which might well be called 'natural' emerge if we study what is in fact involved in any method of social control—rules of games as well as law—which consists primarily of general standards communicated to classes of persons, who are then expected to understand and conform to the rules without further official direction. If social control of this sort is to function, the rules must satisfy certain conditions: they must be intelligible and within the capacity of most to obey, and in general they must not be retrospective, though exceptionally they may be. This means that, for the most part, those who are eventually punished for breach of the rules will have had the ability and opportunity to obey. Plainly these features of control by rule are closely related to the requirements of justice which lawyers term principles of legality. Indeed, one critic of positivism has seen in these aspects of control by rules something amounting to a necessary connection between law and morality, and suggested that they be called 'the inner morality of law'. Again, if this is what the necessary connection between law and morality means, we may accept it. It is unfortunately compatible with great iniquity.[49]

The brevity and dismissiveness of this response naturally did not go unnoticed by Fuller, the unnamed 'critic of positivism' in question. In *The Morality of Law*, Fuller follows his quotation of Hart's point with the comment that 'one could not wish for a more explicit denial of any possible interaction between the internal and external moralities of law than that contained in this last sentence'.[50] Fuller's statement is surely correct, but the effect of Hart's concluding lines has been much greater than this, in terms of how debates about law and morality generally, and analyses of Fuller specifically, developed in the second half of the twentieth century. To see why, it is important to notice the apparent distinction that Hart introduces in the passage between what *might* or *could* be addressed by inquiry

[48] *Concept of Law* (n 41) 117.

[49] ibid 206–7. This passage also receives close attention from Nigel Simmonds, *Law as a Moral Idea* (Oxford, Oxford University Press, 2007) 70–78; Jeremy Waldron, 'Positivism and Legality: Hart's Equivocal Response to Fuller' (2008) 83 *New York University Law Review* 1135, 1145–46; and John Gardner, 'Hart on Legality, Justice, and Morality' (2010) 1 *Jurisprudence* 253, 254–57.

[50] *Morality of Law* (n 1) 154.

into whether there is a necessary connection between law and morality, and what Hart himself evidently thinks should be the subject of any such inquiry, namely, the relationship between observance of Fuller's eight principles and the moral character of legal *ends*. By suggesting that nothing in Fuller's internal morality of law guards against the pursuit of 'great iniquity' through law, the conclusion Hart invites us to take is that nothing salient for debates about the connection between law and morality can flow from jurisprudential inquiry into the features of 'control by rule'. Thus, nothing needs to be said about Fuller's jurisprudential project, at least as a contribution to debates about the necessary connections between law and morality, because those debates pivot on the question of the moral quality of legal ends. As far as Hart is concerned, therefore, the Hart-Fuller debate is over, if indeed it ever got off the ground.

This, indeed, is precisely what happened, and the history of how Fuller's jurisprudence was and largely continues to be received can basically be traced to these lines from *The Concept of Law*, consolidated later by Hart's response that Fuller's 'moral' principles of lawgiving merely make the pursuit of purposes through law more effective, whatever those purposes might be. But in Chapter 4 of *The Morality of Law*, Fuller responds by offering a diagnosis of what, from his perspective, is wrong with Hart's neat conclusion that observance of the principles of the internal morality of law (or what Hart terms 'principles of legality') is 'unfortunately incompatible with very great iniquity'.[51] He suggests that although we might accept that the internal morality of law is essentially 'procedural' in character, such that compliance with it may ostensibly lend efficacy to a wide variety of substantive aims, this need not also mean that '*any* substantive aim might be adopted without compromise of legality'.[52]

Fuller's point here is one about how the form of law may temper its substance; an idea that he suggests Hart is unable to engage with because of how positivists tend to regard the problem of achieving and maintaining legality as deserving 'no more than casual and passing consideration'.[53] The argument Fuller advances in support of this point is somewhat underdeveloped, but we gain insight into its basic thrust when he goes on to note the slender treatment given by Hart in *The Concept of Law* to the philosophical quandaries presented by Nazi legality; a conversation that Fuller is evidently not willing to relinquish.[54] Other instances of oppressive law, he argues, such as South African apartheid laws, seem to point us to the same intuition that he defended in his 1958 reply to Hart; namely, that when pursued through law, oppressive aims tend to be accompanied by a deterioration in the standards of the internal morality of law.[55]

[51] ibid 154.
[52] ibid 153 (Fuller's emphasis).
[53] ibid 154–55.
[54] ibid 155
[55] ibid 160.

C Conception of the Person Implicit in Legality

The analysis just described belongs to Chapter 4 of *The Morality of Law*, a site that for a long time was mostly neglected in scholarly commentary on Fuller's jurisprudence, critical and sympathetic alike. Hart's review of *The Morality of Law*, for instance, makes no meaningful reference to the concerns of the chapter beyond the comment that it contains some interesting observations on 'problems of institutional design'.[56] Ronald Dworkin's review, which I examine in chapter seven, at least contains some passing references to the claims of Chapter 4, but ultimately squares with Hart's in its conclusion that none of those claims conflict with 'the classic or prototypical positivist position that law and morals are conceptually distinct'.[57]

This practice on the part of Fuller's interpreters of basically bypassing the analysis of Chapter 4 of *The Morality of Law* is surely curious, given that the stated aim of the chapter is to defend why the internal morality of law deserves the designation of a 'morality', and thus, by extension, to further defend the case for why this morality threatens the positivist separability thesis. The thrust of this defence, which follows closely from Fuller's claims about how the pursuit of oppressive aims through law tends to be accompanied by a deterioration in the standards of the internal morality of law, is contained in a passage where Fuller elaborates the claim that implicit in the internal morality of law is a conception of the person as a responsible agent:

> I have repeatedly observed that legal morality can be said to be neutral over a wide range of ethical issues. It cannot be neutral in its view of man himself. To embark on the enterprise of subjecting human conduct to the governance of rules involves of necessity a commitment to the view that man is, or can become, a responsible agent, capable of understanding and following rules, and answerable for his defaults.
>
> Every departure from the principles of law's inner morality is an affront to man's dignity as a responsible agent. To judge his actions by unpublished or retrospective laws, or to order him to do an act that is impossible, is to convey to him your indifference to his powers of self-determination. Conversely, when the view is accepted that man is incapable of responsible action, legal morality loses its reason for being. To judge his actions by unpublished or retrospective laws is no longer an affront, for there is nothing left to affront—indeed, even the verb 'to judge' becomes itself incongruous in this context; we no longer judge a man, we act upon him.[58]

[56] HLA Hart, 'Lon L. Fuller: *The Morality of Law*' (1965) 78 *Harvard Law Review* 1281, reprinted in HLA Hart, *Essays in Jurisprudence and Philosophy* (Oxford, Clarendon Press, 1983) 363.

[57] Ronald Dworkin, 'Philosophy, Morality and Law: Observations Prompted by Professor Fuller's Novel Claim' (1965) 113 *University of Pennsylvania Law Review* 668, 671, 673.

[58] ibid 162–63. This passage has recently been brought into focus by scholars sympathetic to Fuller, especially those interested in what it suggests about the connections between Fuller's jurisprudence and an idea of freedom. See especially Colleen Murphy, 'Lon Fuller and the Moral Value of the Rule of Law' (2005) 24 *Law and Philosophy* 239; Jennifer Nadler, 'Hart, Fuller, and the Connection Between Law and Justice' (2007) 27 *Law and Philosophy* 1; Evan Fox-Decent, 'Is the Rule of Law Really Indifferent to Human Rights?' (2008) 27 *Law and Philosophy* 533; as well as Mary Liston, 'Willis, "Theology", and the Rule of Law' (2005) 55 *University of Toronto Law Journal* 767.

The statement of the first paragraph of this passage is certainly boldly put. The subject's capacity for agency is presented to us as part of the very form of law as an enterprise of governance through general rules, constitutive, and not merely contingently connected, to it. The idea hinted at, therefore, is that the subject comes to law in this state, in the sense that her agency is presupposed by the form of law as a mode of governance through general rules. This is the basic message of the first paragraph. But more can still be learned, as is so often the case with Fuller, when we turn to what he has to say in the second part of the passage about the pathological case: that is, about what it means to depart from this presupposition by departing from the requirements of the internal morality of law. As he puts it, every departure from the principles of the internal morality of law is an affront to the legal subject's dignity as a responsible agent.

There is obviously an unfortunate level of hyperbole infecting this statement.[59] If 'every' departure from the principles of the internal morality of law is an affront to the subject's status as an agent, then to depart from those principles in any way would be to automatically disavow law's constitutive conditions and so bring about a condition of non-law. This, we know from both the 1958 essay and his analysis of the story of Rex earlier in *The Morality of Law*, is indeed Fuller's claim about the consequence of significant departures from his principles. But, as I earlier emphasised, it is equally Fuller's claim that certain derogations from the principles of the internal morality of law are permissible (such as occasional instances of retroactivity) *if* those departures serve the particular quality of lawgiver-legal subject relationship that a condition of legality constitutes and maintains.

The subtler aspects of this message reveal themselves in the passage under review when Fuller distinguishes the phenomenon of being subject to law from that of merely being 'acted upon'. Fuller seems to suggest that this shift from 'law' (where the subject is treated as a responsible agent) to 'acting upon' (where she is not) might manifest for one of two reasons. First, it might be because 'the view is accepted' that the subject is 'incapable of responsible action'; presumably, we might extrapolate, for reasons of childhood, or disability, or some other condition that goes to an understanding of legal capacity. Second, a more subtle idea that requires some unpacking, it might be because of what happens when a putative lawgiver interacts with a subject through the medium of formally debased law. Here, the suggestion Fuller makes is that to do this is to degrade, or even dissolve, the subject's status as an agent. As he puts it in the passage, to judge the subject by reference to such debased forms as retroactive or unpublished laws is not to have merely affronted the subject's status as a responsible agent, but to have basically negated it altogether. In such a circumstance, the subject has been rendered into something unintelligible to law, and so is merely acted upon.

Thus, when we read down the hyperbole of Fuller's opening statement that 'every' departure from the principles of the internal morality of law is an affront

[59] Also observed by Leslie Green, 'Law as a Means', in Peter Cane (ed), *The Hart-Fuller Debate in the Twenty-first Century* (Oxford, Hart Publishing, 2009) 182.

to the legal subject's status as an agent, we might restate the message of the passage along the following lines. Departures from the principles of the internal morality of law that maintain the relationship between lawgiving power and the legal subject's status as a responsible agent are permissible, but if either the formal features of law are abused, or the subject is for some reason considered not capable of responsible action, then what purports to be governance through law may slide into something that, in merely acting upon the subject rather than respecting her as an agent, no longer has the character of law.

Read in this way, we can see that the passage actually contains two closely related lines of argument. The dominant one is about the conception of the person implicit in the enterprise of subjecting human conduct to general rules; a message, in sum, about how law and the status of the subject as a responsible agent are symbiotic. But the second line of argument is about how to negate or otherwise fail to respect the subject's agency through (putative) law is not just to negate or fail to respect that subject's status as an agent, but also to fail to engage in the enterprise of law itself. Some other mode of rule or order might be extant in such circumstances, but it cannot be properly understood as the mode of law, because law, for Fuller, is nothing if not a formally distinctive mode of governance that interacts with its subjects in a manner that is qualitatively different than acting upon them. Legal power over subjects is something to be distinguished from mere power over objects, and the question of where a given legal order or relationship can be located on the continuum between these poles is one that goes to the very existence of law.

This appears to be the claim, but we still need more if we are to fully understand its content. In particular, if law's status as such is so critically connected to the status of the legal subject as an agent, what exactly constitutes this status? That is, *who* is Fuller's agent? In chapter one I suggested that, in Fuller's jurisprudence, an agent is a person 'capable of purposive action', in possession of her capacities, and who is to be regarded as an end in herself.[60] She is, as is made clear in the passage under review, more than someone simply capable of responding to a lawgiver's direction, whatever that direction might be. Rather, she is a bearer of 'dignity',[61] with 'powers of self-determination',[62] and so, we can extrapolate, a life to live of her own. But we should also take notice of how, in the passage under review, Fuller speaks of her as a 'responsible' agent; a qualifier that suggests a capacity not only to be answerable to the rules issued by legal officials, and to take responsibility for her conduct vis-à-vis those rules, but also to give regard to how that conduct affects the position of others in society. Fuller's legal subject, we might therefore say, is not just an individual possessed of choices, or a planner

[60] Lon L Fuller, 'Freedom: A Suggested Analysis' (1955) 68 *Harvard Law Review* 1305, 1307.

[61] *Morality of Law* (n 1) 162. See also Lon L Fuller, 'A Reply to Professors Cohen and Dworkin' (1965) 10 *Villanova Law Review* 655, 665.

[62] *Morality of Law* (n 1) 162. There is much in common here with the sense of who the legal subject is and what she possesses conveyed by John Finnis, *Natural Law and Natural Rights* (Oxford, Clarendon Press, 1980) 272–74; and Nigel Simmonds, *Law as a Moral Idea* (Oxford, Oxford University Press, 2007) 111.

with regard only to her own interests, but, akin to the Greek conception of the citizen, is envisaged as an active participant in the legal order.

A further question that is worth raising is how this agency that is possessed by and presupposed in the subject of a legal order becomes visible to us, and indeed, to her. I suggested in chapter two that one way in which this occurs is through her participation in law's procedures.[63] Procedures, though an under-explored site of jurisprudential inquiry, make the agent visible and agency itself tangible because of how they speak directly to that agent and supply conditions for her interaction with law. But perhaps the most important way in which the legal subject's agency becomes visible to us, and through which we also come to appreciate the centrality of her status as such to the possibility of law itself, lies in the clear message we receive from Fuller's jurisprudence about how the possibility of law depends upon its *acceptance* by the subject. In his 1958 reply to Hart, Fuller refers to this acceptance in terms of the 'external' morality which supports law. In *The Morality of Law*, however, he speaks of how law ceases to exist when the subject's fidelity to it is withdrawn, as she is fully entitled to do if the bond of reciprocity between her and the lawgiver 'ruptures' or 'breaks down'. The agent who animates Fuller's conception of the legal subject, therefore, is someone not just capable of following the rules laid down for her by a lawgiver, but someone capable of voluntarily and actively conveying the level of support towards the legal order that is necessary for it to exist.

We can now gain a deeper sense of what Fuller means when, elsewhere in *The Morality of Law*, he speaks of how the principles of the internal morality of law generate a 'relationship with persons generally' that is 'affirmative in nature',[64] and appeal to a 'sense of trusteeship'[65] on the part of a lawgiver. The notions of 'relationship' and 'trusteeship' referred to here are clearly not to be understood in terms of a simple paternalism. While some element of this undoubtedly comes into play—lawgivers have power over people and so they ought to use that power carefully, responsibly—to overstate it would be to contradict the intuition, expressed increasingly forcefully over the course of Fuller's writings, that law is something different to merely acting upon a person and, indeed, something that requires 'more than forbearance' on the part of a lawgiver.[66] It would equally contradict the instruction Fuller gives us, reflecting on the plight of Rex, that fidelity to law is not to be confused with mere deference to established authority. Instead, the responsibility to persons that Fuller argues is envisaged by, internal to, the internal morality of law, is a responsibility to *collaborate* with the legal subject in the creation and maintenance of law. This is the only available reading that is compatible with Fuller's account of the conception of the person implicit in legality.

[63] See chapter two, IIB 'Eunomics "writ small": the models'.
[64] *Morality of Law* (n 1) 42.
[65] ibid 43.
[66] ibid 42.

With these insights in view, it is now possible to return to the frame within which Fuller's claims in the passage are expressed; namely, the frame of a debate about the connections between law and morality, and specifically, one concerned to demonstrate manifestations of that connection that are overlooked in Hart's neat conclusion that the observance of the principles of the internal morality of law is 'unfortunately compatible with great iniquity'. Fuller's response to Hart, we can now see, is to suggest at least two other, interrelated ways in which law's connection to morality might be understood. The first is the moral dimension of law that consists in the *ethos* of the lawgiver's role and the distinctively moral burdens which that role carries. The second is the moral *value* that accrues to the legal subject, as a consequence of law's presupposed acceptance of and respect for her as a responsible agent. The clear suggestion is that these two moral dimensions are not contingently connected to law, in the sense of being that which only a healthy legal order achieves. The point, instead, is that these moral dimensions simply come with the territory of law itself, as an enterprise of governance conducted through the issuance and administration of general rules, irrespective of the specific ends, moral or immoral as they may be, that might be pursued through those rules.

This is the nub of Fuller's conversation-challenging contribution to debates in jurisprudence about the connections between law and morality. But without discounting their importance *as* contributions to those debates, the wider message about the constitutive conditions that make law itself possible must not be missed. These conditions, their distinctiveness *and* their moral dimensions, Fuller tells us, all flow from the particular quality of relationship that is sustained between a site of power and those subject to it through the way that the form of law presupposes and speaks to the status of the legal subject as an agent.

I will show in chapter five how Fuller continued to explore these connections in their own right, as well as part of his response to positivism on the necessary separability of law and morality. But in the 'Reply to Critics', the subject of chapter five, this is done with a very specific conversation in view; namely, a conversation about whether the principles of the internal morality of law are reducible to the value of efficacy, and thus have no intrinsic moral dimension. There are rich questions to be explored in that conversation, as indeed Fuller demonstrates in the 'Reply' and arguably still more so in his working notes for that response. But it is a much narrower and, indeed, a more defensive conversation than others that might have evolved out of the claims presented in *The Morality of Law*, and especially, I would argue, out of his claims about how law presupposes the status of the legal subject as a responsible, self-directed agent. I return to consider just one of the invitations that might otherwise have been taken up here in section III below.

II Hart's Review of *The Morality of Law*

A Mapping Hart's Response

Fuller wrote to Hart in December 1963 enclosing a copy of his soon to be published book, suggesting that since it 'takes positions relevant to those taken in *The Concept of Law*, and sometimes in explicit contradiction to them', Hart might like to have a preview of it. The tone of the letter is collegial and humble. Fuller apologises that he did not have a chance to discuss with Hart some of the questions that he raises, particularly with respect to the rule of recognition, aspects of which Fuller says ('in all candor') he found 'obscure'.[67] He thus invites Hart to 'set me (and your and my readers) straight' in the event that he has erred greatly in his understanding of Hart's project. Congratulating him on his forthcoming publication, Hart's response, a few days later, reassures Fuller that 'of course I shall not complain of misinterpretation'.

The tone of Hart's actual review, however, reads somewhat differently. It is not unfair to say that, from the outset, the review is basically uncompromising in its criticism of Fuller's book, at the level of the project writ large as well as its details. As Hart puts it, it is often difficult amid Fuller's 'firm and clear assertions of what is right and wrong in jurisprudence to identify any equally firm and clear argument in support of these assertions'.[68] He is equally critical of the breadth of Fuller's conception of law (which 'admittedly and unashamedly' includes the rules of clubs, churches, schools 'and a hundred and one other forms of human association');[69] of how Fuller 'speaks throughout as though the notion of a rule were unambiguous and otherwise unproblematic';[70] as well as the polemical way in which Fuller engages with other theories of law. Although these charges come with a caveat ('I am haunted by the fear that our starting-points and interests in jurisprudence are so different that the author and I are fated never to understand each other's works')[71] the thrust of Hart's critique is clear. There is nothing intrinsically moral about Fuller's internal morality of law and no compelling argument can be found in *The Morality of Law* to support the assertion that legal authority necessarily contains implicit moral limitations.

This core objection to Fuller's book is broken down by Hart into three main arguments: first, that Fuller confuses efficacy in the pursuit of purposes with morality; second, that the internal morality of law is neutral with respect to the aims that might be sought through its observance; and third, that the rule of recognition is not as unlimited a limit on lawgiving power as Fuller seems to suggest

[67] Letter from Fuller to HLA Hart, 11 December 1963, The Papers of Lon L Fuller, Harvard Law School Library, Box 3, Folder 14 (correspondence).
[68] Hart, 'Review' (n 56) 343.
[69] ibid.
[70] ibid.
[71] ibid.

it is. Of these three responses, it is the first that has had the most enduring impact on the positivist understanding of Fuller's jurisprudence. This has much to do with how Hart illustrates his point. If Fuller claims that law may be structured by principles of internal morality, he argues, so too might the activity of poisoning contain internal principles relevant to the pursuit of its objects, such as 'avoid poisons however lethal if they cause the victim to vomit', or 'avoid poisons however lethal if their shape, colour, or size is likely to attract notice'.[72] Yet to call these principles of the poisoner's art the 'morality of poisoning', is surely 'to blur the distinction between the notion of efficiency for a purpose and those final judgments about activities and purposes with which morality in its various forms is concerned'.[73]

This is undoubtedly a clever response and, no less, one that kills two birds with one stone. The idea of a 'morality' presents itself as incongruous with a practice (poisoning) that can hardly be thought to be moral in character, which in turn exposes the peculiarity of calling the principles that aid the efficacy of any such practice a 'morality'. It is clear that Hart's objection here is not directed in any way to the *content* of Fuller's eight principles, which he regards as sound and describes as 'principles of good craftsmanship'.[74] The objection, rather, is simply that they are merely principles of the lawgiving craft and nothing more. As Hart puts it at one point, the only way Fuller's principles could be anything more would be if the purpose of subjecting human conduct to the governance of rules was itself considered to be an ultimate value, irrespective of the content of those rules.[75] Otherwise, Hart insists, whether qualified by the term inner or not, Fuller's designation of these principles as a 'morality' perpetrates 'a confusion between two notions that it is vital to hold apart: the notions of purposive activity and morality'.[76]

Hart turns from these foundations to his second argument, which sees him offer his first meaningful engagement with the claims Fuller advanced in 1958 about Nazi law, which are reiterated briefly in *The Morality of Law* through the example of South African apartheid law. But Hart is again unmoved: nothing in Fuller's analysis amounts to a cogent argument in support of his claim that the eight principles 'are not neutral as between good and evil substantive aims',[77] with the example of South African apartheid law illustrating only that the principle that laws must be clearly and intelligibly framed is incompatible with the pursuit of vaguely defined substantive aims, irrespective of whether they are morally good or evil.[78] Hart does, however, allow Fuller a small concession on the issue of Nazi law when he explains that his own 'modest remark' that the internal morality of law is compatible with great iniquity was not intended to suggest that those

[72] ibid.
[73] ibid.
[74] ibid 347.
[75] ibid 351.
[76] ibid 350.
[77] ibid.
[78] ibid 352.

principles are 'compatible with every sort of iniquitous aim, vague or specific',[79] adding that the Nazi government did indeed frequently violate the principles of legality in pursuit of its monstrous aims and that such violations did likely aid the pursuit of those aims. But the jurisprudential point remains clear: none of this is evidence of any 'necessary incompatibility between government according to the principles of legality and wicked ends'. It merely reveals what a government might do to guard itself against the contingent support of its population, or the possibility of external criticism of its actions.[80]

As for Fuller's criticism of the rule of recognition as containing no meaningful limit on legal authority and no provision for revocation in the event of its abuse, Hart responds by suggesting that nothing in his account of the rule places any logical restriction on its possible content, which could provide explicitly or implicitly for such conditions should its designers so choose.[81] But even without such content, Hart suggests that the rule of recognition already includes a limited limit on legal authority in its specification that its authority be accepted by legal officials. This might not be quite the kind of limit that would satisfy Fuller's concerns, but Hart nonetheless insists that if Fuller is to successfully argue that legal authority necessarily contains implicit moral limitations, something much stronger than anything to be found in *The Morality of Law* is needed.[82]

Three brief comments can be made about these three responses to *The Morality of Law*. The first, referring to the point just discussed, is that Hart is mostly right to suggest that Fuller offers little in *The Morality of Law*, or arguably also in his 'Reply to Critics', to meet the challenge that Hart here presents him to formulate an argument to support his intuitions about the moral limits of legal authority, though what we might extrapolate from Fuller's claims in that context is a matter I return to in chapter five, as well as in chapter six when exploring whether Joseph Raz's work on authority might provide a ground for response to Hart on this point.

As for the second argument, Hart's conclusion that Fuller fails to show any 'necessary incompatibility' between observance of the internal morality of law and the pursuit of wicked legal ends, the point to make here is that this criticism arguably misrepresents the claim being contested. Fuller never argues that there is any 'necessary' incompatibility between observance of the principles of the internal morality of law and wicked legal ends. His claim, rather, is that there seems *in practice* to be an affinity between the pursuit of oppressive ends through law and a corresponding deterioration in observance of the principles of the internal morality of law. Fuller's published position on this point is abundantly clear: it is stated explicitly in Chapter 4 of *The Morality of Law* when he suggests that the internal morality of law might be 'neutral over a wide range of ethical issues', but it cannot be so with respect to the view it takes of man himself. It is put still more sharply in

[79] ibid 352.
[80] ibid 353.
[81] ibid 353 and 361–62.
[82] ibid 362.

his response to Marshall Cohen and Ronald Dworkin, published as part of the symposium dedicated to *The Morality of Law*.[83] Facing similar criticisms at the hands of both Cohen and Dworkin as he did from Hart, Fuller responds by stating that he has never asserted 'that there is any logical contradiction in the notion of achieving evil, at least some kinds of evil, through means that fully respect all the demands of legality'.[84] Rather, all that is claimed is the much more humble contention that when we have regard to 'the prosaic facts of human life', we not only see that coherence tends to have 'more affinity with goodness than with evil',[85] but also that in an ordered system of law, formulated and administered conscientiously, there is 'a certain built-in respect for human dignity'.[86] Fuller thus suggests that it is also reasonable to suppose that this 'built in' respect for human dignity, inherent to the internal morality of law, 'will tend to carry over into the substantive ends of law'.[87] But this, and not any claim about 'necessary connections', is all that Fuller argues for in *The Morality of Law*.

As for the main argument of Hart's review—that Fuller confuses mere principles of craftsmanship with something moral—the alleged error here arguably tells us more about Hart's jurisprudence than it does about Fuller's. As Hart explains it, the fundamental difference between Fuller's views and his own lies in how, where Fuller values the principles of legality for their own sake, utilitarians and other positivist thinkers have historically valued those same principles 'only so far as they contribute to human happiness or other substantive moral aims of the law'.[88] This is an uncompromisingly instrumental frame of critique—law is properly regarded as a tool to be used in pursuit of specified ends, and its contribution to moral life is to be assessed exclusively by reference to whether those ends are or prove to be moral in character—and none of its content ought to come as a surprise as a critique emanating from the commitments of legal positivism. Yet, and despite its apparent neatness, a turn to the archival materials suggests that there might be more to be learned from the story of how Hart came to express this critique of Fuller's point in the terms that he did.

B Efficacy and Trusteeship

The contest between Hart and Fuller on this point seems straightforward at first glance: Hart's efficacy argument is a knock-down to Fuller's claims about the internal morality of law, because even the most immoral law can benefit from conditions that make it effective. To read Fuller's jurisprudence in this way is hardly wholly misguided, because there are a number of ways in which Fuller's analysis clearly does invite such an interpretation. Examples include how he tells

[83] Lon L Fuller, 'Reply to Professors Cohen and Dworkin' (1965) 10 *Villanova Law Review* 655.
[84] ibid 664.
[85] ibid.
[86] ibid 665.
[87] ibid 665–66.
[88] Hart, 'Review' (n 56) 357.

the story of King Rex from Rex's (that is, from the lawgiver's) perspective, as well as how he speaks elsewhere in *The Morality of Law* of how the internal morality of law is concerned with the ways in which 'a system of rules for governing human conduct must be constructed and administered *if it is to be efficacious* and at the same time remain what it purports to be'.[89] Fuller also suggests, in the same context, that his internal morality of law is like 'the natural laws of carpentry'; an analogy which seems to suggest that just as the 'laws of carpentry' ostensibly serve the purposes of the builder of the house, the eight desiderata of the internal morality of law serve the efficacy of the lawgiver's pursuits.

But even with these invitations from Fuller himself to view his claims through the prism of efficacy, it is a large step to then suggest that those claims are straightforwardly reducible to such an interpretation, as Hart seeks to argue. The motivations for taking that position might simply be attributed to the polemics of scholarly argument; specifically, to how Fuller's obviously polemical approach to Hart's *The Concept of Law* is met by Hart with an equally polemical response, according to which the internal morality of law has everything to do with efficacy and nothing at all to do with morality. Or there might be more subtle causes, such as how Hart's designation of the internal morality of law as capturing 'principles of craftsmanship' that serve the efficient pursuit of particular legal ends speaks most naturally to how particular laws might be used as instruments for the efficacious pursuit of specific goals, while Fuller, for the most part, was concerned to discern the nature and constitutive features of legal order in a more systemic sense. Indeed, the distance between him and Hart on this point is something that Fuller articulates well in his 'Reply to Critics', when he questions how exactly the criteria of efficacy is to be applied 'to the creation and administration of a thing as complex as a whole legal system'.[90]

But a more likely cause of the impasse between Hart and Fuller on the issue of efficacy is revealed in Fuller and Hart's private correspondence, as indeed elsewhere in Fuller's private papers. When Hart speaks in his review of how the appeal of the internal morality of law is to 'the pride of the craftsman', he omits any reference to the words 'a sense of trusteeship' that precede that phrase in Fuller's formulation.[91] Although the published 'Reply to Critics' makes no reference to this omission, Fuller's private papers suggest that this move on Hart's part was a source of genuine distress, and there is little wonder why. If Fuller's own understanding of the appeal of the internal morality of law is simply to 'the pride of the craftsman', it looks very much like he, with his critics, ultimately understood the idea of the internal morality of law in instrumental terms.

The record of correspondence between Fuller and Hart with respect to this point is instructive. That correspondence begins warmly: upon reading Hart's review in advance of its publication, Fuller expresses delight to see 'such a sharp joinder of issue' reflected in it. While he does note that Hart's treatment of his

[89] *Morality of Law* (n 1) 97 (my emphasis).
[90] 'Reply to Critics' (n 47) 202.
[91] *Morality of Law* (n 1) 43.

claims about the responsibilities of the legislator left out his own reference to the words 'a sense of trusteeship', he adds, collegially, that Hart was 'quite entitled' to do this.[92] But the passage of time seems to have led Fuller to a different view. Several months later, beleaguered by how Hart's review was just one of many which interpreted his claims about the internal morality of law as reducible to the value of efficacy, Fuller wrote to Hart in somewhat more terse terms. Stating that he is 'really puzzled' by the level of criticism that had met his use of the term 'morality', and revealing his sense that this reading of his claims was a direct product of Hart's review, Fuller asks whether 'peace could be had by substituting for "morality" some word like 'trusteeship' (which you so rudely snatched from my mouth in your quotation)'.[93]

It is not a happy scene, and Fuller's working notes for the 'Reply to Critics' reveal the persistence of his worry about this point as well as the sense that he had been unjustly treated. The word 'trusteeship', together with 'Hart', appears frequently throughout these notes, with one including an extended observation about how Hart's partial quotation makes it appear like 'in a moment of absent-mindedness I had tacitly admitted that the question was one of efficacy', and that 'if Hart will let me restore the whole sentence, I would be willing to make my peace with him'.[94] Fuller's candid correspondence with Philip Selznick, whose review of *The Morality of Law* he regarded as 'one of the most perceptive that has appeared',[95] is another site where the efficacy-trusteeship issue is discussed at length. There, Fuller speaks of how 'H.L.A.H. cuts out the reference to trusteeship, and uses my reference to "the pride of the craftsman" as a kind of admission by me that the whole thing is a matter of "efficacy"',[96] and makes it very clear that he 'cannot accept' the view that the internal morality of law is 'simply a prescription for making law effective, law itself being morally neutral'.[97]

Nonetheless, this portrayal of Fuller's own understanding of the internal morality of law in terms of efficacy has been remarkably resilient. This is even the case among some scholars who are openly sympathetic to Fuller's thought and yet who have repeatedly suggested that the primary appeal of the internal morality of law, according to Fuller, 'is to the legislator's pride in his own craft', with no reference to the phrase 'a sense of trusteeship'.[98] It is, of course, highly likely that the roots of Hart's omission of the trusteeship point lay much more in a lack of interest than in any intentional sleight of hand. Indeed, the root cause might simply be

[92] Letter from Fuller to HLA Hart, 3 February 1965, The Papers of Lon L Fuller, Harvard Law School Library, Box 3, Folder 14 (correspondence).

[93] Letter from Fuller to HLA Hart, 18 October 1965, The Papers of Lon L Fuller, Harvard Law School Library, Box 3, Folder 14 (correspondence).

[94] Untitled and undated document, The Papers of Lon L Fuller, Harvard Law School Library, Box 12, Folder 1 (notes for 'Reply to Critics').

[95] Letter from Fuller to Philip Selznick, 18 August 1965, The Papers of Lon L Fuller, Harvard Law School Library, Box 7, Folder 6 (correspondence).

[96] ibid.

[97] ibid.

[98] See, eg Jeremy Waldron, 'Why Law? Efficacy, Freedom or Fidelity?' (1994) 13 *Law and Philosophy* 259, 267, and Waldron, 'Hart's Equivocal Response' (n 49) 1135, 1154.

methodological; that is, to be concerned with the demands of role, of trusteeship, is to be concerned with how a participant in a practice understands that practice and what, by extension, such participation asks of her. Hart's avowedly external and descriptive approach to the study of law, however, generally steers clear of the kind of modes of internal evaluation that exploring questions about the roles necessarily invites.

Still, and whatever the underlying cause, Hart's omission of or refusal to engage with Fuller's claim that the internal morality of law appeals to a sense of 'trusteeship', and has regard to 'the fate of persons', had the effect of misrepresenting Fuller's overall position. An important part of reclaiming Fuller, therefore, is to reclaim the significance of this trusteeship point and to point to the many ways in which its omission has shaped our understanding of what was at stake not only between Hart and Fuller, but between positivists and anti-positivists today. Because what Fuller suggests when he speaks of the appeal of the internal morality of law in terms of both trusteeship *and* craft is an understanding of the value of law as having two centres of gravity: morality *and* efficacy. I return to this issue in chapter seven, where I seek to resituate Fuller's jurisprudence within the landscape of contemporary legal philosophy as it has recently been mapped by Dworkin. But even if Fuller does not mention a word about the 'trusteeship' issue in the 'Reply to Critics', what he does do in the 'Reply' is to situate Hart's efficacy reading of his core claim at the centre of his analysis, seeking to show how and why this reading is much more problematic than Hart and his supporters appreciate, not only with respect to their reading of the value of the internal morality of law, but also for the coherence of some of their own theoretical commitments.

C Lessons from a Tyrant?

It is unsurprising that we should see Fuller's desire to clarify and qualify the efficacy reading of his claims finding expression in the intervening period between Hart's 1965 review and the publication of the second edition of *The Morality of Law* in 1969, to which the 'Reply to Critics' is appended. But what perhaps is surprising is the context in which this takes place: in a 1968 essay titled 'Freedom as a Problem of Allocating Choice', one of a handful of writings on the subject of freedom that Fuller penned over the course of his career.[99]

Fuller's interest in the subject of freedom has a history that shares much in common with his eunomics project, in the sense that although he commenced writing and speaking about the subject in the mid-1950s, and although he intended to elaborate his thinking into a book, the written record reveals at best a preliminary sketch of what might have been developed into a larger project. The freedom writings typically receive little or no attention from his critics, but given the relevance of their animating themes—a desire to challenge the philosophical

[99] Lon L Fuller, 'Freedom as a Problem of Allocating Choice' (1968) 112(2) *Proceedings of the American Philosophical Society* 101.

neglect of affirmative conceptions of freedom, and a concern to theorise about the type of social arrangements that enable freedom to be actually experienced in the world—they are in fact another valuable source through which we can flesh out the parameters of Fuller's jurisprudence.

The interest in 'Freedom as a Problem of Allocating Choice' for present purposes lies in how Fuller introduces us to a hypothetical tyrant who, 'bent entirely on pursuing his own interests' and planning to 'employ his subjects as tools for the realization of his purposes', sets out to become a lawgiver.[100] The tale initially sounds a lot like that of his monarchical predecessor, Rex. But we soon learn that, unlike Rex, the tyrant commences his own lawgiving project with entirely selfish intentions.[101] And this is precisely where things begin to go wrong.

Shortly after beginning his life as a lawgiver, the tyrant arrives at an important insight: that he does not necessarily increase his own freedom of action by restricting that of those under his rule, because he 'cannot effectively use another human being as a tool without according him some power of choice, some opportunity to use his own discretion'.[102] So, he soon realises, from the standpoint of efficiency in achieving a goal, 'some discretion and choice must . . . be accorded the human agent'.[103] Then, on still further experience and reflection, the tyrant comes to realise that beyond minimum choice and discretion, 'a human being will serve as a more effective tool if he is happy and satisfied with his role', and to be happy and satisfied he must feel that he is serving not only the ends of another, but his own as well.[104]

Fuller then suggests some other insights that might yet occur to the tyrant as he continues his adventures with lawgiving, such as how 'it would be well to accord some opportunity to his subjects to enhance and expand their powers beyond those demanded by the immediate job they are doing for their master', because it seems 'reasonable to suppose that they will be happier and more efficient if they are given a chance to improve themselves'.[105] The story then concludes with the following passage:

> Our tyrant, you will observe, has found himself caught in a kind of progression. He started by seeking efficiency, he then moved at least some way toward doing what is essential for human happiness, and he may end by fostering the conditions most conducive to human development. If he traverses fully the three steps of this progression he will, of course, finish by ceasing to be a tyrant. How far he moves toward that outcome will depend less on the balance of good and evil in his soul than it does on the power of his brain to discern the conditions essential for the success of what is, by its very nature, a cooperative enterprise.[106]

[100] ibid 105.
[101] ibid.
[102] ibid.
[103] ibid 105–6.
[104] ibid 106.
[105] ibid.
[106] ibid.

To appreciate how and why this move is so suggestive for our understanding of Fuller's jurisprudence, it is helpful to retrace the key points of development between the three tales of lawgiving that emerge from Fuller's writings over the period of his exchanges with Hart. In his 1958 reply, as will be recalled from chapter two, Fuller's unnamed monarch is 'utterly selfish' and seeks 'solely his own advantage' in purporting to establish a relationship of law with his subjects. This relationship fails, however, because the monarch does not properly understand the responsibilities that attend the creation and administration of a successful legal system. By contrast, King Rex, from *The Morality of Law*, is not portrayed as bent on pursuing his own selfish aims, but rather on making his name in history as a great lawgiver. In Rex, therefore, we meet a lawgiver who meant well (at least until he 'decided to teach them a lesson' for their recalcitrance)[107] but who, like his predecessor, did not fully appreciate that to bring about legal order one must discipline one's activities with an understanding of the principles of lawgiving.

In a number of ways, then, the story of the tyrant completes the message that Fuller sought to convey through the stories of Rex and his unnamed predecessor. The key lies in how the tyrant, being a tyrant, is not benevolently motivated like Rex, and that his story is meant to be different to Rex's is clear from this fact alone. The tyrant's declared attitude towards law is purely instrumental. But once he enters the province of legality, the tyrant finds that his instrumental motivations can occupy only one part of the picture of what he must understand, and do, in order to govern successfully through law. Most importantly, the tyrant finds, to his apparent surprise, that certain demands of treatment, of lawgiver towards subject, qualify the pursuit of his instrumental aims through law. He also learns that these demands ask more and more of him, by way of respect for the agency of his subjects, as his legal order improves its position along the continuum that marks the journey from a minimally constituted legal order to a perfectly constituted one. The suggestion we are left with, therefore, is that a true tyrant cannot be a true lawgiver, precisely because the essence of tyranny consists in an ambition to control human agency in a manner that law, by virtue of the form through which it finds expression, renders unattainable.

Imaginary tales of lawmaking monarchs and tyrants can obviously only take us so far. They are merely stories and, as such, can only be suggestive of what we need to understand about the modes of restraint that attend the enterprise of lawgiving. Still, what makes the story of the tyrant so instructive for the project of reclaiming Fuller is how it seems to recast the lessons learned from Rex about the essential conditions of lawgiving into lessons about the imperatives of treatment, towards the legal subject, that inhere in the form through which legal order is expressed.

The suggestion that the story of the tyrant was motivated by responses to *The Morality of Law* is my own: there is no evidence in Fuller's working papers or

[107] *Morality of Law* (n 1) 36.

indeed in the text of 'Freedom as a Problem of Allocating Choice' that makes the connection explicitly, though the themes, not to mention the date of the essay, are hard to miss. But the value of the story does not rest on the motivations that might account for its inclusion in an ostensibly separate project within Fuller's canon. Its value instead lies in what distinguishes it from, and in many ways makes it richer than, the story of King Rex; namely, the idea that the demands of actually instantiating legal order will qualify the instrumental potential of law when such instrumental intentions do not correspond with the respect for agency upon which a functioning legal order depends.

Moreover, by placing the agency of the legal subject at the centre of a story of how legal ordering arises, is sustained and made successful, Fuller shows us in the tale of the tyrant how we can—and should—look at the phenomenon of legal order from *both* the perspective of the lawgiver and the perspective of the subject. Indeed, one of the most instructive features of the story of the tyrant is precisely the way that the legal subject and her experience of her agency is brought into the centre of the analysis, even though that analysis appears to about the perspective, and the needs, of the lawgiver.[108] The story of the tyrant thus also teaches us much about how we should read the story of Rex.

III A Different Path?

The publication of *The Morality of Law* generated a plethora of reviews as well as a substantial amount of personal correspondence between Fuller and his sympathisers and critics alike. Among these is a series of letters between Fuller and Walter Berns, then of the Department of Government at Cornell University, initiated by Fuller after having read a review of his book by Berns. The motivation for writing, it seems, was Berns' apparent agreement with Hart that a sincere respect for legal morality is compatible with the greatest iniquity, but with Berns also suggesting 'that history has many examples to prove this'.[109] Inviting Berns to provide him with 'just one' such example, Fuller's own position on this alleged historical commonplace is clear:[110]

> Since my book has been out I have discussed this question with a good many people, and I have yet to encounter a single case to prove this point. South Africa is probably as close as any. But as I tried to show in my final chapter, to the extent that an attempt has been made there to write racial prejudice into law, some impairment of legal morality has taken place. So far as South African fidelity to contractual obligation is concerned, this might be quite insulated from racial considerations and I assume the South African

[108] See Jennifer Nadler's astute observations on this point in Nadler, 'Connection Between Law and Justice' (n 58) 25.

[109] Letter from Fuller to Walter F Berns, 21 September 1964, The Papers of Lon L Fuller, Harvard Law School Library, Box 11, Folder 17 ('The Morality of Law').

[110] ibid.

law of contracts is administered with integrity. But the point is that where racial prejudice tries to write itself into law, at that point it and the law both get into trouble.[111]

The four letters back and forth between Fuller and Berns over the course of September and October 1964 are the result of this rather forceful invitation to conversation on Fuller's part. Though at certain points in the correspondence they appear to be at cross-purposes, Fuller's invitation to Berns to extend discussion about his claim that the principles of the internal morality of law are both theoretically and practically compatible with the greatest iniquity sees the two scholars head into an intriguing exchange about the compatibility between the internal morality of law and slavery. As Berns frames it in his letter of reply:

> Would you not regard slavery as a great inquity? Yet slavery has been established by law in regimes characterized by respect for law, or the 'internal morality of law.' Rome, for example, and our own country, until the 13th Amendment. With many Americans it was precisely, indeed, solely, their respect for the 'internal morality of law' that required them to support human slavery. Was this not true of the trial judges in many a fugitive slave case? . . . Racial prejudice was certainly written into our law, by the Founders, and maintained by law-abiding men, yet there was no 'impairment of legal morality'. Or would you argue that there was?[112]

The main argument that Berns seeks to develop is essentially one about moral relativism and public sentiment; that is, he indicates qualified agreement with Fuller's claim that legal morality will be impaired when the law is directed towards iniquity, but only 'as iniquity is understood by the people or regime in question'.[113] Fuller does engage with this suggestion in his letter of reply but, for him, it is clearly not the matter of primary interest. Of interest, rather, is Berns' apparent belief that very great iniquity goes 'arm in arm with the conscientious administration of a legal system by a government willing to accept all of the restraints on its freedom of action implicit in legal morality', such that there is also no reason to see any incompatibility between slavery and observance of the internal morality of law.

Berns' letter is heavily underlined and annotated by Fuller, and his own letter of reply makes apparent that he gave much thought to how best to respond to the arguments raised by his interlocutor. That second letter sees Fuller return to the arguments that he advanced against Hart in *The Morality of Law*, clarifying that he never said there or elsewhere 'that unjust laws are not possible', or that 'that a genuine respect for legal morality guarantees the substantive justice of law'. Fuller also emphasises the distinction he 'tried to drive home' in *The Morality of Law* between deference for established authority and fidelity to law.[114] Berns' apparent assumption, underlying his argument about public sentiment, that the

[111] ibid.

[112] Letter from Walter F Berns to Fuller, 28 September 1964, The Papers of Lon L Fuller, Harvard Law School Library, Box 11, Folder 17 ('The Morality of Law').

[113] ibid.

[114] Letter from Fuller to Walter F Berns, 14 October 1964, The Papers of Lon L Fuller, Harvard Law School Library, Box 11, Folder 17 ('The Morality of Law').

actual task of writing racial prejudice into law is a straightforward one is also noted, with Fuller countering it with the suggestion that such surely involves the task of 'defining the indefinable'.[115] Each of these clarifications and lines of defence, with perhaps only the exception of the last point, are basically familiar. But what is not familiar is what Fuller says in response to Berns' claim that slavery is compatible with observance of the principles of the internal morality of law. The point is teased out as follows:

> I am doubtful of the statement in your letter that 'slavery has been established by law in regimes characterized by respect for law, of the "internal morality of law"'. . . . That laws protected the master's property right in his slave is, of course, true. It is also true that laws gradually gave the slave many rights, including that of protection against abuse, or buying his own freedom, etc. . . . All this is, however, not really relevant to my argument with Hart. I have nowhere said that respect for legal morality inhibits every kind of brutality; I have only argued that it impairs a government's choice of the kinds of brutalities it will impose and the ways it will impose them, and that this impairment serves the cause of justice and humanity.[116]

There are two points being made here, and even if Fuller himself seems here to suggest otherwise, I would argue that they are closely connected. The main point that Fuller is keen to emphasise is one about how to observe the requirements of the internal morality of law is to find oneself limited in the 'kinds' of brutalities that might be pursued through law, as well as the 'ways' available to pursue them. Even if the precise content or scope of that limit is unspecified, Fuller is clear that it nonetheless applies in the direction of means as much as ends. Already, then, we can see that Fuller has here begun to appreciate that, if he is to invite meaningful exchange on the point with Berns or indeed any other interlocutor, his claim that respect for the internal morality of law is incompatible with great iniquity must be reframed into a claim about respect for the internal morality of law being incompatible with certain ways of pursuing iniquity, as well as the kinds of iniquity so pursued.

This intuition links readily to how Fuller then suggests that he doubts that there could be such a thing as slavery that is established through respect for the internal morality of law, even though it is of course true that laws have historically protected a master's property right in his slave. An important distinction is being suggested here, and it is a pity that Fuller did not seek to make more of it. Because the idea alluded to is that the slave was not, as Berns' more casual connection seems to suggest, a legal subject with a relationship to the lawgiver in her own right, but rather was an object of property and regulated by law as such.

Had Fuller sought to make more of this point, and taking a cue from his own observations in Chapter 4 of *The Morality of Law* about the conception of the person implicit in the internal morality of law, he might have reframed his claim about the incompatibility of the internal morality of law with certain kinds of

[115] ibid.
[116] ibid.

iniquity as a claim about respect for the legal subject's status as an agent being incompatible with the pursuit of certain kinds of iniquity, such as the iniquity of designating a person as an object of property and acting upon them accordingly. Because, as Fuller suggests in that passage from Chapter 4, when a person is acted upon in this way, there is 'nothing left to affront': what was formerly a person in possession of a capacity to direct her agency towards the lawgiver's regime of general rules is now an object merely to be acted upon.

This point might be taken further if we extend upon Fuller's other comment in his letter to Berns about how 'laws gradually gave the slave many rights, including that of protection against abuse, or buying his own freedom'. The suggestion here is that, over time, the law began to loosen its constitutive control over the slave's objectified status, presumably until a relationship that was much less like that of master-slave, or property owner-object of property, was extant. If this is the suggestion, there is much in common with the message of the story of the tyrant that I reviewed in the last section. Because in the same way that the tyrant increasingly ceases to be a tyrant the more he lets his subjects express their agency towards their own ends, the more that the law increased her liberties, the more the slave became a legal subject in her own right.

But these implications are only gestured to by Fuller, and so Berns can be forgiven for how in his final letter of reply he holds on to his argument that slavery 'was recognized as legal in the American Constitution' with no mention of the subtleties of Fuller's point about how the legal regime in question was a regime of property rights.[117] Still, Berns makes other points that are problematic for Fuller's jurisprudence and its claims about the status of the legal subject as an agent, as well as about how the form of law limits the kinds of iniquitous substance that can be pursued through it. As Berns puts it, 'I do fear that the law can be used without detriment to itself to disenfranchise negroes or bundle them off to the Bantustans—so long as the whites do not regard it as wrong to do so'.[118] The fact that the law can be used to 'disenfranchise' its subjects is an important rejoinder to Fuller on the question of just how far the legal subject's agency can be debased, and just how serious the claimed corresponding debasement to law's form needs to be, before we have reached the point where law slides into non-law, and thus also where the subject slides into an object. But again, Berns' point here is about public sentiment. The conversation about the relationship between the form of law and the agency of the legal subject is left open.

If Fuller had any inclination to develop his intuitions about law and slavery further, it is likely that they would have been derailed in any event by the intervention of Hart's review and the narrowing of the terms of the debate that this review brought about. But this does not mean that Fuller lost sight of the connections between legal form and the subject's capacity for and ability to express agency that are gestured to suggestively in his correspondence with Berns. To the

[117] Letter from Walter F Berns to Lon L Fuller, 28 October 1964, The Papers of Lon L Fuller, Harvard Law School Library, Box 11, Folder 17 ('The Morality of Law').

[118] ibid.

contrary, what we see in the 'Reply to Critics' is Fuller directing his sense of this fundamental connection towards a different question; namely, what it is that distinguishes law from other modes of ordering, and the moral dimensions, especially pertaining to agency, that give content to this distinction.[119]

IV Conclusion

The primary aim of this chapter has been to clarify and hopefully to resolve some of the confusions that continue to surround Fuller's famous claim that there is something inherently moral about the form through which law finds expression. Fuller's private correspondence seems to strongly suggest that the 'ethos' reading of his morality claim was his own preferred one. Indeed, his correspondence with Philip Selznick makes this explicit, stating that he was actually content (if such could generate meaningful exchange with his ideas) to replace the term 'morality' with 'ethos' or some other term similarly capable of capturing the notion that certain moral responsibilities attend the occupation of certain roles.[120] In a letter to the British philosopher and sociologist Dorothy Emmet whose work on the ethics of roles he admired greatly, Fuller further reports that one of his students had suggested 'that if I could have called my book, instead of *The Morality of Law*, *The Morality of Lawing*, much of the misunderstanding might have been avoided', because while the word law 'calls to mind books lying inertly on shelves', 'lawing' calls to mind 'people in interaction with one another, and that picture in turn would suggest reciprocal responsibilities if the interaction is to proceed properly'.[121] But perhaps most curiously, given that he did not restate the point anywhere in his 'Reply to Critics', Fuller also indicates that he would be willing to replace the term 'morality'—on which, he says, 'nothing hinges'—with 'ethos, conscientious attitude, or trusteeship' in his correspondence with Hart.[122]

It seems, however, that on further thought Fuller came to believe that something did indeed hinge on his use of that term, 'morality'. In the same letter to Hart, as elsewhere in his private papers, Fuller stands firm in his view that the demands of the lawgiving role contain moral dimensions, and that these moral dimensions lend that role a character that simply cannot be captured in the idea of efficaciously discharging a craft.[123] But we need not have recourse to the archival materials to establish this point. A close review of Fuller's main claims in *The*

[119] See chapter five.
[120] Letter from Fuller to Philip Selznick, 18 August 1965, The Papers of Lon L Fuller, Harvard Law School Library, Box 7, Folder 6 (correspondence) (Fuller's emphasis).
[121] Letter from Fuller to Dorothy Emmet, 7 October 1966, The Papers of Lon L Fuller, Harvard Law School Library, Box 2, Folder 16 (correspondence).
[122] Letter from Fuller to HLA Hart, 18 October 1965, The Papers of Lon L Fuller, Harvard Law School Library, Box 3, Folder 14 (correspondence).
[123] Letter from Fuller to HLA Hart, 3 February 1965, The Papers of Lon L Fuller, Harvard Law School Library, Box 3, Folder 14 (correspondence).

Morality of Law, as well as of how they were received and interpreted by his critics, leaves us with a strong sense that the content and orientation of his claims about the internal morality of law simply cannot, as Hart argued, be neatly captured in the idea of a checklist of eight principles that serve the efficacy of a lawgiver's pursuit of ends through law. Fuller's claims about the internal morality of law are clearly larger than this in how they commence from the starting point of the relationship between lawgiver and legal subject, and then seek to show how their potentially opposed positions can be reconciled through a mediating form that expresses the lawgiver's power at the same time as it limits it, and that shapes the legal subject's agency at the same time as it respects it. There is simply more to the picture than efficacy.

Ultimately, then, one leaves a study of *The Morality of Law* with the sense that the book does two closely related things. First, it gives expression to a distinctive jurisprudence which takes the form of law as its starting point, and which explores the relationships, and responsibilities, of both lawgiver and subject within that form. Second, it offers a critique of the undiscriminating character of the positivist account of law that speaks to positivism's apparent failure to make any provision for the withdrawal of lawgiving authority in the event of its abuse. These are two trajectories of jurisprudential analysis in their own right, but when we bring them together we learn much about the meaning of Fuller's claims about law's internal morality. The principles of the internal morality of law are moral, that is, because they articulate an ethos of lawgiving that gives practical expression to the morally significant idea that distinguishes rule through law from rule by men: an internal, non-optional commitment to respecting the legal subject as an agent.

Positivism's general and Hart's particular neglect of the position of the legal subject within an account of law's nature was a matter that came to increasingly preoccupy Fuller as he considered his response to his critics' interpretations of *The Morality of Law* and prepared to pen his 'Reply to Critics'. Much of the preceding analysis has been dedicated to preparing the ground for understanding the basis of this complaint and how it finds expression in that reply. Because contrary to what is suggested in Hart's review, there is nothing at all 'un-Fullerian' about recognising the efficacy value of the eight principles from the point of view of the lawgiver's pursuit of ends through law. But what *is* distinctly un-Fullerian is to suggest that this value of efficacy ought to exhaust our understanding of the value that accrues through observance of the requirements of the internal morality of law. Here, it is helpful to recall the lessons we are invited to take from the story of Rex and his eight failures to make law. Certainly, that story can be read as explaining the ways in which Rex failed to successfully deploy law as an instrument of his own will. But this must be reconciled with how we are told at the outset that Rex wanted to be a great lawgiver and to achieve a legal system that remedied the confusion, expense and corruption of those of his predecessors. He wasn't, it seems, just out for himself.

For Fuller, therefore, law has a dual face: it looks towards the perspective of both lawgiver and legal subject, to efficacy as well as to sites of moral responsibility and value, and seeks to reconcile them through the structural features and relational dynamics of its distinctive form. It takes the 'Reply to Critics' for this position to be articulated and defended more clearly. But what will become apparent in the next chapter is that, to reclaim Fuller's jurisprudence, we need to make much more of the 'morality' side of law's dual face, specifically, of how that dimension arises from the status of the legal subject as an agent. Because, as Fuller's working notes for the 'Reply' make especially clear, the respect conveyed by the lawgiver towards the legal subject's status as an agent, instantiated by the principles of the internal morality of law, is something prior to, and therefore more fundamental than, any instrumental benefits that might accrue to the lawgiver as a consequence of the observance of those principles.

The Morality of Law thus leaves us with a sense of Fuller's objection to legal positivism that might be cast in these terms. The problem with positivism is not that it looks to and takes its cues from the source of lawgiving power. The problem, rather, is how positivism privileges that perspective exclusively, with no meaningful regard for the way that law's distinctive mode of interacting with the legal subject, as an agent, provides the basis for distinguishing law from other modes of rule. To opt into the distinctive enterprise of lawgiving is to take on an obligation to respect the legal subject as an agent, regardless of your motivation for choosing law.[124] For Fuller, as I have expressed it elsewhere, this is simply what comes with the territory of choosing to govern through law in the first place.[125]

[124] For support for this 'no motivation' point, see Waldron, 'Why Law' (n 98) 280–81, and 'Hart's Equivocal Response' (n 49) 1167.

[125] Rundle, 'The Impossibility of an Exterminatory Legality' (n 29) 107.

5

The Reply to Critics

> Naturally you play a prominent role in this *Reply*, and I shall airmail a copy to you as soon as it becomes available. I concentrate on the issue which I consider central to the whole controversy, that is, whether respect for the principles of legality is merely a matter of 'efficacy' or deserves to be regarded as a distinctive morality of law-making and law-administering.[1]

Such are the words that Fuller wrote to Hart in March 1969, notifying him of the new fifth chapter, 'A Reply to Critics', that would be appended to the second edition of *The Morality of Law*. As the letter makes clear, that 'Reply' was to make Hart's and other critics' insistence that the value of observance of the principles of the internal morality of law is merely reducible to efficacy the centrepiece of its response. But it is one thing to diagnose and quite another to develop that diagnosis into a conversation that is more congenial to, or will more congenially serve, one's own contribution. This is the challenge that Fuller faced when he chose to pen the 'Reply'.

Indications of what was to find expression in the 'Reply to Critics' can be found in a number of the writings that Fuller published in the period between the first and second editions of *The Morality of Law*. This, as we saw in chapter four, is clearly so in the case of the story of the tyrant from the 1968 essay 'Freedom as a Problem of Allocating Choice', which appears to test the waters of a response to his critics on the matter of whether the value of legal order is reducible to the efficacy with which it aids a lawgiver's pursuit of his ends through law.[2] But there are several other works, at times equally surprising as contexts for testing his jurisprudential claims, that offer similar insight. One is the introduction that accompanies the republication of Fuller's 1930–31 essays on legal fictions.[3] Penned in 1967, this introduction is intriguing in its own right for how it reflects on the aims of work completed almost forty years earlier. But what is interesting for present purposes is how, at points, that introduction becomes a platform for a polemic against the 'analytical' approach generally, and specifically how legal philosophy, in its disfavour towards 'holistic' and 'comprehensive' views, tends 'to disregard the institutional processes that bring law into existence and produce its efficacy in human affairs'.[4]

[1] Letter from Fuller to HLA Hart, 6 March 1969, The Papers of Lon L Fuller, Harvard Law School Library, Box 3, Folder 14 (correspondence).

[2] See chapter four, IIC 'Lessons from a tyrant?'.

[3] Lon L Fuller, *Legal Fictions* (Stanford, Stanford University Press, 1967).

[4] ibid xi.

Another (intriguing) site for exploring themes that were to become central to the 'Reply' is Fuller's 1965 essay, 'Irrigation and Tyranny'.[5] Fuller himself described 'Tyranny' as 'the best thing I have ever written', by which he meant that 'it compacts into fewer pages more of the basic ideas that I have been struggling with than anything else I have written'.[6] This in itself suggests that there are important cues for us to take from the essay, even if, given its apparent inquiry into the limits of despotic power and the social conditions that foster or inhibit its growth, 'Tyranny' appears at first glance to have little affinity with the criticism of legal positivism found in the 'Reply'. Though it is not clear, from the dates, whether 'Tyranny' was already written when Fuller received Hart's and other reviews of *The Morality of Law*, what is clear is that Fuller seeks in the essay to develop his thinking on the idea that every kind of social power is 'subject to an implicit constitution limiting its exercise',[7] and to link this intuition to his critique of how such is overlooked by those who ordinarily define law 'as consisting of those rules that emanate from some human source that is itself regarded as formally authorized to enact or declare law', and who appear to accept that, in the absence of explicit constitutional limitations, 'this human source can enact anything it sees fit into law', whether or not those laws are 'wise or foolish, intelligible or obscure, just or unjust, prospective or retrospective in effect, general or specific in their coverage, published or unpublished'.[8] In all this variety, Fuller suggests, adherents to source-based conceptions of law assume that there is no structural constancy to law itself, 'except that imposed by the formal rule' which identifies its authorised source.[9]

The working notes for these writings, especially those for what was to be published in 1968 as *Anatomy of the Law*, also tell us much about Fuller's thinking en route to the 'Reply'; indeed, they often speak to a more expansive agenda than that which Fuller ultimately came to address either in that book or in the 'Reply'. Concerns addressed include not only the issue of efficacy and positivism's neglect of the structures and relations antecedent to law, but also how positivism neglects to explore what it means for law to exist in 'a meaningful sense'.[10] There is also a more direct engagement with the question of the connections between law and morality than we see in the 'Reply', such as one note that succinctly explains how the distinction between law and morality 'can touch three different aspects of a legal order: (1) its foundation in public acceptance, (2) the manner in which

[5] Lon L Fuller, 'Irrigation and Tyranny' (1964) 17 *Stanford Law Review* 1021.

[6] Letter from Fuller to Douglas Sturm, 9 August 1965, The Papers of Lon L Fuller, Harvard Law School Library, Box 7, Folder 15 (correspondence). Fuller also described the essay in a letter to Philip Selznick as 'the most explicitly "sociological" thing I have yet written': Letter from Lon L Fuller to Philip Selznick, 24 May 1965, The Papers of Lon L Fuller, Harvard Law School Library, Box 7, Folder 6 (correspondence).

[7] 'Tyranny' (n 5) 1027.

[8] ibid 1029.

[9] ibid.

[10] Untitled notes (paginated in hand as p 33), The Papers of Lon L Fuller, Harvard Law School Library, Box 12, Folder 4 ('Encyclopaedia Britannica').

it is constructed and administered, and (3) the ends at which it is ultimately aimed'.[11]

The writings penned between the first and second editions of *The Morality of Law* thus provide illuminating access into the ideas Fuller was toying with as he considered how to respond to his critics. But the working notes for the 'Reply' itself provide the most telling insight into what he thought to be at stake in the task of responding to those critics. These notes stand as a fascinating record of Fuller's attempt to lay bare the causes of the controversy in which he found his claims embroiled, and the wide array of titles that frame the notes bear witness to this effort. Many are framed around the question 'What is at stake?', or 'Is there such a thing as legal morality?'. Several others are headed with the somewhat more obscure 'Structure of Purpose', or gesture to the content of specific claims that Fuller sought to clarify, such as those given the titles of 'Generality', or 'Reciprocity', or 'Legal morality—identification with subject'. Indeed, one working note is striking simply for how Fuller appears to have changed his mind several times about what exactly its framing idea needed to be. The initial 'What is at stake' is crossed out and replaced with 'Law and Legality', which is then again crossed out again and replaced with 'Legal Morality', before Fuller turns to the point to which this complex attempt at framing is meant to refer; a point, simply and succinctly put, that is itself instructive of what he thought was at stake.[12] As he brought the 11-year trajectory of the Hart-Fuller debate to a close, Fuller evidently still thought that he needed to 'explain fact that legal system always stands between political rule, on one side, and citizen on other'.[13]

The working notes for the 'Reply' play a significant role in the analysis to follow because it is my view that the ideas revealed in them not only help us to clarify the content of Fuller's claims, but also to see their interconnections in ways that we might not necessarily detect through a reading of the 'Reply' standing alone. This is not to suggest that the differences between the published and unpublished texts are radical: all that is said in the working notes is essentially consistent with what is said in the published essay. But, able to record his intuitions without fearing the scrutiny of his critics (scholars who he knew were far more adept at the task of arguing claims of legal philosophy than he was) the working notes are striking for how they frequently see Fuller state his intuitions in much bolder terms. The result is not just a more daring authorial voice. It is also, often, the expression of ideas that either never made it into the published version at all, or which do so only in an implicit way. Had some of these points found inclusion in the published record, especially those which go more deeply into the connection between the generality of law and the status possessed by the legal subject, it is possible that we might have spent much of the last half century debating Fuller's jurisprudential contribution in somewhat different terms than we have.

[11] ibid (paginated in hand as p 24). See further chapter eight, I 'Morality'.

[12] The Papers of Lon L Fuller, Harvard Law School Library, Box 12, Folder 1 (notes for 'Reply to Critics').

[13] ibid.

My aim in this chapter, therefore, is not only to set out what the 'Reply to Critics' does, but also what it could have done, with my suggestions here drawn from how Fuller expressed his intuitions in his working notes and other writings salient to the points at issue. Accordingly, I begin in section I with an account of the key claims of the 'Reply' as published. The strategy through which Fuller develops those claims is noteworthy in its own right. That is, Fuller does not reply to his critics by defending his claims outright, but rather responds by diagnosing the commitments, tacit assumptions and blind spots that might explain the basis of his critics' objections to those claims. Only then does he move to clarify the content of his own position, but again, while always keeping the framework of the debate in view. As a strategy for replying, this entirely makes sense: it is only through exposing the commitments, assumptions and apparent blind spots within his critics' position that Fuller can explain why his own jurisprudential concerns seem to so consistently fall through the gaps of a debate about the connections between law and morality. It is as if, 11 years after his intervention as Hart's respondent in the 1958 *Harvard Law Review*, Fuller finally realised that he had to reclaim the agenda; to expose Hart's terms as riddled with assumptions and ambitions that, independently and severally, invite contestation.

The 'Reply' pursues this objective by elaborating responses to four broad inquiries. The first, under the subheading 'The Structure of Analytical Legal Positivism',[14] sees Fuller seek to diagnose what it is about the starting points of the positivist creed that prevents its adherents from either seeing, or simply taking seriously, the concerns and questions of his own jurisprudential agenda. This diagnosis then forms the background for the central analysis of the 'Reply' that Fuller develops by way of response to two questions: first, 'Is Some Minimum Respect for the Principles of Legality Essential to the Existence of a Legal System?';[15] and second, 'Do the Principles of Legality Constitute an "Internal Morality of Law"?'.[16] The 'Reply' then concludes with a collection of comments, 'Some Implications of the Debate',[17] on matters salient, beyond these key points of impasse, to the close of an exchange that had come to span more than a decade.

The primary focus of my analysis in the pages to follow is on the response that Fuller develops to his stated question of 'Do the Principles of Legality Constitute an "Internal Morality of Law"?', as it this analysis that sees him answer his critics' contention that the principles of his so-called internal morality of law merely serve the lawgiver's efficacy by introducing the distinction between law and managerial direction. Fuller's decision to respond to the efficacy charge by reference to this distinction is significant for a number of reasons. First, it marks an explicit return to the form of law, and how and why that form is to be distinguished from other modes of social ordering, as a means of exposing the difference between the

[14] Lon L Fuller, 'A Reply to Critics', *The Morality of Law* (New Haven, Yale University Press, 1969) 191–97.

[15] ibid 197–200.

[16] ibid 200–24.

[17] ibid 224–42.

basic commitments of positivist jurisprudence and his own. Second, through the process of elaborating the content of the distinction, the analysis sees Fuller go deeper into the connections between his own jurisprudential intuitions at the same time as it sees him refine his critique of positivism.

Precisely what goes on in that analysis needs to be carefully mapped. Certainly, ideas already expressed in the first edition of *The Morality of Law* about the absence of any relationship of reciprocity between lawgiver and legal subject in the positivist account remain central, as do associated intuitions about the failure of positivism to make any meaningful provision for the withdrawal of lawgiving power in the event of its abuse. But the nub of Fuller's argument in the 'Reply', even if ostensibly framed around the reciprocity point, goes to a deeper point that might be summarised as follows. The only way that an efficacy reading of the value served by observance of the internal morality of law can make sense is if it is attached to a top-down, essentially coercive conception of law in which the role of the subject is merely to serve the superior's ends. This, Fuller argues, is precisely what the positivist concept of law boils down to, and this is equally why, at its essence, it is a conception of managerial direction rather than law.

The way that Fuller elaborates this argument, through an intricate analysis of applicability of his eight principles of lawgiving to the form of law as opposed to that of managerial direction, sees him move away from the broad frame of a 'uni-directional versus interactional' contest between competing conceptions of law towards a much more subtle terrain in which the *implications* of certain core commitments of positivism are problematised. I set out how this more subtle terrain emerges in the 'Reply' in section I. But I reserve the majority of my remarks about what we might learn about Fuller's jurisprudence from this analysis for my exploration in section II of the working notes, because it is here that we see Fuller take these points to a still deeper level than what is recorded in the published piece. That is, what is initially framed as an argument about the relationship of reciprocity between lawgiver and legal subject that is constituted and sustained by law's distinctive formal features becomes an argument about the conditions, statuses, modes of responsibility and modes of treatment that underlie that relationship of reciprocity and make it possible. Thus, rather than speak in broad terms about what reciprocity demands, or how it limits lawgiving authority, and, indeed, how the top-down orientation of the positivist conception of law misses all of this, we instead see Fuller inquire specifically into what the principles of generality or congruence require, or what they presuppose, and how, in turn, they make sense of each other as fundamental features of law.

In short, the working notes help us to see how Fuller's jurisprudence fundamentally rests on the idea that the distinctive connections between form and agency are what distinguish law from other modes of ordering, as well as impose limits on the scope and possibilities of lawgiving authority. This is what enables Fuller in the 'Reply to Critics' to take his challenge to positivism to a deeper level. Here, what began as a debate between conceptions of law on the connections between law and morality becomes a more nuanced conversation (or at least provides us with the seeds of such a conversation) about the implications of positivism's commitment to the idea that generality is the fundamental feature of law. Here, crucially, what Fuller identifies is not that the concerns of his own jurisprudence are foreign to positivism, and thus necessarily outside its sphere of concern, but rather that much fits within the implications of positivism's stated commitments. My task in section II, therefore, is to explore whether the fruits taken from a journey into the archive might cast the questions in dispute in the Hart-Fuller debate in a new, or at least a more nuanced, light.

I then turn in section III to reflect on the extent to which the 'Reply' succeeds in reclaiming the stated ambitions of Fuller's jurisprudence and restoring them to the agenda of legal philosophy as it then stood, before concluding the chapter with a brief restatement of Fuller's core claims and their interconnection. The summary I offer here is in much the same terms as that which I provided by way of an introduction to Fuller's core claims in chapter one. But, restated at the close of my exposition of Fuller's primary jurisprudential texts, that summary takes on a new significance for how it is now capable of carrying the meaning of what has been offered in this and the preceding three chapters.

I Mapping the 'Reply to Critics'

Fuller begins his 'Reply' by explaining how, as critical reviews of *The Morality of Law* came in, he 'became increasingly aware of the extent to which the debate did indeed depend on "starting points"—not on what the disputants said, but on what they considered it unnecessary to say, not on articulated principles but on tacit assumptions'.[18] This gesture to 'starting points' is a direct reference to Hart's remark, at the beginning of his review of Fuller's book, that he was 'haunted by the fear that our starting-points and interests in jurisprudence are so different that the author and I are fated never to understand each other's works'.[19] Though Fuller tells us that he deliberated at length about whether he should go in for one more round of debate ('[o]ne has the feeling that at some point such an exchange must terminate'),[20] the 'Reply' stands as proof that he did not wish that debate to draw to a close in the face of the kind of foundational miscommunication to which Hart alludes. And so, Fuller declares, a final round is indeed needed 'to bring these tacit assumptions to more adequate expression than either side has so far been able to do'.[21]

[18] ibid 189.

[19] HLA Hart, 'Lon L. Fuller: *The Morality of Law*' (1965) 78 *Harvard Law Review* 1281, reprinted in HLA Hart, *Essays in Jurisprudence and Philosophy* (Oxford, Clarendon Press, 1983) 343.

[20] 'Reply to Critics' (n 14) 188.

[21] ibid 189.

A 'The Structure of Analytical Legal Positivism'

The first section of his 'Reply' records Fuller's attempt to set out his own understanding of 'the basic intellectual commitments underlying analytical legal positivism',[22] with the term 'analytical' meant here to refer to 'an intellectual mood that finds more satisfaction in taking things apart than in seeing how they fit and function together'.[23] The basic theme of his critique is therefore immediately clear: what's wrong with positivism is its neglect of how law actually works; 'the elements of tacit interrelatedness that infuse—though always somewhat imperfectly—what we call, by no accident, a legal *system*'.

This neglect, Fuller argues, can be traced to the five basic assumptions, or 'starting points', that shape the positivist creed: the way it understands law as a 'one-way projection of authority, emanating from an authorized source and imposing itself on the citizen';[24] the absence within its commitments of any meaningful limit to lawgiving authority;[25] the equal absence of any sense of the distinctiveness of the lawgiver's role; and that this role might require the discharge of a distinctive role morality;[26] and a commitment to the idea that 'clear thinking is impossible unless we effect a neat separation between the purposive effort that goes into the making of law and the law that in fact emerges from that effort'.[27]

Fuller is clear that the first four of these assumptions are interdependent: the positivist 'sees the law at the point of its dispatch by the lawgiver and again at the point of its impact on the legal subject', and so in failing to see that 'the creation of an effective interaction between them is an essential ingredient of the law itself' recognises in the functioning of a legal system 'nothing that can truly be called a *social dimension*'.[28] But it is the fifth starting point that Fuller regards as the 'most central article of faith in the credo of positivism': its apparent intellectual commitment to the idea that 'clear thinking is impossible unless we effect a neat separation between the purposive effort that goes into the making of law and the law that in fact emerges from that effort'.

Two culprits are identified as responsible for this theoretical approach: the influence of ordinary language philosophy (Fuller here reacting to his critics' apparent hysteria in the face of his willingness to apply the term 'law' to the internal regulations of a range of associational forms)[29] and the influence, on Hart's positivism in particular, of utilitarian philosophy. Fuller's elaboration of this last

[22] ibid 191. Fuller suggests that this structure of thought is 'generally shared by Austin, Hart, and Kelsen', and later adds that he also intends it to include the thought of Hart, Ronald Dworkin, Marshall Cohen and Robert Summers, each of whom had written responses to *The Morality of Law* that, Fuller suggests, reveal 'an amazing uniformity'.

[23] ibid.

[24] ibid 192.

[25] ibid.

[26] ibid 193.

[27] ibid.

[28] ibid.

[29] ibid 195–96.

point speaks directly to the critique that has been central to his own jurisprudence since the introduction of his eunomics project; namely, that utilitarian philosophy 'encourages us in the intellectually lazy notion that means are a matter of mere expediency and that nothing of general significance can be said about them'.[30] Later in the 'Reply', however, Fuller takes this point further when he suggests that what is missing in Hart's apparent support for the utilitarians' belief that principles of lawgiving akin to Fuller's eight should be valued only 'in so far as they contribute to human happiness and other substantive aims of the law'[31] is any acknowledgment of how that very position presupposes the kind of 'stability of interactional processes within a society' that is most typically 'furnished by a conscientiously administered legal system'.[32]

But at this early juncture, setting the stage for the arguments of his reply, Fuller's diagnosis of the five starting points of positivism is intended to leave us with this simple message. Legal positivism presents us with a unidirectional conception of law that imposes itself on the legal subject with no apparent limits to its capacity or authority to do so. This in turn explains why such a legal philosophy would be equally unconcerned with the distinctiveness, or the demands, of the lawgiving role, and, indeed, with the means of law more generally.

B 'Is Some Minimum Respect for the Principles of Legality Essential to the Existence of a Legal System?'

These observations form the background for Fuller's examination of the enduring contest between him and his critics on whether some notion of law's purpose ought to find a place within an account of its nature. The context here is Hart's remark, at the close of his review of *The Morality of Law*, that the 'virtues and vices' of the book both spring from how Fuller 'has all his life been in love with the notion of purpose'.[33] Fuller thus takes this apparent impasse about the place of notions of purpose within an account of law as a launching point for exploring the question of whether some minimum respect for the principles of legality is essential to the existence of a legal system.

It seems at first as though Fuller is going to move his reply on this point directly into the territory of debates about the necessary conditions for law to exist; something akin to what he offers in his 1958 reply to Hart, as well as through the tale of Rex in *The Morality of Law*. This, at least, is what seems suggested in how he begins his analysis by noting that he and his critics appear to share the view that the declaration that law exists is contingent, in some important sense, on that putative law conforming to 'certain standards that will enable it to function meaningfully in men's lives', with the standards in view being those he enumer-

[30] ibid 197.
[31] ibid 237–38.
[32] ibid 238.
[33] Hart, 'Review' (n 19) 366, cited in 'Reply to Critics' (n 14) 189.

ates through his model of the internal morality of law.[34] But his desire to respond to Hart on the matter of purpose sees Fuller instead turn to the wider question of the value associated with observance of the internal morality of law. Here he is again simply following Hart's lead. That is, as will be recalled from chapter four, Hart's review of *The Morality of Law* does not engage in any significant way with Fuller's claim that the existence of law is contingent on observance of the principles of the internal morality of law; his attention, rather, is directed primarily to Fuller's claim that these principles are moral in character. What we see in the 'Reply', therefore, is Fuller bringing these elements together to suggest that what divides him and his critics is the question of '*to what end* is law being so defined that it cannot exist without some minimum respect for the principles of legality'.[35] The result is that the existence question joins his response to Hart on the purpose point to become an analysis of why efficacy is an inadequate explanation of the value of the principles of the internal morality of law.

C 'Do the Principles of Legality Constitute an "Internal Morality of Law"?'

The archival materials reviewed in chapter four will have made clear that Fuller was genuinely puzzled by the efficacy reading to which his claims about the internal morality of law were reduced by his critics. Although no mention is made about Hart's omission of the trusteeship point,[36] Fuller sets up a two-track strategy by way of response. He asks, first, how it is that his critics can be so apparently settled in their conviction that there is nothing of inherent moral significance in his eight principles of lawmaking to warrant their designation as a 'morality'; and then, second, why they are so confident that the appropriate designation is instead that which is ostensibly opposed to 'morality'; that is, 'efficacy'.

With respect to the first question, Fuller suggests two possible explanations. One is his critics' apparent belief that the existence or not of law is a matter of neutral moral significance; a view, according to Fuller, that is itself plagued with at least two problems. First, it ignores the relationship between the stability afforded by a legal order and the possibility of moral conduct more generally. To live the good life, Fuller argues, 'requires something more than good intentions, even if they are generally shared'. It also 'requires the support of firm base lines for human interaction, something that—in modern society at least—only a sound legal system can supply.'[37] Second, the idea that the existence or not of law is a matter of neutral moral significance overlooks the special moral responsibilities that attach to the lawgiver's role, and which, Fuller argues, cause that role to deserve 'some more flattering comparison than that offered by the practices of the

[34] 'Reply to Critics' (n 14) 198.
[35] ibid (Fuller's emphasis).
[36] See chapter four, IIB 'Efficacy and trusteeship'.
[37] 'Reply to Critics' (n 14) 205.

thoughtful and conscientious poisoner who never forgets to tear the chemist's label off before he hands the bottle to his victim'.[38]

But the main explanation Fuller offers for why his critics insist that they can trump his 'morality' designation with their 'efficacy' one is rooted in the first of the starting points of positivism that he earlier identified; namely, the assumption 'that law should not be viewed as the product of an interplay of purposive orientations between the citizen and his government but as a one-way projection of authority, originating with government and imposing itself on the citizen'.[39] The argument is about the absence of reciprocity within the positivist account of law, and Fuller elaborates it by introducing the distinction between law and managerial direction.

Distinction between Law and Managerial Direction

According to Fuller, managerial direction and law share much in common. To begin, they share a vocabulary—terms like 'authority', 'orders', 'control', 'jurisdiction', 'obedience', 'compliance' and 'legitimacy' being common to both—and, above all, both are concerned to direct human activity.[40] But they do so in different ways. For instance, the law-abiding citizen does not, like the managerial subordinate, '*apply* legal rules to serve specific ends set by the lawgiver, but rather *follows* them in the conduct of his own affairs'.[41] Or we might say that the directives of a managerial system primarily regulate the relations between the subordinate and his superior, while the rules of a legal system 'normally serve the primary purpose of setting the citizens' relations with each other and only in a collateral manner his relations with the seat of authority from which the rules proceed'.[42]

These are very general statements, but they are given more precise content when Fuller goes on to compare what implications his eight principles of the internal morality of law carry for the mode of managerial direction as opposed to law. The commonalities are highlighted first. Five of the eight principles, Fuller suggests, find a natural home in the managerial context. Some principle of publicity, for instance, must apply to the managerial relation if 'the superior is to secure what he wants through the instrumentality of the subordinate'.[43] The manager's directives must also be reasonably clear, free from contradiction, capable of being executed, and not changed so often as to frustrate the efforts of the subordinate to act on them. Unless these five principles are observed, the efficacy of the managerial enterprise in the pursuit of its desired ends might be seriously impaired.[44]

But, Fuller contends, the same cannot be said for the remaining three principles of generality, non-retroactivity and congruence between official action and

[38] ibid 206.
[39] ibid 204.
[40] ibid 207.
[41] ibid.
[42] ibid 207–08.
[43] ibid 208.
[44] ibid.

declared rule. Generality has no natural place in the managerial context because although this requirement might in some cases serve expediency in certain situations, it is hardly something that can be expected; as Fuller explains, 'the subordinate has no justification for complaint if, in a particular case, the superior directs him to depart from the procedures prescribed by some general order'.[45] This in turn illuminates the equal irrelevance to the managerial mode of the principle that the actions of the superior conform to previously announced rules, because there can be no expectation on the part of the subordinate that the managerial enterprise will be informed by a commitment to governance through general rules in the first place.[46] As for the remaining principle of non-retroactivity, 'no manager retaining a semblance of sanity would direct his subordinate today to do something on his behalf yesterday'; again, it is simply not relevant.[47]

The absence within the managerial form of any constitutive commitment to proceeding through general rules thus generates the corresponding inapplicability of the principles of congruence and non-retroactivity. But this is not just a matter of comparing checklists. The key message is about relationships and the extent to which power is mediated through them; about what one can expect from participation in a managerial order as opposed to a legal one. The point, then, is that what is *not* required to produce a functioning managerial relation on Fuller's account is observance of those of the eight principles of the internal morality of law that protect the expectation on the part of the subordinate that the superior's power over her will be mediated by previously declared intentions that are generally applicable to all. This does not mean that all elements of interaction or reciprocity are necessarily absent within the managerial relation.[48] A superior who overburdens those under his direction, Fuller points out, will likely confront a problem of morale, and so might find it necessary to introduce a certain level of reciprocity between him and his subordinate if he is not to put his goals in jeopardy. But this is merely the 'tacit reciprocity of reasonableness and restraint' that is 'collateral to the basic relation of order-giver and order-executor'.[49] It is something quite different to 'the existence of a relatively stable reciprocity of expectations between lawgiver and subject' that is 'part of the very idea of a functioning legal order'.[50]

Of the eight principles, the latter quality of reciprocity is seen in its most obvious form in the principles of congruence between official action and previously announced rule. Indeed, the commitment by government to faithfully apply those rules that it has previously declared as determinative of the citizen's rights and duties captures the very idea of the 'rule of law'.[51] But Fuller makes clear that it is generality that is the foundational requirement from which all else, including the

[45] ibid.
[46] ibid 208–09.
[47] ibid 209.
[48] ibid.
[49] ibid.
[50] ibid.
[51] ibid 209–10.

principle of congruence, is generated: it is necessary that the rules upon which the whole notion of the rule of law depends take the form of general declarations if they are to do their work in furnishing 'baselines for self-directed action'.[52] The principle of non-retroactivity can equally be traced to generality. This is not only because the expectation that we are bound only by rules that apply to our actions prospectively makes no sense unless there are rules in the first place; the phenomenon of permissible retroactivity (that which is sometimes recruited to remedy 'abuses and mishaps in the operations of a legal system') will also often speak to the implication of the requirement of generality that like cases be treated alike.[53]

This is the stage from which Fuller declares that the positivist vision of law aligns much more closely with the idea of managerial direction than it does with law. The reason is written in positivism's apparent indifference to the 'horizontal element' that characterises the relationships that constitute a functioning legal order.[54] Hart's analysis of the foundations of legal order in *The Concept of Law* is highlighted as a case in point. As Fuller reads it, the only departure from a managerial frame of reference to be found in that analysis lies in Hart's discussion of the 'germ of justice' that attaches to the implication of the requirement of generality that like cases be given like treatment.[55] But even then, there is nothing in Hart's account to suggest that government has any obligation towards the citizen to realise this germ of justice in the way it makes and administers law.[56] Nor is there any suggestion that the citizen's voluntary cooperation in obeying law, readily accepted by Hart as essential to the functionality of a legal system,[57] must be 'matched by a corresponding cooperative effort on the part of government'.[58] The positivist conception of law thus aligns more closely with managerial direction than it does with law because it contains no acknowledgment of how maintaining a legal system in existence 'depends upon the discharge of interlocking responsibilities—of government toward the citizen and of the citizen toward the government'.[59] For positivists, it seems, the lawgiver can do as he likes, while the legal subject can presumably do little other than the lawgiver's bidding.

The argument that this analysis is designed to serve is the contention that the only way a positivist can view the principles of the internal morality of law as mere counsels of expediency is if these last three principles, that necessarily imply a relationship of reciprocity, have no place within the positivist vision of law. But the analysis is equally intended to support the general diagnosis that positivism offers a deficient concept of law because it misses too much about the conditions that make law itself possible. As Fuller sums it up:

[52] ibid 210.
[53] ibid 211.
[54] ibid 233.
[55] ibid 215.
[56] ibid.
[57] ibid 216, citing the first edition of Hart's *The Concept of Law* (Oxford, Oxford University Press, 1961) 196.
[58] ibid.
[59] ibid.

> The commitment implied in lawmaking is not, then, simply an element in someone's 'conceptual model'; it is a part of social reality. I have been emphasizing that obedience to rules loses its point if the man subject to them knows that the rulemaker will himself pay no attention to his own enactments. The converse of this proposition must also be kept in mind, namely, that the rulemaker will lack any incentive to accept for himself the restraints of the Rule of Law if he knows that his subjects have no disposition, or lack the capacity, to abide by his rules; it would serve little purpose, for example, to attempt a juristic ordering of relations among the inmates of a lunatic asylum. It is in this sense that the functioning of a legal system depends upon a cooperative effort—an effective and responsible interaction—between lawgiver and subject.[60]

This passage is meant to be conclusory on the matter of positivism's disregard of the elements of reciprocity that constitute and maintain a legal order. But it clearly also signals to more than this, above all, to *who* the parties to these relationships of reciprocity are, not just in terms of titles or designations, but in terms of statuses, responsibilities and capacities. The reciprocity that Fuller insists is constitutive of the relationships that comprise a functioning legal order will break down if the lawgiver is indifferent in his own observance of the rules that he expects the subject to follow; the subject will likely respond with her own version of indifference. That reciprocity will similarly break down (or will never be constituted at all) if the subject lacks the capacities to interact with the lawgiver's regime of rules in the first place. Fuller's remark that it would be pointless to attempt a 'juristic ordering' among the 'inmates of a lunatic asylum', flippant as it might sound, hardly carries a flippant message. Akin to the passage in Chapter 4 of *The Morality of Law* that explores the conception of the person implicit in the principles of the internal morality of law, what Fuller is suggesting here is that the person capable of being the subject of a legal order, of contributing through her capacities to the creation and maintenance of the relationships of reciprocity with the lawgiver that enable such an order to function, can only be a person possessed of responsible agency. It is this idea that we see developed still further in the working notes for the 'Reply' that I analyse in section II below.

D 'Some Implications of the Debate'

There is not space here to set out the numerous other points that Fuller goes on to make in the aftermath of his analysis of the distinction between law and managerial direction. All, however, go in one way or another to how his critics and the conceptions of law to which they are married overlook the conditions that enable law to function as law. Fuller then moves beyond the parameters of this critique to consider a range of other implications of his ongoing debate with Hart and others. This final section of the 'Reply' contains a range of conversations: interpretation;[61]

[60] ibid 219.
[61] See chapter seven, IV 'Fuller, Dworkin and interpretation'.

morality (and especially the neglected phenomenon of role morality);[62] and, 'with a measure of reluctance', the status of debates between 'positivism' and 'natural law'.[63] Much can also be learned about how Fuller understood his contest with his critics from the way that he brings the 'Reply' to its ultimate close. Akin to how he first articulated his criticisms of positivism in *The Law in Quest of Itself* almost three decades earlier, Fuller brings the 11 years of the Hart-Fuller debate to a close by suggesting that positivism's apparent disregard for 'the general problem of achieving and maintaining legality' can be traced to a methodological outlook.[64] Optimistic that this question might better receive the attention of legal philosophers in future, Fuller concludes the 'Reply' with the hope that:

> Perhaps in time legal philosophers will cease to be preoccupied with building 'conceptual models' to represent legal phenomena, will give up their endless debates about definitions, and will turn instead to an analysis of the social processes that constitute the reality of law.[65]

II Generality, Efficacy and Agency: Insights from the Archive

Fuller invokes the distinction between law and managerial direction in the 'Reply to Critics', above all, to deflect the charge that the value of his principles of the internal morality of law is reducible to the efficacy with which they aid a lawgiver's pursuit of ends through law. But the process of teasing out that point, and the association he sought to make from it about positivism's apparent failure to illuminate what distinguishes law from other modes of ordering, saw Fuller reveal much more by way of a development of his jurisprudence than perhaps even he anticipated. Indeed, the possibility that he did not fully grasp all of the implications of his own reasoning seems to be well-evidenced in how he ultimately chose not to include in the published version of the 'Reply' a number of intuitions that, I would argue, take the richness and coherence of his jurisprudence to a much higher level.

I suggested in the preceding analysis that one of the most interesting and illuminating features of Fuller's analysis of the distinction between law and managerial direction in the 'Reply' is the message it sends about how the subject of a legal order experiences the power of a superior in a manner that is qualitatively different from a similarly positioned subject within a managerial relation. This is the message that we receive from the passages in the 'Reply' where Fuller explains why the principles of generality, non-retroactivity and congruence from his model

[62] 'Reply to Critics' (n 14) 239.
[63] ibid 241.
[64] ibid 242.
[65] ibid.

of the internal morality of law cannot intelligibly be explained through the idea of efficacy. To explore how and why Fuller's working notes on these points offer us richer insight into his position, it is helpful to reproduce those passages in the 'Reply' that link the principle of congruence to that of generality, and then in turn link the phenomenon of generality to the principle of non-retroactivity:

> Applying rules faithfully implies, in turn, that rules will take the form of general declarations . . . Furthermore, if the law is intended to permit a man to conduct his own affairs subject to an obligation to observe certain constraints imposed by superior authority, this implies that he be not told at each turn what to do; law furnishes baselines for self-directed action, not a detailed set of instructions for accomplishing specific objectives.
>
> The twin principles of generality and of faithful adherence by government to its own declared rules cannot be viewed as offering mere counsels of expediency. This follows from the basic difference between law and managerial direction; law is not, like management, a matter of directing other persons how to accomplish tasks set by a superior, but is basically a matter of providing the citizenry with a sound and stable framework for their interactions with one another, the role of government being that of standing as a guardian of the integrity of this system.[66]

Fuller then continues on the next page to explain how:

> under the rule of law, control over the citizen's actions is accomplished, not by specific directions, but by general rules expressing the principle that like cases should be given like treatment. Now abuses and mishaps in the operation of a legal system may impair this principle and require as a cure retrospective legislation. The retrospective statute cannot serve as a baseline for the interactions of citizens with one another, but it can serve to heal infringements of the principle that like cases should receive like treatment.[67]

The message of these paragraphs is that to be a lawgiver is to relinquish any claim to be able to tell the subject at each turn what she should do. To be a lawgiver is instead to convene a relationship with that subject which is mediated by the principles of generality, non-retroactivity and congruence, as well as the other five, and to accept the implications that attach to any attempt to subject human conduct to the governance of general rules.

The working notes for the 'Reply' are valuable for deepening our understanding of these points because they reveal a step-by-step trajectory of reasoning that might be missed in the more narrative and often free-wheeling prose of the 'Reply' as published. The notes begin at the beginning, with what is obviously agreed about the basic nature of law: that it is an enterprise of general rules. In one formulation, this foundational feature is described as a 'fundamental characteristic of a legal system', and contrasted with a system that 'confined itself to specific instructions directed towards individuals', which 'would not be called law, and could not display the characteristic problems of law'.[68] In another, Fuller specu-

[66] ibid 210.

[67] ibid 211.

[68] Undated document titled 'Generality', crossed out and replaced with 'Legal Morality', The Papers of Lon L Fuller, Harvard Law School Library, Box 12, Folder 1 (notes for 'Reply to Critics').

lates on a 'ruler who never issues any general command', and responds that this, 'obviously', cannot be called law.[69]

What is obvious—law's generality—then leads in the notes to what generality means for the kind of position occupied by the lawgiver and legal subject alike. For the lawgiver, Fuller suggests, law's formal feature of generality is hardly something to be embraced for its expediency; that the actions of the government toward the citizen must be by a previously announced general rule is 'plainly not for convenience or efficacy of governmental order, otherwise would be able to depart'.[70] But for the legal subject, generality opens up a particular way of experiencing power, specifically, of experiencing one's own agency in the face of power. As the notes tell it, 'legal rules tell their addressee not what he should do on behalf of someone else, but what he should do for himself', and this is what marks a clear distinction between the servant, for whom, 'where his initiative is invoked, it is ruleless', and the legal subject, who is part of a 'rule-controlled initiative'.[71]

It is his critics' failure to take seriously these modes of limitation and agency that are implicit in law's form that, according to Fuller, causes them to be unable or to refuse to see how 'meeting the demands of legal morality is a *difficult thing*';[72] that it requires 'a real effort at identification with the subject of the law', or, at least, 'sufficient concern for the other fellow to be willing to spend the time and energy necessary to take him into account'.[73] To commit to generality is not to prepare the subject 'to be ready to jump', but to offer 'baselines for the conduct of his own affairs' that enable the subject 'to plan his life, and to adjust his relations with his fellows'.[74] This is not an optional commitment: the legal subject is 'not a man shaped by some outside force', but rather is 'part of the system' itself.[75]

The notes become still more interesting when these thoughts are directed to the issue of efficacy that divides Fuller and his critics on the question of the value of his principles of the internal morality of law. Again, the notes here record Fuller's efforts to develop the implications of this point for both the lawgiver and legal subject. For the lawgiver, Fuller suggests that 'there are moral dilemmas and temptations in creating and the administration of a legal system, and that these hinge about what I have called principles of legality', and that '[t]o try to convert into efficacy is, I submit, a serious and dangerous perversion', with the words 'Trusteeship Hart' appearing next to this point.[76] But the way Fuller develops his

[69] ibid.

[70] Untitled note, The Papers of Lon L Fuller, Harvard Law School Library, Box 12, Folder 1 (notes for 'Reply to Critics').

[71] Untitled note, The Papers of Lon L Fuller, Harvard Law School Library, Box 12, Folder 1 (notes for 'Reply to Critics').

[72] Undated document titled 'External morality relation', The Papers of Lon L Fuller, Harvard Law School Library, Box 12, Folder 1 (notes for 'Reply to Critics').

[73] ibid.

[74] Undated and untitled document, The Papers of Lon L Fuller, Harvard Law School Library, Box 12, Folder 1 (notes for 'Reply to Critics').

[75] Undated note titled 'What is at stake?', The Papers of Lon L Fuller, Harvard Law School Library, Box 12, Folder 1 (notes for 'Reply to Critics').

[76] Undated and untitled note, The Papers of Lon L Fuller, Harvard Law School Library, Box 12, Folder 1 (notes for 'Reply to Critics').

objection to the efficacy point through reference to the status of the legal subject is still more illuminating for our understanding of the core commitments of his jurisprudence. As he explains it:

> A legal system does not succeed or achieve 'efficacy' simply because the citizen is willing to obey orders. It succeeds if it creates a stable order by which the citizen can orient his conduct toward his fellows . . . a functioning society is its goal, *not a subservient populace ready to do what they are told to do*. This is not some extra-legal purpose assigned to law from without; it is intrinsic to the very notion that government should act towards the citizen only in compliance with previously announced general rules.[77]

Over the course of what is recorded in the working notes, therefore, we see Fuller exploring not just what positivists ignore or miss in their account of law, but arguments for why they cannot so readily dismiss the moral value of the principles of the internal morality of law if they accept, as they do, law's fundamental feature of generality. But, above all, the trajectory of reasoning expressed in the notes speaks to how different modes of expressing power are experienced by those who participate within them; how the constraint of one is reflected in the agency of the other. In managerial direction, this constraint is borne largely by the subject, and agency by the manager. But in law, the dynamic works the other way. In law, the lawgiver's constraint is reflected in the subject's agency.

We therefore have a trajectory of reasoning where distinctiveness, the accepted and distinguishing fact of law's generality, leads us to the diagnosis that what is present in law, and absent in managerial direction, is a particular kind of relationship that the formal feature of generality sustains. This in turn causes law to generate a particular kind of value for the lawgiver and legal subject alike—different for each but, crucially, compatible. By proceeding through the medium of previously announced general rules, the lawgiver gains the subject's cooperation, and so finds himself overseeing a system in which the vast majority voluntarily comply with his rules. On the flip-side, under law the subject knows where she stands in relation to the lawgiver and what he asks of her, and so can conduct her own affairs, exercise her agency, as she sees fit. But, and crucially, under law this agency remains her own; she is not a subservient ready to be told what to do. She is a *subject of* law—of the mode of governance that presupposes her as a responsible agent and so must also respect her as such—not *subject to* the direction of another. All of this, Fuller tells us, is 'intrinsic to the very notion that government should act towards the citizen only in compliance with previously announced general rules'.

[77] ibid (emphasis in original).

III Reflections on the 'Reply to Critics'

The 'Reply' is not a point by point debate with positivism, or indeed with Fuller's critics. It is instead, above all, an attempt on Fuller's part to diagnose what is it about the positivist analysis of law that prevents it from seeing what he sees, and from taking seriously what he takes seriously as fundamental concerns of the inquiry of jurisprudence. In his preface to the second edition of *The Morality of Law*, to which the 'Reply' was added as a fifth chapter, Fuller states that the fact that the first four chapters of the book remain virtually unchanged between editions is not to imply his complete satisfaction with either the form or the substance of the presentation achieved in them; he simply had 'not proceeded far enough' in his rethinking of the problems to undertake any substantial reformulation.[78] But he does claim, for the 'Reply', that the essay records 'a better articulation of my own dissatisfactions with analytical legal positivism than I had ever achieved before'.[79]

Certainly, the 'Reply' sees Fuller return to root questions, such as the fundamental differences in starting points between his jurisprudential orientation and that of positivism. But there is at least some case for suggesting that the 'Reply' does not necessarily see Fuller return to some of his own root intuitions about how to navigate and debate those differences. We do not necessarily see in the 'Reply', for instance, despite the 'five starting points' clarification, the kind of diagnosis of positivism's mission to see law examined as a discipline in isolation from others that was the animating theme of *The Law in Quest of Itself*. These thoughts do, however, make their way into the working notes, where Fuller speaks of positivism's efforts at 'retailoring the limits of law to fit the boundaries of some allegedly independent discipline',[80] and its associated disregard for the wisdom to be gained from examining law in practice. Indeed, a concern for the divide between abstract theory versus theory commenced from the internal view of participants looms large in these notes, especially one in which Fuller suggests that '[t]hose of us who have participated directly or vicariously in the law's basic processes . . . should not stand by while others define our tasks, and restate them in terms more congenial to their own predispositions'.[81]

Like so many of Fuller's writings, the 'Reply' suffers for its loosely deployed vocabulary, which in turn puts some of his claims and criticisms at risk of internal contradiction. Fuller's apparent disquiet towards positivism's 'conceptual' approach[82] is expressed alongside his own desire to argue that 'an element of

[78] Lon L Fuller, 'Preface to the Second Edition', in *The Morality of Law*, 2nd edn (New Haven, Yale University Press, 1969) v.

[79] ibid vi.

[80] Undated note titled 'What is at stake', The Papers of Lon L Fuller, Harvard Law School Library, Box 12, Folder 1 (notes for 'Reply to Critics').

[81] ibid.

[82] 'Reply to Critics' (n 14) 219, 242.

commitment by the lawgiver is implicit in the concept of law'.[83] As for more substantive deficiencies, it is notable that little is ultimately made in the 'Reply' of earlier claims about the apparent affinities, in practice, between observance of the internal morality of law and the substantive morality of legal ends, with Fuller here simply referring the reader to his discussion of this point in his much earlier reply to Cohen and Dworkin.[84] Still more importantly for the controversies of his jurisprudence, and as I foreshadowed in section I, the 'Reply' ultimately does not see Fuller return to any direct engagement with the question of the extent to which his principles of the internal morality of law must be observed in order for law to exist. While at points he speaks in one breath of the 'existence and conscientious administration' of a legal system,[85] he does not at any point refine this connection in a way that would see it capable of confronting what has been regarded by most positivists since as a knock-down to Fuller's position; namely, Raz's argument that 'conscientious administration' (ie observance of the principles of the rule of law) is one thing, but the existence of law is another. I return to this point in chapter six.

Clearly, there are a lot of loose ends. Still, there are other ways in which the 'Reply' leaves us in a position to explore an arguably more constructive engagement with positivism on the question of what value (and who) observance of the eight principles of legality serves. On this point, the 'Reply' as published is largely consistent with *The Morality of Law* generally in privileging what I described in chapter four as the 'ethos' reading of the morality of the internal morality of law. Indeed, at one point Fuller explains that 'the onerous and complex responsibility' on the lawgiver to do their job right in the first place is what he has 'tried to describe in the phrase "the internal morality of law"'.[86] At another he is deeply critical of how none of his critics seems willing to pass an adverse moral judgement on the legislator who, through 'indifference to the demands of his role, confuses or misplaces the guideposts by which men coordinate their actions'.[87] The point is again echoed in private correspondence to Hart, where Fuller comments that 'the lawgiver who deliberately uses retroactive laws to frighten the populace into impotence . . . has done something worse than merely impairing the internal morality of law and I would be glad to say that he violated a moral duty.'[88]

The 'ethos' reading, therefore, is clearly dominant, in the sense that it is the meaning possessed most explicitly by Fuller as an explanation for his claims about law's internal morality. But it does not follow from this that the 'moral value' reading (law's value to the subject) is necessarily secondary. The two, in *The Morality of Law* as much as in the 'Reply', stand in a close relationship. Because what these writings make clear is that to say that the lawgiving role implicates a distinctly moral duty is to say that this duty is owed to something, or someone,

[83] ibid 216.
[84] ibid 224.
[85] ibid 206.
[86] ibid.
[87] ibid 238.
[88] Letter from Lon L Fuller to HLA Hart, 3 February 1965, The Papers of Lon L Fuller, Harvard Law School Library, Box 3, Folder 14 (correspondence).

capable of generating the distinctly moral character of that duty. It is a 'relationship with persons';[89] specifically, to the fate of persons in possession of a capacity for responsible agency that entitles them to be respected as such. In Fuller's eyes, therefore, when positivists overlook these interrelated ways in which law incorporates inherent moral dimensions, they are guilty of a double failure: a failure, conceptually, to properly describe legal reality, as well as a failure, morally (as he put it in a working note for the 'Reply) to be 'animated by a concern for the other fellow'.[90]

Finally, is it helpful to reflect on one of the important differences between the 'Reply' as published and the working notes through which Fuller planned its content. Here I have in mind how, for the most part, none of the notes explore the points at issue within the explicit frame of a comparison between law and managerial direction. This analysis might be in view, but the notes themselves speak instead of generality, of the limits that it and the seven other principles that it implies impose on lawgiving, of the status of the legal subject as someone not merely to be acted upon, and of how none of this can be intelligibly explained purely in terms of the efficacy that accrues to a ruler who opts into the enterprise of lawgiving.

In light of these cues from the archive, it seems fitting to suggest that we should not misread, or at least not overstate, the message that Fuller invites us to take by distinguishing law and managerial direction as he does in the 'Reply'. We must be careful, that is, to not make too much of the distinction *as* a distinction, in the sense of assuming that what Fuller is asking us to do is to locate a putative instance of law in one or the other camp (ie if it's not law, it must be managerial direction). There will, of course, be a place for such a declaration. But to jump to it too quickly comes with the risk of disavowing the notion, central to Fuller's jurisprudential approach, that all instances of putative law exist along a continuum; that the existence of law is always a matter of degree.

To overstate the distinction between law and managerial direction might equally be to suggest that the two are intrinsically opposed, and thus that it would be a contradiction in terms to propose that managerial direction might sometimes be called upon in service of a condition of law. But here again, as I explained in chapter four in my discussion of Fuller's position on permissible retroactivity—permissible, that is, if such might serve the cause of legality[91]—the ultimate measure to keep in view is a qualitative one. That is, it is not necessary to realise all eight principles, to perfection, all of the time. Derogations from this or that principle might not always signal the kind of compromise that ought to concern us. Derogations from the quality of relationship between a lawgiver and subject that a legal system constitutes and maintains, however, will. In short, we have to keep the wider point that Fuller was trying to make in view—a point about the

[89] *Morality of Law* (n 14) 42.

[90] Undated note titled 'External morality relation', The Papers of Lon L Fuller, Harvard Law School Library Box 12, Folder 1 (notes for 'Reply to Critics').

[91] See chapter four, IA 'The Story of King Rex'.

qualitative differences, for the position, responsibilities and status of lawgiver and legal subject alike, that are generated by the *formal* differences that distinguish law and managerial direction as modes of social ordering.

But there is a still more important reason why we must not make too much of the distinction between law and managerial direction. It matters little, to our understanding of Fuller's jurisprudence, precisely what law vacates to when it ceases to be law-like. We might call this other mode 'managerial direction', or something else; the designation is not crucial. What is crucial, however, is that we understand why that other mode of ordering is not law-like; why the continuum along which we might chart the recognisably legal has run out, or, as the case may be, how and why the conditions and presuppositions necessary to constitute the recognisably legal were absent in the first place. Here, then, lies the nub of Fuller's distinction between law and managerial direction as a critique of positivism. A philosophy of law that lays claim to being able to tell us what law *is* should be able to tell us what law is *not*. Positivism, however, offers little by way of a meaningful answer to this question, and yet it is one that surely ought to be of fundamental concern to any theory of law.

IV Conclusion

Frederick Schauer has observed that as we look back at the Hart-Fuller debate and recognise who was clearly the better debater and more rigorous philosopher, 'we are far less certain as to who was actually right'.[92] The 'Reply to Critics' does not necessarily leave us with any clear answer to this question. Nor is it necessarily clear that we can agree with Fuller's own view that the 'Reply' achieved 'a better articulation of my own dissatisfactions with analytical legal positivism than I had ever achieved before', or that he had managed to 'lay bare' the basic premises of his thought.[93]

But what we clearly do see in the 'Reply' is Fuller taking up the impasse between him and his critics on the question of whether the principles of his internal morality of law hold any inherent moral significance and attempting to turn that impasse into a conversation about the unintelligibility of his critics' efficacy response as either an evaluation of his own claims or an accurate reflection of their own commitments about law. Because while positivists might think that such can hold as their *own* evaluation of the value to which observance of the principles of Fuller's internal morality of law is reducible, they can only do so, according to Fuller, by holding to a vision of law that downplays the implications of certain features of law's form—above all, its generality—that they apparently

[92] Frederick Schauer, 'A Life of HLA Hart: The Nightmare and the Noble Dream' (2006) 119 *Harvard Law Review* 852, 864–65.

[93] 'Reply to Critics' (n 14) vi.

accept as given. Thus, Fuller tells us, the positivist conception of law is either really better understood as managerial direction (where the requirement of generality is not constitutive) *or* it is a conception of law that is yet to make sense of, or to candidly accept, the implications of its own commitments.

Fuller thus sets himself the task in the 'Reply' of filling these gaps; to lay bare the elements of 'those social processes that constitute the reality of law' that positivists either explicitly or implicitly accept, but which they avoid reconciling with other tenets of their theoretical position. By dwelling in this territory of the implications of avowed or assumed commitments, Fuller brings the legal subject and the lawgiver's responsibilities to her into view; shows how these generality-induced demands introduce limits to lawgiving authority that positivists do not wish to acknowledge; and, in doing so, suggests how we might determine when the form and qualities of a legal order are extant, and when they are not. What began as a conversation about the connections between law and morality, about law's value to a lawgiver's efficacy and no other, is transformed into something much more nuanced, and much more squarely within the territory of Fuller's enduring aim to see debates of jurisprudence attend to the forms, limits, distinctiveness, modes of participation and animating moral values, of the forms through which law finds expression.

We can therefore return to, and review, the statement of the core claims of Fuller's jurisprudence that I offered by way of introduction in chapter one. Fuller's vision of law begins and never sways from the view that to label something as 'law' is to designate a distinctive mode of governance. Law is a formally recognisable alternative to rule by men; a difference that is made especially clear when we consider the status enjoyed by the subjects of a legal as opposed to the kind of order that he associates with the mode of managerial direction. To be a subject of law is to belong to something quite different than 'a subservient populace ready to do what they are told to do'.[94] It is instead to be a participant in a social condition in which one is respected as a responsible agent. This respect, and the social condition that speaks to and constitutes it, arises from the particular way that a legal order creates and communicates its norms; namely, through observance of principles of generality, promulgation, clarity, avoidance of contradiction and of impossibility, constancy through time, non-retroactivity, and the requirement that there be congruence between official action and declared rule.

These eight principles constitute law's 'internal morality' because they speak to a form of governance that imposes moral demands on the task of creating and administering it that are a direct reflection of the status that the principles must envisage the subject as possessing if they are to have any meaning in practice. These morally significant commitments distinguish rule through law from rule by men, and they are internal to law because they arise from the presuppositions that must be in place in order to support law as a distinctive form of social ordering. If the necessary reciprocity between lawgiver and subject that is generated by these

[94] Undated and untitled document, The Papers of Lon L Fuller, Harvard Law School Library, Box 12, Folder 1 (notes for 'Reply to Critics').

commitments and presuppositions collapses, law disappears. The relationship between the source of power and those subject to it will have been reconvened—if, indeed, it was ever constituted in the first place—along some other, non-law, lines.

According to Fuller's jurisprudence, then, there can be no meaningful concept of law that does not include a meaningful limitation of the lawgiver's power in favour of the agency of the legal subject. This much we might already have taken, joining its claims and criticisms together, from *The Morality of Law*. But what we learn from the 'Reply' is that these limitations do not depend on any benevolent disposition, any desire that law be a certain kind of good, on the part of the source of power that finds itself so limited. Such is not a moral objective that is imposed on the enterprise of lawgiving from without. It is, rather, simply something that follows from the formal distinctiveness of law as the enterprise of subjecting human conduct to the governance of general rules.

6

Resituating Fuller I: Raz

My aim in the foregoing chapters has been to illuminate the content of Fuller's central jurisprudential claims by giving close attention to the writings in which those claims are articulated, as well as the debates to which those writings understood themselves as answerable.

With this purpose in view, in chapter two I emphasised the instructiveness of Fuller's unfinished eunomics project for highlighting the dual centres of gravity of his jurisprudential vision—moral value for the subject, and efficacy for the lawgiver—and for revealing his intention to correct the tendency of legal philosophers to neglect the form of law as a site of theoretical investigation in its own right. Against this background, in chapter three I steered the standard reading of the 1958 Hart-Fuller debate towards Fuller, highlighting how and why we should read his contribution to that exchange as an attack on the incompleteness of the positivist account generally, and as a challenge to Hart to explain the positivist understanding of the relationship of law to legality in particular. The debate about Nazi law within that exchange was emphasised for how it challenged Hart to explain the nature of and reasons for a legal subject's obligation to obey law when the form of the latter falls so far short of conditions in which the subject can possess and express agency in her interaction with that form.

I then turned in chapter four to consider how Fuller developed these ideas in *The Morality of Law*. Here I suggested that the message that Fuller sought to convey through his model of the internal morality of law is distorted if we regard it as a checklist of criteria that require exhaustive and inflexible application. Instead, I argued, the eight principles must be understood as larger than the sum of their parts, a representation of how governance through general rules works and is made possible and which includes respect for the legal subject as an agent as part of law's form. I also paid particular attention to both the 'ethos' and 'moral value' readings of the 'morality' aspect of Fuller's internal morality of law, and highlighted Fuller's apparent distress in the face of Hart's interpretation of his claims about the internal morality of law as reducible to conclusions about instrumentality and efficacy, with no regard for the dimension of trusteeship for the fate of persons as agents that Fuller saw as built-in to law's form. Finally, in chapter five, I explored Fuller's attempt to explain and address this impasse about efficacy through the introduction of a distinction between law and managerial direction; a distinction that saw him return not only to the kinds of questions about law's formal distinctiveness that animated his eunomics project, but also opened up new ways of exploring the necessary modes of treatment of lawgiver to legal subject that attend law's defining formal feature of generality.

There is much in this analysis of Fuller's jurisprudential writings that is very likely to be congenial to, or at least to share strong affinities with, a number of previous engagements with his thought. This is especially so in relation to those who have suggested that Fuller's jurisprudence is underscored by an idea of freedom as 'independence from the power of others';[1] that Fuller understands the rule of law as 'inherently respectful of people's autonomy';[2] that a concern for agency and specifically for a view of the person as a free and responsible 'centre of action' animates his theoretical claims;[3] and, more generally, that if we are to gain the best understanding of Fuller's position on the moral dimensions of law, and especially its generality, we must shift our perspective from that of the lawgiver to that of the legal subject.[4] There are equally many commonalities between the reading I have offered in the preceding chapters and studies of Fuller's thought by such scholars as Kenneth Winston,[5] Robert Summers,[6] David Dyzenhaus,[7] David Luban,[8] Gerald Postema,[9] and Nicola Lacey,[10] as well as scholars including Roderick Macdonald,[11] Wibren van der Berg, and Willem Witteveen,[12] who have turned to Fuller for his insights into the possibilities and limits of institutional design. Much is also likely to be acknowledged as sound by scholars who, though generally not concerned to engage squarely with Fuller's jurisprudence, have nonetheless expressed considerable sympathy for his underrecognised insights. Neil MacCormick's suggestion that Hart's interpretation of the principles of Fuller's internal morality of law as 'merely technical requirements of legal efficiency' overlooks the possibility that 'some ways of organising human affairs can have a positive value, and that this value can be moral value, even in situations in which there

[1] Nigel Simmonds, *Law as a Moral Idea* (Oxford, Oxford University Press, 2007) 100, 101.

[2] Colleen Murphy, 'Lon Fuller and the Moral Value of the Rule of Law' (2005) 24 *Law and Philosophy* 239, 250.

[3] Evan Fox-Decent, 'Is the Rule of Law Really Indifferent to Human Rights?' (2008) 27 *Law and Philosophy* 533, 535–36, 552.

[4] Jennifer Nadler, 'Hart, Fuller, and the Connection Between Law and Justice' (2007) 27 *Law and Philosophy* 1, 25.

[5] Kenneth I Winston, *The Principles of Social Order: Selected Essays of Lon L. Fuller*, revised edn (Portland, Hart Publishing, 2001).

[6] Robert S Summers, *Lon L. Fuller* (Stanford, Stanford University Press, 1984).

[7] David Dyzenhaus, 'The Grudge Informer Case Revisited' (2008) 83 *New York University Law Review* 1000.

[8] David Luban, 'Natural Law as Professional Ethics: A Reading of Fuller' (2001) 18 *Social Philosophy and Policy* 176.

[9] Gerald Postema, 'Implicit Law' (1994) 13 *Law and Philosophy* 361.

[10] Nicola Lacey, 'Philosophy, Political Morality, and History: Explaining the Enduring Resonance of the Hart-Fuller Debate' (2008) 83 *New York University Law Review* 1059; 'Out of the "Witches" Cauldron?: Reinterpreting the Context and Reassessing the Significance of the Hart-Fuller Debate', in Peter Cane (ed), *The Hart-Fuller Debate in the Twenty-first Century* (Oxford, Hart Publishing, 2009).

[11] Roderick Macdonald, 'Legislation and Governance', in Willem J Witteveen and Wibren van der Burg (eds), *Rediscovering Fuller: Essays on Implicit Law and Institutional Design* (Amsterdam, Amsterdam University Press, 1999).

[12] See Witteveen and van der Burg, *Rediscovering Fuller* (n 11).

are countervailing moral values, even overriding ones', is one striking example.[13] John Finnis' comment that Fuller's reaction to the efficacy interpretation of his principles of the internal morality of law contained 'more underlying sense than his critics were willing to allow' is another.[14]

It is Jeremy Waldron's engagement with Fuller, however, that I wish to focus on as a frame for guiding the ambition of the next two chapters to resituate Fuller within the landscape of contemporary legal philosophy. Waldron's recent works suggest that he wishes to lead fellow legal philosophers towards a reconsideration of some of the narrower, conversation-closing commitments within the project of Anglo-American legal philosophy as it has taken shape since the ascendancy of Hart's project. 'The Concept and the Rule of Law' puts this appeal starkly, with Waldron there inviting legal philosophers to seek a more 'discriminating' or 'less casual and accommodating'[15] concept of law than the prevailing positivist one derived from Hart, which basically boils down to the view that whatever legal officials declare to be such is law. He has also suggested that, as we pursue this task, a better understanding of why having a legal system 'matters to us' might be 'our best guide to what is distinctive about legal as opposed to non-legal modes of governance'.[16]

Within this agenda, Waldron has long reserved a sympathetic ear for Fuller's contributions in the same vein, and has been a key voice among those who question the correctness of reducing the message of Fuller's jurisprudence to one about efficacy. Moreover, as I highlighted in chapter one, since his 1994 essay, 'Why Law? Efficacy, Freedom or Fidelity',[17] Waldron has signalled to form as the site from where Fuller's jurisprudence and his challenge to positivism proceeds, suggesting that Fuller's writings might be regarded as the initiation of a research programme to explain what the connection between legal forms and fidelity to law might actually consist in.[18] Significantly for the concerns of this book, the occasion of the 50th anniversary of the Hart-Fuller debate in 2008 has seen Waldron develop this interest in Fuller's claims about the form of law towards the status, within that form, that is occupied by the legal subject. Capturing Fuller's challenge to Hart succinctly, and positioning that challenge as a standing question for contemporary legal philosophy, he has proposed that:

> Law itself may be an enterprise unintelligible apart from the function of treating humans as dignified and responsible agents capable of self-control; unscrupulous rulers must make what they can of that fact when they decide, for reasons of their own, to buy into the 'legal' way of doing things.[19]

[13] See Neil MacCormick, 'The Separation of Law and Morals', in Robert P George (ed), *Natural Law Theory: Contemporary Essays* (Oxford, Oxford University Press, 1992) 122–23.

[14] See John Finnis, *Natural Law and Natural Rights* (Oxford, Clarendon Press, 1980) 272–74.

[15] Jeremy Waldron, 'The Concept and the Rule of Law' (2008) 43 *Georgia Law Review* 1, 13, 15.

[16] ibid 61.

[17] Jeremy Waldron, 'Why Law: Efficacy, Freedom or Fidelity?' (1994) 13 *Law and Philosophy* 259.

[18] ibid 276.

[19] Jeremy Waldron, 'Positivism and Legality: Hart's Equivocal Response to Fuller' (2008) 83 *New York University Law Review* 1135, 1167.

This convergence of Waldron's and other scholars' concerns with Fuller's jurisprudential agenda suggests something significant for the state of contemporary legal philosophy. Certainly, we are witnessing an increasing and increasingly sophisticated interest in Fuller's thought generally, and especially with respect to his intuitions about the connections between the form of law and human agency. But, I would suggest, we are also witnessing a new willingness to explore those sites within the field that open up the possibility of bridge-building conversations; to candidly examine the merits of mapping the field along the kind of polarised lines that constituted jurisprudential debate for much of the twentieth century. In short, there seems to be a new willingness to consider whether old enemies might become new friends.

This sense of a common programme provides an appropriate background against which to turn to the task that I have assigned to this and the following chapter; namely, to study the two major projects of contemporary legal philosophy which, by all appearances, have *not* 'signed on' to Fuller in any meaningful way in the past, even if they have turned to address, or to embrace, some of his most central concerns. I refer here to the legal positivism of Joseph Raz, and the anti-positivist jurisprudence of Ronald Dworkin.

To attempt to resituate Fuller's jurisprudence by reference to these two projects is not to suggest that such is all that is required to meet that task. But because Raz and Dworkin, like Hart, have each contributed to the marginalised position that Fuller's jurisprudence has occupied for much of the second half of the twentieth century, part of the project of reclaiming Fuller must surely be to examine how and why this came to pass. I turn now, then, to explore how we might resituate Fuller within the landscape of contemporary legal philosophy as it has been constituted by Raz, before moving in chapter seven to attempt the same with respect to Dworkin.

I Fuller and Raz

The version of positivism against which Fuller's contribution to jurisprudence has mostly been judged so far is obviously Hart's, and within this, Hart's claims about the separation of law and morality. The thrust of Hart's response to Fuller on this point can be recalled from chapter four: that the principles of the internal morality of law do not evidence any necessary connection between law and morality because they are merely principles of efficacy and, as such, are indifferent to the moral quality of the ends pursued through law. Thus, Hart concluded, because the principles of legality are unfortunately compatible with great iniquity, and indeed may assist the pursuit of such iniquity, Fuller failed to demonstrate any necessary connection between law and morality.[20]

[20] See chapter four, II 'Hart's review'.

Hart held solidly to this position in both sites where he engaged directly with Fuller's work, *The Concept of Law*,[21] and his 1965 review of *The Morality of Law*.[22] If there are points in those writings where Hart suggests that there might be other ways in which law contains important connections to morality, such as the moral significance that consists in how law must address its subjects as rational agents,[23] they are, at best, only gestures. Otherwise, Hart's approach was to actively assign the concerns of Fuller's jurisprudence to a marginal place within debates about law and morality, and indeed about the nature of law generally. In doing so, a scholarly memory of the Hart-Fuller debate was created which turns upon questions that, as we have seen in the preceding chapters, were never in fact Fuller's primary concern or, at least, which obscured the expression of other concerns.

Part of the project of reclaiming Fuller must therefore be to scrutinise what Hart's positivism actually boils down to with respect to the questions that were of much larger significance to Fuller. One way of stating this—indeed, to evoke a prism of analysis that is central to my discussion in chapter seven—is to suggest that Fuller was challenging Hart to explain the positivist's understanding of law's answerability to legality, as well as the value that Hart understood the latter to stand for.

On the first point, Jeremy Waldron has done much to show how Hart's answer to Fuller oscillates between ostensibly opposed positions on both questions.[24] But on the matter of value, Hart's 'efficacy' answer, as I emphasised at length in chapter five, was a source of enduring frustration to Fuller. Quite apart from the trouble it caused for how his own claims were received, Fuller was also insistent that 'efficacy' was an unintelligible answer for positivists to offer when the question at issue was the value that emanates from the existence of legal order, writ large, rather than from a particular end pursued through a particular law.

The debate about efficacy is clearly crucial to diagnosing and resolving the impasse between Fuller and positivism, but in many ways it shields us from Fuller's still greater concern to criticise positivism's expansive, permissive concept of law on the ground that such a concept offers no meaningful limits on lawgiving power. On this point of criticism, Hart's commitments are arguably nowhere better captured than in the passage in *The Concept of Law* where he elaborates the idea that only officials, not subjects, need hold a critical reflective attitude towards the rule of recognition in order for a legal system to exist. Such a society 'might be deplorably sheeplike', and the 'sheep might end in the slaughterhouse'. According to Hart, however, so long as the legal officials of such a system adopt an internal point of view towards the rules, there is no grounds for denying it the title of a legal system.[25]

[21] HLA Hart, *The Concept of Law*, 2nd edn (Oxford, Oxford University Press, 1994).

[22] HLA Hart, 'Lon L. Fuller: *The Morality of Law*' (1965) 78 *Harvard Law Review* 1281, reprinted in HLA Hart, *Essays in Jurisprudence and Philosophy* (Oxford, Clarendon Press, 1983) 343.

[23] Hart, *Concept of Law* (n 21) 206–07, when discussing the 'minimum form of justice' arising from law's generality.

[24] Waldron, 'Hart's Equivocal Response' (n 19) 1135.

[25] Hart, *Concept of Law* (n 21) 117.

A generous reading might be given to Hart here if we pay closer attention to the context in which the 'slaughterhouse' comment arises: his discussion of the two 'minimum conditions necessary and sufficient for the existence of a legal system'.[26] Official acceptance of the rule of recognition that specifies the criteria of legal validity for the given system is obviously one of these minimum conditions, but so too, Hart argues, is the requirement that rules of behaviour singled out by that rule as valid law be 'generally obeyed'.[27] The latter, he explains, is the only condition 'which private citizens need satisfy', even if 'they may obey each "for his part only" and from any motive whatever', and even though in a healthy society 'they will in fact often accept these rules as common standards of behaviour and acknowledge an obligation to obey them'.[28] The assertion that a legal system exists, Hart therefore concludes, is 'a Janus-faced statement looking both towards obedience by ordinary citizens and to the acceptance by officials of secondary rules as critical common standards of official behaviour'.[29]

All of this, Hart says, is simply a reflection of the 'composite character' of a sophisticated modern legal system: a system that involves the union of both primary and secondary rules.[30] But he nonetheless goes on to argue that in an 'extreme case'—a case, we might safely extrapolate, that is not excluded by the criteria for legal validity supplied by a theory of legal positivism—the internal point of view towards the rule of recognition and its criteria of legal validity 'might be confined to the official world', with only officials, not subjects, accepting this criteria.[31] It is in these circumstances that those subjects, now sheep, may end up in the slaughterhouse.

It thus seems clear that Hart envisages a healthy (or perhaps even standard) legal system as one which takes the attitude of the legal subject toward the legal order seriously. Nonetheless, it seems fair to suggest, as both a practical and theoretical matter, that the 'sheep to the slaughterhouse' comment, and the conception of the legal subject which it speaks to, invites a 'when it all boils down' understanding of Hart's positivism that might be stated as follows. Law is a framework for instrumental legal activity which will still properly be regarded as law even when the legal subject is envisaged, or treated, as no more capacious an agent than a mere sheep to the slaughterhouse.

This point offers a fitting segue to the question that lies at the heart of any engagement between Fuller's jurisprudence and the positivism of Joseph Raz. The question is whether, given the concerns of his own project, Fuller would see in Raz's legal positivism as undiscriminating a concept of law as he saw in Hart's: one, that is, which ranges with no apparent conceptual trouble between witless sheep to the slaughterhouse and a fully realised relationship of respect and reciprocity between lawgiver and subject. My purpose in the analysis to follow is to

[26] ibid 116–17.
[27] ibid 116.
[28] ibid.
[29] ibid 117.
[30] ibid.
[31] ibid.

suggest that the answer to that question would have likely been in the negative. Before explaining why, however, the qualifying words in this statement require emphasis. On virtually all points that are crucial to assessing the distance between Raz and Fuller on the question of law's answerability to conditions beyond the positivist formula for legal validity, including conditions which seem to reveal law's inherent moral dimensions, Raz's position is ambiguous or, at the very least, non-committal. This is why, in what I offer below, I do not seek to attribute any conclusive position to Raz that his own writings are not capable of yielding. But I do seek to suggest that Raz *can* be read as not advancing as undiscriminating a concept of law as Hart did, and, if accepted as viable, that this reading throws up a new site of engagement between positivism and Fuller on questions that were far from exhausted in the Hart-Fuller debate.

To illuminate the bases of these claims, the shape of Raz's distinctive positivist project must be briefly sketched. Raz adopts the foundations of Hart's positivism that a given norm is a *legal* norm due to its membership of a system that is given its systemic quality from the social practice in which officials recognise some norms as law and others not. But the idea of law that then emerges from Raz's project is quite different to Hart's. Law, in Raz's account, is a special kind of exclusionary reason for action. This way of envisaging law arises from how Raz's legal philosophy is an outgrowth of his wider philosophical inquiry into the phenomenon of practical reason in combination with how he considers law's defining structural feature to be its claim to possess authority. To understand law through the prism of authority, therefore, is to understand it as something which speaks to our capacity for practical reason in a distinctive way.

I attempt to show below that, when the key commitments of Raz's legal philosophy are mapped out, some important points of commonality between his jurisprudence and Fuller's emerge that have likely gone overlooked while positivists have held confidently to their conclusion that Fuller contributed little by way of a compelling critique of Hart's project. On the question of whether this conclusion ought to be revisited, I expect to meet resistance from those who, even if willing to acknowledge certain points of commonality, will still likely insist that an intractable difference between Raz and Fuller lies in Raz's foundational positivist commitment to the view that moral questions have no place in determining the existence conditions for law.[32] Still, the claim that I wish to defend is that these and other differences, intractable as they might appear, are problematised by what seems to be a commitment shared by Raz and Fuller: a commitment to understanding law as a phenomenon that is fundamentally linked to respect for the legal subject as an agent.

In Raz's case, this apparent commitment, and why it is problematic, emerges from what I read to be a disconnect between his assessment of the relationship

[32] I make no attempt here to engage Fuller with the apparently revised version of Hart's position that is reflected in the proposed postscript to *The Concept of Law* that was published after Hart's death, as this version is salient to intramural debates among legal positivists much more than it is to an engagement between Fuller and Razian positivism.

between law and the principles of the rule of law, and his work on the authority of law. As far as I am aware, these two sites of inquiry are generally not brought into close contact with each other for the purpose of analysing the content and features of Raz's legal philosophy. This makes sense in so far that Raz's analysis of the connection between law and the rule of law appears to have much narrower aims than his work on authority. Moreover, as I will explain below, this analysis of law and the rule of law represents little departure from Hart's responses to Fuller with respect to the same question in the Hart-Fuller debate. Raz's work on authority, however, *does* mark an important departure from Hart's positivism. And, in terms of differentiating its concerns from Raz's analysis of law and the rule of law, this work on the nature of legal authority bears little relationship to the more familiar contest about law's connection to morality beyond a common defence of the 'sources thesis'—the argument that the existence and content of law can be identified by reference to social facts alone, without resort to evaluative argument[33]—as the correct positivist account of the existence conditions for law.

Yet it is precisely because the brilliance and interest of Raz's project lies in the interdependence of its elements that it is appropriate to examine how and whether these different sites of his legal philosophy can be reconciled with each other. My starting point for undertaking this examination is to ask how, or whether, Raz's legal positivism regards the status of the legal subject as that of a robust and responsible agent. While I reserve any final conclusion on this point in favour of the ambiguities of Raz's analysis, I nonetheless hope to show that Raz's interest in the legal subject as an agent invites those committed to his project to differentiate the fundamental commitments of their legal philosophy from Fuller's idea that law is defined by a distinctive form that engenders respect for the legal subject as an agent, and which is morally valuable for this reason, more clearly than is presently the case.

II Raz on the Rule of Law

If Raz's project in legal philosophy is to offer an interdependent analysis of central characteristics of law, the message conveyed by his essay 'The Rule of Law and its Virtue' is that the concept of the rule of law is not among those central characteristics.[34] 'The rule of law', as Raz explains it, designates the idea that people should be ruled by and obey law, and that the law should be such that people will be able to be guided by it.[35] It is a concept that derives from the insight that if law is to be obeyed, it must be capable of guiding the behaviour of its subjects, and can be

[33] Joseph Raz, 'Authority, Law, and Morality', in *Ethics in the Public Domain: Essays in the Morality of Law and Politics* (Oxford, Clarendon Press, 1994) 211.

[34] Joseph Raz, 'The Rule of Law and its Virtue', in *The Authority of Law: Essays on Law and Morality* (Oxford, Clarendon Press, 1979) ch 11.

[35] ibid 213.

expressed through eight principles akin to what Fuller offers through his model of the internal morality of law (which, indeed, Raz describes as 'full of good sense').[36]

Although celebrated as a virtuous moral ideal that is of moral value to those subject to it, the crux of Raz's analysis of the rule of law comes down to a strong defence of three points: that the rule of law is merely a political ideal that a given legal system may possess to a greater or lesser degree; that its observance is not necessary to secure the existence of a legal system; and that any such observance is ultimately of neutral instrumental moral value, in the sense that any moral value that accrues to law through a connection to the rule of law is contingent on the morality of the ends being pursued through law.

Each of these claims is, in different ways, relevant to an engagement between Raz and Fuller. Certainly, the 'Virtue' essay is widely regarded by legal philosophers as having completed Hart's work in disposing of Fuller's argument that observance of his internal morality of law is necessary for the existence of law, *and* necessarily reveals a connection between law and morality. But because it was published shortly after his death, Fuller himself never had the opportunity to respond to Raz's declaration that his claims about law's internal morality were a failure, although a number of scholars sympathetic to Fuller have attempted to do so.[37]

My own contribution to that attempt differs from its predecessors in that my primary aim is to clarify where the most important site of jurisprudential contest between Raz and Fuller actually lies. This site is not, as is often thought, to be found in Raz's claim that observance of the rule of law is of neutral instrumental moral value. Indeed, the architecture of Raz's essay strongly suggests that he also did not regard the issue of law's instrumental value to be the salient point of controversy, as his famous argument that the rule of law is to law what sharpness is to a knife—that which gives the instrument its virtue of efficiency—is not introduced until *after* he has dismissed Fuller's claims about the connections between law and morality. The real controversy between Raz and Fuller lies, instead, in Raz's insistence that law's existence does not, as Fuller thought, depend on the observance of the rule of law, and that, by extension, no moral value attaches to the mere existence of law through any constitutive connection to the rule of law.[38]

To navigate what is going on in this conclusion, it is helpful to divide Raz's analysis into those of his arguments which explore the idea of the rule of law in its own right, and those which, in their defence of the idea that moral criteria do not figure in the question of law's existence, address its connection to the concept and existence of law. To begin with the idea of the rule of law in its own right, the first point to note is that Raz readily accepts that the rule of law has moral value when its principles are observed. This moral value derives from how the rule of law stands in contrast to and curbs arbitrary power;[39] from how it facilitates agency

[36] ibid 214, 218 note 7.

[37] See, eg Murphy, 'Moral Value of the Rule of Law' (n 2).

[38] Raz, 'Virtue' (n 34) 210.

[39] ibid 219–20.

through stabilising social relationships and providing a basis for individual planning;[40] from how the predictability that the rule of law nurtures increases the individual's power of action[41] and minimises frustrated expectations.[42] Moreover, and perhaps most importantly for present purposes, Raz also suggests that observance of the rule of law is morally valuable for how it instantiates respect for human dignity by treating the legal subject as a rational autonomous creature who is capable of planning and plotting her future.[43] Thus, Raz concludes, to violate the rule of law is to offend human dignity by expressing disrespect for the autonomy of the legal subject.[44]

But, Raz goes on to argue that in other ways the rule of law can lack moral value because, as the good-making quality of law, it is ultimately of neutral instrumental value with respect to the moral quality of the ends that are pursued through it.[45] He then also suggests that the rule of law can lack moral value through how laws that radically subvert the moral value of autonomy can still be produced in compliance with its principles. Law may, Raz argues, 'institute slavery without violating the rule of law', because this is entirely compatible with the business of law 'to guide human action by affecting people's options'.[46] The rule of law is thus not only contingently connected to the realisation of autonomy (as is so starkly suggested by its apparent compatibility with legalised slavery) but should also not be over-estimated in the extent to which it serves the value of autonomy generally. Securing the condition of autonomy, he suggests, might in fact require the pursuit of political and social conditions that implicate a lesser degree of conformity with the rule of law.[47]

These comments, which belong to different parts of the 'Virtue' essay, form the backdrop to Raz's assessment of the place occupied by the concept of the rule of law within the commitments of positivist legal philosophy. The key point here is that, unlike Hart, Raz sees that the question of the relationship between the concept of law and the rule of law has important implications for the positivist claim that there is no conceptual connection between law and morality, and that this in turn requires that positivists answer Fuller's objections on this point. As Raz states the problem:

> Lon Fuller has claimed that the principles of the rule of law which he enumerated are essential for the existence of law. This claim if true is crucial to our understanding not only of the rule of law but also of the relation of law and morality. I have been treating the rule of law as an ideal, as a standard to which the law ought to conform but which it can and sometimes does violate most radically and systematically. Fuller, while allowing that deviations from the ideal of the rule of law can occur, denies that they can be radi-

40 ibid 220.
41 ibid.
42 ibid 222.
43 ibid 221–22.
44 ibid 222.
45 ibid 225.
46 ibid 221.
47 ibid 227–28.

cal or total. A legal system must of necessity conform to the rule of law to a certain degree, he claims. From this claim he concludes that there is an essential link between law and morality. Law is necessarily moral, at least in some respects.[48]

This passage offers an accurate portrayal of Fuller's claims, and of the nature of their challenge to a positivist account of law: a challenge directed to the positivist separability thesis as well as to the positivist understanding of the relationship between the concept of law and the concept of the rule of law. Rejecting the idea that a real challenge to positivism exists on either front, Raz begins by expressing basic agreement with Fuller that the principles of the rule of law 'cannot be violated altogether by any legal system'.[49] This is because legal systems require an institutional framework that must itself be established through rules that possess certain attributes of the rule of law, including that those rules be general rather than particular so as to facilitate the establishment of institutions of adjudication, and prospective so as to provide instruction with respect to how the law is to be enforced.[50]

This apparent common ground with Fuller takes a sharp turn, however, when Raz goes on to insist that the *extent* to which principles such as generality, clarity, prospectivity, and so forth, are essential to the existence of law 'is both minimal and consistent with gross violations of the rule of law'.[51] Raz quickly anticipates the objection to this argument that comes from the law and morality point: that if he concedes that there must be *some* observance of the principles of the rule of law in order for a legal system to come into existence, then does it not follow that there is necessarily at least some moral value in every legal system?[52] His response to this objection is to argue that the rule of law is 'an essentially negative value': designed to minimise 'the great danger of arbitrary power' that is 'inevitably' created by the law itself.[53] The rule of law is equally a negative value in how it may prevent the dangers to freedom and dignity that are instantiated through laws that are unstable, obscure and retrospective.[54] The rule of law, Raz thus concludes, is a negative virtue in two senses: 'conformity to it does not cause good except through avoiding evil and the evil which is avoided is evil which could only have been caused by the law itself'.[55] Fuller's attempt to establish a necessary connection between law and morality through law's connection to the rule of law must therefore fail.[56]

It is easy to accept this conclusion without further argument, as, indeed, it seems that positivists have done since 'Virtue' was published several decades ago.

[48] ibid 223.
[49] ibid.
[50] ibid.
[51] ibid 223–24.
[52] ibid 224.
[53] ibid.
[54] ibid.
[55] ibid.
[56] ibid.

I wish to suggest, however, that it is a conclusion that contains certain ambiguities. To illuminate these, it is helpful to break Raz's engagement with Fuller down into its two main elements. First, Raz argues that the principles of the rule of law need only be minimally observed to bring legal order into existence, with the implication here being that any moral value otherwise realised through observance of the rule of law does not accrue to a legal system simply by virtue of it being a legal system. This argument enables Raz to keep the question of the conditions for law's existence—and, crucially, any connection these might have to moral value—insulated from the concept of the rule of law. The central positivist tenet that moral considerations do not figure in the determination of law's existence is thus preserved. The function of the second argument, the negative virtue claim, is to contend that there can also be no inherent moral value to any and every legal system through its connection to the rule of law because the virtues of the rule of law merely correct the evils that law itself creates. Even if this might mean, as Raz puts it, that 'there are some kinds of evil which cannot be brought about by law', this is hardly to law's moral credit.[57]

I am not the first commentator to suggest that these two arguments are problematic, at least as a response to Fuller and as grounds for concluding that his claims about the connections between law and morality must fail.[58] To begin, by speaking of the danger of arbitrary power that only law itself can create, Raz basically denies the very basis of Fuller's understanding of what makes law a distinctive form of ordering; namely, how it curtails arbitrary power through its particular way of creating and communicating norms to the agent legal subject. For Fuller, law cannot be a cause of the problem of arbitrary power, as Raz suggests, because law, properly understood, is an answer to precisely this pathology. To suggest otherwise would, amongst other things, be to disavow law's distinguishing feature of generality.

Raz might, however, have a response here; indeed, one that is essentially consistent with claims I have made elsewhere about the moral work done by the way that the form of law (or in Raz's terms, the 'rule of law') treats the legal subject, even while, or despite the fact that, the ends being pursued through that form are oppressive or unjust in their intentions.[59] Any moral value realised in this way is obviously different to that which manifests through the pursuit of a just legal end, and it is likely that the *dis*value that accrues from an *un*just legal end will not be cancelled out by it.[60] Still, the species of moral value in question very arguably remains independent of any disvalue that might accrue through the injustice of a given legal end.

[57] ibid.

[58] See, eg Jeremy Waldron, 'The Concept and the Rule of Law' (2008) 43 *Georgia Law Review* 1, 11; Martin Krygier, 'The Hart-Fuller Debate, Transitional Societies and the Rule of Law', in Peter Cane (ed), *The Hart-Fuller Debate in the Twenty-first Century* (Oxford, Hart Publishing, 2009) 117; and Nigel Simmonds, 'Reply: The Nature and Virtue of Law' (2010) 1 *Jurisprudence* 277, 285.

[59] Kristen Rundle, 'The Impossibility of an Exterminatory Legality: Law and the Holocaust' (2009) 59 *University of Toronto Law Journal* 65.

[60] Simmonds makes this point well in 'Reply: The Nature and Virtue of Law' (n 58) 286.

Bringing this point to the 'negative virtue' claim, then, the suggestion might be made that the 'evil which could only have been caused by the law itself' of which Raz speaks is the arbitrariness, or perhaps better stated as the oppressiveness, of law's content, and not of its form.[61] If this is the right reading of the 'negative virtue' argument, however, it is one that does not appear to harm Fuller. Rather, it suggests that we ought to look more closely at Fuller's argument about the independent moral work done by law's form when we assess the claims he advanced in the Hart-Fuller debate.

In any event, the more important engagement between Fuller and Raz, as I read it, does not ultimately rest on the 'negative virtue' issue, but rather on Raz's claim that there need only be minimal compliance with the principles of the rule of law in order for a legal system to be both brought into existence and maintained. The way that Raz addresses this point in 'Virtue' makes apparent that he considers this argument to be a knock-down response to Fuller on both the question of law's connection to legality, as well as the question of whether law is intrinsically morally valuable.[62] On a closer look, however, the real difference between Raz and Fuller on this point seems to be less one of directly opposed positions than a question of degree.

That is, we know that, if law is to exist on Fuller's account, there must be substantial compliance with the principles of the internal morality of law. This is the message we receive from how, even though he accepts that his model of the internal morality of law is an aspirational ideal, Fuller refuses to designate as a legal order any attempt at lawgiving that derogates too greatly from that ideal. For him, such a derogation would represent a movement from law to non-law. Still, *just how great* a departure from one or other of the principles of the internal morality of law can be tolerated within an order that makes claim to the quality of legality—and, correspondingly, *just how much* respect for agency needs to be instantiated through law in order for it to be law-like—is unclear.

This site of underspecification within Fuller's jurisprudence has always been a problem for his critics. But it is hardly a reason to conclude that his claims must necessarily fail and Raz's must succeed. Raz's own analysis of the connections between law and the rule of law is equally vague about how much compliance with the principles of the rule of law is necessary to bring legal order into existence. His designation of *minimal* compliance arguably tells us little or nothing more, in terms of a clear position, than the substantial designation we ordinarily attribute to Fuller. Ultimately then, it might be said that Fuller and Raz (and the opposed jurisprudential camps that we might regard them as representing) simply disagree on the degree of compliance with the principles of the rule of law that is necessary to create and sustain a legal system.

This might be right, but a deeper and more novel response needs to be considered, one that invites a closer examination of the place within Raz's legal philosophy of what, for Fuller, is presupposed by the form of law; namely, a conception

[61] I am grateful to Joseph Raz for an informal discussion of this point.
[62] Raz, 'Virtue' (n 34) 223–24.

of the legal subject as an agent. Clearly, Raz's analysis of the moral value of a substantially realised rule of law is significantly informed by such an idea. It might even be said that he speaks in a decidedly Fullerian voice when he states that the most important sense in which the rule of law has moral value lies in how its observance instantiates respect for human dignity by treating the legal subject as a rational autonomous creature who is capable of planning and plotting her future.[63] But, as I have highlighted, Raz then goes on to argue that not just law but observance of the rule of law is compatible with gross disrespect for that same capacity for agency, exemplified by how the law may institute slavery without any violation of the rule of law.[64]

It is difficult to assess what to make of these apparent tensions within Raz's analysis or, at least, whether we can undertake that assessment in the isolated context of the 'Virtue' essay where little seems available to the reader to resolve them. But, as foreshadowed earlier, the claim I wish to elaborate is that the complexity of these tensions and their implications for the overall coherence of Raz's project can be brought into sharper focus if we move beyond 'Virtue' to Raz's work on authority. From this wider view, I wish to suggest that it becomes viable to ask whether Raz's jurisprudence really *does* accept what Fuller's cannot; namely, that a putative legal order can still be regarded as such despite being plagued with systematic legal arbitrariness, instability, obscurity, retroactivity, drastic disrespect for the legal subject to the point of their enslavement, and other pathologies.

III Raz on Authority

Raz's inquiry into the nature of authority seeks to meet two main goals.[65] The first is to offer an account of authority in its own right, and which provides a means for distinguishing mere de facto authority, the capacity to effect authority in fact, from de jure or 'legitimate' authority; that is, authority capable of justifying law's claim over a community of moral agents. The second is to illuminate why the claim to possess authority is a structural feature of law; an argument which, if made out, fits neatly with the conclusion that the sources thesis (that the existence and content of law can be identified by reference to social facts alone) is the best available account of the existence conditions for law. It is, however, not possible to fully appreciate this last claim in the absence of an explanation of Raz's analysis of authority, the work that authority does in relation to our capacity for practical reason, and the ways in which the concept of authority connects to the concept of law.

Beginning with the first of these points, Raz's subject might be the idea of authority in general, but his writings make clear that his main concern is with the

[63] ibid 221–22.
[64] ibid 221.
[65] Raz, 'Authority, Law, and Morality' (n 33).

phenomenon of legitimate authority. To this end he offers an account (which he labels the 'service conception') that is constituted by three theses. The 'dependence thesis' states that in order to be authoritative in character, all directives from a putative authority should be based on reasons that apply to the subjects of those directives; for example, a child welfare law must address issues of child welfare.[66] The work done by the dependence thesis is thus to link Raz's idea of authority to conditions of intelligibility and compatibility with the subject matter over which such authority is claimed. The 'pre-emption thesis' performs a different function, standing for the idea that a putative authoritative directive becomes authoritative in fact when it provides content-independent reasons for action that displace, or 'pre-empt', the reasons for action that an individual might otherwise have.[67] Put at its most succinct, the fact that an authority requires performance of an action is, of itself, a reason for the performance of that action. The 'normal justification thesis' then stands for the idea that a person should acknowledge another as an authority when that person is more likely to comply with reasons that apply to her if she accepts the directives of the authority rather than if she were left to decide for herself.[68] The normal justification thesis thus offers an account of how a practical reasoner assesses her relationship to authority, and what goes into the balance when she accepts that authority's regulation of her circumstances as legitimate.

This general account of the conditions of legitimate authority ties into Raz's legal philosophy through the argument that a fundamental feature of law is that it claims not just de facto, but legitimate, authority. As Raz explains it, this claim on law's part will not always be well-founded. There will be instances, likely many, where claim and reality are some distance apart, and precisely how often or to what extent this can occur in a system that nonetheless lays claim to be law is a question left unanswered in Raz's writings. Still, what is clear is that Raz insists that it is part of the nature of law that it claims the authority to comprehensively regulate the lives of its subjects, *and* that it must be capable of possessing this legitimate authority which it claims for itself.[69] Law, as he puts it, 'must be a system of a kind which is capable in principle of possessing the requisite moral properties of authority',[70] with the consequence, for legal philosophers, that 'we can learn from the doctrine of authority something about the nature of law'.[71]

Raz then completes the circle back to his defence of the positivist sources thesis as the best account of the existence conditions for law by making the following argument. If law is to possess authority of any kind—that is, if it is to be capable of displacing the reasons for action that its subjects otherwise might have by offering a reason for action in its own right—then law itself must first be identifiable. This capacity to be identified is best supported by the sources thesis because of

66 ibid 214.
67 ibid.
68 ibid.
69 ibid 211, 215.
70 ibid 215.
71 ibid.

how that thesis insulates the question of law's existence from the contests about morality that an exercise of authority is meant to resolve.

Scholars who have engaged with Raz's service conception of authority have done so from multiple angles, and their responses include raising objections to the idea that an account of legitimacy can be offered solely in terms of superior reasons for action as opposed to the process through which such reasons (superior or not) are arrived at,[72] as well as challenging whether the service conception is as compatible with the sources thesis as Raz claims it to be.[73] However, my own interest in turning to Raz's account of authority lies primarily in how his attempt to understand law's claim to authority involves a turn towards the legal subject and, specifically, towards evaluating law's claim to authority from her perspective.

The key action of this move, which has no equivalent in previous versions of legal positivism, takes place within the architecture of the normal justification thesis. To see how, the first point to highlight about the normal justification thesis is how it is of a qualitatively different character to the dependence and pre-emption theses. The dependence and pre-emption theses arguably could, under certain institutional conditions, be wholly compatible with tyranny (although even here a problem might be presented by how the dependence thesis seems to demand conformity with, or at least intelligibility between, the authority's reasons and the subject's circumstances, something likely to prove inconvenient to a tyrant). The normal justification thesis, by contrast, heads in a very different direction. By making the subject's reasons for action its centre and starting point, the normal justification thesis introduces the figure of a robust agent who is capable of assessing the extent to which her own reasons for action will be better realised if she follows the authority's reasons rather than her own. Indeed, and quite contrary to the act of deference that one might ordinarily think constitutes the relationship of subject to authority, Raz suggests at one point that the subject's turn to authority *supports* her realisation as a moral agent in a world in which she otherwise lacks expertise and the capacity for coordinating social life.[74]

Therefore, the figure that animates Raz's service conception of authority is clearly not someone who submits to authority in the manner of Hart's sheep to the slaughterhouse. She is someone, rather, who has reasoned her way to authority and has invited the authority's reasons to displace her own, whatever her justification might be for doing so. The conclusion that seems readily on offer, then, is that a condition of authority that meets the three theses of the service conception is one in which the authority's reasons are intelligible to the subject's self-understanding as an agent.

None of this is likely to be controversial to the proponents of Raz's legal philosophy because, it seems safe to say, the service conception of authority simply

[72] Kenneth Himma, 'Just 'Cause You're Smarter than Me Doesn't Give You a Right to Tell Me What to Do!' (2007) 27 *Oxford Journal of Legal Studies* 121.

[73] Margaret Martin, 'Raz's *The Morality of Freedom*: Two Models of Authority' (2010) 1 *Jurisprudence* 63.

[74] See Joseph Raz, 'The Problem of Authority: Revisiting the Service Conception' (2005) 90 *Minnesota Law Review* 1003, 1018.

could not work unless it were grounded in a conception of the practical reasoner as a robust agent.[75] Moreover, Raz makes no explicit commitment to the idea that *law itself* instantiates a relationship between lawgiver and subject of this quality. Law, rather, merely possesses authority of some kind (possibly de facto, possibly legitimate) and we can make no theoretical generalisation as to which one. It thus follows, the response might continue, that although Raz goes to great lengths through the service conception to elaborate an account of legitimate authority which is compatible with his positivist legal philosophy, the only conceptual claim that is and can be made about law itself is that it *makes claim* to possess this quality of authority, and that it must therefore be capable, in-principle, of possessing what it claims for itself.

Yet it is precisely here, on my reading, that a site of controversy does seem to arise. That is, because Raz makes such an effort to elaborate an account of the conditions of legitimate authority as part of his legal philosophy (incomparably more effort than he does with respect to explaining the nature of de facto authority), it is worth considering whether his philosophy of law really is capable of remaining so non-committal on the question of which kind of authority is a structural feature of law.

To suggest that there is a question here seems fitting not least because Raz himself never declares outright that law can possess nothing but de facto authority and still be law. He also suggests that even when law *is* reducible to a mere expression of de facto authority, this condition is qualified by the requirement that law still claims to possess legitimate authority and must be capable in-principle of making such a claim. From the perspective of the legal subject, then, the notable point is this. There are signs that Raz might not fully embrace the undiscriminating concept of law that animated Hart's positivism, because his turn towards the subject at least raises the question of whether he is fully on board with the view that anything that officials declare to be such should be regarded as law.

If this is a viable reading, and in offering it I by no means intend to discount the ambiguities in Raz's writings that are salient to it, it is worth considering what implications it might present for other elements of Raz's legal philosophy. To this end, it is fitting to return to the 'Virtue' essay, and to how the reasoning and evaluating figure who is so central to Raz's account of legitimate authority is decidedly at odds with the much more minimally, indeed barely, realised agent that animates Raz's account of the relationship between the concepts of law and the rule of law.

This tension is illuminated strongly when we consider Raz's contention, in 'Virtue', that laws of slavery could be made in full compliance not only with the concept of law, but the concept of the rule of law. There is a tension here because

[75] I do not seek here to resolve the question of whether Fuller's 'responsible agent' can be readily equated with Raz's practical reasoner. In chapter four, however, I suggested that Fuller advances a fairly robust moral conception of the agent as a bearer of dignity, thus ostensibly a richer moral conception than what we know of Raz's practical reasoner. See chapter four, IC 'Conception of the person implicit in legality'.

even if we accept Raz's non-committal stance on the character of law's authority, it is difficult to see how the precondition of the capacity for legitimate authority over a subject that is, on Raz's own terms, part of the nature of law and something which it cannot fundamentally fail to possess, could possibly be satisfied if the law designated that subject as a slave. Moreover, the main conceptual claim in 'Virtue' that only a very minimal compliance with the principles of the rule of law is necessary to secure the existence of a legal system (a view of legal order which ostensibly permits secrecy, arbitrariness and all other manner of ills in the name of legality) hardly seems to amount to a legal order in which any agent could function; or, that is, not unless Raz's positivism effectively disassociates itself from the commitment to law's generality that is otherwise a shared starting point of all positivist theories of law.

This question of whether an agent could function within the quality of legal order presented in the 'Virtue' essay is one that invites direct consideration of where Raz's commitments lie, not only on the matter of agency, but also on the matter of form. The salient question here is whether Raz's positivism can actually be read as rejecting the idea that the intelligibility of the concept of law is linked to the idea that law has a distinctive form. This, indeed, seems to be the conclusion we are directed to take from 'Virtue'. But strong hints run against it elsewhere. One example is when Raz suggests that in order to actually be an authority, a putative authority must communicate law's demands effectively to its intended subjects, because 'what cannot communicate with people cannot have authority over them'.[76] This comment that clearly resonates with Fuller's sense that law's normative status is closely linked to the distinctive form through which its claim to authority finds expression, and which in turn generates a responsibility to uphold standards of right communication if such authority is to be effective.

The presence of an agent subject in Raz's service conception of authority also seems to complicate his otherwise apparent disregard for the law's form as both distinctive and as necessitating respect for agency. The interactional character of the authority-subject relationship in Raz's account, where the authority's reasons must be intelligible not just to the subject's self-understanding as an agent but also to her circumstances, has, in its shape, much in common with the necessary interaction between lawgiver and subject which Fuller claims makes the enterprise of law an inherently collaborative one. For authority to actually manifest, on Raz's account, there must be a meaningful interaction not just between authority and subject, but authority and agent as well.

Again, therefore, if we accept this, we must ask whether it is really possible for Raz to allow law to be anything that a lawgiver declares it to be, irrespective of the form through which it finds expression. Even in the face of the ambiguities in his position with respect to law's relationship to legitimate as opposed to de facto authority, it seems that a valid question can be raised as to whether Raz can be as indifferent to law's form as his arguments in 'Virtue' suggest, no matter how

[76] Raz, 'Authority, Law, and Morality' (n 33) 217.

drastically that form falls short of the standards that operate to support the legal subject's status as someone actually capable of relating to authority in the first place. It is equally valid to ask whether the answer to this question would actually be fundamentally different if law is understood as a vehicle of legitimate authority or, as seems most true to Raz's writings, it is understood as a vehicle of de facto authority that must nonetheless possess the capability to claim the legitimate authority that it claims for itself.

IV Conclusion: Form, Agency and Authority

It is worth repeating at this point that these arguments are offered as part of a reading that supporters of Raz's version of positivism are likely to dispute. Raz's non-committal stance on the points of ultimate interest clearly gives Razians the liberty to respond that his account of authority does nothing to disturb the immunity of his positivist legal philosophy from Fuller's challenges. The agent, they are likely to argue, rather than being a constitutive presence, essentially disappears in those moments when law's claim to, and capability to claim, legitimate authority is tenuous. Law *as* law makes no commitment to the security of agency, just as law *as* law makes no claim to satisfy the morally valuable principles of the rule of law. On the matter of authority and the rule of law alike, they are likely to conclude, there is nothing inherently morally significant about law, and there is no meaningful conceptual connection between law and morality that goes to the question of law's existence. The sources thesis prevails as the beginning and end-point of Razian positivism, undisturbed by anything just said.

My challenge to these conclusions might be one that comes from the margins of jurisprudential inquiry as it is presently conceived, and, indeed, might be one ultimately kept adrift by the ambiguities in Raz's position. But if it is to be ignored, those who regard Raz's account of authority as consistent with the other commitments of his legal philosophy are obliged to overlook how Raz himself never suggests that law is ultimately reducible to, constituted purely and consistently by, nothing other than de facto authority. They would also be obliged to downplay his argument that even if law does not always possess legitimate authority, it must always be capable of claiming it. The essential point, therefore, is this: as the agent creeps in to Raz's legal philosophy, so too, it seems, does a presumptive concern for law's integrity as a distinctive form capable of sustaining the relationship within which authority is effected. This suggests the possibility of a concern for form, and for presupposing and maintaining the status of the legal subject as an agent, that is closer to the core commitments of Fuller's jurisprudence than the apparently settled contest between Fuller and positivism would have it.

To read Raz in this way also seems consistent with how he has sought to distance his own version of legal positivism from accounts based in coercion, or mere habits of obedience, or in an idea of the subject's relationship to law as the mere doing of official bidding, or, I would argue, in the idea that law is still law (and compatible with the principles of the rule of law) when it enslaves those who are subject to it. Indeed, Raz's understanding of the nature of de facto authority seems to be too rich to support this last idea. When he speaks of what it means to possess de facto authority, that is, Raz suggests that such a person needs more than mere power or influence. There is 'an important difference', he argues, 'between the brute use of force to get one's way and the same done with a claim of right'; the latter being what effective de facto authority claims for itself when it claims that it in fact possesses, or is held by others to possess, legitimate authority.[77]

But, as I have suggested here, if Raz's positivism is to be distinguished from the projects of his predecessors for its interest in the legal subject's experience of law's normativity, other implications must also follow. First, it might be necessary for those committed to Raz's positivism to revisit those claims in 'Virtue' which appear to accommodate, as fully law-like, an enslaved legal subject and a formless law. Because even if efficacy continues to hold its place as the value to which law and the rule of law are reducible within positivist legal philosophy, what is also present in Raz's distinctive contribution to that project is the idea that, in order to discharge its power effectively, a lawgiver must exist in a relationship with the legal subject in which that power is limited, to some unspecified but not negligible extent, by the demand of meaningful communication with her. How this is to be reconciled with other declared commitments of Raz's positivism is something for those who support Raz's project to answer. In my view, however, these turns on Raz's part can be seen as an encouraging indication of how positivism may in fact offer a more discriminating concept of law—and one that shares Fuller's concern for the form of law and the agency of the legal subject—than settled beliefs presently indicate.

It seems, then, that there is work to be done by those committed to Raz's positivism if, to borrow Waldron's words, they are to dismiss the idea that law itself may be an enterprise unintelligible apart from the function of treating humans as dignified and responsible agents.[78] Reclaiming Fuller's focus on law's form and its relationship to a conception of the legal subject as an agent is, in my view, among the catalysts needed to reinstate this question squarely onto the agenda of twenty-first century legal philosophy. I will now turn to how the project of reclaiming Fuller also offers something new and distinctive to the way that this question figures in Ronald Dworkin's anti-positivist jurisprudence.

[77] Joseph Raz, *The Authority of Law* (Oxford, Oxford University Press, 2009) 9.
[78] Waldron, 'Hart's Equivocal Response' (n 19) 1167.

7

Resituating Fuller II: Dworkin

I Fuller and Dworkin

Ronald Dworkin's writings have long given the impression that there is little in common between the method and claims of Fuller's jurisprudential project and his own. *Taking Rights Seriously*, for instance, dismisses Fuller within its first few pages, joining him to other 'processual' theorists, such as Hart and Sacks, who all 'insisted on the importance of regarding law as an instrument for moving society toward certain large goals'.[1] *Law's Empire* makes no mention of Fuller at all.[2] The implication we might take from this is that Dworkin has long suggested that Fuller offers little to enlighten the pressing questions of jurisprudence generally, including inquiries, like Dworkin's own, that challenge the positivist account of law on the grounds of its incompleteness.

If this is an accurate assessment of Dworkin's sense of Fuller's contribution to jurisprudence generally, and of the connections between that contribution and Dworkin's own project in particular, it is one that operates to obscure several stark commonalities that cut across the two theorists' projects. For present purposes, however, it is not my aim to highlight these commonalities as mere points of interest, or, indeed, with a view to illuminating how many aspects of Fuller's jurisprudence offer a preliminary and sometimes quite substantial sketch of some of Dworkin's most important challenges to positivism. My aim, rather, is one that seeks to keep a wider and more contemporary conversation in view: one concerned to uncover the ways in which Fuller's jurisprudence speaks to a conception of legality that can be distinguished from Dworkin's own in important ways.

The context of this conversation must be briefly sketched. It arises from how Dworkin now sees the idea of 'legality' as the 'foundation' for inquiry in legal philosophy and the reference point against which the field itself might be remapped.[3] As Dworkin now sees it, contemporary legal philosophy might best be understood as populated not only by competing conceptions of law (legal pragmatism, legal positivism and his own 'law as integrity')[4] but also by the competing conceptions

[1] Ronald Dworkin, *Taking Rights Seriously* (London, Duckworth, 1977) 4.

[2] Ronald Dworkin, *Law's Empire* (Oxford, Hart Publishing, 1998).

[3] Ronald Dworkin, 'Hart's Postscript and the Character of Political Philosophy' (2004) 24 *Oxford Journal of Legal Studies* 1, 23–35.

[4] See generally *Law's Empire* (n 2).

of the value of legality that lie at the heart of those conceptions and which account for the differences between them.

Fuller's place on Dworkin's landscape is notable for its absence. When we review the history of Dworkin's engagement with Fuller, as I do below, this should not come as a surprise. But this history of disengagement seems to have reached a kind of natural end in Dworkin's recent turn to the idea of legality, an idea that can in many ways be seen as Fuller's starting point while, for Dworkin, marks a destination only recently arrived at. The question for now, then, is whether Fuller's contribution to understanding the meaning and value of legality can be collapsed into one or other of the conceptions that Dworkin identifies, thus justifying its omission from Dworkin's map, or whether it should be recognised as standing for something distinctive in its own right.

In what follows I seek to answer this question by undertaking a trajectory of inquiry that begins with an account of Dworkin's earliest and most sustained engagement with Fuller: essays published in 1965 in response to *The Morality of Law.* I then move through both Fuller and Dworkin's thinking on matters of interpretation and the proper methodology for jurisprudence before engaging squarely with the question of whether Fuller's conception of legality ought to be given its own place on the landscape of contemporary legal philosophy as Dworkin has recently mapped it.

The elements of this trajectory each serve a dual purpose. With respect to Dworkin's earliest and most sustained engagement with Fuller, the 1965 essays, my aim is both to chart the basic concerns and character of Dworkin's engagement, as well as to analyse how that engagement seems to overlook those elements (and, indeed, occasionally explicit arguments) of Fuller's position that invite Dworkin and fellow critics to recognise the *formal* basis of Fuller's claims about law's inner morality. With respect to my analysis of, and comparison between, Dworkin and Fuller's thinking on interpretation and methodology, here again the aim is not comparison for comparison's sake, but rather an attempt to reveal how and why Fuller's own critique of positivism on these points speaks to a distinctive conception of legality that deserves an equally distinctive place within Dworkin's map of the current state of legal philosophy.

It follows from this that my aim in undertaking the analysis to follow is not to appraise Dworkin's project generally, or, indeed, to evaluate the success of its attack on the viability and coherence of positivist jurisprudence. This stands on its own terms. Still, because Dworkin has in many ways been as much a catalyst as Hart and Raz for the marginal place that Fuller's jurisprudence currently occupies within contemporary legal philosophy, the project of reclaiming Fuller necessarily involves a detailed exploration of the claims, assumptions and manoeuvres of argumentation that lie behind that marginalisation, as it has been secured, or at least assisted, by Dworkin's hand.

II The 1965 Essays

Dworkin published two essays in response to *The Morality of Law*, one a substantial article in the *University of Pennsylvania Law Review*,[5] and the other a contribution to a panel discussion with Marshall Cohen and Fuller himself, reproduced in the *Villanova Law Review*.[6] I referred to the latter essay briefly in chapter four, as it contains some of Fuller's most explicit statements about the 'built-in' respect for human dignity contained in his idea of the internal morality of law. For present purposes, however, the point of reviewing these essays is to ascertain how Dworkin understood and represented Fuller's position in this moment when he engaged with that position directly, because, I will suggest, this goes some way towards explaining why Dworkin has subsequently showed so little interest in Fuller's jurisprudence, or has regarded it as having no apparent relevance to his own enterprise.

Anyone who has followed the course of jurisprudential debates in the last half century will surely be surprised, on review of the 1965 essays, to witness how closely the substance of Dworkin's criticisms of *The Morality of Law* aligns with Hart's review of Fuller's book.[7] The sensible conclusion to draw from Fuller's claims about the internal morality of law, according to Dworkin, is that these eight desiderata are 'not only criterial, in the sense that some compliance is necessary to make law, but also strategic in the sense that some level of compliance is necessary to achieve whatever governmental purpose a legislator might have in mind'.[8] The crucial alignment with Hart thus lies in how Dworkin's primary charge against Fuller's claims is the instrumental one: that Fuller fails to demonstrate why the value of the internal morality of law is anything other than that of efficacy, as that value accrues to the lawgiver's pursuit of ends by virtue of complying with the principles of the internal morality of law. Like Hart, Dworkin also charges Fuller with offering no compelling argument in support of his contention that observance of the internal morality of law is somehow connected to the moral quality of legal ends.

These two claims constitute the essence of Dworkin's response to Fuller, and provide the basis for his declaration that *The Morality of Law* is 'an unsuccessful attempt to establish a novel claim about law and morality'.[9] Yet it would be wrong to suggest that Dworkin's engagement with Fuller, especially in his *Pennsylvania Law Review* essay, is as dismissive as this conclusion seems to suggest. When turning to sketch his own project for legal philosophy concerning the logic of legal

[5] Ronald Dworkin, 'Philosophy, Morality and Law: Observations Prompted by Professor Fuller's Novel Claim' (1965) 113 *University of Pennsylvania Law Review* 668.

[6] Ronald Dworkin, 'The Elusive Morality of Law' (1965) 10 *Villanova Law Review* 631.

[7] See chapter four, II 'Hart's review'.

[8] Dworkin, 'Elusive Morality' (n 6) 632.

[9] Dworkin, 'Observations' (n 5) 668.

standards—a 'different connection'[10] between law and morality to which I will return below—Dworkin suggests that some 'sound and important perceptions' underlie Fuller's 'mistaken' claims about the moral dimensions of law. These perceptions include Fuller's suggestion that 'the so-called formal features of law, bound up in the very meaning of the term and worked out by jurisprudential exercises, are indeed of normative as well as linguistic or strategic importance'.[11] Still, the overriding message remains one about how Fuller is 'mistaken' in his idea that law is somehow 'internally' moral, with Dworkin here understanding Fuller as claiming that the principles of the internal morality of law are moral principles, in the 'perfectly conventional sense' of the term 'moral'.[12]

This is an important point, as it provides the foundation of Dworkin's assessment of Fuller's morality claim. As Dworkin understands it, Fuller's argument runs something like this. Because some of the most notorious examples of political immorality in history, such as in Nazi Germany or South Africa, involved gross violations of one or more of Fuller's eight canons, we can conclude that the canons themselves are moral principles.[13] Dworkin's charge is that this claim on Fuller's part contains two related mistakes. The first lies in drawing conclusions, as Fuller apparently does, from examples of political immorality which involve violations of the eight canons that these canons state moral principles. Dworkin rejects this: if a badly drafted statute is to have terrible moral implications, this will be a result of 'the uses to which the statute is put, not merely the bad drafting'.[14] But everyday cases of poor legislative draftsmanship, he argues, involve 'no particular immorality', but rather just a 'considerable degree of confusion and inefficiency'.[15]

Fuller's second mistake lies in his misunderstanding of what moral principles actually are. According to Dworkin, if we accept the idea that the eight canons of the internal morality of law state moral principles, it does not follow from that a lawgiver who observed these canons only to the extent necessary to make bad law (illustrated by Dworkin through the figure of a tyrant lawgiver, 'Tex') 'could be said to be complying with moral principles'.[16] When taking care to frame his discriminatory legislation with 'some consideration for the physical capacities of his subjects', Tex can in fact only be said to be complying with strategic, not moral, principles. Fuller's canons, he therefore concludes, do not offer even a shade of moral argument in favour of what Tex is doing.[17]

Dworkin thus concludes that Fuller's attempt to establish a novel claim about law and morality must fail. But there is more to the *Pennsylvania Law Review* essay than this. Most notably, unlike Hart, Dworkin does take space in the essay to

[10] ibid 678.
[11] ibid.
[12] Dworkin, 'Elusive Morality' (n 6) 632.
[13] ibid 633.
[14] ibid.
[15] ibid.
[16] ibid.
[17] ibid 634.

mention Fuller's argument that implicit in the internal morality of law is a conception of the person as a 'responsible agent', and that, for Fuller, this relates closely to how observance of the requirements of the internal morality of law upholds 'man's dignity as a responsible agent'.[18] But it is clear that these ideas strike no significant chord for Dworkin, because he immediately dismisses them as involving no assertion of a necessary connection between law and substantive morality, and thus also no important conflict with 'the classic or prototypical positivist position that law and morals are conceptually distinct'.[19]

The point I wish to develop is that what is missing here, and what could have generated a much more nuanced appraisal of Fuller's position on Dworkin's part (and indeed one more readily distinguishable from Hart's) is any apparent sensitivity to the formal basis of Fuller's claims. A combination of the archival materials and Dworkin's published response to Fuller in the *Villanova Law Review* sheds light on this, as does Fuller's own published reply.[20] These remarks make clear that, faced with Dworkin's interpretation of his claims, Fuller made a concerted effort, orally, to clarify how and why that interpretation had missed his point. Most strikingly, the remarks make apparent that Fuller had directed Dworkin to the morally significant idea of ethos that Fuller himself saw as animating the lawgiver's role, and which the principles of the internal morality of law capture as principles of law*giving*.[21] Dworkin's response to this suggestion that law's inner morality might be tied up with 'a special set of moral duties or ideals binding upon those who accept certain offices or participate in certain enterprises' is instructive.[22] Acknowledging the interest of the idea in principle, he nonetheless marginalises its significance to debates about law and morality, properly understood, by designating it as 'not part of my argument':[23]

> I do not quarrel . . . with the proposition that there is a morality particularly concerned with law and its enforcement. I differ with Professor Fuller *only* in denying that the very abstract canons he has produced (that law should be clear, public, prospective, enforced as written, etc.) in and of themselves are principles of this morality.[24]

This is the thrust of the historical record, but the question for now is whether this and other aspects of Dworkin's engagement with *The Morality of Law* offers any instruction on how we might assess the relationship between his and Fuller's jurisprudential projects today. To this end, the point that must be emphasised, as just foreshadowed, is the absence in Dworkin's analysis of any meaningful interest in the *form* of law in the sense that Fuller intended—and, indeed, attempted to clarify—beyond the passing reference to how the 'so-called formal features of law'

[18] Dworkin 'Observations' (n 5) 672.
[19] ibid 671, 673.
[20] Lon L Fuller, 'Reply to Professors Cohen and Dworkin' (1965) 10 *Villanova Law Review* 655.
[21] A distinction emphasised by David Luban in 'Natural Law as Professional Ethics: A Reading of Fuller' (2001) 18 *Social Philosophy and Policy* 176.
[22] Dworkin, 'Elusive Morality' (n 6) 636.
[23] ibid 636.
[24] ibid 637.

are 'indeed of normative as well as linguistic or strategic importance'[25] when Dworkin moves to his own concern for how legal standards are generated and what role they play in legal reasoning. But nowhere in either the *Pennsylvania* or *Villanova* essays is there any serious engagement with the possibility that this 'normative importance' might arise from the nature and presuppositions of these formal features. Instead, Dworkin's published response to Fuller is striking for how he holds to his initial assessment of Fuller's claims, despite clarifications about ethos, and despite Fuller's presentation of the idea (echoing closely what he says in chapter four of *The Morality of Law*) that in an ordered system of law there is 'a certain built-in respect for human dignity'.[26]

Fuller's private papers suggest that Dworkin's interventions on these points arguably struck a greater chord of irritation than did Hart's review of *The Morality of Law*. In addition to the hot-headed first footnote that is attached to his 'Reply to Professors Cohen and Dworkin',[27] Fuller's archival notes wish Dworkin 'God speed in his labours' and hope that, when those labours are done, 'he will do me the elementary courtesy of recognizing the highly qualified sense in which I can be said to be a "naturalist"'.[28] The notes also speak with frustration of Dworkin's 'romantic figure of the fiendishly evil legislator, who delights to chase down his victims with an exquisite respect for legality', but which, in Fuller's eyes, cause Dworkin to overlook 'what everyone knows', namely, 'that bosses or order-givers everywhere can have an interest in leeway, in obscurity, along with desire that everyone know the rules, a countervailing desire that a convenient flexibility be retained'.[29]

These reactions, irritated and combative as they are, might simply be an indication of the wearying effect on Fuller of having heard similar interpretations of his claims so many times before. But they also suggest that Dworkin became, for Fuller, something of a symbol for how far off the mark his critics' understanding of his account of the internal morality of law could be, and how, despite his attempts to clarify the sources of the impasse, these misunderstandings could remain so evidently undisturbed.

III Dworkin's Project

Dworkin, as is well known, moved on to develop his own distinctive challenge to the completeness of the positivist account of law. The thrust of the jurisprudence

[25] Dworkin, 'Observations' (n 5) 678.

[26] 'Reply to Cohen and Dworkin' (n 20) 665.

[27] ibid 655, where Fuller expresses his grievance about how the comments Dworkin submitted for publication in the *Villanova* symposium included remarks additional to those in the piece to which Fuller's reply responds.

[28] Document titled 'Natural law', The Papers of Lon L Fuller, Harvard Law School Library, Box 12, Folder 1 (notes for 'Reply to Critics').

[29] Document titled 'Legal morality', The Papers of Lon L Fuller, Harvard Law School Library, Box 12, Folder 1 (notes for 'Reply to Critics').

that has emerged from this challenge might be stated in the following terms. When confronted with a case in which the relevant legal rule is either unclear or incomplete, judges resolve that case not only by reference to established, posited legal rules, but also by recourse to moral principles which show the meaning of the law in its morally best light. These moral principles are equally legal principles because they are immanent in, or internal to, existing legal materials. When the judge brings these diverse materials to bear on the interpretive practice, she engages not in an exercise of extra-legal discretion but rather in a constructive process in which she aims for the best possible interpretation of what the law is, in light of the values of the legal order as a whole.[30] The tests that guide this interpretive process, 'fit' and 'appeal', or 'justification', speak to both a concern for the coherence of the proposed interpretation vis-à-vis the wider legal order within which it will take its place, as well as a basis for justifying the state's use of coercion towards the subject, through law, in the given instance.

When we see how declarations of what the law is involve an interpretive endeavour along these lines, Dworkin argues, we come to see that law itself is necessarily an interpretive practice. That is, law is a practice that is understood by its participants as having point, or purpose, that serves something valuable in our lives. Legal participants, when engaging with law, thus aim to interpret that practice in a manner that reveals it in its best possible light. Recognising this enables us to explain the phenomenon of theoretical disagreement in law. When judges disagree about the meaning of a given law, their disagreements reflect fundamental contests not only about the meaning of the individual law in question, but also the point of legal practice as a whole.

This line of argument leads Dworkin to the claim that not just the practice of adjudication, but all exercises in legal theory, are necessarily interpretive: jurisprudence is 'the silent prologue to any decision at law'.[31] It follows from this that what is really at stake in jurisprudence is a contest between different conceptions of the concept of law.[32] Dworkin's own conception of 'law as integrity' can be distinguished from his interpretive version of legal positivism, 'conventionalism'[33] and legal pragmatism,[34] because:

> it supposes that law's constraints benefit society not just by providing predictability or procedural fairness, or in some other instrumental way, but by securing a kind of equality among citizens that makes their community more genuine and improves its moral justification for exercising the power it does . . . rights and responsibilities flow from past decisions and so count as legal, not just when they are explicit in these decisions but also when they follow from the principles of personal and political morality the explicit decisions presuppose by way of justification.[35]

[30] See 'Jurisprudence Revisited', in Dworkin, *Law's Empire* (n 2) ch 3.
[31] ibid 90.
[32] See generally ibid chs 3–6.
[33] ibid ch 4.
[34] ibid ch 5.
[35] ibid 95–96, see also 93.

Law as integrity, then, 'is both the product of and the inspiration for comprehensive interpretation of legal practice'.[36] It speaks not only to how judges are guided in their task when faced with determining the meaning of a law that is unclear, but also to the content of the political obligation to obey law of the community that accepts integrity as a political virtue.[37]

Dworkin's sense of the *point* of law under law as integrity—the way that it secures 'a kind of equality among citizens'—becomes critical to my discussion later in this chapter of how his account of the value served by legality might be distinguished from Fuller's. For now, with a view to the engagement between Dworkin and Fuller on the issue of interpretation that I sketch below, two interrelated points invite emphasis. The first is how Dworkin insists that the interpretive process is highly bounded and constrained. That is, when the applicability of extant positive law runs out in a hard case, judges do not, as Hart argued, exercise discretion and make new law, but rather continue to be constrained by law that is here understood to also include the moral principles that bear upon the correct resolution of the case by showing the law in its best light. The second point, which flows from the first, is how this account is underscored by a clear sense of the judge as a bearer of responsibility. When making her interpretive decisions, the Dworkinian judge is acutely aware that, to discharge her interpretive task, she is responsible for meeting the demands of coherence and justification that sustain the legal order as a single, interconnected order. She is also aware of how this responsibility constrains the interpretive choices available to her.

IV Fuller, Dworkin and Interpretation

As he developed his debate with positivism about interpretation and its implications for rethinking the traditional contests of jurisprudence, it seems that Dworkin saw no cause to engage with Fuller's observations on the same issues. Yet these observations feature in virtually all of Fuller's jurisprudential writings, from *The Law in Quest of Itself* in 1941 and 'Reason and Fiat in Case Law' in 1946 through to his exchange with Hart in the 1958 *Harvard Law Review* and the two editions of *The Morality of Law* that represent a continuation of that exchange. As I foreshadowed in the preceding chapters, I have reserved a sustained engagement with Fuller's understanding of interpretation until now because Fuller's views on this subject are often presented as separate to, or at least separable from, his other claims. This is especially apparent in the 'Reply to Critics', where Fuller's (re) engagement with the subject of interpretation forms one of several 'implications' of his ongoing debate with his critics on the matter of law's morality and the inad-

[36] ibid 226.
[37] ibid 188.

equacy of the efficacy interpretation of his claims to this effect. He does not seek to position it as a key part of that debate.[38]

Yet, when we examine Fuller's main statements about interpretation in succession, we can readily see the error of separating his thinking on interpretation too strictly from his wider jurisprudential concerns. The connection, above all, lies in how Fuller consistently speaks of the interpretative task as charged with a distinctive kind of responsibility: responsibility for 'maintaining legality'. The point to explore for present purposes, then, is what we might learn from bringing this understanding of interpretation into conversation with how Dworkin's more recent work has also arrived at 'legality' as the animating value of interpretation and, indeed, of law.

Fuller's approach to interpretation starts from the idea, expressed consistently throughout his writings, that those who participate in the practice of interpreting law must orient their task to understanding its purposes, small and large. A judge is charged not only with the task of interpreting individual laws in light of the purposes that the legislator had in mind when enacting them, but also in light of how those laws belong to a wider legal order which makes demands on the interpretive task in its own right.

Proceeding chronologically, Fuller first articulates this view in *The Law in Quest of Itself*.[39] There, his target is the narrow range of vision of (pre-Hart) legal positivism generally, but also specifically in the positivist's resistance to 'the looser and freer ways characteristic of ethical thinking'.[40] In the concluding pages of *Law in Quest*, Fuller develops this point into an account of the judge's role when he suggests that the judge, when deciding cases, 'is not merely laying down a system of minimum restraints designed to keep the bad man in check, but is in fact helping to create a body of common morality which will define the good man'. In doing so, the judge discharges a responsibility 'to the future' that is otherwise evaded if she were to adopt 'a passive and positivistic attitude towards "the existing law"'.[41]

Published five years after *Law in Quest*, 'Reason and Fiat in Case Law' reveals more about Fuller's sense of interpretation as a bounded enterprise, in which 'the nature of the task' imposes certain limitations on the judge.[42] Among the many responsibilities that interpretation requires a judge to discharge, Fuller identifies a responsibility towards 'rightness'. That is, a judge faced with the interpretive task understands that it is her responsibility to see that her decisions are 'right', meaning, 'right for the group, right in the light of the group's purposes and the things that its members sought to achieve through common effort'.[43]

The next site where interpretation occupies a central place in Fuller's writings is his response to Hart in the 1958 *Harvard Law Review*. Despite being by far the most famous statement of his approach to interpretation and its relationship to

[38] See chapter five, I 'Mapping the "Reply to Critics"'.

[39] Lon L Fuller, *The Law in Quest of Itself* (Chicago, The Foundation Press, 1940).

[40] ibid 130.

[41] ibid 137–38.

[42] Lon L Fuller, 'Reason and Fiat in Case Law' (1946) 59 *Harvard Law Review* 376, 378.

[43] ibid.

his rejection of the positivist separability thesis, Fuller's treatment of interpretation in this context in many ways detracts from, rather than profitably develops, the line of thinking that animates the earlier writings. Moreover, as Frederick Schauer has suggested, valid questions can be raised about the extent to which Fuller properly appreciated the content of Hart's arguments about interpretation to which he was responding.[44] Still, what is clear is that Fuller's 1958 reply to Hart sees his avowedly purposive approach to interpretation coming through, and most apparently in his rejection of Hart's idea that when the clarity of 'core' extant law runs out, judges exercise discretion and make new law in a 'penumbral' space. His commitment to the view that the interpretive task is bounded by the judge's larger responsibilities towards the demands of maintaining the coherence of legal order, writ large, is equally clear.

This last point emerges strongly in Fuller's objections to Hart's view that the majority of cases see judges remaining within the core of accepted legal meanings rather than in the penumbra of uncertainty. Here, Fuller raises the question of how one might characterise the interpretive task in relation to a hypothetical statute 'which prohibits anyone to sleep in any railway station', and which sees two men brought before a judge, the first for falling asleep while sitting waiting for a train, and the second for settling down for the night but who was not yet asleep when he was arrested.[45] The question Fuller puts to Hart is to explain 'how the ideal of fidelity to law is served if the judge fines the second man and sets free the first'.[46] The question is intended to support his argument that interpretation is, above all, not about individual words, but about taking on responsibility for maintaining laws that are intelligible to their subjects' situation:

> The judge does not discharge his responsibility when he pins an apt diagnostic label on the case. *He has to do something about it, to treat it, if you will.* It is this larger responsibility which explains why interpretive problems almost never turn on a single word, and also why lawyers for generations have found the putting of imaginary borderline cases useful, not only 'on the penumbra', but in order to know where the penumbra begins.[47]

Little changes in this understanding of the task of interpretation from Fuller's first response to Hart in 1958 to his final 'Reply to Critics' in 1969, including his statement in *The Morality of Law* that 'legality' requires that legal officials 'apply statutory law, not according to their fancy or with crabbed literalness, but in accordance with principles of interpretation that are appropriate to their position in the whole legal order'.[48] Like the passage just quoted, the Dworkinian flavour of this statement is hard to miss. But it is in the 'Reply' where Fuller takes his account

[44] See Frederick Schauer, 'A Critical Guide to Vehicles in the Park' (2008) 83 *New York University Law Review* 1109.

[45] Lon L Fuller 'Positivism and Fidelity to Law: A Reply to Professor Hart' (1958) 71 *Harvard Law Review* 630, 666. Fuller revisits this example in *Anatomy of the Law* (New York, Frederick A Praeger, 1968) 17.

[46] 'Positivism and Fidelity' (n 45).

[47] ibid.

[48] Lon L Fuller, *The Morality of Law* (New Haven, Yale University Press, 1969) 82.

of interpretation in the direction of the theme of the 'cooperative nature of maintaining legality' that is central to his attempt to clarify the impasse between him and his critics in that 'last word' of the Hart-Fuller debate:[49]

> If we discern, as a basic element of law, a commitment by government to abide by its own law in judging the acts of its subjects, then interpretation will occupy in theory the central place it has always occupied in our everyday thinking about law. This emphatically does not mean that the problem will become simple; on the contrary its hidden complexities will come to light and we shall no longer be able to pretend that it is a peripheral matter to be left to unreflective common sense.[50]

To see what Fuller has in mind here, the content of his 'legality' orientation, it is helpful to look elsewhere in the 'Reply' where he observes that his analysis may seem on its face to suggest that what is demanded of an interpreting agency is simply that 'it achieve a balance of restraint and initiative in correcting the errors and oversights of superior authority'.[51] In his explanation of why the demands of interpretation are in fact much more complex than this, Fuller observes that an interpreting agency must always bear in mind that its perceived standards of interpretation 'are likely to create expectations among those affected by them and that sudden shifts in those standards may impair the collaboration essential for achieving and maintaining legality'.[52]

This point invites emphasis for how it sees Fuller return to his sense that the form of law is a structure that facilitates the collaborative activity, between mutually responsible agents, necessary for law to work. Among his writings, the site where this concern for responsiveness not only speaks to the adjudicator's role morality, but also to the institutional form of adjudication, is 'The Forms and Limits of Adjudication'.[53] I explained in chapter two, when considering the contribution of this essay to Fuller's eunomics inquiry, that the thrust of 'Forms and Limits' is the argument that the distinctive feature of adjudication, as an institutional form, is not the office of the judge but the peculiar form of participation that adjudication affords to its participants: the opportunity to present proofs and reasoned arguments before an impartial observer.[54] Adjudication, on Fuller's account, is a 'device which gives formal and institutional expression to the influence of reasoned argument in human affairs', thus respecting not only its participants' capacity for reason but also, in affording them an equality of participation through its form, their equality as agents.[55]

'Forms and Limits' gives little direct attention to the interpretive task that attaches to the adjudicator's role as that form's third participant, because its primary focus is on the institutional shape of adjudication as a distinctive form of social ordering. But nothing in 'Forms and Limits' contradicts anything said so

[49] Lon L Fuller, 'A Reply to Critics', *The Morality of Law* (n 48) 224.
[50] ibid 227.
[51] ibid 231.
[52] ibid.
[53] Lon L Fuller, 'The Forms and Limits of Adjudication' (1978) 92 *Harvard Law Review* 353.
[54] ibid 364. See further chapter two, IIB 'Eunomics "writ small": the models'.
[55] ibid 366.

far about Fuller's understanding of interpretation as a task directed to bringing disputes to a workable resolution at the same time as it is directed to maintaining the coherence and declared values of the wider social order within which that interpretive task takes place. The message of 'Forms and Limits' is wholly in keeping with the abiding message of Fuller's jurisprudence that all forms of social ordering are animated and given their distinctive shape by how their participants maintain the responsiveness of those forms towards the situation of those whose lives are shaped by them. Indeed, examining Fuller's analysis in 'Forms and Limits' enables us to notice his sense that adjudication is an enterprise which gives regard to the agency and interests of both parties, with the judge's role being not just to mediate this interaction, but to do so in light of the relevant law.

This attention to the agent's role in the adjudicative process also finds expression in Fuller's 1968 book, *Anatomy of the Law*. There Fuller diverts briefly from a long analysis of the common law to comment on the need to 'view adjudication as a collaborative process of decision in which the litigant plays an essential role'.[56] He also appeals to an understanding of an adjudicative judgment as 'not simply a determination made by an official bound by oath to act impartially' but as 'the product of a procedure in which the litigant is assured of an opportunity to present proofs and arguments for a decision in his favour'.[57]

Within the foregoing overview of Fuller's thinking on adjudication, the judicial role and interpretation, I have left to last the lengthy analysis of the common law that is set out in *Anatomy*, and in which this gesture to the thesis of 'Forms and Limits' finds its place. *Anatomy*, as I explained elsewhere, is a somewhat unusual text among Fuller's writings. Though it has many 'Fullerian' attributes in its style, scope and content, the fact that it was written for the *Encyclopaedia Britannica* gives it a decidedly different character than the kind of deliberately polemical writings we see in Fuller's exchanges with Hart. This is a helpful point to bear in mind when approaching the analysis of the common law that Fuller presents in *Anatomy*. Here, some of the rougher edges of those debates are notable in their absence, as if *Anatomy* provides Fuller with a more capacious space for articulating his thinking about law, on his own terms.

The analysis itself might best be described as a both an explanation, and a defence, of the nature and methods of the common law as contrasted with 'codified law' and all of its 'specious geometry'.[58] Adjudicative law, Fuller explains, is the product of the resolution of actual disputes which come before a court.[59] This means that the court has a double task: 'it *must* decide the case before it and do the best it can in finding some intelligible principle on which to rest its decision';[60] and it is precisely the double nature of this task that brings a range of complexities and compromises into the territory of adjudicative lawmaking.

[56] *Anatomy of the Law* (n 45) 101.
[57] ibid.
[58] ibid 106.
[59] See generally ibid 89–91.
[60] ibid 89.

The most notable ill here is the fact of the common law's retrospective application. There is nothing in Fuller's account to suggest that he seeks to hide from the ostensibly negative impact of this feature of the common law in action, one that 'is at times capable of working serious hardship', though he does point out that the doctrine of precedent 'derives much of its motivation from a desire not to impose the hardships that would result from retrospective changes in the law'.[61] He is equally candid about how the decisions from which adjudicative law flows might leave the 'exact nature of the citizen's duties' in some obscurity.[62] But here again, Fuller suggests that 'in theory at least, when the obscurity has been lifted what will be revealed is an assertion about what the citizen must do or refrain from doing and not a command to think right or to work cooperatively towards some imposed goal'.[63] The message, it seems, and consistent with the distinction between law and managerial direction that he draws in the 'Reply to Critics' shortly after, is that law, including adjudicative law, does not instrumentalise the agency of its subjects for another's ends, in the manner characteristic of a managerial relation.

For present purposes, however, the most striking aspect of Fuller's analysis of the common law in *Anatomy* is what we would today regard as its distinctly 'Dworkinian' flavour. This should already be apparent from his observations about the common law court's dual task, and how courts 'have always taken it for granted that they must demonstrate that the rules they apply are "grounded in principle"'.[64] There is also a strong nod towards Dworkin's idea of 'coherence', or 'fit', when Fuller suggests that:

> Those responsible for creating and administering a body of legal rules will always be confronted by a *problem of system*. The rules applied to the decision of individual controversies cannot simply be isolated exercises of judicial wisdom. They must be brought into, and maintained in, some systematic interrelationship; they must display some coherent internal structure.

Thus, it is the judge's task to:

> discern what is essential and what is not in the decision of a case, one must have resort to principles that transcend their immediate application; it is these general principles that bind the elements of law into a coherent system of thought.[65]

In short, the similarities between Fuller and Dworkin on these points are hard to miss. But the purpose of the foregoing has not been to draw comparisons for comparison's sake. My aim, rather, has been to provide a foundation for assessing how Fuller's understanding of interpretation contributes to his understanding of the nature and value of legality, and how, in turn, the constitutive features of this account might be distinguished from that offered by Dworkin.

[61] ibid 103.
[62] ibid 91.
[63] ibid.
[64] ibid (the quotation marks are Fuller's).
[65] ibid 94. See also 96.

To this end, some preliminary comments can be offered. For Fuller, the demands of the task of interpretation are part and parcel of a judge's responsibility for maintaining an order that honours and realises a moral valuably phenomenon—respect for persons as dignified agents—that is internal to the form of law. Whether specifically reflected in his thinking on interpretation, or in the broader themes of his jurisprudence, the centrality that the responsible agent occupies within Fuller's thinking obviously has much in common with how Dworkin's law as integrity speaks to a community of principle in which individuals are to be treated with equal concern and respect. Thus, even at the level of interpretation, there is clearly much similarity between Fuller and Dworkin on the question of the value that legality realises and, indeed, *who* such a condition serves.

But there is, I think, an important difference. In the way that it derives its substance from the character of the form of law, and in how this form presupposes the collaboration of mutually responsible agents, the political morality salient in Fuller's jurisprudence—respect for persons as dignified agents—simply comes with the territory of choosing to govern through law in the first place. Certainly, this status of responsible agency is also nurtured and sustained by law. But Fuller's key insight is that it is a status that is prior to law, in a way that does not seem to be the case (or at best is only implicitly the case) for Dworkin. I will return to this crucial point of distinction below. First, however, it is necessary to chart Fuller's and Dworkin's common thinking on the issue of methodology in jurisprudence, as this, too, is an important site for understanding the features of Fuller's conception of legality and its place within Dworkin's schema.

V Fuller, Dworkin and Methodology

The question of the appropriate methodology for doing jurisprudence occupies a central place within Dworkin's challenge to Hart's positivist theory of law. Dworkin's point here can be neatly summarised. Not only is a descriptive approach to illuminating the concept of law in Hart's vein unfruitful, it is not possible. All attempts to illuminate the concept of law invariably involve a process of evaluation, in one form or another. Thus, if jurisprudence is to be an illuminating exercise, we must abandon any pretence of merely observing the subject of our study from a position of objective neutrality. We must instead approach law from the inside-out, taking seriously its participants' understanding of what it means to make a declaration of what the law is, and what it means to keep a legal order in a workable and coherent shape.[66] Only then, Dworkin insists, can the enterprise of jurisprudence be rescued from the 'Archimedean obscurantism'[67] that has plagued the discipline at the behest of the dominant positivist approach.

[66] See generally Dworkin, *Law's Empire* (n 2) chs 1–3, and 'Hart's Postscript' (n 3) 1–5.
[67] ibid 25.

This is a conversation between Dworkin and Hart. But anyone familiar with Fuller's exchanges with Hart, not least the 'Reply to Critics', will identify the ways in which Fuller attempted to have the same kind of conversation about the proper, or most illuminating, methodology for jurisprudence. The 'Reply', as I emphasised in chapter five, not only identifies an absence of shared 'starting points' as the source of the impasse between Fuller and his critics,[68] but also makes an unequivocal appeal to legal philosophers, if the issues of concern to his own jurisprudential agenda are to be addressed, to adopt what contemporary debates refer to as an 'internal' perspective: the viewpoint of participants to the actual enterprise of lawgiving, law-applying and law-obeying.

This matter of perspective is well indicated in the closing words of the 'Reply', which express hope for a future in which 'legal philosophers will cease to be preoccupied with building "conceptual models" to represent legal phenomena, will give up their endless debates about definitions, and will turn instead to an analysis of the social processes that constitute the reality of law'.[69] It is still better indicated in some of the working notes for that reply that I examined in chapter five, especially those which see Fuller chastise positivists for failing to meaningfully identify with the subject of the law, or, as one note puts it, to show 'sufficient concern for the other fellow to be willing to spend the time and energy necessary to take him into account'.[70]

Whether we regard these lines, published and unpublished, as merely a final expression of frustration on Fuller's part, or a prophecy of important debates to come, what is abundantly clear is that Fuller recognised the extent to which contests about methodology were crucial to defining and, indeed, explaining the character of the contests of jurisprudence. This is an insight which can be readily traced to his eunomics project and its criticism of the significant distance between the concerns of legal philosophy and the actual work of lawyers,[71] as well as other earlier writings that I reviewed in chapter two. But what is also clear here, in terms of the Hart-Fuller debate and the shape which that debate ultimately took, is that Fuller saw the question of perspective—of the methodological demands of jurisprudential inquiry—to be crucial to his effort to break down, diagnose and address the obstacles which plagued his efforts to instantiate, let alone sustain, a constructive conversation with his critics.

There is no question that Dworkin's own methodological critique of Hart's project surpasses Fuller's in both its clarity and its depth. Again, my purpose in bringing them into the same space has not merely been to illuminate a compelling similarity, but rather to add further dimensions to the conversation to which this chapter is heading. The immediate question, then, is how Fuller's methodological commitments contribute to his conception of legality, and whether any such

[68] 'Reply to Critics' (n 49) 189.

[69] ibid 242.

[70] Undated document titled 'External morality relation', The Papers of Lon L Fuller, Harvard Law School Library, Box 12, Folder 1 (notes for 'Reply to Critics').

[71] Lon L Fuller, 'American Legal Philosophy at Mid-Century: A Review of Edwin W. Patterson's *Jurisprudence, Men and Ideas of the Law*' (1954) 6(4) *Journal of Legal Education* 457, 476–77.

contribution aids the case for giving that conception its own place on the landscape of contemporary legal philosophy as Dworkin has mapped it.

To this end, two points need to be emphasised. The first has already been foreshadowed in my reference to Fuller's working papers for the 'Reply'; namely, the orientation of his jurisprudence towards the legal subject, and the way that he insists she not fall from view while the theorist, as positivists since Hobbes had done, exhausts all intellectual energies on interrogating the features of law's source.[72] Taken as a methodological point, it might be said that the contribution here lies in Fuller's awareness of how choices about methodology determine *which* legal participants—lawgiver or legal subject—a given account of law keeps within its gaze. As he puts it in another working note for the 'Reply':

> This mistake comes from posture of their minds . . . they stand off pasting verbal tickets on what they observe, instead of participating vicariously in the activities, seeking to understand what it means to engage in them. Get some sense of what it is like to assume legislative responsibilities . . . If instead of standing off, watching the scene, and debating what words to use in describing it, they would only move a little closer and participate vicariously in the task of lawmaking.[73]

Participating vicariously in the task of lawmaking, and by doing so giving attention to the position of both lawgiver and legal subject, is very clearly Fuller's own methodological stance on the question of the proper mode for doing jurisprudence. His conception of what a legal order actually is absorbs this commitment in how it brings those two participants, otherwise so widely regarded as standing at two poles, into its domain, as equals. The two, as I will shortly elaborate further, constitute dual centres of gravity within Fuller's jurisprudence.

The second observation on how Fuller's methodological commitments contribute materially to the distinctiveness of his conception of legality comes from a different angle—through the idea of form. This is a more subtle point, but an important one. If legal philosophers are to take the form through which law finds expression seriously, as Fuller believed they should, an internal methodological perspective is immediately implied because form is ultimately only something that is intelligible in practice. We cannot understand the legal form unless we have an eye to legal practice, its dynamics and its patterns. When we keep this idea in view, Fuller's appeal to legal philosophers at the end of the 'Reply' that they dispense with their 'conceptual models' and turn instead to an analysis of 'the *social processes that constitute the reality of law*'[74] can be seen as part and parcel of his enduring ambition to take form seriously as the site which not only distinguishes law from other forms of ordering, in a structural sense, but which possibly also provides a different answer to the perennial question of the source of law's normativity. This last point is crucial to the engagement between Fuller and Dworkin on the meaning and value of legality to which I now turn.

[72] See chapter two, I 'The early Fuller: positivism and natural law at mid-century'.

[73] Document titled 'Internal *morality*', The Papers of Lon L Fuller, Harvard Law School Library, Box 12, Folder 1 (notes for 'Reply to Critics').

[74] 'Reply to Critics' (n 49) 242 (emphasis added).

VI Fuller, Dworkin and the Value of Legality

Although its foundation lay in a contest about methodology, the destination of Dworkin's 2004 essay, 'Hart's Postscript and the Character of Political Philosophy', is the idea of legality and its contribution to reassessing the calcified debates that populated the landscape of jurisprudence at the close of the twentieth century. 'Legality', Dworkin declares, is not only the 'distinctly legal value' that lies at the heart of 'what claims of law mean and what makes them true or false',[75] but is also the 'foundation' from which future inquiries of legal philosophy must proceed.[76]

Dworkin's claim is that we can understand legal practice better, and make more intelligible propositions of law, if we pursue 'an explicitly normative and political enterprise: refining and defending conceptions of legality and drawing tests for concrete claims of law from favoured conceptions'.[77] As Dworkin presents it, this is not a sudden or unexpected turn within his own project, though he does concede that an emphasis on legality, as such, has previously lurked in the background of his work. But the idea that 'a philosophical theory of law must begin in some understanding of the point of legal practice as a whole', he insists, has always been at the forefront of that project.[78] The 'Postscript' essay therefore stands as a statement of intent to relocate those underlying preoccupations to a far more central place.

The working definition that Dworkin attributes to the term 'legality' is that such stands for the enterprise of constraining the use of political power in accordance with standards established in advance of its deployment.[79] But this is a working definition, a starting point only, and Dworkin immediately distances himself from such a descriptive statement because it is 'entirely uninformative' with respect to the question of what kinds of standards satisfy legality's demands or, indeed, the question of why legality is valuable to us.[80] This—legality's value, and the demands it places on its participants,—is Dworkin's primary sphere of concern, and his main purpose is to investigate how this idea constitutes the 'centres of gravity' of different understandings of law. For natural law thinking, this centre of gravity is the value of 'accuracy'; for legal positivism, the value of 'efficiency'; and for his own conception of law as integrity, the value of 'political integrity'.

'Accuracy', Dworkin suggests, is a way of distilling the concern of the natural law tradition to see the state's coercive power exercised 'in a substantively just and wise way'.[81] Thus, one might argue that legality promotes the value of accuracy if

[75] Dworkin, 'Hart's Postscript' (n 3) 24.
[76] ibid 35.
[77] ibid 25.
[78] ibid. *Law's Empire* uses the term 'the rule of law' only once; see Dworkin, *Law's Empire* (n 2) 93.
[79] Dworkin, 'Hart's Postscript' (n 3) 26, 35.
[80] ibid 24.
[81] ibid 26.

one believes that 'official acts are more likely to be wise or just if they are governed by established standards than if they represent merely the contemporary judgment of some official',[82] or if one holds a conception of legality 'that allows the tests for established standards to promote or even guarantee' accuracy.[83] The heart of accuracy, then, lies in how the exercise of the state's coercive power in accordance with pre-established standards might contribute to the substantive justice of legal outcomes.

In contrast, the centre of gravity for legal positivism, 'efficiency', speaks to law's value in substituting 'crisp direction for the uncertainties of customary or moral imprecation',[84] or providing 'fair warning' about when the state will intervene in its subjects' affairs or punish them.[85] The emphasis here is on how the exercise of the state's coercive power in accordance with pre-established standards contributes to the efficiency and clarity of state action and, reciprocally, to the clarity with which the subject of that action can foresee and accommodate its impact on her life.

While it is apparent that Dworkin seeks to distinguish the centre of gravity of his own jurisprudence from positivism's value of efficiency, it is equally clear that he is at pains *not* to suggest that efficiency is an unimportant legal value. In a way that echoes the lessons of Fuller's story of the tyrant lawmaker that I set out in chapter four, Dworkin readily agrees that no ruler, even a tyrant, survives for long or achieves his goals, even very bad ones, 'if he altogether abandons legality for whimsy or terror'.[86] Thus, Dworkin does not seek to place his own understanding of the value of legality ('political integrity') in direct competition with positivism's efficiency. Rather, what he seeks to suggest is that political integrity, alternately explained as the value of 'fairness' and 'equality before the law', is sufficiently independent of 'efficiency' to provide the basis of a distinctive conception of the animating point of legal order.[87]

The distinguishing feature of Dworkin's 'political integrity' is 'equality before the law' in the sense that 'government must govern under a set of principles in principle applicable to all'.[88] It is not my aim here to appraise the success of Dworkin's effort to argue for the superiority of this conception of legality, his own, among the three which he sets out. My interest lies instead with his wider effort to remap the landscape of jurisprudence in terms of the value served, or represented, by the idea of legality within different understandings of law. If Fuller is to be (re)situated on this map as a central rather than marginal player, it is necessary that his understanding of the value of legality be distinguished from Dworkin's 'political integrity' and, indeed, from 'efficiency' and 'accuracy', rather than collapsed into one or other of these ideas.

[82] ibid 26–27.
[83] ibid 27.
[84] ibid 28.
[85] ibid.
[86] ibid 29. See chapter four, IIC 'Lessons from a tyrant?'.
[87] Dworkin, 'Hart's Postscript' (n 3) 29.
[88] ibid.

Because he addressed these issues at multiple points in his writings, distinguishing Fuller's understanding of the value of legality from accuracy is a relatively straightforward task. As I explained in some detail in chapters two and three, Fuller identified himself with the natural law tradition in a number of ways: in its commitment to the role of reason in human affairs, its interest in matters of institutional design, and its concern for the substantive justice of legal ends.[89] He expressed considerable sympathy for projects, like Radbruch's, that sought to sustain a conceptual connection between law and substantive justice within twentieth century jurisprudence.[90] Yet, the preceding chapters should have made clear, detailed especially in chapter three, that beyond these sympathies Fuller insisted that his own was a natural law jurisprudence in only a very qualified sense. Fuller saw observance of his principles of the internal morality of law as having, in practice, some affinity with the realisation of substantively just legal ends, but he equally and repeatedly made clear that he had never claimed that observance of those principles necessarily guaranteed moral outcomes. Indeed, as I explained in chapter four, the moral thrust of Fuller's jurisprudence—the idea that the form of law expresses respect for agency—remains intact even when the aims pursued through law are unjust.[91] On this argument alone, therefore, the value of legality for Fuller can be distinguished from 'accuracy' and its concern for the substantive justice of legal ends.

However, distinguishing how Fuller understands the value of legality from positivism's 'efficiency' is a more complex exercise. This is because, as I have emphasised repeatedly, attentiveness to law's efficacy for the lawgiver is a substantial part of Fuller's understanding of law. It is, one might say, borrowing Dworkin's language, one of the dual 'centres of gravity' of his jurisprudence.[92] To straddle dichotomies in this way is precisely what we should expect from Fuller. Across the entire course of his writings, he consistently rejected the idea that the nature of law could be grasped by a singular theoretical viewpoint; Marshall Cohen's 'principle of polarity' informed Fuller's theoretical approach very strongly.[93] Yet to accept that Fuller's jurisprudence incorporates dual centres of gravity is not necessarily also to accept that the efficacy commitment within that duality should suddenly be collapsed into, rather than distinguished from, the positivist's sense of the value of legality. This is not least because Fuller gave us the means, in the 'Reply', to see how the positivist answer to what is good about law—how it realises the value of efficacy—struggles to reveal an argument for *why* this value is better realised through law than through direct rule by men.[94]

Fuller's jurisprudence thus clearly embraces aspects of both 'accuracy' and 'efficiency' as Dworkin represents them. But it is equally clear that Fuller's position is

[89] See chapter two, III 'Navigating the labels'.
[90] See chapter three, IIC 'Analysis: Reading Fuller through the Nazi law debate'.
[91] See chapter three, IV 'Conclusion'.
[92] Dworkin, 'Hart's Postscript' (n 3) 26.
[93] See chapter two, I 'The early Fuller: positivism and natural law at mid-century'.
[94] See David Dyzenhaus, 'The Grudge Informer Case Revisited' (2008) 83 *New York University Law Review* 1000, 1030 on this point.

not reducible to either as its dominant value. In suggesting this, I do not mean to overlook how Dworkin himself makes clear that no tradition 'has chosen one of these three values as the exclusive key to legality, and disparaged or neglected all others'.[95] If not a duality, there is at least some kind of shared landscape on which accuracy, efficiency and political integrity each find their place. But this last point is the one to emphasise: Dworkin clearly does commit to the idea that a theory of law is naturally dominated by a single value, and to the idea that the main ground of difference between conceptions of law lies in the importance each assigns to that value.[96]

It is precisely this idea that a given jurisprudential position can be distinguished on the basis of its commitment to a dominant value that is complicated by a turn to Fuller. Again, this is because of how Fuller's starting point is the idea of law as a distinctive form which generates a distinctive kind of relationship between lawgiver and legal subject. For him, legality is a condition that speaks meaningfully in two directions: to the moral value it represents for the position of the legal subject, as well as to its efficacy for the lawgiver. Attention to form brings both of these values into focus.

Form, then, must also be the site on which to distinguish Fuller's conception of the value of legality from Dworkin's. To illuminate the source of this claim, it is helpful to restate Fuller's understanding of form, as I have interpreted his jurisprudence. For Fuller, to speak of the form of law is to speak of law's distinctive shape. It is an enterprise of general rules, and this starting point of generality generates a number of implications, including the other seven principles which together make such a mode of governance both possible and effective. But these principles of lawgiving, which enable law's distinctive formal attributes to manifest, speak not only to the set of responsibilities that burden the lawgiving task, but also to a particular conception of the legal subject as someone who must possess the necessary responsible agency to be able to interact with general rules and the conditions through which they are communicated. For Fuller, then, there is not a form called law that acts upon the legal subject, but rather, the form of law includes the legal subject's capacity for agency within it, precisely because it presupposes that capacity.

With this reminder of Fuller's understanding of law's form in view, we are able to see why the task of distinguishing Fuller's conception of legality and its value from Dworkin's 'political integrity' is more nuanced than was the case for 'accuracy' and 'efficiency'. This nuance arises from Fuller and Dworkin's apparent shared concern for the quality of the space that the legal subject occupies under law. Dworkin is regarded as one of the leading liberal theorists of our era for good reason—his consistent, animating concern for how political power is experienced by those who are subject to it. These subjects, in his view, are bearers of inherent dignity who are entitled to be treated with equal concern and respect. Indeed, Dworkin seems to offer us something richer here than Fuller's conception of

[95] Dworkin, 'Hart's Postscript' (n 3) 26.
[96] ibid.

the legal subject as a responsible agent who is a bearer of dignity and an end in herself.

Still, the argument I will offer to meet my claim that Dworkin and Fuller can nonetheless be distinguished on this point can be stated as follows. It is only in Fuller's jurisprudence that law's respect for the agency of persons, for the distinctive space occupied by the legal subject, is inherent to law's distinctive form. The implication of this for distinguishing his sense of the value of legality from Dworkin's is this. The value of legality, for Fuller, is *prior* to Dworkin's value of political integrity, or equality before the law, because Fuller's speaks to the shape, content and presuppositions of law's distinctive form in a way that Dworkin's does not. This does not mean that Fuller's conception of the value of legality is in any way incompatible with Dworkin's. On the contrary, Fuller's provides foundations on which a conception such as Dworkin's rests or, at the very least, can be supported. This, indeed, leads to a further point: Fuller's understanding of the value of legality can also be distinguished from Dworkin's in how it directly addresses a question that Dworkin's jurisprudence neglects; namely, the question of what it is that makes law distinctive, that sets it apart from, other modes of ordering.

There is much going on in this argument that needs to be unpacked. To do so, I will focus on two related points: first, how we might distinguish the moral significance of law's generality in Dworkin and Fuller; and second, how we might understand the place of form within Dworkin's understanding of the competing concepts of law that populate the landscape of jurisprudence.

A Content and Moral Significance of Generality

As I explained in chapter five, understanding the content and moral significance of generality in Fuller's jurisprudence is a central part of understanding the place of the legal subject within that jurisprudence—her status as an agent, as someone capable of responding to, and thus enlivening, a regime of governance through general rules. It is also key to distinguishing Fuller's sense of the value of legality respect for agency—from positivist theories of law.[97] But it is much less clear whether this point of distinction has any salience for the project of distinguishing Fuller's understanding of the value of legality from Dworkin's.

Dworkin's 'law as integrity' expresses its orientation towards the position of the legal subject in terms of law's respect for the community of equals to whom the state's coercive power must be justified. This community, the citizenry, the subjects of law, are the end-concern of Dworkin's project. Securing equality before the law for the legal subject is, for Dworkin, the 'putative point served' by legality in its constraint of political power in accordance with standards established in advance of its deployment.[98] What is 'good' about legality, then, is how it

[97] With the possible exception of the reading of Raz that I offered in chapter six.

[98] Dworkin, 'Hart's Postscript' (n 3) 26.

promotes the value of equality, or equal concern and respect, for the subject as a subject of law. But it is nonetheless the case, even with this community in view, that the most explicitly important figure in that project, its primary site of jurisprudential exploration, is the judge. On Dworkin's account, this judge must, through the discharge of his interpretive responsibilities, see to it that law's meaning manifests in accordance with a political morality of equal concern and respect. We might say, therefore, that to the extent that we come to know the legal subject within Dworkin's jurisprudence, it is principally through the vision of the judge, discharging his role morality *as* a judge.

Dworkin's engagement with the *formal* dimensions of this idea is primarily a negative one, in the sense that he makes explicit that the 'force' of the claim of integrity lies *not* with a claim of formal equality before the law, as might be captured in the idea that 'law should be enforced against everyone according to its terms'.[99] Instead, what integrity has in view is a 'substantial' idea of equality before the law, such as that reflected in Dicey's formulation that 'every man, whatever his rank or condition, be subject to the ordinary law of the realm', or Hayek's emphasis on generality as 'the most important aspect of that attribute of law which we have called its "abstractedness"'.[100] This idea of legality, something beyond what Dworkin regards as a mere 'formal' idea, is what would be breached by laws that 'applied only to the poor, or exempted the privileged'.[101] Its opposite is the arbitrary exercise of coercion or infliction of punishment, or, as he puts it at one point, 'whimsy' and 'terror'.[102]

In analysing what we might take from this picture, and its relevance to the task of resituating Fuller, I do not wish to rest my claims in a narcissism of small differences. In its orientation to the subject as someone to be recognised as an equal before the law, the picture painted by Dworkin has much in common with Fuller's. Still, it is distinguishable in key ways. The first relates to where the legal subject is positioned within Dworkin's vision. As counter-intuitive as this might seem, there are ways in which Dworkin's understanding of equality before the law finds expression in an essentially 'top-down' way; that is, in a way that examines how, mediated through law, power *acts upon* those who are subject to it. The idea is that the legal subject gains her status, as an equal to fellow citizens, *if* law meets the conditions of integrity. What is absent is any meaningful sense, or recognition, of the subject and her capacity for agency as having a morally significant status *within* law, as is the case for Fuller.

This might be a subtle point, but it is an important one. For Fuller, what is 'good' about legality is not just how it brings rules and principles, lawgivers and legal subjects, into a coherent and sustainable relation. It is equally not just how law treats those subjects as responsible agents. What is good about legality, for Fuller, derives above all from what law simply is; how it is a mode of governance

99 ibid 30.
100 ibid.
101 ibid.
102 ibid 29.

that finds expression through a distinctive form which presupposes, and thus must also respect, the status of the legal subject as an agent. In Fuller, the place we look to illuminate this presupposition, and its implications for imposing meaningful limitations on a lawgiver's power in favour of that subject's agency is, or starts with, law's defining feature of generality.

The direction of Fuller's thinking on this point, as I highlighted in chapter five, is suggestively expressed in one of his working notes for the 'Reply to Critics', where he explicitly links the feature of generality which distinguishes law from managerial direction to the status of the legal subject as an agent. As he puts it there, generality is not about giving the legal subject advance notice 'to be ready to jump', or about 'reducing the superior's work load'. It is something beyond this. The key point, then, is that Fuller's understanding of generality recognises the subject as an actor, or agent, independent from law, even if that same understanding then envisages how law contributes to the maintenance of this status through its provision of baselines for the conduct of the agent's own affairs. To commit to generality is to commit to a conception of the legal subject in these terms—to accept what is prior to the law, but which law also constitutes and sustains.

This understanding of the moral fibre of generality is obviously much richer than anything offered by Hart's suggestion that a 'germ of justice' consists in law's generality in so far as this feature of law necessitates like treatment in like cases.[103] For Hart, this was something worth noting, but hardly of serious concern to debates about the connections between law and morality, because laws which met this standard could apply 'with the most pedantic impartiality as between persons affected, laws which were hideously oppressive', and 'might deny to a vast rightless slave population the minimum benefits of protection from violence and theft'.[104]

I bring Hart back into the picture here because his stance on the moral significance of generality provides a contrast, within a taxonomy of sorts, that assists us to chart the differences between Dworkin and Fuller on the same point. Like Fuller, Dworkin also understands the idea of treating like cases alike in much more morally robust terms than Hart, insisting that his own idea of integrity speaks to more than consistency by another name.[105] Integrity, rather, demands that the public standards of the community be both made and seen, so far as this is possible, 'to express a single, coherent scheme of justice and fairness in the right relation'.[106] This, indeed, is a richer understanding of generality than Fuller's, in the sense that Dworkin's appears to be burdened with a more demanding moral task.

Still, even if richer in this sense, there are other ways in which the implications of Dworkin's conception of generality and his sense of the work that it does basically bypass Fuller's. This happens mainly through how, as it stands, Dworkin's

103 See chapter three, IA 'Setting the agenda: Hart's claims'.

104 HLA Hart, 'Positivism and the Separation of Law and Morals' (1958) 71 *Harvard Law Review* 593, 624.

105 Dworkin, *Law's Empire* (n 2) 219.

106 ibid.

project gives little apparent attention to the way that respect for the equality of persons—the nub of political integrity—is supported by law's general form. It overlooks how law's distinctive form is valuable not just for how it *serves* the agent, but for how it *presupposes* her status as such.

One might respond here by saying that Dworkin's conception of legality is basically a development of Fuller's. That is, while both theorists begin with a shared concern for the moral value of law's form, it is Dworkin who moves that concern in the direction of an explicitly normative agenda that gravitates around a particularly rich understanding of political morality. Dworkin and Fuller are thus sufficiently close, the intervention might continue, as to be indistinguishable or, at least, to override the case for Fuller's conception being given its own place on Dworkin's map.

There is much merit to this response, but a number of factors counsel against accepting it unqualifiedly. The first might be approached by recalling how Dworkin opposes his idea of political integrity to ideas of arbitrariness, or 'whimsy' and 'terror'. This seems sound in principle, but what is missing is a deeper consideration of how whimsy and terror actually manifest; that is, how they are effected not only by disavowing the idea of governing through standards established in advance, but also by disavowing the idea that there be any such standards at all.

A second point brings this idea into connection with the observations I made earlier about Dworkin's neglect of the moral dimensions of form in the early essays where he engaged with Fuller's arguments in *The Morality of Law*. There, Dworkin's intuitive sense was that the moral claims of Fuller's jurisprudence spoke only to the extremes of formal legal pathology, pathologies that very often also transgress 'conventional' moral principles in a flagrant manner. As he then understood it, Fuller's jurisprudence otherwise said nothing to suggest why there is anything of moral significance to more 'everyday instances' of legal pathology, such as poor legislative draftsmanship.[107]

It is instructive to return to this point for what it reveals about how Dworkin did not appear to appreciate that, for Fuller, the moral dimensions of law's form arise from the particular quality of relationship that that form constitutes between lawgiver and legal subject, one that goes directly to law's respect for the dignity of the subject as a responsible agent. For Fuller, these moral dimensions are extant in, and have implications for, all instances of legal pathology, drastic and minor alike. In his more recent work, however, there are signs that Dworkin might be coming around to Fuller's point; that is, to Fuller's attention to the form of law as a site of moral interest in its own right.

Certainly, in his most recent book, *Justice for Hedgehogs*, Dworkin revisits his earliest approaches to debates about law and morality in an effort to redraw them in a manner more congenial to his contemporary concerns.[108] Opportunities for

[107] See section II 'The 1965 essays' above.

[108] Ronald Dworkin, 'Law', in *Justice for Hedgehogs* (Cambridge, The Belknap Press of Harvard University Press, 2011) ch 19, 402, but with Dworkin here referring to the views expressed in *Taking Rights Seriously* (n 1), rather than in the 1965 essays.

new conversations are thus potentially opened up here. But, more saliently, it is his 2004 'Postscript' essay which sees Dworkin speak of how his turn towards legality as the foundational idea of jurisprudence might serve to illuminate issues 'underdeveloped or ignored' in his previous work.[109] One such issue is the need to inquire more deeply into the attributes of procedural fairness as the 'nerve centre' of his interpretive criteria of 'fit'.[110] Still, beyond these gestures, all evidence indicates that Dworkin's understanding of form is much narrower than Fuller's. Indeed, the boundaries of that understanding seem to be more in the vein of traditional readings of Fuller; readings, that is, which interpret Fuller's jurisprudence as speaking to 'procedural' desiderata or 'requirements of process', rather than as speaking to the form required to instantiate a particular quality of relationship between a lawgiver and an agent legal subject. This last observation is not intended to downplay the importance of renewed attention to process and procedures within legal philosophy, as indeed Jeremy Waldron presently advocates.[111] It is simply to point out that, for Fuller at least, there is much more to be said, foundationally, about why such procedures matter; that is, what work, and contribution to legality, they fulfil.

The key points, therefore, might be summarised thus. The value that Dworkin attributes to legality—political integrity—might speak to the legal subject in the sense that it has her status as a person deserving of equal concern and respect squarely within its gaze. At the same time, within Dworkin's jurisprudence, that subject exists within a somewhat abstract space. The reason for this can be neatly stated. At least as it stands, Dworkin's project has not taken the idea of legality as a relationship between a subject who is presupposed to possess the status of an agent, and a lawgiver whose efficacious pursuit of ends must account to that status, as far as Fuller did, precisely because Dworkin has not taken form seriously as a site of jurisprudential inquiry.

For Fuller, law's moral project arises at least in part from the features of this site, which are in turn what distinguish law from other modes of ordering. They are equally congenial to supporting the kind of political morality, and conception of the judicial role, that Dworkin's jurisprudence argues for. But Fuller need not go down Dworkin's path to make a case for the moral work that is done by law's form. This is why, in my view, Fuller seems to offer us a conception of the meaning and value of legality in his own right, one that not only straddles morality and efficacy, accuracy and efficiency, but, in its adjudicative guise, aspects of Dworkin's domains of 'fit' and 'justification' as well.

109 Dworkin, 'Hart's Postscript' (n 3) 25.

110 ibid.

111 See, eg Jeremy Waldron, 'The Rule of Law and the Importance of Procedure', in James E Fleming (ed), *Getting to the Rule of Law* (2011) *Nomos* vol L.

B Form and Concepts of Law

A complaint that Dworkin makes about understanding legality merely in terms of the practice of governance in accordance with pre-established standards is that they offer nothing by way of illumination of 'what kinds of standards satisfy legality's demands', nor of 'what counts as a standard having been established in the right way in advance'.[112] The place of this complaint within the trajectory of Dworkin's project is illuminated if we turn to his 2006 book, *Justice in Robes.*[113] Dworkin's objective in the first chapter of that book is to arrange extant concepts of law into a taxonomy of 'doctrinal', 'sociological', 'taxonomic' and 'aspirational' categories, with different jurisprudential writers assigned to their appropriate place within each.[114]

A striking feature of this taxonomy is that Fuller *does* find a home within it, assigned to those who investigate law in the 'sociological sense'. As Dworkin explains it, scholars who approach the concept of law in this vein employ the term 'law' to name 'a particular type of institutional social structure', as Fuller did when he suggested that 'there is no law unless certain minimal requirements of procedural justice are met'.[115] Dworkin readily concedes that there might be a range of reasons why one might seek to stipulate 'a precise definition of what kind of social structure counts as a legal system'.[116] But he is clear that, in his opinion, little ultimately turns on the precise boundaries of such an exercise: we normally think it unnecessary 'to ask for a more precise definition of a "legal system" than our rough working ideas provide'.[117]

It is therefore obvious that Dworkin does not conceive of his own jurisprudential enterprise as having any meaningful affinity with the efforts of those in the 'sociological' camp. That enterprise, rather, can first be identified with 'doctrinal' inquiry into what *the* law of a given place or time actually is, and the role of moral criteria in determining the truth conditions of these propositions of law.[118] But the way that Dworkin understands his main jurisprudential contribution, his conception of 'law as integrity', is as an 'aspirational' concept of law—concepts which are concerned with the 'ideal of legality or the rule of law'.[119] Unlike sociological concepts, he argues, a great deal does turn on the boundaries we give to these aspirational concepts.

I do not here seek to question the designations that Dworkin has chosen for his own jurisprudence within the taxonomy that he constructs. His designation of Fuller's in the 'sociological' camp, however, is something that needs to be

112 Dworkin, 'Hart's Postscript' (n 3) 24.
113 Ronald Dworkin, *Justice in Robes* (Cambridge, The Belknap Press of Harvard University Press, 2006).
114 ibid 1–5.
115 ibid 3.
116 ibid.
117 ibid.
118 ibid 2.
119 ibid 5.

questioned. This placement is, of course, not entirely mysterious. Fuller's project was undoubtedly engaged with the question of what counts as a legal system, and was also, to a certain extent, concerned with advancing criteria against which such an assessment might be made. Still, as I have emphasised and especially in chapter four, Fuller's jurisprudence is a very long way from one that aims to stipulate 'a precise definition' of law. Moreover, and in no way inconsistent with its 'sociological' concern to offer standards for assessing what counts as a legal system, Fuller's jurisprudence is wholly animated by an 'aspirational' project; one which understands the existence conditions for law as something to be achieved, rather than simply to be taken for granted or measured against some kind of concrete, factual test.

The problem with Dworkin's designation of Fuller's jurisprudence as 'sociological', therefore, is not that it wholly misses the mark, but that it leaves no room for the possibility that Fuller's 'minimal requirements of procedural justice' should be understood as more than a checklist of static criteria.[120] This is precisely the reading of the internal morality of law that I rejected in chapter four. Instead, a far more accurate designation of Fuller's jurisprudence would be to situate it at the interface between the 'aspirational' and 'sociological' conceptions as Dworkin presents them. Indeed, as should be apparent from the preceding analysis of Fuller and interpretation, there is also very arguably a place for Fuller in the 'doctrinal' camp of Dworkin's map, too.

Precisely where Fuller ought to be placed on that map, however, is not the most important point. The more important point is what is revealed by Dworkin's choice to place Fuller in the sociological camp in the first place; how it suggests a failure, with respect to the task that Dworkin sets for himself, to consider what *qualities* a putative legal order must have in order to 'count', formally, as a recognisable legal system. There are suggestions in his 2004 'Postscript' essay that Dworkin does think that the question of what qualities a putative legal order must have in order to formally 'count' as a recognisable legal system is one worth thinking about. There, he refers to inquiries into 'what kinds of standards satisfy legality's demands', and, 'what counts as a standard having been established in the right way in advance' as among the most underdeveloped questions in jurisprudence.[121] But, at least as it stands, Dworkin's project does not offer anything to address this question by way of attention to law's form and the quality of relationship between lawgiver and subject that this form presupposes and speaks to. Here again, therefore, Dworkin might be counselled to turn to a resituated Fuller to illuminate the basis of such an inquiry, because, at least as things stand, Fuller continues to take us further towards a specification of the 'kinds of standards which satisfy legality's demands' than Dworkin does.

[120] ibid 3.
[121] Dworkin, 'Hart's Postscript' (n 3) 24.

VII Conclusion: Taking Form Seriously

I hope that it has remained clear throughout the foregoing analysis that it has not been my intention to overstate the points of difference between the end-concerns of Dworkin's and Fuller's respective projects. The commonalities between the two are abundant, and go much further, and are much broader, than the similarity of approach on matters of interpretation and methodology and, indeed, the nature and value of legality, discussed above. Dworkin and Fuller unite, for instance, in rejecting the view that even if the phenomenon of legality is sensitive in its application to the history and standing practices of a given community, 'nothing of importance can or should be done to explore the value at a philosophical level that transcends most details of place'.[122] The two also share a strong belief, again to borrow Dworkin's words, in the need for jurisprudence to engage in an 'explicitly normative' exercise of 'refining and defending conceptions of legality and drawing tests for concrete claims of law from favoured conceptions'.[123] Indeed, such a project was, in all relevant respects, the one with which Fuller confronted Hart when he challenged him to explain why certain instances of Nazi law, when measured against its presuppositions as a distinctive and workable form, ought to be anointed with the status of law. From the perspective of the task of resituating Fuller, then, it might be said that yet another site of value that arises from Dworkin's restated jurisprudential agenda is how it helps us to see what was always going on, from Fuller's perspective, in the Hart-Fuller debate.

But in terms of my own project of resituating Fuller in the landscape of twenty-first century legal philosophy, the most crucial aspect of Dworkin's turn towards legality is his appeal to legal philosophers to take seriously the idea that law has value for us, and to see how the answers we give as to why this is the case animate our understandings of law and the purposes that it serves. From a different end of the jurisprudential landscape, as I noted at the beginning of chapter six, Jeremy Waldron has also recently suggested that we must engage seriously with this question of what we think is 'good' about legality. Waldron, however, goes in a more Fullerian direction in this appeal than does Dworkin because, for him, 'a full understanding of how the legal system matters to us is our best guide to what is *distinctive* about legal as opposed to non-legal modes of governance'.[124]

This is a 'Fullerian' direction because Fuller's own answer to the question of what is 'good' about legality, or 'why law matters to us', was always inseparable from his sense of what distinguished law from other modes of ordering—the way that the respect for agency that is inherent in law's form imposes meaningful limits on lawgiving power. This answer can be distinguished from that offered by the

[122] ibid 35–36.

[123] ibid 25.

[124] Jeremy Waldron, 'The Concept and the Rule of Law' (2008) 43 *Georgia Law Review* 1, 61 (emphasis added).

natural law tradition because it makes no claim to law's necessary relationship to the substantive justice of legal ends. It can also be distinguished from the positivist answer, not because it rejects the value of efficacy as a key legal value, but because this positivist response both ignores the important moral work done by law's distinctive form, and struggles to reveal an argument for why efficacy is better realised through law than through some other mode of rule. Finally, Fuller's answer to what is good about legality can be distinguished from Dworkin's because the former speaks to a phenomenon—the moral work done by the relationship between legal form and agency—which presupposes the legal subject as an agent prior to the point at which law constitutes respect for that status. Though certainly supportive of the kind of normative political project that is stated in Dworkin's conception of law as integrity, this kind of work, what Fuller sees law as doing, need not rely upon, nor be developed into, such a conception.

So though I do not wish to overstate the differences between Fuller's and Dworkin's particular versions of anti-positivist jurisprudence, it remains the business of legal philosophy to draw distinctions precisely because there are things that we can learn by doing so. In legal philosophy, this distinction-making enterprise very often seems to dissolve into contests about mere questions of degree. For instance, it might be said that Dworkin's claims face the same challenge of just how much 'political integrity' or 'equal treatment' must be secured for law to be law, in the same way that Fuller's jurisprudence faces the question of just how great a level of disrespect for agency, occasioned through derogation from the principles of the internal morality of law, will cause any meaningful conception of law to disintegrate. In both cases, the questions admit of no certain answer. But perhaps it is time that legal philosophers began to accept such a compromise of clarity rather than resisting it as some kind of admission of theoretical failure. As I also concluded in chapter six in my analysis of Fuller and Raz on the relationship between law and the rule of law, these questions of degree cannot be the basis on which one jurisprudential project is declared to triumph over another because, when it all boils down, all are victim to the same affliction or, depending on one's perspective, the same promise.

This is why my purpose in these two 'resituating' chapters has ultimately been a dual one. My primary goal has been to resituate Fuller's distinctive contribution, as I have interpreted it, in the landscape of contemporary legal philosophy. But it has equally been my aim to illuminate how that landscape is defined by more shared commitments, inquiries and, indeed, challenges, than I suspect tend to be recognised; or, one might suggest, that have tended to be recognised in the absence of Fuller being part of that landscape.

8

Three Conversations

The 'forms liberate' image that has provided the frame for the inquiry of this book is perhaps the most intriguing of the many working notes that comprise the archive of Fuller's private papers. But there is another, also to be found in the file relating to his 'Reply to Critics', that is quite peculiar, in so far as it says little else beyond its title, 'Words We Use'.[1]

The words we use did indeed matter greatly to how Fuller's claims were both expressed and received. Just how acutely he appreciated this is made clear throughout his working papers, but perhaps nowhere more so than in the correspondence with Hart on the issue of trusteeship that I discussed in chapter four, where Fuller indicates a willingness to 'make peace' in his debate with Hart by 'substituting some term like "ethos" or "trusteeship" for "morality"'.[2] That term, 'morality', continued to be the main cause of Fuller's troubles in his exchanges with Hart about the unconventional claims of his jurisprudence. Still, despite the concessionary gestures we see in his letters to Hart, Fuller ultimately chose not to concede the term in any of his published writings. The 'Reply to Critics', for instance, sees no mention of the terms 'trusteeship' or 'ethos', although Fuller does introduce the idea of 'role morality'. Here again, however, the point to note is that the term 'morality' stays in, and the reason why is surely not hard to find. Had Fuller have conceded the term 'morality', he would almost certainly have found himself out of a conversation that was, for better or for worse, about the connections between law and morality.

Fuller thus clearly had occasion to think deeply and often about the words we use and how decisive they are not just to the possibility of meaningful conversation, but to any conversation at all. His working papers for the 'Reply to Critics' make it abundantly clear that the only way Fuller thought he would be able to make it through the conversation about the connections between law and morality that he found himself in was simply to try to change the conversation itself. Whether or not he succeeded in this effort is not the main point. After all, it takes two. The more salient point for present purposes is that Fuller's participation in that debate stands as a record of one scholar's attempt to expand the boundaries of extant conversations so as to accommodate questions and perspectives that he thought were being unduly neglected.

[1] Undated document titled 'Words We Use', The Papers of Lon L Fuller, Harvard Law School Library, Box 12, Folder 1 (notes for 'Reply to Critics').

[2] See chapter four, IIB 'Efficacy and trusteeship'.

It is thus fitting to close this book by exploring what can be learned from the project of reclaiming and resituating Fuller for the kinds of conversations we might now have about the animating themes of his jurisprudence; specifically how that jurisprudence engages with, interrupts, or gives content to the ideas of morality, instrumentalism and legality.

I Morality

One of the main purposes of this book has been to clarify the confusions that endure with respect to what Fuller meant when he spoke of the internal 'morality' of law. This question would not have cast the shadow that it has over the reception of Fuller's jurisprudence if it were not a valid one. Still, as I mentioned in chapter one, what has received too little acknowledgment in the history of the Hart-Fuller debate is how Fuller himself saw this confusion as actually running two ways. As he put it to Hart in 1958, positivists might be confident in their claim that there is no necessary connection between law and morality, but what exactly do they seek to exclude when they exclude 'morality' from the domain of law?

Fuller saw at least three kinds of connections between law and morality that he thought were salient to debates in jurisprudence. The first was a connection that goes to the question of why we accept law, why it has its normative force. As I have made clear in the preceding chapters, Fuller came to trace this connection to the special features that characterise the position of the legal subject as an agent who understands the crucial importance of the support she conveys towards the legal order to the very possibility of its existence. The second connection that Fuller identified was that which goes to the demands that burden the lawgiving task: the moral dimensions of the ethos of lawgiving, which are in turn referable to the lawgiver's responsibilities to the responsible agents who comprise the subjects of law. Third, Fuller also identified a connection between law and morality that goes to the moral quality of legal ends, and thus which concerns the moral substance of law.

The order of priority here is instructive.[3] Fuller's jurisprudence was always much more squarely concerned with the first and second of these connections between law and morality than with the third. Indeed, any turn that he did make towards the third was inseparable from the claims he advanced about the others. That is, at one level, Fuller was interested to explore the extent to which the moral quality of legal ends seemed to be conditioned, in practice, by the moral dimensions of law's distinctive form. We see this point expressed not only by reference

[3] Fuller lists the points in this order in an untitled and undated working note, paginated in handwriting as p 24, The Papers of Lon L Fuller, Harvard Law School Library, Box 12, Folder 4 ('Encyclopaedia Britannica').

to the pathological case in his discussion of Nazi law,[4] but also in his reply to Cohen and Dworkin, where he speaks of how it is reasonable to suppose that the 'built-in' respect for human dignity, inherent to the principles of the internal morality of law, 'will tend to carry over into the substantive ends of law'.[5] At another level, attaching to his thinking about how the moral foundations of a legal order lie in the acceptance conveyed by the legal subject toward that order, Fuller's working notes for *Anatomy of the Law* also speak of how this acceptance 'may be destroyed if there is no integrity in the manner of its internal construction and administration' or 'if the ends sought by the legal system are sufficiently odious to the population as a whole'.[6]

Fuller's contribution to debates about the connections between law and morality, and indeed the interconnections between these connections, thus was and remains a rich one. But throughout this book I have also suggested that to reclaim Fuller's jurisprudence is, to a certain extent, to accept the need to distance ourselves from the prism of those debates; or, at least, there is in my view a need not to grant that prism the hegemony that it has occupied so far in interpretations and appraisals of Fuller's claims. The battle of the schools between positivism and natural law, and specifically the debate about the connections between law and morality, was the context that Fuller entered when he commenced his foray into questions of jurisprudence. It was still more squarely the conversation that he entered when he chose to respond to Hart's defence of legal positivism on these grounds. But, as I suggested in chapter one, and which I trust has become clear over the course of the foregoing chapters, that debate was hardly the most congenial forum within which Fuller could elaborate his intuitions about what was so independently interesting and important about the form of law. Certainly, as it has been one of the purposes of this book to show, Fuller's intuitions about the distinctive form of law were and are capable of being translated into debates about the connections between law and morality. This is clear in his claims about why the legal form is valuable to and indeed presupposes the legal subject as an agent, how it instantiates an ethos of responsibility to persons on the part of the lawgiver, and how, in practice, there appears to be an affinity between the health of that form and the justice of legal ends. Still, the point I have sought to make is that the law and morality conversation has often obscured these jurisprudential insights as much as it has illuminated them.

As we move forward with Fuller, therefore, we might think about some of the reclaimed ways of conversing about the connections between law and morality that I have sketched at different points in the foregoing chapters. For instance, and to recall the analysis I undertook in chapter six, those who defend the position of exclusive positivism and its commitment to the idea that moral criteria do

[4] See chapter three, IIB 'Nazi law according to Fuller'.

[5] Lon L Fuller, 'Reply to Professors Cohen and Dworkin' (1965) 10 *Villanova Law Review* 655, 665–66.

[6] Untitled and undated working note, paginated in handwriting as p 35, The Papers of Lon L Fuller, Harvard Law School Library, Box 12, Folder 4 ('Encyclopaedia Britannica').

not figure in the determination of what law is might consider whether the apparently robust conception of the legal subject as an agent that is implicit in Raz's positivist account of authority complicates that commitment in any way; or, to gesture to an argument I have made elsewhere, conversations about the connections between law and morality carried primarily through a conversation about the connections between the form of law and human agency might further explore the instances of apparent correspondence between legal substance that seeks to oppress or to destroy the subject's agency and debasement to the formal features of law.[7] Indeed, in that vein, fresh conversations might still be developed out of Fuller's intuition that, as a matter of conceptual coherence, it is not possible to accept the idea that the institution of slavery is compatible with the institution of law.[8]

Above all, however, to reconvene conversations about the connections between law and morality in Fuller's name requires a clarified sense of the ways in which he is to be regarded as a natural law participant in those conversations. It should be clear by now that Fuller was a natural lawyer in the sense that he thought a conversation could be had about the connections between law and morality that is internal to conversations about the concept of law. He was equally a natural lawyer in the sense that he had faith in the role of reason in human affairs. And he was a natural lawyer in the sense that he thought that our obligation to obey law could not be meaningfully explicated in the absence of a sense of law's purpose. But the twist, or distinctiveness, of Fuller's contribution lies in what his jurisprudence suggests to us about how law's purpose is, to a significant extent, revealed in its form. Any conversation about law and morality convened in Fuller's name, therefore, must keep this insight in view.

II Instrumentalism

My analysis in chapter five will have made particularly clear just how much the task of clarifying, reclaiming and resituating Fuller's understanding of the moral dimensions of law requires that his claims be extracted from a conversation that, at Hart's lead, pitted claims about law's 'morality' against ones about its 'efficacy'. But for Fuller, as I have explained at length, the values of morality and efficacy were not incompatible but rather closely intertwined; together part and parcel of the demands as well as the promise, to both lawgiver and legal subject, of law's distinctive form. This was, and remained, Fuller's intuition from the early articulations of his eunomics project through to the close of his exchanges with Hart.

[7] Kristen Rundle, 'The Impossibility of an Exterminatory Legality: Law and the Holocaust' (2009) 59 *University of Toronto Law Journal* 65.

[8] See chapter four, III 'A different path?'.

One of my aims in this book has thus been to dissolve the unhelpful polarity of this 'morality versus efficacy' conversation, or at least as it applies to a proper reading of Fuller's jurisprudence. But this does not mean that all that is to be said about Fuller's sense of when and how it is appropriate to adopt an instrumental understanding of law has been said. It is fruitful to explore this question further, not least because it is increasingly the case that any theoretical conversation about law, whether commenced in the name of analytic legal philosophy, or in service of the agenda of legal pragmatism, the economic analysis of law, or some other project, seems to commence from the uncontested assumption that law's basic nature is that of an instrument. Can anything further be taken from a reclaimed Fuller to assist those who seek to unsettle, or at least to add nuance, to this increasingly colonised theoretical space?

To begin answering this question it is helpful to recall what Fuller's complaint against an instrumentalist understanding of law actually was. Above all, this complaint was directed to the shape of instrumentalist analysis: how its inherently one-way, top-down character blinds us to the interactions, interrelationships and sites of mutual adjustment that constitute the workings of a legal system in practice. In his specific responses to Hart, Fuller identified this shape as being part and parcel of the utilitarian genealogy of Hart's positivist project, where law was regarded as an instrument through which the political objectives of the utilitarian political and moral project could be met.[9] But, ultimately, the utilitarian frame that Fuller identified as underlying Hart's position is not crucial to the thrust of his critique. What's wrong with an instrumental conception of law, according to Fuller, is that because it takes the source of legal power as its starting point, and then understands the business of that source to be the pursuit of specified ends through the instrument of law, the conditions antecedent and] subsequent to the very possibility of law being used in this way simply fall from view.

For positivists, these conditions are undoubtedly accepted as *practically* important, but their consideration as a subject of jurisprudential inquiry is consistently regarded as external, rather than internal, to conversations we might have about the basic truths of the nature of law. Fuller identified this tendency very clearly in Hart's positivism, where 'the problem of achieving and maintaining legality' seemed worthy of 'no more than casual and passing consideration'.[10] He might equally have identified it in Raz's concern to show that whatever can be said about the nature of law's authority is still compatible with the sources thesis and its associated commitment to a conception of law as an instrument.[11] Seeking, like Raz, to defend the core commitments of legal positivism but while developing the project in different directions than did Hart, Scott Shapiro's recent work equally assumes the basic truth of law's status as an instrument in its suggestion that it might be fruitful to turn our jurisprudential attention to the task of compil-

[9] See chapter five, IA 'The Structure of Analytical Legal Positivism'.
[10] Lon L Fuller, *The Morality of Law* (New Haven, Yale University Press, 1964) 153.
[11] See chapter six, III 'Raz on authority'.

ing a 'user manual' to guide our understanding as well as our use of that instrument.[12]

But there are other conversations about law's instrumentality within which Fuller might also take a place. Fuller, for instance, had things to say about our instrumental reasons, whether objective or motivating, for obeying law; or, in Fuller's terminology, for why we might give our 'fidelity' to law.[13] We might say that in this kind of conversation Fuller would propose that the legal subject has 'instrumental' reasons for giving her fidelity to law that are derived from how the form of law respects her as a responsible agent. This, indeed, is a point well captured by Waldron when he reads Fuller as suggesting that it is what law *is*, quite apart from what it *does*, that generates our fidelity to it.[14]

This, however, is a different kind of conversation still from one which might attribute to Fuller the view that we have instrumental reasons for supporting one theoretical conception of law over another. This is what is suggested by Liam Murphy in his diagnosis of the Hart-Fuller debate as a contest about the best way, on political grounds, to understand the nature of law.[15] Certainly, there are strong hints of an argument in this vein in Fuller's writings, especially in his 1958 reply to Hart where, in the context of broadly defending Radbruch, he suggests that there are real world reasons not to commit to the philosophy of legal positivism.[16] But the core of Fuller's jurisprudence ultimately does not pivot around such an overt political agenda for legal theory. The core of that jurisprudence is instead arrived at through a process that begins with an account of the distinctive features of law as a form, and then uncovers the implications, including the normative implications, of what has been so described.

Put succinctly, the normative impulse of Fuller's jurisprudence follows from his diagnosis of law as a distinctive form, rather than from any separate or consciously over-arching normative agenda. To understand the difference, it is helpful to compare Fuller's jurisprudence with Dworkin's. For Fuller, law is not, in the first instance, distinctive because it gives expression to a distinctive political morality, as Dworkin would suggest. Rather, law is distinctive because it finds expression through a distinctive form, and even if this form has moral significance, which Fuller of course insists it does, the point is that the form itself comes first. It is only after establishing this—the distinctiveness point—that Fuller moves to the normative significance of the responsibilities and benefits, for lawgiver and legal subject alike, that are generated out of that form.

But the main conversation about instrumentalism in which Fuller actively participated, of which his critique of the unidirectional shape of the positivist analysis of law is part, was one that sought to unsettle the classically instrumentalist assumption that law is purely a means to an end. That conversation, at least if it

[12] Scott Shapiro, *Legality* (Cambridge, The Belknap Press of Harvard University Press, 2011) 399.

[13] See Liam Murphy, 'The Normativity of Law', draft manuscript on file with the author.

[14] Jeremy Waldron, 'Why Law? Efficacy, Freedom or Fidelity?' (1994) 13 *Law and Philosophy* 259.

[15] Liam Murphy, 'Better to See Law This Way' (2008) 83 *New York University Law Review* 1088.

[16] See chapter three, IIB 'Nazi law according to Fuller'.

does not entertain the moral dimensions and value of the 'instrument' itself, provides essentially no place for the concerns (indeed, often explicitly in response to such a view) that Fuller's jurisprudence sought to centrally situate. These concerns include considerations relating to creating and maintaining law as a distinctive form, the distinctive burdens of lawgiving responsibility, and the status that must be possessed by the legal subject. This is why it is helpful to return to Shapiro's point about our need for a 'user manual' to illuminate and guide the ways we deploy the instrument of law if we are to understand the particular challenge that Fuller presented to instrumentalist legal thinking. In multiple ways, Fuller's jurisprudence supplies us with precisely the kind of 'user manual' that Shapiro appeals for. But his way of offering us that manual is likely to be unwelcome to a positivist, for obvious reasons: it is designed to interrupt the very notion that we can start with settled convictions about law's status as an instrument. In Fuller's jurisprudence, that is, the 'user manual' to which Shapiro gestures is something we must already have in hand before we can bring the instrument itself into being, mediating not just the uses to which it can be put, but its very status as a mere instrument at all.

III Legality

In much the same way that a conversation about Fuller's sense of the connections between law and morality segues readily into a conversation about how his jurisprudence unsettles dominant assumptions about law's status as an instrument, a conversation about Fuller's understanding of the instrumentality of law segues readily into one about his views on the meaning and value of legality. It is thus fitting to close this book with that conversation, because Fuller's repeatedly expressed objections to positivism's disregard for the conditions antecedent to law is one now increasingly shared by contemporary legal theorists, but more commonly framed in terms of the threat posed by instrumentalist legal thinking to 'the rule of law', or 'legality'.[17]

Fuller himself used both the terms 'legality' and 'the rule of law', with the latter finding expression at various points in *The Morality of Law*, especially in connection with his eighth principle of congruence between official action and declared rule ('if the Rule of Law does not mean this, it means nothing').[18] But he was much more likely to use the term 'legality', which indeed made increasingly regular appearances as his exchanges with Hart progressed. Throughout *The Morality of Law* and the 'Reply to Critics' can be found arguments about how the fact of reciprocity between lawgiver and legal subject illuminates 'the cooperative task

[17] Exemplified in Brian Tamanaha, *Law as a Means to an End: Threat to the Rule of Law* (New York, Cambridge University Press, 2006).

[18] Lon L Fuller, 'A Reply to Critics', *The Morality of Law* (New Haven, Yale University Press, 1969) 209–10.

of maintaining legality', and analyses of the judicial role as one that implicates a 'responsibility for maintaining legality'.[19]

Though it is open to debate whether it is ultimately a distinction without a difference, there is perhaps something more capacious about the term 'legality' as an access point to the message of Fuller's jurisprudence than 'the rule of law'. The latter phrase tends (though it need not) to lend itself to a more criterial understanding of the phenomenon that Fuller sought to capture through his eight principles of the internal morality of law; indeed, something akin to the 'checklist' reading of those principles that I have rejected.[20] 'Legality', by contrast, arguably opens up a conversation that speaks more readily to the quality of the conditions generated by a legal system, above all, the distinctive quality of relationship between lawgiver and subject that is constituted by law's distinctive way of communicating norms through general rules.

Whichever term is preferred, what matters for any discussion we might have about how Fuller's jurisprudence contributes to conversations about the nature and value of legality is that we represent his sense of that idea in a manner true to his writings; that is, in terms closely connected to the intuition that law is a distinctive form of social ordering that constitutes and sustains a distinctive quality of relationship between the lawgiver and the legal subject. For Fuller, as I have stated repeatedly, there can be no meaningful concept of law that does not include a meaningful limitation of the lawgiver's power in favour of the agency of the legal subject. Any idea of 'legality' that we choose to attribute to Fuller must, I argue, reflect this insight.

But this also means that if we are to raise an objection against instrumentalist accounts of law in Fuller's name, we must do so in a manner that sustains this message at the same time as it does not oppose one, instrumentalism, directly against the other, legality. Certainly, Fuller viewed instrumentalist accounts of law as fundamentally incomplete, and he diagnosed this incompleteness as arising above all from how such accounts ignore or neglect the conditions that underscore the very possibility of governance through general rules. But he saw no contradiction in the idea that legality is a condition capable of absorbing *both* the considerations of efficacy in the pursuit of ends that are the primary focus of instrumentalist conceptions of law, *and* the rich moral territory that we associate with the idea of the rule of law in view of the benefits that accrue to the legal subject through its observance. On Fuller's account, from his earliest writings, the right way to theorise about law is to see these elements brought into a proper relation rather than to oppose them.[21] Thus, we might say, what is wrong with source-based conceptions of law is their failure to qualify their instrumentalism with the legality of law.

[19] *The Morality of Law* (n 9) 91, 224.
[20] See chapter four, IA 'The story of King Rex'.
[21] See, eg Lon L Fuller, 'Reason and Fiat in Case Law' (1946) 59 *Harvard Law Review* 376, 388.

Fuller and Shapiro: A New Conversation?

The possibility of new conversations about how Fuller might be resituated within contemporary discussions about the nature and value of legality has recently moved to a promising level at the hands of a legal positivist who, parting ways with his predecessors to a considerable extent, has willingly invited Fuller into the space. To bring this book to a close, therefore, it is fitting to respond to Scott Shapiro's important new contribution to legal philosophy on these questions: a work, titled simply, *Legality.*

The centrepiece of *Legality* is the 'Planning Thesis': Shapiro's claim that legal activity is best understood as an activity of social planning.[22] Legal institutions, he explains, 'plan for the communities over whom they claim authority, both in the sense of telling their members what they may or may not do, as well as authorizing some of these members to plan for others.'[23] Legal activity is not simply the creation and application of rules, but is rather 'an incremental process whose function is to guide, organize, and monitor behaviour through the settling of normative questions and which disposes its addressees to comply under normal conditions.'[24] Shapiro's 'Planning Thesis' is thus a significant contribution to legal philosophy in its attempt to move the theory of legal positivism away from Hart's model of rules and towards the idea of plans; a move that also sees Shapiro readily accept the idea that law serves the morally valuable purpose of settling normative questions for us.

Shapiro's book is recent, and so we are yet to see precisely what kind of general challenge to his central claims, as well as to his account of the meaning and value of legality, anti-positivists might seek to articulate. There is, at the very least, likely to be much sympathy for Shapiro's embrace of the idea that law serves an important moral purpose. Still, we can surely expect anti-positivists to insist that Shapiro's idea of law as a planning activity is incomplete in crucial ways; indeed, in precisely the kind of ways that Fuller complained of when he responded to Hart's positivism. To recall the spirit of those complaints, the kinds of questions that an anti-positivist in this vein might present to Shapiro include the following.[25] What are the conditions antecedent to law as a planning activity, and how does the Planning Thesis account for them? Who are Shapiro's 'planners', and what capacities and responsibilities are they assumed to have in order to be capable of participation in the enterprise of law as a planning activity? What relationships are Shapiro's planners in with each other, and how are those relationships constituted? How does the Planning Thesis comprehend the distinctiveness of law as a form?

[22] Shapiro, *Legality* (n 11) 195.
[23] ibid.
[24] ibid 203.
[25] Jeremy Waldron has offered some of his own in 'Planning for Legality' (2011) 109 *Michigan Law Review* 883, 896–99.

The interconnections between what might at first glance here present as a somewhat unruly catalogue of questions become apparent when we turn to the engagement with Fuller on the question of the meaning and value of legality that brings Shapiro's book to a close.[26] Its focus is the familiar site of Fuller's tale of King Rex and his eight failures to make law; a story that Shapiro reads as intended to illuminate 'the various steps one can take to *destroy* a legal system'.[27] This seems mostly right, though it is perhaps more accurate to say that Rex's ultimate failure was not that he *destroyed* a legal system, but that his failure to never in any meaningful sense *attain* one frustrated Rex's own aims as much as it did those of his personally loyal but ultimately disaffected subjects On Shapiro's reading, however, the nature of the relationship between Rex and his (putative) legal subjects is a relationship between planners, with Rex's hopeless efforts making 'the task of everyday planning extremely difficult' for his subjects.[28] Here, then, is the source of Rex's failure in a nutshell. Rex did not destroy his legal system because he was unjust in his planning, even if he surely did cause significant injustice through his incompetence. Nor was his failure, at least straightforwardly, one of inefficiency. Rex, rather, was unsuccessful 'because *he was not a social planner at all*': he simply lacked the kinds of dispositions that a true planning agent possesses, and needs to possess, to be a planner.

Shapiro readily concedes that Fuller's own intentions for the tale might have been quite different to this account of Rex as a failed social planner, even if he is otherwise on board with Fuller's claim that 'the wholesale flouting of any particular principle results in the failure to create or maintain an existing legal system'.[29] In terms of claims that can be made from the story of Rex, therefore, Shapiro is in agreement with half of Fuller's position. But what he cannot accept is the other half: that is, Fuller's 'notorious' claim that the eight principles derived from Rex's failures constitute a 'morality'.

As Shapiro understands it, what Fuller means by 'morality' in this context is the following:

> It is a *morality* on his view because its observance necessarily generates certain moral goods: any system that obeys the Rule of Law will provide its citizenry with a fair opportunity to obey the law. The morality is *internal*, rather than external, in that its value does not depend on the substantive political objectives that the rules aim to serve.[30]

This interpretation of Fuller's 'morality' claim, identifying law's intrinsic moral dimension as arising from how observance of the principles of the internal morality of law gives the subject a fair opportunity to obey the law, is essentially the same as that offered by Robert Summers in his 1985 book on Fuller.[31] There is much to support this reading, and there is no reason why Fuller would have

[26] Shapiro, *Legality* (n 11) 392–400.
[27] Shapiro calls this a 'Deconstructivist Strategy': ibid 392.
[28] ibid 393.
[29] ibid 394.
[30] ibid.
[31] Robert S Summers, *Lon L. Fuller* (Stanford, Stanford University Press, 1984) 37–38.

objected to it outright. Nonetheless, the 'fair opportunity to obey' reading is a much thinner and weaker understanding of Fuller's claims about the internal morality of law than that conveyed through the idea that law envisages the legal subject as a responsible agent; someone who comes to law in this state, and who is capable, on Fuller's view, of much more than simply obeying the law. By opting for the 'fair opportunity to obey' reading, therefore, Shapiro keeps his account of Fuller's claims at the usual positivist distance from the legal subject. But, more significantly, Shapiro's reliance on this reading also keeps his analysis at some distance from an adequate explanation of what, for Fuller, makes the morality in question an 'internal' one.

That is, like Raz in 'The Rule of Law and its Virtue', Shapiro is on the right track in how he sees Fuller's morality claim as lying not, or at least not squarely, with the moral quality of legal ends.[32] We therefore know what Fuller's internal morality is *not* derived from: namely, external, substantive political objectives. But what is much less clear is whether we have illuminated from where that morality *is* derived, and the cause of this deficiency is not hard to find. In short, Shapiro's understanding of the 'internality' of Fuller's internal morality remains essentially formless: absent in his analysis is any meaningful appreciation of how this 'internal' morality arises from the distinctive character, presuppositions and relationships generated by law's form. Moreover, there is little indication that Shapiro is prompted to move his engagement with Fuller in this direction. The meaning of Fuller's use of the term 'morality', he says, is a 'thorny question' about which much ink has been spilled and that he personally wishes to 'sidestep'.[33]

Shapiro's reason for taking this sidestep, however, is not because he wishes to relegate the whole 'morality' question to the too-hard basket. Rather, the reason for the sidestep is because Shapiro thinks that there is nothing in the message of Fuller's jurisprudence, of what we are meant to learn from Rex's failures as a lawgiver, that cannot be adequately represented by (indeed, fully absorbed into) his own account of law as a planning activity. This, Shapiro suggests, is actually an 'easy' move.[34] The desiderata of generality and publicity, he explains, 'follow immediately' from the claim that legal planning is 'social' in nature; that it regulates 'the bulk of communal activity via general, publicly accessible policies.'[35] The requirements that law must be 'clear, consistent, prospective, satisfiable, and stable' each also follow from the dispositions that a lawgiver, as a social planner, must possess in order to see plans filled in over time, rendered consistent with one another, and so forth.[36] Finally, Shapiro suggests that the planning thesis can also readily account for the eighth requirement of congruence between official action and declared rule, because this simply follows from how the 'composite' character of planning involves 'not only plan adoption, but plan application'.[37]

[32] See chapter six, II 'Raz on the rule of law'.
[33] Shapiro, *Legality* (n 11) 394.
[34] ibid.
[35] ibid 394, 395.
[36] ibid 395.
[37] ibid.

These are the foundations from which Shapiro is able to make an important jurisprudential claim that, in his view, settles the debate between positivism and Fuller on the question of the meaning and value of legality. To understand Rex's failure in terms of his failure to be a social planner is not only to illuminate the message of Fuller's tale in its best light, but also to answer the anti-positivist challenge. That is, positivists can 'agree with Fuller that observance of his eight principles is necessary for the existence of a legal system and yet deny that the existence of law depends on moral facts'.[38]

This is a compelling claim, and indeed an intriguing and innovative reading of Fuller's famous allegory. But when tested against the interpretation of Fuller's jurisprudence that I have offered in this book, Shapiro's position is highly problematic. The root of the problem might be stated thus. The key point is not whether the best understanding of Rex's failed attempt at lawgiving ought to be cast in terms of a failure to succeed in the enterprise of subjecting human conduct to the governance of rules, or a failure of social planning. The key point, instead, is whether we can accept how Shapiro understands the *content* of Rex's failure; an understanding that in turn leads to his neat dismissal of Fuller's claims about the connections between law and morality.[39] Because even if plans serve, for Shapiro, the morally valuable purpose of settling our normative questions for us, his version of positivism still seems ultimately indifferent to the quality of those plans *as* plans, and, crucially, to what they ask of and presuppose about their participants. *This* is the territory of Fuller's morality claims, and of why he regards that morality to be internal to law. Shapiro's engagement, however, appears to miss this in much the same way as Hart's engagement with Fuller did. The consequence, again like Hart, is that Shapiro does not understand the source of Rex's failure in the way that Fuller did: that is, that Rex failed as a lawgiver because of what he neglected to notice about how the forms through which he sought to govern related to the capacities and circumstances of his subjects.

The message of Fuller's jurisprudence is that when you try to be a lawgiver, to pursue your ends efficaciously through the form of law, you find out fairly quickly that the choice to govern through this mode carries its own presuppositions and demands. You find out, in particular, that there are ways that you must treat your subject if you are to proceed through this as opposed to some other means of rule. This imperative of treatment arises from how the subject who you seek to address through law possesses the status of an agent. She comes to law in this way, and the features of law's form, not least its generality, speak to, rely upon, and ultimately respect her status as such. All of this, you soon find out, has crucial implications for the task of lawgiving. To communicate with your subject, to bring her around as your collaborator in the enterprise of creating and maintaining a legal order, it is necessary to mediate your aims as well as your desire for their efficacious realisation through a system of communicative commitments of the kind that Fuller

[38] ibid.
[39] Waldron has made a similar point: see 'Planning for Legality', (n 25) 898–99.

captured in his model of eight principles. Fail to heed these fundamentals about how to collaborate with the agents who you are trying to address, and you are likely, indeed, very likely, to fail as a lawgiver. You might achieve something else (some other mode of rule, perhaps, such as managerial order) but it is highly improbable that you will be able to lay claim to being a lawgiver, and to having created a legal system.

These, as I argued in chapter four, are the lessons we are invited to take from Fuller's story of the tyrant lawmaker, and they are equally the lessons that we ought to take from the tale of Rex.[40] Because what Rex thought was a toolkit, an easy to follow template for governing, turned out to be a system of relational and communicative commitments within which the lawgiver's desire to pursue ends efficaciously through law coexists with the need for the legal subject to be, and to be respected as, a responsible agent. This made Rex much more than a failed social planner. He was also, much more gravely, a failed trustee for the fate of persons.

There might, of course, be a number of possible bridges between this reading of the causes of Rex's failure and Shapiro's analysis as it currently stands. We might speculate, for instance, about whether Shapiro's 'planner' (or at least the version of her that is a legal subject as opposed to a lawgiver) might be conflated with Fuller's 'responsible agent', or vice versa. Similarly, we might consider whether the form-agency connection that I have teased out as the unifying theme of Fuller's jurisprudence might find a place within Shapiro's idea that law is a planning activity which has its own internal rationality. Perhaps it can, but a longer conversation than the present is needed if we are to explore such questions.

Still, and this is surely the key point, it will remain difficult to see how any such conversation could see Shapiro remaining firm in his support for the jurisprudential verdict that his engagement with Fuller is intended to yield: namely, that positivists can 'agree with Fuller that observance of his eight principles is necessary for the existence of a legal system and yet deny that the existence of law depends on moral facts'.[41] In this sense, therefore, the question I have presented to Shapiro is essentially the same as that which I presented to Raz at the conclusion of chapter six. The bridge-building moves towards what would traditionally have been regarded as anti-positivist positions that contemporary positivists seem increasingly willing to make do not come without complications. One such complication relates to how a positivist account of law understands the status of and position occupied by the legal subject within a legal order. To this end, Shapiro, like Raz, seems to have a case to answer. Because once we begin to ask questions about the nature of law that take the perspective, position and intuitions of the legal subject seriously, and once, in doing so, we let the status of that legal subject as an agent creep in, we are likely to find that the positivist commitment to the idea that the existence of law depends in no meaningful way on moral facts is, at the very least, problematised.

[40] See chapter four, IIC 'Lessons from a tyrant?'.
[41] Shapiro, *Legality* (n 11) 395.

IV Conclusion

In light of the preceding analysis, it is fitting to bring this book to a final close by referring back to Jeremy Waldron's appeal to legal philosophers to search for a more 'discriminating', or 'less casual and accommodating', concept of law than that offered by legal positivism as it presently stands.[42] It is surely no accident that Waldron made his appeal in the aftermath of the occasion of the fiftieth anniversary of the 1958 Hart-Fuller debate; an occasion that saw contemporary legal philosophers begin to think more seriously about how Fuller's contribution to debates about law and morality might be equally, or in fact better, appraised as an important contribution to debates about the relationship between law and the rule of law, or legality.

Waldron's appeal is identical in all material respects to that which Fuller presented to Hart, but with Fuller doing so at a time when legal philosophers were far less receptive to the idea of turning inquiry into the nature of law towards the territory of legality. That time, however, appears now to have passed, and so it is surely now also time to return to the conversations that Fuller tried to initiate and to ask their core questions again.

Whatever those conversations might ultimately come to be, if they are to include Fuller as a meaningful participant it is essential that the insights and intuitions of his jurisprudence that I have brought together in this book be taken much more seriously than, for the most part, they have been to date. Above all, these include the under-explored relationship between the form of law and human agency; the conditions that limit lawgiving authority; the phenomenon of legal pathology; and the neglected status of the legal subject. If Fuller is truly to be reclaimed, these themes must be taken seriously not only within interpretations of his writings, but as part of the agenda of twenty-first century legal theory more generally, whether that agenda announces itself in the name of legal philosophy, legal theory or, indeed, jurisprudence.

[42] Jeremy Waldron, 'The Concept and the Rule of Law' (2008) 43 *Georgia Law Review* 1, 13, 15.

Index

Please note: Fuller's *writings* are listed under his main entry

www.ingramcontent.com/pod-product-compliance
Ingram Content Group UK Ltd.
Pitfield, Milton Keynes, MK11 3LW, UK
UKHW020813290926
15470UKWH00003B/1139